# Neuropsychodynamic Psychiatry

Heinz Boeker • Peter Hartwich
Georg Northoff
Editors

# Neuropsychodynamic Psychiatry

Springer

*Editors*
Heinz Boeker
Psychiatric University Hospital Zurich
Zurich
Switzerland

Georg Northoff
Mind, Brain Imaging, and Neuroethics
Institute of Mental Health Research
University of Ottawa
Ottawa
ON, Canada

Peter Hartwich
Hospital of Psychiatry-Psychotherapy-Psychosomatic
General Hospital Frankfurt
Teaching Hospital of the University
Frankfurt
Germany

ISBN 978-3-319-75111-5 ISBN 978-3-319-75112-2 (eBook)
https://doi.org/10.1007/978-3-319-75112-2

Library of Congress Control Number: 2018948668



This Springer imprint is published by the registered company Springer Nature Switzerland AG
The registered company address is: Gewerbestrasse 11, 6330 Cham, Switzerland

# Preface

When reading the title of our book, *Neuropsychodynamic Psychiatry*, the reader may well feel curious, but also somewhat surprised: Is this simply another attempt to reduce complex phenomena like subjective experience, psychopathology, and psychodynamics to neuronal activation patterns and dysfunctions?

One of this book's key issues is in fact to encourage and promote the critical discussion surrounding neuroscientific knowledge in the context of psychiatry, psychotherapy, and particularly of psychodynamic psychiatry. We particularly wanted to address that psychodynamic models and mechanisms can well be integrated with neurobiological approaches without reducing the former to the latter. For that, as we claim, one needs to take a slightly different and novel look on both psychodynamics and neurobiology though. One needs to consider psychodynamic processes in the larger social context in concrete relation to the world—psychodynamic mechanisms like defense mechanisms are psychosocial configurations of our relation to the world. At the same time, one needs to free the brain from its encapsulation within the skull and put it into the context of the world—neuroscience thus needs to extend beyond the brain to the world (of which the body is part). This points out the world as common reference for both psychodynamic mechanisms and brain (Northoff 2011)—that, in turn, makes possible to easily integrate psychodynamics and neurobiology and to thereby account for the complexities one can see in daily clinical reality.

The book focuses on mental functions such as the self, the ego, the conscious and the unconscious, and the psychological mechanisms and functions underlying these. The self in psychiatric patients is considered in a relational perspective, and the relational connecting processes between the brain and the environment are examined. The empirical plausibility of this relational definition of the self and the brain presents a particular challenge.

As neurobiological scientific knowledge on psychiatric illness grows, numerous therapeutic implications will certainly come into play. In view of this, fundamental, epistemologically oriented research, for instance, whether evidence-based research can give adequate answers to important questions in clinical practice, should always be kept in mind.

Considering how neurobiological knowledge can potentially benefit psychiatry and psychotherapy (for instance, a stronger position among the general public and in government healthcare policies, breaking down stigmas and taboos about mental illness, a better understanding of the etiology involved, and of how physiological and mental aspects of psychiatric disorders are linked, aiming at neurobiological variables in psychotherapeutic treatment, a better understanding of the interaction between various brain regions, and developing selective rules for indications using neurobiological predictors), the limits and possible risks that may be involved in neuroscientific approaches and, in the worst case, their one-dimensional application should not be overlooked. A neurobiological view of mental disorders and their treatment can always become problematic, as Fuchs (2006) puts it, if neurobiological approaches are no longer considered an alternative and complement to psychological models and procedures, but claim to have the absolute sovereignty of interpretation in human science.

Recent studies have shown a relationship between such reductionistic biological interpretations of psychiatric illness and the empathy shown by the clinician concerned: for example, Lebowitz and Ahn (2014) reported on unintentional negative consequences of exclusively biological models of illness depending on how the therapeutical relationship was built, on how the symptoms were perceived, interpreted, and dealt with, and also particularly depending on the therapist's empathy.

Psychiatrists, psychotherapists, and patients encounter one another as individual subjects or persons. The brain, however, is an "object." This obvious difference has considerable implications, in both a conceptual and an empirical perspective. Bennet and Hacker (2003) warn, for instance, about confusing individual subjects with their respective brains, as this would not consider the fundamental difference between persons and objects. Neuronal processes and mechanisms concern the brain and can be viewed as a necessary, but not sufficient condition for psychiatric treatment, and psychotherapy in particular, to be effective. This is because there are also other factors, such as interpersonal constellations and the cultural environment, that have to be considered. On the other hand, the effectiveness of psychiatric therapy and psychotherapy has to do with the personal level, which is implicitly related to the brain, but ultimately may not be identified with it in a conceptual perspective. Considering all these points, this book faces the particular challenge of trying to bridge the gap between the principal differences of a person's individual level and the brain's general level. One of the greatest methodological challenges of the future will be to develop experimental designs and analytical methods, which will make it possible to link individual and general features on a neuronal level. "First-person science" may thus be defined as a methodological strategy endeavoring to systematically link subjective experience with the observation of neuronal states (from a third-person perspective) (Northoff et al. 2006; Boeker et al. 2013). Like this, first-person neuroscience is different from general neuroscience, which is based on the observation of neuronal states, more or less independently of subjective experience. The complex task of investigating neuronal effects of psychiatric treatments, and especially those of psychotherapy, reflects in an almost paradigmatical way the complexity of the brain, resulting in greater insight and a better understanding of the general principles of neuronal organization (Boeker and Northoff 2010).

This book's chapters have been written by internationally renowned clinicians and researchers in the fields of psychiatry, psychotherapy, psychoanalysis, neuroscience, and other related disciplines. It is their contributions that have made it possible to adequately consider the broad spectrum of clinical-psychiatric, psychotherapeutic, psychoanalytical, and neuroscientific perspectives. Paradigms and methodological procedures in neuroscientific studies and results from fundamental research, as well as questions relevant for clinicians concerning the limited application of neuroscientific findings in clinical practice, are considered. This is presented within the broader framework of historical, epistemological, and philosophical discussions. This book has partly resulted from numerous discussions over many years with various professionals in psychiatric clinics and institutions. On behalf of everyone involved in these discussions, we would like to thank our colleagues from the "Therapy and Process Research" group at the University Hospital of Psychiatry, Zurich/Switzerland, from the Psychiatric University Hospital in Frankfurt/Main/Germany, from the Psychiatric University Hospital Charité, Humboldt University Berlin/Germany, Giessen/Germany, and Freiburg/Breisgau/Germany, and from the Hospital for Psychiatry and Psychotherapy in Göttingen-Tiefenbrunnen/Germany. Our cooperation with Professor Marianne Leuzinger-Bohleber's research group (Freud Institute Frankfurt/Germany and the Psychological Institute at the University of Kassel/Germany) was particularly fruitful. Many of the important findings discussed in this book were generated in close cooperation with the research groups at the ETH and University of Zurich/Switzerland (Professor Dr.rer.nat. P. Boesiger), the University Hospital of Psychiatry, Zurich, and the Humboldt University Berlin (Frau Dr. rer. nat. S. Grimm) and the University of Ottawa. Our stimulating discussions with lecturers and participants of various training courses in psychotherapy and psychoanalysis (Further Education in Psychiatry in North-East Switzerland, University of Zurich, Freud Institute Zurich, International Psychoanalytical University Berlin/IPU/Germany, Further Education in Analytical Therapy of Psychoses at the Academy of Psychoanalysis and Psychotherapy in Munich/Germany) and in various US-American, Canadian, and Chinese training and research institutes gave the editors of this book great inspiration.

Furthermore, we would like to thank Springer Publishers and especially Renate Scheddin for her encouragement right from the start, and Debika Bose (Project Coordinator for Springer Science) and Ejaz Ahmad (Assistant Project Coordinator) for their excellent, constructive cooperation.

We would particularly like to express our gratitude to Dawn Eckelhart for her valuable assistance in the preparation, coordination, and organization of the book's material and for making use of her talent to create order in chaos.

Zurich, Switzerland — Heinz Boeker
Frankfurt, Germany — Peter Hartwich
Ottawa, ON, Canada — Georg Northoff
May 2017

## References

Bennet MR, Hacker PMS. Philosophical foundations of neuroscience. Oxford, UK: Blackwell; 2003.

Boeker H, Northoff G. Die Entkopplung des Selbst in der Depression: Empirische Befunde und neuropsychodynamische Hypothesen. Psyche – Z Psychoanal. 2010;64:934–76.

Boeker H, Richter R, Himmighoffen H, et al. Essentials of psychoanalytic process and change: how can we investigate the neural effects of psychodynamic psychotherapy in individualized neuro-imaging? Front Hum Neurosci. 2013;7:355.

Fuchs T. Difficult issues in neuroscience. Curr Opin Psychiatry. 2006;19:600–7.

Lebowitz MS, Ahn W. Effects of biological explanations for mental disorders on clinicians' empathy. PNAS. 2014;111(50):17786–90.

Northoff G. Neuropsychoanalysis in practice. Brain, self, and objects. Oxford, New York: Oxford University Press; 2011.

Northoff G, Boeker H, Bogert P. Subjektives Erleben und neuronale Integration im Gehirn: Benötigen wir eine Erste-Person-Neurowissenschaft? Fortschr Neurol Psychiat. 2006;74:627–33.

# Contents

# Introduction

# 1

Heinz Boeker, Peter Hartwich, and Georg Northoff

**Abstract**

Please check and confirm the hierarchy of the section headings are appropriate.

**What Is "Neuropsychodynamic Psychiatry"?**

This book, "Neuropsychodynamic Psychiatry", follows the tradition of psychodynamic psychiatry (Gabbard 2005, 2014), and discusses its fundamental principles, linking them with recent findings in neuroscientific research. The guiding principle is the concept of the "socially embedded brain" and the "relational self". Both psychopathological phenomena and psychic experience are considered in the context of social and interpersonal experiences. Hence, psychodynamic relationships are discussed in a neurobiological context. We are interested in the neuropsychosocial mechanisms underlying psychic experiences and social relationship as knowledge of the former might strongly impact how we can approach the latter in therapeutic terms. Hence "mechanism-based approach" integrating neuronal and psychic mechanisms is of central importance.

H. Boeker (✉)
Department of Psychiatry, Psychotherapy, and Psychosomatics,
Psychiatric University Hospital Zurich, University of Zurich, Zurich, Switzerland
e-mail: heinz.boeker@bli.uzh.ch

P. Hartwich
Department of Psychiatry, Psychotherapy, Psychosomatics,
General Hospital of Frankfurt am Main, Teaching Hospital University Frankfurt,
Frankfurt, Germany
e-mail: Prof.Hartwich@casytec.de

G. Northoff
Mind, Brain Imaging, and Neuroethics,
Institute of Mental Health Research, University of Ottawa, Ottawa, ON, Canada
e-mail: georg.northoff@theroyal.ca; www.georgnorthoff.com



H. Boeker et al. (eds.), *Neuropsychodynamic Psychiatry*,
https://doi.org/10.1007/978-3-319-75112-2_1

Accordingly, the essentials of neuropsychodynamic psychiatry may be defined as follows:

Neuropsychodynamic psychiatry is a diagnostic and therapeutic approach as well as a scientific model, encompassing unconscious conflicts and dilemmata, as well as distortions in intrapsychic structures and internalized object relationships, with a view to explaining, understanding, investigating, diagnosing and treating psychopathological phenomena. It focuses on the functionality and dysfunctionality of psychic and neuronal mechanisms and integrates these elements into the context of recent findings in neuroscience with a specific focus on the relational-social dimension of the brain and our self (cf. Böker et al. 2015).

As there is widespread literature in psychodynamics itself, we here define our psychodynamic starting point in clear terms. The starting point in the psychodynamic perspective is the three-dimensional model drawn up by Mentzos (1991, 2009). The three dimensions comprise defence mechanisms, personality structure and conflict. In this model, conflict in various stages of development should be understood as resulting from the interaction between defence mechanisms and structure. We aim to develop a model equivalent to the one designed by Mentzos on the neuronal level. Here, defence and compensation mechanisms on a psychic level correspond to neuronal mechanisms. These focus mainly on the interaction between intrinsic and extrinsic activity, and thus on how the brain and the outside world interplay with each other.

We here aim to extend Mentos' three-dimensional model from the psychological level to the neuronal and, more precisely, neuro-ecological level, that is, our brain's relation to and integration within the world, the world–brain relation as we will term it (Chap. 5 and Northoff 2016, 2017). Like Mentzos' three-dimensional model which was linked with various mental and psychiatric-syndromal disorders, different psychiatric syndromes are integrated into the three-dimensional neuropsychodynamic model. This neuropsychodynamic model thus forms the foundation of our book. As it forms the foundation of this very book, we illustrate our three-dimensional neuropsychodynamic model by the following figure (Fig. 1.1) to which we add a short explanation as given below which will be detailed and extended in Chap. 5.

This marble sculpture can be seen as a symbol of the spatial-temporal structure of the self, which represents what is described as self-continuity and can be found in the relationship between brain, body and environment as a continuous process in balancing out dynamic powers and structuring order.

Specifically, we suggest that the dimension of structure on the psychological level corresponds to the spatiotemporal organisation of the brain's spontaneous activity and its different layers. As it will become clear throughout the book, changes in different layers of the spontaneous activity's spatiotemporal features correspond to different psychiatric disorders. Conflict describes the encounter between internal and external demands on the psychological level which, as we assume, can be traced to the interaction between brain, i.e., its internal spontaneous activity, and external events in the world. As well manifest in the various symptoms, that very same interaction, i.e., rest–stimulus interaction (Northoff et al. 2010; Huang et al. 2017) is altered in several psychiatric disorders. Finally, defence mechanisms on the

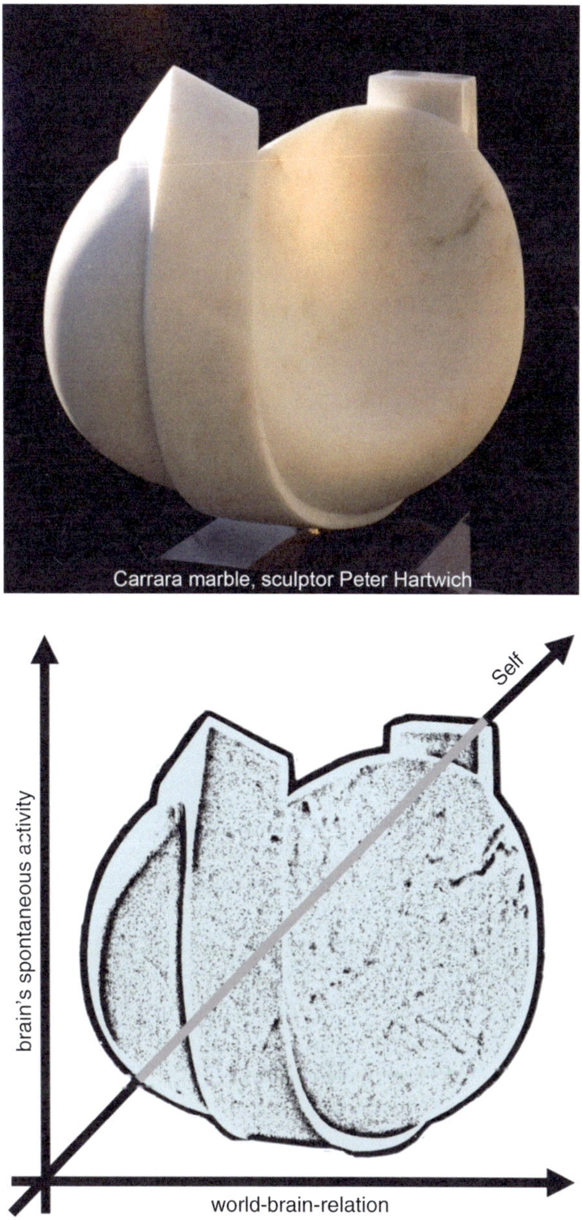


**Fig. 1.1** Balance of dynamics and structure (Carrara-marble, sculptor Peter Hartwich, 2001)

psychological level supposedly correspond to the brain and its spontaneous activity's processing and elaboration of the various information about events in the world it receives. Hence, defence mechanisms reflect the spontaneous activity's capacity for dynamic change in response to internal and external information and events (Northoff 2011)—we will see that that very same capacity or predisposition is seriously altered in many psychiatric disorders.

## 1.1 The First Part of the Book: Neuropsychodynamic Foundations

In the book's first section, the neuropsychodynamic essentials outlined above are presented and discussed. As well as the historical development of fundamental psychoanalytical concepts (the unconscious, the self, psychic development and mental structure, emotions, conflicts and dilemmata, transference and counter-transference, attachment, mentalisation, defence and compensation mechanisms) and what role they play in modern neuroscience, the concept of the "socially embedded brain", the "relational self" and personality structure are addressed.

## 1.2 The Second Part of the Book: Neuropsychodynamics of Psychiatric Disorders

The book's second section focuses on the neuropsychodynamics of psychiatric disorders using a syndromal approach, including chapters on schizophrenia and other psychoses, depressive syndromes, manic and bipolar syndromes, anxiety disorders, obsessive-compulsive disorders, somatisation and body distress disorder, anorexia and bulimia, traumatogenic disturbances, personality disorders, dissociative syndromes, and last but not least, addiction.

Each chapter covers the following aspects:

- Clinical observations
- Psychodynamics
- Scientific results
- Neuropsychodynamic hypotheses
- Clinical and therapeutic implications

In specially marked boxes (with a grey background) specific viewpoints are discussed:

- Historical background
- Recent and current discussions
- Critical reflection
- Ideas for future research.

## 1.3 The Third Part of the Book: Neuropsychodynamic Perspectives

The book's third section focuses on neuropsychodynamic therapy, its established principles and their application in different therapeutic settings (in-patient, out patient and day clinic). Each treatment intervention is put into the context of the neuropsychodynamic mechanisms mentioned above. Comparing the current state of the art with neuropsychodynamic mechanisms, future therapeutic perspectives may be developed. The five aspects mentioned above (clinical observations, psychodynamics, scientific results, neuropsychodynamic hypotheses, clinical and therapeutic implication) are also covered in these chapters.

The third section of the book also presents future neuropsychodynamic perspectives: psychotherapy research in the context of neuroscience and ethical and neurophilosophical aspects are considered.

## References

Böker H, Hartwich P, Northof G, editors. Neuropsychodynamische psychiatrie. Heidelberg: Springer; 2015.

Gabbard GO. Psychodynamic psychiatry in clinical practice. Washington DC: American Psychiatric Publishing Inc; 2005.

Gabbard GO. Psychodynamic psychiatry in clinical practice. 5th ed. Washington DC: American Psychiatric Publishing; 2014.

Huang Z, Zhang J, Longtin A, Dumont G, Duncan NW, Pokorny J, Qin P, Dai R, Ferri F, Weng X, Northoff G. Is there a nonadditive interaction between spontaneous and evoked activity? Phase-dependence and its relation to the temporal structure of scale-free brain activity. Cereb Cortex. 2017;27(2):1037–59. https://doi.org/10.1093/cercor/bhv288.

Mentzos S. Psychodynamische modelle in der psychiatrie. Göttingen: Vandenhoeck und Ruprecht; 1991.

Mentzos S. Lehrbuch der Psychodynamik. Die Funktion der Dysfunktionalität psychischer Störungen. Göttingen: Vandenhoeck und Ruprecht; 2009.

Northoff G. Neuropsychoanalysis in practice, self, object and brain. Oxford: Oxford University Press; 2011.

Northoff G, Qin P, Nakao T. Rest-stimulus interaction in the brain: a review. Trends Neurosci. 2010;33(6):277–84. https://doi.org/10.1016/j.tins.2010.02.006.

Northoff G. How do resting state changes in depression translate into psychopathological symptoms? From 'spatiotemporal correspondence' to 'spatiotemporal psychopathology'. Curr Opin Psychiatry. 2016;29:18–24.

Northoff G. The brain's spontaneous activity and its psychopathological symptoms–spatiotemporal binding and integration. Prog Neuro-Psychopharmacol Biol Psychiatry. 2017;80(Pt B):81–90. https://doi.org/10.1016/j.pnpbp.2017.03.019.

# Part I

# Neuropsychodynamic Foundations

# Why Do We Need Psychopathology? From the Brain's Spontaneous Activity to "Spatiotemporal Psychopathology"

2

Georg Northoff

**Abstract**

The resurgence of biological psychiatry raises the question how and why we need psychopathology. Psychopathology has been well developed in the time where the brain was not yet explored; it has brought forth psychological approaches like cognitive psychopathology and experiential approaches like phenomenological psychopathology. Both psychological and experiential approaches suffer from a divide to the brain though, the divide between the brain and cognition as well as the divide between the brain and experience. I here suggest a novel form of psychopathology that focuses on spatiotemporal rather than cognitive or experiential features, i.e., spatiotemporal psychopathology. Thereby the brain's spontaneous activity plays a central role since it provides and shows various kinds of spatial and temporal features which, as I suppose, are organized cognition and are transformed into experience. I illustrate such spatiotemporal approach to psychopathological symptoms by the examples of depression and mania in bipolar disorder. I conclude that spatiotemporal psychopathology holds the promise to bridge the gap between the brain and symptoms including the divides between the brain and cognition/experience. Taken in this sense, spatiotemporal psychopathology will also be able to trace both psychological and experiential approaches to psychopathology to a commonly underlying basis, i.e., the spatiotemporal structure and features of the brain's spontaneous activity. Accordingly, we need psychopathology and, more specifically, "spatiotemporal psychopathology" to understand both the brain's neural activity and psychopathological symptoms and how the former translates into the latter.

G. Northoff
Mind, Brain Imaging, and Neuroethics,
Institute of Mental Health Research, University of Ottawa, Ottawa, ON, Canada
e-mail: georg.northoff@theroyal.ca; www.georgnorthoff.com



H. Boeker et al. (eds.), *Neuropsychodynamic Psychiatry*,
https://doi.org/10.1007/978-3-319-75112-2_2

This, in turn, opens a new understanding of psychodynamic mechanisms like defense mechanisms that can then be considered as different and specific "spatiotemporal configurations" as engineered on the basis of the brain' spontaneous activity and its spatiotemporal structure.

Neither the "brainless" psychiatry of the middle of the 20th century, nor the "mindless" variety of the past 30 years should be taken to represent the most we can achieve. The future should yield a synthesis. Panksepp (2004, p. 17)

## 2.1 Introduction

Neuroscience has made enormous progress in the last 20–30 years on all levels ranging from the genetic over the molecular to the regional and network level of neural activity. This has also affected psychiatry as in biological psychiatry. Various psychiatric disorders including schizophrenia, major depressive disorder (MDD), and bipolar disorder (BP) as well as others like addiction, personality disorders, etc. show molecular, genetic, regional, and network abnormalities in the brain. However, despite all progress in biological psychiatry, we still fall short in explaining the exact neuronal mechanisms of the various psychopathological symptoms. Specifically, biological psychiatry cannot yet explain how the brain's neuronal changes transform into the mind's alterations, the psychopathological symptoms.

Traditionally, the explanation and understanding of psychopathological symptoms has been the focus of psychopathology. Put in a nutshell, psychopathology concerns the empirical and theoretical framework in which symptoms, behavior, and experiences in psychiatric patients can be described, categorized, and classified (Parnas et al. 2008, 2013; Stanghellini 2009a, b; Stanghellini and Broome 2014). Different empirical and theoretical frameworks have been suggested in past and present approaches to psychopathology. However, how the different approaches to psychopathology (see below for details) are linked to the brain and its various neuronal mechanisms remains unclear.

Taken all together, we are facing a divide between biological psychiatry and psychopathology. The advocates of biological psychiatry tend to claim that all we need is the brain: the more we understand the brain and its abnormal changes in psychiatric disorders, the better we will understand and explain the psychopathological symptoms. This makes psychopathology as separate scientific discipline (Stanghellini and Broome 2014) meaningless and senseless and thus superfluous. Conversely advocates of psychopathology resist such interpretation. There is "more" to psychopathological symptoms than just the brain, and this "more" consists in the central role of experience or consciousness, i.e., the mind (Parnas et al. 2008, 2013; Stanghellini 2009a, b; Stanghellini and Broome 2014). Taken in this view, biological psychiatry remains as "mindless" as psychopathology is "brainless," to pick up our initial quote.

How can we resolve the divide between the brain and mind/symptoms and thus biological psychiatry and psychopathology? The aim in this paper is to show that a novel approach, "spatiotemporal psychopathology," can bridge this divide by providing a "common currency" between the brain and symptoms—that "common currency" is supposed to consist in spatiotemporal features that transform abnormal neuronal activity in psychopathological symptoms.

## 2.2 "Spatiotemporal Psychopathology": Determination and Distinction

### 2.2.1 Psychological Approaches to Psychopathology

Roughly, one may want to divide between psychological and experiential approaches to psychopathology. Psychological approaches focus on specific psychological functions like cognitive or affective cognitive as in cognitive psychopathology (David and Halligan 2000; Halligan and David 2001), affective psychopathology (Panksepp 2004). With the development of neuroimaging, these approaches are now able to link the objectified changes in cognitive and affective functions onto the brain. However, such "mapping" of cognitive into neural functions leaves open how and why abnormal changes in the brain's neural activity are transformed into psychopathological symptoms.

How can cognitive and affective and cognitive functions more strongly link to the brain and its neuronal mechanisms? We would need to unravel a yet unclear "common currency" that allows to transform neuronal into psychological activity. To be clear, I am not raising the question which regions or networks in the brain are related to cognitive functions. Such "cognitive-neural mapping" has been well established in cognitive neuroscience that showed how cognitive functions like executive functions, attention, memory, etc. are related to specific regions or networks in the brain. This and the respective changes in those regions and networks have been well researched intensely over the last 10–20 years. Instead, I am raising the question why and how the brain's neuronal activity in those regions and networks transforms into cognitive (and affective functions) rather than remaining merely neuronal (and non-cognitive).

What is needed is a "common currency" between neural and cognitive functions—due to such yet unclear "common currency," neural activity translates into cognitive function basically by default. And it is this transformation or translation that seems to be altered in psychiatric disorders that can indeed be characterized by numerous cognitive deficits. I postulate that the spatial and temporal features of the brain's spontaneous activity provide such "common currency"—cognitive symptoms are spatiotemporal symptoms for which reason I speak of "spatiotemporal psychopathology."

### 2.2.2 "Common Currency" Between the Brain and Cognition

We are confronted with a divide between the brain on the one hand and cognition on the other. Biological psychiatry focuses on the brain while leaving out the mind and its experience. While psychological approaches to psychopathology focus on cognitive functions and the relation of their contents to the brain, neither has yet provided a full-fledged explanation and understanding of psychopathological symptoms though. We are thus confronted with a divide between the brain and cognition.

Psychological approaches to psychopathology focus on contents, i.e., cognitive, affective, sensorimotor, and social contents and their related functions. The cognitive, affective, sensorimotor, and social contents are then "mapped" upon the brain and its various regions and networks—this is where psychological approaches to psychopathology converge with biological psychiatry. This neglects one central dimension of the contents though. The contents are organized and structured in a particular way, and this organization is mainly spatial and temporal. Spatiotemporal psychopathology as suggested here focuses on the temporal and spatial organization of the contents rather than the nature of the contents themselves, i.e., cognitive, affective, sensorimotor, or social.

Spatiotemporal psychopathology aims to unravel the spatiotemporal organization and structure within which the various kinds of contents are embedded hence the name "spatiotemporal psychopathology." Alterations in cognition in psychiatric disorders are consequently not related to specific contents, i.e., cognitive, affective, etc. Instead, abnormal cognition is related and traced to abnormal spatial and temporal organization within which the contents are embedded.

Let us give an empirical example. Duncan et al. (2015) recently demonstrated that early childhood traumatic experience is manifested in adulthood in the spatiotemporal patterns of the brain's spontaneous activity (as indexed by entropy) which, in turn, impacts subsequent stimulus-induced activity in relation to aversive stimuli. The early childhood traumata were thus encoded in terms of spatiotemporal features, i.e., entropy, rather than in terms of specific contents and cognitions. Sure, the very same spatiotemporal pattern impacts the contents and their subsequent cognition—however, it is clear that the latter has a spatiotemporal basis in the spatiotemporal features of the brain's spontaneous activity. Hence early childhood traumas are primarily a matter of spatiotemporal organization of the contents, i.e., life events, rather than being directly related one-to-one to the life event and its content itself.

Taken together, the spatiotemporal organization of the brain's spontaneous activity may provide the "currency" that translates directly into the cognitive level with the cognition of contents. Spatial and temporal features as manifest in both the brain's spontaneous activity and our cognition of contents may consequently provide the "common currency" between the brain and cognition. Changes in cognition as in psychiatric disorders may then be traced to alterations in the resting state's spatial and temporal features. This would link the psychological approaches to psychopathology even more tightly to the brain while, at the same time, providing a new view on the brain and especially its resting state, a spatiotemporal rather than cognitive view (Northoff 2016a).

### 2.2.3 Experiential Approaches to Psychopathology

In contrast to psychological approaches to psychopathology, experiential approaches focus on the subject's experience, i.e., subjective experience, of self, world, and body rather than on objectified cognitive and affective functions. The hallmark experiential approach is phenomenological psychopathology takes the subject's experience of self, body, and world and thus the structure of its consciousness as explanatory framework for psychopathological symptoms (Fuchs 2007, 2013; Northoff 2016b; Parnas et al. 2008, 2013; Stanghellini 2009a, b; Stanghellini and Broome 2014).

Siblings of phenomenological psychopathology include existential psychopathological, which focuses on the existence as the deeper layer underlying experience, and the hermeneutical psychopathology that emphasizes the meaning of symptoms in a wider biographical and environmental context (Stanghellini 2009a, b). Despite the difference in focus or emphasis, the overall explanatory framework in all three approaches consists in experience or consciousness for which I reason I subsume under the "experiential approaches" to psychopathology.

Phenomenological psychiatry takes experience or consciousness itself as starting point and focuses on exploring first-person experiences in detail (Parnas et al. 2008, 2013; Stanghellini 2009a, b). Specifically, the focus is on the first-person experience of time and space as well as body, self, and world. The brain, in contrast, nowhere surfaces in experience, in particular, and phenomenology, in general, since it cannot be accessed in experience in first-person perspective but only in observation as in third-person perspective. The brain is thus excluded in experience of the own self, body, and world including time and space in particular and phenomenology in general. Such exclusion of the brain in experience or consciousness occurs by default, e.g., on methodological grounds, since the brain cannot be accessed in experience in first-person perspective. Importantly, this leaves the link to the brain open and renders the experiential approaches to psychopathology ultimately as "brainless" (as picking up our initial quote).

### 2.2.4 "Common Currency" Between the Brain and Experience

How can we close the gap between experience and brain? Closing this gap is central for psychiatry since we need to understand the processes that transform abnormal neuronal into phenomenal states which psychiatric patients experience in first-person perspective. How can we apprehend these transformative processes, e.g., neuronal-phenomenal transformation? For that we may want to search for a shared overlap or "common currency" between neuronal and phenomenal states that drives the transformation of the former into the latter.

The shared overlap or common currency between neuronal and phenomenal states, e.g., the brain and experience, may consist in spatiotemporal features. On the side of the brain, it is the spontaneous activity (rather than its stimulus-induced or task-evoked activity) that may be central in providing or constituting such spatiotemporal structure (see below for details). The brain's spontaneous activity shows certain spatiotemporal features, a particular spatial and temporal structure in its neural activity that surfaces in and is transformed into phenomenal state, e.g., experience (see (Northoff 2016a) for many examples). One would consequently expect a common, similar, or analogous spatiotemporal structure between the brain's spontaneous activity and the phenomenal features of experience.

Such common, similar, or analogous spatiotemporal structure between the brain and experience amounts to what I describe as "spatiotemporal correspondence." The concept of spatiotemporal correspondence means that the brain's spontaneous activity and the phenomenal features of experience show corresponding or analogous spatial and temporal features: the spatial and temporal configuration or structure of the neural activity in the brain's spontaneous activity surface in the spatial and temporal features within which the contents of experience (like specific objects or events including body, self, and world) are integrated and thus structured and organized.

For instance, a recent study of ours demonstrated that private self-consciousness is directly related to the temporal patterns of spontaneous or resting state activity across different frequency ranges (as indexed by what is described as "power law") (Huang et al. 2016). This suggests that mental features like self may be rooted in spatiotemporal features of the brain's spontaneous activity. The self as mental feature may then be characterized in spatiotemporal terms, that is, by specific spatiotemporal schemata or structure rather than by cognition of particular contents (see Fig. 4.1).

Unlike biological psychiatry that focuses on the brain itself independent of its respective ecological context, phenomenological psychiatry emphasizes the integration of experience including the subject of experience within the ecological context of the world. There is continuity between experience and world with such continuity often assumed to be mediated by the body, e.g., experience of the body as lived body (see, for instance, Northoff and Stanghellini 2016). Such continuity between subject and world is deemed central for making experience including the first-person perspective itself first and foremost possible.

## 2.3 Spatiotemporal Psychopathology: Depression and Bipolar Disorder

### 2.3.1 Spatiotemporal Psychopathology: Bipolar Disorder and Neuronal Variability

How about spatiotemporal changes in the resting state in bipolar disorder (BP)? Several resting state investigations observed changes in functional connectivity in the default mode network in bipolar disorder though the phases, i.e., depressed, euthymic, and manic, are rarely specified. Going beyond functional connectivity, we investigated neuronal variability in different resting state network in manic, euthymic, and depressed phases of BP as well as healthy subjects. Neuronal variability is measured by the root means square of the amplitude, in that it reflects the change in the amplitude over time and the degree to which these changes vary over time. Taken in this sense, neuronal variability can be considered a measure of the temporal structure and, more specifically, the temporal dynamics of the ongoing spontaneous activity.

We focused on neuronal variability (SD) in the main neural networks, default mode network (DMN), central executive network (CEN), salience network (SN), and sensorimotor network (SMN). Depressed BP patients showed significantly decreased SD in the sensorimotor network, while their SD was significantly increased in the DMN. The other neural networks like SN and CEN did not show any SD changes. We then calculated the ratio or balance between DMN SD and SMN SD; this was tilted significantly toward the DMN SD at the expense of the SMN SD.

What does this mean? Neuronal variability may be linked to the initiation of internally directed cognition in DMN and movements/actions in SMN. The more often the neuronal variability change surpasses a certain threshold, the more often the respective regions internally, i.e., by itself independent of external stimuli, initiate either cognition or action. Let us be more specific regionally. The DMN has been associated with internally directed cognition as in spontaneous cognition and mind wandering. If now neuronal variability is abnormally high in the DMN, there is a higher likelihood that spontaneous thoughts will be initiated. This is exactly what one can observe in depressed BP where the patients suffer from increased spontaneous thought which are described as rumination.

How about the SMN? In that case, neuronal variability may be related to the spontaneous or internal initiation of movements and actions. If now neuronal variability in SMN is decreased, one would expect decreased internal initiation of movements and actions. This, again, is exactly what can be observed in depressed BP where patients often suffer from psychomotor retardation. Most interestingly, it seems that the balance between DMN SD and SMN SD is central since the balance correlated significantly positively with depressive symptoms (as measured with Hamilton depression scale): the more the SD balance was shifted toward the DMN at the expense of the SMN, the more and stronger depressive symptoms.

The reverse could be observed in the manic phase. Here SD was significantly lower in the DMN and abnormally high in the SMN; the balance between DMN SD and SMN SD is consequently tilted toward the SMN at the expense of the DMN. This is symptomatically manifested in increased internal initiation of movement/action as it is reflected in the well-known psychomotor agitation in mania. In contrast, internally directed cognition like spontaneous thought is no longer initiated internally as much—this is reflected in the fact that many manic patients say "that they do not think much or not at all" in the manic episode.

### 2.3.2 Spatiotemporal Psychopathology: From Neuronal Variability to Cognition and Experience

What do these findings tell us about the nature of psychopathological symptoms? There is still internal initiation of movements as related to SMN and internally directed cognition, i.e., spontaneous thought as based on DMN. However, the neuronal mechanism potentially underlying such internal initiation, i.e., neuronal variability, is expressed to an abnormal degree. It is either too high or too low which leads to either increased or decreased internal initiation of the respective function. That very same neuronal mechanism is temporal, i.e., SD, and spatial, i.e., in different networks like DMN and SMN, and can therefore be considered "spatiotemporal mechanism" as I say (see Fig. 4.2).

Let me be more precise. The function, i.e., internal initiation of movements/action and internally directed cognition, is still intact by itself—the bipolar patients are still able to internally initiate them. This distinguishes psychiatric patients from neurological patients. In the latter, the region itself is lesioned which makes impossible the internal initiation of, for instance, movement and action as in Parkinson's disease. However, the function of internal initiation of movement/action and internally directed cognition is expressed in an abnormally high or low way due to some spatial and temporal abnormalities in the brain's spontaneous activity, i.e., SD in SMN and DMN. The resulting abnormalities in movement/action and internally directed cognition, i.e., the psychopathological symptoms, are thus based on and can be traced to spatiotemporal changes in the brain's spontaneous activity. Rather than being primarily motor, as in Parkinson's disease, the psychomotor changes in mania and depression are thus spatiotemporal at their very basis.

The same holds analogously for internally guided cognition. Unlike in neurological lesion patients, the bipolar patients can still initiate internally directed cognition like spontaneous thought. However, that very same internal initiation is temporally disorganized by the abnormal high neuronal variability in DMN in depression and the low SD in DMN in mania. The cognitive symptoms like rumination (or decreased thought) are consequently not primarily cognitive but rather spatiotemporal as they are related to spatiotemporal changes in the brain's spontaneous activity.

Taken together, the example of BP nicely demonstrates that cognitive and motor symptoms in both depression and mania are not related to primary dysfunction in either cognitive or motor functions. Instead, the basic function, i.e., cognitive or motor, is preserved by itself but abnormally organized in spatial and temporal terms. Therefore, the symptoms are spatiotemporal rather than cognitive and motor. What is described as cognitive in cognitive psychopathology is based on and can be traced to spatiotemporal abnormalities in the brain's spontaneous activity and thus spatiotemporal psychopathology.

The same holds for experience and phenomenological psychopathology. Depressed patients often experience their "inner time," i.e., the time of their own self, as extremely slow which, when taken as reference, lets them perceive the "outer time," i.e., the time in the environment, as extremely fast (Fuchs 2013). We measured neuronal variability in the neural network underlying "inner time," i.e., the somatosensory network (SS), and the one related to "outer time," i.e., primary sensory regions like visual cortex (VS). This yielded decreased SD in the SS and increased SD.

How are these findings related to the experience or perception of time? Neuronal variability indicates change in neural activity, and the more change there is, the faster the time. Decreased SD in SS thus indicates slower "inner time," while increased SD in VC reflects faster "outer time"—this corresponds exactly to the experience of time depressed patients report. The opposite SD pattern with increased SD in SS and decreased SD in VC was observed in manic patients—this corresponds well to their experience of faster "inner time" and slower "outer time."

Taken together, these findings indicate how a temporal measure like neuronal variability is translated into experience or perception, i.e., the experience of the speed of time. Hence, experience of the speed of time may be traced to and be based on a corresponding neuronal measure that indicates the speed of the brain's time, i.e., neuronal variability. Hence, the change in the brain's time speed, i.e., its neuronal time as indexed by neuronal variability, is transformed into corresponding experience or perception, i.e., the experience of the speed of time. Experience of time and experience in general is thus spatiotemporal by itself and thereby based on the spatiotemporal features of the brain's spontaneous activity. Experiential approaches like phenomenological psychopathology are thus ultimately based on and can be traced to spatiotemporal features and hence spatiotemporal psychopathology.

**Conclusion**

How can we bridge the divide between the brain and cognition and hence between biological psychiatry and cognitive psychopathology? I demonstrated how cognitive changes like rumination in depression and decreased cognition in mania are related to abnormal expression of spatial and temporal mechanisms of the brain's spontaneous activity. Hence, I postulate that what is described as abnormal cognition in cognitive psychopathology is based on and can be traced to abnormal spatial and temporal organization of cognitive functions—this entails what I describe as "spatiotemporal psychopathology." Accordingly, I postulate that the spontaneous activity's spatial and temporal features provide the bridge between the brain and cognition. Therefore, spatiotemporal psychopathology provides the bridge between biological psychiatry on the one hand and cognitive psychopathology on the other.

How about the divide between the brain and experience and hence between biological psychiatry and phenomenological psychopathology? I showed how the abnormal experience of time in depression and mania may be based on abnormal temporal features like neuronal variability in the brain's spontaneous activity. Experience is thus based on spatiotemporal features—the spatiotemporal features of the brain's spontaneous activity transform into experience which thereby can be characterized as spatiotemporal. Hence, the spontaneous activity's spatiotemporal structure allows linking the brain and experience and can therefore bridge the divide between biological psychiatry and phenomenological psychopathology.

The initial question and title in this paper is: Why Do We Need Psychopathology? We need psychopathology to bridge the gap between the brain and cognition as well as the one between the brain and experience. This does not only provide common link between biological, cognitive, and experiential forms of psychopathology but also a novel, i.e., spatiotemporal, understanding of both the brain and symptoms. I postulate that spatiotemporal psychopathology as sketched here provides exactly that form of psychopathology that allows us to understand the brain and how its neural activity transform into cognition and experience and subsequently the kind of symptoms we observe in our patients.

Why is spatiotemporal psychopathology relevant for neuropsychodynamic psychiatry? First and foremost, it opens a novel way of understanding psychodynamic mechanisms in a neurobiological context. For instance, defense mechanisms can then be understood in a spatiotemporal way as specific constellations of how time and space can (or cannot) be constructed by the brain's spontaneous activity in its relation to both body and world. In other terms, defense mechanisms are spatiotemporal mechanisms—this sheds a novel light on the defense mechanisms and other psychodynamic features. And, at the same time, the elaborate description of different defense mechanisms may also provide insight into the different kinds of possible spatiotemporal configurations as constructed on the basis of the brain's spontaneous activity. Hence, there may be true bilateral exchange between neuropsychodynamic psychiatry and spatiotemporal psychopathology.

## References


David AS, Halligan PW. Cognitive neuropsychiatry: potential for progress. J Neuropsychiatry Clin Neurosci. 2000;12(4):506.

Duncan NW, Hayes DJ, Wiebking C, Brice T, Pietruska K, Chen D, et al. Negative childhood experiences alter a prefrontal-insular-motor cortical network in healthy adults: a preliminary multimodal rsfMRI-fMRI-MRS-dMRI study. Hum Brain Mapp. 2015;36:4622–37.

Fuchs T. The temporal structure of intentionality and its disturbance in schizophrenia. Psychopathology. 2007;40(4):229–35.

Fuchs T. Temporality and psychopathology. Phenomenol Cogn Sci. 2013;12(1):75–104.

Halligan PW, David AS. Cognitive neuropsychiatry: towards a scientific psychopathology. Nat Rev Neurosci. 2001;2(3):209–15.

Huang Z, Obara N, Davis HH IV, Pokorny J, Northoff G. The temporal structure of resting-state brain activity in the medial prefrontal cortex predicts self-consciousness. Neuropsychologia. 2016;82:161–70.

Northoff G, Stanghellini G. How to link brain and experience? Spatiotemporal psychopathology of the lived body. Front Hum Neurosci. 2016;10:172.

Northoff G. Spatiotemporal psychopathology I: no rest for the brain's resting state activity in depression? Spatiotemporal psychopathology of depressive symptoms. J Affect Disord. 2016a;190:854–66. https://doi.org/10.1016/j.jad.2015.05.007.

Northoff G. Spatiotemporal psychopathology II: how does a psychopathology of the brain's resting state look like? Spatiotemporal approach and the history of psychopathology. J Affect Disord. 2016b;190:867–79. https://doi.org/10.1016/j.jad.2015.05.008.

Panksepp J. Textbook of biological psychiatry. New York: Wiley; 2004.

Parnas J, Sass LA, Zahavi D. Recent developments in philosophy of psychopathology. Curr Opin Psychiatry. 2008;21(6):578–84.

Parnas J, Sass LA, Zahavi D. Rediscovering psychopathology: the epistemology and phenomenology of the psychiatric object. Schizophr Bull. 2013;39(2):270–7.

Stanghellini G. A hermeneutic framework for psychopathology. Psychopathology. 2009a;43(5):319–26.

Stanghellini G. The meanings of psychopathology. Curr Opin Psychiatry. 2009b;22(6):559–64.

Stanghellini G, Broome MR. Psychopathology as the basic science of psychiatry. Br J Psychiatry. 2014;205(3):169–70.

# 3 Psychoanalysis and Neuroscience: The Development of Neuropsychoanalysis

Heinz Boeker

**Abstract**

In his *Project for a Scientific Psychology*, published in 1895, Freud tried to connect psychoanalysis with the neuroscience of his day. Finally, he was forced to give up this endeavour because of the lack of diagnostic possibilities and empirical facts. Freud's paradigmatic change to psychoanalysis came about through his work on the mind-body problem and its central question of how the brain is able to generate subjective experience (consciousness) by means of existing anatomical structures and physiological functions.

This chapter focuses on the development of modern neuroscience with its diagnostic-technical repertoire and the possibility to gain insight into neuronal processing of mental processes such as emotional-cognitive interaction, which enabled to look for connections between what is known in both disciplines.

Lurija's neurodynamic approach, Kaplan-Solms and Solms' neuroanatomical methods, Damasio's *The Feeling of What Happens*, Panksepp's *Affective Neuroscience* and the discovery of the so-called mirror neurons (Rizzolatti, Galese) are milestones of the development of neuropsychoanalysis.

Finally, the further development of a "first-person neuroscience" (Northoff) may contribute suitable strategies to answer core questions on the relation between subjective experience and neuronal integration in the brain.

The apportioning of the determining factors of our life between the 'necessities' of our constitution and the 'chances' of our childhood may still be uncertain in detail; but in general it is no longer possible to doubt the importance precisely of the first years of our

H. Boeker
Department of Psychiatry, Psychotherapy, and Psychosomatics, Psychiatric Hospital, University of Zurich, Zurich, Switzerland
e-mail: heinz.boeker@bli.uzh.ch



H. Boeker et al. (eds.), *Neuropsychodynamic Psychiatry*,
https://doi.org/10.1007/978-3-319-75112-2_3

> childhood. We still show too little respect for Nature which (in the obscure words of Leonardo which recall Hamlet's lines) 'is full of countless causes ('ragioni') that never enter experience'.
>
> Everyone of us human beings corresponds to one of the countless experiments in which these 'ragioni' of nature force their way into experience.
>
> (Freud 1990, p. 137)

## 3.1 Introduction

This chapter will focus on challenges which ultimately led to the development of a "theory of the unconscious" by Freud and later psychoanalysts from a clinical perspective. The beginnings of the dialogue between psychoanalysis and neuroscience will be presented, starting with Freud's *Project for a Scientific Psychology*. Furthermore, the important role of the unconscious and the unconscious phantasy in modern-day psychoanalysis will be examined. In this context, connections with neuroscientific research strategies and concepts of "cognitive neuroscience" concerning unconscious information processing will also be mentioned. This historical review of how a "theory of the unconscious" was developed and rejected will show just how relevant this issue remains today.

## 3.2 Freud and the Demystification of the Unconscious

The concept of the unconscious and the theory of instincts are the cornerstones of a psychoanalytical understanding of man's psychic reality. They go hand in hand with each other, are dependent on each other and are merely two different viewpoints from which psychoanalysis considers and examines psychic reality. Together they form a theory of man's first and most urgent wishes, his phantasies and conflicts, expressed above all in his sexuality, aggression and striving for self-assertion (Müller-Pozzi 2002).

Freud is often credited as having discovered the unconscious: this is an honour he neither merited nor claimed for himself or for psychoanalysis. On the contrary, when Freud began exploring the unconscious origins of mental disorders, the consideration of the unconscious forces of the mind was a popular subject in literature and philosophy. To counterpoint one-sided rationalism, the romantic transfiguration and mystification of the dark and mysterious unconscious side of the psyche reached its late climax at the end of the nineteenth century (for instance, in the works of the philosophers Arthur Schopenhauer, Friedrich Nietzsche and Ludwig Klages and the authors Hugo von Hofmannsthal and Arthur Schnitzler).

Freud kept a critical distance from these cultural schools of thought. He denied himself, as he put it (Freud 1914b, pp. 15–16), "…the very great pleasure of reading the works of Nietzsche, with the deliberate object of not being hampered in working out the impressions received in psycho-analysis by any sort of anticipatory ideas. I had therefore to be prepared–and I am so, gladly—to forgo all claims to priority in the many instances in which laborious psychoanalytic investigation can merely confirm the truths which the philosopher recognized by intuition".

Prior to his psychoanalytical discoveries, Freud's career was apparently that of a young scientific researcher in the late nineteenth century. However, Freud lacked the financial means needed to pursue a purely scientific career. Yet he may well have succeeded in this endeavour, had he not fallen in love. Today, in the light of his partly published letters to his fiancé, it may inevitably be concluded—as Israël (1983) wrote—that his work and his private life were closely connected and that, indeed, without passion nothing can be discovered, invented or captured.

These circumstances in Freud's private life are worth mentioning, since, as we will see later, his examination of the unconscious was preceded by a far-reaching professional and, thus for Freud, personal crisis, namely, when he discovered that what some of his female patients told him did not concern the actual trauma itself but was an expression of their unconscious phantasies.

As a neurologist, Freud was confronted on a daily basis with the symptoms of hysteria, a common illness at that time. Hysteria was manifest externally by striking functional disturbances, such as paralysis, difficulties swallowing, visual disturbances, pain and impaired consciousness. For medical professionals in those days, a condition which could not be somatically explained and treated was considered a nuisance and a contradiction, so that the patient was often suspected of faking the illness.

Freud did not close his eyes to the "neurotic suffering" of these female patients. Based on knowledge from doctors who had at times used hypnosis to cure hysterical symptoms, he began, together with Josef Breuer, to investigate hysterical patients' subjective experience.

He looked for the causes of hysterical symptoms not in organic factors but in the patient's subjective experience and was finally convinced that hysterical symptoms should be viewed as the physical manifestation of mental reality. Thus, the symptoms have meaning, but are expressed in body language which cannot be directly understood. Freud believed that this strangely distorted language of the symptoms was connected to the unconscious dimension, which not even the patient herself can access. Freud realized that the hysterical patient suffers from her reminiscences. These are often memories of emotionally meaningful experiences involving significant persons from the patient's childhood. Reflected in these reminiscent memories are intensive wishes and phantasies which could not be lived out in earlier relationships and have been pushed aside into unconscious phantasies. The "unconscious phantasy"—in sharp contrast to the psychology of consciousness—became one of the most important concepts in psychoanalysis.

The foundation for the psychoanalytical concept of the unconscious was thus laid: what previously seemed incomprehensible and nonsensical now seemed to make sense as a derivative and as the manifestation of the unconscious. This is not only the case for psychopathology in the narrow sense but also for the "psychopathology of everyday life", i.e. the Freudian slips and symptomatic actions which, like wit, dreams, daydreams, phantasies and even being artistically creative, can be seen as the manifestation or as deriving from unconscious phantasies. With this interpretation, Freud "challenged the whole world" and summoned up "the greatest critical minds against psychoanalysis" (Freud 1915a, b, p. 287, 294).

Freud's rational definition of the unconscious, which comprised definite facts in a scientific concept, contributed to the demystification of the romantic transfiguration of the unconscious by the philosophy and literature of the time. Müller-Pozzi (2002, p. 55) rightly underlined: "It was not the fact that Freud had started to speak about the unconscious, but how he did it, which challenged the whole world. The fact that he lifted the unconscious out of non-committal aesthetic speculation and literary description, substantiated and conflictualised it, making it the subject of a psychological-scientific discourse, earned him the criticism and disapproval of writers and philosophers on the one hand, and physicians and psychologists on the other" (Translation by H.B.).

## 3.3 Freud's Voyage of Discovery

Now back to the beginning of Freud's voyage of discovery. As a neurologist, Freud had always been confronted with hysterical patients: with hysterical patients, who, to quote Israël (1983), "for thousands of years had simply served as a screen for projecting male medicine" (p. 243) (translation by H.B.). Freud discovered that the theatrical picture of these women was only supposed to cover up something and that what people saw of hysteria "was only a testing out". He stopped just looking at these women and started listening to them. By using language as the basis of medical practice, Freud was doing exactly the opposite of Charcot and other renowned clinicians in the nineteenth century, who described their hysterical patients impressively but totally excluded what they actually said.

Freud himself was astonished to find that his patients began to speak while under hypnosis—a method he initially used—and that they were not just talking about anything, but often specifically about their sexuality. Finally, he was also surprised that his patients talked even when they were not being hypnotized and that the less he looked at them, touched them or tried to supervise them, the more they talked to him. In the end, there was no longer any doubt in his mind about just how important what he was doing and discovering was, which made him very unpopular with his medical colleagues.[1]

On 21 May 1894, he wrote a letter to his friend, Fliess: "…There are still a hundred gaps, large and small, in my ideas about the neuroses; but I am getting closer to a comprehensive view and to some general lines of approach. I know three mechanisms: transformation of affect (conversion, hysteria), displacement of affect (obsessions) (3) exchange of affect (anxiety neurosis and melancholia). In every case what seems to undergo these alterations is sexual excitation, but the impetus to them is not in every case something sexual…" (Freud 1915a, p. 188).

[1] Freud (1895b) wrote in this context in a letter to Fliess: "...ich bin hier ziemlich allein mit der Aufklärung der Neurosen. Sie betrachten mich so ziemlich als einen Monomanen, und ich habe die deutliche Empfindung, an eines der grossen Geheimnisse der Natur gerührt zu haben..." (This text in the original German version was not integrated in the English Standard Edition.)

Today, in our more or less emancipated world, it is not difficult to understand what was wrong with these women. What they wanted to say "could not and should not be said. That would have been an attack on male majesty" (Israël 1983, p. 244; translation by H.B.). Above all, it was particularly important for these women to come across a man who actually dared to listen to them. By doing so, the contents of her message was revealed to a hysterical female precisely at the moment she put it into words: and so psychoanalysis was born, with all its paradoxical conditions, that is, the need for the analyst to remain silent and to interpret, but also the importance of his not knowing: it is the patient who knows.

At first, Freud thought that his patients' complaints implied that someone was involved who had done something bad to them. He formulated his theory of sexual trauma as the first draft for a pathogenetical theory, devised from the hysterical discourse: hysterical patients had been overtaken by a sexual happening which they could not make sense of and which aroused emotions in them which they could not cope with. Sexual trauma implied that the patient was confronted unexpectedly and prematurely with her sexuality, as Freud described in his *Studies on Hysteria* (Freud 1895a, pp. 75–312) using the example of his patient, "Katharina": Katharina suffered from various throat complaints which the doctors could not find a reason for. Freud let her talk and discovered that her symptoms had developed after her uncle had attempted to rape her (a later footnote states that it was her father).

However, his theory of sexual trauma soon seemed insufficient to Freud. His new understanding can be seen in his letter (dated 21.09.1897) to his friend, Fliess:

> …I will confide you at once the great secret that has been slowly dawning on me in the last few months. I no longer believe in my *neurotica* (theory of neuroses)… So I will begin historically from the question of the origin of my reasons for disbelief… Then came surprise at the fact that in every case the father, not excluding my own, had to be blamed as a pervert – the realization of the unexpected frequency of hysteria, in which the same determinant is invariably established, though such a widespread extent of perversity towards children is, after all, not very probable… Then, thirdly the certain discovery that there are no indications of reality in the unconscious, so that one cannot distinguish between the truth and fiction that is cathected with affect… Fourthly, the reflection that in the most deep-going psychosis the unconscious memory does not break through, so that the secret of the childhood experiences is not betrayed even in the most confused delirium…Can it be that this doubt merely represents an episode in advance towards further knowledge?…. (Freud 1897, pp. 259–260)

Freud, several years later, saw it retrospectively in this way:

> The same clarification (which corrected the most important of my early mistakes) also made it necessary to modify my view of the mechanism of hysterical symptoms. They were no longer to be regarded as derivatives of the repressed memories of childhood experience; but between the symptoms and the childish impressions there were inserted the patient's *phantasies* (or imaginary memories), mostly produced during the years of puberty which on the one side were built up out of and over the childhood memories, and on the other side, were transformed directly into the symptoms. It was only after the introduction of this element of hysterical phantasies that the texture of the neurosis and its relations to the patient's life became intelligible. (Freud 1906, p. 274)

It would be a mistake to speak in terms of an "abandonment" of the theory of infantile seduction by Freud; we could think more precisely of a deepening of the concept of infantile sexual trauma that recognizes phantasy life, that is, psychic reality, as the protagonist as compared with the truly experienced events and the pathogenic effects that we can confirm: "Freud's disillusionment with the protagonistic role of seduction opens the way to a more complex theory of trauma that emphasizes its internal aspect, but that at the same time does not renounce the 'real' basis of the traumatic sexual situations, now seen in the form of universal and paradigmatic situations. On the other hand, ulterior analytic experience confirms the very great frequency of fully verifiable infantile seductions in analysands, hysterics as well as others" (Baranger et al. 1988).

## 3.4 The Discovery of the Phantasm

This change in Freud's interpretation of hysterical symptoms, connected with the realization that much of what his patients told him was not based on real events, but rather was a way of expressing unconscious wishes and phantasies, also marks a significant crisis in his life. In retrospect, it may be said that the discovery of the phantasm was almost certainly bound to lead to an examination of his own unconsciousness and to a deeper understanding of his own self, i.e. to self-analysis (for further details, see Gay's biography of Freud published in 1988).

The rhetoric in Freud's letter mentioned above, often called the "letter of revocation" (Grubrich-Simitis 1998), hinted at a certain, simplified interpretation: it was assumed that Freud was revoking his so-called seduction theory once and for all, that is, his understanding that hysterical symptoms could be explained by a case of sexual seduction during childhood and thus by a real trauma afflicted on the subject by the outside world. Furthermore, it was speculated that Freud's doubts about his seduction theory and his suggestion that sexual phantasies may possibly be of causal significance in the development of neuroses led him to discover the psychic conflict and the unconscious inner world and thus opened up the dimension of real psychoanalytical thinking.

In contrast to this dichotomizing view of trauma and conflict, Grubrich-Simitis (1998) was able to show that Freud's thoughts had much earlier turned to the intrapsychic mechanisms in the development of symptoms and particularly to the dynamics of unconscious conflicts. Furthermore, he had certainly not lost sight of the traumatic aspects in the aetiology, even in his later works. Grubrich-Simitis (1998) was indeed convinced that at no time did Freud understand trauma and conflict as aetiological factors excluding each other, but rather in the sense of a complex and causal supplementary sequence.

What exactly is the phantasm indicated in the letter to Fliess? The phantasm is what underlies human actions, plans, intentions and wishes. As Israël (1983, p. 249) underlines, it is "the origin of what someone can discover from his unconscious" (translation by H.B.). This also includes phantastic ideas, which may become apparent in the course of psychoanalytical treatment. Image presentations, as described by

Freud (1919), for example, in his work *A Child is Being Beaten*, recur in associations, without the subject being able to give any details about how they came about. Starting with this first image, numerous transformations can be found, leading to the centre of the phantasm, to the basic phantasm which is at the basis of all the others after they have been reduced. The subject's object is a separate part of him/her, which can therefore be considered lost. Thus, the wish, the desire ("désir"), may be understood in the context of an assumed object loss. The phantasm can be considered an organizing element of the unconscious; it is connected with someone's earliest experiences with significant others and, in the further course, structures the personality. The phantasm's elements are partly derived from the unconscious of the subject's parents. Let us imagine, for instance, a child who is confronted with expressions of motherly love: it is easy to understand why the child's question: "What does mother expect from me" "makes the phantasms grow rampantly" (Israël 1983).

Following this characterization of the phantasm, once again the question of trauma comes up. If the traumatic event involves the possible persons in the phantasm, it goes without saying "that the latter suddenly becomes obvious to the subject; thus, confronted with this discovery which is basically the discovery of the subject's own wish, he/she has no choice but to react with denial" (Israël 1983, p. 251).

Even if it may be assumed that incest is more prevalent than generally assumed, we have to agree with Freud when he declares in his letter to Fliess (dated 21 September 1897) that hysteria is such a common illness that it can certainly not be explained by "trauma". In this context, Israël summarizes what psychoanalysis in hysteria basically amounts to: what a hysterical patient is searching for is a father image which is supposed to fill the gaps left by her real father. Therefore, the psychoanalytical treatment should help the hysterical patient to stop trying to make her real father perfect and to accept that he can be imperfect with failures and weaknesses. From then on, she should also be willing to accept the same things about herself:

> That is, she will give herself permission to make use of what she has at the present time, instead of missing out on the present and focussing on a future which is full of empty hopes and keeps being pushed further away. (Israël 1983, p. 252; translation by H.B.)

## 3.5 The Topographical Model

Freud's topographical model, his concept of the preconscious and the dynamic unconscious, will be outlined in the following: so according to Freud, man's mental organization includes a memory with a practically unlimited capacity. The greater part of this is always available and retrievable. Freud calls this available memory "preconscious" (Pcs). According to him, this preconscious is unconscious but only in a "descriptive sense", i.e. it is potentially conscious, acts in accordance with the rules and principals of the conscious and is accessible to language, rational thinking and decision-making.

In contrast to this, the actual unconscious itself is only experienced by the conscious in the incomprehensible language of substitutes and symptoms and is completely different from the subconscious. Unconscious ideas and emotions are extremely virulent; somehow they find their way into our experience, but cannot at all be processed or integrated in a conscious way. It is as if an important message is being transmitted to the subject in a foreign language. Freud called these effective and unconscious forces the "dynamic unconscious":

> We have learnt from psycho-analysis that the essence of the process of repression lies, not in putting an end to, in annihilating, the idea which represents an instinct, but in preventing it from becoming conscious. When this happens we say of the idea that it is in a state of being 'unconscious', and we can produce good evidence to show that even when it is unconscious it can produce effects, even including some which finally reach consciousness. Everything that is repressed must remain unconscious; but let us state at the very outset that the repressed does not cover everything that is unconscious. The unconscious has the wider compass: the repressed is a part of the unconscious. (Freud 1913, p. 166)

The ideas, phantasies and emotions of the dynamic unconscious revolve around earlier wishes and their later derivatives which have been removed from the conscious and "repressed" (cf. Fig. 3.1).

There is strong censorship between the system ucs (unconscious) and the system cs (conscious) or pcs (preconscious), described by Freud as defence in general and repression in particular.

Freud called the countermovement to repression, which causes ideas which occurred unconsciously to become conscious in dreams, phantasies, symptoms and transference scenarios, the "recurrence of the repressed". However, when repressed

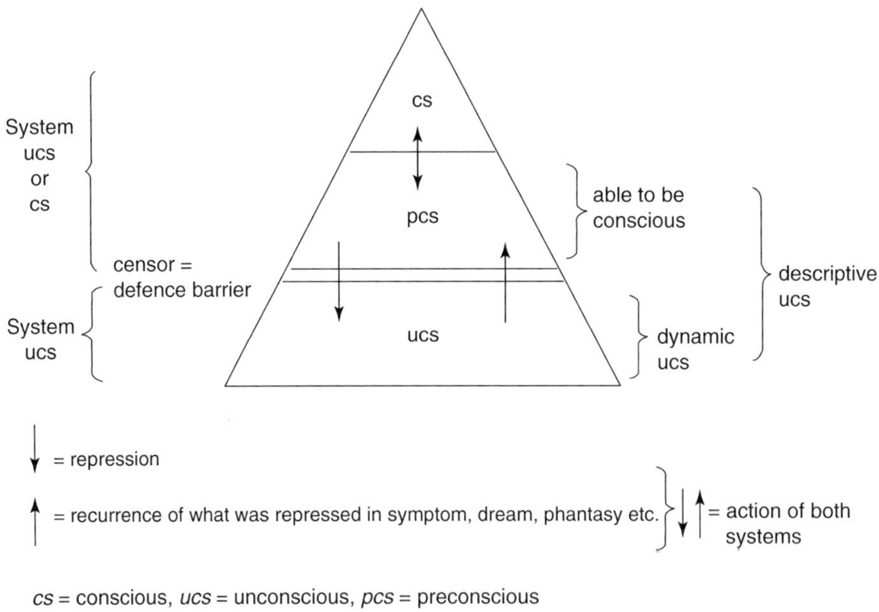


**Fig. 3.1** Freud's topographical model. *cs* conscious, *ucs* unconscious, *pcs* preconscious

ideas are made conscious, they become distorted and unrecognizable and may not arouse the same libidinous emotions they originally evoked.

Here a possible misunderstanding should be mentioned, namely, the idea that it is at all possible to actually locate the unconscious and the conscious, as understood by Freud. Freud's topographical model is not the anatomy of the psyche, and the unconscious is not some kind of mental trash can. Rather, the topographical model points to the significance of resistance to transference which is a core issue of psychoanalytic therapy. This considers human defence strategies as a way of mentally dealing with conflicts and their significance for the development of mental structures. Freud believed that the unconscious contains and protects human passions which can be neither lost nor given up, meaning, however, that they can never be lived out in real life. Therefore, mental conflicts are conflicts of passion, conflicts full of love and hate. Psychoanalysis is thus the "attempt to understand the conflicts causing suffering" (Müller-Pozzi 2002, p. 57).

When psychoanalysis creates a connection between unconscious wishes and wishes involving conflict, the question of how the unconscious came about and how it is different from consciousness gains central importance. In his essay "The unconscious" (Freud 1915a, p. 159 ff.), Freud postulated that man's mental world is made up of two fundamentally different systems involving memory and imagination, which he called *thing-presentation* and *word-presentation*. Conscious imagination consists of the presentation of the thing and its corresponding presentation of the word. Unconscious imagination is when the connection between the presentation of the thing and its corresponding representation of the word is torn apart. The systematic difference between these two systems lies in the language, i.e. a system of symbols (Lorenzer 1970). The unconscious does not speak. Therefore, repression means making speechless.

In the sense of Freud, thing-presentations are traces of memories of direct experiences, which a person stores independently of language and long before that person has learned to speak. Thing- presentations are sensual and pictorial, tangible and associative and are not rationally or logically connected to each other. They reflect the primary processing system (Freud also talks about "object-associations"); without symbolic representation, the self stays totally cut off from them and cannot access them.

Word-presentations, on the other hand, are colloquial signs which have been established by convention and which attain symbolic significance through their connection with certain thing representations:

> What we have permissibly called the conscious presentation of the object can now be split up into the presentation of the *world* and the presentation of the *thing*; the latter consists in the cathexis, if not of the direct memory-images of the thing, at least of remoter memory-traces derived from these. We now seem to know all at once what the difference is between a conscious and an unconscious presentation (…). The two are not, as we supposed, different registrations of the same content in different psychical localities, nor yet different functional states of cathexis in the same locality; but the conscious presentation comprises the presentation of the thing plus the presentation of the word belonging to it, while the unconscious presentation is the presentation of the thing alone. The system *Ucs.* contains the thing-

> cachexes of the objects, the first and true object-cathexes; the system *Pcs.* Comes about by this thing-presentation being hypercathected through being linked with the word-presentations corresponding to it. It is these hypercatexes, we may suppose, that bring about a higher psychical organization and make it possible for the primary process to be succeeded by the secondary process which is dominant in the *Pcs.* (Freud 1915a, pp. 201–202)

Language specifically enables humans to store their subjective experiences in their memory not just in pictures, and to reproduce them in their behaviour, but also to create from these subjective experiences an inner space of symbolic representations and to present it in an intersubjective "objective" code (Müller-Pozzi 2002).

An idea becomes unconscious if a word-presentation is separated from the thing-presentation or if they are not connected at all in the first place. Freud described the latter as "primal repression". From the perspective of making the unconscious conscious, interpretation in the analytic situation focuses on restoring the torn connection between word- and thing-presentation.

In contrast to this, a hallucination is a form of consciousness which is not dependent on word-presentations. In a hallucination, when the trace of a memory is activated, it does not follow the path of word-presentations and word-associations but that of "object-associations". From a psychoanalytical perspective, it is a mode of experience and functioning of the psyche, in which self and object, inside and outside, phantasy and reality and ideas and emotions cannot be distinguished.

The unconscious mode of functioning was described by Freud as a primary process and that of the preconscious and conscious as secondary processes.

**Significant Features of Unconscious Processing in a Psychoanalytical Perspective**

- The dynamic unconscious follows the primary process: unconscious ideas and emotions are connected by means of compression and displacement.
- The unconscious lacks discursive logic: in the unconscious there are no contradictions, wishes or ideas which are incompatible for the conscious; they exist alongside each other in the unconscious and have no influence on each other.
- Abstractions have no place in the unconscious: the unconscious is pictorial, concrete and "sensual".
- In the unconscious there is no concept of time, space and causality: the unconscious is timeless.
- Unconscious reality is psychic reality alone: unconscious wishes and impulses follow the pleasure principal.
- The primary process is closely connected to "body language" and "organ language": unconscious processes have a plastic influence on body processes, e.g. in conversion symptoms and somatization (Müller-Pozzi 2002, p. 65 ff.).

## 3.6 Freud's Second Topography and Concepts of Instincts

In the development of the psychoanalytical theory, the transition from the topographical model of mental processing to the so-called structural model played a major role in the new understanding of unconscious mental processes. With the structural model's broader perspective, psychic conflict takes on a special form without detracting from the significance of the topographical perspective (Sandler and Freud 1985). However, with the topographical model, it was not possible to conceptualize adequately the active side of internal conflict processing, as represented, for instance, in transference resistance. Freud finally came to believe that not only what is repressed but also the act of repression itself is unconscious. Thus, the dynamic unconscious could no longer be considered equivalent to the system ucs; the unconscious ego was part of the dynamic unconscious as well. In his "New series of lectures on an introduction to psychoanalysis", Freud wrote (Freud 1933, p. 71 ff.):

> You will not expect me to have much to tell you that are new about the id apart from its new name. It is the dark, inaccessible part of our personality; what little we know of it we have learnt from our study of the dream-work and of the construction of neurotic symptoms, and most of that is of a negative character and can be described only as a contrast to the ego. We approach the id with analogies: we call it a chaos, a cauldron full of seething excitations. We picture it as being open at its end to somatic influences, and as there taking up into itself instinctual needs which find their psychical expression in it, but we cannot say in what substratum. It is filled with energy reaching it from the instincts, but it has no organization, produces no collective will, but only a striving to bring about the satisfaction of the instinctual needs subject to the observance of the pleasure principle. The logical laws of thought do not apply in the id, and this is true above all of the law of contradiction. Contrary impulses exist side by side, without cancelling each other out or diminishing each other: at the most they may converge to form compromises under the dominating economic pressure towards the discharge of energy. There is nothing in the id that could be compared with negation; and we perceive with surprise an exception to the philosophical theorem that space and time are necessary forms of our mental acts.
>
> There is nothing in the id that corresponds to the idea of time; there is no recognition of the passage of time, and - a thing that is most remarkable and awaits consideration in philosophical thought - no alteration in its mental processes is produced by the passage of time. Wishful impulses which have never passed beyond the id, but impressions, too, which have been sunk into the id by repression, are virtually immortal; after the passage of decades they behave as though they had just occurred. They can only be recognized as belonging to the past, can only lose their importance and be deprived of their cathexis of energy, when they have been made conscious by the work of analysis, and it is on this that the therapeutic effect of analytic treatment rests to no small extent.

Furthermore, it should be considered that deprivation creates conflict. The conflict leading to neurosis is never the direct "external" conflict with a person who is loved, but rejects this love. In order to protect the relationship which the child is absolutely dependent on, it internalizes the conflict and unconsciously takes over the responsibility for it. It identifies with the rejective aspects of the significant other person. Freud conceptualized the internalization of these rejective aspects of the relationship in the super-ego. Id, ego and super-ego thus became three parts of the psychic apparatus. This new perspective later came to be called the second topography, the tripartite or structural model (see Fig. 3.2). Freud himself merely referred to the "structural conditions of the psychic apparatus" (Freud 1933, p. 85).

Some further approaches at tackling the problem of the unconscious in the era of post-Freudian psychoanalysis will briefly be outlined in the following. For example, Sandler and Sandler (1994) attempted to link the psychoanalytical theory of the

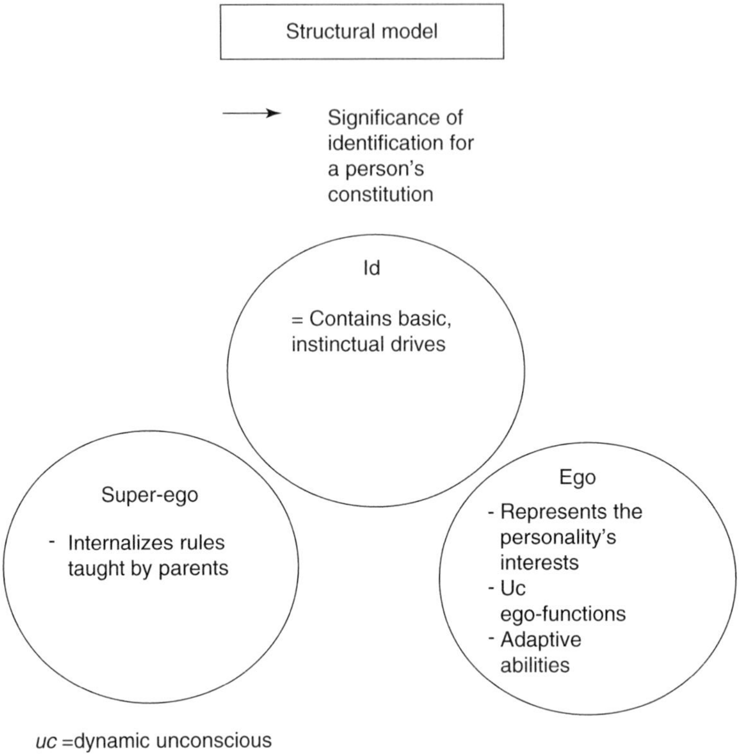


**Fig. 3.2** Freud's second topography: the structural model. *uc* dynamic unconscious

unconscious with clinical experience. It was particularly their concept of the present and past unconscious that made this endeavour easier, namely, they connected the present unconscious with the unconscious ego and the past unconscious with the early origins of psychic development, i.e. with the infantile unconscious. The present unconscious can be seen as coming from the current interaction in the analytical relationship. In contrast, the past unconscious is only accessible via psychoanalytical intervention and reconstruction.

According to Klein's theory, the psychoanalytical perspective of understanding focuses on early oral-introjective interactions between mother and child. Early development has been understood in the context of psychic survival and the integration of ambivalent relationship wishes (Stein 1999). Klein's idea of the schizo-paranoid and depressive position may be understood as a widely diversified dual model of anxiety and guilt.

In Freud's structural model, final psychic integration is achieved by means of identification and internalization after the demise of the oedipal complex. The ego develops a feeling of signal anxiety, in order to control the unconscious mental process on a new level of autonomy (Freud 1926).

Freud continued to elaborate his second topology and to deepen his study of the sources of anxiety. "Inhibitions, symptoms and anxiety" (Freud 1926) mark the final restructuring of the concept of trauma as related to anxiety and the definitive replacement of this concept for that of traumatic situation, which includes both the interaction of internal and external situations and also the interstructural nature of all traumatic situations.

In Freud's view, but also later on, the perceptory-sensory aspect took on great importance for a deeper understanding of elementary drive phenomena. This is not just about feelings to do with pleasure and unpleasure in the unconscious, but it is also about perception and emotional reactions connected with imaginary ideas generated by instinct. Freud described this kind of perception which is never registered on a conscious level, as id-perception:

> The id, cut off from the external world, has a world of perception of its own. It detects with extraordinary acuteness certain changes in its interior, especially oscillations in the tension of its instinctual needs, and these changes become conscious as feelings in the pleasure-unpleasure series. It is hard to say, to be sure, by what means and with the help of what sensory terminal organs these perceptions come about. But it is an established fact that self-perceptions—coenaesthetic feelings and feelings of pleasure-unpleasure—govern the passage of events in the id with despotic force. The id obeys the inexorable pleasure principle. But not the id alone. It seems that the activity of the other psychical agencies too is able only to modify the pleasure principle but not to nullify it; and it remains a question of the highest theoretical importance, and one that has not yet been answered, when and how it is ever possible for the pleasure principle to be overcome. The consideration that the pleasure principle demands a reduction, at bottom the extinction perhaps, of the tensions of instinctual needs (that is, *Nirvana*) leads to the still unassessed relations between the pleasure principle and the two primal forces, Eros and the death instinct.
>
> The other agency of the mind, which we believe we know best and in which we recognize ourselves most easily—what is known as the *ego*—has been developed out of the id's cortical layer, which, through being adapted to the reception and exclusion of stimuli, is in direct contact with the external world (*reality*). (Freud 1937–1939, p. 197)

The dynamic unconscious cannot be considered out of the context of instinct, and vice versa, the concept of instinct cannot be properly understood without insight into the way unconscious processes work (see Müller-Pozzi 2002, p. 72 ff. for details on the concept of instinct in psychoanalysis). Wanting to identify with a significant other person (the object), and at the same time wanting to have a relationship with them, marks a fundamental dualism in the subject's development. Here, the relationship with the primary person represents the social core as a basis for the emotional interaction with the significant other person, around which the child builds its own inner world. The core of such primary relationships and their significance for future psychic development is conceptualized by means of the concepts of instinct (libido and aggression) or wish, the concept of object relationships (and their internalization) and the concept of the self. Specific human development starts with the ability to put off satisfying a desire. It is this decoupling of a desire from its fulfilment which creates a mental space, that is, space in which mental experience including ideas, phantasies and emotions can develop.

Freud (1915b) defined the psychoanalytic concept of instincts as a "concept on the frontier between the mental and the somatic":

> If now we apply ourselves to considering mental life from a *biological* point of view, an 'instinct' appears to us as a concept on the frontier between the mental and the somatic, as the psychical representative of the stimuli originating from within the organism and reaching the mind, as a measure of the demand made upon the mind for work in consequence of its connection with the body. (Freud 1915b, pp. 121–122)

According to Freud, instinct originates from its own physical momentum; this determines the dynamic-emotional quality of the wish and connects the wish with sensuality. The wish's dynamism only becomes accessible psychoanalytically on the level of experience when it is felt as an emotional urge and can be satisfied through the relationship with the object. In this context, "object relationship" describes the interaction with the significant other person from the perspective of the subject's experience.

With the focus on instinct as a phenomenon somewhere between the somatic and the mental, it may be said that the object of psychoanalysis is the subject's sensual experience and the way instinct reveals itself consciously and unconsciously. Listen to what Freud has to say:

> Psycho-analysts never forget that the mental is based on the organic, although their work can only carry them as far as this basis and not beyond it. (Freud 1910, p. 217)

Then he continues:

> An instinct can never become an object of consciousness – only the idea that represents the instinct can, Even in the unconscious, moreover, an instinct cannot be represented otherwise than by an idea. If the instinct did not itself attach to an idea or manifest itself as an affective state, we could know nothing about it. (Freud 1914a, p. 177)

A wish is thus the mental representation of the instinct. Accordingly, modern psychoanalysis interprets the theory of instinct as a concept of wish and as theory of emotion.

Every experience, from when the wish first comes about, right until it is fulfilled through the object, with all the mental processes which happen in between, is deposited as a trace of memory and so ultimately contributes to the generation of mental structure.

Freud explained the intensity, that is, the quantitative momentum of the drive using the metaphor of energy, and originally described its quantity as an "amount of emotion" or the "sum of excitation". From today's point of view, the energy metaphor and the energy model of excitation and removal do not adequately conceptualize clinical observations. It should, however, be stressed that this economic view, i.e. viewing mental processes from the perspective of the "energy balance", was the basis of the later development of a more comprehensive theory of affect. Only after internalization and cathexis of the representation of the significant other person can intense feelings be experienced. On the other hand, any emotion will become traumatic and will have to be warded off once a certain intensity is exceeded and if it cannot be integrated into an object relationship.

With the dogma of representation—although it seems rather mechanistic in its terminology—psychoanalysis points to the fact that the subject's inner image of the object is not a photographic reproduction but an inner "phantasmatic" creation determined by the subject's wishes, attitudes and phantasies about the longing for the other person. Freud (1926, p. 171) described the qualitatively differentiated affective aim also as "high and unsatisfiable cathexis of longing which is concentrated on the object…".

## 3.7 The Mysterious Message of the Other and Its Consequences for the Concept of the Unconscious

How does contemporary psychoanalysis understand the development of the repressed unconscious?

Laplanche (2004) describes in this context the "fundamental anthropological situation": this consists of the relationship between adult and infant, an adult with an unconscious, that is, a sexual unconscious mainly made up of infantile remains, and a child without hormones to trigger sexuality, and originally without sexual phantasies. What Freud originally described as "self-maintenance", based on empirical results from modern research on babies and infants, is today described as "attachment" with all its implications. On a clearly instinctual, genetic basis, a dialogue between adult and infant develops at a very early stage, practically right from the start. The former theory of "symbiosis" has vanished due to the observation of differentiated, early relationships that were mutual right from the start, in which the non-ego is always discriminated from what belongs to the person himself/herself. In this context Laplanche points out a flaw in modern attachment and affect theories, namely, the failure to consider asymmetry on the sexual level. It should also be taken into consideration that right from the start the dialogue between adult and infant is "disturbed by something else". The message is not clear. From the adult's side, interference from the unconscious takes place unilaterally. We could even say: "from the adult's infantile unconscious …" (Laplanche 2004, p. 900; translation by H.B.).

The other person's "mysterious message" is thus a message which has been interfered with by the unconscious and has been "developed through compromise". Laplanche goes on to explain that what counts in the end in this situation is what the recipient makes of it, namely, the attempt at translation and its inevitable failure. Given the fundamental difference between sexual instinct in childhood and that in adolescence, it should be assumed that during adolescence, sexual instinct catches up with the distinct which originated intersubjectively and has developed independently over the years; thus, a great problem concerning the coherence and connection of their contents arises. The adult message in its contradictory totality cannot be comprehended in the original communication between adult and infant:

> In the typical model of breast feeding, e.g. love and hate, fulfillment and arousal, milk and breast, the 'containing' and the sexually aroused breast, etc. become intermingled …. (Laplanche 2004, p. 902; translation by H.B.)

The translation of this mysterious adult message does not happen at once but in two steps and by reverting to schemes from the cultural surroundings.

Freud commented "The interest of psychoanalysis from a developmental point of view":

> The content of the *Ucs.* may be compared with an aboriginal population in the mind. If inherited mental formations exist in the human being—something analogous to instinct in animals - these constitute the nucleus of the *Ucs.* Later there is added to them what is discarded during childhood development as unserviceable; and this need not differ in its nature from what is inherited. A sharp and final division between the content of the two systems does not, as a rule, take place till puberty. (Freud 1914a, b, c, p. 194)

In Laplanche's view, the translation of the adult's mysterious message is always incomplete, leaving behind *remains*. Being the opposite of the subconscious ego, these remains form the unconscious in the Freudian sense of the word.

Freud's model of the psychical apparatus ("Seelenapparat") is a neurotic-normal model. However, what about the great number of cases, e.g. in psychiatric-psychotherapeutic practice, which do not fit into this model (psychoses, borderline cases, severe personality disorders, etc.)? Here, many psychoanalysts have laid Freud's concept, which was based on the repression of the unconscious, to rest and have developed new models. These models are mostly desexualized and have little to do with the concept of the unconscious. Laplanche (2004) compares these models with a "general seduction theory", enabling one standard overview of the supposedly separate models, the neurotic-normal and the psychoses-borderline model. Starting from the basic anthropological situation, it is assumed that every person has a certain amount of "untranslated messages". This "trapped unconscious" is at a standstill, but at the same time there is expectation, that is, expectation that these untranslated messages can possibly be translated. Remember that Freud had already spoken of two mechanisms working together in one and the same individual: the neurotic mechanisms of repression and the psychotic/perverse of denial. The vertical boundary of division in relation to the horizontal barrier of repression does not cause conflict, but marks—as with Freud—the separation of two different defence processes. This barrier can be overcome, e.g. if a new translation process is started—in the course of psychotherapy.

In the case of repression (especially primaeval repression), the other person's messages are initially encoded into the "trapped unconscious" or preconscious. Then they can later be taken up again, translated and consequently divided between subconscious translation and unconscious remains.

## 3.8 Unconscious Phantasy: Psychoanalysis and "Embodied Cognitive Science"

Today, the concept of the unconscious remains a central feature of modern psychoanalysis. It has opened the way for a constructive dialogue between the various streams in modern psychoanalysis (e.g. psychoanalytic structural theory, Kleinian and object relation theoretical approaches) and also for a dialogue with contemporary science of cognition and neuroscience. Decoding unconscious traces in physical reactions can open the door to unconscious phantasies about prelingual experiences (Leuzinger-Bohleber 2004). Communication between the analyst's/therapist's unconscious and that of the patient is mostly mediated by projective and introjective identification processes. The concept of "embodied cognitive science" (Edelman 1987, 1989, 1992) or "embodied remembering" can then raise the awareness for unconscious traces in the patients' physical reactions. The concept of embodiments suggests that early, preverbal experiences play a decisive role in the unconscious throughout someone's whole life. Memory and recall are functions of the entire organism, i.e. functions of a complex, dynamic, re-categorizing and interactive process which is always "embodied", that is, depending on sensory motor processes manifest in the whole organism. To recall something in an "embodied" way does not simply mean that it takes place "non-verbally", and neither is it synonymous with the "descriptive unconscious", but rather it is to a high degree constructive, dynamic and historically determined. For example, if certain experiences, such as expressing anger and aggression towards a significant person who is depressive (particularly the mother), are not permitted, they contribute to feelings of resignation when interacting with the significant other, that is, feeling that nothing can be achieved and that there can be no emotional resonance with the other person. These experiences are naturally reflected in the neuronal network and unconsciously determine perception and information processing in later situations. For instance, it may be assumed that the same experiences later harden such early relationship experience patterns (e.g. interactionally conveying that it is "forbidden" to show strong emotions like love and/or anger towards the mother), because the experiences are constantly being transcribed, resulting in a more active way in the banishing of the corresponding wishes, emotions and phantasies from the conscious. Interactionally communicating that it is forbidden for someone to show significant others their own desires and intense emotions may best be worked on with a psychotherapist/psychoanalyst (as a "thematic organizational point").

Edelman's central hypothesis that the neuronal network is from the beginning developed through the interaction between genetically determined biological factors on the one hand and environmental influences on the other hand fits in well with the psychoanalytical concept of the dynamic unconscious. The continual modification of early body experiences through later experiences corresponds with Freud's concept of "afterwardsness". Here, Freud's famous words are recalled: the ego is originally somatic. This also means that historical truths can never be reconstructed in the sense of "one-to-one observation", and neither should they be dismissed as the basis of later adaptations.

## 3.9 Beginnings of the Dialogue Between Psychoanalysis and Neuroscience

In his essay *Project for a Scientific Psychology*, published in 1896, Freud attempted to connect psychoanalysis with the neuroscience of his day. Finally, he was forced to give up this endeavour because of the lack of diagnostic possibilities and empirical facts. It was only with the development of modern neuroscience with its diagnostic-technical repertoire and the possibility to gain insight into neuronal processing of mental processes such as emotional-cognitive interaction that it once again became possible to look for connections between what was known in both disciplines. Freud's paradigmatic change to psychoanalysis came about through his work on the mind-body problem and its central question of how the brain is able to generate subjective experience (consciousness) by means of existing anatomical structures and physiological functions. For 20 years, Freud had already been considering neurophysiological questions about the nervous system's structure and functions when he was finally confronted with the central problem in neuropsychology at that time, the cerebral localization of language and a possible explanation for aphasia (Kaplan-Solms and Solms 2000, 2003). Ultimately, Freud tackled one of the greatest challenges in medicine of his time, the physiologically unexplainable symptoms of neuroses, particularly of hysteria. Between 1895 and 1900, his interest in both of these problems made him give up neuroscientific methods in favour of psychoanalytic ones.

Before Freud's ambivalence and desire to find a scientific explanation for psychoanalysis can be discussed, the historical roots and forerunners of neuropsychoanalysis should first be considered. When Freud completed his training, neurology was still a very new field, based mainly on a specific scientific method, namely, the method of *clinical-anatomical correlation*. At that time, this method was mostly taught by physicians from the field of internal medicine. Considering clinical experience, answers were sought to the question of how certain clinical lifetime abnormalities correlate with certain pathological autopsy results. Thus, pathognomonic symptom constellations were discovered, making it possible to almost exactly predict the localization of the underlying disorder and to initiate appropriate treatment (*concept of clinical syndromes*). Neurology finally became a specialized field of internal medicine in its own right. When Freud began his training in clinical neurology in 1980, he set out to learn the concept of clinical syndromes, based on the clinical-anatomical correlation method in diagnosis and treatment planning.

### Clinical-Anatomical Correlation Methods

The clinical-anatomical correlation method in neurology differed from that in internal medicine in one very important aspect: defects in the brain contribute not only to localized pathophysiological changes but also have an immediate and direct effect on a person's mind and possibly also on a person's personality. In this context, the classic case of Phineas Gage is often quoted: an iron bar penetrated this railroad worker's left frontal lobe. The resulting effects and personality changes after this accident were described by Harlow (1868) as follows:

> ...the subject of it was Phin. P. Gage, a perfectly healthy, strong and active young man, 25 years of age, ..., five feet six inches in height, average weight 150 pounds, possessing an iron will as well as an iron frame; muscular system unusually well developed – having had scarcely a day's illness from his childhood to the date of his injury ...

> ... his physical health is good, and I am inclined to say that he has recovered. Has no pain in head, but says it has a queer feeling which he is not able to describe. Applied for his situation as foreman, but is undecided whether to work or travel. His contractors, who regarded him as the most efficient and capable foreman in their employ previous to this injury, considered the change in his mind so marked that they could not give him his place again. The equilibrium or balance, so to speak, between his intellectual faculties and animal propensities, seems to have been destroyed. He is fitful, irreverent, indulging at times in the grossest profanity (which was not previously his custom), manifesting but little deference for his fellows, impatient of restraint or advice when it conflicts with his desires, at times obstinate, and oscillating, devising many plans of future operation, which are no sooner arranged than they are abandoned in turn for others appearing more feasible. A child in his intellectual capacity and manifestations, he has the animal passions of a strong man. Previous to his injury, though untrained in the schools, he possessed a well-balanced mind, and was looked upon by those who knew him as a shrewd, smart businessman, very energetic and persistent in executing all his plans of operation. In this regard, his mind was radically changed, so decidedly that his plans and acquaintances said he was "no longer Gage". ...

This example of a personality change induced by brain injury underlined that damage in certain brain regions cause specific mental changes. Accordingly, there came to be a fundamentally new understanding of the clinical-anatomical method in neurology, namely, as a *localization of mental functions*.

In the early 1860s, the French neurologist, Paul Brugger, demonstrated that damage in the brain region known today as the "Brugger region" induced a specific syndrome characterized by loss of speech although the speech organs had not been impaired. Thus, with the aid of clinical-anatomical observational methods, a mental function—symbolic vocalization—was identified for the first time.

Not long after this, the German neurologist, Carl Wernicke, showed that damage in another part of the brain known as the "Wernicke region" leads to the inability to understand what someone is saying, even though hearing ability is intact (Wernicke aphasia). Hence, the brain region responsible for understanding speech was localized. Further clinical-anatomical correlations concerning fundamental mental functions such as reading, writing, fine motor activity and object recognition finally formed the basis for what was to become a specialist field of neuroscience, namely, behavioural neurology.

As seen from his writings during this period, Freud was familiar with the methods of and differences between the then prevalent German and French school of classical neurology. The main issue in classical German neurology was developing anatomical and physiological theories in the Helmholz tradition ("The only forces operating in an organism are the common physical and chemical ones", Du Bois-Reymond 1842, 1927, translation by H.B.). The main issue in the French school of neurology, however, was identifying, classifying and describing different clinical pictures. Freud remarked that Charcot never got tired of defending the importance of purely clinical work, that is, viewing it and putting it into order, against the assaults from theoretical medicine. Freud became more and more impressed by Charcot and wrote "... la théorie, c'est. bon, mais ca n'empêche pas d'exister". If one only knew what exists (Freud 1893, p. 139).

In spite of the, for the most part, complementary nature of the German and French school of classical neurology, there was a group of disorders which particularly underlined the conflicting views of the two schools: the group of neuroses, in particular hysteria and neurasthenia. After autopsy, no injuries to the nervous system could be found to account for the clinical symptoms observed while the person was still alive. While Charcot continued to describe the pathognomonic symptoms of hysteria and neurasthenia, the problem of the lack of evidence for anatomical lesions in these disorders proved to be almost unsolvable for the German school of neurology. Meynert published a psychiatric textbook entitled *Psychiatry: A Clinical Treatise on Diseases of the Fore-Brain* and wrote in his preface: "The historical name, psychiatry, as 'treatment of the mind', promises something that simply cannot be achieved and takes off leaving natural science behind" (Meynert 1884) (translated by H.B.).

Since Freud was particularly interested in the complex field of neurotic disorders, his growing preoccupation with Charcot while studying at the Salpetrière (1885–1886) can be well understood. However, in view of this change in direction, it should be stressed that this did not amount to a transfer from neurology to psychology. Freud did not regard neuroses as non-physical, that is, mental disorders, but rather as non-anatomical or physiological disorders that cannot be localized (Kaplan-Solms and Solms 2000, 2003). On this subject, Freud wrote: "Neurasthenia is not an illness in the sense of the text books exclusively based on pathological anatomy, but should rather be described as a way that the nervous system reacts" (Freud 1887, p. 65 ff., translated by H.B.). On the subject of hysteria, he added: "In hysteria, physiological modifications of the nervous system are involved, and its nature could be described by a formula taking into account the excitability conditions of the different parts of the nervous system" (Freud 1888, p. 72; translation by H.B.).

Freud's criticism of classical localization methods and the development of a functional mechanism and ultimately of a structure-based model is already hinted at in these quotations from Freud. This development was started through discussions with the English neurologist, Jackson. Jackson rejected the idea that complex mental abilities could be limited to localized brain regions. Freud understood hysterical paralysis as a mental disorder, describing it in functional terms, and he explained its physiological correlates as "associations" existing between the anatomical elements of the nervous system: "… by way of summary, we may say that hysteria is an anomaly of the nervous system which is based on a different distribution of excitations, probably accompanied by a surplus of stimuli in the organ of the mind. Its symptomatology shows that this surplus is distributed by means of conscious or unconscious ideas. Anything that alters the distribution of the excitations in the nervous system may cure hysterical disorders: such effects are in part of a physical and in part of a directly psychic nature" (Freud 1888).

While considering the subject of aphasia, Freud decided to describe mental syndromes using their own psychological terms. He considered the classical clinical-anatomical method which showed mental functions as a mosaic of regions in the brain's hemispheres as completely unsuitable for bringing the most important features of mental activity in line with one another:

> I will add the further comment that the psychical topography that I have developed here has nothing to do with the anatomy of the brain, and actually only touches it at one point. What is unsatisfactory in this picture—and I am aware of it as clearly as anyone—is due to our complete ignorance of the *dynamic* nature of the mental processes. We tell ourselves that what distinguishes a conscious idea from a preconscious one, and the latter from an unconscious one, can only be a modification, or perhaps a different distribution, of psychical energy. We talk of cathexes and hypercathexes, but beyond this we are without any knowledge on the subject or even any starting-point for a serviceable working hypothesis. Of the phenomenon of consciousness we can at least say that it was originally attached to perception. All sensations which originate from the perception of painful, tactile, auditory or visual stimuli are what are most readily conscious. Thought-processes, and whatever may be analogous to them in the id, are in themselves unconscious and obtain access to consciousness by becoming linked to the mnemic residues of visual and auditory perceptions along the path of the function of speech. (Freud 1939, p. 96)

It was not until 1893–1900 that Freud began to apply the dynamic and development-related principles of the concept which he had initially developed while investigating aphasia and various movement disorders, to psychopathology as well. This is how psychoanalysis came to be a new field of science: the interpretation of dreams ultimately marked the crossroads between psychoanalysis and neuroscience.

In his further research, Freud described the psychical apparatus, which he used when trying to understand his clinical observations, as a provisional design and as a system of functional relationships somehow reflected in the brain tissue. Therefore, he insisted that “We are justified, in my view, in giving free rein to our speculations so long as we retain the coolness of our judgement and do not mistake the scaffolding for the building. And since at our first approach to something unknown all that we need is the assistance of provisional ideas, I shall give preference in the first instance to hypotheses of the crudest and most concrete description” (Freud 1900a, p. 535).

Referring to the spatio-temporal organization of the mental apparatus Freud continued:

> Accordingly, we will picture the mental apparatus as a compound instrument, to the components of which we will give the name of ‘agencies’, or (for the sake of greater clarity) ‘systems’. It is to be anticipated, in the next place, that these systems may perhaps stand in a regular spatial relation to one another, in the same kind of way in which the various systems of lenses in a telescope are arranged behind one another. Strictly speaking, there is no need for the hypothesis that the psychical systems are actually arranged in a *spatial* order. It would be sufficient if a fixed order were established by the fact that in a given psychical process the excitation passes through the systems in a particular *temporal* sequence. In other processes the sequence may perhaps be a different one; that is a possibility that we shall leave open. For the sake of brevity we will in future speak of the components of the apparatus as ‘ψ-systems’.
>
> The first thing that strikes us is that this apparatus, compounded of ψ-systems, has a sense or direction. All our psychical activity starts from stimuli (whether internal or external) and ends in innervations. Accordingly, we shall ascribe a sensory and a motor end to the apparatus. At the sensory end there lies a system which receives perceptions; at the motor end there lies another, which opens the gateway to motor activity. Psychical processes advance in general from the perceptual end to the motor end. Thus the most general schematic picture of the psychical apparatus may be represented thus. (Freud 1900a, p. 536)

In many of his writings, Freud underlined that psychoanalysis and neuroscience would sooner or later have to be brought together: "… we must recollect that all our provisional ideas in psychology will presumably someday be based on an organic substructure" (Freud 1914c, p. 78). He put the shortcomings in his metapsychology largely down to not enough being known about the energetical procedures:

> On the other hand it should be made quite clear that the uncertainty of our speculation has been greatly increased by the necessity for borrowing from the science of biology. Biology is truly a land of unlimited possibilities. We may expect it to give us the most surprising information and we cannot guess what answers it will return in a few dozen years to the questions we have put to it. They may be of a kind which will blow away the whole of our artificial structure of hypotheses. If so, it may be asked why I have embarked upon such a line of thought as the present one, and in particular why I have decided to make it public. Well—I cannot deny that some of the analogies, correlations and connections which it contains seemed to me to deserve consideration. (Freud 1920, p. 59)

Freud considered these two steps to be key requirements if future neuroscience was to be able to explain the neuronal organization of mental processes: firstly, complete psychological analysis aiming at understanding the internal structures of the relevant functional systems, irrespective of their cerebral organization, and secondly, valid identification of the brain correlates of the psychic process concerned.

## 3.10 Lurija's Neurodynamic Approach

Lurija's dynamic neuropsychology is a method enabling the combination of psychoanalysis and neuroscience and is at the same time in accord with Freud's central hypotheses mentioned above (Kaplan-Solms and Solms 2000, 2003). The principles of Lurija's approach can be summarized as follows:

1. Psychological analysis should be a priority in mental disorders (depending on the aetiology).
2. Methods of the individual case report, emphasizing the qualitative-descriptive examination (particularly syndrome analysis).
3. Appreciation of the dynamic nature of mental experience (both pathological and normal mental phenomena should be considered the result of functional interaction between elementary components of the psychic apparatus).
4. The developmental-dynamic and hierarchical model of the psychic apparatus (conceptualized as a functional system).

In his search for a psychology which could connect descriptive (idiographic) and explanatory (nomothetic) science, Lurija felt at an early stage an affinity to psychoanalysis (Lurija 1979, pp. 21–23). His varying attitude to psychoanalysis, clearly due to political and ideological pressure, will be discussed elsewhere (Kaplan-Solms and Solms 2000, 2003, p. 33 ff.).

Lurija thought that following brain damage, mental functions are not simply "lost" but dynamically distorted in many different ways (Lurija 1947, 1970). He pointed out that every system serves a psychic function. Even though this is largely in agreement with Freud's aphasia concept, there is one noteworthy difference between the two models: whereas Freud has assumed that only primary sensory motor functions situated on the periphery of the speech apparatus could be localized, Lurija was convinced that, taking into consideration the dynamic nature of the complex, psychological process of speech, every stage could be localized, including those generated by structures situated further down. Lurija's work was based on a differentiated psychology of speech and particularly on a modification of the classical clinical-anatomical method. His "dynamic localization" method comprises two key steps: firstly, qualification of the symptom (including careful psychological analysis of the defects) and, secondly, analysis of the syndrome. The latter focuses on a thorough description of the symptom complex or the behavioural changes resulting from local brain damage (Lurija 1973, p. 38). In a further step, the question of what effect damage in various brain regions has on functional systems is addressed. Step by step, the diverse fundamental factors making up the functional system are identified, and the various functions of the individual brain regions are defined at the same time. Like this, the components of each functional system can be defined and localized in the brain cortex. The "dynamic localization" does not focus on the localization of the function as such but on the localization of the subcomponents of the psychic apparatus supporting the function.

Lurija warned about being tempted "to localize the mental process directly in the cortex". Rather, he said it was a question of "the particular effects of mental activity resulting from differently localized brain damage, and what factors are involved in making up the structure and complex forms of mental activity in each brain system" (Lurija 1973, p. 42).

Charcot's clinical-descriptive method (syndrome analysis) and Lurija's method of dynamic localization became important fundaments of Kaplan-Solms and Solms' later neuro-psychoanalytical theory (Kaplan-Solms and Solms 2000, 2003). With this approach, both authors consider the requirements necessary to bring psychoanalysis and neuroscience back together again fulfilled. For this, they first tackled the problem of the neuronal organization of dreaming. They specified the varying severity of dream disorders by identifying the damaged brain regions and recording their effects on dreaming. These dream disorders subsequently underwent a detailed psychological analysis. This study on dreaming ability in patients with focal brain damage resulted in the observation of the following six syndromes:

1. Loss of dreaming ability (damage in the left parietal lobe)
2. Loss of dreaming ability (damage in the right parietal lobe)
3. Loss of dreaming (damage in the bifrontal region located further down)
4. Non-visual dreaming (damage in the occipital and temporal lobe)
5. Blurring of dream and reality (damage in the frontal limbic system)
6. Recurring nightmares (temporal lobe seizures)

## 3.11 Kaplan-Solms and Solms' Neuroanatomical Methods

Kaplan-Solms and Solms (2000, 2003) describe the neuro-psychodynamic structure of dreaming. They point out that no function associated with dreaming could be localized in one of the regions involved, but rather the dynamic process develops as an interaction between different subcomponents of the whole functional system. Here, they refer to Lurija's conclusions that the process of dreaming extends over a functional system consisting of six basic subcomponents (cf. Sect. 3.10). Lurija assumed that these processes which are regulated from deeper lying ventromedial frontal brain regions form the end of the path which initiated the dream process (the "dreamwork" in the sense of Freud 1900b).

Kaplan-Solms and Solms (2000, 2003) describe dreaming as a "motivated process" induced by the same forces that regulate thinking and spontaneous behaviour in the waking state. According to them, dreaming only occurs when a stimulus arouses motivational interest during sleep. The question of why it is particularly the white matter in the ventromedial frontal lobe that forms the basis of this motivational factor can be answered by referring to the neuronal networks and neurotransmitter systems which are involved: the white matter in the ventromedial frontal lobe largely consists of fibres connecting the ventral midbrain nuclei with the limbic system (the gyrus cinguli and nucleus accumbens) and the frontal cortex. Moreover, these fibres form part of the ascending dopaminergic system, which is a component of the brain's "seeking systems and anticipatory states" (Panksepp 1985, 1998). These systems induce the organism's goal-oriented behaviour and appetence interactions with the environment.

As shown in the example of dreams, Kaplan-Solms and Solms (2000, 2003) view Lurija's method of syndrome analysis as the appropriate platform for psychoanalysis to meet neuroscience. The authors hope that this approach would make it possible to integrate psychoanalysis into all areas of neuroscience, that is, not only to connect psychoanalysis with anatomy but also with chemistry and molecular biology. In a first step, personality, motivational and emotional changes in patients with various forms of brain damage who were treated with psychoanalysis or psychoanalytical psychotherapy were investigated. Their clinical observations led Kaplan-Solms and Solms to the conclusion that the right perisylvic convexity is a key component in the neuroanatomical substrate of the representation of the whole object and is therefore significant as a neurophysiological pathway for object cathexis. If this part of the brain is damaged, the result is the inability to develop a relationship with objects in a mature and adequate way. Symbolic word representations are encoded in the left perisylvic cortex; narcissistic object representations therefore largely depend on physiological modifications in the ventromedial frontal lobe. Furthermore, these are connected with disturbances in the interior core of the self-regulating functions of the ego and super-ego.

Based on the results from 35 patients with delimited brain damage who were treated with psychoanalysis or psychoanalytical psychotherapy, the authors have drawn up a general theory focussing on the question of how the mental apparatus as a whole may be represented in the brain. The ventromedial frontal brain region hypothetically forms the basic, economic transformation processes, which are responsible for the inhibition of mental primary processes. The economic transformation in the energy balance is then identical with the process which Freud called "attachment". If this economic function is disturbed, all the secondary functions connected with it are also inevitably disturbed. The ego and the super-ego are described as a set of memory systems, as a set of structured internalizations. The applied method of dynamic localization enables a relatively comprehensive picture of neuronal organization and also of the mental functions (e.g. repression, attention and reality check) important for psychoanalysis. It may be shown how psychoanalysis can empirically lead the way to neuroscience and how a neuroscientific approach can be added to the investigation of "non-recognizable reality" in the sense of Freud.

## 3.12 The Neuroscience of Subjective Experience

Mark Solms' and Oliver Turnbull's book *The Brain and the Inner World* (2002) gives an overview on the "neuroscience of subjective experience". A broad spectrum of results (on emotion, motivation, memory, phantasy, dreams and hallucinations, differential and complementary functions) of the left and right hemisphere is presented. Further hypotheses focus on the possible basis of the psychoanalytical "talking cure", on the nature of unconscious and conscious processes and on the basis of subjectivity, consciousness and the self. This somewhat ambitious approach involves translating Freud's theory into a series of verifiable hypotheses on the functional organization of the brain.

Not least, this book also pays tribute to the outstanding pioneering work of Antonio Damasio: Damasio (1994, 1999) underlined that consciousness does not only depend on the perception of internal conditions but on a fluctuating connection of current states of the self with those of the object world. Each unit of consciousness is accompanied by a connection between the self and the objects. The title of Damasio's book, *The Feeling of What Happens*, underlines that consciousness consists of emotions which are projected onto what is happening in the outside world. Damasio described the attachment mechanism for linking the various channels of consciousness as "core consciousness". Access to episodic memory, i.e. remembering past circumstances in the self-object-relationship, contributes to the development of the "autobiographical self". Put in psychoanalytical terms, the "core self" may be described as the perception of the current state of the "id", whereas the extended, autobiographical self is synonymous with the "ego".

## 3.13 Affective Neuroscience

Further pioneering work significant for the dialogue between neuroscience and psychoanalysis was done by Jaak Panksepp (1998, 2009). It was Panksepp who coined the term "affective neuroscience". He describes seven fundamental emotional systems defined as activation and orientation systems (with neuronal forces common across all species). According to Panksepp, in the course of evolution, these systems were able to prevail because of the necessary survival advantages, and they use genetically determined brain regions. When one of these systems becomes active, the others usually become inhibited (Le Doux 1996, 2006). These systems are needed for activation, evaluation and preparation of actions. A physiological adaptation towards a greater willingness to learn is also connected with this.

Panksepp's basic emotions (Panksepp 1998, 2009) each have an evolutionary significance. The first two basic emotions are used for survival. They enable adequate learning time (this is particularly important to open the brain's construction programmes).

**Basic Emotional and Motivational Processes (According to Panksepp):**

- *Care:* attachment, nurturing love. Intrinsic brain systems that promote nurture and behaviour of mothers, and occasionally fathers, towards their offspring. Oxytocin mediates maternal behaviour. Neural circuitry for maternal behaviour (and oestrogen-responsive populations of oxytocin cells) is situated within the dorsal preoptical area (POA) just above the brain areas that elaborate male sexuality.
- *Panic/separation distress:* panic, separation anxiety. The panic circuit becomes aroused when young animals are separated from their social support systems. "Since opioid systems had already evolved to modulate the intensity of physical pain, it is not surprising that these same neural chemistries can sooth the pain evoked by social isolation" (Panksepp 1998, p. 261). Dopamine and opioid systems interact in several ways, including through the arousal of brain dopamine by opioid receptors within the ventral tegmental area (VTA) and opioid inhibition of dopamine activity in the terminal area of the striatum. The panic system appears to arise from the midbrain PAG, very close to where one can generate physical pain responses. Anatomically, it almost seems that separation has emerged from more basic pain systems during brain evolution. Separation distress is related to perceptions of pain. The panic system is also well represented in the medial thalamus, especially in the dorsomedial thalamus. There is a remarkable resemblance with the neural anatomy of this behavioural control system and those for corticotropin-releasing factor (CRF) and beta-endorphin systems.
- *Seeking systems and anticipatory states of the nervous system:* the seeking system is sensitized by (1) regulatory imbalances to general arousal and persistent forward processing and (2) external stimuli that can either have strong or weak interactions with this emotional system, and (3) it helps mediate appetitive learning so that animals will become eager and exhibit expectancies in response to cues that have been previously associated with arousal and disarousal of this system.
- *Play:* joy promotes social learning ("play circuits allow other emotional operating systems, especially social ones, to be exercised in the relative safety of one's home environment. Play may help animals project their behavioural potentials

joyously to the very perimeter of their knowledge and social realities, to a point where true emotional states begin to intervene. Thus, in the midst of play, an animal may gradually reach a point where true anger, fear, separation distress or sexuality is aroused …" (Panksepp 1998, p. 283)).
- *Fear:* anxiety, used to ward off danger. The basic fear system extends from the temporal lobe (from central and lateral areas of the amygdala) through the anterior and medial hypothalamus to the lower brain stem, through the periventricular grey matter of the brain, and then down to specific autonomic and behavioural output components of the lower brain stem and spinal cord, which control the physiological symptoms of fear (including increases in heart rate, blood pressure, vegetative response, elimination and perspiration).
- *Rage:* Anger promotes the ability to be assertive and supports access to resources. The rage circuits run from medial areas of the amygdala, through the hypothalamus and down into the periaqueductal grey matter of the midbrain. These areas are hierarchically arranged so that higher functions are dependent on the integrity of lower ones (see Panksepp 1998, p. 187).
- *Lust:* Lust, sexuality (survival of the species). Different neuropeptides control sexuality differentially between the sexes.

This approach is potentially valuable for generating hypotheses which could enable a better understanding and preciser description of the specificity and relevance of emotional factors in change processes (e.g. in psychotherapy).

**Mirror Neurons**

Another important neuroscientific contribution to this transdisciplinary discussion has been made by Rizzolatti's research group: they discovered so-called mirror neurons, localized on the surface of the frontal and parietal lobe (Gallese et al. 1996; Rizzolatti and Arbib 1998; Rizzolatti et al. 1999). Their experiments with monkeys showed that the motor neurons of those monkeys who merely passively observed the behaviour of the other monkeys, fire with the same pattern as those who took action. Although more experimental data are needed, it may nonetheless be assumed that mirror neurons form the neuronal basis for empathy and may be viewed as the physiological basis of internalization: by means of mirror neurons, executive programmes are established through repeated activation following observation, passive behaviour is transformed into active behaviour and activity into thought processes.

## 3.14 Concluding Remarks

Because of methodological limitations, Freud had to give up his project in 1895 to bring together psychoanalysis and the neuroscience of that time. Core questions concerning neuroscience in our time focus, among other things, on the relation between subjective experience and neuronal integration in the brain. In this way, serious and important matters concerning psychoanalysis can also be opened up for neuroscientific research. For example, Georg Northoff has raised the question of whether we need first-person neuroscience (Northoff and Boeker 2006; Northoff et al. 2006). This first-person perspective could be a suitable methodological strategy for a better understanding of neuronal processes (e.g. of defence mechanisms) and their modulation in psychoanalytical psychotherapy (Northoff et al. 2007; cf. Boeker et al. 2013) and for the development of paradigms which take subjective experience and first-person perspective into account (e.g. Boeker et al. 2013).

## References


Baranger M, Baranger W, Mom JM. The infantile psychic trauma from us to Freud: pure trauma, retroactivity, and reconstruction. Int J Psychoanal. 1988;69:113–28.

Boeker H, Richter A, Himmighoffen H, Ernst J, Bohleber L, Hofmann E, Vetter J, Northoff G. Centres of psychoanalytic process and change: how can we investigate the neural effects of psychodynamic psychotherapy in individualized neuro-imaging? Front Hum Neurosci. 2013;7:355.

Damasio A. Descartes' error. New York: Grosset/Putnam; 1994.

Damasio A. The feeling of what happens. London: Heinemann; 1999.

Du Bois-Reymond E. Zwei grosse Naturforscher des 19. Jahrhunderts: Ein Briefwechsel zwischen Emil Du Bois-Reymond und Karl Ludwig, Leipzig: J.A. Barth; 1842; 1927.

Edelman GM. Neural Darwinism. The theory of neural group selection. New York: Basic Books; 1987.

Edelman GM. The remembered present: a biological theory of consciousness. New York: Basic Books; 1989.

Edelman GM. Bright air, brilliant fire: on the matter of the mind. New York: Basic Books; 1992.

Freud S. Besprechung von "Die akute Neurasthenie", Wiener Medizinische Wochenschrift, Band 37, S. 138., vol. 1, Standard ed. (1886–1899). Pre-Psychoanalytic publications and unpublished drafts; 1887. p. 35.

Freud S. Hysteria, vol. 1, Standard ed.; 1888. p. 57.

Freud S. Footnotes to Charcot, vol. 1, Standard ed.; 1893. p. 139.

Freud S. Studies on Hysteria, vol. 2, Standard ed. (including Breuer's contributions); 1895a. p. 1–312.

Freud S. A project for a scientific psychology, vol. 1, Standard ed.; [1895b] (1950). p. 283–94.

Freud S. Extracts from the Fliess papers, vol. 1, Standard ed.; 1897. p. 259–60.

Freud S. The interpretation of dreams (First Part), vol. 4, Standard ed.; 1900a. p. ix-627.

Freud S. The interpretations of dreams (Second part) and on dreams, vol. 5, Standard ed.; 1900b. p. 339–721.

Freud S. My view on the part played by sexuality in the etiology of neuroses, vol. 7, Standard ed.; 1906. p. 274–82.

Freud S. The psycho-analytic view of psychogenic disturbance of vision, vol. 11, Standard ed.; 1910. p. 209–18.

Freud S. The claims of psycho-analysis to scientific interest, vol. 23, Standard ed.; 1913. p. 163–200.

Freud S. Remembering, repeating and working-through (Further recommendations on the technique of Psycho-analysis II), vol. 12, Standard ed.; 1914a. p. 145–56.

Freud S. On the history of the psycho-analytic movement, vol. 14, Standard ed.; 1914b. p. 1–66.

Freud S. On narcissism: an introduction, Standard ed.; 1914c. p. 78.

Freud S. The unconscious, vol. 14, Standard ed.; 1915a. p. 159–215.

Freud S. Instincts and their vicissitudes, vol. 14, Standard ed.; 1915b. p. 109–40.

Freud S. A child is being beaten: A contribution to the study of thr origin of sexual perversions, vol. 17, Standard ed.; 1919. p. 175–204.

Freud S. Beyond the pleasure principle. Group psychology and other works, vol. 28, Standard ed.; 1920. p. 1–283.
Freud S. Inhibition, symptom and anxiety, vol. 20, Standard ed.; 1926. p. 75–176.
Freud S. New introductory lectures on psychoanalysis, vol. 22, Standard ed.; 1933. p. 1–182.
Freud S. An outline of psycho-analysis and other works, vol. 23, Standard ed.; 1937–1939. p. 141–96.
Freud S. Moses and Monotheism: Three essays (1934–38), vol. 23, Standard ed.; 1939. p. 3–140.
Freud S. Leonardo da Vinci and a memory of his childhood, vol. 14, Standard ed.; 1990. p. 57–137.
Gallese V, Fadiga L, Fogassi L, Rizzolatti G. Action recognition in the premotor cortex. Brain. 1996;119:593–609.
Gay P. Freud; A life for our time. New York: W. W. Norton; 1988.
Grubrich-Simitis J. Es war nicht der "Sturz aller Werte". Gewichtungen in Freuds atiologischer Theorie. In: Schlosser AM, Hohfeld K, editors. Trauma und Konflikt. Giessen: Psychosozial-Verlag; 1998. p. 97–112.
Harlow J. Recovery from passage of an iron bar through the head. Publ Mass Med Soc. 1868;2:327–47.
Israël L. Die unerhörte Botschaft der Hysterie. Aus dem französischen von P. Müller und P. Posch. Reinhardt, München, Basel; 1983.
Kaplan-Solms K, Solms M. Neuro-Psychoanalyse. Eine Einführung mit Fallstudien. Klett-Cotta, Stuttgart. Clinical Studies in Neuro-Psychoanalysis (2000). Madison: International Universities Press; 2003.
Laplanche J. Die rätselhaften Botschaften des anderen und ihre Konsequenzen für den Begriff des "Unbewussten" im Rahmen der allgemeinen Verführungstheorie. Psyche. 2004;58(9/10):898–913.
Le Doux J. The Emotional Brain, London: Weidenfeld and Nicholson. Deutsch: Das Netz der Gefühle, München: Deutscher Taschenbuch-Verlag; 1996; 2001.
Le Doux J. Das Netz der Persönlichkeit. München: Deutscher TaschenbuchVerlag; 2006.
Leuzinger-Bohleber M. Die unbewusste Phantasie: Klinische, konzeptuelle und interdisziplinäre Perspektiven. Europäische Psychoanalytische Föderation, Bulletin. 2004;58:49–68.
Lorenzer A. Sprachzerstörung und Rekonstruktion. Frankfurt: Suhrkamp; 1970.
Lurija AR. Traumatic aphasia. its syndromes, psychology and treatment, Den Haag: Mouton; 1947; 1970.
Lurija AR. The working brain: an introduction to neuropsychology, New York: Basic Books. Deutsch: Das Gehirn in Aktion. Einführung in die Neuropsychologie, Reinbek: Rowohlt; 1973; 1996.
Lurija AR. The making of mind: a personal account of Soviet psychology. Cambridge: Harvard University Press; 1979.
Meynert T. Psychiatry: clinical treatise on the diseases of the fore-brains. Translated Sachs B. New York, London: G.P. Putnam; 1884.
Müller-Pozzi H. Psychoanalytisches Denken: Eine Einführung, vol. 3, erw. Aufl., Bern, Göttingen, Toronto, Seattle: Huber; 2002.
Northoff G, Boeker H. Principals of neuronal integration and defence mechanisms: neuropsychoanalytic hypotheses. Neuro-psychoanalysis. 2006;8(1):69–84.
Northoff G, Böker H, Bogerts B. Subjektives Erleben und neuronale Integration im Gehirn: Benötigen wir eine Erste-Person-Neurowissenschaft? Fortschr Neurol Psychiatr. 2006;74:627–33.
Northoff G, Bermpohl F, Schöneich F, Boeker H. How does our brain constitute defence mechanisms? First-person-neuroscience and psychoanalysis. Psychother Psychosom. 2007;76:141–53.
Panksepp J. Mood disorders. In: Winken P, Bruyn G, Klawans H, Fredericks J, editors. Handbook of clinical neurology, vol. 45. Amsterdam: Elsevier; 1985. p. 271–85.
Panksepp J. Affective neuroscience: the foundations of human and animal emotions. New York: Oxford University Press; 1998.

Panksepp J. Brain emotional systems and qualities of mental life. In: Fosha D, Ziegel DJ, Solomon M, editors. The healing power of emotion. Affective neuroscience, development and clinical practice. New York, London: Norton; 2009.
Rizzolatti GB, Arbib MA. Language within our grasp. Trends Neurosci. 1998;21(5):188–94.
Rizzolatti GB, Fadiga L, Fogassi L, Gallese V. Resonance behaviors and mirror neurons. Arch Ital Biol. 1999;137:85–100.
Sandler J, Freud A. Die analysis of defence: the ego and the mechanisms of defence revisited. New York: International University Press; 1985.
Sandler AM, Sandler J. The past unconscious and the present unconscious. A contribution to a technical frame of reference. Psychoanal Study Child. 1994;49:278–92.
Stein R. Psychoanalytic theories of affect. London: Karnack Books; 1999.

# 4 Self Between Brain and World: Neuropsychodynamic Approach, Social Embedded Brain and Relational Self

Georg Northoff

**Abstract**

The present chapter aims to target yet another central feature of the mind, the self as the subject of all our experience and hence of consciousness. More specifically, the focus is on different concepts of the self and how they are related to recent findings about neural mechanisms related to the self-reference of stimuli. I first introduce different basic concepts of the self as they are currently discussed in philosophy. The first concept of self is the self as mental substance, which was introduced originally by Descartes. This is rejected by current and more empirically oriented concepts of the self where the idea of a mental substance is replaced by assuming specific self-representational capacities. These self-representational capacities represent the body's and brain's physical, neuronal states in a summarized, coordinated, and integrated way. As such, the self-representational concept of the self must be distinguished from the phenomenological concept of self that is supposed to be an integral part of the experience and thus of consciousness. This phenomenal self resurfaces in the current debate as the "minimal self"—a basic sense of self in our experience that is supposed to be closely related to both brain and body. Current neuroscience investigates the spatial and temporal neural mechanisms underlying those stimuli that are closely related to the self when compared to the stimuli that show no relation or reference to the self. This is described as the self-reference effect. When comparing self- versus non-self-specific stimuli, neural activity in the middle regions of the brain, the so-called cortical midline structures, is increased. Moreover, increased neuronal synchronization in the gamma frequency domain can be observed. The question is how specific these findings are for the concept of self as discussed in philosophy. Neuronal specificity describes the specific and exclusive association

G. Northoff
Mind, Brain Imaging, and Neuroethics,
Institute of Mental Health Research, University of Ottawa, Ottawa, ON, Canada
e-mail: georg.northoff@theroyal.ca; www.georgnorthoff.com



H. Boeker et al. (eds.), *Neuropsychodynamic Psychiatry*,
https://doi.org/10.1007/978-3-319-75112-2_4

of the midline regions with the self. This is not the case since the same regions are also associated with a variety of other functions. This goes along with the quest for the psychological and experimental specificity of psychological functions and experimental paradigms and measures used to test for the self. One may also raise the issue of phenomenal specificity: the concept of phenomenal specificity refers to whether the phenomenal features of the self, that is, minessness and belongingness, are distinguished from other phenomenal features like intentionality or qualia. Finally, one may discuss the question of conceptual specificity that targets the distinction between the concepts of self-reference and self.

## 4.1 Background Ia: Conceptual Determination of the Self in Philosophy

You read these lines. You're winning a game of tennis while your girlfriend is watching. You feel pride. Who experiences that pride? You. You are the subject of the experience of boredom. Without you as subject of this experience, you could not experience anything at all, not even boredom. This subject of experience has been described as the "self." Your "self" makes it possible for you to experience things. In other words, it is a necessary condition for experience and thus also for consciousness. It is clear, therefore, that there is much at stake when it comes to the self.

The concept of self has been subject to intense philosophical discussion over the centuries. Different philosophers have suggested different concepts of self. Because of time and space constraints, we will only focus on those that are relevant in the attempt to map the interface between philosophical and neuroscientific accounts of the self.

There are four main different concepts of self discussed in current philosophy. First is "the mental self," which is based on our thoughts and a specific mental substance. Second is the "empirical self"—this concept of the self represents and reflects the biological processes in one's body and brain.

Third is the "phenomenal self," from which originates our experience in consciousness. Our consciousness is accompanied by an awareness of our self, referred to as pre-reflective self-awareness or phenomenal self. Finally, and most recently, philosophers speak of a "minimal self." This concept of the self is based on our body and its physiological processes. I will discuss each of these different concepts and how they relate to the brain in this chapter.

Before we do this, I have to shed some light on several related concepts. We experience our self in daily life during, for example, the act of perceiving certain objects, persons, or events in our environment. For example, while making a list of all the things you have to do today, you experience not only the act of thinking and writing but an awareness and experience of your own self. Hence, your self as the very subject of experience seems to be part of that experience. In other words, your self is a content of your consciousness. This is described as self-consciousness.

However, there is more to the self than the self itself and our experience of it in self-consciousness. You wake up every morning, every day, every week, and every year. Your body changes. You become older. You get wrinkled and your hair turns white. Despite all these bodily changes, you nevertheless have the feeling that you are the same self. You still experience your self as being the same self of 20 years ago.

You are one and the same person. There is thus a temporal dimension to your self that seems to be coherent and persistent over time. You and your self are continuous across time. The temporal dimension of your self has consequently been discussed under the umbrella of what is called "personal identity" in philosophy. While our discussion will touch upon the temporal dimension of the self and thus upon personal identity, we will not explicitly discuss it.

In a world of over seven billion people, there are many, many selves: you, your friends, your family, etc. Most interestingly, you can relate to them—you can communicate with other selves and sometimes even feel their emotions as in, for instance, the grief someone might feel when they lose a loved one. Or you might experience pain when your boyfriend's arm is broken. How is this possible? In philosophy, this is called "intersubjectivity." Finally, your self is not isolated from the rest of the world. You can share others' experiences and feel connected to the world. The world and its specific objects, persons, and events have meaning to you—you can relate to it more or less and can appropriate it for your own self. How is such basic integration of your own self within the world possible? And how is that related to your brain and its neuronal mechanisms? That shall be the focus in the following.

## 4.2 Background Ib: Empirical Investigation of the Self in Neuroscience

How can we investigate the self? In order to experimentally address the self, we need some quantifiable and objective measures that can be observed from third-person perspective. How can we obtain such measures? Psychologists focusing on memory observed that items related to ourselves were better remembered than those unrelated (see Northoff et al. 2006). For example, as a resident of Ottawa, I recall the recent thunderstorm that wiped away several houses locally much better than a person who, perhaps living in Germany, just heard about it in the news.

There is thus superiority in the recollections of those items and stimuli that are related to one's self. This is described as the self-reference effect (SRE). The SRE has been well validated in several psychological studies. Most interestingly, it has been shown to operate in different domains, not only in respect to memory but also in relation to emotions, sensorimotor functions, faces, words, etc. In all these different domains (see below for details), stimuli related to one's own self, known as self-specific stimuli, are recalled much better than those that are unrelated to one's own self, known as non-self-specific stimuli.

How is the SRE possible? Numerous investigations (see, e.g., Klein 2012; Klein and Gangi 2010 for summaries) show that the SRE is mediated by different psychological functions. These range from personal memories including autobiographical memories over memories of facts (semantic memories) to those cognitive capacities that allow for self-reflection and self-representation. Hence, the SRE is by itself not a unitary function but rather a complex multifaceted psychological composite of functions and processes.

How can we link the SRE to the brain? Before the introduction of functional imaging techniques such as fMRI at the beginning of the 1990s, most studies conducted focused on the effect of dysfunction or lesions in specific brain regions caused by brain tumors or stroke. These revealed that lesions in medial temporal regions that are central in memory recall, such as the hippocampus, change and ultimately abolish the SRE effect.

With the introduction of brain imaging techniques such as fMRI, we could then transfer the experimental paradigms of comparing self- and non-self-specific stimuli to the scanner and investigate the underlying brain regions. The basic premise here is that if self-specific stimuli are recalled better than non-self-specific stimuli, they must be processed by the brain in a different way. This might be, for instance, by higher degrees of neural activity and/or different regions.

This led to the investigation of numerous experimental designs of SRE-like paradigms in the fMRI scanner. For example, subjects were presented trait adjectives that were either related to themselves (such as for me, my hometown, Ottawa) as opposed to (Sydney, an unrelated city for me). In other tests, subjects were presented with images of their own face, and these were compared with faces of other people. Also autobiographical events from the subject's past were compared with those from other people. One's own movements and actions could also be compared with those of other people, implying what is called ownership (e.g., my movements) and agency ("I myself caused that action").

The stimuli belonged to different domains such as memory, faces, emotions, verbal, spatial, motor, or social. Most of the stimuli were presented either visually or auditorily, and the presentation of these stimuli was usually accompanied by an online judgment about whether the stimuli are related and personally meaningful or not to the research subject.

On the whole we can see that current neuroscience can investigate the self in various experimental ways using mainly functional brain imaging. However, any empirical research relies on certain presuppositions. This also holds true for current neuroscientific research on the self, which aims to reveal the neuronal mechanisms underlying our experience or sense of self. However, before examining the neuroscientific findings, we need to briefly shed some light on the concept of the self and how it has been defined in philosophical discussions.

## 4.3 Philosophical Concepts Ia: Mental Self

What is the self? What must it look like in order to presuppose experience and be the subject of our experience? The self has often been viewed as a specific "thing." Stones are things; the table on which your laptop stands is a thing. And in the same way the table makes it possible for the laptop to stand on it, the self may be a thing that makes experience and consciousness possible. In other words, metaphorically speaking, experience and consciousness stand on the shoulders of the self.

However, another question is whether the self is a thing or, as philosophers such as René Descartes suggest, a substance? A substance is a specific entity or material that serves as a basis for something like a self. For instance, the body can be considered a physical substance, while the self can be associated with a mental substance.

Is our self real and thus does it exist? Or is it just an illusion? Let us compare the situation to perception. When we perceive something in our environment, we sometimes perceive it not a real thing but an illusion that in reality does not exist. The question of what exists and is real is what philosophers call a metaphysical question. Earlier philosophers, such as René Descartes, assumed that the self is real and exists.

However, Descartes also assumed that the self is different from the body. Hence, self and body exist but differ in their existence and reality. Thus, from this perspective, the self cannot be a physical substance and is a mental substance instead. It is a feature not of the body but of the mind.

However, the characterization of the self as a mental entity has been questioned. For example, the Scottish philosopher David Hume argued that there is no self as a mental entity. There is only a complex set or "bundle" of perceptions of interrelated events that reflect the world in its entirety. There is no additional self in the world; instead there is nothing but the events we perceive. Everything else, such as the assumption of a self as mental entity, is an illusion. The self as mental entity and thus as a mental substance does not exist and is therefore not real.

To reject the idea of self as mental substance and to dismiss it as mere illusion are currently popular. One major proponent of this view today is the German philosopher Thomas Metzinger (Metzinger 2004). In a nutshell, he argues that through our experience, we develop models of the self, so-called self-models. These self-models are nothing but information processes in our brain. However, since we do not have direct access to these neuronal processes (e.g., all those processes and activities of the cells, neurons, in the brain), we tend to assume the presence of an entity that must underlie our own self-model. This entity is then characterized as the self (Fig. 4.1a).

According to Metzinger, the assumption of the self as a mental entity results from an erroneous inference from our experience. We cannot experience the neuronal processes in our brain as such. Nobody has ever experienced their own brain and its neuronal processes. Therefore, the outcome of our brain's neuronal processes, the self, cannot be traced back to its original basis, the brain, in our experience.

Where then does the self come from? We assume that it must be traced back to a special instance different from the brain. This leads us to assume that the mind and the self are mental entities rather than a physical, neuronal entities originating in the brain itself. Metzinger argues that the self as a mental entity simply does not exist. Therefore, Metzinger (2004) concludes selves do not really exist. Hence, the title of his book *Being No One*.

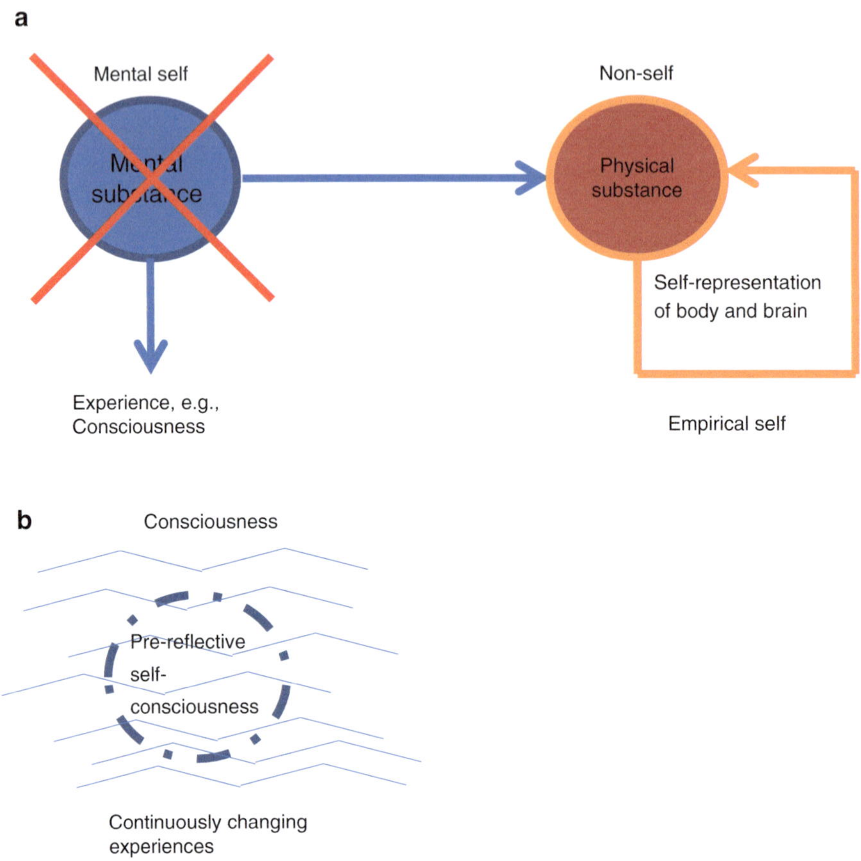


**Fig. 4.1** The figure schematically illustrates different concepts of self, the self as mental substance (a) and the phenomenal self (b). (**a**) The self is determined as mental substance (left) that is distinguished from the body (and brain) as mere physical substance (right). Thereby the self as mental self control and directs the body following the earlier french philosopher Descartes. This is denied in current empirical approaches to the self (e.g., vertical red lines). They reject the notion of the self as mental substance and claim that such mental self does not exist. All there is is the body as physical substance with the brain allowing for the representation of both body and brain in the brain's neural activity. Such self-representation may then amount to what can be described empirical self. (**b**) The phenomenal self no longer claims to be outside and prior to any experience. Instead, the phenomenal self is supposed to be 'located' or part of the experience itself in the gestalt of pre-reflective self-consciousness. This is indicated by the insertion of the circle within the midst of the experience, e.g., consciousness, itself

## 4.4 Philosophical Concepts Ib: From the Metaphysical to the Empirical Self

What is the self if not a mental entity? Current authors, such as Metzinger (2004) and Churchland (2002), argue that the self as mental substance or entity does not exist. How do we come up with the idea of a self or the self-model as Metzinger calls it? The model of our own self is based on summarizing, integrating, and coordinating all the information from our own body and own brain.

What does such integration look like? Take all that information together, coordinate, and integrate it, and then you have a self-model of your own brain and body and their respective processes. In more technical terms, our own brain and body are represented in the neuronal activity of the brain. Such representation of the own brain and body amounts then to a model of your self. The self-model is therefore nothing but an inner model of the integrated and summarized version of your own brain and body's information processing. The self is thus a mere model of one's own body's and brain's processes.

The original mental self, the self as mental substance or entity, is in this line of thinking replaced by a self-model. This implies a shift from a metaphysical discussion of the existence and reality of self to the processes that underlie the representation of the body and brain as a self-model. Since this representation is based on the coordination and integration of the various ongoing processes in the brain and body, it is associated with specific higher-order cognitive functions such as working memory, attention, executive function, and memory, among others.

What does this imply for the characterization of the self (presupposing a broader concept of self beyond the self as mental substance)? The self is no longer characterized as a mental substance but as a cognitive function. Methodologically, this implies that the self should be investigated empirically rather than metaphysically.

We therefore need to search for the cognitive processes underlying the special self-representation. The self is consequently no longer an issue of philosophy but rather one of cognitive psychology and ultimately of cognitive neuroscience. According to this model, the self is no longer a metaphysical matter but a possible subject of empirical investigation.

## 4.5 Philosophical Concepts IIa: Phenomenal Self

One of the problems one encounters is that such substance or meta-representation cannot be experienced as such. Nobody ever experienced a mental substance or a meta-representation in consciousness. We are not conscious of any such substance or meta-representation. Therefore, instead of speculating about something that lies beyond the scope of our experience, why not start with experience itself and thus with consciousness. Rather than looking at what lies "outside" our consciousness, like a substance or meta-representation, the self may be found within that very consciousness itself.

However, this localization is denied in phenomenological philosophy precisely because it focuses on consciousness itself and what lies "inside" our experience. More specifically, phenomenological philosophy is interested in investigating the structure and organization of our experience and thus of consciousness. It focuses on how our experience is structured and organized and reveals phenomenal features as we experience them from the first-person perspective.

How does the phenomenal approach determine the self? Currently, it is argued that the self is an integral part of experience itself (Northoff 2012).The self is always present and manifested in the phenomenal features of our experience such as intentionality (e.g., the directedness of our consciousness toward specific contents), qualia (e.g., the qualitative character of our experience; what it is like), etc. Without these features, the self would remain impossible.

Consequently, phenomenological philosophers such as Zahavi (2005) consider the self to be an inherent part of consciousness itself. Here, the self is supposed to be always already accompanied by some kind of consciousness of the external world, even if we are not aware of the self being part of that experience. Phenomenological philosophers therefore speak of what they call pre-reflective self-awareness (or pre-reflective self-consciousness).

The concept of pre-reflective self-consciousness contains two main terms, "pre-reflective" and "self-consciousness." "Pre-reflective" means that the experience of the self does not stem from any reflection or cognitive operation. Instead it is already always there as an unavoidable part of our experience as such. The self is thus pre-reflective. It is simultaneously an inherent part of our experience and thus of our consciousness. The self is consequently no longer outside of our consciousness, but an integral part of it, hence the second term, "self-consciousness." Such an approach suggests an intimate and intrinsic link between self and consciousness (Fig. 4.1b).

Characterizing the self in terms of self-consciousness implies a significant shift. The self is no longer metaphysical as Descartes proposes nor is it empirical as advocated by Hume and others such as Metzinger and Churchland. Instead, the self is part of experience and of consciousness itself and can therefore be characterized as the "phenomenal self." Such a phenomenal self is open to systematic investigation of the phenomenal features of our experience, which would complement the metaphysical, empirical, and logical approaches to the self.

## 4.6 Philosophical Concepts IIb: Minimal Self

How can we describe the pre-reflective self-consciousness in more detail? It is always already there in every experience so that we cannot avoid it or separate it from the experience. The self is always present in our consciousness and thus in our subjective experience. Even if we do not focus on the self as such, we cannot avoid or remove its presence. Hence, the term pre-reflective self-conscious describes an implicit or tacit experience of our self in our consciousness.

Since the self as pre-reflectively experienced is the basis of all phenomenal features of our experience, it must be considered as essential for any subsequent cognitive activity. Such a basic and fundamental self occurs in our experience before any reflection. For instance, when reading the lines of this book, you experience the contents, and, in addition, you also experience your self as reading these lines.

Hence, your immediate experience and consciousness comes with both the content and your own self, since the experience of such self occurs prior to any reflection and recruitment of higher-order cognitive functions. This is why this concept of self is sort of a minimal version of the self. Current phenomenological philosophers such as Gallagher (2000) or Zahavi (2005) speak therefore of a "minimal self" when referring to the self as implicitly, tacitly, and immediately experienced in consciousness.

How can we describe the concept of the "minimal self"? The minimal self refers to a basic form of self that is part of any experience. As such, it is not extended across time like it is in the experience of the self as a continuity across time in personal identity. Instead the minimal self describes a basic sense of self at any particular given moment in time, but does not yet provide a link between different moments in time and thus continuity across time.

How can such continuity across time be constituted? Cognitive functions such as memories and autobiographical memories in particular may be central. In this model, the self may become more complex. One might speak of a cognitive, extended, or autobiographical self, as does, for example, the Portuguese-American neuroscientist, Damasio (see, e.g., Damasio 1999, 2010).

Another important feature of the minimal self is that although we experience it, we may not be aware of it as such. This means that we might not be able to reflect upon it in order to gain knowledge of it. We are, to put it in technical terms, only pre-reflectively aware of the minimal self. In contrast to such pre-reflective awareness, there is no reflective awareness of the minimal self. How can we become reflectively aware of the minimal self? For that to be possible, the different moments or points in time need to be integrated and, as philosophers say, represented. For such representation to occur, cognitive functions are needed which make it possible to put and link together the different time points.

Finally, the minimal self may also occur prior to verbalization and thus linguistic expression. Rather than being tied to specific linguistic concepts, as is the case with more cognitive concepts of the self, the minimal self must be considered prelinguistic. It is an experience, a sense of self that can barely be put into concepts. We can experience it as self but are not really able to describe these experiences in terms of concepts and thus articulate them in a linguistic way.

Thus, the minimal self is prelinguistic and preconceptual and will therefore, speculatively, not be affected by second language acquisition. It is the kind of experience, an implicit sense of self, which most likely subjects will take with them as more or less stable when moving to a new country where they have to acquire a new language. However, at the same time, the minimal self provides the essential basis upon which more cognitive forms of self are developed. These are then central and instrumental in providing the ability to learn a second language.

## 4.7 Philosophical Concepts IIc: Social Self

How does the self interact with other selves? So far we described the self in an isolated and purely intraindividual way. However, in daily life, the self is not isolated from others but always related to other selves. This is called inter-individualism rather than intra-individualism. This raises questions about what is described as the "problem of other minds" or, more generally, questions concerning intersubjectivity. Here we will give a brief description of the problem of intersubjectivity.

How can we make the assumption of attributing mental states and thus self and mind to other people? Philosophy has long relied on what is called the "inference by analogy." What is the "inference by analogy"? "Inference by analogy" goes like this. We observe person A to show the behavior of type X. And we know that in our own case the same behavior X goes along with the mental state type M. Since our own behavior and that of the person A are similar, we assume the other person A to show the same mental state type M we experience when exhibiting behavior X.

What kind of inference do we draw here? There is similar or analogous behavior between ourselves and the other person. In addition, my own behavior is associated with a particular mental state. Since now the other person shows the same behavior, I infer that she also show the same mental state as it is associated with my own behavior. Hence, by indirect inference and analogy via our own case, we claim to obtain knowledge of the other person's mental state. How can we make such inference? We may make it on the basis of our own mental states and their associated behavior. And what we do may also hold true for the other person who in the same way attributes mental states to us by inferring them from the comparison between our behavior and their own mental states.

Why do we make such inferences? Because it seems to be the easiest and best way for us to explain the other people's behavior. The assumption of mental states thus seems to be the best explanation for your behavior. The "inference by analogy" may thus be considered an inference to the best possible explanation.

The inference by analogy describes intersubjectivity in a very cognitive and ultimately linguistic way when attributing mental states and a self to other persons. There might be, however, a deeper level of intersubjectivity. We also feel the other persons' mental states when sharing the emotional pain one's spouse experiences when her father died. Such sharing of feeling is described as empathy and sheds light on a deeper precognitive and preverbal dimension of intersubjectivity. This has been emphasized especially in phenomenological philosophy (see, for instance, Metzinger 2004).

However, both empathy and the attribution of mental states to another person are puzzling: despite the fact that we do not experience the other's mental states and consciousness, we nevertheless either share them (as in empathy) or infer them (as in inference by analogy). We have no direct access to other persons' experience of a self and its mental states in first-person perspective and nevertheless share their mental states and assume that they have a self. How is that possible?

This is where we need to introduce yet another perspective. There is first-person perspective—tied to the self itself and its experience or consciousness of objects, events, or persons in the environment. Then there is the third-person perspective—this perspective allows us to observe the objects, events, or persons in the environment from the outside, rather than from the inside. The picture is not complete.

What is the second-person perspective? The second-person perspective has initially been associated in philosophy with the introspection of one's own mental states. Rather than actually experiencing one's own mental states in first-person perspective, the second-person perspective makes possible to reflect and introspect about one's own mental states. An example of this is when you ask yourself whether the voice you heard was really the voice of your good friend (see also Schilbach et al. 2013).

The second-person perspective thus allows us to put the contents of our consciousness as experienced in first-person perspective into a wider context, the context of oneself as related to the environment. In other words, the second-person perspective makes it possible to situate and integrate the purely intraindividual self with its first-person perspective into a social context. This transforms the intraindividual self into an interindividual self. Another way of thinking of second-person perspective is to call this concept of the self the "social self."

How can we define the concept of the social self? The concept of the social self describes the linkage and integration of the self into the social context of other selves. This shifts the focus from experience or consciousness in the first-person perspective to the various kinds of interactions between different selves as associated with the second-person perspective. As we already indicated, there may be different kinds of social interactions including affective precognitive and more cognitive ones that involve meta-representation as described above.

## 4.8 Neuroscientific Findings Ia: Spatial Patterns of Neural Activity During Self-Specific Stimuli

How can we relate the various philosophical concepts of the self to the neuroscientific findings of self-reference? Above, we discussed that psychology, and later neuroscience, quantified the self in terms of the self-reference effect (SRE). The SRE describes the different impacts of self-referential and non-self-referential stimuli on psychological (e.g., reaction time, recall, etc., see above) and neural (e.g., degree of activity, regions, etc. see below) measures. Below we want to briefly highlight some of the main findings of recent imaging studies on the self-reference effect.

What results did the various imaging studies yield in the fMRI? Two different kinds of regions showed up. First, one could see that the regions specific for the respective domains like emotions or faces were recruited. For instance, there is a region in the back of the brain that processes specifically faces (as distinguished from, say, houses); this is called the fusiform face area. This region is obviously active during the presentation of faces, no matter whether it is one's own face or another person's face. Importantly, clear differences between self- and non-self-specific stimuli could not be observed in these domain-specific regions in most studies (see Northoff et al. 2006).

What about other regions that are not specific to particular domains (also known as domain-independent regions) involved in the neural processing of the self? Meta-analyses of the various studies demonstrated the involvement of a particular set of regions in the middle of the brain. These regions include the perigenual anterior cingulate cortex (PACC), the ventro- and dorsomedial prefrontal cortex (VMPFC, DMPFC), the supragenual anterior cingulate cortex (SACC), the posterior cingulate cortex (PCC), and the precuneus. Since they are all located in the midline of the brain, they have been coined "cortical midline structures" (CMS).

The self-specific stimuli—those that were personally relevant for the subjects—induced higher neural activity in these regions than non-self-specific stimuli or those that remained irrelevant und unrelated to the person. This was observed in the various domains for faces, trait adjectives, movements/actions, memories, and social communication. Therefore, the CMS seem to show a special significance to the self and self-reference.

However, there is also some differentiation within the CMS. The self-specific stimuli may be presented in different ways to the subject in the scanner. If subjects have to make judgments requiring cognitive involvement, the dorsal and posterior regions such as the SACC, DMPFC, and PCC are recruited to a stronger degree. If, in contrast, stimuli are merely perceived without any judgment, and thus without any cognitive component, the ventral and anterior regions such as the VMPFC and PACC were highly involved (Fig. 4.2a, b).

This led to the assumption that the different regions mediate different aspects of self-reference. The ventral and anterior regions, such as the PACC and VMPFC, may be more involved in the representation of the degree of self-reference in the stimulus. However, dorsal regions, such as the SACC and the DMPFC, may be related to monitoring and reflection of the stimulus and its self-reference when we become aware of the stimulus as self-specific.

Finally, the posterior regions, such as the PCC, may be implicated in integrating the stimulus and its degree of self-reference into the autobiographical memory of the respective person. These regions seem to be implicated in the recall and retrieval of especially personally relevant and autobiographical information from the past of that person. Thus, it can be concluded that specific regions in the midline of the brain, the cortical midline structures, seem to be involved in the neural processing of self-reference or attributing personal relevance or self-relevance to stimuli.

## 4.9 Neuroscientific Findings Ib: Temporal Patterns of Neural Activity During Self-Specific Stimuli

In addition to the spatial patterns of self-reference, its temporal patterns have also been investigated using the EEG. Again self-specific and non-self-specific stimuli have been compared with each other while the subjects undergo EEG measurement. This revealed early changes during self-specific stimuli at around 100–150 ms after stimulus onset.

More specifically, self-specific stimuli induced different electrical activity changes already at 130–200 ms after their onset when compared to non-self-specific stimuli. This was accompanied by later changes at around 300–500 ms. Hence, the temporal pattern between self- and non-self-specific stimuli shows both early and late differences.

In addition, different frequencies of neural activity were investigated. The neural activity oscillates rhythmically in different frequency ranges in the fluctuations of the neuronal activity.

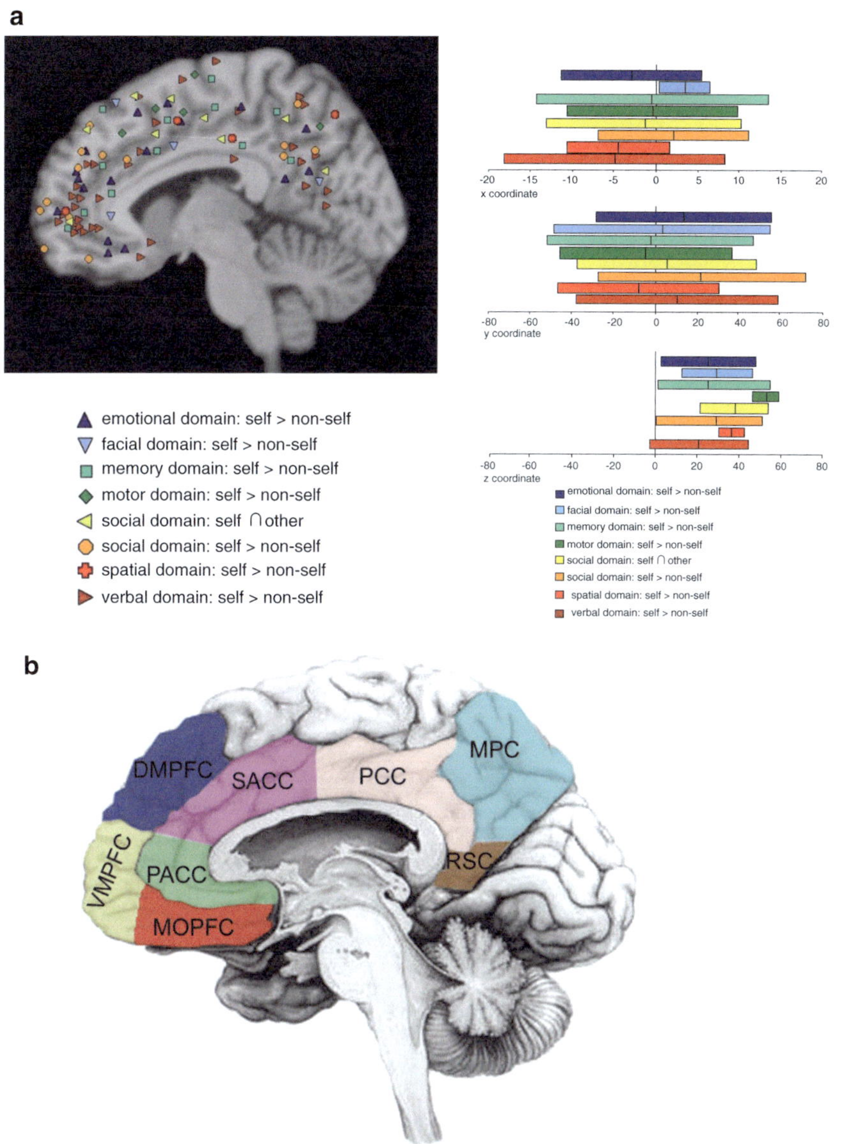


**Fig. 4.2** The figure demonstrates the results of a meta-analysis on imaging studies of self-reference (Northoff et al. 2006) (a) and anatomical illustration of the midline regions (b). (**a**) The figure on the left depicts all the imaging studies on the self as plotted in their obtained location on one brain. This includes self-referential stimuli in various domains or functions like memory, social, spatial, etc as indicated in the lower text with the colours as shown above and on the right. On the right three different coordinates (x, y, z) are shown that determine the direction (medial-lateral, inferior-superior) of the location in the brain. One can see that all studies locate in the midline regions of the brain (left image) as seen in the x-coordinates that describe the medial-lateral location (right image). (**b**) The figure shows the anatomical regions in the midline of the brain. MOPFC = Medial orbital prefrontal cortex, PACC = Perigenual anterior cingulate cortex, VMPFC, DMPFC = Ventro- and dorsomedial prefrontal cortex, SACC = Supragneual anterior cingulate cortex, PCC = Posterior cingulate cortex, MPC = medial parietal cortex, RSC = Retrosplenial cortex

One frequency often induced by stimuli are gamma frequencies in the range of 30–40 Hz. Interestingly, some EEG (and MEG) studies observed higher power in the gamma range in anterior and posterior midline regions during self-specific stimuli than non-self-specific stimuli. The question though is whether such increase in gamma power is specific to self-specific stimuli since it can also be observed in other functions independent of self-reference (see below).

## 4.10 Neuroscientific Findings Ic Social Patterns of Neural Activity During Self-Reference

How can we investigate the earlier described social nature of the self? Various studies have been conducted to investigate different kinds of interaction between different selves. Pfeiffer et al. (2013) and Schilbach et al. (2012) distinguish two different methodological approaches. One investigates social cognition, the cognition of mental states in other people, from third-person perspective. Here, social cognition is investigated in an "offline" mode. More recently this "offline" methodological strategy has been complemented by an "online" mode. In the "online mode," social interaction is investigated from the "inside," by taking on the perspective of the interacting selves (rather than the observer's point of view).

Besides conducting several studies, the same group has recently investigated the neural overlap between emotional processing, resting-state activity, and social-cognitive processing (Schilbach et al. 2012). They conducted a meta-analysis including imaging studies from all three kinds of investigations, resting-state, emotional, and social-cognitive. In a first step, they analyzed the regions implicated in each of the three tasks. This yielded significant recruitment of neural activity in especially the midline regions like the ventro- and dorsomedial prefrontal cortex and the posterior cingulate cortex (bordering to the precuneus). In addition, neural activity in the temporoparietal junction and the middle temporal gyrus was observed.

In a second step, they overlaid the three tasks, emotional, social-cognitive, and resting-state, in order to detect commonly underlying areas. This indeed revealed the midline regions, the dorsomedial prefrontal cortex and the posterior cingulate cortex, to be commonly shared among emotional and social-cognitive tasks and resting-state activity. Based on this neural overlap, the authors concluded that there may be an intrinsically social dimension in our neural activity which might be essential for consciousness of both our own self and other selves. If this is true, it will have radical consequences, not only for the concept of the self but also for consciousness in general.

## 4.11 Conclusion: Self as Brain-Based Neurosocial Structure and Organization

What does this imply for the self? Our self may be considered as intrinsically linked to the body. This is called embodied self. Furthermore, since it is based on self-reference, our self may also be intrinsically linked to the environment. This is called the embedded and social self. Our self cannot consequently be regarded as an entity located somewhere in the brain and isolated from both body and environment. Instead our self seems to be intrinsically social, as suggested by the advocates of the concept of a social self (see above).

What does this intrinsically bodily and social nature imply for the conceptual characterization of the self? Our self may be described as structure and organization, rather than as an entity—be it mental or physical. Such structure and organization need to develop through childhood and adolescence with persistent changes even throughout adulthood. Despite all the changes, there may also be persistence and continuity across time, which then accounts for what can be described as identity. Identity may describe the persistence and continuity of self over time which, in an exploratory study, has recently been associated with the midline structures and their high intrinsic activity.

We can also see that this concept of self as structure and organization is embodied and embedded. Hence, the virtual structure of the self spans across the brain, body, and environment. At the same time, that very same virtual structure is dependent upon the respective environmental context. Freud's characterization of the ego as structure and organization surfaces here in a more specific way as being integrated in the body and environment, that is, embodied and embedded. Put differently, the ego consists in a relation, the one between the brain, body, and environment, and can thus be determined in an intrinsically relational way. Future investigation might link the different features Freud attributed to the ego to the self.

What however do we mean by the concepts of structure and organization? The structure must be virtual in that it spans across the physical boundaries of the brain, body, and environment. Does this mean that we have to revert to a mental structure and organization as distinct from the physical structure and organization of the brain? No! The results from neuroscience clearly link the self with neuronal processes related to both intraindividual experiences and interindividual interaction. There is thus a neuronal basis for the distinct aspects of the self within the context of the brain, body, and environment. We therefore reject the mental characterization of the structure and organization that is supposed to define the self.

How can we define the concepts of structure and organization in a more positive way? One way is to characterize structure and organization as social. This distinguishes it from mental or physical features. The social characterization would then be the underlying basis that links and integrates between the purely physical and the purely mental. The self would then be based on the brain but would also extend beyond it to the body and the environment. This means that conceptually, we need to characterize the concept of the self as brain-based, rather than brain-reductive (as the proponents of the empirical self tend to do). The brain-based nature of the self also excludes both mind- and consciousness-based approaches to the self.

If the social characterization of the structure and organization as related to the self is indeed basic and fundamental, one would assume that our brain's neural activity is intrinsically neurosocial: the brain cannot avoid including the social environmental context in the encoding of stimuli into its own neural activity. The neural activity is thus by default neurosocial rather than merely neuronal. This is supported by the above-described neural overlap between resting-state activity and the neural activity changes during emotional and social-cognitive tasks.

Whether the brain encodes its neural activity in an intrinsically neurosocial way remains unclear at this point. What is clear is that the exact characterization of the brain's neural activity will be essential if we are to develop a truly neurophilosophical, brain-based (rather than brain-reductive) and neurosocial (rather than merely neuronal) concept of the self.

## References

Churchland PS. Self-representation in nervous systems. Science. 2002;296(5566):308–10. https://doi.org/10.1126/science.1070564.
Damasio AR. How the brain creates the mind. Sci Am. 1999;281(6):112–7.
Damasio A. Self comes to mind: constructing the conscious mind. New York: Pantheon; 2010.
Gallagher II. Philosophical conceptions of the self: implications for cognitive science. Trends Cogn Sci. 2000;4(1):14–21.
Klein SB. Self, memory, and the self-reference effect: an examination of conceptual and methodological issues. Personal Soc Psychol Rev. 2012;16(3):283–300. https://doi.org/10.1177/1088868311434214.
Klein SB, Gangi CE. The multiplicity of self: neuropsychological evidence and its implications for the self as a construct in psychological research. Ann N Y Acad Sci. 2010;1191:1–15. https://doi.org/10.1111/j.1749-6632.2010.05441.x.
Metzinger T. Being no one: the self-model theory of subjectivity. Cambridge: MIT Press; 2004.
Northoff G. Immanuel Kant's mind and the brain's resting state. Trends Cogn Sci. 2012;16(7):356–9. https://doi.org/10.1016/j.tics.2012.06.001.
Northoff G, Heinzel A, de Greck M, Bermpohl F, Dobrowolny H, Panksepp J. Self-referential processing in our brain—a meta-analysis of imaging studies on the self. NeuroImage. 2006;31(1):440–57.
Pfeiffer UJ, Timmermans B, Vogeley K, Frith CD, Schilbach L. Towards a neuroscience of social interaction. Front Hum Neurosci. 2013;7:22. https://doi.org/10.3389/fnhum.2013.00022.
Schilbach L, Bzdok D, Timmermans B, Fox PT, Laird AR, Vogeley K, Eickhoff SB. Introspective minds: using ALE meta-analyses to study commonalities in the neural correlates of emotional processing, social & unconstrained cognition. PLoS One. 2012;7(2):e30920. https://doi.org/10.1371/journal.pone.0030920.
Schilbach L, Eickhoff SB, Schultze T, Mojzisch A, Vogeley K. To you I am listening: perceived competence of advisors influences judgment and decision-making via recruitment of the amygdala. Soc Neurosci. 2013;8(3):189–202. https://doi.org/10.1080/17470919.2013.775967.
Zahavi D. Subjectivity and selfhood: investigating the first-person perspective. Cambridge: MIT Press; 2005.

# Three-Dimensional Neuropsychodynamic Model of Mental Disorders and Their Defence Mechanisms

5

Heinz Boeker and Georg Northoff

**Abstract**

On the basis of a three-dimensional model encompassing the psychodynamic dimensions defence mechanisms, conflict, and structure (Mentzos, Lehrbuch der Psychodynamik. Die Funktion der Dysfunktionalität psychischer Störungen. Vandenhoeck und Ruprecht, 2009), an analogous model was developed on the neuronal level. It is shown how this original, three-dimensional model of psychic illness can be translated into a neuropsychodynamic context.

The first dimension of defence and compensation describes certain modes of intrinsically predisposed mechanisms for processing extrinsic life events. From a neuronal perspective, the different modes of defence and compensation of conflicts and/or traumata are analogous to the resting state spatiotemporal structures' ability to modulate and adapt as well as to the resting state's self-specific organisation.

The second dimension (conflict) in a neuropsychodynamic perspective may be characterised as an interaction between the intrinsic resting state activity and the extrinsic event or stimuli, which has been operationalised as rest-stimulus interaction.

The third dimension (structure) focusses on the organisation of the personality and the psychic structure of the subject including her/his relationship to the object and to the self. The intrinsic activity and its spatiotemporal structure would appear to fulfil the criteria for self- and object-specificity analogous to the

H. Boeker (✉)
Department of Psychiatry, Psychotherapy, and Psychosomatics,
Psychiatric University Hospital Zurich, University of Zurich, Zurich, Switzerland
e-mail: heinz.boeker@bli.uzh.ch

G. Northoff
Mind, Brain Imaging, and Neuroethics,
Institute of Mental Health Research, University of Ottawa, Ottawa, ON, Canada
e-mail: georg.northoff@theroyal.ca; www.georgnorthoff.com



H. Boeker et al. (eds.), *Neuropsychodynamic Psychiatry*,
https://doi.org/10.1007/978-3-319-75112-2_5

structure in the psychodynamic context. This spatiotemporal structure is most probably constituted by the interaction between different networks and regions and the interaction between different fluctuations in different frequencies.

Thereby, self- and object-specificity may be understood as two extreme poles of an underlying commonly shared continuum which runs along the relationship between the brain and world, i.e. their respective spatiotemporal structures.

The three-dimensional neuropsychodynamic model of psychic disorders—as a core element of neuropsychodynamic psychiatry—may contribute to the development of future diagnostic classification systems focussing on functional mechanisms in psychopathology and their subject-related application. Furthermore, it is a diagnostic instrument for the planning, application, and evaluation of psychotherapy.

## 5.1 Introduction

The neuropsychodynamic approach aims at understanding the dynamics of the different psychiatric disorders. Given that the suffering in each single case is always something very particular, for an orientation, it is necessary to comprehend the common ground between the single cases and. If possible, do develop and make subtypes visible. This approach and search for general features should by no means be limited to the description of psychiatric disorders on a phenomenological level but especially focus on the variations of the psychodynamics. Whereas the descriptive-phenomenological approach of the international diagnostical classification systems (DSM-5, ICD-10/ICD-11) enables a first orientation, this is mostly not sufficient, if one tries to comprehend the specific problems in the individual subject.

Contrary to descriptive psychiatry, neuropsychodynamic psychiatry focusses on the specific importance of subjective experience. Whereas the descriptive-phenomenological approach mainly focusses on the question, whether a patient corresponds to others, who show congruent features, neuropsychodynamic psychiatry essentially focusses on the history of subjective experiences and the development of the symptoms trying to understand how the individual patient differs in his particular life history from others. Symptoms and behaviours are viewed, as Gabbard (2005/2014) underlines, "only as the final common pathways of highly personalized subjective experiences that filter the biological and environmental determinants of illness" (p. 8). Furthermore, neuropsychodynamic psychiatry—as an actual development of the former psychodynamic psychiatry—places paramount value on the patients' internal world: fantasies, dreams, hopes, wishes, fears, impulses, self-images, perception of others, and psychological reactions to symptoms. Neuropsychodynamic psychiatrists are curious about the neurobiological and psychological, i.e. neurobiopsychological or neuropsychodynamic mechanisms, which lead to the "final common pathway" of symptoms and behaviours; central here is to address the question, how these mechanisms were necessary to protect the interior world and how the "exterior" world reflects the inner contents of the respective person.

It had been consequent to look for the necessary supplement of descriptive psychiatry in psychoanalysis 100 years ago, which had dealt with the psychic drives and energies and unconscious motivations beyond the exterior phenomena.

Freud hoped to be able to develop a systematology according to the traditional paradigm of medicine. He started from the concept of nosological entities, which get along with certain and always equal causes, pathogeneses, symptoms, courses, and therapeutic specificities. In view of a big group of mental disorders, Freud (1926) assumed a specific intrapsychic conflict which would underlie the respective disorder. These disorders on the basis of unconscious inner conflicts were called neuroses. Freud regarded the correlated symptoms and syndromes as parts of mechanisms of defence and compensation and therefore also as—inadequate—compromising trials to solve conflicts.

In this group of neurotic disorders, a greater subgroup of psychoneuroses was separated. It was assumed that their phenomenology serves the symbolic expression of psychic conflicts (Laplanche and Pontalis 1989). The origins of these conflicts were localised in childhood.

Mentzos (2009) underlined that the emphasis on the symbolisation as most decisive differential criterion of the psychoneuroses compared to other forms of neuroses is applicate, but not sufficient. Finally, symbolic patterns of expression also get along with other psychic disturbances, e.g. anorexia, psychosomatic disorders of the skin or with psychoses. In order to be able to differentiate psychoneuroses sufficiently, rather the different maturational levels of symbolism have to be taken into account, e.g. the maturational level of the personality organisation and of the defence mechanisms; furthermore also the kind of the conflict has to be considered. Subsequent to the drive conflicts (especially the triadic or oedipal conflict), which were focussed on by the early psychoanalysis, the later self-psychological, object relational, and intersubjective developments of psychoanalysis during the last decades point to further relevant dimensions far beyond the original conflict.

The assumption of a drive conflict following Freud's tripartite structural theory of ego, id, and superego—e.g. drive wishes vs. superego—mostly is not sufficient to adequately comprehend the multidimensionality of the single case though Böker 1979; Böker et al. 2000). For instance, an intensive bonding between father and daughter or between mother and son (as well as its neurotic equivalences in later relationships in adulthood), which seems to be incestuous from the outside perspective, may not at all based on a libidinal wish and a forbidding incestuous taboo but may rather be the consequence of the emotional dependency from the primary object and/or the wish to stabilise one's own self by means of identification (Mentzos 2009, p. 85).

The disadvantages of the former classifying and categorising procedure in psychoanalysis, especially the assumed specificity of conflicts in psychic disorders, are especially obvious in the context of the discussion on hysteria: the nosological entity of "hysteria", in which—according to the drive-theoretical conceptualisation—always an oedipal conflict was involved, proved to be wrong by means of empirical research and could no longer be maintained. By the way, this also concerns the former assumptions of specific constellations in psychosomatic disorders (e.g. "pensee operatoire" or "psychosomatic phenomenon", Marty and de M'Uzan 1963; Böker 1979). The overcoming of the former concepts concerning hypothesised conflict-specific conditions in psychoneuroses or finally also in obsessive-compulsive disorders, the neurotic depression, and phobia got necessary: the expected constellations of conflict, defence, and symptomatology—according to the former concept of nosological entities—could not be validated empirically.

In view of a new orientation, which was necessary not only because of theoretical reasons but also because of practical reasons, Mentzos (1982, 2009) proposed a three-dimensional diagnostics, which enable to characterise each individual case with regard to three psychodynamic dimensions or criteria (cf. Fig. 5.1).

The first dimension refers to the mode of defence and compensation of the conflict and/or the trauma. The second dimension concerns the specific features and the level of the conflict or dilemma (e.g. self-identity vs. symbiosis as an "immature" conflict contrary to the "mature" oedipal conflict). The third dimension refers to the level of the personality organisation, especially to the relationship to significant

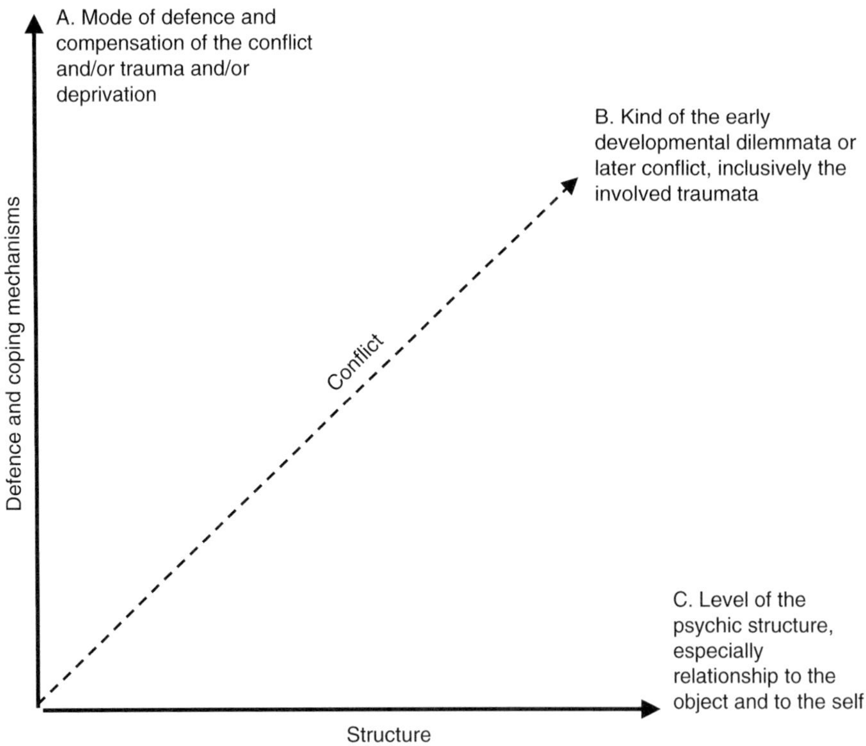


**Fig. 5.1** Three-dimensional diagnostics of mental disorders. (**a**) Mode of defence and compensation of the conflict and/or trauma and/or deprivation. (**b**) Kind of the (early) developmental dilemma or later conflict, inclusively the involved traumata. (**c**) Level of the psychic structure, especially relationship to the object and to the self

others or to one's own self. Focussing on this dimension, Rudolf (2006) proposed in his concept of a structure-related psychotherapy a gradation in four steps: well integrated, moderately integrated, poorly integrated, and disintegrated (Rudolf 2006, cf. 4.4, Operationalised Psychodynamic Psychotherapy, OPD 1996).

In summary, the development of a three-dimensional diagnostics for psychiatric disorders was necessary, because similar psychic disorders differ from one another in regard to psychogenesis, conflict, and developmental level, and no conflict specificity referring to the occurrence of psychic disorders could be stated. There is often an overlap between the dimensions of conflict and structure. For instance, an adolescent may develop an identity crisis in the course of the separation situation of adolescence, who may had reached a relative stable level of personality organisation before, but was, for example, confronted with extremely unfavourable conditions in his/her environment: this crisis may get along with momentary psychotic symptoms and a reactualisation of an earlier conflict or dilemma. A possible diagnosis of such a case only on the basis of the actual and momentary psychotic symptoms—without considering the relative stable personality structure—would contribute to a distorted diagnostic assessment and to the assumption of a more unfavourable prognosis.

On the basis of this three-dimensional model encompassing the psychodynamic dimensions defence mechanisms, structure, and conflict (cf. Mentzos 2009), an analogous model was developed on the neuronal level. How can this original, three-dimensional model of psychic illness be translated into a neuropsychodynamic context?

## 5.2 First Dimension of the Three-Dimensional Neuropsychodynamic Model: Mode of Defence and Compensation

Also in this adapted and extended three-dimensional neuropsychodynamic model, which will be outlined in the following, the first dimension—the mode of defence and compensation—is of outstanding relevance. Mentzos (2009) already had pointed out in regard to the clinical reality and the intended congruence with descriptive approaches that the phenomenology and the symptoms of psychic disorders mainly correlate with the respective mode of defence and compensation. The intrapsychic constellation of defence as well as the interpersonal coping mechanisms especially represent the relationship of the person to one's own self and to significant others. Furthermore, clinical experiences and empirical data underline that there is often a statistical cluster of single cases which correlates with a certain mode on a certain level, though there may be a relative broad distribution of the single modes within the organisational levels. Within the mode of defence and compensation, a hierarchy of more mature (see Table 5.1) and less mature (see Table 5.2) mechanisms of defence and compensation can be assumed.

One especially significant group of coping mechanisms, in the neuropsychodynamic context apostrophised as mechanisms of defence and compensation, does not refer any longer only to intrapsychic processes but to the development of

**Table 5.1** Hierarchy of mature mechanisms of defence

| Defence mechanism | Function |
|---|---|
| Intellectualisation | Using excessive and abstract ideation to avoid difficult feelings |
| Rationalisation | Justification of unacceptable attitudes, beliefs, or behaviours to make the tolerable to oneself |
| Isolation of affect | Separating an idea from its associated affect state to avoid emotional turmoil |
| Undoing | Attempting to negate sexual, aggressive, or shameful implications from a previous comment or behaviour by elaborating, clarifying, or doing the opposite |
| Reaction formation | Transforming an unacceptable wish or impulse into its opposite. This is often a permanent and habitual process which gets along with the development of respective character traits |
| Displacement | Shifting feelings associated with one idea or object to another that resembles the original in some way |
| | Shifting aggressive feelings which originally were directed to another person against one's own self (autoaggression) |
| Identification | Internalising the qualities of another person by becoming like the person |
| | This is an essential process in the development of a person, by which intrapsychic structures emerge (mature forms of internalisation). Identification may be used as a defence (e.g. identification with the aggressor or development of conversion as a defence against psychic pain after loss and separation) |
| Introjection | Internalising aspects of a significant person as a way of dealing with the loss of that person and/or the anxieties resulting from separation and the differentiation between subject and object. Introjection leads to an internalised representation of the object ("in-toto- internalisation"), whereas identification is experienced as part of the self ("selective internalisation") |
| Repression | Essential psychic process which is connected with the development of the unconscious |
| | Expelling unacceptable ideas or impulses or blocking them from entering consciousness (amnesia, scotomising certain contents). The above-described defence mechanisms serve repression in a wider sense, i.e. blocking them from entering consciousness |
| Sublimation | Transforming socially objectionable or internally unacceptable aims into socially acceptable ones. This is a controversial term, especially because of Freud's assumption that cultural achievements necessarily presuppose to give up drive wishes |

unconscious interpersonal constellations, which confirm and justify the intrapsychic changes and let them appear to be real. These specific mechanisms of interpersonal defence and compensation open up a more person perspective: these interpersonal mechanisms were characterised by Mentzos (1982) as "**psychosocial arrangements**", i.e. unconscious constellations of relationships, which either happen by the selection of an adequate partner (who has complementary "neurotic"

**Table 5.2** Hierarchy of immature (early) mechanisms of defence

| Defence mechanism | Function |
|---|---|
| Idealisation | Attributing perfect qualities to others as a way of avoiding anxiety or negative feelings, such as contempt, envy, or anger |
| Psychotic introjection | Internalising aspects of a significant person as a way of dealing with the loss of that person and/or the differentiation between subject and object. Introjection leads to an internalised representation experienced as "other" ("in toto internalisation") |
| Projection | Perceiving and reacting to unacceptable inner impulses and their derivatives as though they were outside the self. Results in a distorted perception of reality (e.g. delusion) |
| Projective identification | Both an intrapsychic defence mechanism and an interpersonal mode of communication and interaction. Behaving in such a way that subtle interpersonal pressure is placed on another person to take on characteristics of an aspect of the self or an internal object that is projected into that person. The person who is the target of that projection then begins to behave, think, and feel in keeping what has been projected |
| Splitting | Compartmentalising experiences of self and other. When the individual is confronted with the contradictions in behaviour, thought, or affect, he or she reacts with denial or indifference. Contrary to repression incompatible contents are in principle conscious or subconscious |
| Denial | Avoiding awareness of aspects of external reality that are difficult to face by disregarding sensory data. The individual avoids to perceive aspects that may be conflictuous or traumatising |
| Dissociation | Disrupting one's sense of continuity in the areas of identity, memory, consciousness, or perception as a way of retaining an illusion of psychological control in the face of helplessness and loss of control. Contrary to splitting, dissociation may involve alteration of memory of events because of the disconnection of the self from the event |
| Acting out | Enacting an unconscious wish or fantasy impulsively as a way of avoiding painful affect. Through this unconscious actualisation of the past in the present, the origin and the repetitive character of the enactment are unrecognised |
| Somatisation | Converting emotional pain or other affect states into physical symptoms and focussing one's attention on somatic (rather than intrapsychic) concerns. Somatisation involves different functional modes: somatisation by means of "histrionic" identification (conversion), somatisation as an emotional correlate (in somato-psychosomatic processes), and projective somatisation (externalisation of unbearable affects and pains into one's own body (hypochondria) |
| Regression | Returning to an earlier phase of development or psychic functioning to avoid the conflicts, pains, and tensions associated with one's present level of development or actual situation |
| Autism | Returning to one's inner world, often connected with schizoid fantasies, to avoid anxiety in interpersonal situations |

desires) or by means of role attribution (i.e. specific roles of children in the family through the parents) or, furthermore, by means of manipulation, seduction, or influence on the partner into a specific direction (Richter 1967; Mentzos 1982). These psychosocial or interpersonal mechanisms of defence (or arrangements) are widespread. The unconscious interaction between partners was also described as "collusion" (Willi 1975): "love as symbiosis" in the narcissistic collusion, "love as care for one another" in the oral collusion, "love as completely belonging to one another" in the anal-sadistic collusion, and "love as male confirmation" in the phallic-oedipal collusion.

A further variant refers to sado-masochistic relationships, in which the masochism of the one serves the sadism of the other and vice versa.

Beyond the intrapsychic and the interpersonal level institutions as a whole are also predestined—besides other defined functions—to represent and carry over functions of psychosocial defence ("**institutionalised defence**", Mentzos 1988). Accordingly, structures and principles of institutions do not only serve rational aims and tasks; moreover they are based on common values, attitudes, and emotion, often not clearly recognisable undefined motivations. Institutions enable, besides their defined tasks, to satisfy regressive drive desires and to develop defensive and protective behaviour against irrational, fantasised, and infantile anxieties, depression, shame, and guilt feelings (Mentzos 1988: 88ff.)

## 5.3 Second Dimension of the Three-Dimensional Neuropsychodynamic Model: Mode of Conflict

Clinical experiences underline that often not unspecific stress induces the development of psychic disorders but rather inner contradictions and conflictuous interpersonal relationships. On this background conflicts also play an important role in psychodynamics and in neuropsychodynamics as well.

Evolutionary and cultural theoretical aspects have to be considered when answering the question, why the mode of conflict is of great significance in neuropsychodynamic psychiatry: the development of the individual represents a **dialectic process**, as Mentzos (2009) emphasises. Potentially incompatible contradictions or "**bipolarities**" are involved in this dialectic process and are balanced time after time. It may be assumed that this dynamic pattern was successful during evolution, because on this way renewal, dynamics, progress, and differentiation were guaranteed. Such dynamic processes do not develop in a linear way, but rather risks are implicated, which may involve a possible blockade of the dialectic principle through a dichotomising "either-or" in the solution of conflicts and/or the generation of permanent rigid structures. Contrary to the instinct-based behaviour of animals, in whom the integration of contradictory tendencies is regulated and determined by means of innate instinct behaviour (e.g. separation from the mother animal), a great variety of possible solutions are available to humans, who were "thrown into freedom", on the background of the acquired ability to symbolise. Culturally grounded normative rituals assist the members of the respective culture to solve such ambivalence conflicts. Of course, there are quantitative and qualitative differences between cultures and within one culture (e.g. differences from one family to the other), which have to be taken into account.

In principle **outer conflicts** (constellations which are located on a real, superindividual level) have to be discriminated from **inner conflicts**. Inner conflicts are based on different contradictory desires of an individual. Often they are not conscious but result from strong unconscious energies, which strive to express themselves and which continuously have to be controlled by opposing tendencies. These interacting, opposing energies may be understood as wish and as defence mechanism: as different intrapsychic parts with different aims and as an impulse, which is contrary to the internalised perception of the demands of external reality.

Conflicts belong to healthy psychic life. The psychic disorders caused by them may only be expected in those in whom intrapsychic contradictions are hardened in rigid and absolute "of either-or patterns" on the background of unfavourable experiences and through specific defence reactions. On this way a further development of the individual is stopped.

The psychophysical organism is continuously occupied with the satisfaction of desires and interests and the protection against dangers as well. These functions had been described by Freud (1926) with the terms "lust-unlust principle" and "reality principle". Experiences of "lust" and "unlust" and the according further differentiated emotions (e.g. joy, well-being, elation on the one side and anxiety, mourning, shame, and guilt on the other side) may be functionally interpreted as indicators within a complex regulation system. In this process the maintenance of an **optimal level of tension** is of great importance. In case of overstimulation by outer or inner stimuli, regulatory mechanisms are automatically activated, to overcome an unpleasant or dangerous condition by decreasing the level of tension and generating an optimal level.

On the contrary, an intrapsychic level of tension, which is too low, contributes to an unfavourable psychic development of the person (increasing apathy and extreme psychic retreat and connected loss of "input"). If a compensation of extreme fluctuations of the tension is not possible, then emergency mechanisms may be activated (the development of psychopathologic symptoms may also be understood in this sense). A loss of psychic functions may result (or a loss of the psychosocial functional level, respectively). The contradictory and gradually petrified motivational tendencies, which are involved in the conflict, finally induce an increased tension. This increasingly unbearable situation may get along with dysfunctional adaptation processes and reactions of the stress aches, by which further neurobiological-psychosocial vicious circles are induced.

Given the great number of biological and psychic opposites, one pair of opposites is of specific relevance, namely, the conflict or dilemma between **self-oriented and object-oriented tendencies** of the human, i.e. the opposite between desires and tendencies of autonomous identity, autarky, and independence on the one hand and the wishes for bonding, communication, solidarity, and emotional near to the other on the other hand. Whereas this bipolarity contributes under favourable conditions to an undisturbed dialectical elevation and balancing of the opposites and

**Table 5.3** Conflict/dilemma and corresponding emotions

| | Conflict | Fear of |
|---|---|---|
| I | Autistic retreat versus fusion with the object | |
| | Object | Loss of self through loss of object or fusion with the object |
| II | Absolute autonomous self-worth versus absolute dependent self-worth | Loss of self-value through devaluation of the self or through devaluation of the idealised object |
| III | Separation versus dependence or individuation versus relationship | Threatening of the self through loss of object or clinging to the object |
| IV | Autarky versus submission | Rejection, not being loved, separation, or humiliating dependence |
| V | Identification with male role versus identification with the female role | Total loss of female role or total loss of male role |
| VI | Loyalty conflicts | Loss or need to betray the one or the other |
| VII | Triadic "oedipal" conflicts | Being expelled from parents, threatening of one's own integrity and safety, "castration anxiety" |

consequently to an enrichment and differentiation, the development of psychic disorders may be induced by an obstruction of this developmental process (Mentzos 2009, p. 30ff.). Therefore the psychogenesis of such psychic disorders either depends on psychosocial conditions as on specific biological determinations as well. With it compensating mechanisms are involved, which may, for instance, contribute to compensate the deficient consequences of biological determination by means of a favourable psychosocial environment, at least partly.

Starting from the existence of normal bipolarities and the resulting basic conflict or dilemma, Mentzos (1982, 2009) described the emerging conflicts in the disturbed psychic development as **variations of the basic conflict or dilemma** (see Table 5.3).

The conflict between self-oriented and object-oriented tendencies results during the course of development—dependent on the primary neurobiological disposition, temperament, personality of the person, and the respective psychosocial constellations—in different conflict constellations. Autistic retreat versus fusion/symbiosis with the object, marked "autonomous" self-worth versus object-dependent self-worth, marked separation versus dependency, and marked autarky versus submission represent the respective extreme poles of the same basic conflict.

The conflicts concerning psychosexual identity and loyalty mostly represent contradictions within the self-pole (do I experience myself more as a woman than a man?); the same conflict may be connected with the object pole (love to men or women; relationship to near relatives or even collectives, e.g. "my home country", etc.) or also in a form of loyalty to superiors or solidarity with colleagues. Especially in case of loyalty conflicts, more self-oriented (e.g. focussing on one's career) versus object-oriented tendencies (which result from emotional relationships and responsibility) may be differentiated. This differentiation also concerns the oedipal conflict.

Originally Freud understood the term "oedipal conflict" as the rivalry conflict of the boy with his father concerning the relationship to the mother (or the conflict of the girl concerning the relationship to her father). The oedipal conflict, which develops in the course of the fourth or fifth year of life, was supposed to be an important phase in the development of sexuality and used as description of the respective developmental phase of the child: oedipal or genital phase of psychological development.

Freud himself stressed in the later editions of his early works that the oedipal condition does not only implicate rivalry but also concern the love to mother and father. The boy gets involved in the relationship with his father within the oedipal constellation into a deep and nearly unsolvable intrapsychic conflict, wishing to overcome somebody whom he especially loves and whom he needs for the development of his male identity. This correlation of love and rivalry analogously also concerns the girl in relationship with her mother.

It had gotten necessary to overcome reductionistic assumptions resulting from drive theory (e.g. identification of the boy with his father or analogously identification of the girl with her mother, as consequence of homosexuality). Nevertheless, these necessary changes of the conceptualisation should not lead to a desexualising of psychoanalysis and its theories. The sexual drive represents one of the most important motivational sources of human life. The later described aspects of the self-psychology, the object relation theory, the intersubjective psychoanalysis, and the affect theory widen our understanding of the dyadic pregenital relationships and the oedipal constellation. The integration of the dyadic and the triadic relationship world represents one of the greatest challenges in human life.

## 5.4 Third Dimension of the Three-Dimensional Neuropsychodynamic Model: Mode of Structure

The third dimension of the neuropsychodynamic model concerns the level of the personality organisation (cf. structural dimension in the operationalised psychodynamic diagnostics/OPD, Working Group OPD 1998; OPD Task Force 2009). The maturational level of the psychic structure of a person is closely connected with the relationship to the object and to the self. In a psychogenetic perspective, structure and function are elements of a dialectical relationship and essentially may not be divided from one another.

Subsequent to the characterisation of a systematology of the essential conflicts and their corresponding mechanisms of defence and compensation, the question is whether and how such a similar systematology may also be formulated for the psychic structure (encompassing maturational levels, variations of structural deficits, and self-pathology).

Rudolf (2006) proposed a gradation with four different levels: well integrated, moderately integrated, poorly integrated, and disintegrated.

The assumption that psychic disorders are of conflictuous nature is not at all questioned by clinical observations which underline that a subgroup is not characterised by conflicts themselves but is based on the pathological processing of very early conflicts and the resulting structural deficits. This subgroup was described by Kohut (1971, 1977) with the term "self-psychology". Mentzos (1982, 2009) understood psychoses, borderline personality disorder, and narcissistic personality disorders as variations of structural deficits.

**Operationalised Psychodynamic Diagnostics (OPD)**

The working group operationalised psychodynamic diagnostics (OPD) aimed at supplementing the descriptive-phenomenological diagnostics (DSM, ICD) by means of a psychodynamic-psychogenetic-oriented operationalisation.

Psychodynamic elements were selected which are relevant for the understanding of the psychodynamics of patients on the one hand and moreover may be sufficiently be operationalised for empirical validation on the other hand. This selection resulted in four axes:

- Experience of the disorder
- Relationships
- Conflict
- Structure

A comprehensive, unique, and precise language and terminology was strived for. The following table (Table 5.4) enables an overview on the possible use and the aims of the operationalised psychodynamic diagnostics.

The necessary data for the OPD diagnosis are gained by means of a specific interview, which is open in the beginning and which focusses in the further course on certain dimensions aiming at assessing the respective OPD axes (cf. interview guideline, Janssen et al. 1996). The interview guideline represents a combination of

**Table 5.4** Possible use and aims of the operationalised psychodynamic diagnostics (OPD)

| | |
|---|---|
| – | Teaching of clinical diagnostic guidelines |
| | Training the application of psychodynamic and descriptive-phenomenological classifications in the course of the education in psychodynamic psychotherapy |
| – | Improvement of communication |
| | Interchange about the concepts of psychodynamic theory in the scientific community |
| – | Research |
| | Attributing to a greater homogeneity of random samples in studies by means of clearly defined operationalised diagnostic criteria |
| – | Baseline data |
| | Phenomenologic and psychodynamic assessment of clinically relevant data before the beginning of a planned psychotherapy |
| – | Assessment of the illness course |
| – | Assessment of the indication and differential indication for psychotherapy |
| – | Assessment of the efficacy of psychotherapy |

the classic psychoanalytic first interview, of the structural first interview, of the psychodynamic-biographic case history, and of psychiatric exploration.

The interrater reliability of the OPD is good (kappa value for conflict, 0.61; kappa value for structure, 0.71). The results of the reliability studies underline the importance and necessity of intensive and systematic teaching and training of the OPD diagnostics.

OPD is an internationally acknowledged diagnostic instrument for the assessment of operationalised psychodynamic diagnostics. A further developed second version of OPD is available in the meanwhile (OPD-2).

The essential descriptions of the first four axes (axis I, experience of the disorder and preconditions of the treatment; axes II, relationship; axis III, conflict; axis IV, structure) are in accordance with psychoanalytic concepts (of personality structure, intrapsychic conflict, and transference). Axis V adopts the established descriptive-phenomenological diagnostics (ICD-10), which underlines the necessity of a more detailed assessment of psychopathologic features also in the framework of psychodynamic diagnostics. The structure and contents of the OPD axes are elucidated in the following:

- **Axis I: Experience of the Disorder and Preconditions of the Treatment**

Different sections of feature were defined enabling the differential indication of psychotherapy and assessing the severity of the symptoms, suffering and mental trauma, treatment expectations of the patients, and their psychosocial resources. The assessment is based on five categories, for which respective anchor examples are presented.

**Axis II: Relationship**

The habitual and dominant relationship pattern of the patient is analysed. This axis represents a categorical system of behaviours, which can be observed by means of a list of 30 items. In a first step the relationship pattern is described in the emotional perspective of the patient, basing on the characterisation of relationships in the interview. Then, in a second step the relationship pattern is described in the emotional perspective of the other (including the psychotherapist). This is also based on the observed behaviour of the patient during the interview and, furthermore, the countertransference of the psychotherapist (or researcher). Finally, in a third step both emotional perspectives are integrated into a relationship-dynamic formulation. Within this procedure, especially the differences between the experiences of the patients and the experience of their partners may be focussed on (cf. Grande et al. 1997).

- **Axis III: Conflict**

The central role of inner conflicts is comprehended on this axis. Determining, internalised conflicts may be confronted with their respective actual, externally conflictuous situations. Different types of conflicts are defined, which have a predominant role in life:

- Dependence versus autonomy
- Submission versus control
- Care versus autarky
- Self-worth conflicts (self-worth versus object worth)
- Superego and guilt conflicts
- Oedipal-sexual conflicts
- Identity conflicts
- Conflictuous external life conditions
- Mode of compensation

As a further category, the clinical impression of a lacking perception of conflicts and emotions is described.

The manual defines criteria for the assessment of these conflicts in following domains: choose of partner, attachment behaviour, family relationships, family of origin, work/profession, relationships in the social environment, and illness behaviour.

The existence and significance of a conflict is rated on a four-stage scale; furthermore, it has to be decided, which two conflicts are the most significant ones for the patient. In a summarising decision the question is answered, whether the patient uses more a passive or an active mode coping the conflicts (cf. Schüssler et al. 1996). The main affects (e.g. rage after narcissistic hurts), which are often connected with long-lasting conflicts, have to be considered.

- **Axis IV: Structure**

The psychic structure represents the background, on which conflicts with their more or less well-adapted or dysfunctional solution patterns take place. Axis IV comprehends the functional level of the patients in the sense of their structural capabilities and vulnerabilities by means of six dimensions:

- Self perception
- Self control
- Defence
- Object perception
- Communication
- Bonding

The integration level is evaluated in regard of these dimensions by means of four different levels: "well integrated", "moderately integrated", "less integrated", and "disintegrated" (cf. Rudolf 2006). The global structure level is evaluated in a summarising decision also on a four-graded scale. This procedure is oriented on the structure levels defined by Kernberg (1975, 1976, 1981). Each of the six structural dimensions includes several subordinated aspects, e.g. the ability to self-control contains aspects of affect tolerance, self-worth regulation, impulse control, and anticipation.

– **Axis V: Psychic and Somatic Disorders**

In the framework of the diagnostic evaluation according to ICD-10/ICD-11 (research criteria) and DSM-5, only one main diagnosis should be given to each patient. This main diagnosis is the one with the highest importance for the actual treatment. Furthermore, three additional diagnoses may be coded. And last not least, personality disorders may be coded according DSM/ICD on axis V. Diagnostic criteria for the narcissistic personality disorder, which is not contained in the original version of ICD-10, are supplemented.

## 5.5 Specific Neuropsychodynamic Features of Psychiatric Disorders in the Three-Dimensional Model

- **Which are the specific neuropsychodynamic extensions of this originally three-dimensional psychodynamic model of psychic disorder?**
- **How can this three-dimensional model be transformed into a neuropsychodynamic context?**

In a first step, it will be demonstrated, how the three dimensions—mode of defence and compensation, conflict, and structure—may be put into the context of the brain. Yet it has to be stressed that this is not a reductionistic connection but a corresponding one, i.e. the psychic level of defence and compensation corresponds on the neuronal level with certain functions and mechanisms (see Fig. 5.2).

### 5.5.1 Mode of Defence and Compensation: Neuropsychodynamic Specification

The dimension of defence and compensation describes certain modes of intrinsically predisposed mechanisms for processing extrinsic life events. In the context of the brain, the following question arises: what intrinsic predispositions does the brain use as mechanisms of defence and compensation?

The intrinsic activity of the brain may be characterised by different neuronal networks, for instance, the default-mode network, the executive network, the sensorimotor network, the attention network, and the language network. These different networks are found in various regions. They are related to one another in a specific way: the default-mode network (DMN) may be characterised particularly by the midline structures of the brain, whereas the executive network (EN) covers lateral prefrontal and parietal regions. Functional imaging studies showed that both these networks are related to one another in a negative way, anti-correlating with one

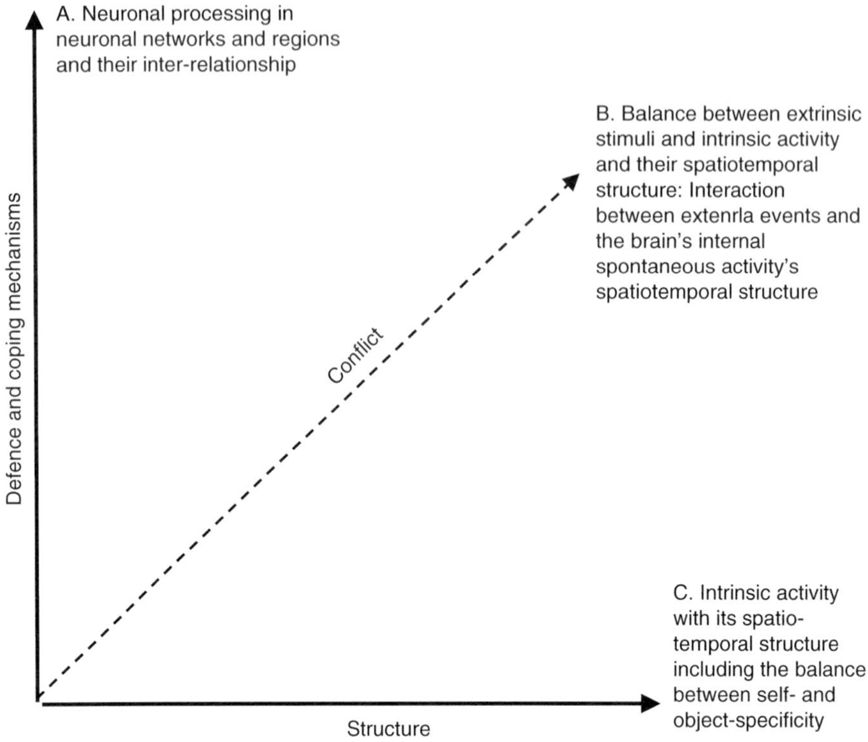


**Fig. 5.2** Three-dimensional neuropsychodynamic model of mental illness. (**a**) Neuronal processing in networks and regions and their interconnectivity. (**b**) Balance between extrinsic stimuli and intrinsic activity and their spatiotemporal structure. (**c**) Intrinsic activity and its spatiotemporal structure including the balance between self- and object-specificity

another. For instance, the activation of the DMN leads to a reduced activation of the EN and vice versa. If the EN is strongly activated, for instance, by cognitive processes, the activity in the DMN is reduced, and the associated emotions or self-specific processes may be recruited in only a limited way. Because of the balance between different networks, there are predetermined mechanisms, by means of which extrinsic stimuli and tasks may be processed.

Hence, the different networks form different layers of spatial structure. Analogousl so on the temporal side. Different regions/networks show different distributions of faster and slower frequencies, which together amount to a complex temporal structure with different layers, i.e. from very slow to very fast. We now suppose that these different layers in the spatial and temporal organisation of the brains' neural activity correspond to the different layers of defence mechanisms. From a neuronal perspective, the different modes of defence and compensation of conflicts and/or traumata are analogous to the resting state spatiotemporal structures' ability to modulate and adapt as well as to the resting state's self-specific organisation.

For instance, more mature defence mechanisms may concern the upper neural layers like sensory networks and faster frequencies. In contrast, more immature defence mechanisms may shift towards corresponding changes in the most basic and fundamental layers like the default-mode network and the very slow frequencies which both provide the "spatial and temporal basement" of the brain. That is analogous to the immature defence mechanisms that provide the most basic mechanisms of homeostasis within the brain and allostasis of the brain within the environment.

Two different spatiotemporal axes are important: firstly, a longitudinal axis with a very long spatiotemporal scale, possibly reaching back into early childhood. Secondly, the mode of processing of the conflict or trauma depends on the interaction between a certain life event at a specific time with the spatiotemporal structures of the involved conflicts or traumata. In the future, different mechanisms of neuronal processing which are analogous to different mechanisms of defence in a neuropsychodynamic context may be investigated.

### 5.5.2 Mode of Conflict: Neuropsychodynamic Specification

How can the extrinsic-intrinsic interaction between the external reality and the internal psychic reality be translated into the neuronal context of the brain? The psychic conflict in a neuropsychodynamic perspective may be characterised as an interaction between the intrinsic resting state activity and the extrinsic event or stimuli, which has been operationalised as rest-stimulus interaction.

One limitation should be mentioned here: resting stimulus and stimulus-resting state interaction describe specific mechanisms resulting from the interaction between intrinsic activity and an extrinsic stimulus. In contrast, conflict in a psychodynamic perspective is more a result than a mechanism. Focussing on the question of whether the resting state-stimulus and stimulus-resting state interaction and the interaction between intrinsic activity and an extrinsic stimulus result in different psychic conflicts will be highly significant for future neuropsychodynamic research on acute symptoms in psychiatric disorders. Ideally, different types of conflicts may correspond to different modes or types of interaction between external events and intrinsic resting states or spontaneous activity. That remains to be the subject of future research though.

### 5.5.3 Mode of Structure: Neuropsychodynamic Specification

The dimension of structure focusses on the organisation of the personality and the psychic structure of the subject including her/his relationship to the object and to the self (Kernberg 1975, 1976). Where can such a structure relating to the self and objects be found in the neuronal activity of the brain? The extrinsic activity of the brain focusses on the neuronal activity, which is triggered by certain extrinsic stimuli or tasks. This extrinsic activity should be differentiated from the brain's own activity which develops in the brain independently of extrinsic stimuli. This intrinsic activity is paradoxically called resting state activity—paradox because the brain is not continuously in the resting state. This can be seen, for instance, during sleep and dreams when the brain is still active despite a lack of certain stimuli or tasks.

The intrinsic activity has a specific spatiotemporal structure, but this should not be understood in a purely physical sense but rather in a virtual sense or in a statistically based way (for instance, like a computer programme which remains invisible to a third party). This spatiotemporal structure is most probably constituted by the interaction between different networks and regions and the interaction between different fluctuations in different frequencies (0.001–180 Hz) (see above). Recent studies showed that there is a continuous adaptation of a continuous variability, which is an essential characteristic of the intrinsic activity and its spatiotemporal organisation.

According to our own studies, it may be hypothesised that this spatiotemporal structure is essential for the transformation of purely neuronal processes into subjective experiences and phenomenal processes. Only if a certain extrinsic stimulus or task is integrated into this intrinsic activity can something be consciously experienced (Northoff 2014a, 2014b). Hence we assume correspondence between the structural organisation of the personality in its virtual spatiotemporal terms and the spatiotemporal structure of the brain's spontaneous activity. For instance, certain personality types like anxiousness may be characterised by predominance of faster frequencies relative to slower one in their brain's spontaneous activity; the continuous predominance of faster frequencies make them anxious as they lack the emotionally calming and soothing presence of slower frequencies.

A further analogy between the psychodynamic and the neuronal level concerns the self- and object-specificity. Intrinsic activity and its spatiotemporal structure show a strong overlap with the neuronal processes which underlie self-experience and are also at the same time greatly influenced by the outside world and its extrinsic stimuli. There is much evidence that particularly the midline regions show a strong overlap between intrinsic activity and activity by means of self-related stimuli (for instance, by saying one's own name). This is only possible if specific self-related information is decoded by the intrinsic activity and its spatiotemporal structure. On the other hand, neuroimaging studies have shown that the intrinsic activity and its spatiotemporal structure may be strongly influenced by the outside world, especially by stressful life events. Thus, intrinsic activity and its spatiotemporal structure would appear to fulfil the criteria for self- and object-specificity analogous to the structure in the psychodynamic context. Thereby, we ultimately want to understand self- and object-specificity as two extreme poles of an underlying commonly shared continuum which runs along the relationship between the brain and world, i.e. their respective spatiotemporal structures and how the brain is aligned to and integrated within the world—this amounts to what one of us calls world-brain relation (Northoff 2016, 2017).

**Case Study: Mrs. Rose**

*The imago of the dead mother*

One of the authors (HB) knows Mrs. Rose since 20 years.

At the time of her first admission to the Psychiatric University Hospital, she was 48 years old. She then suffered from a severe depression with psychotic features (delusion of guilt), showed stuporous symptoms, was mutistic, and had had suicidal ideas. She had not been able to work as an architect for several months. Obviously her former psychotherapist had not recognised the psychotic features of her depression and had interpreted her behaviour as a rejection.

This first manifestation of her depression was triggered off by a deep crisis in her relationship with her husband. The year before, she had felt more and more distant to him, and she was disappointed because of her impression that he seemed only to be interested in his career. Finally, she had fallen in love with another man. From that moment on, she was confronted with feelings of guilt and suffered from a loyalty conflict (between her responsibility for her 10-year-old daughter and her "selfish" emotional and erotic wishes). Mrs. Rose described this situation later on in the course of psychotherapy: "My sensuality is used against me and I have to pay for it".

During the inpatient treatment, Mrs. Rose was treated with antidepressants and lithium. She remitted after 10 weeks; afterwards she started with a psychoanalytic outpatient psychotherapy: she was very much motivated to tackle with her biography, the background of her guilty feelings, and her self-worth problems.

Soon after the beginning of the psychotherapy Mrs. Rose felt sad and remembered the death of her father, who had surprisingly died, when the patient was 11 years old. Mrs. Rose remembered the shock after the sudden death of her father and her "ocean of tears, as it was yesterday".

Mrs. Rose, the youngest of three children (sister plus 7 years, brother plus 12 years) had had a very deep relationship with her father, whom she had loved and admired. She still idealised him and started crying during many sessions of the psychotherapy.

On the other hand, the relationship to her mother was more negative and ambivalent. The patient described her disappointment, e.g. she often had felt that she was not able to get near to her and to find her resonance. These feelings of "blankness" increased after the dead of the patient's father. Mrs. Rose described the "petrification" of her mother, who suffered from a long-lasting depression (as she had done also before her husband's death). The patient often got the feeling being bounced off her mothers' "critical wall of silence".

At the time of adolescence, Mrs. Rose did not find the wished help and support of her mother and no orientation for her psychosexual development.

In spite of the traumatic death of the father and the long-lasting pathological mourning of her mother, the further development of the patient was favourable. She was successful in school and at university; Mrs. Rose was well integrated in her peer group and had girlfriends and some intense relationships with boyfriends. She married in the age of 29 years, a young academic, whom she admired because of his intelligence and knowledge. She kept on working as an architect till the birth of her only son (10 years later, when she was 39 years old).

Transference and countertransference: while the patient developed an idealising, at times sexualising relationship, the therapist felt the need to protect her because of her fragility.

One main focus was the traumatic loss of her father during the early adolescence: the psychotherapy enabled—as a "second chance"—a deep mourning, which had not been possible for the patient during adolescence because of the emotional "petrification" of her mother. Therefore this focus was soon connected with another essential focus, the "imago of the dead mother" in the sense of A. Green (2002): the living mother who is already "dead", not being able to react positively and empathetically onto the child's wishes to be near.

Another focus concerned the sequential process of getting into a state of exhaustion and increasing "petrification" (on the background of actual and former relationship and conflict patterns). This included the development of new strategies to overcome ruminations and negative beliefs.

In the transference-countertransference relationship, the dilemmatic connection of wishes for bonding, idealising love, and erotic wishes on the one hand and loss anxieties and guilty feelings on the other hand could be worked through. The understanding of illness-relevant psychodynamic constellations and maladaptive coping mechanisms supported the further stabilisation of the patient and contributed to the prophylaxis against recidivism.

Nevertheless, in spite of the positive development of the patient, a second depressive episode occurred 2 years later, after the husband had decided to divorce and had confronted the patient with his wish to live together with another woman.

This second depressive episode was characterised by an intense inhibition, ruminations, and cognitive symptoms. A long inpatient treatment (9 months) was necessary, before the patient gradually recovered. Mrs. Rose was again treated with a combination of different antidepressants and lithium. Finally, lithium had to be stopped because of side effects. Trileptal was given as an alternate mood stabiliser, but unfortunately the sodium level was reduced dangerously under this treatment, so that the patient had to be treated in an intensive care unit.

Three weeks after recovery, manic symptoms were observed for the first time. They lasted for several weeks; the diagnosis was changed (bipolar affective disorder).

The psychotherapy was continued in the following years. Mrs. Rose remitted and stayed in a stable condition for several years.

It was to the psychotherapist's astonishment, when he was informed after his return from a stay in a foreign country that the patient had been admitted again to the psychiatric hospital during his absence. This third inpatient treatment had been necessary because of a changed symptomatology, namely, a psychotic episode: she was in a highly agitated and irritated psychotic state with paranoid anxieties (e.g. to lose her job, mistrustful, feeling betrayed by her colleagues and even by her best friends). She was not at all compliant, claimed the injustice of the school system and of psychiatry, and hoped that the psychotherapist could free her out of the "prison". The diagnosis "schizoaffective disorder" was discussed.

Mrs. Rose wished to be in contact with the psychotherapist during this crisis, and she calmed down in the therapeutic sessions. Several weeks later the outpatient psychotherapy was continued, after most of the psychotic symptoms (besides her mistrust against some of her friends) had remitted. The triggers, which had led to this psychotic episode, could be worked through. These stressors concerned disappointing experiences in her job and in her new partnership.

The low-frequency psychoanalytic psychotherapy (combined with low-dose antidepressant treatment and lithium, which was tolerated now) was continued in the following years. Mrs. Rose continued her work successfully until she retired (at the age of 64).

The remission continues, and Mrs. Rose looks confidently into the future, though she sometimes fears that "the door back to depression could be opened again". In these situations, it is important for her to hear the voice of the therapist, and she gives him a call.

She knows that sleeping disturbances are possible "alarm symptoms". But in the meanwhile Mrs. Rose is able to cope with her vulnerability and certain triggering events, without further depressive or psychotic episodes.

**Comment on the Case from the Psychotherapist's Perspective**

Mrs. Rose is to be characterised by a mature personality, which enables her to live a rich life with an interesting profession, many cultural interests, and deep interpersonal relationships.

But there are some constellations, which threw a shadow on her life and which led to a mixed biopsychosocial vulnerability.

The somatic-genetic component of this vulnerability for psychiatric disorders may be related to the history of depression of near relatives (recurrent depression of the mother, bipolar disorder of the uncle).

The psychosocial components of the vulnerability concern the emotional deprivation in early childhood and adolescence, which resulted from the lack of empathy of her mother, who could not react empathetically on the emotional wishes of the patient to be near (because of her own chronic depression). This lack of resonance was only partly compensated by the warm father, who was very interested in the development of the patient. All the worse, the sudden, unexpected death of her father in her early adolescence got the quality of a traumatic loss experience. Mourning did not happen, not at least because of the "petrification" and the long-lasting retreat of the mother.

The dyadic relationship of the patient with her mother may be characterised by the emotional and relational constellation, which was described by A. Green (2002) as "imago of the dead mother", which is connected with feelings of "blankness" in the attachment situation and strongly and negatively influence the further development of identity and self-worth.

Another essential component of the superego structure of the patient is a consequence of the experienced "deadness" of her mother: the development of a very rigid, "motherly" superego on the background of the identification with the strict and always silent mother. This rigid "motherly" superego is one of the psychodynamic components of the later psychotic depression (delusion of guilt).

Furthermore, the integration of the dyadic and the triadic relationship was not possible for Mrs. Rose. She kept on idealising her father and had not the chance to internalise the imago of both her parents' relationship. The solution of the genital-oedipal conflict (love to mother and father and overcoming the ambivalence) was blocked. Sensuality and erotic wishes were connected with guilty feelings and a loyalty conflict. This psychodynamic constellation played an important role in the first depression of the patient. On a deeper, dyadic level, loss anxieties contributed to the manifestation of the depressive episodes and had an essential importance in the further life of Mr. Rose.

The mature personality of Mrs. Rose enabled her to use the regressive process in the psychoanalytic psychotherapy for her own development (including the use of more mature mechanisms of defence and compensation as intellectualisation, rationalisation, identification, displacement, and repression).

On the other hand, the self-pathology components of Mrs. Rose resulting from the emotional deficits in the early relationship with her mother may have contributed to the later psychotic episode (which occurred in the absence of the psychotherapist). A psychotic level of the psychic development was reactualised contemporarily (including immature mechanisms of defence as splitting, psychotic projection, denial, acting out, and regression).

The course of the illness underlines the significance of the three psychodynamic dimensions concerning the different symptomatologies (syndrome change): one may say that Mrs. Rose left her "defence lines" in specific crisis situations (loss of self-worth and feelings of guilt in depression, loss of identity and reality testing in psychoses on the background deep anxieties of loss). Her mature personality structure (and the reliability of the therapeutic relationship) enabled her to return to her former level of psychosocial functioning. The mode of interpersonal defence and compensation is still in her own focus (experiencing the reactualisation of early relationship pattern in the actual partnership, e.g. idealising tendencies connected with masochistic submission).

Some more general comments:

In the long course of the psychotherapeutic relationship, different therapy-technical problems were obvious: The psychotherapy of patients with major depression is often confronted with a partial temporary decoupling of the psychic process and the actual symptoms (in spite of a relative positive and successful psychic process in psychotherapy, relapses occur; kindling).

Different levels of symbolisation may be obvious, within which somatopsychic processes are involved.

Countertransference feelings are sometimes influenced by the depressive affect in the sense of "presymbolic affect contagion". The process of resymbolisation then involves the sharing of a somatic processed sensorimotor affect, by means of a transmodal change, which enables mourning and separation in the further course.

A stepwise use of different psychotherapeutic interventions is necessary regarding state and trait variables. The "now moments" in the therapeutic relationship are of outstanding significance. Moreover, time factor, network dynamics, and gene expression afford a sufficiently long-lasting treatment, including psychotherapeutic "maintenance strategies".

**Comment on the Case from the Neuroscientist's Perspective**

The biographical history of the patient with her several episodes of childhood trauma must have certainly impacted the spatiotemporal structure of her brain's spontaneous activity. It is known that early childhood trauma leads to instability, i.e. disordered spatiotemporal structure that provides less temporal continuity across time. This is, for instance, reflected in decreased power in the slow frequencies (Nakao et al. 2013) and unstable pattern in the anterior midline structures (Duncan et al. 2015). One may now suspect that our patients' spontaneous activity is characterised by exactly that decreased power in slow frequencies which makes her inner world experience unstable and temporally less continuous. While the spatial instability leads to abnormal development of an inner experiential space, her inner space is most likely not only unstable but smaller and shrinked in its extension which becomes rather virulent in acute depression when she feels "locked in". These spatiotemporal abnormalities concern her personality organisation, the third dimension in our neuropsychodynamic model.

Our patient showed good cognitive development with realistic life and some mature structure. This more mature "structural overlay" superpositions her early immature and unstable spatiotemporal structure. One would, for instance, assume that cognitive networks like the central-executive network and corresponding faster frequencies are well developed—there is somewhat mature cognitive functioning as it is well developed in her biography.

However, such mature cognitive functioning in central-executive network and faster frequencies is based on a somewhat immature unstable earlier spatiotemporal structure as related to the default-mode network and slower frequencies—this is like well-developed upper floors of a house built on an unstable basement or fundament.

How is that manifest in her biography and behaviour? Life events that are meaningful to her earlier biography reactivate the instabilities of her early immature spatiotemporal structure in DMN and slower frequencies—unlike in nonmeaningful life events, the immature state of the earlier spatiotemporal structure can no longer be compensated for in the case of meaningful life events. The early immaturities break through and become manifest in depression and psychosis as realisation of the earlier inner world experience as infant. One would thus suspect that during depression and psychosis shows a spatiotemporal pattern in her spontaneous activity that resembles the earlier one, strong predominance of DMN and slower frequencies (as manifest in the increased self-focus and detachment from the world in depression). While her central executive network and the faster frequencies vanish in the background and decrease—this is what one observes in acute depression (Chap. 11).

The immature spatiotemporal structure with DMN and slow frequencies prevents mature defence mechanisms; her scope of possible reaction to both outer and inner experience is limited and extremely restricted; her defence level regresses to an immature level in the acute depressed state, while it is somewhat mature in her stable intervals. This covers the first dimension of our three-dimensional model, the defence mechanisms.

How about the second dimension, the abnormal reactivity to external life events, i.e. conflict? That is paradigmatically reflected in the life events triggering her depressive episodes. Due to its earlier immaturities, the brain's spontaneous activity reacts abnormally to specifically meaningful, i.e. self-related, events—they abnormally decrease the central-executive network and its faster frequencies which shift the predominance back to the default-mode network and slower frequencies. It is like if one puts strong power on the floor in the upper basement of a house—if the foundation or basement is unstable, there is the likelihood that the upper floor will collapse completely and crumble. That is exactly what happens in her case—the more mature layers of her spontaneous activity' spatiotemporal structure crumble and revert to the immature ones. Hence, her second dimension, the interaction between external life events and her inner brain's spontaneous activity is highly volatile and thus prone to crumble her back into her earlier immature life as child.

### 5.5.4 Neuropsychodynamic Model: Schizophrenia and Depression

#### 5.5.4.1 Structure Level

Both disorders, depression and schizophrenia, concern basic disturbances, changes in the neurobiological and psychic structure, and in psychodynamic mechanisms as well. Correspondingly can be assumed on the basis of the neuropsychodynamic model that modifications in the resting state activity or intrinsic activity of the brain and especially its spatiotemporal structure are concerned. Exactly this is the case. Actual neuroimaging studies showed strong changes of the intrinsic activity and its spatiotemporal structure. For instance, the different neuronal networks, the default-mode network (DMN) and the executive network (EN), have significant alterations, e.g. the frequency fluctuations—especially in the higher gamma frequency between 40 and 180 Hz—are significantly decreased.

#### 5.5.4.2 Schizophrenia

We remember that we connected the level of processing with the relation between different regions of the brain and neuronal networks. In schizophrenia a very interesting result was found. Normally, there is a negative or anticorrelating relation between the DMN and the EN. That is, when the DMN is increased, then the EM is decreased and vice versa. Interestingly, also different mental contents are connected with these networks: the DMN most of all mediates internal mental contents, e.g. one's own thoughts, whereas the EN is more directed to external thoughts and actions. On this background the anticorrelation is very useful. We experience that on ourselves always again: when the internal thoughts are strong, the external-oriented actions are much less and vice versa. In schizophrenic patients, it is especially this balance, this negative or reciprocal balance between internal mental contents (the DMN) and the external-oriented actions (the EN), which seems to be contrary, namely, more positive than negative. This ultimately means that internal and external mental contents are mixed with one another, instead of being reciprocally modulated. This is just the fact, which also can be observed in the clinical routine from the symptoms, namely, that schizophrenic patients are not able any longer to differentiate internal psychic contents and external events.

Let us go back to the dimension of conflict. This depends on the interaction of resting state and stimulus, the resting state-stimulus interaction. Because of the deficits as well in the structure as on the level of processing, in schizophrenic patients, necessary deficits of the resting state-stimulus interaction result. Typically is the example of the acoustic hallucinations. Here we find clear indications that the auditory cortex shows an abnormal resting state activity and possibly also an abnormal increased variability. These spontaneous variations of the variability occur with such a strong intensity that they may correspond exactly to the difference between the resting state activity and an extrinsic stimulus in the healthy brain. Consequently, the schizophrenic patient does perceive acoustic hallucinations not in his head but in the outside. This results in a conflict by hearing internal voices, which the patient does not perceive externally. This exactly corresponds to the conflict, which the schizophrenic patient is confronted with on the psychodynamic level in the interaction with the social environment.

#### 5.5.4.3 Depression

In depression we find multiple indications for an increased resting state activity especially in the DMN. This causes a change of the complete spatiotemporal structure of the intrinsic activity and to a disbalance between DMN and EN. But here the reciprocal or negative relation between DMN and EN is shifted abnormally in the direction of the DMN. That is, the DMN is abnormally strong, whereas the EN is abnormally low. This disbalance results in an abnormal increase of internal mental contents, whereas the external-oriented actions are abnormally decreased. Exactly this can be observed in depression, namely, an increased self-focus with strong ruminations and a deceased relation to the environment, a decreased outside focus. Therefore depressive patients feel isolated from the social environment, but instead—quasi in a circular way—an abnormal strong focus on their own self. Each interaction with external stimuli—now we are moving to the level of conflict—results in an abnormal integration of this external stimulus into one's own self, and finally the differentiation between external stimuli and one's own self is not possible any longer. This sequence leads to conflicts between the self and the outside, which are increased in a circular way through the abnormally increased resting state activity in the DMN, which cannot be modulated by external stimuli any longer. Finally, this sequence results into a deficient differentiation between the self and the outside world.

## 5.6 Summary

The three-dimensional neuropsychodynamic model of psychic disorders enables a better understanding of the link between emotional experience and neuronal mechanisms concerning the essential dimensions of the human development, the mode of defence and compensation, the mode of conflict, and the mode of structure. Therefore it may be used for the conceptualisation of psychiatric disorders and a better understanding of psychiatric patients. Furthermore, it is a diagnostic instrument for the planning, application, and evaluation of psychotherapy.

Last but not least, the three-dimensional neuropsychodynamic model of psychic disorders—as a core element of neuropsychodynamic psychiatry—may further contribute to the development of future diagnostic classification systems focussing on functional mechanisms in psychopathology and their subject-related application. Neuropsychodynamic psychiatry can thus contribute to a better understanding of irregular, turbulent, and dynamic processes, which has long been standard practice in scientific models of complex system.

## References

Böker H. Sekundary amenorrhoea and psychosomatic phenomenon: Empirical studies of the personality structure and the object relations of patients with amenorrhoea. Medical Dissertation, Faculty of Medicine, Justus-Liebig-University, Giessen/Germany; 1979.

Böker H, Hell D, Budischewski K, Eppel A, Härtling F, Rinnert H, von Schmeling C, Will H, Schoeneich F, Northoff G. Personality and object relations in patients with affective disorders: idiographic research by means of the repertory grid-technique. J Affect Disord. 2000;60:53–60.

Duncan NW, Hayes DJ, Wiebking C, Tiret B, Pietruska K, Chen DQ, Rainville P, Marjańska M, Ayad O, Doyon J, Hodaie M, Northoff G. Negative childhood experiences alter a prefrontal-insular-motor cortical network in healthy adults: a preliminary multimodal rsfMRI-fMRI-MRS-dMRI study. Hum Brain Mapp. 2015;36(11):4622–37.

Freud S. Inhibition, symptom and anxiety. Standard Edition, Volume XX, 1926. pp. 75–176.

Gabbard GO. Psychodynamic psychiatry in clinical practice. Washington, DC: American Psychiatric Publishing Inc; 2005/2014.

Grande T, Burkmayer-Lose M, Cierpka M, Galwender RW, Davis-Osterkram S, Vriesert G, Jurasky PO, Brach C, Schauenburg H, Strak N, Strauss W. Die Beziehungsachse der Operationalisierten Psychodynamischen Diagnostik (OPD) – Konzept und klinische Anwendungen. Z Psychosom Med Psychoanal. 1997;43:280–96.

Green A. The imago of the "dead mother". Psyche Z Psychoanal. 2002.

Janssen PL, Dalbender RW, Freiberger HJ, Holst G, Mans EJ, Rudolg G, Schneider W, Seitler GH. Leitfaden zur Psychodynamisch-diagnostischen Untersuchung. Psychotherapeut. 1996;41:297–304.

Kernberg OF. Borderline conditions and pathological Narcissism. New York: Jason Aronson; 1975.

Kernberg OF. Object relations theory and clinical psychoanalysis. New York: Jason Aronson; 1976.

Kernberg O. Structural interviewing. Psychiat Clin N Am. 1981;4:169–95.

Kohut H. The analysis of the self: a systematic approach to the psychoanalytic treatment of narcissistic personality disorders. New York: International Universities Press; 1971.

Kohut H. The restoration of the self. New York: International Universities Press; 1977.

Laplanche J, Pontalis J. Das Vokabular der Psychoanalyse. Frankfurt: Suhrkamp-Taschenbuch; 1989.

Marty P, de M'Uzan M. La pensée operatoire. Rev franc psychosomat 27:345 ff; 1963.

Mentzos S. Neurotische Konfliktverarbeitung. Einführung in die psychoanalytische Neurosenlehre unter Berücksichtigung neuer Perspektiven. München: Kindler Verlag; 1982.

Mentzos S. Interpersonale und institutionalisierte Abwehr. Frankfurt: Suhrkamp; 1988.

Mentzos S. Lehrbuch der Psychodynamik. Die Funktion der Dysfunktionalität psychischer Störungen. Göttingen: Vandenhoeck und Ruprecht; 2009.

Nakao T, Bai Y, Nashiwa H, Northoff G. Resting state EEG power predicts conflict-related brain activity in internally guided but not in externally guided decision-making. NeuroImage. 2013;66:9–21.

Northoff G. How is our self altered in psychiatric disorders? A neurophenomenal approach to psychopathological symptoms. Psychopathology. 2014a;47(6):365–76.

Northoff G. Unlocking the brain, Coding, vol. 1. Oxford: Oxford University Press; 2014b.
Northoff G. How do resting state changes in depression translate into psychopathological symptoms? From 'spatiotemporal correspondence' to 'spatiotemporal psychopathology'. Curr Opin Psychiatry. 2016;29:18–24.
Northoff G. The brain's spontaneous activity and its psychopathological symptoms–spatiotemporal binding and integration. Prog Neuro-Psychopharmacol Biol Psychiatry. 2017;80(Pt B): 81–90. https://doi.org/10.1016/j.pnpbp.2017.03.019.
OPD. Operationalisierte Psychodynamische Diagnostik. Bern: Grundlagen und Manual; 1996.
OPD (1998) Operationalisierte Psychodynamische Diagnostik. Grundlagen und Manual. 2. Korrigierte Auflage. Hans Huber, Bern.
OPD Task Force. Operationalized psychodynamic diagnostics OPD-2 – manual of diagnosis and treatment planning. Göttingen: Hogrefe Publishing GmbH; 2009.
Richter HE. Eltern, Kind und Neurose. Rheinbeck: Rowohlt; 1967.
Rudolf G. Strukturbezogene Psychotherapie. Leitfaden zur psychodynamischen Therapie struktureller Störungen. Stuttgart: Schattauer; 2006.
Schüssler G, Holst G, Hoffmann SO, Mans E, Mentzos S. Operationalisierte Psychodynamische Diagnostik (OPD): Konfliktdiagnostik. In: Buchheim P, Cierpka M, Seifert D, editors. Lindauer Texte. Wien: Springer; 1996.
Willi J. Die zweier Beziehung. Spannungsursachen, Störungsmuster, Klärungsprozesse, Lösungsmodelle. Rowohlt, Rheinbeck; 1975.

# 6 Unconscious: Psychoanalytic Perspective

Marianne Leuzinger-Bohleber

**Abstract**

Neuropsychoanalysis is a young discipline that developed in the last 20 years. One leading proponent was Mark Solms followed by others like Eric Kandel, Heinz Boeker, and Georg Northoff. A central issue in neuropsychoanalysis, as in psychoanalysis, is the concept of the unconscious. This can be understood in various ways of cognition and, relying on Jaak Panksepp, affect and emotion. The unconscious has been associated with traumatic memories in psychoanalysis; neuropsychoanalysis extends this by associating unconscious traumatic memories with memories in the sensorimotor system of the body rather than the cognitive functions of the brain. This suggests convergence between neuropsychoanalysis and embodied cognitive science as it is also illustrated by a case report whose implications for the transference between analysand and therapist are pointed out. It is concluded that neuropsychoanalysis can draw on many fields including neuroscience and embodied cognitive science to sharpen and more detail the concept of the psychoanalytic concept of the unconscious.

> As you know, we have never prided ourselves on the completeness and finality of our knowledge and capacity. We are just as ready now as we were earlier to admit the imperfections of our understanding, to learn new things and to alter our methods in any way that can improve them. (Freud, SE. XVII, p. 159)

M. Leuzinger-Bohleber
Sigmund-Freud-Institut, Frankfurt am Main, Germany
e-mail: Leuzinger-Bohleber@sigmund-freud-institut.de; M.Leuzinger-Bohleber@gmx.de



H. Boeker et al. (eds.), *Neuropsychodynamic Psychiatry*,
https://doi.org/10.1007/978-3-319-75112-2_6

## 6.1 Personal Introductory Remarks

The unconscious is still the central concept of psychoanalysis, one reason why the fifteenth Joseph Sandler Research Conference in Frankfurt was devoted to the topic: *The Unconscious: A bridge between psychoanalysis and cognitive neuroscience*. The publication of the main papers (Leuzinger-Bohleber et al. 2017) is one source of the following chapter (Sect. 6.2). Mark Solms, an expert on Freud's neuroscientific papers and founder of the International Society for Neuropsychoanalysis, summarized the state of the art of the dialogue between psychoanalysis and the neurosciences on "the unconscious" and its consequences for psychoanalysis. Therefore, I will summarize his main theses in Sect. 6.4. A second source for discussing a contemporary understanding of the unconscious was the conference of the IPA "Exploring Core Concepts: Sexuality, Dreams, and the Unconscious" in 2011 (Sect. 6.3). Finally I have been involved in the interdisciplinary dialogue with Rolf Pfeifer, an expert of embodied cognitive science for almost 30 years. Therefore, I will illustrate that the concept of embodied memories might enrich the ongoing struggling for understanding the "unrepresented" in contemporary psychoanalysis (Sects. 6.5 and 6.6).

## 6.2 The Unconscious in Dialogue Between Psychoanalysis and the Neurosciences[1]

As is well-known, in Freud's time psychoanalysis was characterized as "the science of the unconscious mind." In the last 100 years, many other disciplines, among them cognitive science, have studied nonconscious mental functions. What are the differences between the conceptualization of "the unconscious" in psychoanalysis and in cognitive science? Is the core thesis of psychoanalysis still plausible, namely, that unbearable impulses and fantasies from the past and present are banished into the unconscious, from whence they continue to shape feelings, thoughts and behaviors in unknown ways? And is such an understanding of the unconscious still central for helping patients in contemporary psychotherapy?

For many authors, like the Nobel Prize winner, Eric Kandel, Sigmund Freud's dream has become a reality in recent decades: he never gave up the hope that developments in the neurosciences might someday contribute to a "scientific foundation" of psychoanalysis. He abandoned this attempt, his "Project for a Scientific Psychology" (1895), due to the obvious limitations of the neurosciences of his time (see Kaplan-Solms and Solms 2000), subsequently defining psychoanalysis as a "pure 'psychology' of the unconscious." As Kandel (1998, 1999, 2006) and many others point out, however, the developments in the neurosciences, such as neuroimaging techniques (MEG, EKP, PET, fMRI, etc.), open a new window for psychoanalysis to the non-psychoanalytic scientific world. Kandel is passionate about this vision:

[1] The following chapter is based on a former publication by Leuzinger-Bohleber and Solms (2017).

### 6.2.1 The Necessity to Investigate Psychoanalytic Treatments by Neuroscientific Methods

He is convinced that psychoanalytic treatments must show their effectiveness also in studies applying methods of contemporary neurosciences. He certainly is right in one respect: if psychoanalysis could show that its treatments influence the brain's functioning, this would tremendously heighten its acceptance as a treatment method in medicine and the mental health systems. Several groups of researchers presently engage in such studies (see Chap. 20).

### 6.2.2 Neuropsychoanalysis

Kaplan-Solms and Solms (2000, 2003) have developed the so-called clinico-anatomical research method for investigating patients with brain lesions using clinical psychoanalytical techniques. In different countries interdisciplinary research groups using this method work systematically with patients with localized brain lesions (see Röckerath et al. 2009; Leuzinger-Bohleber et al. 2015; Leuzinger-Bohleber and Kächele 2015). Their findings are broadly relevant for studying the ancient mind-body problem in new and fascinating ways (see, e.g., Damasio 1999; Sacks 2007 and many others).

The first volume of the international journal *Neuropsychoanalysis* was published in 1999; leading psychoanalysts and neuroscientists published their studies on emotion and affect, memory, sleep and dreams, conflict and trauma, conscious and unconscious problem-solving, etc. The International Society for Neuropsychoanalysis, founded in 2000 by Mark Solms and others, organizes annual congresses of the Society on such topics around the world.

It seems undeniable that an exchange between psychoanalysis and neuroscience is most promising for both parties. The neurosciences have developed objective and exact methods to verify complex hypotheses on human behavior, while psychoanalysis can contribute the necessary rich knowledge concerning the meanings and motivations of psychic processes and can therefore direct interesting questions at the neurosciences.

### 6.2.3 Psychoanalytical Conceptual Research and Some Epistemological Remarks

Another field of research, mentioned by Eric Kandel, is psychoanalytical conceptual research, a specific and genuine psychoanalytical research field. As was discussed in several papers at the fifteenth Joseph Sandler Research Conference in Frankfurt in March 2014 mentioned above, the interdisciplinary dialogue fertilizes clinical psychoanalytical work in an indirect way (see Leuzinger-Bohleber et al. 2017). Results from neuroscientific studies will never be able to tell a clinical psychoanalyst how to cope with a certain analysand in a certain psychoanalytical situation. The psychoanalytical treatment technique and intuition are something fundamentally different than the further development of concepts and theories. Therefore, the exchange between the knowledge base of psychoanalysis and the neurosciences always takes place at the level of concepts and theories, never at the level of concrete clinical interactions (see, e.g., Leuzinger-Bohleber 2015). Nevertheless for the further development of psychoanalysis as a scientific discipline, an openness and an attempt to achieve "external coherence" (C. Strenger) of psychoanalytical concepts with the knowledge of neighboring disciplines are inevitable. Psychoanalytical concepts and theories should not be in uncritical contradiction with the current knowledge in other scientific disciplines. Perhaps surprisingly many of the central psychoanalytical concepts of Sigmund Freud have indeed proven to be "externally coherent" with modern neuroscientific understanding of the same complex psychic processes. Some of them can even be understood more precisely and deeply. On the other hand, we also have to discuss some critical points in our theorizing and modify or even rethink of some of our psychoanalytical concepts and even central concepts as "the unconscious", in the light of modern neuroscience (see, e.g., Solms 2013, Northoff in this volume).

From an epistemological point of view, we have to take into consideration that the dialogue between psychoanalysis and the neurosciences is still relatively new. Therefore, some of the possibilities of this dialogue are overestimated or even idealized. Besides, according to personal experiences in a common endeavor of 20 psychoanalysts and neuroscientists, studying memory, dreams, and cognitive and affective problem-solving in a joint research project in 1992–1998 (supported by the Köhler Foundation, Darmstadt, Germany), while fascinating and innovative, is challenging and complicated for both sides (see Koukkou et al. 1998; Leuzinger-Bohleber et al. 1998).

We often don't speak the same language, apply different concepts in analogous terms, and often identify with divergent traditions in science and in philosophy of science. Much tolerance and stamina is needed to achieve an intensive exchange of ideas enabling us to reach new intellectual frontiers: to crack up former understandings and concepts and resist idealizing tendencies to expect "solutions" for unsolved problems in our own discipline from the other foreign one, which—like a white screen—attracts projections and projective identifications. To take new findings of the other discipline means to undergo uncertainty and unease; it is painful to leave aside "certainties" and false beliefs developed in your own field. Going through a period of uncertainty and unease is inevitable, a must for a productive and constructive dialogue reaching beyond a rediscovery of already established disciplinary knowledge. The comparison of models developed by both disciplines in order to explain their specific data collected by specific (and very different) research methods is linked to complex and sophisticated problems of philosophy of science and epistemology. The well-known danger of the eliminative reductionism of psychic processes onto neurobiological processes or the consequences of a transfer of concepts, methods, and interpretations from one scientific discipline onto another, without reflecting them, need to be prevented.

Therefore, we cannot agree completely with the passionate conviction of Eric Kandel that modern neurosciences really can save the future of psychoanalysis. On the one hand, we share his view that curiosity and openness toward scientific developments, neurosciences included, are a must for innovation and creativity. In order to remain a "Wissenschaft" of the mind, psychoanalysis must refresh and further develop its concepts and theories, showing again and again that psychoanalytic theories are "externally coherent" (Strenger 1991) with the state of art of other disciplines, e.g., the neurosciences. In this sense, the future of psychoanalysis as a productive "Wissenschaft" depends on openness toward contemporary neurosciences.

However, we must carefully avoid "categorical mistakes" (see, e.g., Leuzinger-Bohleber and Pfeifer 2002; Leuzinger-Bohleber 2015). The data of the neurosciences are on a completely different level than those of psychoanalysis, aiming to understand and decode meanings of unconscious psychic functioning of human beings. Epistemologically, psychoanalysis is a "specific science" (Wissenschaft) with a specific methodology suited to investigate its specific research object (unconscious conflicts and fantasies) and its specific scientific quality criteria. Psychoanalytic research method has contributed a large body of knowledge and cannot be replaced by any other one, including neuroscientific ones (see Chap. 20 in this volume). Michael Hagner (2008), a philosopher of science and historian, who investigated how the visualization of processes which takes place in the interior of our bodies and brains is influencing our thoughts, fantasies, and emotions as well as our culture in general, stated:

"There is a distinction (in studies on imaging techniques) between disordered thinking from mathematical problem-solving, … those first memories of childhood experiences, of the last quarrel with one's life partner or the conflicts with parents, of erotic dreams about the most exciting love relationship. As is well-known, in the twentieth century it was primarily psychoanalysis which was to first single out such phenomena for research. The biographical detail, intimacies and concealed layers this discipline retrieved will doubtless never be matched by screening the brain…

This shift [from psychoanalysis to neuroimaging, L-B] could lead to circumstances in which the multiplicity and relevance of the life of the mind are measured primarily by the methods of visualization. The price for such a development consists in the fact that: the investigation into the deeper connections, the explanation, listing, narration, and evaluation, in short, historical, scientific textual linear thought is displaced by a new, image-based, "superficial way of thinking" (ebd.). The consequence of this shift with respect to the sciences of man is that the analytic depth of former forms of thought, for which psychoanalysis may be considered representative, will be replaced by the superficial insight of neuroimaging. Human understanding is thus relegated to the status of an excrescence of material forms of representation.… (p. 278f.) (See also Chap. 20 in this volume)

## 6.3 Conceptualization of the Unconscious in Contemporary, Pluralistic Psychoanalysis

As we all know, Freud effectively contributed the third largest insult to mankind by discovering "the dynamic unconscious." He shocked us with the insight that none of us are "masters of our own mental houses"; we are driven by libidinal and aggressive drive impulses and unconscious fantasies derived from them. In every one of his works, Freud warned not to deny these unconscious powers. Only in acknowledging their effectiveness can we guarantee a wise handling of them. Turning our backs and negating the unconscious not only leads to psychic illness, it also enhances the danger of uncontrolled outbreaks of drives and threatens human cohabitation and culture.

Throughout its history of more than a century, psychoanalysis has differentiated itself as a science with 12,000 members of the International Psychoanalytical Association in terms of its central concepts such as "the unconscious" to the point of a "plurality of theories," and the question arises: Does "a psychoanalysis" exist at all? Don't we have to speak of "many psychoanalyses?" While modern ego psychology-oriented psychoanalysts such as New York's Fred Pine (2011) still refer to the "dynamic unconscious" as the product of fended off impulses and drives, which are to be examined by psychoanalysis, others (e.g., Giuseppe Civitarese (2011) from Pavia) define a continuum of the conscious and unconscious, with reference to Bion. The unconscious does not protrude via, for instance, slips of the tongue into the symptoms of the conscious; rather every conscious process is accompanied by an unconscious process. Based on neuroscientific findings and experimental psychological research on unconscious forms of information processing, Werner Bohleber (2010) also goes by the notion of a *non-repressed unconscious* and differentiates it from the "dynamic unconscious" and a "creative unconscious." Jorge Luis Maldonado from Brazil (2011) on the other hand firmly believes in the concept of the dynamic unconscious and the psychoanalytic structural theory, which distinguishes psychoanalysis from other disciplines examining hidden, nonconscious information processing. Finally, based on Jacques Lacan, Miguel Kolteniuk Krauze (2011) from Mexico City advocates two dimensions of the unconscious as a system of "primary repression, which is characterized by its inertia and lack of symbolization capabilities and a secondary repression which is characterized by the primary process and its fate. Hence Andrè Greens approach concerning the preservation of the drive dimension" (p. 2).

All of these authors were keynote speakers at the IPA Conference 2011 in Mexico City, themed "Exploring Core Concepts: Sexuality, Dreams, and the Unconscious." This short summary of the diverging views is able to illustrate how the plurality of theories is a characteristic of the prosperity of modern, international psychoanalysis as a discipline which has always been concerned with highly complex clinical phenomena and has strived to decode conscious, preconscious, and unconscious inner workings in joint efforts with patients. When referring to psychoanalysis as a scientific discipline, which just like any other science puts its findings up for a critical discussion in the non-psychoanalytic community, we must always continuously refurbish the lenses of this kaleidoscope in order to recognize commonalities as well as differences to individual conceptualizations of the unconscious and to enable fruitful discussions. This is a prerogative for any innovative advancement in psychoanalysis as an internationally acclaimed science.

## 6.4 Which Concepts of "the Unconscious" in Psychoanalysis May Remain Unchanged, Which Have to Be Modified or Even Been Dropped by Neuroscientific Finding?

Mark Solms (2017) discussed these questions extensively and even tried to approach a first integration of contemporary psychoanalytical knowledge on unconscious conflicts and fantasies on the one hand and neuroscientific findings on the other hand. Therefore, I am referring to his considerations in the following section:

### 6.4.1 Most Mental Processes Are Unconscious

Since Freud's description of unconscious mental functioning, many studies from experimental and development psychology have strengthened his conviction:

> that at any given moment consciousness includes only a small content, so that the greater part of what we call conscious knowledge must in any case be for very considerable periods of time in a state of latency, that is to say, of being psychically unconscious. (Freud 1915, p. 167, quoted in Solms (2017), p. 17)

Bargh and Chartrand (1999) estimate that consciousness plays a causal role in only 5% of cognition. Intensively discussed were, e.g., the experimental neurophysiological studies by Libet (1985) which demonstrated that voluntary motor acts are initiated at the supplement motor area before a subject becomes aware of the decision to move. This discovery has initiated an intensive and controversial discussion on the "free will," in German philosophy. This debate takes up the abovementioned irritation due to the "third largest insult to mankind" postulated by Freud that mankind not even is the master of his own home but determined in all his (conscious) decision by unconscious motivations, drives, longings, and unsolved conflicts of his past.

### 6.4.2 Unconscious Processes Are Automatized Cognition

In the 1920 Freud mentioned in several papers that not only psychic conflicts are kept in the so-called dynamic unconscious but that many secondary processes are unconscious as well (see, e.g., Freud 1923). However, up to present, controversial discussions focus the concept of the dynamic unconscious, in other words, the psychoanalytical core thesis, that unsolved psychic conflicts and fantasies determine—without the knowledge and control of the human subject—emotions, thoughts, and actions in the present. Only very few cognitive scientists (e.g., Ramachandran 1994; Anderson et al. 2004 and Pfeifer and Bogard 2007) try to connect Freud's dynamic unconscious with contemporary neuroscientific findings. Most cognitive psychologists and scientists postulate that the unconscious is a repository of automatic and automatized information processing and behavioral capacities (see, e.g., Kihlstrom 1996).

### 6.4.3 Conscious Is Endegenous

In contrast to cognitive neuroscientists, many researchers in the field of affective neuroscience as, e.g., Jaak Panksepp (1998), don't limit their studies on the investigation of cognition but also include affective processes associated with them. "What Freud called the "id" is the principal object of study in affective neuroscience (Solms and Panksepp 2012)" (Solms 2017, p. 18).On the other hand, these researchers disagree that the "id" is mainly connected with the unconscious localized in basic structures of the brain (brainstem). Therefore, Solms talks about the "conscious id."

> On this view, consciousness derives from the deepest strata of the mind, it is inherently affective, and it is only secondarily "extended" (to use Damasios's term) upwards to the higher perceptual and cognitive mechanisms that Freud described as the systems Pcpt.-C., and Pcs. In other words, it is the *higher* systems that are unconscious in themselves. They borrow consciousness via associative links from the lower system, not the other way round...... There is a clear contradiction here. The pleasure principle cannot simultaneously be a bottom-up force and a top-down sensory óffshoot of the cortical layer. (Solms 2017, pp. 21, 22).

### 6.4.4 Affect Is Always Conscious

Solms illustrates this thesis, e.g., by the famous example of a child which was born without neocortex and thus blind and deaf (Shewmon et al. 1999). However, these children are capable to show (and probably experience) feelings. Therefore, affects are always conscious.

As we have discussed in several papers, affects—according to concepts of "embodied cognitive science"—have the function to simplify complex (embodied) perceptions and enable a spontaneous (unconscious) evaluation of these perceptions in order to initiate prompt reactions of the subject. To give just an example, if a teacher of a class with small children wants to cross a street with a lot of traffic and sees a boy waving with a fascinating toy on the other side of the street, he immediately feels intensive fear or anxiety because he unconsciously recognizes immediately the danger that children could run over the street; the (conscious) emotional reaction, fear, enables him to react promptly and to hold back the children from running over the street! His emotional reaction (fear) is thus an immediate evaluation of a complex situation of acute danger which enables a functional reaction much earlier than a cognitive analysis of the situation would allow (see, e.g., Pfeifer and Leuzinger-Bohleber 1986; see Sect. 6.4).

### 6.4.5 Not All Consciousness Is Declarative

Solms is referring to the work of Edelman (1993) and his differentiation between primary and secondary consciousness.

Freud's usage of the word "consciousness" typically refers to secondary consciousness, that is to awareness given various namely by different theorists, such as "declarative" consciousness, "reflective'consciousness," "áccess'consciousness," "áutnoetic'consciousness," "éxtended'consciousness," "higher-order" thought, etc. Primary consciousness, by contrast, refers to the indirect, concrete, phenomenal stuff of sentience. As we have seen, Freud was dimly aware of this distinction, but he did not think through the implications. (Solms 2017, p. 26)

Analogous arguments can be derived from the dialogue with embodied cognitive science, as we have discussed in several papers (Leuzinger-Bohleber and Pfeifer 2013; Leuzinger-Bohleber 2015, also see Sect. 6.5). All these just mentioned different categories of memory (declarative, reflective, extended, higher-order thought, etc.) are focusing a so-called "descriptive level" of memory functions. They do not describe the "explanatory level" which means the mechanisms of the brain which determine memories. In contrast embodied cognitive sciences have developed a very different conceptualization of memory which is essential for understanding unconscious and conscious memory processes. All experiences, from the very beginning on, are influencing sensomotoric coordinations and thus unconscious memories. Therefore, e.g., traumatic events are unconsciously remembered already in the very first months of life not only when declarative memory is determined. Susan Coats and Theodore J. Gaensbauer (2009) illustrated this thesis by impressive case examples with children who had experienced severe trauma already during the first year of life (see Sect. 6.5). These experiences have an unconscious influence on later affects, behavior, and fantasies and only may become conscious by the interpretative work of child psychoanalyst as, e.g., Agneta Sandell, illustrated in a case example with a 2-year-old child (see Sandell 2014).

### 6.4.6 The Systems, CS, and PCS Are Unconscious in Themselves

As Solms elaborates, these new conceptualizations of mental functioning have important implications for psychoanalysis. As mentioned above most of the psychic processes are unconscious.

> Thinking is necessary only when problems arise. This (the problem) generates the conscious `presence'of affect and, thereby, attention to the object of perception and cognition. (or embodied memories as mentioned above, MLB). However, the whole purpose of the reality principle (of learning by experience) is to improve one's predictive model: that is, to minimize the chances of surprise - to solve problems - and thereby to minimize the need for consciousness. The classical model, therefore, is again turned on its head. (Solms 2017, p. 28)

### 6.4.7 Repression Is Premature Automatization

Solms tries to integrate all this new knowledge into a "metapsychology of repression" (p. 29 ff.). He refers to the neural process of reconsolidation by which previously consolidated memories are made labile again through reactivation of the consolidated memory traces. For Solms repressed memories are prematurely consolidated solutions:

"that is, non-solutions- predictions that constantly give rise to prediction errors. Hence the ever-present threat of a "return of the repressed" which gives rise to neurotic symptom formation…. The tragedy of repression (or premature automatisation) is that it renders childish solutions immune to updating, Hence the central task in psychoanalytic therapy is to de-automatise, to render conscious once more, to permit reconciliation to take place, and then to automatise better solutions." (Solms 2017, p. 30) (see also Solms and Friston 2014).

### 6.4.8 Conclusions

Solms (2017) summarizes his attempt to integrate psychoanalytical and neuroscientific findings on the unconscious in the following conclusion:

This review of Freud's metapsychology of 'The Unconscious' in relation to some findings of contemporary cognitive and affective neuroscience suggests that his model is in need of major revision:

1. The core processes of the system *Ucs.* (the processes that Freud later called 'id') are not unconscious. *The id is the fount of consciousness*, and consciousness is primarily affective. I therefore propose that *the Ucs. and the id are different mental systems, and that they should be located separately.*
2. The primary consciousness generated in the id is of a different kind to that generated in Freud's system *Cs.* Freud's systems *Pcpt.-Cs.* And *Pcs.* are concerned primarily with what is now called secondary or 'declarative' consciousness.
3. The systems *Pcpt.-Cs.* and *Pcs.* (the systems that Freud later called 'ego') are *unconscious in themselves*; and by inhibiting the id *they aspire to remain so.* They inhibit the id in order to supplement stereotyped instincts with learning from experience. Unsuccessful instinctual predictions generate affective consciousness (prediction error; free energy) which can only be tamed through thinking (problem solving).
4. The ego systems borrow consciousness as a compromise measure, they *tolerate* consciousness, in order to solve problems and resolve uncertainties (to bind affect). Once a realistic solution is found for an id demand, however, the *raison d'être* of consciousness disappears. *Then a memory-trace arises instead of consciousness.* This is 'Nirvana'.
5. The system *Ucs.* includes all such automatized predictions. This system is not the id; the *Ucs.* is hived off from the ego. The 'dynamic' part of the *Ucs.* is simply the part of it that malfunctions, that causes prediction errors (causes affect; re-awakens the id). The dynamic ('repressed') part of the *Ucs.* therefore *tends to re-attract consciousness.* This is the threat of the 'return of the repressed.'
6. The task of psychoanalytic therapy is to connect the affect (the 'free energy' of the id) generated by prediction errors (by 'surprises' in reality) with the illegiti-

mately automatized predictions that gave rise to it (the 'repressed' in the *Ucs.*). This enables the individual (the conscious ego) to *think* its way through an unsolved problem once more, and then to *reconsolidate* (to re-automatize in the unconscious ego) the memory traces in question. Conscious thinking is thus a temporary state, located half way between affect on the one hand (problems) and automatized behaviour (solutions) on the other. (pp. 31, 32).

In the last two sections of this paper, I would like to mention another attempt to use knowledge from cognitive neurosciences for widening the psychoanalytical understanding of unconscious mental functioning. It is the struggle for a new understanding of the "unrepresented" by applying concepts from the field of so-called embodied cognitive science.

## 6.5 The "Dynamic Unconscious" and the "Unrepresented": Embodied Memories and the Unconscious

Generations of psychoanalysts since Freud have concerned themselves with the way in which repetition of unconscious fantasies and conflicts in transference can be rendered a healing process of remembering. This primarily involves symbolically (unconsciously) represented and repressed memories or relationship patterns. However, psychoanalytic theories on the unconscious have focused for quite some time on psychic material present in the analytical relationship in other ways. Levine et al. entitled their anthology "Unrepresented States and the Construction of Meaning" (2013, in honor of André Green) and focus on the question of the search for meaning in the unrepresented from a contemporary psychoanalytical perspective. With his broadly received concept of "dead mother," Green (2007) described the early identification with an absent mother leading to a withdrawal cathexis and thus to a disappearance of the inner representation which, in the transference relationship, can be perceived by the analyst as an empty, negative hallucination of the object, "a representation of the absence of representation" (Green 1999, p. 196, quoted from Reed 2013, p. 39). Reed (2013, p. 29 ff.) points out that this negative hallucination of the object leads to an emptiness rather than a representation of the lost object—an empty mirror, which, with these patients, is always there—but which is frequently observed in the analysand's extreme reactions to separation from the analyst.

Green is concerned with the process of de-objectification, namely, the obliteration of representation. Other psychoanalysts, by contrast, focused on the psychic material of patients, which had only insufficiently, if at all, gone through the processes of symbolization. Dominique Scarfone (2013, 2015) presented a conceptual integration of different forms of psychic representation and their various psychoanalytic conceptualizations. He compared Pierce's sign theory to Freud's conception of primary and secondary processes; Lacan's theory of the real, the imaginary, and the symbolic; Wilfred Bion's beta and alpha elements; Jean Laplanche's infantile sexual theories and their decoding in analytic discourse; and Pierra Aulangier's concept of the primary, such as "primary violence," which entered the stage ("mise-en-scène") and that could ultimately open up the discourse on secondary processes: a brilliant example of contemporary conceptual research.

In several papers we have pursued another path by drawing on several studies in the field of basic research, more specifically, embodied cognitive science and the cognitive neurosciences, so as to show that these disciplines offer first explanations for this clinically important phenomenon, such as the analyst's spontaneous inspiration, which represents an initial central step to understanding hitherto unrepresented psychic material, and which is capable of making psychoanalytic processing accessible (Leuzinger-Bohleber 2015). Hence, this should provide new perspectives on familiar concepts, such as "scenic understanding" (Argelander, Lorenzer), "hearing with the third ear" (Reik), "cracking up" (Bollas), or the "now moments" by the Boston Change Process Study Group. Furthermore, aspects of current discourse on intersubjective psychoanalysis and on enactment are touched on, as well as further understanding of countertransference around the bodily sensations of the analyst.

In the psychoanalytic model of representation and in the computer metaphor derived from "classic cognitive science," memory and recollection were for a long time understood as processes whereby (statically) retained knowledge was transformed from long-term memory to short-time memory and called up into a current problem-solving situation. We still find comparable thinking in some textbooks in clinical psychology. Aristotle's famous example comparing memory to a wax tablet into which experiences etch themselves appears to live on. This (erroneous) idea of memory has also entered popular language usage: "we call up saved knowledge" or "we search for forgotten names in memory" (much like the search for an object in a wardrobe).

According to various views in embodied cognitive science, today's memory can no longer be understood as comparable to a computer, as storage disk with statically stored content from which information can be "retrieved" in a present situation. What patients expected from the analyst are new, existential, and important relationship, not an unconscious "statically entrenched" representation of former relationship to traumatizing object unconsciously reactivated, as had been understood, for example, in reference to the model of representation in classical psychoanalysis (cf., e.g., Karl Menninger's triangle of insight 1958). Memory is a function of the entire organism, the product of complex, dynamic, recategorizing, and interactive processes, which are invariably "embodied." "Embodied" not only means "non-verbal": memory arises by way of a "coupling" of reciprocally influential sensoric and motoric processes. This "coupling" is biologically implemented through neuronal maps embedded in the organism's sensomotoric system. Thus, Clancey (1993) defined memory as the ability to coordinate neurological process and to categorize sensoric and motoric processes, as these occurred in an analogous earlier situation.

To summarize the essential theses of embodied cognitive science on unconscious and conscious mental processes:

1. Biological systems are self-organized and develop "intelligent" bodies, namely, structures in which they interact with the environment by way of sensomotoric coordinations without central regulation.

2. Learning always simultaneously occurs sensomotorically (in the body) and in the brain (in neuronal networks). Thus, learning always is mostly unconscious—only a very small portion of the learning processes becomes conscious.
3. Learning, problem-solving, and memory are thus no longer functions of a "saving in the brain" but invariably the product of complex, self-regulated, and sensomotoric coordination.
4. Psychic processes, such as "unconscious memories" or affects and fantasies evoked in a certain situation, are "constructed" between subject and environment in the here and now of a current interaction: consequently, thinking, feeling, and action thus arise only interactively; the subject cannot learn in an insular quasi autistic capsule and further develop itself; it requires interaction with the environment.
5. Similarly, such categories that constitute the basis of all learning and understanding do not develop by retrieval or modification of stored knowledge. They are automatically brought forth by sensomotoric coordination (spontaneously "constructed").

   Since this is decisive for our subject of understanding that which is non-represented, "unconscious" in contrast to "conscious," one experiment should be cited: if we give a 1-year-old child a red rubber ball in one hand and in the other a brown chocolate bar, it will put both in his mouth several times though prefers the chocolate bar no later than after two or three attempts; through sensomotoric coordinations—the learning by doing—it has formed categories without an adult having to explain it to him, namely, without the aid of cognitive schema: the brown, long-shaped object tastes good, one can eat it, and although one can bite the round object, it does not taste good and one cannot eat it! And yes, at some point the mother will remark: "and, does the chocolate taste good?" from which point on the child also associates the linguistic concept with his self-constructed categories. As this example indicates, the concept of embodiment provides a solution for one of the central problems of developmental psychology, namely, the early prelinguistic, (unconscious) acquisition of categories and, finally, also symbols and language.
6. The concept of "embodiment" is thus radically "historical," as psychic processes in the present always take place as the product of sensomotoric coordinations analogous to those in the subject's idiosyncratic past: the past inevitably impacts the present and future—that is, for the most part, unconsciously.
7. In that each new experience further develops sensomotoric coordinations, earlier experiences are permanently rewritten. Hence, the "historic truth" can never be reconstructed "one to one" on the basis of specific behavior in the present. Put more bluntly: this is the subjective part of all psychic experience. And yet, in the sensomotoric coordinations, past real experiences are retained "objectively" ("embodied") and can be measured, in principle, with the aid of neurobiological methods. For this reason, psychic experience, such as memory, always receives a "subjective" as well as an "objective" side. (For further details, see Leuzinger-Bohleber et al. 2017, p. 145 ff.)

## 6.6 Present in the Body, But Not Represented: Embodied Memories and Trauma—A Case Example[2]

A further controversy concerns memories of very early, traumatic experiences. As Fonagy (2010) emphasizes, implicit memory assumes a key role in the mediation of post-traumatic symptoms. Relatively primitive structures of the nervous system, such as amygdala and the hippocampus, presumably participate in the mediation of the memory of these experiences. According to Fonagy, traumatic memories are decontextualized via the sensory system in the form of synesthetic perceptions, smells, tastes, or visual images and cannot be conscious in cases in which they are not provided with new significance. From a psychoanalytic perspective, initially it may well be useful that a traumatic experience is not in consciousness. However, it continues to exert an effect unconsciously and thus determines current thought, feelings, and actions undetected.

According to Olds and Cooper (1997), the 2-year-old human hippocampus is immature in contrast to amygdala, which is completely developed by this time. Hence, very early childhood anxieties are stored in the "emotional memory" of the "immature" amygdala-integrated circuits and are barely accessible to (adult) conscious verbal narrative. These theses would appear to contradict the findings of Rovee-Collier (1997, 1999) and Rovee-Collier and Cuevas (2009), namely, that infants from 23 months on can already form a declarative explicit memory. Hence, there was no developmental phase in which only procedural implicit memory emerged. The formation of memory is a very diverse, complex, and variable process including feelings, motifs (one's own and foreign), anxieties, and conflicts and which takes place very early on in life. Gaensbauer (2002, 2011) holds a comparable view, showing, by way of impressive clinical examples, that at the age of 2 and 3, children already remember extremely traumatic events that took place in their first year of life (e.g., the shooting of their father). With the aid of the concept of "embodied memories," the Freudian thesis can be supported that early and earliest memories deposit themselves "in the body" (cf. also, among others, Leuzinger-Bohleber 2008; Leuzinger-Bohleber et al. 2014).

Therefore, as discussed in Sect. 6.5, I am postulating that embodied memories are always "kept in the body" and are often unconscious sources of "irrational," "inadequate" feelings, thoughts, and actions of patients who are seeking treatment. To understand these unconscious embodied memories often proves to be indispensable for achieving a psychic transformation of severely traumatized patients, as I would like to illustrate by the following case example.

As outlined above, in many cases traumatic experiences can only be fragmentarily recollected or else dissociated entirely from current consciousness. In psychoanalytic therapy, they repeat themselves in enactments and other manifestations of transference. Formerly, this memory of traumatic experience has been explained by way of a model of representation in which, due to excessive arousal, traumatic experiences are not integrated but incompletely represented or even only registered. Contemporary

[2] The case example was published in Bohleber and Leuzinger-Bohleber (2016).

interdisciplinary research results are now available following radical rethinking on the conception of memory and recollection. In this section I will try to illustrate how this new conception is able to offer an alternative explanation of the way traumatic memories function and their understanding in psychoanalytic treatment.

As summarized in Sect. 6.5, inspired by biology and the life sciences, embodied cognitive science currently understands human organism—and the human psyche—as being in an ongoing (embodied) state of change involved in constant dynamic processes of interaction with the environment in which a continuous process of recategorizing experiences occurs. Memories of earlier situations unconsciously determine present thought, feelings, and action, though not in the sense of stored knowledge in analogy to a computer or static memory traces. In contrast memories are products of dynamic, complex constructions in the here and now. In the sense of embodiment, sensorimotor coordinations in the present always operate in an analogue manner as was the case in earlier situations. The similarities between a current and a past situation are not perceived cognitively, e.g., by cognitive pattern matching, but by similar complex information gained by different senses (auditory, visual, olfactory, touch, smell, etc.) and actions of the body (characterized as sensorimotor coordination in embodied cognitive science). Through such sensorimotor coordination, memories and categories are constructed automatically as self-regulating process of learning by doing (Dewey), in other words, by means through coordinating information from sensory channels and connected (motor) actions of the body. Memories resulting sensorimotor coordination thus provide orientation in a new situation.

Another field of research is important for understanding social interaction in general and transference relationships in particular. Recent studies have illustrated the decisive role of the so-called mirror neuron system, which enables human beings to identify immediately with the observed behavior and the mental state of others (see, for instance, Gallese (2013)). In the analytic context, this means that during interaction with the analysand, analogue sensorimotor coordinations take place within the analyst as in the analysand implying that unconscious processes of immediate identification are occurring. These processes bring forth categories of understanding—automatically, spontaneously, and unconsciously—which are connected with the analysand's unconsciously occurring memory processes from earlier, important relationship experiences. In the case of traumatized patients, these are recurring memories of psychically unbearable experiences of over-flooding, extreme powerlessness, desperation, pain, panic, and fear of death. By identifying with the analysand's ongoing sensorimotor coordination and the construction of memories of the traumatic experiences, the analyst immediately (unconsciously) understands the traumatic psychic reality of the patient. And yet, at the same time, the extreme quality of traumatic experiences mobilizes his own spontaneous defense, thereby hindering becoming conscious of that which is perceived.

The following example serves to illustrate these conceptualizations:

> Hardly had I opened the door before Ms. M. stormed in across the threshold. She clasped my hand feverishly, pressing it between hers in a peculiar and sexually stimulating manner while at the same time stepping up very close to me, thus

encroaching on my normal sense of bodily intimacy: "Well, hi there. . . . I'm so glad to have the opportunity to speak with you." Intuitively, I took two steps back, immediately perceiving a forceful, negatively emotional reaction combined with an aversive physical response: "What an overwhelming woman! I find this too much. She's really coming too close for comfort. . . . Why did I offer her an appointment? Will I ever be able to send her away? Evidently, she is very needy."

Then she asked for the toilet and left the door wide open. Only once she was seated in the chair opposite me did I become aware of her pretty, girlish face as it strove to maintain a permanent social smile and of her beautiful female form, which she apparently sought to conceal beneath loose-fitting jeans and a frayed, plain pullover. Though in her mid-40s, she rather looked like a 60-year-old woman. She had already informed me over the telephone that her family doctor had recommended she seek out psychotherapeutic help. She was ill and suffered from burnout syndrome with attendant heavy depression.

As mentioned, the first (conscious) thoughts occurring to me, "What an overwhelming woman," and "I find this too much!" clearly contained both the perception of an overpowering quality of the patient's trauma-induced psychic reality, as well as my own defensive reactions.

When treating traumatized patients, it often takes considerable time before the traumas suffered can be understood in greater depth and observed in detail in the transference relationship itself. Over the course of therapeutic interaction, (new) sensorimotor coordination develops which, in patients, successively constructs memory of the holding function of the new analytic object. Among the well-known, most enduring experiences of severe traumatization is the complete breakdown in the basic trust of a helping object. As is generally known, connected with this are the unconscious convictions and phantasies wherewith the affected person ascribes guilt to himself for the traumatic event to which it is connected. Thus, initially, traumatized patients will once again continually repeat this inner truth in the transference, before gradually limiting its validity following alternative relationship experiences in the psychoanalytic treatment. The unconscious memories of the traumatic relationship experiences cannot be erased and are thus repeated time and again in the analytic relationship. At the same time, however, alternative sensorimotor coordinations (i.e., expressed metaphorically, alternative neuronal paths) can be constructed successively which is, in turn, connected to the (new) categories, security, reliability, understanding, and survival which characterize the analytic relationship. The old recategorization processes (of a basic mistrust in the object and the self) run parallel and disconnected alongside the new ones, which form in the analytic relationship for considerable time. Only once new recategorizations have led to more or less stable categories, such as trust, security, etc., do the two paths in sensorimotor coordination (the neuronal maps according to Edelman 1987) connect with one another. This is the precondition for the possibility of traumatic experience being reexperienced directly in the analytic relationship. This is also a precondition for creating significant associations in the analyst's mind as initial keys for finally understanding the specific detail of the trauma. These associations may initiate a therapeutic process of recollection and of understanding the trauma and ultimately open a process for working through these complex processes which may be illustrated briefly by the following clinical example:

Only in the third year of psychoanalysis did the significance of the aforementioned scene in the first interview reveal itself. The intrusive behavior outlined repeated itself in the analytic treatment in numerous variations. One day prior to the following psychoanalytic session, I found Ms. M.'s behavior intensely irritating. She appeared unannounced at one of my lectures, seating herself in the first row. During the subsequent session, I had been listening to Ms. M. for about 10 min: when explaining how, prior to his death, her uncle had recounted her youthful impatience when waiting for him in front of his studio, my immediate association was that her uncle had sexually abused her.

MLB: Could it be that what you remember is that the impatience and the visits to your uncle were yours, indeed, that you actively sought to be close to him, because it was, perhaps, too painful for you to think that your uncle abused your yearning for your father and had thereby transgressed the borders of intimacy? [To my great astonishment, she replied:]

M: Naturally, we shared affections—but I enjoyed it. When he touched my breasts, I finally felt myself to be an attractive young woman.

This example illustrates the way in which the first spontaneous and still theory-free association of sexual abuse in the analyst can only form once some degree of trust has developed in the analytical relationship.

The subject sexual abuse disappeared from the sessions for some considerable time, though brutally sexualized scenes increasingly appeared in Mrs. M's dreams. I again sought to establish a connection to this.

MLB: You were already in your adolescence when you visited your uncle and can probably remember the experiences. Some time ago, you explained how you and your uncle shared affections. Could it be that you are reluctant to think about further details of what had occurred between you and your uncle?

Ms. M. reacted vehemently to this question. She went to the toilette and vomited. In subsequent sessions, it gradually became possible for her to talk about memories of the coitus experiences with her uncle that had been marked by violence. Disgust, loathing, and repugnance showed themselves: the acting out of the overwhelming, traumatic experiences gave way to a successive memory and verbalization.

Ms. M. blamed herself for these events: "I was so in need of love and affection. Little wonder my uncle responded to this." Only gradually was she able to admit that this really concerned a matter sexual abuse, which had exercised a huge influence on her sexuality as a woman. "When visiting my uncle as a

13-year-old, I would always rush into the studio and initiate our sexual adventures: It was me who wished to be the emancipated, unconventional person, not him. I found it good."

Only then did I understand that the scenes during the first interview outlined previously contained unconscious embodied memories of her traumatic experiences with her uncle: she had also literally overrun and overwhelmed me in the first interview and had "come too close for comfort." However, although at the time of the assessment interview I had unconsciously understood the traumatic psychic reality of the patient (by means of the aforementioned identification processes), at that time it was not possible to decipher precisely these unconscious memories of Ms. M. in enactment: only once I came to know the analysand much better, and would frequently experience the intrusive infringements in the transference relationship directly, and had established a sustained analytic relationship to her, did the decisive association (sexual abuse) occur to me.

I had clearly hit the mark. It now transpired that Ms. M. had been sexually abused by her uncle between the ages of 13 and 20. And yet it was only during the sustained psychoanalytic relationship, by way of new memories of brutal scenes that she was finally able to admit that these were sexual assaults and that it was not a case of voluntary "emancipated," "happy" affairs, which she had initiated. Only the secure and empathetic analytic relationship enabled her to gain painful insight into how destructive the effects of these experiences had been for her and that they had contributed substantially to the fact that she had until now never been able to allow herself a constant, affectionate, as well as passionate love relationship.

My association (sexual abuse) facilitated, for the first time, the expression in language of that which had hitherto been present in her body but not been represented and to thus initiate a process of working through in the transference relationship. Although the limits of this article place constraints on a more detailed discussion, it ought to be mentioned here that—as the concept of embodied memories postulates—Ms. M's traumatic experiences were repeatedly overwritten. Thus, over the course of the fourth year, the dreams ultimately led to a further, unexpected discovery: with her adolescent experiences of sexual abuse, Ms. M. had been subject to additional unconscious embodied memories of the brutal rape of her mother by Soviet troops and which Ms. M. had observed as a 3-year-old; These were traumatic memories that, in late adolescence, had also unconsciously induced her to engage in several dangerous sexual adventures that led to seven abortions within the space of 10 years. The unconscious feelings of guilt triggered by this, among others, determined her depressive breakdown, as became clear only later during psychoanalysis.

Ultimately, these embodied memories were also closely connected with traumatic experiences of separation from her mother, who suffered from severe postpartum depression after she had received news that her husband was missing at the Russian front. She was no longer able to care for her baby and was obliged to put her in care for several months. Incidentally, embodied memories of these early separations were likewise contained in the initial scenes mentioned at the outset: the manner in which Ms. M. had pressed my hand between hers not only had a sexually stimulating character but, as we later came to understand, were a way of literally holding on to me and of not losing me. "Will I ever be able to send her away? She seems to be so needy," were, at that time, the categories of understanding that spontaneously occurred and formed within me, and with that—in retrospect—I already then understood the early trauma of separation, but—also due to my own defense reactions—was not yet able to decipher in detail.

With respect to the theory of treatment, the focus on new theories of memory as based on research in the life sciences may influence psychoanalytic attitudes insofar as they sensitize one's own sensitive (embodied) bodily responses. Furthermore, the aforementioned insights in psychoanalysis over foregoing decades—which lead to both an exclusive work with transference and to an exclusive (mostly intellectual) reconstruction of the (traumatic) life history of the analysand, which do not lead to a sustained therapeutic change—have experienced a new interdisciplinary support by way of the concept of embodiment (see section 6.5....). On the one hand, the psychic processes of the patient invariably depend on his current interaction with meaningful others (e.g., in the transference) and are thus consequently invariably intersubjective. On the other hand, current experiences are always determined by sensorimotor coordinations formed in the subject's idiosyncratic (biographic) past. In this sense, the individual's distinctive history is embodied because sensorimotor coordinations emerge in the earliest relationship experiences and, as mentioned, continually (causally) determine presently occurring psychic processes in relationships. Above all, in the psychoanalyses of severely traumatized analysands, it is indispensable to approach the distinctive life and trauma history, the historical reality of the trauma, in spite of the fact that it is never possible to discover the historical truth of the trauma in a one-to-one sense. Although life-historical events—retrospectively speaking—are repeatedly rewritten and adapted to the present, these rewritings still retain the core of their historical truth.

Thus, the process of discovery and of understanding embodied memories in the psychoanalytical relationship and the working through of the traumatic experiences in transference enable analysands to overcome dissociative states and fragmentations of the self and the inner objects and to regain some basic trust in a helping object. For analysands, as for Mrs. M., this means improved psychic integration and delimitation of the destructiveness of trauma.

The following graph illustrates an embodied, "unconscious" interaction between analysand and psychoanalyst as discussed above.

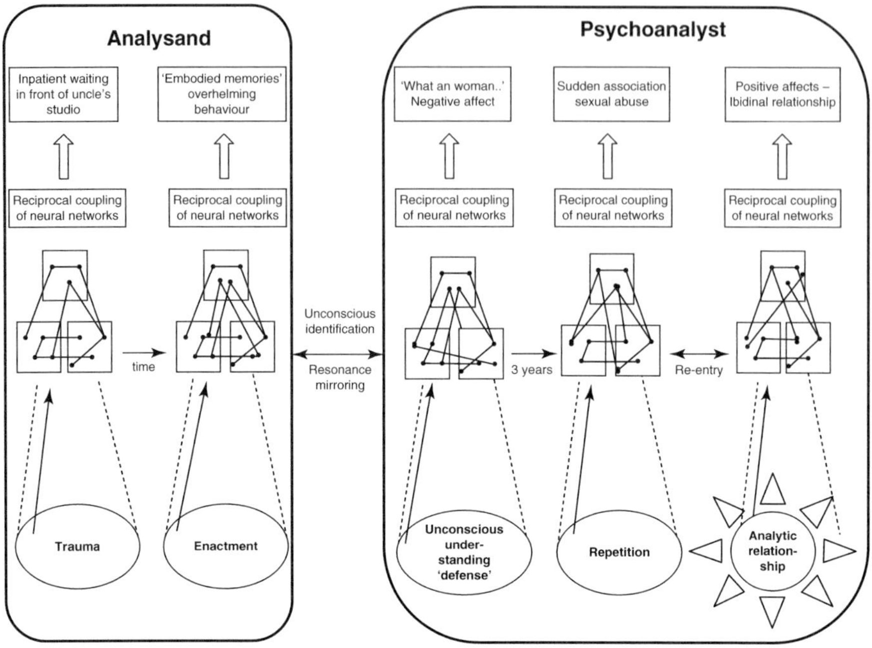


## 6.7 Summary and Outlook

Contemporary psychoanalysis is in a state of pluralism of concepts and research approaches. Therefore, a great intellectual effort is necessary to discuss which parts of central psychoanalytical concepts, as the unconscious, are in need for changes and modifications and which ones of them still can be considered to be plausible and "true." In the limited frame of this chapter, only a few of the ongoing discourses have been summarized focusing particularly the field of neuropsychoanalysis and embodied cognitive science. I am convinced that these interdisciplinary exchanges have opened new doors for the conceptual development of psychoanalysis as well as for clinical practice as was illustrated with an extensive case example.

However, due to epistemological and methodological arguments, I cannot agree completely with the passionate conviction of Eric Kandel (2009) that modern neurosciences really can save the future of psychoanalysis. But I am unambivalently sharing his view that curiosity and openness toward scientific developments, neurosciences included, are a must for innovation and creativity. In order to remain a "Wissenschaft" of the mind, psychoanalysis must refresh and further develop its concepts and theories, showing again and again that psychoanalytic theories are "externally coherent" (Strenger 1991) with the state of art of other disciplines, e.g., the neurosciences (see e.g. Böker and Seifritz 2012; Leuzinger-Bohleber and Weiss 2014). In this sense, the future of psychoanalysis as a productive "Wissenschaft," in my eyes, depends on openness toward contemporary neurosciences.

## References


Anderson M, Ochsner K, Kuhl B, Cooper J, Robertson E, Gabrieli S, Glover G, Gabrieli J. Neural systems underlying the suppression of unwanted memories. Science. 2004;303:232–5.

Bargh J, Chartrand T. The unbearable automaticity of being. Am Psychol. 1999;54:462–79.

Bohleber W. Destructiveness, intersubjectivity, and trauma. The identity crisis of modern psychoanalysis. London: Karnac; 2010.

Bohleber W, Leuzinger-Bohleber M. The special problem of interpretation in the treatment of traumatized patients. Psychoanal Inq. 2016;36:60–76.

Böker H, Seifritz E. Psychotherapie und Neurowissenschaften. Ein Blick in die Zukunft. Psychotherapie und Neurowissenschaften. Integration – Kritik – Zukunftsaussichten. Bern: Huber; 2012.

Civitarese G. Exploring core concepts: sexuality, dreams and the unconscious. Int J Psychoanal. 2011;92:277–80.

Clancey WJ. Situated cognition: how representations are created and given meaning. In: Lewis R, Mendelsohn P, editors. Lessons from learning. Amsterdam: North-Holland; 1993. p. 231–42.

Coates SW, Gaensbauer TJ. Event trauma in early childhood: symptoms, assessment, intervention. Child Adolesc Psychiatr Clin N Am. 2009;18(3):611–26.

Damasio A. The feeling of what happens. New York: Harvest; 1999.

Edelman G. Neural Darwinism: the theory of neural group selection. New York: Basic Books; 1987.

Edelman G. Bright air, brilliant fire. New York: Basic; 1993.

Fonagy P. Attachment, trauma and psychoanalysis: where psychoanalysis meets neuroscience. In: Leuzinger-Bohleber M, Canestri J, Target M, editors. Early development and its disturbances: clinical, conceptual and empirical research on ADHD and other psychopathologies and its epistemological reflections. London: Karnac; 2010. p. 53–75.

Freud S. Project for a scientific psychology. SE. 1895;1:281–397.

Freud S. The unconscious. SE. 1915;14:166–204.

Freud S. The ego and the id. SE. 1923;19:12–59.

Gaensbauer TJ. Representations of trauma in infancy: clinical and theoretical implications for the understanding of early memory. Infant Ment Health J. 2002;23.259–77.

Gaensbauer TJ. Embodied simulation, mirror neurons, and the reenactment of trauma in early childhood. Neuropsychoanalysis. 2011;13:91–107.

Gallese V. Mirror neurons, embodied simulation, and the neural basis of social identification. Psychoanal Dialogues. 2013;19:519–36.

Green A. The work of the negative. London: Free Association; 1999.

Green A. Resonance of suffering. Coutertransference in non-neurotic structure. London: International Psychoanalytic Association; 2007.

Hagner M. Der Geist bei der Arbeit. Die visuelle Repräsentation zerebraler Prozesse. In: Leuzinger-Bohleber M, Roth G, Buchheim A (Hrsg.). Psychoanalyse – Neurobiologie – Trauma, Stuttgart: Schattauer; 2008.

Kandel ER. A new intellectual framework for psychiatry. Am J Psychiatr. 1998;155(4):457–69.

Kandel ER. Biology and the future of psychoanalysis: a new intellectual framework for psychiatry revisited. Am J Psychiatr. 1999;156(4):505–24.

Kandel ER. Psychiatrie, Psychoanalyse und die neue Biologie des Geistes. Frankfurt am Main: Suhrkamp; 2006.

Kandel ER. The biology of memory: a forty-year perspective. J Neurosci. 2009;29(41):12748–56.

Kaplan-Solms K, Solms M. Clinical studies in neuropsychoanalysis. London: Karnac; 2000.

Kaplan-Solms K, Solms M. Neuro-Psychoanalyse. Eine Einführung mit Fallstudien. Stuttgart: Klett-Cotta; 2003.

Kihlstrom J. Perception without awareness of what is perceived, learning without awareness of what is learned. In: Velmans M, editor. The science of consciousness: psychological, neuropsychological and clinical reviews. London: Routledge; 1996. p. 23–46.

Koukkou M, Leuzinger-Bohleber M, Mertens W (Hrsg.). Erinnerung von Wirklichkeiten. Psychoanalyse und Neurowissenschaften im Dialog. Bd. 1: Bestandsaufnahme. Stuttgart: Internationale Psychoanalyse; 1998.
Krauze MK. Exploring core concepts: sexuality, dreams and the unconscious. Int J Psychoanal. 2011;92:283–5.
Leuzinger-Bohleber M. Biographical truths and their clinical consequences. Understanding "embodied memories" in a third psychoanalysis with a traumatized patient recovered from severe poliomyelitis. Int J Psychoanal. 2008;89:1165–87.
Leuzinger-Bohleber M. Finding the body in the mind – embodied memories, trauma, and depression, International Psychoanalytical Association. London: Karnac; 2015.
Leuzinger-Bohleber M, Arnold S, Solms M, editors. The unconscious: a bridge between psychoanalysis and cognitive neuroscience. London: Routledge; 2017.
Leuzinger-Bohleber M, Emde RN, Pfeifer R (Hrsg.). Embodiment. Ein innovatives Konzept für Entwicklungsforschung und Psychoanalyse. Göttingen: Vanenhoeck & Ruprecht; 2014.
Leuzinger-Bohleber M, Fischmann T, Böker T, Northoff G, Solms M (Hrsg.). Psychoanalyse und Neurowissenschaften. Chancen – Grenzen – Kontroversen. Reihe Psychoanalyse im 21. Jahrhundert. Stuttgart: Kohlhammer; 2015.
Leuzinger-BohleberM, Kächele H, editors. An open door review of outcome and process studies in psychoanalysis, 3rd ed. London: International Psychoanalytical Association; 2015. Zugriff am 09.02.2016 unter http://www.ipa.org.uk/en/What_is_Psychoanalysis/Open_Door_Review/en/Psychoanalytic_Theory/Research/open_door.aspx?hkey=fade476a-ae81–48b3–8fab-280070a69bf6
Leuzinger-Bohleber M, Mertens W, Koukkou M (Hrsg.). Erinnerung von Wirklichkeiten. Psychoanalyse und Neurowissenschaften im Dialog. Bd. 2: Folgerungen für die psychoanalytische Praxis. Stuttgart: Internationale Psychoanalyse; 1998.
Leuzinger-Bohleber M, Pfeifer R. Remembering a depressive primary object. Memory in the dialogue between psychoanalysis and cognitive science. Int J Psychoanal. 2002;83:3–33.
Leuzinger-Bohleber M, Pfeifer R. Embodiment: Den Körper in der Seele entdecken - Ein altes Problem und ein revolutionäres Konzept. In: Leuzinger-Bohleber M, Emde RN, Pfeifer R, editors. Embodiment - ein innovatives Konzept für Entwicklungsforschung und Psychoanalyse. Göttingen: Vandenhoeck & Ruprecht; 2013. p. 14–38.
Leuzinger-Bohleber M, Weiss H. Psychoanalyse – Die Lehre vom Unbewussten. Geschichte, Klinik und Praxis. Stuttgart: Kohlhammer; 2014.
Levine HB, Reed GS, Scarfone D. Unrepresented states and the construction of meaning: clinical and theoretical contributions. London: Karnac; 2013.
Libet B. Unconscious cerebral initiative and the role of conscious will in voluntary action. Behav Brain Sci. 1985;8:529–39.
Olds D, Cooper AM. Dialogue with other sciences: opportunity for mutual gain. Int J Psychoanal. 1997;78:219–25.
Panksepp J. Affective neuroscience. New York: Oxford University Press; 1998.
Pfeifer R, Bogard J. How the body shapes the way we think. A new view of intelligence. Cambirdge: MIT Press; 2007.
Pfeifer R, Leuzinger-Bohleber M. Applications of cognitive science methods to psychoanalysis. A case study and some theory. Int Rev Psychoanal. 1986;13(2):221–40.
Pine F. Beyond pluralism: psychoanalysis and the workings of mind. Psychoanal Q. 2011;80:823–56.
Ramachandran V. Phantom limbs, neglect syndromes, repressed memories, and Freudian psychology. Int Rev Neurobiol. 1994;37:291–333.
Reed G. And empty mirror: reflections on nonrepresentation. In: Levine HB, Reed GS, Scarfone D, editors. Unrepresented states and the construction of meaning. Clinical and theoretical contributions. London: Karnac; 2013. p. 18–41.
Röckerath K, Strauss LV, Leuzinger-Bohleber M (eds.). Verletztes Gehirn—Verletztes Ich. Göttingen: Vandenhoeck & Ruprecht; 2009.

Rovee-Collier C, Cuevas K. The development of infant memory. In: Courage ML, Cowan N, editors. The development of memory in infancy and childhood. Hove: Psychology Press; 2009. p. 11–41.

Rovee-Collier CC. Dissociations in infant memory: rethinking the development of implicit and explicit memory. Psychol Rev. 1997;104(3):467–98.

Rovee-Collier CC. The development of infant memory. Curr Dir Psychol Sci. 1999;8(3):80–5.

Sacks O. Musicophilia: tales of music and the brain. New York: Random House; 2007.

Sandell A. Vom namenlosen Grauen zu ertragbarer Furcht. Die psychoanalytische Behandlung eines 22 Monate alten Kindes. In: Leuzinger-Bohleber M, Emde R, Pfeifer R (Hg). Embodiment. Ein innovatives Konzept für Entwicklungsforschung und Psychoanalyse. Göttingen: Vandenhoeck & Ruprecht; 2014, p. 367–85.

Scarfone D. From traces to signs. Presenting and representing. In: Levine HB, Reed GS, Scarfone D, editors. Unrepresented states and the construction of meaning: clinical and theoretical contributions. London: Karnac; 2013. p. 75–95.

Scarfone D. Conceptual research in psychoanalysis. In: Leuzinger-Bohleber M, Kächele H (eds.). An open door review of outcome and process studies in psychoanalysis, 3rd ed. London: International Psychoanalytical Association; 2015. Zugriff am 09.02.2016 unter http://www.ipa.org.uk/en/What_is_Psychoanalysis/Open_Door_Review/en/Psychoanalytic_Theory/Research/open_door.aspx?hkey=fade476a-ae81–48b3–8fab-280070a69bf6 (written version: p. 33–7).

Shewmon D, Holmse D, Byrne P. Consciousness in congenitally decorticate children: developmental vegetative state as a self-fulfilling prophecy. Dev Med Child Neurol. 1999;41:364–74.

Solms M. The conscious id. Neuropsychoanalysis. 2013;15(1):5–19.

Solms M. "The unconscious" in psychoanalysis and neuroscience: an integrated approach to the cognitive unconscious. In: Leuzinger-Bohleber M, Arnold S, Solms M, editors. The unconscious. A bridge between psychoanalysis and cognitive neuroscience. London: Routledge; 2017. p. 16–37.

Solms M, Friston K. Consciousness by surprise. In: Oral presentation (Solms) and discussion (Friston) at the international psychoanalytical association research conference. Frankfurt: Sigmund Freud Institute; 2014. https://www.youtube.com/watch?v=xP8Y2f1I0jE.

Solms M, Panksepp J. The id knows more than the ego admits. Brain Sci. 2012;2:147–75.

Strenger C. Between hermeneutics and science. An essay on the epistemiology of psychoanalysis. New York: International Universities Press; 1991.

# What Is the Unconscious? A Novel Taxonomy of Psychoanalytic, Psychological, Neuroscientific, and Philosophical Concepts

7

Georg Northoff

**Abstract**

The concepts of consciousness and unconscious have been widely debated in neuroscience, psychoanalysis, and philosophy. Thereby, three different lines of thoughts often get confused. One the hand, consciousness is distinguished from the unconscious; this is, for instance, the case in psychoanalysis and more specifically Freud when he distinguishes psychological material and contents in consciousness from the ones remaining preconscious or dynamic unconscious. Current neuroscience and psychology associate the distinction between consciousness and unconscious rather different modes in which the same functions, affective, cognitive, etc. can appear and thus be quasi duplicated. Finally, philosophy associates the distinction between consciousness and unconscious with a principal difference between mind and brain and thus mental and neuronal states. To bridge the gap between these three lines, I here suggest a novel conceptual characterization. I distinguish the principal conscious from the principal unconscious: The principal conscious describes those states that in principle can become conscious and thus have the potential for consciousness independent of whether they are actually conscious, preconscious, or dynamically unconscious, while the principal unconscious refers to the principal impossibility of a state becoming conscious because it may simply be coded in the wrong format. I show that this conceptual distinction between principal conscious and principal unconscious carries important implications for empirical, i.e., neuronal matters and that both concepts may be associated with different neuronal mechanisms. I here refer to the brain's intrinsic activity, its resting state activity, and how that impacts subsequent stimulus-induced activity as mediated by a particular set of regions in the brain's midline, the cortical midline structures. Finally, I indicate that this

G. Northoff
Mind, Brain Imaging, and Neuroethics,
Institute of Mental Health Research, University of Ottawa, Ottawa, ON, Canada
e-mail: georg.northoff@theroyal.ca; www.georgnorthoff.com



H. Boeker et al. (eds.), *Neuropsychodynamic Psychiatry*,
https://doi.org/10.1007/978-3-319-75112-2_7

distinction may also be relevant to psychoanalysis and philosophy alike in opening the door to bridge the gap between mind and brain and thus between mental and neuronal states.

## 7.1 Introduction

What is the unconscious? Notions of the unconscious can be traced back as far as to Plato and Aristotle and have been elaborated since then in philosophical and later in psychological thought. Whatever frameworks have been presupposed, unconscious states have been characterized by hidden characteristics of a person's self (fate, temperament, soul, character) that need to be inferred and cannot be accessed directly. Such hidden characteristics were distinguished from those that were believed to be transparent, experienced directly, open to introspection, and thus accessible to consciousness (Uleman 2005). Building on descriptions of such hidden and inaccessible characteristics in philosophy (Schopenhauer, Nietzsche) and literature (Dostojewsky), Freud was probably the first trying to give a systematic empirically based account of the unconscious (see also for an overview, as well as Northoff (2011) for the broader context of neuropsychoanalysis). According to Freud, the unconscious in a psychoanalytic sense, the dynamic or repressed unconscious, does not only operate outside awareness but is extremely complex including distinct aspects of the persons' self. It includes innate and inherently sexual and aggressive blind drives (i.e., the id), most of its conscience and ego ideals (i.e., the superego), and processes (perception, action, etc.) that deal with reality (i.e., the ego). The dynamic or repressed unconscious mediates wishful, associative, instinctual primary processes. The dynamic or repressed unconscious must be distinguished from what Freud called system preconscious that includes mental contents which are descriptively but not dynamically unconscious because, as non-offensive to the person's self, they are readily accessible to consciousness. The system preconscious is thus more rational, disciplined, reality-oriented, and energetically "bound" than the dynamic unconscious.

Originally, Freud described the contents or material of the unconscious by predominantly primary processes like wishes, drives, instincts, and associative "blind" nonrational processes.[1] Current accounts include both rational and nonrational contents or materials in the unconscious. The content or material of the unconscious may concern cognition, emotions, movements, perceptions, behavior, etc. Consequently, terms like the "cognitive unconscious" (Kihlstrom 1987), "emotional unconscious" (Kihlstrom 1987, 1999; Berridge and Winkielman 2003), "behavioral unconscious" (Uleman 2005), and "procedural unconscious" (Schuessler 2002) have been introduced to characterize different contents or material of the unconscious.

[1] In his later work, he however refrained, at least partially, from exhaustive and exclusive definition of the dynamic unconscious by primary and primitive processes by including more rational processes associated with the ego and superego into the unconscious (see Westen 1999).

The variety of the processes (memory, motor, perception, etc.) and functions associated with the unconscious suggests that one can no longer speak of "the unconscious" as though it were a unified and isolated sector of the mind—the concept of "the unconscious" has thus outlived its usefulness (Westen 1999, p. 1064). It is clear that here the focus is on different contents; unconsciousness in this sense mirrors cognition content. One characterizes unconsciousness in this sense as "psychological unconscious" (see Kihlstrom et al. 1992). This means that the unconscious can occur in a variety of different psychological functions processing different types of materials or contents. Unlike in the early time of the "cognitive unconscious," the psychological unconscious includes not only cognitions but also affect, motivation, goals, motives, self-regulation, interpersonal encounters, and the causes of phenomenal experience like the free will; all of which are empirically studied with a variety of different methods—this has recently been called the "new unconscious" (Uleman 2005, p. 6).

I here want to characterize unconsciousness in further detail. First, I discuss the concept of unconsciousness in neurophilosophy as suggested by Searle (2004). This is followed by a characterization of the unconscious as "principle unconscious" that must be distinguished from the various forms of the unconscious in either psychodynamic or psychological gestalt, as put forward by Freud and current cognitive psychology. Such taxonomy of the unconscious (see Fig. 7.1) provides the conceptual ground for future research on the empirical conditions of psychological and psychodynamic forms of unconsciousness as well as for better understanding of the neural mechanisms underlying consciousness.

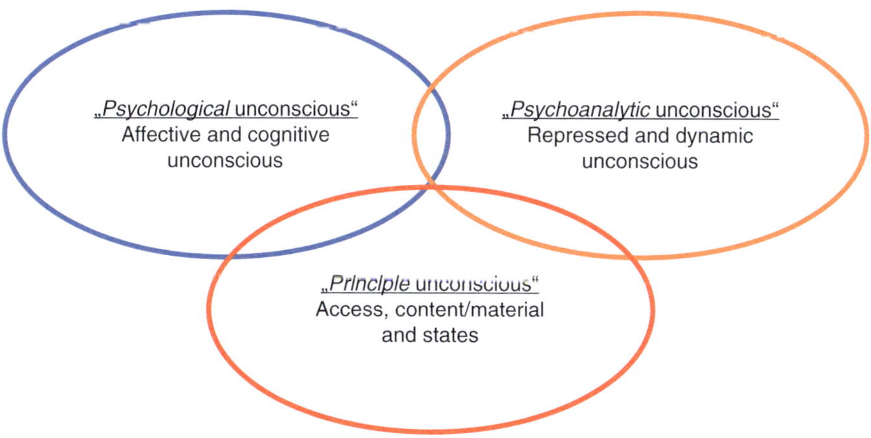


**Fig. 7.1** Taxonnomy of the unconscious in psychology psychoanalysis and philosophy. The figure shows the relationship between difference of the unconscious as it is conceptualized in cognitive neuroscience and psychology ("psychological unconscious"), psychoanalysis ("psychoanalytic unconscious"), and philosophy ("principle unconscious"). Since the different disciplines presuppose different referents (affective/cognitive functions, suppressed/repressed material, access to knowledge), the unconscious is defined and determined differently in the context of different disciplines

## 7.2 Neurophilosophical Characterization of the Unconscious

Searle (2004, pp. 165–72) distinguishes between different types of unconsciousness. He speaks of "preconscious" which refers to more or less the same as Freud's notion of system preconscious. Then there is the "dynamic unconscious" that describes cases where the "unconscious mental states functions causally, even when unconscious" (Searle 2004, p. 167); this mirrors more or less what Freud described as dynamic or repressed unconscious. The third type is what Searle describes as "deep unconscious"; here the unconscious mental state is not factually brought into consciousness, as in the dynamic unconscious, but it also remains principally impossible to do so. Following Searle, this is so because what is unconscious here is not "the sort of thing that can form the content of a conscious intentional state" (Searle 2004, p. 168). He underlines this by the example of the computational rules that we follow unconsciously in acquiring language. While we can be preconscious or dynamic unconscious about the language and its letters, we remain deeply unconscious about the rules and principles of its universal grammar that guide our learning of the language. Rules that guide the acquisition of language and our construction of perception in the retina and the visual cortex are simply not the sort of things we can become conscious of at all.

Finally, there is the "nonconscious" that concerns neurobiological phenomena that remain nonconscious and cannot become cases of consciousness at all:

There are all sorts of things going on in the brain, many of which function crucially in controlling our mental lives but that are not cases of mental phenomena at all. So, for example, the secretion of serotonin at the synaptic cleft is simply not a mental phenomenon. Serotonin is important for several kinds of mental phenomena, and indeed some important drugs, such as Prozac, are used specifically to influence serotonin, but there is no mental reality to the behaviour of serotonin as such. Let us call these sorts of cases the 'nonconscious' (see Oppenheim, this volume for further discussion on this concept). There are other examples of the nonconscious that are more problematic. So, for example, when I am totally unconscious, the medulla will still control my breathing. This is why I do not die when I am unconscious or in a sound sleep. But there is no mental reality to the events in the medulla that keep me breathing even when unconscious. I am not unconsciously following the rule 'Keep breathing'; rather, the medulla is just functioning in a nonmental fashion, in the same way that the stomach functions in a nonmental fashion when I am digesting food. (Searle 2004, p. 168)

Where does this leave us with regard to the distinction between cognition form/mode and content? If consciousness is the cognition form/mode, all possible states, all possible cognition should be able to become conscious in principle, be they preconscious, dynamic unconscious, or nonconscious. All three forms of unconscious reflect nothing but the continuum of consciousness. Even nonconscious states, neurobiological states, behave as if they were intentional and conscious, as Searle (2004, pp. 171, 173) says who speaks of so-called background capacities. The problematic case are those unconscious states that I left out here, what Searle called the "deep unconscious" that describes states of which we cannot become conscious in principle. The assumption of "deep unconscious" contradicts the characterization of consciousness being the form or mode of consciousness since that, as mentioned above, does not leave any room for any states other than being principally accessible to consciousness. The "deep unconscious" claims however exactly that, possible states that remain principally inaccessible to consciousness. Why could they possibly remain principally inaccessible to consciousness?

The format or form of these states may remain incompatible with the form of consciousness; this is, for instance, illustrated by an analogous example, the DNA, by Revensuo (2006, p. 63): "There is biological information coded in the DNA of our brain cells, but that type of information is in a totally nonconscious format and we will never be able to read it out just by reaching into our own minds and trying to retrieve it into consciousness. It is in a format unreadable at the phenomenal level." Similar so in the case of the "deep unconscious," this indicates a format that remains principally inaccessible to the one employed by consciousness. Since however we can cognize only in the conscious format, all our cognition being conscious or principally accessible to consciousness, we remain principally unable to cognize and thus to assume a "deep unconscious." In short, the concept of the "deep unconscious" must be rejected since it is contradictory to our cognition being principally conscious.

Due to the fact that all our cognition is in principally conscious, consciousness and unconsciousness must be logically connected to each other. Searle calls this the "connection principle" which states that the notion of unconscious is logically connected to the notion of consciousness with an unconscious mental state being "the kind of thing that could be a conscious mental state" (Searle 2004, p. 171; see also Strawson 1994). Presupposing the "connection principle," one may regard consciousness indeed a form or mode of cognition that concerns all our possible cognition and distinguish it from impossible cognition. For that purpose I want to speak of the "principle conscious" that encompasses all states that are conscious or can become in principle conscious including the preconscious, the dynamic unconscious, the psychological unconscious as the "new unconsciousness", and the nonconscious as variants of the "principle conscious." The "principle conscious" must be distinguished from the "principle unconscious" that includes all states that in principle cannot become conscious like Searle's notion of the "deep unconscious."

The distinction between "principle conscious" and "principle unconscious" is prior and more basic than the one between different types of the "principle conscious" like consciousness, preconscious, dynamic unconscious, etc. This means that consciousness, preconscious, and dynamic unconscious must be regarded as specifications of the "principle conscious" and must thereby all be distinguished from the "principle unconscious." Such double distinction has important implications for both domains empirically and conceptually.

Let me start with the empirical domain. When raising the question for the neural correlates of the unconscious, we must first address which distinction we refer to: Do we want to search for neural processes that distinguish consciousness, preconscious, and dynamic unconscious from each other? Or do we take those neural processes into our view that allow for the more basic distinction between "principle conscious" and "principle unconscious?" This is important since both may refer to distinct kinds of neural processes. Consciousness, preconscious, and dynamic unconscious may be distinguished by the degree of neural activity and/or the degree to which certain neural networks are recruited. The distinction between the "principle unconscious" and the "principle conscious" may refer to a more basic neural mechanism like the abovementioned format which may be rephrased as the search for the neural code. Rather than the degree of neural activity or neural network recruitment, the neural code describes how the brain's neural activity as such is generated and formatted in relation to the respective stimulus. Hence, we may need to target different neuronal mechanisms, degree of neural activity versus format, or code of neural activity, which makes this conceptual distinction also empirically relevant for neuroscience itself.

In addition to its empirical relevance, the double distinction carries important conceptual and thus philosophical relevance. Distinguished between different forms of consciousness: phenomenal and access consciousness with the former describing the subjective (or phenomenal) experience, while the latter refers to the way we can become aware and thus access the former. This distinction clearly supersedes and specifies the concept of the consciousness as distinguished from the preconscious and the dynamic unconscious. One may consequently extend the principle conscious by adding different forms of consciousness (access and reflective consciousness and probably also other forms of consciousness) to the here put forward distinction between consciousness, dynamic unconscious, and preconscious. This makes it also clear that the current neuroscientific and philosophical discussion about consciousness takes a rather limited and restricted focus by often only considering the tip of the iceberg, consciousness, and how the tip itself may show different layers, i.e., access and phenomenal consciousness.

While neuroscience often focuses only on the tip of the iceberg, i.e., consciousness, philosophy of mind and especially tackle the distinction between the tip of the iceberg and the iceberg itself. Chalmers distinguishes between an "easy problem" and a "hard problem": The "easy problem" describes the mechanisms underlying the distinction between consciousness on the one hand and that which I here called the unconscious in its empirical gestalt, i.e., preconscious and dynamic unconscious, while what he calls the "hard problem" raises the question why there is consciousness at all rather than nonconsciousness. This clearly pertains to the distinction between the "principle conscious" and the "principle unconscious" as put forward here: Why is there a "principle conscious" at all rather than just "principle unconscious?" This means that he raises the question why there is an iceberg at all that shows a tip that is conscious and a body that remains unconscious but principally accessible to consciousness so that we could, if we want, always climb from the iceberg's body to its top (and vice versa of course as especially pointed out by Freud).

In the following, I want to characterize the "principle unconscious" in further detail along three distinct dimensions, accessibility, material/content and states. In addition to characterize the "principle unconscious" in this way, I discuss the epistemic and empirical conditions that make the "principle unconscious" possible for us. This in turn will shed some light on those conditions that make the "principle conscious" possible for us.

## 7.3 Dimension of Accessibility

The *dimension of accessibility* refers to the ability or inability to gain access to unconscious states and to bring them into consciousness. This is definitely given in what Freud called the system preconscious and what I will simply call the "preconscious." It is also principally given in the dynamic or repressed unconscious whose primary processes can be accessed in principle though in fact they remain suppressed because of their offensive character to the self. I call this the "repressed unconscious," whereas accessibility is no longer given in what I call the "principle unconscious" whose processes (like defense mechanisms) remain principally inaccessible and can thus not be brought to consciousness in principle "because it is not the sort of thing that can form the content of a conscious intentional states" (see above Searle 2004, p. 168).

Why are certain processes "not the sort of thing that can form the content of a conscious intentional state?" I argue that there are two possibilities. A specific process may not be the "not the sort of thing that can form the content of a conscious intentional states" because we lack the means to access it as such. The process may principally be accessible to consciousness, but we lack the means to do so in our epistemic equipment. I call this type of unconscious the "principle unconscious (a)" with (a) standing for lack of access. A condition for the "principle unconscious (a)" is what I call phenomenal autoepistemic limitation (Northoff 2004, 2011): Due to the fact that the brain lacks an interoceptive sensory system, we do not possess the means or tools to access the own brain and to bring it into (phenomenal experience and hence to) consciousness; the own brain remains therefore principally unconscious (a).

This lack of means to access principle unconscious states must be distinguished from the case where these states simply show the wrong format or code. Even if we have the means to access certain processes, they may nevertheless remain principally inaccessible because they may be characterized by a format or code that we remain unable to grasp. For example, running a windows program on an Apple platform or vice versa remains impossible because of their principal differences in code or format so that the one cannot recognize and decipher the signal and processes of the respective other. The same is true of our body's and brains' physical processes as it is nicely expressed by Revonsuo with regard to DNA (2006, p. 63): "There is biological information coded in the DNA of our brain cells, but that type of information is in a totally nonconscious format and we will never be able to read it out just by reaching into our own minds and trying to retrieve it into consciousness. It is in a format unreadable at the phenomenal level." I call this type of principle unconscious the "principle unconscious (c)" with (c) standing for wrong code or format. A necessary condition of such "principle unconscious (c)" is what I call physical autoepistemic limitation (physical AL; Northoff 2004, 2011): What can be brought to consciousness in principle must necessarily be coded or formatted in phenomenal (or mental) rather than physical terms thereby making phenomenal (or mental) states and hence consciousness possible. In contrast, experience of states in physical terms in consciousness remains impossible. Physical states thus show the wrong format or code to be brought into consciousness so that they remain principally unconscious (c).[2]

[2] One could argue that what I here described as "principally unconscious (c)" should rather be regarded as what has been called nonconsciousness (see Strawson 1994; Revonsuo 2006; Searle 2004). Neurophysiological and thus physical processes are considered as nonconscious by these and other authors because they simply have the wrong format to be ever brought into consciousness. Though "principal unconscious (c)" amounts to the same as nonconscious in terms of the result, the difference may not only be terminological. "Principal unconscious (c)" concerns the format or code of a specific material or content (see below), whereas it does not concern the material or content itself. It is this what Searle seems to have in mind when he characterizes physical brain processes as nonconscious: "There are all sorts of things going on in the brain, many of which function crucially in controlling our mental lives but that are not cases of mental phenomena at all. So, for example, the secretion of serotonin at the synaptic cleft is simply not a mental phenomenon. Serotonin is important for several kinds of mental phenomena, and indeed some important drugs, such as Prozac, are used specifically to influence serotonin, but there is no mental reality to the behavior of serotonin as such." Let us call these sorts of cases the "nonconscious." There are other examples of the nonconscious that are more problematic. So, for example, when I am totally unconscious, the medulla will still control my breathing. This is why I do not die when I am unconscious or in a sound sleep. But there is no mental reality to the events in the medulla that keep me breathing even when unconscious. I am not unconsciously following the rule "keep breathing"; rather, the medulla is just functioning in a nonmental fashion, in the same way that the stomach functions in a nonmental fashion when I am digesting food" (see also this quote above in Searle (2004), p. 168). Searle seems to refer to the material or content in this quote rather than, as Revonsuo does (see above), to the code or format of a certain material or content. If the material or content itself cannot become conscious as such at all, one may indeed speak of nonconscious. If, however, the material or content cannot be brought into consciousness not because of its specification as such and such material or content but rather because of its format or code, I prefer to speak of "principal unconscious (c)" to distinguish this case from the one of "nonconscious (n)" with (n) standing for narrow sense referring only to the material/content itself but not to its format/code. If, in contrast, the term nonconscious is used in a wide sense, it includes both material/format itself and its code or format as possible underlying reasons for preventing access to consciousness. It seems to me that neither Revonsuo (2006, p. 63) nor Searle (2004, pp. 167–9) distinguish between both cases since both speak of nonconscious in either case. If he would have distinguished both cases, Revonsuo (2006, p. 63) meaning "principal unconscious (c)" rather than nonconscious (see his quote above) would have not stated that he relies on Searle in his account of the unconscious. Though Searle (2004) at first distinguishes between what he calls the "deep unconscious," as being unconscious in principle, and the nonconscious, he later considers the former as case of the latter. Considering the difference between material/content and code/format, this however remains impossible.

## 7.4 Dimension of Material or Content

Originally, Freud described the contents or material of the unconscious by predominantly primary processes that describe the principles according to which it operates; this in turn may result in wishes, drives, instincts, and associative "blind" nonrational processes.[3] Current accounts include both rational and nonrational contents or materials in the unconscious. The content or material of the unconscious may concern all different types like cognition, emotions, movements, perceptions, behavior, etc. Consequently, terms like the "cognitive unconscious" (Kihlstrom 1987), "emotional unconscious" (Kihlstrom 1999), "behavioral unconscious" (Uleman 2005), and "procedural unconscious" (Schuessler 2002) have been introduced to characterize the content or material that remains unconscious. Some of this material/content may remain "principally unconscious (a)," like, for example, of our relationships and the world in the first months (or even first 2 years) of our life to which, due to lack of access,[4] we seem to remain principally unable to bring into consciousness. Once being accessible, later parts (after the first months or 2 years of our life) of these materials/contents may become either repressed unconscious or preconscious depending on its offensive or non-offensive character for the self.

Unconscious material or contents may also include specific processes associated with a particular material or content. For example, self-related stimuli (like a piano for a pianist or a brain for a neuroscientist) reflecting a certain material or content and their processing, i.e., self-related processing, may remain unconscious in most parts. One could however imagine that material/content and processing of that material/content may dissociate from each other; this means that the process itself may not be accessible to consciousness at all thus being principally unconscious. In contrast, the result of that process and the degree of self-relatedness assigned to that stimulus may be principally accessible to consciousness thus being characterized by the principle conscious rather than the principle unconscious.

Based on these considerations, I characterize the referent of the unconscious in more detail with (p) standing for process and (m) standing for content/material (see also Fig. 7.2). For example, self-related processing that describes the processing of stimuli in relation to the organism and its needs, goals, etc. (Northoff et al. 2006; Northoff and Bermpohl 2004) may remain "principally unconscious (c) and (p)" by itself, whereas self-related stimuli, i.e., the material or content of self-related processing, may become either repressed conscious (m) or even preconscious (m). The

[3] In his later work, he however refrained, at least partially, from exhaustive and exclusive definition of the dynamic unconscious by primary and primitive processes by including more rational processes associated with the ego and superego into the unconscious (see Westen 1999).

[4] One could, of course, argue that the material/content of the first months or first 2 years of our life may be principally unconscious (c) rather than principally unconscious (a) thus showing the wrong format or code. Though this is at least logically possible, it seems rather unlikely to be the case under natural conditions and thus in our case, since this would presuppose that the format or code of the early material/content is different from the later material/content. The same argument could be applied if one argues that the early material/content remains nonconscious (n); it is not plausible to assume that different types of material/content are used in early and later life.

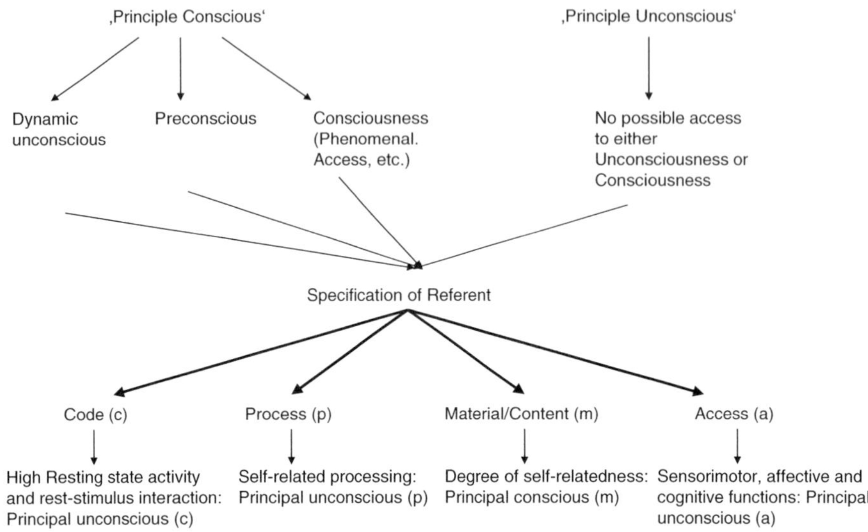


**Fig. 7.2** Taxonomy of consciousness and unconsciousness. The figure shows the different concepts of the consciousness and the unconscious (upper part) and how they are related to possible neuronal processes as indicated in the lower part of the figure

assumption of self-related stimuli being repressed conscious or preconscious is well compatible with the concept of the relational self as put forward by Anderson (Anderson and Chen 2002; Anderson et al. 2005; Northoff et al. 2006; Northoff and Panksepp 2008; Panksepp and Northoff 2009).

The concept of the relational self as put forward by Anderson and Chen (2002) refers to the fact that we experience our own self in relation to the self of another person; what is central for such interaction is transference by means of which we can connect and apprehend the other person's self which in turn yields a sense of our own person's self. This is a purely psychological determination of the relation between the own and the other person's self. The term self-related processing refers to the process that is supposed to underlie such psychological relationship between different persons' selves. As such self-related processing can be understood in either a purely functional sense and thus as software without designating its underlying hardware. Or self-related processing can be specified further with regard to the underlying hardware when it is understood as a neural process. This is, for instance, the case when self-related processing is associated with neural activity in particular parts or networks of the brain like the cortical midline structures (CMS). The CMS describe a set of regions that are particularly located in the midline of the brain; they include the ventro- and dorsomedial prefrontal cortex (VMPFC, DMPFC), the anterior and posterior cingulate cortex (ACC, PCC), and the medial parietal cortex (MPC) (Northoff et al. 2006). Interestingly, this set of regions has been shown to be implicated particularly in the processing of highly self-related stimuli and thus in self-related processing, though the exact nature of this relationship between neural processing and functional anatomy remains unclear.

How now is self-related processing in this sense related to the distinction between the "principle unconscious" and "principle conscious" with regard to processes (p) and content (c)? For that we may need to go a little more into the physiology of the brain. The abovementioned cortical midline structures are not only related to self-related processing but do also show high neural activity in the resting state since they are part of the so-called default mode network (DMN). This has led to the assumption that our sense of self and thus self-related processing may be closely related to the brain's intrinsic activity and thus its high level of resting state activity which is further supported by overlap and modulation of resting state activity by self-related stimuli.

What though does the brain with such high resting state activity? Why is it there and what is it for? Recent studies show that it strongly impacts the neural processing of any stimulus and the degree of neural activity it induces in the brain; this has been described as rest-stimulus interaction. Most importantly, the degree to which such rest-stimulus interaction takes place determines in the degree to which a stimulus becomes conscious or not. This means that rest-stimulus interaction may be central in determining whether a stimulus can become conscious or remains rather preconscious or dynamically unconscious. This means that I hypothesize that the degree of rest-stimulus interaction is directly relevant for the principle conscious itself and its different manifestations.

How is that related to self-related processing? I hypothesize that the degree of rest-stimulus interaction reflects the degree to which a stimulus is related to the own person. The higher the degree of rest-stimulus interaction, the higher the degree of self-related processing because then the stimulus is more intimately related to the brain and its intrinsic resting state activity. And the more the stimulus is related to the brain itself, the more it becomes linked and integrated into the organism which then can experience the respective stimulus as highly self-related. And this in turn may then determine the degree of consciousness, i.e., whether the stimulus becomes conscious or remains rather preconscious or dynamically unconscious.

Why though is there such high level of resting state activity? Imagine the case of there were no intrinsic activity and thus no high resting state activity in the brain. Would we still be able to develop consciousness, preconscious, or a dynamic unconscious? My answer is clear and straightforward: No! This though means that the brain's high resting state activity may exert or better impose a certain format or neural code on any kind of neural processing in the brain during stimulus-induced activity. What this format or neural code is remains unclear at this point. But what can at least be hypothesized now is that this format or neural code required and imposed by the brain's intrinsic activity may be principally different from the neural code of a brain that would not show any intrinsic activity.

How does that stand to the here suggested distinction between the principle conscious and the principle unconscious? If the degree of self-related processing predicts or determines the degree of consciousness and thus the different forms of consciousness in the principle consciousness, its presence or absence should decide upon whether something can become principally conscious or not. This means that the presence or absence of self-related processing and thus the brain's resting state may be crucial for distinguishing the principle conscious from the principle unconscious.

What is the "correct" code or format to make material or contents principally accessible to consciousness? I claim that the "correct" format or code consists in relation between organism and world and thus in what I call self-relatedness. Once there is self-relatedness, as in the case of self-related material/content, the material becomes principally accessible to consciousness, i.e., becoming repressed unconscious (m) or even preconscious (m). If, in contrast, there is no self-relatedness, as in the case of non-self-related material/content, access to consciousness remains principally blocked because of the "wrong" format or code resulting in "principally unconscious (c) and (m)."

I claim that self-related processing provides the transition between two different codes or formats with respect to the same material or contents—the transition from non-self-related material/content, lacking any self-relatedness and therefore remaining principally unconscious (c), to self-related material/content as characterized by self-relatedness and thus by repressed unconscious (m) or even preconscious (m). This implies that what is generally subsumed under the term unconscious, the repressed unconscious, and the preconscious necessarily presuppose self-related processing since without it the material/content has not the "correct" format or code. The repressed unconscious and the preconscious are thus essentially relational, reflecting the self-related relation between organism and environment. However, the unconscious cannot be restricted to such self-relatedness since that would mean to neglect the "principal unconscious." This implies that neuronally, the brain's intrinsic activity or its resting state activity may be central in providing the "correct" format for the neural processing of stimuli which in turn makes it possible that they can become principally conscious.

How about self-related processing itself? Can we experience the process of self-related processing as such by itself, i.e., as process? No, we can only experience and access its result; the degree of self-relatedness a particular stimulus is assigned, while the process yielding such result remains inaccessible and blocked to us. I henceforth claim that self-related processing remains "principally" unconscious (c). This is so I assume because it provides the transition from the "wrong" format or code to the "correct" one, i.e., the one that can be brought into consciousness and that can be read out by our means. If however self-related processing provides the transition from an unreadable to a readable code or format, self-related processing must be considered a working function and must itself therefore remain necessarily unreadable and thus "principally unconscious (c) and (p)".[5]

[5] Note that I here apply the same argumentative structure, what I call the "working function" argument, for self-related processing as Freud does with respect to defense mechanisms (see above). This will be further strengthened later when I consider defense mechanisms as manifestations of self-related processing.

Self-related processing may thus dissociate from self-related material or content that indeed can principally be brought into consciousness by being characterized as either "repressed unconscious (m)" or even preconscious (m).[6] For example, one's own body can apparently be self-related processed so that it becomes self-related material thus becoming principally accessible to consciousness with the absence of phenomenal AL of one's body. This contrasts with non-self-related material/ contents which, as I claim, can principally not be brought into consciousness because they have the "wrong" format or code thus remaining "principally unconscious (c) and (m)." An instance of such non-self-related material or content is, for example, one's brain which, unlike one's body (see above), cannot be self-related processed resulting in phenomenal AL.

## 7.5 Dimension of States

The dimension of state describes the type of state the unconscious is supposed to refer to. In philosophy of mind, the unconscious is often considered to refer to mental states so that one speaks of unconscious mental states. This however seems to be problematic. Since Descartes, the notion of mental states is used to characterize conscious states and thus consciousness. The idea of unconscious mental states would thus amount to an unconscious consciousness that however must be regarded as self-contradictory (see also Searle 2004, pp. 165–6). An unconscious mental state would then be nothing but a conscious mental state minus the consciousness. This makes it clear that the characterization of the unconscious in terms of mental states is deeply problematic. One way out is to define the unconscious in terms of consciousness by characterizing the former at least as principally and potentially accessible to the latter. In this case, the definition of mental states in terms of (potential) consciousness can remain so that the assumption of unconscious mental states is no longer contradictory.[7] This however would mean to neglect the "principal unconscious" including the distinction between "repressed unconscious" and "principle

[6] However, one may also logically imagine the reverse case with self-related processing being suppressed but principally being accessible to consciousness and thus being "repressed unconscious (p)" or even preconscious (p), whereas self-related material or contents may remain principally unconscious (c) or (a) and (m). In this case, however, self-related processing could have no longer the crucial function as I claim in the present book. Such characterization of self-related processing and material would thus presuppose a totally different setup and design of the human mind and brain.

[7] See, for example, Searle (2004, pp. 165–78) who follows this way by denying the possibility of "principal unconscious" what he calls "deep unconscious" which he subsumes under nonconscious. He can thus define the unconscious, the repressed unconscious, and the preconscious, in terms of potential consciousness and thus mental states. He calls this the "connection principle" which states that the notion of unconscious is logically connected to the notion of consciousness with an unconscious mental state being "the kind of thing that could be a conscious mental state" (Searle 2004, p. 171).

unconscious" with its substantial epistemic implications reaching beyond mere terminological differentiation.[8]

What are these substantial epistemic implications that make the distinction between "repressed unconscious" and "principle unconscious" necessary rather than being a mere terminological subtlety? The seemingly contradictory linkage between the unconscious and mental states, as discussed in the philosophy of mind, is replaced in cognitive neuroscience by association with the unconscious with various psychological states like cognition and emotions. As mentioned in the introduction, Kihlstrom et al. (1992), who also introduced the notion of the "cognitive unconscious," speaks therefore of the "psychological unconscious." Philosophy in general and philosophy of mind in particular presuppose the concept of mental states as hallmark feature of the mind as distinguished from the merely physical states of the brain. As such mental states were linked and associated with the concept of consciousness that was supposed to be regarded as *the* feature of the human mind. Mental states = consciousness. The discovery of the unconscious by Freud and its seemingly mental gestalt and its huge impact on mental states put this association between mental states and consciousness and thus the whole characterization of the mind by consciousness into doubt. Hence philosophy encounters a real problem to its concept of mind when mental states can no longer be identified with consciousness.

[8] Once one denies the possibility of "principle unconscious" by subsuming it under the nonconscious, the assumption of phenomenal autoepistemic limitation (phenomenal AL) must be considered contradictory. This is so because something, i.e., phenomenal AL, that presupposes the principal absence of a principally possible experience and thus the "principal unconscious" cannot be characterized by the principal absence of a principally impossible experience and thus the nonconscious (n). It seems that the principal absence, as characterizing the "principal unconscious," is confused with the principal impossibility of experience as reflecting the nonconscious (n). It is true that once one claims for the principal absence of principally impossible experience, one does not even need to consider phenomenal AL. However, this claim does not necessarily imply the impossibility of the first claim, the principal absence of a principally possible experience and thus of the "principal unconscious". Once however one acknowledges the difference between both claims, one cannot do otherwise than to postulate phenomenal AL as epistemic manifestation of the "principal unconscious." This demonstrates that phenomenal AL extends the boundaries of the notion of the unconscious by showing new aspects like the "principal unconscious (c) and (a)" and its necessary distinction from the "nonconscious (n)." Most importantly, phenomenal AL makes a redefinition of the concept of mental states necessary. Phenomenal AL as the absence of mental states with respect to one's brain can only be understood if the concept of mental states becomes detached from the concept of (potential) consciousness. Mental states may then be characterized as a dimension of principally possible experience being independent and separate from its actual absence, as in the unconscious, or its actual presence, as in consciousness. The presence of mental states and their concurrent absence in conscious experience may then be considered the defining feature of the unconscious as it is understood here—the unconscious can thus be regarded as nonconsciously experienced mental states.

What to do? Cognitive neuroscience and psychology associate consciousness no longer with mental states but with psychological functions like affect, emotions, cognition, etc. Since they are no longer tied to mental states, these psychological functions can then also be associated with the unconscious thus being duplicated, if one wants to say, in two modes, conscious and unconscious. This move in turn makes the unconscious accessible to empirical research and thus to its neural exploration.

What does this entail for the philosophical concept of mental states? One may either abandon this concept altogether as it is indeed done in neuroscience. Or one may modify and extend it to describe any kind of non-neuronal state independent of whether it is experienced in a conscious or unconscious mode. As such the concept of mental state becomes detached from the one of consciousness. Rather than signifying the difference between consciousness, preconscious and dynamic unconscious all three reflecting the principal conscious, it may then rather describe the principal conscious as such as distinguished from the principal unconscious.

One may finally ask why this is relevant at all for both neuroscience and psychoanalysis? Let the philosophers deal with these conceptual matters but give us peace from it. Psychoanalysis has demonstrated that the domain of the mental, as used in an extended sense, reaches deep into the unconscious, e.g., preconscious and dynamic unconscious. Even objects and events that remain unconscious are already non-neuronal and thus mental. This though means that in these deep layers of neuronal processing that do not become conscious, the same kind of coding or format as in the conscious mode must already be at work and thus be operative. To put it into more simple terms, the same neural code may be operative in all three forms of the principle conscious, consciousness, preconscious, and dynamic unconscious. These different forms of the principle conscious may then represent different degrees of the same neuronal mechanisms, as, for instance, rest-stimulus interaction, rather than being associated with principally different neural codes and modes of neural functioning (as it is often assumed these days).

This though entails a different research strategy than presupposed currently. As said above, there is currently a tendency to duplicate affective and cognitive functions in both unconscious and conscious modes. Rather than duplicating them and searching for the neuronal differences of the duplicates, I here suggest to search for neuronal commonalities in their underlying mechanisms; on the basis of these neuronal commonalities, we may then be able to better figure out what their difference consists in. Hence, though looking rather complicated, my conceptual distinctions carry important methodological and neuroscientific load.

Finally, the philosopher may ask why I force him to give up his so beloved characterization of mental states by consciousness. Do I not commit then what the philosophers call "category error," to confuse the category of the mind with the one of the brain? This though is not my intention. Rather it is to escape from the dichotomy between mind and brain and thus the dichotomy between mental and physical states. By anchoring the concept of mental states deep down in the unconscious while at the same time not designating it as merely physical, such extended concept of mental states may give us some conceptual light how it is possible that the brain's neuronal states transform via the unconscious into the mind's mental states in consciousness. In other terms, the extended concept of mental states may provide a conceptual bridge between brain and mind and may thereby contribute (from the conceptual and also the epistemological[9] side) to solve the above described hard problem, why there is principal conscious rather than principal unconscious. This in

[9] Philosophy of mind presupposes the unconscious with reference to mental states which captures what I called phenomenal AL. In contrast, cognitive neuroscience refers to psychological states when it speaks of the unconscious. This however does not yet capture what I called physical AL. Physical AL presupposes reference of the unconscious to physical states. Unlike in phenomenal AL, where mental state of one's brain remain "principally unconscious," or in cognitive neuroscience, where psychological states are considered unconscious, the case of physical AL implies that physical states remain unconscious. Accordingly, physical AL makes the extension of the unconscious to physical states necessary even if they remain "principally unconscious."

If, in contrast, one neglects such extension, the reference of the unconscious to physical states may appear paradox if not contradictory. The unconscious is usually regarded to refer to material or contents that are either psychological or mental, whereas they are supposed to be caused by non-mental and non-psychological material/contents, i.e., the physiological processes. It is generally regarded that the unconscious can only refer to the former but not to the latter which are therefore characterized as nonconscious (see above). The claim that the unconscious, i.e., the "principal unconscious," may refer to physical states, as it is implied in physical AL, must thus appear contradictory implying confusion between cause and effect.

What however is neglected are two possible scenarios. One is at least a logical possibility that physiological processes and thus physical states, as caused by other nonconscious (n) remaining physiological processes, may possibly become unconscious by means of some change in material or content. Another one is a natural possibility that the physical material/content is generally suited for the unconscious and the consciousness but that it possesses the "wrong" code or format by itself. In this case physical states can no longer be considered "nonconscious (n)" but rather "principally unconscious (c)." I claim that the latter case, physical states being the "correct" material/content though with the "wrong" code or format, is the case which applies to physical AL. If this is true, the reference of the unconscious can no longer be restricted to mental and psychological states only but should include physical states as well. Accordingly, physical AL stretches the notion of the unconscious beyond its own boundaries by revealing novel references like physical states that make the distinction between "repressed unconscious" and "principle unconscious" necessary for epistemic rather than mere terminological reasons.

turn may then pave the way for neuroscience to tackle the question of the neural code which may contribute an empirical, i.e., neuronal answer to the hard problem.

## 7.6 Conclusion

I here discussed the concept of the unconscious in a rather broad way including psychodynamic, psychological, and philosophical notions. Though there is a certain overlap between the different concepts of the unconscious in all three disciplines, philosophy, psychoanalysis, and cognitive neuroscience, I also showed substantial differences (see Fig. 7.1). These concerned the dimensions of accessibility, material/content, and states that make the distinction between the unconscious as dealt with in psychoanalysis and cognitive neuroscience and the unconscious as presupposed in philosophy. Hence the "repressed unconscious" and the "psychological unconscious" must be distinguished from what I here call "principle unconscious." One may argue that such distinction may be of mere conceptual-philosophical relevance and does not contribute to better empirical, psychodynamic, psychological, and neuronal understanding of how unconscious and conscious states are generated. However, by revealing the conditions of the possible "principle unconscious," we may learn something about what makes both unconscious and conscious states possible. I here argue that the process of self-related processing that neuronally seems to be associated with rest-stimulus interaction and neural activity in the subcortical-cortical midline structures (Northoff et al. 2006; Northoff and Panksepp 2008; Panksepp and Northoff 2009) may make the "principle unconscious" impossible for us as conscious (human) beings. This, in turn, implies that the brain's resting state activity, the rest-stimulus interaction, and the self-related processing may be necessary conditions to constitute principally conscious states (including dynamic unconscious and preconscious) so well described in psychoanalysis, psychology, and cognitive neuroscience. Such hypothesis about the relationship between resting state/rest-stimulus interaction, self-relatedness, and unconscious/conscious may stipulate different and novel approaches to investigate the psychological and neuronal basis of unconscious and conscious states. As such it may contribute to neuroscience in stipulating novel questions like the one for the neural code, to psychoanalysis in opening the door of the brain, and to philosophy in allowing to develop conceptual bridges to solve the hard problem.

## 7.7 Definitions

**Psychoanalytic Unconscious**

According to Freud, the dynamic or repressed unconscious does not only operate outside awareness but is extremely complex including distinct aspects of the persons' self. It includes innate and inherently sexual and aggressive blind drives (i.e., the id), most of its conscience and ego ideals (i.e., the superego), and processes (perception, action, etc.) that deal with reality (i.e., the ego). The dynamic or repressed unconscious mediates wishful, associative, and instinctual primary processes. The dynamic or repressed unconscious must be distinguished from what Freud called system preconscious that includes mental contents which are descriptively but not dynamically unconscious because, as non-offensive to the person's self, they are readily accessible to consciousness. The system preconscious is thus more rational, disciplined, reality-oriented, and energetically "bound" than the dynamic unconscious. Originally, Freud described the contents or material of the unconscious by predominantly primary processes like wishes, drives, instincts, and associative "blind" nonrational processes. In his later work, he however refrained, at least partially, from exhaustive and exclusive definition of the dynamic unconscious by primary and primitive processes by including more rational processes associated with the ego and superego into the unconscious (see Westen 1999). In this latter sense, the psychoanalytic unconscious resembles the "psychological unconscious."

**Psychological Unconscious**

Psychological accounts include both rational and nonrational contents or materials in the unconscious. The content or material of the unconscious may concern cognition, emotions, movements, perceptions, behavior, etc. Consequently, terms like the "cognitive unconscious" (Kihlstrom 1987), "emotional unconscious" (Kihlstrom 1987, 1999; Berridge and Winkielman 2003), "behavioral unconscious" (Uleman 2005), and "procedural unconscious" (Schuessler 2002) have been introduced to characterize different contents or material of the unconscious. It is clear that here the focus is on different contents; unconsciousness in this sense mirrors cognition content. One characterizes unconsciousness in this sense as "psychological unconscious" (see Kihlstrom et al. 1992). This means that the unconscious can occur in a variety of different psychological functions processing different types of materials or contents. Unlike in the early time of the "cognitive unconscious," the psychological unconscious includes not only cognitions but also affect, motivation, goals, motives, self-regulation, interpersonal encounters, and the causes of phenomenal experience like the free will, all of which are empirically studied with a variety of different methods—this has recently been called the "new unconscious" (Uleman 2005, p. 6).

**(Neuro)philosophical Unconscious**

In the context of philosophy, the unconscious denotes the kind of knowledge we can and cannot acquire when accessing information originating in either our own mental states, our body, or other bodies and the rest of the world. The notion of unconscious is taken here no longer in a psychodynamic or psychological context and hence in an empirical context but rather in an epistemological context. The epistemological context distinguishes itself from the empirical context in that it no longer asks for the observable mechanisms but rather for the kind of principal knowledge we can and cannot acquire about the information related to the unconscious. Within such epistemological context, one may want to distinguish between unconscious information that can principally be accessed by us and thus know and the kind of information that remains principally inaccessible by us; the first denotes what I determine as principal conscious, while the second describes what I define as principal unconscious (see below). Finally, the prefix "neuro" to the term of the (neuro)philosophical unconscious derives from the search for the kind of neural mechanisms that allow as necessary (and/or sufficient) conditions of the possible distinction between principal conscious and principle unconscious; as I hypothesize these neural mechanisms pertain to the kind of neural coding the brain employs and imposes upon all incoming stimuli which in turn makes them accessible or inaccessible to our possible knowledge of them (see Northoff (2011) for details).

**Principle Conscious**

Encompasses all states that are conscious or can become in principle conscious including the preconscious, dynamic unconscious, and the psychological/new unconscious.

**Principle Unconscious**

Includes all states that in principle cannot become conscious, like Searle's notion of the "deep unconscious," i.e., states that remain principally inaccessible to consciousness that cannot be cognized. Consciousness, preconscious, and dynamic unconscious must be regarded as specifications of the "principle conscious" and must thereby all be distinguished from the "principle unconscious." The principle unconscious has two types, the "principle unconscious (a)" with (a) standing for lack of access and the "principle unconscious (c)" with (c) standing for wrong code or format. We cannot have conscious access to the "principle unconscious (a)" because we are not epistemically able to have access to it. We cannot access to the "principle unconscious (c)" because it is not coded or formatted in phenomenal (or mental) terms, thereby making phenomenal (or mental) states and hence consciousness impossible.

**Questions for Research**

1. What are the neuronal mechanisms and thus the neural code underlying the possible distinction between principal unconscious and principal conscious?
2. Are the here made conceptual distinctions between different forms of the unconscious empirically plausible meaning do they correspond to distinct neuronal mechanisms?
3. What are the exact neuronal mechanisms gating the transition from the psychodynamic unconscious to the consciousness?

## References


Anderson S, Chen S. The relational self: an interpersonal social-cognitive theory. Psychol Rev. 2002;109(4):619–45.

Anderson S, Reznik I, Glassman N. The unconscious relational self. In: Hassin Ran R, et al., editors. The new unconscious. New York: Oxford University Press; 2005.

Berridge KC, Winkielman P. What is an unconscious emotion? (the case for unconscious "liking"). Cognit Emot. 2003;17(2):181–2111.

Kihlstrom J. The cognitive unconscious. Science. 1987;237:1445–52.

Kihlstrom J, Barnhardt TM, Tataryn D. The psychological unconscious. Am Psychol. 1992;47(6):788–91.

Kihlstrom JF. The psychological unconscious. In: John OP, Pervin LA, editors. Handbook of personality: theory and research. New York: Guilford; 1999.

Northoff G. Philosophy of the brain. The brain problem. Amsterdam: John Benjamins; 2004.

Northoff G. Neuropsychoanalysis in practice. Brain, self and objects. New York: Oxford University Press; 2011.

Northoff G, Bermpohl F. Cortical midline structures and the self. Trends Cogn Sci. 2004;8(3):02–107.

Northoff G, Heinzel A, de Greck M, Bermpohl F, Dobrowolny H, Panksepp J. Self-referential processing in our brain – a meta-analysis of imaging studies on the self. NeuroImage. 2006;31(1):440–57.

Northoff G, Panksepp J. The trans-species concept of self and the subcortical-cortical midline system. Trends Cogn Sci. 2008;12(7):259–64.

Panksepp J, Northoff G. The trans-species core SELF: the emergence of active cultural and neuro-ecological agents through self-related processing within subcortical-cortical midline networks. Conscious Cogn. 2009;18:193.

Revonsuo A. Inner presence. Cambridge, MA: MIT Press; 2006.

Schuessler G. Aktuelle Konzepte des Unbewussten. Z Psychosom Med Psychother. 2002;48:192–214.

Searle J. Mind. A brief introduction. New York: Oxford University Press; 2004.

Strawson G. Mental reality. Cambridge, MA: MIT Press; 1994.

Uleman JS. Introduction: becoming aware of the new unconscious. In: Hassin Ran R, et al., editors. The new unconscious. New York: Oxford University Press; 2005.

Westen D. The scientific status of unconscious processes: is Freud really dead? J Am Psychoanal Assoc. 1999;47(4):1061–106.

# Dreams

# 8

Tamara Fischmann and Marianne Leuzinger-Bohleber

**Abstract**

Dreams have been one core element in psychoanalysis and have been emphasized already by Freud as central therapeutic elements. Here we review recent findings on the neurobiology of sleep in general and dreams in particular. We here consider dreams as indicators of inner transformation processes in the structure of the ego—the vivid experiences of dreams are thus supposed to index change in our self and its structure. This is based on various findings from psychoanalytical dream research and, more specifically, the contents of dreams. That is complemented by recent psychological-cognitive and neurobiological findings of dreams—considered in a cognitive perspective, dreams can be regarded as 'embodied memories'.

Different theories of dreams are compared and discussed. Taken in this sense, dreams have a central function for the ego in allowing for transforming previous memories into the present and current state of the self—they thus serve for the self to adapt itself to its changing environmental contexts without losing itself.

## 8.1 Introduction: Neuropsychoanalytical Dream Research[1]

Sigmund Freud has chosen the beginning of the twentieth century, 1900, as the publishing date for *The Interpretation of Dreams* because he was convinced that this book would create a new scientific discipline: psychoanalysis. Indeed, dreams are still seen as the 'via regia to the unconscious' by many contemporary psychoanalysts.

[1] The introduction is based on an unpublished text by Mark Solms (2017).

T. Fischmann (✉) · M. Leuzinger-Bohleber
Sigmund-Freud-Institut, Frankfurt, Germany
e-mail: Dr.Fischmann@sigmund-freud-institut.de; m.leuzinger-bohleber@gmx.de



H. Boeker et al. (eds.), *Neuropsychodynamic Psychiatry*,
https://doi.org/10.1007/978-3-319-75112-2_8

In the meantime, dreams have been studied in many fields, experimental dream-sleep research, neurology, psychiatry and neuropsychoanalysis (see, e.g. Solms 1997; Leuzinger-Bohleber 2015). Just a few introductory remarks on some of the findings (see footnote 1):

Humans spend almost a third of each day asleep and about one-fifth of their sleep time dreaming. This means that we spend approximately 90 min of every day in the state of dreaming. However, sleep science and neuroscience have been unable to convincingly elucidate the biological function of dreaming.

There is every reason to believe that dreaming is not a uniquely human function. Dreaming is highly correlated with REM sleep, and REM sleep is a ubiquitous mammalian state. However, dreaming is not synonymous with REM sleep; in fact, the two processes are doubly dissociable (Solms 2000). Moreover, it is impossible to access dream reports in any species other than humans. To address the biological function of *dreaming* (as opposed to REM sleep), therefore, it is necessary to study humans and humans only. Therefore, several ongoing neuropsychoanalytical studies are focusing this research question (see, e.g. Solms 2017).

On Freud's view, sleep was characterized by disinhibited endogenous drives which activate (or are activated by) volitions which would normally provoke motor activity. For this reason, dreams were considered as a mode of 'diverting' potentially sleep-disturbing impulses, through hallucinatory fulfilment of the impulses in question. Recent research strongly suggests that sleep is indeed punctuated by potentially disturbing endogenous arousal and motivational events (see Perogamvros and Schwartz 2012 for review). Not only is sleep characterized by the cyclical arousal state of REM itself (Aserinsky and Kleitman 2003), but it is also disturbed by surges of midbrain dopaminergic activation in the mesocortical-mesolimbic circuit responsible for 'the most vigorous exploratory search activity an animal is capable of' (Panksepp 1998, p. 145; see Dahan et al. 2007; Léna et al. 2005). How then does the animal remain asleep? On the Freudian view, it does so by dreaming.

The clear experimental prediction from this hypothesis, which has never been tested (and, indeed, was famously described as untestable; Popper 1963), is that *non-dreaming patients with posterior cortical lesions should display poor-quality sleep*. Preliminary work suggests that this is indeed the case. Bischof and Bassetti (2004) reported a single case of acute thrombotic infarction of the occipital lobe in the region of the posterior cerebral arteries with cessation of dreaming but preservation of REM sleep. The authors noted, apparently incidentally and without realizing the theoretical significance, that their patient suffered sleep-maintenance insomnia. Solms and his research group have since confirmed this observation in five further cases (see Solms 2017), but still further research must be done.

One other theory in the biological function of dreaming concerns the role that dreaming plays in memory processes during sleep. In recent years, there has been increasing interest in the hypothesis that sleep contributes to and influences memory processing in a significant manner. This hypothesis is widely accepted, and focus has now shifted to the relative contributions of various memory processes during sleep. This includes, for example, memory consolidation (stabilization, enhancement and reconsolidation) in both REM and NREM sleep (Rasch and Born 2013).

In this theoretical context, dreaming is hypothesized to play a specific role in memory processing during sleep. Dreaming is said to play an especially important role in emotional declarative memory[2] during REM sleep. The neurobiological association between REM sleep and emotional declarative memory has been made due to the neurochemical changes that take place in the brain during this sleep stage (e.g. increased acetylcholine release), as well as the activation of certain brain regions (e.g. increased activity in the amygdala and cingulate cortex) that make the brain amenable to affect-related memory consolidation (Stickgold et al. 2001; Hu et al. 2006; Nishida et al. 2009).

This neurobiological association between emotional declarative memory and REM sleep is widely supported in the literature. For example, a study by Wagner et al. (2001) found that memory for emotional material was significantly enhanced after a period of late sleep in which REM sleep was predominant. The same was not true for memory related to emotionally neutral stimuli and for early sleep (predominated by NREM sleep). Similar results have been obtained by Nishida et al. (2009) who found a correlation between offline emotional memory enhancement and amount of REM sleep, while no correlation was found regarding emotionally neutral material.

Due to the strong correlation between REM sleep and dreaming, several theorists speculate that dreaming—being a conscious state—plays a special role in emotional declarative memory consolidation (Stickgold et al. 2001; Nielsen and Stenstrom 2005). However, any role for dreaming in memory consolidation remains to be empirically demonstrated, due to the difficulty in experimentally distinguishing dreaming per se from REM sleep.

There are several hypotheses regarding the role of dreaming in memory consolidation. For example, Stickgold et al. (2001) note that during REM sleep, limbic forebrain structures along with the amygdala are activated, while there is also an inhibition of hippocampal outflow that presumably prevents the reactivation of episodic memories. Consequently, dreams would be constructed mainly from weak neocortical associations that are available during REM sleep. Dreams are thus typically unpredictable, bizarre, and emotion-laden. The authors hypothesize that these features reflect the brain's attempt to recognize and assess novel cortical associations in the context of emotions mediated by limbic structures. They propose that one functional consequence of REM dreaming is the strengthening or weakening of specific activated associations, with regression to pictographic imagination providing compensation for the relative loss of motor function during sleep.

There are several other theories regarding the function of dreaming in relation to memory processes. For example, firstly, it has been proposed that the appearance of memories in dreams promotes learning by reactivating those elements in their original (perception-like) state; secondly, that the binding of various memory elements (especially around emotionally relevant themes) strengthens and consolidates those elements; and thirdly, that dreaming about newly learned material enhances subsequent recall of that material (for reviews, see Payne and Nadel 2004; Nielsen and Stenstrom 2005).

However, it is important to reiterate that although the majority of dreams occur during REM sleep, multiple lines of evidence demonstrate that REM sleep and

[2] Memory for events associated with intense and salient emotions.

dreaming are in fact dissociable states, governed by different mechanisms in the brain (Solms 2000). This evidence suggests that REM sleep is controlled mainly by cholinergic brainstem mechanisms (McCarley et al. 1995), while dreaming is controlled mainly by dopaminergic limbic and other forebrain mechanisms (Perogamvros and Schwartz 2012). Thus, dreaming can occur outside of REM sleep, and REM sleep can occur in the absence of dreaming.

As this short summary might have illustrated: Dream research is a fascinating field in contemporary neuropsychoanalysis. We already have finished a pilot study on the biological function of dreaming and hope to be able to start the main study soon (see Fischmann and Leuzinger-Bohleber 2017).

In this chapter, we will focalize another psychoanalytical research project focusing on dreams *Changes of dreams as indicators of sustaining inner transformation processes in psychoanalysis.* It is a sub-study of the large ongoing LAC depression study comparing the outcome of psychoanalytical and cognitive-behavioural long-term psychotherapies. In this study, we are taking dreams as an indicator for transformations of the inner object world of the patient, in other words indicators of so-called structural changes in psychoanalysis. These transformation processes are connected to symptomatic change but go much beyond. Structural change refers to uncovering unconscious psychic structures and mechanisms which determine current inadequate mental functioning, problem-solving, human relationships, affects and satisfaction in life. Structural changes enable patients to resolve psychic and psychosocial conflicts by stressors in current life situations in a more adequate manner and thus not only influence its psychopathological symptoms but also his capability 'to work, to love and to enjoy life'—the well-known aims of psychoanalyses.

Structural changes are difficult to operationalized. In another publication, we have reported on findings concerning structural changes based on the so-called Operationalized Psychodynamic Diagnostics (OPD) and the Heidelberger Umstrukturierungsskala (HUS) (see Leuzinger-Bohleber et al. forthcoming). In this chapter, we are investigating systematic changes of the manifest (and latent) dreams in a psychoanalysis. We are referring to a single case study with a chronic depressed patient which is summarized in Chap. 30. The theoretical background of the content analysis of the manifest dreams is a model by Moser and von Zeppelin (1996) integrating a broad knowledge base of psychoanalytical, neuroscientific and empirical knowledge on the generation of dreams (see Sect. 8.2).

## 8.2 Changes of Dreams as Indicators of Sustaining Inner Transformation Processes in Psychoanalysis

### 8.2.1 Some Findings from Psychoanalytical Psychotherapy Research

Marianne Leuzinger-Bohleber (Leuzinger-Bohleber 1987, 1989) compared manifest and latent dream content of the first and last 100 sessions in a total of 5 psychoanalyses and could show systematic changes in these aggregated single case studies. In the 'successful psychoanalyses' (defined by the analysands, their analysts and

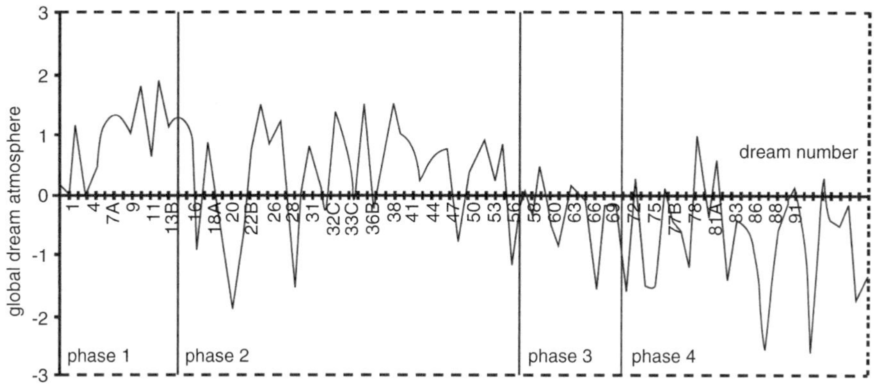


**Fig. 8.1** Changes in the dream atmosphere during psychoanalyses

independent observers), the following changes were found in the dreams of the last 100 psychoanalytic sessions:

- Atmosphere of the manifest dream is more frequently positive.
- More successful problem-solving.
- Broader spectrum of affects (in contrast to the domination of one single affect (most frequently panic) in the dreams of the first 100 sessions.
- Dreamer was in an active position (not in the position of the observer).
- More intensive and satisfying human relationships.
- More human subjects, less animals.

These findings have been replicated by Kächele et al. (2015). In the frame of the LAC study, another replication study is ongoing (Figs. 8.1 and 8.2).

### 8.2.2 The 'Dream Generation Model' by Moser and von Zeppelin: An Attempt to Integrate Psychoanalytical and Interdisciplinary Knowledge on Dreams

In modern dream theories, dreaming is described as a thought-process engaging our inner system to process information (Dewan 1970). Inner (cognitive) models are constantly being modified in coordination with what is perceived. In contrast to a dreaming state, the reactions to our environment are immediate while we are awake, thus enabling information consolidation into memory with one constraint, namely, that consolidation processes are not always possible due to capacity restrictions of the system. Consolidation processes do continue though during sleep in an 'offline' modus, thus enabling integration to long-term memory.

According to Moser and von Zeppelin (1996)[3]—psychoanalysts and dream researchers at the same time—so-called dream complexes—activated by current

[3] Ulrich Moser and Ilka von Zeppelin are full-trained psychoanalysts engaged in interdisciplinary research for decades. Ulrich Moser was professor for clinical psychology at the University of

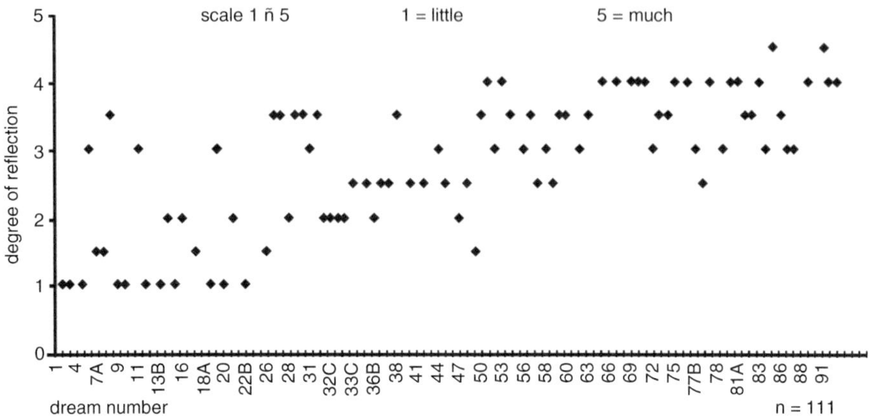


**Fig. 8.2** Changes in problem-solving during psychoanalyses

events process the entirety of information from unsolved conflicts and traumatic situations while dreaming. The dream searches solutions or rather best possible adaptations for these dream complexes. A dream, which is usually pictorial, consists of at least one situation produced by a 'dream organizer'. Dream organization may be considered, according to Moser, as a bundle of affective-cognitive procedures, generating a microworld—the dream—and controlling its course of action. Within this system, the 'dream complex' is considered to be a template facilitating dream organization.

Thus, it may be assumed that a 'dream complex' originates from one or more complexes stored in long-term-memory, rooted in conflictuous and/or traumatizing experiences, which found their condensates in *introjects*. These conflictuous or traumatic dream complexes are easily triggered by stimuli from the outside world, which are structurally similar to stored situations of these complexes. The search for solution of the complex is governed by the need for security and wish for involvement, i.e. the *security principle* and the *involvement principle* which govern dream organization.

Wishes within these complexes are links between self- and object-models and RIGs (i.e. representation interaction generalized), which are accompanied by convictions and a hope for wish fulfilment. Conflictuous complexes are areas of bundled wishes, RIGs and self- and object-models with a repetitive character, thus creating areas of unbound affective information. Affects within such an area are interconnected by k-lines, which are blocked and thus not localized. To solve these conflictuous or traumatic complexes, it is necessary to retrieve this affective

Zurich. Already in the 1960s and 1970s, he was involved in modelling parts of psychoanalytic theories. By the means of computer simulation, he tested the logical and terminological consistency of psychoanalytic theories of defence and the generation of dreams. Based on this basic research on dreams, he developed an own model of the generation of dreaming as well as a coding system for investigating the manifest dreams. In this chapter as well as in the paper by Varvin et al. (2012), the dream model and the coding system by Moser and v. Zeppelin are applied.

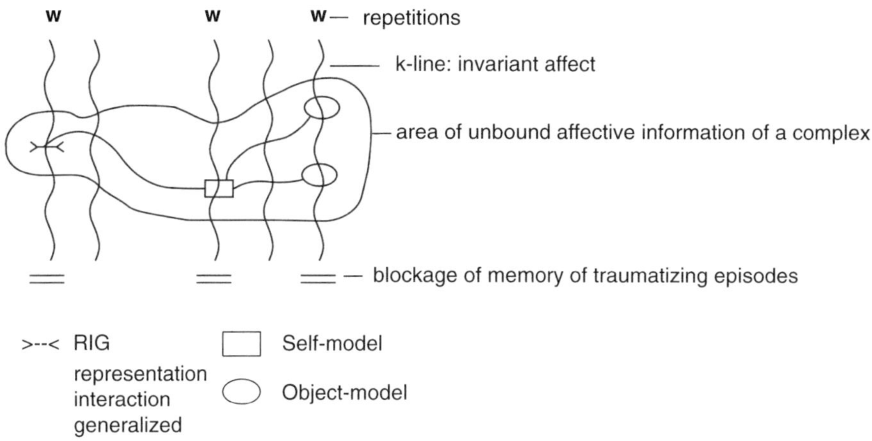


**Fig. 8.3** Memory model of conflictuous complexes according to Moser and von Zeppelin (1996)

information and reintegrate it into a relational reality to make them come alive (cf. Fig. 8.3). This is being attempted in dreams whose function is to search for a solution of the complex. This search for a solution within a dream is governed by the above-mentioned need for security and wish for involvement, i.e. the *security principle* and the *involvement principle*. The following illustration may serve as an elucidation of this model.

Based on this sophisticated model, the research group in Zurich has developed a coding system for analysing the manifest dream contents in a very valid and reliable way—the Zurich Dream Process Coding System (ZDPCS). We have applied this system in the LAC study in order to investigate systematic changes in the manifest dream content of analysands during their psychoanalyses. Some of these analysands—due to their severe sleep disturbances—also agreed to undergo a medical examination of their sleep including three nights in the sleeping laboratory. This gave us the possibility to compare their dreams after the REM phases in the laboratory with the dreams reported in the psychoanalytical sessions with very interesting findings (see Fischmann et al. 2013). Even though the content of the manifest dreams showed some differences, the structure of the laboratory and the dreams reported during psychoanalytic sessions were identical—a finding which was important for the systematic investigation of the changes of the manifest dreams during the long-term treatments.

### 8.2.3 Dreaming and 'Embodied Memories'

Another interesting new perspective on dreams was developed in the dialogue between psychoanalysis and the so-called embodied cognitive science (see, e.g. Leuzinger-Bohleber 2015) and Chap. 6 in this volume. In the psychoanalytic model of representation and in the computer metaphor derived from 'classic cognitive science', memory and recollection were for a long time understood as processes whereby (statically) retained knowledge was transformed from long-term memory to short-time memory and called up into a current problem-solving situation. We still find comparable thinking in some textbooks in clinical psychology. Aristotle's famous example comparing memory to a wax tablet into which experiences etch themselves onto appears to live on.

But according to various views in embodied cognitive science today, memory can no longer be understood as comparable to a computer, as storage disk with statically stored content from which information can be 'retrieved' in a present situation. What analysands experience in their important and for them existential relationship to their analysts are not unconscious 'statically entrenched' representations of their past relationships, e.g. to their primary objects which are unconsciously reactivated, as had been understood, for example, in reference to the model of representation in classical psychoanalysis (cf., e.g. Karl Menninger's Triangle of Insight, Menninger 1958). Memory—in contrast to this 'classical conceptualization'—is a function of the entire organism, the product of complex, dynamic, recategorizing and interactive processes, which are invariably 'embodied'[4].

According to this new understanding of memory, 'embodied memories' have often been triggered in dreams, e.g. by a current interactional experience in the psychoanalytical situation, as is discussed in the chapter by Leuzinger-Bohleber and Fischmann in this volume. Therefore, dreams often are keys for discovering central 'embodied memories' of the (traumatized) patients. Nightmares, e.g. contain some of the embodied experiences of a self which was totally helpless, impotent and lonely confronted with unbearable pain, scared to death without a holding or containing primary object, a helpful other. The passive, frozen bodily state of the dreaming subject thus often uncovers unconscious 'embodied memories' of former traumatization. Often movements of the dream subject and first solutions in a terrifying situation indicate turning points in psychoanalyses as will be illustrated with the extended case example in this paper (see dream of the 'odd bird' in the third year of treatment; publications concerning this issue: see Leuzinger-Bohleber 2015; Leuzinger-Bohleber and Pfeifer 2017).

### 8.2.4 Conceptual Considerations Based on the Interdisciplinary Dialogue with Experimental Dream Researchers

Weinstein and Ellman (2012) have published an innovative neuropsychoanalytical dream model integrating Freud's dream theory with object-relational theories. They

[4] 'Embodied' not only means 'non-verbal': memory arises by way of a 'coupling' of reciprocally influential sensoric and motoric processes. This 'coupling' is biologically implemented through neuronal maps embedded in the organism's sensomotoric system. Thus, Clancey (1993) defined memory as the ability to coordinate neurological process.

are defining the drive theory of Freud in a new way making the connection to endogenous stimulations of the brain:

> 'Following Freud, we assumed that there is at least one (…) neurophysiological system(s) that provide mammals endogenous stimulation. Freud saw the dream as a type of safety valve; in terms of modern psychobiology this led us to think about the REM sleep state as a type of regulator for what he called drive or cognitively, unconscious phantasy.' (p. 111) Part of Freud's conception was that, at least periodically, endogenous stimulation builds up, and motivates the person to engage in action to reduce that stimulation. Reduction of stimulation was also how he defined pleasure.

Before we describe the regulatory function of REM sleep we should spell out another assumption about REM mechanisms. REM, in our view, is a manifestation of the basic rest activity cycle (BRAC) (Kleitman 1963). In Kleitman's view the 24-h or circadian cycle is a series of alternating states of resting and activity. In our theory REM mechanisms fire periodically to create activity in mammals. This activity is a manifestation of drive or endogenous mechanisms[5] where thresholds for centers that are critical to the animal's survival are regulated. Thus, during this activity phase the animal is primed to perform behaviors crucial to its survival or its group's survival such as food seeking, courting or nest building. REM sleep is part of BRAC and is an activity period during sleep. This activation permits an unbroken period of sleep while also allowing for periodic activation of survival mechanisms in case the animal encounters an emergency, for example, a predator. REM sleep also alters waking thresholds; if, for example, waking thresholds are high for drive behaviors (as in depression), there will tend to be less REM sleep or less discharge during sleep in an attempt to lower waking thresholds. In this way, one can view REM sleep as performing a regulatory function. Alternatively, one can say that REM sleep is part of the basic rest activity cycle and is affected by waking thresholds and experiences. Implied in both statements is the idea that the same mechanisms that fire during REM sleep are also activated periodically during wakefulness. (Ellman and Weinstein 1991; Weinstein and Ellman 2012, p. 4).

The research group investigated these hypotheses in different animal experiments. To summarize in a simplified way: The animals are seeking different tasks to get positive stimulations:

> To look at the physiological correlates of pleasure we studied positive reward systems in a variety of animals. These are sites that Olds first discovered (Olds 1956, 1962) when he found that if he delivered electrical stimulation to the hypothalamus, rats would learn a variety of tasks to obtain this stimulation. When the animal works to obtain this stimulation, this behavior has been labeled intracranial self-stimulation (ICSS). ICSS sites are usually in midbrain areas implicated in the control of behaviors such as eating, drinking, sex, and aggression, all sites involved in essential survival for an animal… (ibid., p. 6)

The researchers could show the close connection between REM sleep and the ICSS: If REM sleep was prevented, the self-stimulating behaviour increased. If

[5] For psychoanalytic readers, you can use the term drive instead of endogenous stimulation.

ICSS was prevented, the REM sleep increased! To summarize it again in a simplified manner: The activation of the ICSS system relates to the activation of neuronal circuits which activate eating, sexuality and aggression (in analogy to Freud's unconscious wishes, the latent dream content, which are activating the dream):

> Putting together to results of our human and animal studies, we reasoned that during REM sleep the firing of the ICSS system increases the probability of activating the neurophysiological substratum of a drive system(s) – that is the activation of neural networks involved in eating, sex and aggression. In humans, the activation of ICSS pathways frequently triggers memory systems that involve conflict. Typically, the mentation present in REM sleep contains material about the issues that are most relevant and/or threatening to the individual at any given point in time. This is not necessarily always true; one might dream about a not conflictual pleasurable situation but in our view for most adults (and most children) the important situations in their lives involves some conflict. Under optimal circumstances the dream provides a way of resolving the conflict, as the literature on the emotion regulation function of REM sleep suggests (see Nielsen and Lara-Carrasco 2007 for a review). If this is the case, the dream is forgotten. On the other end of the continuum are traumatic dreams, those in which the dreamer cannot envision a pain free resolution, but can only imagine an outcome that would result in injury or death. Traumatic dreams are ones that are frequently repeated: unfortunately, they express a threat of survival that the dreamer feels they can neither avoid nor resolve. (ibid., p. 13)

These theories have, in our view, a high relevance for the clinical understanding of dreams of chronic traumatized, depressed patients. One first point: the intracranial self-stimulation (ICSS) is only experienced as a positive reward if the subject can evoke it actively by himself. The rewarding seems to relate to an elementary experience of autonomy and self-agency. As we know from psychoanalytical trauma theories: One of the main characteristics of a traumatic event is that the individual is suddenly and unexpectedly confronted with an extreme situation of total helplessness and impotence in relation with extreme pain and the threat of death *without* getting any help from another person, thus losing a basic feeling of self-agency. Therefore in a traumatic situation, the basic trust in a helping 'other' and an active self is destroyed with sustaining consequences.

### 8.2.5 Taking Conceptual Considerations to the Experimental Lab

Based on these theoretical thoughts, we analysed the changes of dreams of a severely traumatized, chronic depressed patient (Mr. X, cf. Chap. 30 for more details) during his therapy. For one we looked at dreams he reported in the clinical situation and compared them to dreams of the same time elicited at the sleeping laboratory using the above-mentioned method of Moser and von Zeppelin, to see whether we could discern changes of dream atmosphere, relational capacities and problem-solving capabilities (i.e. self-agency). We were also interested to see if those changes occurred in both types of dreams—the laboratory dreams and the ones reported in the clinical situation.

Mr. X's dreams reported in the first 6 months of psychoanalysis were characterized by unbearable, traumatic situations and may be considered to be nightmares. In these dreams his dream self is captured in extremely dangerous, life-threatening situations usually flooded with panic and anxiety and lacking any capability to liberate himself from these situations.

The following dream, taken from the first 6 months of therapy, may serve as an exemplary specimen of such a nightmare:

> I am in a narrow tunnel, kind of a tube. Behind me my brother is crawling. We cannot go backwards – behind us is the stormy sea. The tunnel becomes narrower and narrower. I wake up in panic.

In comparison, his laboratory dreams of the same time during therapy are blander although palpably full of anxiety and the feeling of being left alone helplessly:

> I am walking through a building – a residential building. I don't know to which destination. Down the stairs – there are elevators. I walk through a door, behind it there are my parents, my brother. I try to talk to them. Then there is a fellow [female] student – her face is alienated. I am surprised and bewildered, happy to see her again. I ride elevators up or down. During the ride the floor of the elevator suddenly drops underneath me and is gone. I look outside – there are gigantic hangars with tools. They are deserted. I gaze for several minutes. There is an underground passageway. I am scared. Where does it go to? I am uncertain. A ride into the unknown?

The Zurich Dream Process Coding System (ZDPCS) of the two dreams reveals interesting facts. The dream from the clinical situation is shorter and ends in panic. The laboratory dream, though longer, has more interrupts—a sign for unbearable accumulation of affects, which must be interrupted—but has more distancing and failing interactions. The latter may be interpreted as a sign of the dreamers' lack of problem-solving capacities; he feels helplessly extradited to the situation he is in (Table 8.1).

As is discussed in detail in Chap. 30, the manifest dreams of this severely traumatized patient changed obviously during psychoanalysis. In the frame of this chapter, we can only refer to one other example in the third year of psychoanalysis:

> I played with the famous jazz guitarist Ralf Towner. It went quite well and it was fun. I didn't fail and the neck of the guitar was not soft[6] (laughs). The guitarist played along with my improvisations and held back. Of course, I knew that he is better than me, but this did not matter – it was just great fun…

The corresponding laboratory dream goes as follows:

> I was on the way with someone, whom I cannot name. He was familiar, but I cannot put a name to him. And we had a strange substance with us. In the beginning, it was a lump of earth or clay and he showed me how to make new forms out of it. In fact, not by processing

[6]The patient refers to another 'funny' dream. Before the dream he had a conflict with his wife which wasn't treated openly. Instead the conflict led to an erectile dysfunction. Then he dreamed that he played on a guitar which had a very soft neck….

**Table 8.1** ZPCS dream coding comparing clinical dreams with laboratory dream of the first 6 months of psychoanalysis

| Dream from clinical situation | | | | | Laboratory dream | | | | |
|---|---|---|---|---|---|---|---|---|---|
| Dream narrative | Sit | PF | LTM | IAF | Dream narrative | Sit | PF | LTM | IAF |
| I am in a narrow tunnel, kind of a tube | S1 | SP (dreamer)<br>Place (tunnel)<br>ATTR (narrow)<br>ATTR (tube) | | | *I am walking through a building—a residential building* | S1 | SP (dreamer)<br>Place (building)<br>ATTR (residential) | LTM | |
| Behind me my brother is crawling | S2 | SP (dreamer)<br>$OP_1$ BEK (brother)<br>POS REL | LTM | | *I don't know to which destination* | C.P. | | | |
| We cannot go backwards—behind us is the stormy sea | S3 | SP (dreamer)<br>$OP_1$ BEK (brother)<br>Place (sea)<br>ATTR (stormy)<br>POS REL | | IR.C RES<br>LTM<br>Fail (cannot go backwards) | *Down the stairs—there are elevators* | S1 | SP (dreamer)<br>CEU (stairs)<br>CEU MULT (elevators)<br>POS REL | | |
| The tunnel becomes narrower and narrower | S4 | SP (dreamer)<br>Place (tunnel)<br>ATTR (narrow) | | IR.D (IR.S) | *I walk through a door, behind it there are my parents, my brother* | S2 | SP (dreamer)<br>CEU (door)<br>CEU (door)<br>POS REL<br>OP BEK (parents)<br>OP BEK (brother) | LTM | |
| I wake up in panic | EX AFF R | | | | *I try to talk to them* | S3 | SP (dreamer)<br>OP BEK MULT (them) | | IR.C int |

| | | | | | | | | | |
|---|---|---|---|---|---|---|---|---|---|
| | | | | | *Then there is a fellow [female] student—her face is alienated. I am surprised and bewildered, happy to see her again* | S4 | SP (dreamer)<br>EX AFF R (surprised…)<br>OP BEK (student)<br>ATTR<br>OP PART OF (face)<br>ATTR (alienated) | | |
| | | | | | *I ride elevators up or down* | S5 | SP (dreamer)<br>CEU (elevators) | LTM | |
| | | | | | *During the ride the floor of the elevator suddenly drops underneath me and is gone* | S6 | SP (dreamer)<br>CEU (elevator)<br>CEU PART OF (floor)<br>POS REL | | IR.D (IR.S) |
| | | | | | *I look outside—there are gigantic hangars with tools. They are deserted. I gaze for several minutes* | S7 | SP (dreamer)<br>CEU MULT (hangars)<br>ATTR (gigantic, deserted) | LTM perc | |
| | | | | | *There is an underground passageway* | S8 | SP (dreamer)<br>CEU (passageway)<br>ATTR (underground) | | |
| | | | | | *I am scared. Where does it go to? I am uncertain* | EX AFF R | | | |

*Sit* situation, *PF* positioning field, *LTM* locotime motion, *IAF* interaction field

> it but by crumbling it. And this was interesting; so, he crumbled this thing and it became fine flakes, fell and when it reached the ground new forms emerged. I tried to do it too but did not manage in the beginning – I thought it did not really work for me. But he said: "no, this is quite good already. It's not perfect yet, but I should keep on trying and it will get better and it will work. And I still thought it didn't really work for me and that the product was baddish…

The dreams from the third year of psychoanalysis reveal a very different picture in the ZDPCS. In his dream from the clinical situation, he again wakes from his dream highly affectively aroused, but this time he feels elated, and during most of the dream, he is in responsive interaction with a 'helping' object. The laboratory dream is in line with this insofar as that the dreamer is in a positively responsive interaction with a 'helping' other. He is still full of doubt if this 'good' 'relationship' will carry and help him to become 'better'.

These dreams may have illustrated to a degree the way Mr. X's early traumatization become observable in his manifest dreams and how this changed during the treatment. The underlying traumatic complex that governed the dream organization at the beginning of treatment was successively better integrated in the psychic functioning of the patient. The dream coding showed how the dreamer established an increasing feeling of self-agency, control and basic trust in a helping other (Table 8.2).

## 8.3 Summary and Discussion

Different contemporary interdisciplinary dream models, integrating knowledge from diverse scientific disciplines, agree that nightmares of patients may be connected to (early) traumatization of the analysand. In analogy to the traumatic situation, the dream subject repeats being in a traumatic situation of extreme helplessness, impotence and unbearable negative affects like panic, despair, rage and death anxiety again and again—also while dreaming (i.e. underlying traumatic complex). The dream self has lost any control of the situation and is confronted with annihilation anxiety and the threat of death. As Weinstein and Ellman (2012) discuss: nightmares are not only triggered by extreme overflooding anxiety but at the same time by the missing of a holding, containing object. As is well known: trauma is defined as a situation in which the basic trust in a helping 'other' and the self-agency is destroyed—an experience with sustaining consequences (see, e.g. Bohleber 2010). Moser and von Zeppelin (1996) postulate in their model of the generation of dreams that traumatic complexes can be characterized by the fact that extreme affects are not 'bonded' (integrated) in a structure of human relationships. Finally, we have referred to the concept of 'embodied memories' (Leuzinger-Bohleber and Pfeifer 2012; Leuzinger-Bohleber 2015, 2016). As is illustrated in the case example in Chap. 30 in detail, we can find indicators of 'embodied memories', which means bodily experiences in early relationships, in the manifest dreams which can often be used as indicators for a successive understanding of the (unconscious) 'history of the trauma' of the analysand in the psychoanalytic situation.

**Table 8.2** ZPCS dream coding comparing clinical dreams with laboratory dream of the third year of psychoanalysis

| Dream from clinical situation | | | | | Laboratory dream | | | | |
|---|---|---|---|---|---|---|---|---|---|
| Dream narrative | Sit | PF | LTM | IAF | Dream narrative | Sit | PF | LTM | IAF |
| I play with the famous jazz guitarist Ralf Towner. It goes quite well | S1 | SP (dreamer) OP BEK (towner) ATTR (famous) ATTR (jazz g.) | | IR. C RES | *I was on the way with someone, whom I cannot name. He was familiar* | S1 | SP (dreamer) OP (someone) ATTR BEK | | IR.C RES LTM |
| It is fun | EX AFF R | | | | *But I cannot put a name to him* | C.C. | | | |
| I don't fail, and the neck of the guitar is not soft | S2 | SP (dreamer) CEU (guitar) PART OF (neck) ATTR (not soft) | | | *We have a strange substance with us. In the beginning a lump of earth or clay* | S1 | ATTR (have) CEU (substance) ATTR (strange, lump, like clay) | | (s. above) |
| The guitarist plays along with my improvisations and holds back | S3 | SP (dreamer) OP BEK (guitarist) ATTR (held back) | | IR.C RES | *He shows me how to make new forms out of it. In fact, not by processing it but by crumbling it* | S2 | SP (dreamer) OP (he) CEU (substance) CEU mult (forms) | | IR.C RESP (shows) |
| Of course, I know that he is better than me, but this does not matter— | C.P. | | | | *This is interesting* | C.C. | | | |

(continued)

**Table 8.2** (continued)

| Dream from clinical situation | | | | | Laboratory dream | | | | |
|---|---|---|---|---|---|---|---|---|---|
| Dream narrative | Sit | PF | LTM | IAF | Dream narrative | Sit | PF | LTM | IAF |
| it is just great fun… | EX AFF R | | | | *He crumbles this thing, and it becomes fine flakes* | S3 | SP (dreamer) OP (he) CEU (thing) PART OF (flake) ATTR (fine) | | IR.D (IR.C KIN (crumble) IR.D PHYS (become flakes) |
| | | | | | *Falls down* | S4 | SP (dreamer) CEU (substance) | LTM | |
| | | | | | *And when it reaches the ground, new forms emerge* | S5 | SP (dreamer) CEU (substance) CEU (forms) MULT ATTR (new) | | IR.C PHYS (new emerge) |
| | | | | | *I then also try to do it too but do not manage in the beginning* | S6 | SP (dreamer) | | IR. C KIN (try) FAIL |
| | | | | | *I think it does not really work for me* | C.C. | | | |
| | | | | | *He then says: no, this is quite good already. It's not perfect yet, but keep on trying and it will get better and it will work* | S7 | SP (dreamer) OP (he) | | V.R. (says) |

*Sit* situation, *PF* positioning field, *LTM* locotime motion, *IAF* interaction field

All these different conceptualizations of dreams agree that the frequent nightmares, which are often told in the initial phases of psychoanalysis (also by patients in the LAC study), may hint at cumulative (early) traumatizations of the analysands which have finally lead into chronic depression. As is tried to illustrate in the clinical case: Changes in the manifest dream contents as well as in the association of the dream (indicators for the latent dream content) thus can be used as signs for often hidden transformations during psychoanalyses. If the dream subject gains a more active stance and control over dangerous situation and no longer exclusively is passive, lonely victim but in company with helping others often means that there are 'turning points' in the psychoanalytical process. Another indicator is the systematic change of affects in the manifest dreams: the spectrum is enlarged. Not one single affect (like panic) is dominating the dream plot anymore as in the initial phase of the treatment. In the frame of this paper, we only could illustrate, but not systematically show or theoretically discuss in detail, that in 'successful' psychoanalysis, the analyst achieves to bring the (split-off) trauma with its unbearable affects and unconscious beliefs back into the psychoanalytical relationship. This may lead to a modification of unconscious convictions that 'no-one- but no-one - is interested in me when I am in a unbearable, life threatening situation with complete helplessness and impotence, without any self-agency'. Of course, the traumatic experiences cannot be deleted by such experiences in the transference/countertransference of the psychoanalytical relationship but may lose its quality of the unbearable horror as well as the psychic quality of nightmares.

Therefore, changes of the manifest dreams as well as the working with dreams in the psychoanalytical situation still seems a 'via regia to the unconscious' for us. They are key to the understanding of unconscious conflicts and fantasies and possible transformations of psychic functioning. Of course we don't want to simplify these processes: they are never unilineal but very complex. Often transformations take place in a very hidden way and are characterized by ups and downs (as in real life!). But, as we tried to show in several extensive case summaries: a systematic clinical and extra-clinical investigation of the changes of dreams could be a more psychoanalytic way to study changes in psychoanalyses or psychoanalytic long-term treatments than an exclusive investigation of symptoms (see also Leuzinger-Bohleber 2015, 2016). At least we do hope that we can add some of such more complex, and in our view, *psychoanalytical* perspectives, to the comparative outcome studies as the LAC Study. Another hope is that we could illustrate that the dialogue between psychoanalysis and the neurosciences may even be fruitful in the field of comparative psychotherapy research.

## References

Aserinsky E, Kleitman N. Regularly occurring periods of eye motility, and concomitant phenomena, during sleep. J Neuropsychiatr Clin Neurosci. 2003;15(4):454–5.

Bischof M, Bassetti CL. Total dream loss: a distinct neuropsychological dysfunction after bilateral PCA stroke. Ann Neurol. 2004;56:583–6.

Bohleber W. Destructiveness, intersubjectivity, and trauma: the identity crisis of modern psychoanalysis. London: Karnac Books; 2010.
Clancey WJ. Situated action: a neuropsychological interpretation (Response to Vera and Simon). Cogn Sci. 1993;17(1):87–116.
Dahan L, Astier B, Vautrelle N, Urbain N, Koscis B, Chouvet G. Prominent burst firing of dopaminergic neurons in the ventral tegmental area during paradoxical sleep. Neuropsychopharmacology. 2007;32:1232–41.
Dewan EM. The programing (P) hypothesis for REM sleep. Int Psychiatry Clin. 1970;7(2):295–307.
Ellman SJ, Weinstein LN. REM sleep and dream formation: a theoretical integration. In: Ellman SJ, Antrobus JS, editors. The mind in sleep: psychology and psychophysiology, vol. 166. New York: Wiley; 1991. p. 466–88.
Fischmann T, Leuzinger-Bohleber M. Veränderungen von Träumen als Indikatoren für Therapieerfolg. Trauma—Zeitschrift Für Psychotraumatologie Und Ihre Anwendungen. 2017;15(2):80–9.
Fischmann T, Russ MO, Leuzinger-Bohleber M. Trauma, dream, and psychic change in psychoanalyses: a dialog between psychoanalysis and the neurosciences. Front Hum Neurosci. 2013;7:877. https://doi.org/10.3389/fnhum.2013.00877.
Hu P, Stylos-Allan M, Walker MP. Sleep facilitates consolidation of emotional declarative memory. Psychol Sci. 2006;17(10):891–8.
Kächele H, Albani C, Pokorny D. From a psychoanalytic narrative case study to quantitative single-case research. In: Gelo OCG, Pritz A, Riegen B, editors. Psychotherapy research: foundations, process and outcome. Vienna: Springer; 2015. p. 367–79.
Kleitman N. Sleep and wakefulness. Chicago: University of Chicago Press; 1963.
Léna I, Parrot S, Deschaux O, Muffat-Joly S, Sauvinet V, Renaud B, Suaud-Chagny MF, et al. Variations in extracellular levels of dopamine, noradrenaline, glutamate, and aspartate across the sleep-wake cycle in the medial prefrontal cortex and nucleus accumbens of freely moving rats. J Neurosci Res. 2005;81:891–9.
Leuzinger-Bohleber M. Veränderung kognitiver Prozesse in Psychoanalysen. Bd. 1: Eine hypothesengenerierende Einzelfallstudie. Berlin: Springer; 1987.
Leuzinger-Bohleber M. Veränderung kognitiver Prozesse in Psychoanalysen. Bd. 2: Fünf aggregierte Einzelfallstudien. Berlin: Springer; 1989.
Leuzinger-Bohleber M. Working with severely traumatized, chronically depressed analysands. Int J Psychoanal. 2015;96(3):611–36.
Leuzinger-Bohleber M. Enactments in transference: embodiment, trauma and depression. What have psychoanalysis and the neurosciences to offer to each other. In: Weigel S, Scharbert G, editors. A neuro-psychoanalytical dialogue for bridging Freud and the neurosciences. Cham: Springer; 2016. p. 33–46.
Leuzinger-Bohleber M, Pfeifer R. Paper given at the 12th international neuropsychoanalysis congress "neuropsychoanalysis: minding the body", Berlin, 25 June 2011; 2012.
Leuzinger-Bohleber M, Pfeifer R. Struggling with unconscious, embodied memories in a third psychoanalysis with a traumatized patient recovered from severe poliomyelitis: a dialogue between psychoanalysis and embodied cognitive science. In: Leuzinger-Bohleber M, Arnold SE, Solms M, editors. The unconscious: a bridge between psychoanalysis and cognitive neuroscience. New York: Taylor & Francis; 2017. p. 138–63.
Leuzinger-Bohleber M, Hautzinger M, Fiedler G, Keller W, Bahrke U, Kallenbach L, Kaufhold J, Ernst M, Negele A, Schoett M, Kuechenhoff H, Guenther F, Rueger B, Beutel M. Outcome of psychoanalytic and cognitive-behavioral therapy with chronic depressed patients. A controlled trial with preferential and randomized allocation. In review.
McCarley RW, Greene RW, Rainnie D, Portas CM. Brainstem neuromodulation and REM sleep. Semin Neurosci. 1995;7(5):341–54.
Menninger K. Theory of psychoanalytic technique. London: Imago; 1958.
Moser U, von Zeppelin I. Der geträumte Traum. Wie Träume entstehen und sich verändern. Stuttgart: Kohlhammer; 1996.

Nielsen T, Lara-Carrasco J. Nightmares, dreaming, and emotion regulation: a review. In: Barrett D, McNamara P, Barrett D, McNamara P, editors. The new science of dreaming: volume 2. Content, recall, and personality correlates. Westport: Praeger Publishers/Greenwood Publishing Group; 2007. p. 253–84.

Nielsen TA, Stenstrom P. What are the memory sources of dreaming? Nature. 2005;437(7063):1286–9.

Nishida M, Pearsall J, Buckner RL, Walker MP. REM sleep, prefrontal theta, and the consolidation of human emotional memory. Cereb Cortex. 2009;19(5):1158–66.

Olds J. A preliminary mapping of electrical reinforcing effects in the brain. J Comp Physiol Psychol. 1956;49:281–5.

Olds J. Hypothalmic substrates of reward. Physiol Rev. 1962;42:554–604.

Panksepp J. Affective neuroscience. New York: Oxford University Press; 1998.

Payne JD, Nadel L. Sleep, dreams, and memory consolidation: the role of the stress hormone cortisol. Learn Mem. 2004;11(6):671–8.

Perogamvros L, Schwartz S. The roles of the reward system in sleep and dreaming. Neurosci Biobehav Rev. 2012;36(8):1934–51.

Popper K. Conjectures and refutations: the growth of scientific knowledge. London: Routledge and Kegan Paul; 1963.

Rasch B, Born J. About sleep's role in memory. Physiol Rev. 2013;93:681–766.

Solms M. The neuropsychology of dreams: a clinico-anatomical study. Mahwah: L. Erlbaum Associates; 1997.

Solms M. Dreaming and REM sleep are controlled by different brain mechanisms. Behav Brain Sci. 2000;23(6):843–50. Discussion 904–1121.

Solms M. The scientific standing of psychoanalysis. Unpublished paper; 2017.

Stickgold R, Hobson JA, Fosse R, Fosse M. Sleep, learning, and dreams: off-line memory reprocessing. Science. 2001;294(5544):1052–7.

Varvin S, Fischmann T, Jović V, Rosenbaum B, Hau S. Traumatische Träume: Streben nach Beziehung. Psyche—Z Psychoanal. 2012;66(9–10):937–67.

Wagner U, Gais S, Born J. Emotional memory formation is enhanced across sleep intervals with high amounts of rapid eye movement sleep. Learn Mem. 2001;8(2):112–9.

Weinstein L, Ellman SJ. Die Bedeutung der endogenen Stimulation für das Träumen und für die Entwicklung: Ein Versuch der Integration und Neuformulierung (The role of endogenous stimulation in dreaming and development: an attempted integration and reformulation). Psyche—Z Psychoanal. 2012;66(9–10):862–88.

# Transference and Countertransference 9

Peter Hartwich and Heinz Boeker

**Abstract**

The discovery of the phenomenon of transference as well as later of the countertransference has led to a paradigm shift in the treatment of mentally ill patients by significantly broadening the understanding of the relationship between patient and therapist as well as the range of psychotherapeutic possibilities. The various variants of this phenomenon are exemplified in depression, mania, narcissistic personality disorders, suicidal patients, and schizophrenic and schizoaffective psychoses. This also involves the variants of countertransference and how they can be used as an effective therapeutic instrument. Since these phenomena are mostly unconsciously effective in all therapeutic encounters, it is also necessary to make them aware, even, e.g., in prescribing psychopharmaceuticals.

## 9.1 History and Definition of Transference and Countertransference

### 9.1.1 Transference

The transference is a central event which determines the relationship between the patient and the therapist in every psychotherapeutic encounter. Regardless of whether psychodynamic or classical psychoanalytic treatments are concerned,

P. Hartwich (✉)
Department of Psychiatry, Psychotherapy, Psychosomatics, General Hospital of Frankfurt am Main, Teaching Hospital University Frankfurt, Frankfurt, Germany
e-mail: Prof.Hartwich@casytec.de

H. Boeker
Department of Psychiatry, Psychotherapy, and Psychosomatics, Psychiatric University Hospital Zurich, University of Zurich, CH, Zurich, Switzerland
e-mail: heinz.boeker@bli.uzh.ch



H. Boeker et al. (eds.), *Neuropsychodynamic Psychiatry*,
https://doi.org/10.1007/978-3-319-75112-2_9

there is always an unconscious transference. Even with behavioral therapy techniques, transference phenomena occur, even if they are named differently. Transference is, for example, positively or negatively noticeable when expectations, desires, feelings, and behavior manifest themselves in a form which does not correspond to the real relationship in an encounter situation and cannot be explained from it. Historically, this is a discovery of early psychoanalysis, meaning in the narrower sense emotional attitudes of the patient to the therapist or to other persons in the present which were originally referring to childhood but which are unconsciously transferred to the present. These include, for example, early childhood-acquired love, desire, hate, expectation, and rejection, which are applied in the treatment situation. It is an unconscious event, the realization and awareness of which is a core of psychodynamic and analytical therapies.

Freud (1904–1905, pp. 279, 280): "What are the transferences? These are reprints, imitations of the impulses and phantasies that are to be aroused and made aware during the advancement of the analysis with a person's characteristic replacement of a former person by the physician. To put it another way: A whole series of earlier psychological experiences is not experienced as a past, but as a current relationship to the person of the physician".

Numerous authors, among them Laplanche and Pontalis (1972), have extended the concept further, so that the transference encompasses almost all aspects emotions and attitudes that constitute the patient's relationship with the therapist and the therapeutic environment. In this extended use, we also consider it useful to use the term "transference" in all mental disorders including the psychoses.

### 9.1.2 Countertransference

The countertransference is the complementary reaction of the therapist, which should also be understood in a broader sense, by subsuming all the emotional reactions which the patient causes. That is, in the broader sense, the countertransference is the total unconscious emotional response of the treating person and a therapist team to the behavior of a patient, which includes his reactions and attitudes resulting from the transference.

Freud (1909–1913, p. 108): "We have become aware of the countertransference which comes to the physician by the influence of the patient on the unconscious feeling of the physician, and are not far from raising the demand that the physician must recognize and cope with countertransference".

Today, many authors, by contrast, understand everything that can interfere with the therapist's personality in the treatment as countertransference; this is also a concept extension, which Laplanche and Pontalis (1972) say that there is a wide range of variations in the definition of the concept.

In all mental disorders treated psychodynamically, the countertransference, as far as the therapist becomes aware, is regarded as an important diagnostic tool for the patient's unconscious aspects.

Hering (2004, pp. 33–34) proposes to apply the global understanding of the countertransference not only for psychoreactive disorders and personality disorders but also for psychodynamic treatment of psychoses: "With an increasing view of psychoanalysis as an intersubjective process between two persons involved, the application of empathy and countertransference as a therapeutic tool, and the increased use of psychoanalysis in borderline personality disorder, other personality disorders and psychotic patients, concepts of a "self-presentation" of the therapist have become of considerable significance".

Maier (2001, p. 120) emphasizes that in the practical therapeutic situation with psychotic patients, the countertransference is the most valuable source of information for the therapist.

## 9.2 Transference and Countertransference in Depressive Disorders

The facets of the transference aspects as well as the corresponding countertransference are so varied and individually nuanced that we can only restrict ourselves to a few characteristic examples.

### 9.2.1 Transference

A typical transference constellation, which is to be observed with depressive patients, often unfolds in the first meeting. The patient feels weak and helpless, speaks quietly, and is unconsciously seeking maternal protection, someone who understands him and who does everything for him. If his depression, as is often experienced, has been associated with early loss experiences in his childhood, his unconscious transference serves the desire to restore the situation before the loss. This may be associated with idealizing hopes, so far the transference aspects to the therapist.

In a more detailed history, this transference situation is often found in the relationship with the partner, whether female or male. The partner has unconsciously assumed the protective and often dominant role.

### 9.2.2 Countertransference

In the therapist maternal-protective and patient-oriented feelings are triggered. If these complementary reactions are not reflected, the therapist quickly favors the strong, helping, and active role. The patient first unconsciously rewards the therapist by signaling a slight improvement. This, in turn, perpetuates the therapist's behavior. When experiences of pain and loss are then treated in the therapy, resistance in patient becomes a problem; he persists in the depressed plaintive attitude, in order not to have to revive the loss. As a resistance, which is also unconscious, the depressive symptoms are used against psychodynamic attempts of interpretation, which may take a long time. Since the therapy process no longer proceeds, as it seemed initially "ideal," anger comes up within the therapist. However, his increasing aggressive feelings are tied to the patient's suffering and helplessness. If the therapist nevertheless realizes that this is his countertransference reaction he can experience his own feelings as a **diagnostic tool**. And he can become aware of the unconscious aggression of the patient, who is upset, angry and disappointed because of previous "undigested" experiences of loss that is so great that he directs the aggression not only angains himself but also unconsciously against the therapist. The therapist realizes that it is not primarily his own anger but that he feels it as a substitute for the patient. In this way the path becomes free for working through these unconscious emotional constellations, which so far have played a major part in the disease of the depressed person. Determinant causes can now gradually be worked through and slowly remedied. If the countertransference is not reflected, the psychotherapy would be somehow similar to the interaction in the partnership.

### 9.2.3 Depression and Partnership

When the depressive illness lasts long time, it is not uncommon to see the change of the attitude in the family to an overly caring behavior. But after a while when they experience that well-meaning assistance and interventions are not fruitful, but are rejected by the depressive person with constant strength, this usually encourages their anger. Psychodynamically speaking, the unconscious aggressive effect of the patient is implanted in this way into his relatives. However, they slow down to express their anger adequately, as the depressive is constantly showing his excessive vulnerability. This in turn leads to an attitude among the family members, in no way to be aggressive against the depressive, since otherwise they would destroy him. It is typical that in such relations, whether family or partnership, adequate conflicts are suppressed from the outset and are therefore missing. Instead, he continually implants feelings of guilt among the relatives, which they carry with them and whose origin remains unconscious.

The two psychodynamic components are reflected in the interaction pattern of the relationships in which one partner is chronically deep depressive or even suffers from a psychotic depression. The healthy partner is constantly busy repressing the implanted anger, which comes originally from the depressive person, and has to devote a great deal of his life's energy. Freud (1900–1901, p. 610) described the energy as the "anticathexis." In his distorted perception the depressive person experiences the resulting emotional behavior of his partner as emotional coldness and rejection. In some cases, he experiences some kind of a reproduction of his early loss in his childhood, but he is not aware of this.

## 9.3 Countertransference in Suicidal Patients

According to the many different facets of suicidal behavior, there are also a number of different countertransference variants. We would like to mention a particularly difficult situation as a clinical example. There are additional narcissistic insults to the treating persons, who may also be exposed to sense of guilt attributions by outsiders.

In the case of a long-lasting, repeatedly expressed serious chronic suicidal behavior of a patient who is treated, for example, at a protected inpatient unit, it is possible that unconsciously it comes to a mental state which Gabbard (2014) calls "countertransference hate." Gabbard (2014, p. 247): "Therapists who treat seriously suicidal patients will eventually begin to feel tormented by the repeated negation of their efforts. Countertransference hate is likely to develop at such times, and treaters will often harbor an unconscious wish for the patient to die so that the torment will end. … The inability to tolerate their own sadistic wishes toward such patients may lead treaters to act out countertransference feelings. … Countertransference hatred must be accepted as part of the experience of treating suicidal patients. It often arises in direct response to the patient's mythical sword of Damocles, tormenting and controlling the therapist night and day."

In the case of the inability to become aware of one's own destructive impulses, this form of countertransference can lead to action with lethal consequences. In addition, Milch (1994) points to the danger of the possible suicidal impulses in the therapist himself, which may be unconsciously reversed to the contrary and brings the treater to actions which can have deleterious effects because of the jointly shared death wishes. These statements, derived from the psychodynamic view, seem to contradict the professional self-understanding of therapist and the helpful treatment team. In this conflict between professional roles and the abovementioned countertransference reactions, supervision can be helpful in which the destructive desires are made conscious and expressed as "permitted feelings," which others also have in such situations.

## 9.4 Transference and Countertransference in Narcissistic Personality Disorders

For the narcissistic personality variant, Kohut (1973) describes two main forms of transference: the idealizing transference and the mirror transference.

### 9.4.1 Idealizing Transference

In this form of transference, it is assumed that an early reference person is idealized in the narcissistic context. It is a matter of occupational energy, the content of which refers, for example, to idealized objects of the inner parents. An idealizing transference can then be applied to the therapist if a corrective change through experience has not taken place in the psychic development and maturation of the person concerned. Kohut (1973, p. 67) describes what happens in the psychotherapy: "The patient revives the need for an archaic self-object, narcissistically experienced, which preceded the formation of psychical structure in a specific sector of the psychic apparatus. From the object sought (i.e. the psychoanalyst), the patient expects the fulfillment of certain crucial tasks in the field of narcissistic equilibrium, which cannot take over his own psyche". So it becomes that the therapist is experienced as particularly great and omnipotent. Thus the therapist is not only attributed to a positive transference, positive qualities of early reference persons, but rather to a kind of unlimited power.

### 9.4.2 Mirror Transference

The mirror transference goes even further; here in the idealized transference, the megalo-self is reactivated. Kohut (1973, p. 130) describes this phenomenon as follows: that the mirror transference and its precursors represent the therapeutic revival of that aspect of a developmental phase in which the child tries to preserve the original all-encompassing narcissism by bringing perfection and power into the self—here called the great-self—and turned away contemptuously from an external world to which all imperfections are attributed.

This kind of transference is especially found in people with narcissistic personality structures who are continually trying to adapt their actual abilities and knowledge with their unreal increased expectations of themselves. They are also not in a position to admit the gap in his knowledge in order not to expose themselves to the shame of being unmasked.

## 9.5 Countertransference in Idealization and Mirroring

If the therapist is flattered by the transference that he is particularly capable and potent, his therapy is special, and he is the only one who really helps, then two countertransference reactions are characteristic.

On the one hand, the therapist's feeling can be unpleasant and he feels rejection tendencies. On the other hand, the therapist can feel confirmed in his own unconscious narcissism. There is a narcissistic seduction, in which the therapist's great-self is "tapped." In both cases, the therapist's awareness of the countertransference is crucial for the progress of the therapy, otherwise a psychotherapeutic process stand still or does not even begin.

## 9.6 Transference and Countertransference in Maniform States

Even more pronounced is the transference-countertransference constellation in manic patients, which are still in the hypomanic state at the beginning of a developing manic psychosis. In the transference, the patients consider themselves either as equated with the idealized therapist or superior to him. They fascinate the therapist with their wealth of ideas. He is entangled in their world and, in his countertransference, seduced by the genius-like creativity. If this gets out of hand, and the seduction becomes conscious, the need for structuring and confinement increases.

## 9.7 Transference and Countertransference in Schizophrenic Disorders

From the development of psychoanalysis, it should be pointed out that Paul Federn (1956) did not agree with Freud, who argued that patients who suffer from psychosis are not capable of transference. Federn (1956/1978, p. 108) reports on his experience that a strong transfer to the therapist can occur in psychotic patients: "It was generally assumed that for this reason no psychoanalysis was possible. Nowadays, many authors know that both the findings and the conclusions were wrong. There is, however, something true, the transference of psychotics is quite unstable and does not justify the use of the same psychoanalytic method as in neurotic patients".

In the historical controversy, whether or not psychotic patients are capable of transference has been strengthened by the fact that the different investigators have seen different psychopathological manifestations in their patients and then have generalized their viewpoints according to the selection of their patient group.

**Historical Controversy**
- Freud: A patient who suffers from psychosis is not capable of transference and therefore unsuitable for psychoanalysis.
- Federn: Psychotic patients are capable of strong transference and can be treated psychoanalytically but methodically different.

### 9.7.1 Closeness and Distance in the Transference-Countertransference Pattern in the Treatment of Schizophrenics

Schwarz (2001) has pointed out the particular danger that weak self-borders in close and too intensive therapeutic contacts are threatened by dissolution, which can occur in psychodynamic therapy. Pao (1979) and Volkan (1994) speak here of "organismic panic," the psychopathology of disintegration and disorganization, which make up the lack of ego strength. Disorganizing impulses, emotions, etc. can no longer be modulated and regulated. These states may have their neurobiological correlate in the enhancement of low-frequency fluctuations in the central nervous system, as well as in the EEG as delta-waves parenrhythmias (see Chap. 4). Therein, the high degree of disturbance in the relationship, when too close, may be seen with schizophrenic and schizoaffective patients. In countertransference, it is of crucial importance for the success of a therapy that the psychotherapist never suppresses this particular vulnerability of the patient, which he usually cannot have learned from his own experience and training analysis, but that he becomes aware of it and has the inherent fragmentation risk of the psychotic patient "at the back" during each therapy session.

Schwarz (2001) pointed out the danger of a symbiotic fusing in this context. This means that in the case of schizophrenics, their ego-demarcation and their ability to distinguish between the ego and the external world are not any more possible. To see the therapist's position as an object, as something external, outside himself, is made more difficult. However, this pathologically distorted close-distance regulation still has an opposite pole, namely, autistic isolation with an extreme form of demarcation. The threatening symbiotic fusion can lead to an abrupt withdrawal from a too close constellation. Then the schizophrenic patient ends the psychotherapy abruptly. Therefore the development of a positive transference should be encouraged at the beginning of the therapy, as emphasized by Benedetti (1987).

**Transference Problems**

- Schwarz: the transference facet of the threatening symbiotic fusion means that the patient feels that the therapist is too close to him. He withdraws abruptly, stops the therapy, or does not even begin.
- Benedetti: actively promote positive transference.

### 9.7.2 Rapid Change of Transference Facets in Psychoses

The resistance of the psychotic patient to a positive transference can be that fear of threatening self-fragmentation, of organismic panic and destruction of trust is given. Maier (2001, p. 119) points out that one form of expression of the resistance is a rapid change of transference facets.

Psychotherapist's interpretations would not be appropriate but rather harmful. We see this type of transference most frequently in young patients who suffer from hebephrenia when they constantly change their conversation contents and jump from one subject to another with the shortening of the "intentional arc" (Beringer 1927), which is additionally characterized by a flat and inadequate affect. A distinction must be drawn here between associative loosening of thoughts with increased distraction on the one hand and, on the other hand, a protective mechanism for the hebephrenics, not to become determined. Since many of these patients must be characterized by a sustained self-fragmentation, they are not able to be determined to one of their self-fragments. The patient is in a polylemma (Hartwich and Grube 2015). In the countertransference, the therapist should be able to withstand violent irritation and the disappointment of the permanent slipping of the relationship. The neurobiological disturbances described by Northoff (2011) following Freud as decathexis in the sense of the withdrawal of occupational energy correspond to these flattening of the affects in hebephrenics.

**Change of Transference Facets**

- Resistance of the psychotic patient by rapid change of transference facets
- Frequently with hebephrenic symptoms
- Question of the fixation: to which fragment?
- Often non-fixation and polylemma

### 9.7.3 Transference Psychosis

When psychotic manifestations occur in transference, one speaks of transference psychosis (Rosenfeld 1966). Little (1958) speaks of delusional transference. If the psychotic patient inserts the therapist into his delusional construct, this can block the therapy. Either the therapist is experienced as positive, and then he becomes an available figure for the patient in his delusional construction, or if he is negatively occupied, he becomes the object of the negative projections. In both cases, the therapist becomes the creature of the patient. There are states in which the patient, in the sense of a projective defense, finds the more rigidly the transference psychosis, the more the therapist tries to change the transference facet. In the countertransference emerge feelings of anger and a strong rejection against the patient. Often the therapist is not any more able to work together with the patient.

### 9.7.4 Participating Countertransference in Schizoaffective Psychoses

Hering (2004, 2006) described the procedure for the therapist to participate in the countertransference for the treatment of schizoaffective psychoses in certain situations and gives an example by telling the patient: *One feels your fear that you may lose your own control over things.* The patient is thereby informed that his disintegration anxiety and fragmentation can be perceived and shared by the therapist which is very important for the progress of a helpful therapy.

### 9.7.5 Identical Countertransference, Countertransference Resistance, and Action Dialogue

If the therapist perceives the risk of fragmentation of the psychosis, it is obvious that he develops a resistance to become aware of such a corresponding experience. It is essential for the progress of the therapeutic process that the therapist's unconsciously arising countertransference resistance is made conscious and processed positively. In this context presumably the praecox feeling, which Rümke (1941) described as a diagnostic sign for schizophrenia, can be classified.

If the therapist identifies himself with the experience of the patient, he may come to the conclusion that he should offer more structure for the self-fragmented patient in the therapeutic session. This need for structuring first applies to the therapist himself. This structuring countertransference needs the therapist not to be carried away by the pull of chaos. In the sense of an action dialogue, the aim is to offer the patient structuring and strengthening in verbal form or as a medium. The term "action dialogue" was first introduced by Klüwer (1983), and the terms "enactment" (Jacobs 1986) and "staging" (Streeck 2000) are used in this sense as a sort of coaching response of the therapist.

## 9.8 Transference and Countertransference in the Treatment with Antipsychotic Medication

There are a number of studies showing that there can be structural abnormalities in the brain in people with the diagnosis of schizophrenia, which can be increased by taking antipsychotics, depending on the level of dosage. This is shown to be clinically associated with cognitive disturbances (Aderhold et al. 2015). The discussion of whether or not psychotic patients should be given antipsychotic medication in long-term or shorter-term therapy *or* psychotherapeutic treatment has been concluded in favor of both as well as in the previous century. Today, it is important to take even greater account of the findings on the side effects on the brain, by using the principle of "as little as possible and only as much as absolutely necessary." As a result, other therapeutic methods, in particular psychotherapeutic ones, should be used more. In the case of an apportionment of the therapy, in which the psychotherapist assumes exclusively psychotherapeutic treatment and the psychiatrist exclusively takes the pharmacological part of the treatment, a transference split in the patient may occur. That means positive transference wishes occur to the psychotherapist with sometimes unreal healing expectations. In contrast there occur rejecting feelings to the psychiatrist, because the patient does not like to accept the somatic aspects of the etiology and the treatment of the disease. Then within schizophrenic patients develop the danger that this kind of separation may enhance the inner tendency to psychotic ego splitting.

The countertransference of the psychotherapist can creep in the feeling to be the "better healer," and the psychiatrist experiences himself as "stamped" for the mere prescription of medication. It is best that the two therapists be aware of their countertransference and they should communicate about the patient with each other continually. Then in addition, the schizophrenic tendencies of inner splitting of the ego can be alleviated.

## 9.9 Neuropsychodynamic and Neurobiological Aspects

The question of whether transference and countertransference are associated with neurobiological correlations is still largely unexplained. In the study by Perez et al. (2016) about transference-focused psychotherapy (TFP), they found that the association between clinical improvement, low pretreatment, and relatively elevated posttreatment anterior dorsal ACC (anterior cingulate cortex) activation suggests that TFP may potentially modulate neural activity in this region to improve behavioral restraint. Activation in the anterior dorsal ACC and amygdala-hippocampus was associated with improvements in behavioral constraint and emotional regulation and/or aggression in patients with BPD. It could be, but remains still unclear whether the transference as an isolated item was investigated because (TFP) of an evidence-based treatment for BPD, developed by Kernberg and colleagues that relies on techniques of clarification, confrontation, and interpretation of affect-laden themes that emerge within the transference relationship.

Loughead et al. (2010) combined the core conflictual relationship theme (CCRT) method and functional magnetic resonance imaging (fMRI) to identify brain regions involved in recall of autobiographical relationship episodes. They found that in an exploratory analysis, higher CCRT scores correlated with increased brain activation in the left hippocampus, parahippocampal gyrus, and middle occipital gyrus. Their interpretation is that brain systems subserving memory processes are more active when recalling relationship episodes with greater CCRT content.

Vivona (2009) describes the mirror neuron discussion, particularly regarding countertransference, as a critical field. Within neuroscience there are competing interpretations of mirror neuron findings, with diverse implications for psychoanalysis, but there is still a lack of empirical validation.

In so far the neuronal activation in several regions of the brain including the research about mirror neurons on one side and transference and countertransference on the other side may have correlative relationship, but this has still to be investigated more detailed in the future.

## References


Aderhold V, Weinmann S, Hägele C, Heinz A. Frontale Hirnvolumenminderung durch Antipsychotika? Nervenarzt. 2015;86(3):302–23.

Benedetti G. Psychotherapeutische Behandlungsmethoden. In: Kisker KP, et al., editors. Psychiatrie der Gegenwart. Schizophrenien, Bd 4. Berlin: Springer; 1987.

Beringer K. Der Meskalinrausch. Berlin: Springer; 1927.

Federn P. Ichpsychologie und die Psychosen. Frankfurt am Mai and Bern-Stuttgart: Huber and Suhrkamp; 1956.

Freud S. Die Traumdeutung. Gesammelte Werke Bd 2/3; S Fischer, Frankfurt/M, S 1–642, 1942; 1900–1901.

Freud S. Bruchstücke einer Hysterie-Analyse. GW Bd 5, S Fischer, Frankfurt/M, S 161–286, 1981; 1904–1905.

Freud S. Die zukünftigen Chancen der psychoanalytischen Therapie, Bd 8. S Fischer, Frankfurt/M, S 103–115, 1973; 1909–1913.

Gabbard GO. Psychodynamic psychiatry in clinical practice. 5th ed. Washington, DC: American Psychiatric Press; 2014.

Hartwich P, Grube M. Psychotherapie bei Psychosen. 3rd ed. Heidelberg: Springer; 2015.

Hering W. Schizoaffektive Psychose. Psychodynamik und Behandlungstechnik. Göttingen: Vandenhoeck & Ruprecht; 2004.

Hering W. Psychodynamische Aspekte der schizoaffektiven Psychosen. In: Böker H, editor. Psychoanalyse und Psychiatrie. Heidelberg: Springer; 2006.

Jacobs T. On countertransference enactments. J Am Psychoanal Assoc. 1986;34(2):289–307.

Klüwer R. Agieren und Mitagieren. Psyche. 1983;37:828–40.

Kohut H. Narzißmus. Frankfurt/M: Suhrkamp; 1973.

Laplanche J, Pontalis JB. Das Vokabular der Psychoanalyse. Frankfurt/M: Suhrkamp; 1972.

Little M. Über wahnhafte Übertragung (Übertragungspsychose). Psyche. 1958;12:258–69.

Loughead JW, Luborsky L, Weingarten CP, Krause ED, German RE, Kirk D, Gur RC. Brain activation during autobiographical relationship episode narratives: a core conflictual relationship theme approach. Psychother Res. 2010;20(3):321–36. https://doi.org/10.1080/10503300903470735.

Maier C. Deutung und Handlungsdialog. In: Schwarz F, Maier C, editors. Psychotherapie der Psychosen. Stuttgart: Thieme; 2001. p. S 117–27.

Milch W. Gegenübertragungsprobleme bei suizidalen Patienten unter stationärer psychiatrischer Behandlung. Psychiatr Prax. 1994;21:221–5.

Northoff G. Neuropsychoanalysis in practice. New York: Oxford University Press; 2011.

Pao PN. Schizophrenic disorders. Theory and treatment from a psychodynamic point of view. New York: International University Press; 1979.

Perez DL, Vago DR, Pan H, Root J, Tuescher O, Fuchs BH, Leung L, Epstein J, Cain NM, Clarkin JF, Lenzenweger MF, Kernberg OF, Levy KN, Silbersweig DA, Stern E. Frontolimbic neural circuit changes in emotional processing and inhibitory control associated with clinical improvement following transference-focused psychotherapy in borderline personality disorder. Psychiatry Clin Neurosci. 2016 Jan;70(1):51–61. https://doi.org/10.1111/pcn.12357.

Rosenfeld H. Psychotic states. A psychoanalytical approach. New York: International University Press; 1966.

Rümke HC. Das Kernsyndrom der Schizophrenie und das 'Praecox-Gefühl'. Zentralbl Neurol Psychiatrie. 1941;102:168–9.

Schwarz F. Übertragung und Gegenübertragung bei der Psychotherapie schizophrener Patienten. In: Schwarz F, Maier C, editors. Psychotherapie der Psychosen. Stuttgart: Thieme; 2001. p. S 127–35.

Streeck U. Szenische Darstellung, nichtsprachliche Interaktion und Enactments im therapeutischen Prozess. Göttingen: Vanderhoeck & Ruprecht; 2000.

Vivona JM. Leaping from brain to mind: a critique of mirror neuron explanations of countertransference. J Am Psychoanal Assoc. 2009;57(3):525–50. https://doi.org/10.1177/0003065109336443.

Volkan VD. Identification with the therapist's function and ego-building in the treatment of schizophrenia. Brit J Psychiatry. 1994;164(23):77–82.

# Part II

# Neuropsychodynamics of Psychiatric Disorders

# Schizophrenia and Other Psychoses

# 10

Peter Hartwich and Georg Northoff

**Abstract**

Schizophrenia is a complex disorder that shows various symptoms which, as emphasized in older and recent psychopathological approaches, can be traced to a basic disturbance of the self with "self-fragmentation" (Kohut) of its spatiotemporal structure. Our neuropsychodynamic perspective focuses on the basic disturbance of self and its underlying neuropsychodynamic mechanisms—the latter are considered spatiotemporal at their very core as they focus on the spatiotemporal relation or alignment of the brain's spontaneous activity to the respective environmental context the world.

We will explain how findings of neurobiological research of abnormalities in the brain open a new understanding of the etiopathogenesis and psychodynamic of psychotic diseases.

An abnormal resting state activity and imbalance between anterior and posterior midline regions (hyperconnectivity) correspond with disturbances of the self. The disbalance between default mode network and central executive network, lateral regions of the prefrontal and parietal cortex on the neural level, seems to mirror the confusion of external and internal mental content, which is on the neuropsychodynamic level a "self-environment blurring." Such "self-environment blurring" leads to what psychodynamically has been described as loss of ego boundary (Federn, Freud) with "abnormal cathexis." We consecutively describe examples of the *investment energy "cathexis"* in different

P. Hartwich (✉)
Department of Psychiatry, Psychotherapy, Psychosomatics, General Hospital of Frankfurt am Main, Teaching Hospital University Frankfurt, Frankfurt, Germany
e-mail: Prof.Hartwich@casytec.de

G. Northoff
Mind, Brain Imaging, and Neuroethics,
Institute of Mental Health Research, University of Ottawa, Ottawa, ON, Canada
e-mail: georg.northoff@theroyal.ca; www.georgnorthoff.com



H. Boeker et al. (eds.), *Neuropsychodynamic Psychiatry*,
https://doi.org/10.1007/978-3-319-75112-2_10

psychotic qualities: hypocathexis, oscillating cathexis, decathexis, anticathexis, paracathexis, and hypercathexis.

How is such abnormal cathexis related to the different symptoms? On the clinical *neuropsychodynamic* level, we see the psychotic patient's counter-regulation to restore the spatiotemporal structure of the self as a *paraconstruction*. Many symptoms are the result of different qualities of paraconstruction, which we describe in examples for particular cases, e.g., delusion of pregnancy, delusion of love, Othello syndrome, catatonic and coenesthetic syndrome, etc.

These compensatory mechanisms in psychosis do not reach the structural level of defense mechanisms as it was described in neuroses; therefore, it is necessary to create the concept of *neuropsychodynamic paraconstruction* which is to be understood in an unfolding matrix which is biological and psychodynamic as well.

One of the results is a paradigm shift of the therapist, who understands paraconstruction as a necessary protection for the self, which is in danger to fragment, and who should use treatment methods which are helpful beyond the symptoms in order to strengthen the spatiotemporal structure of the self.

## 10.1 About Etiopathogenesis

The psychodynamic understanding and the psychodynamic action are always linked to the question of the history of psychiatry and especially schizophrenia. The theories, which have grown over decades, have moved (switched) in the medical history between the poles of biological-somatic and psychodynamic orientation. An overwhelming number of experts favored the extreme positions. In particular, what is called "mainstream" has led nowadays to a great influx among those who represent the biological and somatic aspect. Consequently, the *psychodynamically* active psychiatrists and psychotherapists are now represented only in quite a small number. A question may occur, why the interpretation of the etiopathogenesis of psychotic illnesses changed in the long history of psychiatry from one extreme position to the other nd back again. This recalls the *Structure of the Scientific Revolution* by the philosopher Kuhn (1976), who points out that every period of science also contains a dogmatic element. This applies to both the biological-somatic as well as the psychoanalytic-psychodynamic representatives of the subject. Since dogmatism is used to suppress inappropriate alternatives, a saturation effect ensues after a while, which prepares the basis for a threshold situation in which alternatives can no longer be excluded. Then there is a period of failure which leads to a paradigm shift to the disadvantage of the hypotheses and models that were previously so surely represented.

We assume that the paradigm shift in the understanding of schizophrenic disorders will not change from the current biological-somatic domination back to a one-sided and narrowly understood psychodynamic dominance. Instead, a new dimension will be added which offers the opportunity to connect the two previously seen as alternative approaches. These are the well-developed and evolving findings and research results of neurobiological scientists. It seems to provide a common basis for the two mentioned and often competing principles. Consequently, we are talking today of Böker and Northoff's (2010) *neuropsychodynamic* (see Chap. 2) understanding of schizophrenic and schizoaffective disorders, both in etiopathogenetic understanding and in therapeutic concepts. This is to be understood in the sense of Solms and Turnbull (2002), who have introduced the term neuropsychoanalysis. In his foreword of the mentioned book, Oliver Sacks writes: "Neuro-psychoanalysis, one feels, is spreading its wings, but always remaining, as it must, firmly grounded on the demonstrable and the testable." We see this in the same way for the *neuropsychodynamic* approach; this term is superordinate (overarching) and encompasses psychoanalysis as well as all insight-oriented concepts.

## 10.2 Self and Psychosis

If, for the first time in an individual's life, a schizophrenic episode breaks out, whether slowly or suddenly, then the affected person experiences the previously naturally self-evident reality of his inner and outer life as alienated and his experience of the uncanny spreads. The connection between thoughts, feelings, and their contents no longer fit together. Foreign powers may influence him, and his consciousness of his self fades into helplessness and fear for his existence. From the perspective of descriptive psychopathology, it is a disintegration and disorganization of the order of his psychic structures. In a *neuropsychopathological* perspective, we speak of loss of the connection of the spatiotemporal continuity of the structure of the self.

If one includes the personal experience of the subject, the psychodynamic view of understanding opens up. It is then the aberrant change of experience interpreted as a fragmentation of the self (Kohut 1973), in which the coherence of the ego experience is lost. Scharfetter (2003, pp. 42–60) reports some examples of the loss of self-coherence in schizophrenic patients:

> I was no longer sure I was myself, I was no longer master of my own thoughts and actions; it was as if I were spellbound. (p. 54)

> I am changed; my ego doesn't exist anymore. (p. 52)

> Only half my opinion is my own, the other half belongs to other people. (p. 56)

> I didn't know for sure any more that I was I. My relatives had changed too. (p. 60)

> The most uncanny thing was that one didn't know oneself any more, didn't know that one was oneself. One can't control one's own action any more, one's thoughts. (p. 60)

**Characteristics of Psychosis**

- Descriptive psychopathology: disintegration and disorganization.
- Psychodynamic understanding: fragmentation of the self and loss of ego coherence.
- The patient says in retrospect: "My ego has been destroyed."

Northoff (2016, p. 160) remarks:

> It is as if the experience of the inner and outer world is no longer their experience. Because their very Self is absent in their experience of the world, patients with schizophrenia become detached, alienated and estranged from that very experience. This detachment from their own Selfhood makes it impossible for them to feel their experience as subjective. The experiencing Self is consequently no longer affected by its own experiences.

And in page 171, he writes "Schizophrenia can therefore be conceived as an existential disorder in the same way the world-brain relation can be described as an existential relation."

If one sees health and illness as opposed pairs, then the healthy ego and the destroyed ego represent the inherent opposite (I and non-I) in an antithesis. The ancient philosopher Heraclitus would presume a unity behind this (Stemich-Huber 1996). We try to go one step into the direction of this idea and may find it rooted in our relation to the world which our brain integrates and aligns us to the wider context of the world. "Only on the basis of this word-brain relation can we experience ourselves as part of the world and capable of developing a sense of the Self (as it is lost, for instance, in schizophrenia)" (Northoff 2016, p. 172). It is such integration and alignment of the brain into the world that seem to be lost in psychosis and therefore lead to the loss of self-coherence in these patients which underlies as "basic disturbance" their various symptoms (see below). Where is this coming from? How is it generated? We will review the various theories and approaches and, at the end, suggest a truly neuropsychodynamic approach.

## 10.3 Neurobiological Findings

Connectivity represents the relation between the neuronal activities in different brain regions and should be differentiated: between anatomical connectivity, which describes the relation between anatomical substrates of the nervous system of the brain, and functional connectivity. Functional connectivity describes the correlation between neuronal events, which are more or less distant, and effective connectivity describes the direct neuronal interaction between brain regions (see Chap. 11 for details).

From a neurobiological point of view, disorganization and the experience of self-fragmentation are correlated to functional *hyperconnectivity*, that is, connectivity between anterior and posterior median regions of the brain becomes abnormally strong. The abovementioned core phenomena of schizophrenia have weaker and stronger degrees of intensity not only in psychopathology and in the experience of the person concerned but also in their neurobiological correlations. Thus, Northoff (2012) states that the increase in low-frequency fluctuations in the midline regions with the frequency ranges of 0.01–0.1 Hz is abnormally strong in the case of delusion and pronounced self-disturbances. The stronger the low-frequency fluctuations, the more intense the symptoms are pronounced. Symptoms such as hallucinations, delusion, and ego disturbances are directly dependent on the phases of the low-frequency fluctuations. It is thus clear that the strength and intensity of these core symptoms are associated with measurable brain activities and their changes.

This is also shown by EEG examinations by Doege et al. (2010). In this context, it is worth mentioning that Huber and Penin (1968) already described abnormal rhythmizations in the form of alpha-, theta-, and delta-parenrhythmias in the EEG, *in correlation* with the intensity of expression of delusion, cognitive sliding, and coenesthetic and vegetative syndromes.

The crucial thing is that the brain activities are no longer modulated accordingly and adapted to new contexts. This is called phase resetting. The schizophrenic symptoms therefore seem to be attributable to altered temporal processes, the phases that describe the ongoing cycles of the fluctuations in the different frequencies (Northoff 2012, p. 245).

The research results confirm that certain amplitudes of MMN (mismatch negativity) correlate with the severity of the disease and the severity of cognitive disorders, which is sometimes referred to as so-called negative symptoms, in schizophrenic patients. However, we see in the coarse division of negative and positive symptoms as an arbitrary simplification, presumably resulting from reductionism of psychopharmaceutical research strategies.

**Some Neurobiological Findings in Schizophrenia**

- Hyperconnectivity: connectivity between the posterior and anterior midline regions of the brain is abnormally strong.
- The more intense the symptoms, the stronger the low-frequency fluctuations and parenrhythmias in the EEG.
- The severity of the disease and the extent of the cognitive disorders correlate with the amplitudes of MMN (mismatch negativity).

Mulert et al. (2010) observed a significant correlation between the severity of acoustic hallucinations and the phase synchronization in the gamma frequency range. With regard to the bilateral primary auditory cortex, the higher the gamma-phase synchronization between left and right, the more pronounced the acoustic hallucinations. The phenomenon of voice hearing occurs not only in paranoid hallucinatory schizophrenia but also in a number of other psychotic states. More specifically, the increase in resting state activity in the auditory cortex appears to be associated with severe fluctuations (Quin and Northoff 2011). These fluctuations of the intrinsic resting state activity itself seem to be as strong as the extrinsically stimulated activity change. Quin and Northoff (2011) suggest that encoded voices are activated when the intrinsic activity exhibits strong fluctuations similar to those of the extrinsic stimuli.

From the multitude of studies showing aberrant neuronal functions, some of the most important ones will be mentioned here. Whitfield-Gabrieli et al. (2009) demonstrated that the anterior cortical midline structures (CMS) and the posterior CMS, such as the posterior cingulate cortex (PCC) and the precuneus, no longer show as much activity change in response to short-term memory tasks. This finding suggests that the resting state is no longer as reactive to stimuli and can therefore no longer induce proper activity changes in response to cognitive tasks. This decreased activity change was observed in both patients with schizophrenic disorders and their first-degree relatives when compared with healthy individuals.

Holt et al. (2011) showed that abnormal anterior-to-posterior midline connectivity is related to self-specificity processing in patients with schizophrenia. An abnormal resting state activity and imbalance between anterior and posterior midline regions (hyperconnectivity) were also observed in many other studies. The resting state spatiotemporal structure seems not to operate properly in schizophrenics; therefore, they are not able to modulate and adapt the pathologically increased hyperconnectivity. The result is that we see a precondition of a possibility of what is to be observed on the phenomenal level: the change or even deterioration of the consciousness of the self. The idea behind this is that the neuronal (prephenomenal) virtual spatiotemporal structure does not correspond properly with the phenomenal spatiotemporal structure of the self-experience, so that from the psychoanalytic point of view, schizophrenics suffer from self-fragmentation (Kohut 1973; Kernberg 1978) which may be seen as a trans-phenomenal concept. The "trans-phenomenal level targets the implicit yet operative matrix that underlies these anomalous subjective experiences" (Northoff 2016).

The anticorrelation between DMN and CEN (central executive network, lateral regions of the prefrontal and parietal cortex) in healthy persons is mentioned above (see the details in Chap. 11): The crucial point in schizophrenics is that the balance of the connectivity between DMN and CEN changes, depending on the amount of the abnormally high resting state activity and hyperconnectivity in the CMS (Northoff et al. 2004; Vanhaudenhuyse et al. 2011; Wiebking et al. 2014a, b; Northoff 2015). The result is that external mental content is now no longer reduced when internal mental content is strong and schizophrenics lose the clear distinction between internal and external mental content. Northoff describes this confusion as "self-environment blurring."

Psychodynamically speaking, such "self-environment blurring" may well correspond to the loss of *ego boundary cathexis*, when the barrier between what is experienced inside and what is experienced outside cannot be distinguished. The often used term "ego boundary" was described by Federn (1956/1978). Cathexis (Besetzungsenergie) describes the *investment energy* in something in general and in another person, event, or idea in particular. Comparing the phenomenological (psychopathological) and the pre-phenomenological (neuronal) concept of cathexis, Northoff (2011, p. 93, 264) describes investment of energy into objects parallel to the investment of the brain's intrinsic activity into its own neural processing of stimuli: rest-rest, rest-stimulus, and stimulus-rest interaction.

Also the term of the "schizophrenic *dilemma*" between self-relation and object-relation, as referred to by Mentzos (2011), could also apply here, when the mutually dependent co-occurrence between self and objects is no longer balanced out in some acute schizophrenic states, and it is impossible for the patient to distinguish between the inner self and the outer object. The processing mode could be such that there was a total withdrawal to the self-pole in autism or, on the other hand, dissolve of the ego boundaries into a fusion. Mixed states are frequent in which both components oscillate.

When all relations are lost and in an acute psychotic state the choice between objects and subjects is no longer possible and the person concerned sinks into chaos, we speak of an *existential polylemma* of the schizophrenic because the cathexis is shattered and cannot be invested somewhere properly. The schizophrenic patient does not only need to choose between object and subject relation but is also confronted with a multitude of different relations, which she/he remains unable to classify and categorize and consequently unable to make fast decisions which result in the well-known decision ambiguity in these patients with long decision or reaction times or even no decision.

Depending on whether the accentuation of a disorder is more in the direction of DMN or CEN, we suppose psychopathological symptoms change that is true for attentional dysfunctions (Hartwich 1980) for delusion and for a strange sense of agency. The connectivity between other regions of the brain, however, should also be taken into account, e.g., orbitofrontal dysfunction and catatonia (Northoff and Boeker 2003). Some studies demonstrated abnormally high resting state activity in the auditory cortex during *auditory hallucinations*. This may be related to the DMN and CEN and their relation with the auditory cortex. In the resting state, the DMN seems to be less connected to the auditory cortex, which, in contrast, is rather strongly connected to CEN. Such disengagement of DMN functional connectivity from auditory cortex and the latter's association with CEN may account for the assignment of an external origin to hallucinated voices, rather than relating them back to an internal origin (Northoff and Qin 2011). However, in the case of schizophrenia, in contrast to toxic and organic disturbances, the experience of the *subjective certainty* of the hallucinations is a crucial phenomenon. For this phenomenon, there are still no considerations regarding correlations or even connections with neuronal anomalies.

Robinson et al. (2015) referred to the sense of agency (SoA), which means the self as the subjective experience that oneself is the agent of perception, action, cognition, and emotion. Older German psychiatrists described this as "Meinhaftigkeit" (Schneider 1962), "consciousness of the ego as opposed to the outside" (Jaspers 1953), and "ego activity" (Scharfetter 1980, 1986) which means functioning as a self-directing unity that is disrupted in individuals with schizophrenia. That may be related on the neuronal level to the disbalance between resting state hyperactivity in medial prefrontal regions as part of the DMN and resting state hypoactivity in lateral prefrontal regions as part of the CEN. Patients may then *misattribute the source that generates a stimulus on the phenomenal level*. In *psychopathology*, the corresponding result is the *loss of the self-reference of the sense of agency*: they then experience the origin of their own thoughts, actions, etc. as coming from the outside of another person's self rather than the inside with their own self. In some cases is the psychopathological experience that ideas, voices, orders, and other influences are forced into the patient's head by some powerful authority from the outside.

Not only the correlation but also the more direct relation between bilogical findings in schizophrenics, their psychopathology, and psychodynamics are described in what we call *neuropsychodynamic approach* and thus represent an overarching concept that leads to new hypotheses and models.

**Coenesthesia and Catatonia** Resting state abnormalities and abnormal body experience, e.g., coenesthesia, are taken all together; this suggests that we view abnormalities of bodily experience in terms of their underlying abnormal spatiotemporal features which, as we suppose, can be traced back to the spatiotemporal features of the brain's spontaneous activity.

"The interaction of the intero- and proprioceptive input from the body into the brain's resting state activity entails that these inputs are set and integrated within the ongoing spatial and temporal dynamics, i.e., structure of the resting state. They are integrated and encoded within the resting state's internally-directed processing and its functional connectivity (as shown above) and possibly also into its ongoing frequency fluctuations (as it remains to be demonstrated). More generally, this means that the incoming input from the body, intero- and proprioceptive, becomes integrated within a larger spatiotemporal framework of the resting state's internally-directed processing." (Northoff and Stanghellini 2016, p. 11) "Coenesthesia is the name for the integration of the manifold of impressions coming from one's own body, including phenomena related intero- and exteroceptive functions. The origin of these symptoms could be searched for at the basic level where the temporal coherence of conscious awareness is constituted. A failure of the constitutive temporal synthesis may create micro-gaps of conscious experience." … "The coupling between bodily sensations and life situation needs the integrity of TT (transcendental temporality, trans-phenomenal temporality) as these are two moments in the spatiotemporal Self-word relation." … "We argue that the fragmentation of bodily experience taking place on the phenomenal level is originated by a fragmentation of TT—that is, of the pre-reflexive synthesis of impression retention-protention. The latter takes place on the transphenomenal, non- experienced level and is closely related to anomalies of the resting state, which in their own turn takes place on the pre-phenomenal level."

From this point of view, psychoanalytic theories, the somatic etiological components and their interactions with psychic factors from the historical beginning period until today, did not meet all the actual needs of the persons who suffer from schizophrenia. It is the same with purely biological-somatic models, in which the treatment was almost exclusively based on psychopharmaceuticals. Some years ago, we spoke of somatopsychodynamics (Hartwich 2006a), in the sense of the somatopsychic-psychosomatic concept of Mentzos (1991, 1996, 2000) which dates back to the neurophysiologist Jung (1967, 1980, p. 769) who had already described in detail the somato-psychological and psycho-somatic interactions. Today we call it from a comprehensive perspective *neuropsychodynamics* (Northoff 2011; Böker and Northoff 2010).

## 10.4 From Conflict-Versus-Deficit Controversy to Psychodynamic Interpretation of Schizophrenia

The path from the initial period up to the new concept presented above has been intertwined and will be briefly traced here. One basic assumption was that of a *somatic deficit*, which leads to disintegration, thus establishing the defect hypothesis. The other basic assumption was the *conflict*, which in most cases leads to the fragmentation of the self in the early but also, in many cases, in the later psychical development.

The *defect hypothesis* was based on the findings of somatic, genetic, biochemical, and imaging research including twin and adoption studies. Thus, the genetic influence in an individual life could, under certain further unfavorable conditions, pave the way for psychopathological deficits which led to the development of a schizophrenic and schizoaffective psychosis.

*Conflict hypothesis* is related to traumatic environmental influences, especially in early childhood development, mostly caused by the problematic behavior of the close relatives. Freud (1909–1913, 1920–1924, 1925–1931), Fromm-Reichmann (1940), Bateson et al. (1956, 1978), Lidz et al. (1965), Fleck (1992), Federn (1978), Klein (1956), Mahler (1972), and Kohut (1973) stand for different variations of psychological concepts of understanding and treating schizophrenia.

Some authors, beginning with Jung (1979), Benedetti (1979), and Mentzos (2000, 2001, 2011), have already taken into account the conflict-deficit interaction, which was later underpinned by a lot of researchers, e.g., by Tienari (1991) and Tienari et al. (1994). They found out that positive rearing experience can protect high-risk children and adolescents against future development of schizophrenia, whereas the genetically vulnerable individuals who grew up in a disturbed adoptive family tended significantly to develop schizophrenia, and this was not present in those at low genetic risk. This supports an interactive effect of genetic risk and rearing environment.

What is the interaction of the somatic-genetic and psychological factors in the overall group of psychotic disorders and, in particular, in the personal developmental history of a single individual?

The interaction involves the weighting of individual components within a network of influencing factors. The weightings shift in the course of the development of a person. If one wants to analyze the special network of the conditions at a certain point in time, that is, in the cross-section, a circular interaction process can be said. If, however, the factor *time* is taken into account, the conditional network moves three-dimensionally and can be seen as a *spiral*.

The somatopsychic interaction in the etiological view of schizophrenic and schizoaffective diseases is nowadays acknowledged. The genetic (e.g., Maier and Hawallek 2004, pp. 63–72, Schmitt et al. 2015), the neurobiological (e.g., Northoff 2011), and the psychodynamic (e.g., Tienari 1991; Tienari et al. 1994) research has provided many weighty arguments.

Today's neuropsychodynamic understanding of schizophrenia and other psychoses replaces the alternative models of the previous century and represents a meaningful and scientifically founded combination of those components which have proven themselves and survived the previously discussed alternative concepts.

Starting from the researches of Tienari et al. (1994), the sensitive genotypes have a greater instability against environmental influences than nonsensitive genotypes when stress-triggering factors are equally present. Thus, inversely genetically predisposed children have better protection from a later psychosis when they grow up in a stable family with healthy parents. To what extent the genetic disposition has a significant influence on the neuronal organizations and structural orders of the brain with its functional characteristics is individually weighted on a case-by-case basis. The interaction with early and also later relationships to experienced objects (father, mother, and other reference persons) as well as traumatizing or also stabilizing environmental conditions has an influence on the neuronal and functional somatic processes, whereby they can be modified in the development in turn to contribute to the personal processing of environmental influences. The respective behaviors of a person in the early or later development are, on the one hand, caused by the sensitivity of the psychical structure formed up to then, and on the other hand, the psychic structure is further modified by the reactions of the environment to the behavior which can be more or less stabilizing or traumatizing. In the history of psychiatry, types and groups of psychotic persons have been distinguished. Examples are hebephrenia, paranoid schizophrenia, schizoaffective psychoses, delusional diseases, psychotic reactions, and further subgroups.

Today we see the etiopathogenesis of the heterogeneous illness commonly called schizophrenia as a complex interaction between somatic-genetic factors, the psychological environment, and the individual personality structure. From the viewpoint of descriptive symptomatology and *experimental psychopathology*, patients can be differentiated into groups, e.g., a paranoid hallucinatory group with "positive" symptoms, such as hallucinations, catatonia, and delusions; a nonparanoid group with symptoms such as attention deficit, restricted affect, anhedonia, and social withdrawal; a coenesthetic schizophrenia group with aberrant symptoms of the experienced body (Leiberleben); and more groups as well.

## 10.5 Toward a Neuropsychopathology

In the second edition of his textbook *Psychiatrie*, Ziehen (1902) describes the disturbance of the connection of the ideas and their associations as incoherence and dissociation. Dominant ideas, through which normal thinking creates a connection, are lost in psychotic patients. The connection of a whole series of ideas loosens, and no or only distant relations are recognizable:

> Where this anomaly occurs, it is called dissociation or incoherence of the idea association. A severe case of such dissociation is as follows: I ask a patient for the present year, and she answers me "blue." The notion of "blue" has no recognizable relation to the concept of "present year," which I have suggested by my question. It is the language, which reflects the various degrees of incoherence most faithfully. (Ziehen 1902, pp. 96–98)

The overall state of "disorientation" and "incoherence" is also referred to as confusion. Even Ideler (1847) called severe states of incoherence the "chaotic devastation of consciousness," and Kraepelin (1889) described it as "destruction and paralysis of superior institutions." In analogy to neurological disorders, Stransky (1914) saw in this disorder an "intrapsychic ataxia" and Bleuler (1911) the splitting of the connection between thinking, perception, memory, and feeling. The described phenomena were later referred to as "disintegration" in the sense of the essential core of the psychotic disorder (Conrad 1952, 1958; Heimann 1957; Petrilowitsch 1958). The imaginative language of Jung, who compared this phenomenon with a "splintered mirror," is similar to the expression "rubble field" from Wellek (1953). These are characteristic features which are to be found on schizophrenic patients especially with the nonparanoid subtype, e.g., hebephrenia, early-onset schizophrenia, and also chronic states. Besides the attentional deficit and the cognitive disturbances (Hartwich 1980, 1987), hebephrenics are often characterized by inadequate affects, grimacing and *mirror* gazing, often active but in an aimless, nonconstructive manner. One problem in experimental psychopathology is that perception and emotion in studies usually are separated, but this can only be done theoretically because perception and emotion are normally connected intensely (Fig. 10.1).

**Fig. 10.1** The shattered mirror. As early as the 1930s, French psychiatrists described young schizophrenic patients who saw themselves in the mirror, did not recognize themselves, and then smashed the mirror. In the picture you see an example of a shattered mirror (Fig 10.1); the acute and aroused psychotic patient no longer recognizes himself; he also sees a "stranger," but as a dangerous opponent who runs to him threateningly; he cannot avoid him because he moves in the same way; he can only attack him; and thereby the mirror is destroyed. Later, when things got better, the patient embedded the fragments of the mirror in plaster and created such a picture in art therapy

In experimental psychopathology, the heterogeneous construct, what is called schizophrenia, was divided into subgroups. Three dimensions have proved its worth:

1. The *paranoid-nonparanoid dimension* has proved especially useful.
2. The acute-chronic dimension is defined in terms of length of hospitalization; however, this has its difficulties. The time at which psychotic symptoms began would often be a matter of subjective judgment by unsophisticated observers.
3. Good-poor premorbid adjustment, although this has proved useful, is not without problems.

It depends on the severity of the disease and how severely the patients are disorganized. If one asks such patients what they really experience, they sometimes report mostly in retrospect, since they cannot describe the experience themselves in the acute state: "I disintegrate internally, … everything flows out of me, … I dissolve, … everything is a chaotic confusion … I am no longer me, my I is extinguished."

The experiences have their psychopathological descriptions:

- Lack of attention and disturbance of concentration, signifies the inability to focus on an object and to fixate attention, a marked distractibility, and fluctuations in the duration of attention
- Inhibition of thought, thought blocking, break in thought, narrowing of thought, paralogical thinking, and incoherent thinking
- Poor motivation, inappropriate affect (parathymia), poverty of affect, and emotionally flat

Broadbent has presented a model for the explanation of normal information processes (Broadbent 1971) and described a "filter mechanism" which is situated between short-term memory and transition system for the further processing of memory and perceptive contents. Many results from the field of experimental psychopathology of schizophrenics were subsequently interpreted as a disturbance of the selective filter mechanism. The schizophrenic disorder was seen in such a way that internal and external stimuli can flood the organism without the corresponding inhibition and filtering, so that selection and limitation of the essential stimuli are not possible for a reality-appropriate processing. In modern neuropsychopathology, Northoff (2011, 2012) picks up the filter disorder and names experiments by Javitt (2009), according to which schizophrenics can no longer filter when it comes to differentiating important stimuli from unimportant input stimuli. Instead of the aberrant filtering, Northoff speaks nowadays from a neuronal perspective of aberrant difference-based coding, which includes encoding and decoding. That would, e.g., mean that the encoding of stimuli from the environment does not work properly in schizophrenics. But this does not explain all psychotic symptoms.

All this is a description in the *third-person* perspective.

If in the *first-person* perspective one asks patients what they really experience, they often report mostly in retrospect, since they cannot describe the experience themselves in the acute state: the abrupt blocking of the stream of thought. He ceases to speak in the middle of a sentence. When he is questioned about that, he is likely to report that he had the physical sensation of somebody's taking his thoughts out of his head.

It is to be mentioned that there is not only an inability to focus on attention, to select between items of sensory input, but also impressions of abnormalities in the *quality* of the sensory input. For example, "Noises seem to be louder to me than they were before; I noticed it with background noises."

"Colors seem to be brighter now almost as if they are luminous." (Venables 1977, p. 5). The patient suffers from abrupt blocking of the stream of thought. He ceases to speak in the middle of a sentence.

About hypercathexis (Mc Ghie 1977, p. 62), "My senses were sharpened. I became fascinated by the little insignificant things around me—sights and sounds possessed a keenness that I never had experienced before—my senses were sharpened, sounds were more intense and I could see with great clarity everything seemed very clear to me. Even my sense of taste seemed more acute."

**About Subjective Space and Time Experience** Jaspers (1953) and Sartre (1956) describe the temporal-spatial object in the world as an essential structure of a temporal-spatial situation in the world. We will quote how schizophrenic patients describe their experience of time. Venables (1977, p. 6): "When I move quickly it's a strain on me. Things go too quick for my mind. They get blurred and it's like being blind. It's as if you were seeing a picture one moment and another picture the next." "Things are coming too fast I lose my grip on it and get lost. I am attending to everything at once and as a result I do not really attend to anything" (Venables 1977, p. 5).

Jaspers differentiates between loss of subjective consciousness of time, loss of reality of time experience, and the experience that the time stands totally still. Klages (1974) describes the para-order in the room where they live. Objects seem not to have any relation to each other.

One of our schizophrenic inpatients is standing at the window and looking down at the street; he is obviously confused. He explains to us after a while: "The cars are driving and standing at the same time. The car's doors are open and closed at the same time." This patient has a psychopathological disorder, in which the experience of the flow of time, the succession of things, is changed and fuses to one now.

On the neuronal level, Northoff (2016, p. 168) emphasizes that the mutual matching process between the spatiotemporal structures of extrinsic stimuli and the resting state seems to no longer operate properly in schizophrenia. If schizophrenia is a disorder of the resting state and its spatial and temporal structures, various symptoms seem to be primarily spatiotemporal and based on abnormalities in the resting state activity. "This endeavor could be called a spatial-temporal approach to psychiatric disorders such as schizophrenia" (Northoff 2016, p. 168).

## 10.6 How and Why Do Schizophrenics Create Such Symptoms? Concept of Cathexis and Paraconstruction

When there are states of losing the ego boundary and symptoms such as delusion, signs of catatonia or perceptions of coenesthesia, etc. and the correlative (or sometimes the underlying) neuronal aberration and differences in neuronal synchronization, we will now ask the question: *How* do the schizophrenics create their symptoms?

### 10.6.1 Concept of Cathexis

Freud used the term cathexis as a psychological representation of the drives. These are occupied with certain energies (Freud 1925–1931). Some variants of the cathexis concept can be found in the English translations of the Freudian writings, where they are related to defense mechanisms in neuroses.

We use variants of the cathexis in a different way, namely, neuropsychodynamically, with regard to the pathologically changed spatiotemporal structure of the self in schizophrenic, schizoaffective, and some other psychoses.

As Northoff explains (Northoff 2011), cathexis means the non-specified invested energy in objects (and the transformation into self-objects) and the binding of energy to specific objects lead to certain organization of objects. Cathexis is both static and dynamic at the same time; it can stream from one object to another and shifts the organizational structure of the objects. The different movements and shifts of energy have been described by the terms anticathexis, decathexis, and hypercathexis. On the neuronal level, the predisposition of the investment of energy into objects is the investment of the brain's intrinsic activity into its own neural processing of stimuli. This leads to a neural structure and organization. Northoff postulates that the brain's intrinsic activity may enable and predispose to what Freud called cathexis.

On the use of the term cathexis, some critical remarks: On the one hand, the term is seen more mechanistic, as an invested energy without direct content, which would be logical. On the other hand, the self-brain-environment is a unit that does not always follow our logical categories. As a result, the investment energy could be controlled by content somehow. But how this step works is still unknown.

**In the Following, We Describe Different Variations of Cathexis in Schizophrenia:**

- Hypocathexis
- Oscillating cathexis
- Decathexis
- Anticathexis
- Paracathexis
- Hypercathexis

#### 10.6.1.1 Hypocathexis

Hypocathexis means the diminished invested energy into the self/ego. In consequence, the ego boundaries (Federn 1978) are weakened, incomplete, and can even break down. Federn describes in schizophrenic patients a diminution of the investment energy of the ego: Depending on the degree of diminution of the investment energy of the ego, inner and outer objects threaten to cross the ego boundary and even flood the ego consciousness. Depending on how strongly or how little the hypocathexis is pronounced, it corresponds to the range between the low structure level of disorganized schizophrenia and the simple forms which manifest themselves as derealization and depersonalization with a relatively firm structure.

#### 10.6.1.2 Oscillating Cathexis

A crucial disturbance is most pronounced among the group of hebephrenia; they suffer from the fact that there is a continuous shift of investment energy from one object to another. The main problem with these psychotic disorders is that the balance between static and dynamic investment energy is disturbed. The static component is more or less switched off. This refers not only to external objects but also to the self. However, if the patients experience their self as fragmented, they have no chance to invest their energy on a particular self-fragment or even an ego core because there is a constant change between the different self-fragments. As a result, the investment energy oscillates back and forth between the ego fragments. This is particularly noticeable in the transference to the therapist, as there are permanent new attempts to invest. In the countertransference, this can be very irritating if the therapist has not learned how to deal with it (see Chap. 9).

#### 10.6.1.3 Decathexis

A crass example is the totally missing relationship to the newborn with a mother suffering from postpartum psychosis. This psychodynamic peculiarity is a core symptom in some postpartum psychoses, which hardly occurs in other psychiatric disorders. Normally, the emotions of joy and love for a newborn child are very strong. Here, however, they cannot be experienced for the mother. The natural and self-evident investment energy, which is also anchored in animals as instincts, is completely absent in these patients. The reaction is incomprehension of oneself and guilt feelings. Deficits in the care of the child and suicidal behavior could be the result. Family members also react with helplessness. In the meantime, there are neurobiological findings that could explain this form of decathexis. In the context of the postnatal decrease in the concentration of GABA neuroactive steroids, the findings of Chase et al. (2014) of resting BOLD fMRI show in unmedicated women with postpartum depression that the posterior cingulate cortical connection with the right amygdala is disturbed. One might assume that the interruption of connectivity is accompanied by a lack of investment energy in this area of the brain.

#### 10.6.1.4 Anticathexis

At a level on which the investment energy is not yet linked to the content, there is a tendency for the opposing object to be automatically selected. We do not know how the opposite constellation occurs, since, according to the definition of the cathexis, the content is not implied. Perhaps our clean separation of investment energy from content is only in our logic, but not in the functions of our brain.

A clinical example is the nihilistic delusion. We speak of anticathexis, since in the nihilistic delusion something contrary is actively invested. The anticathexis may also be associated with a reversal of attention with simultaneous contrast association or converse association. This could be shown in an experimental study (Hartwich 1980) the group of paranoid schizophrenics reacted differently than the nonparanoids in a stimulus reaction experiment. Unlike the nonparanoids and the controls, the paranoids in an acoustic stimulus did not associate with the corresponding picture, which was shown in a tachistoscope. They named the other picture, which did not fit. The behavior can be associated with an inhibition of the nearby stimulus or with a preference of non-nearby. This can be related to the mistrust of the paranoids. This deformation of the associations to the contrary can in the extreme case be a pathological negativism. In this anticathexis, we see a cornerstone of some delusion with all the paradox and systematization of the inner opposite. In case of an imminent loss of self-cohesion, *anticohesion* (Hartwich 2004) occurs. This is not a non-cohesion but a special constellation of self-fragments in paradoxical relation. This may serve to prevent a greater loss of self-cohesion (see later).

#### 10.6.1.5 Paracathexis

We have described above that the investment energy leads to certain *organization* of objects. In schizophrenics, the organization can be disrupted. Thus, false attributions occur which correspond to false realities. This disturbance of the organization of the investment energy could be called *false cathexis*, but we prefer the term *paracathexis*, because we see a protective attempt in the wrong attributions (see below the explanation for the paraconstruction). Patients experience a "new reality" that runs like film scenes, without having the opportunity to check the reality. This is mostly found in paranoid hallucinatory schizophrenia.

#### 10.6.1.6 Hypercathexis

In contrast to the oscillating cathexis, hypercathexis is characterized by the fact that the balance between static and dynamic investment energy is disturbed. The dynamic component is more or less switched off. Instead, the static component is overly enhanced.

As a clinical example, the exaggerated occupation can manifest itself in a psychotic conviction that is not correctable. We see this in the Othello syndrome (see 10.7.2.3), the delusion of jealousy, and other systematized, overvalued ideas.

### 10.6.2 Concept of Paraconstruction

Now we will come to the question: *Why* do schizophrenics and some other psychoses create such symptoms?

Ideler (1847, p. 11) was one of the first to attempt psychodynamic interpretations of schizophrenic symptoms: "Psychoses show the stressful effort of consciousness to reorganize the Self." He understood delusions as a defense against unbearable situations. Freud, who may have known Ideler's writings, mentioned in the Schreber's case: "The delusional formation, which we take to be the pathological product, is in reality an attempt at recovery, a process of reconstruction. Such a reconstruction after the catastrophe is successful to a greater or lesser extent …" (Freud 1911, p. 74).

Bleuler (1911), Benedetti (1987), Scharfetter (1999, 2003, 2012), Mentzos (2011), and others followed this line and also regarded psychotic symptoms as compensatory attempts to restructure and reorganize the constitution and construction of self and objects, including self-object differentiation. In a neuropsychodynamic context, these compensatory mechanisms cannot be regarded as defense mechanisms in the proper sense of the term. Therefore, today we speak of *neuropsychodynamic paraconstructions*. Because due to the lack of psychological structure and organization, the rather amorphous compensatory mechanisms in psychosis do not reach the level of defense mechanisms in a proper or a narrow sense and may thus be conceptualized as "paraconstructions" (Hartwich 1997, 2006a, b). In connection with the view of neuronal mechanisms and their abnormalities on the prephenomenal level, we speak of *neuropsychodynamic paraconstructions on a transphenomenal level*. That means also their pathogenesis is to be understood in an unfolding matrix which is biological *and* psychodynamic as well.

Therefore, the conception of *defense mechanisms*, which originates from the neuroses, as many mental diseases were named in earlier times and which are defined from only a psychic (and not somatic) point of view, has to be *changed* when we speak about *severe psychotic symptoms*. There is evidence in the direct link between psychodynamic and neuronal mechanisms. In the history of psychiatry and psychoanalysis, defense mechanisms were originally taken up by Freud, and his daughter, Anna Freud (1966), described in detail individual defense mechanisms (such as introjection, projection, etc.), and also all psychoanalytic schools represent and emphasize the psychological origin of the defense mechanisms. Even the term primitive defense mechanisms (e.g., splitting, denial, etc.), which is often used in psychoses, does, in this kind of interpretation, not contain a somatic component. However, if we are to take the somatic, especially the neuronal, viewpoint and emphasize their interaction with the psychodynamic approach to psychoses, the term defense mechanisms could not be used without misunderstanding.

But if, in the post-acute state, the psychotic symptoms decrease and the spatiotemporal structure of the self becomes more solid, like the level of (which was formerly called) neuroses, then one could sometimes speak of defensive mechanisms. In the intermediate stage, there sometimes can be paraconstructions and defense mechanisms next to each other. Thus, we see that the terms *neuro*psychodynamic paraconstruction and *psycho*dynamic defense mechanism depend on the strength of the structure of the personality and the weakness of the spatiotemporal structure of the self and how far neuronal and other somatic anomalies also contribute to the psychopathological symptoms. This can change intraindividually in the course of a psychotic illness together with the change of the severity of the symptoms.

### 10.6.3 Counter-Regulation

We see neuropsychodynamic paraconstructions as counter-regulations against the danger of breakdown of the spatiotemporal structure of the self and thus as expressions of a modification of trans-phenomenal matrix. We see these counter-regulations as very basic mechanisms and see an analogy with the somatic area. For example, if the body is injured and bleeds, blood immediately coagulates to protect the body; another example is the forming of antibodies. We suppose that there are also comparable protection mechanisms in our psyche in interaction with our brain on hormonal and neuronal levels. Schizophrenics may use this kind of protection against the danger of the deterioration of the spatiotemporal structure of the self. Because they are not able to build real reconstructions, they create instead neuropsychodynamic paraconstructions which, on the phenomenal level, form delusions, catatonic and coenesthetic symptoms, etc. (Fig. 10.2).

The concept of *neuropsychodynamic paraconstruction* can be explained using the example of delusion (delusion of persecution, delusion of pregnancy, delusion of jealousy, etc. see later). Our hypothesis is that the disturbed balance between DMN and CEN may be a neuronal precondition which corresponds with the beginning of the

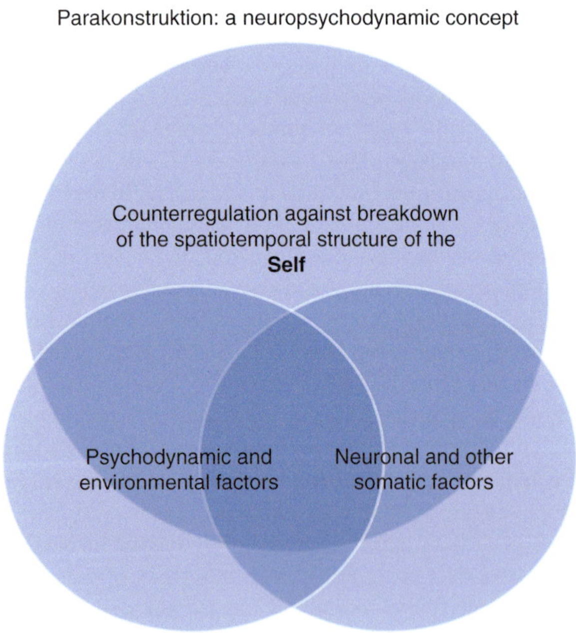


**Fig. 10.2** The neuropsychodynamic paraconstruction. The concept of paraconstruction is explained by the interaction of three components: the counter-regulation forces against the danger of the breakdown of the spatiotemporal structure of the self, the influence of traumatic psychodynamic and environmental factors, and the aberration of neuronal and other somatic components in the brain. The weighting of each factor differs from individual to individual and depends from the severity of the psychotic illness

confusion between inside and outside experiences. This causes a dangerous insecurity for the consciousness of the self. Depending on the strength of the personality structure, psychic and neuronal systems try to protect the self by creating, e.g., delusional ideas which are subsequently followed by an individual delusional system.

Such a paraconstruction of delusion is able to strengthen the self and give back some kind of order and stability of the consciousness of the self. Now it is understandable that the psychotic patient is not able to question or even give up the new system of his view of the world, because he needs this kind of protection; otherwise the new spatiotemporal paraconstruction of the self and subjective world would deteriorate rapidly in a state of complete dissolution of the self.

Why is it not possible to influence the symptom effectively, e.g., delusion by psychic means? In every generation, the psychiatrist experiences that his efforts to change or even heal a solid delusional system by means of arguments is an unsuccessful attempt and it is totally in vain. The reason is that somatic neuronal factors are mixed with psychological factors. This means that in this example, there is a strong weight on the somatic and neuronal side in the interactional relationship between psychodynamics and somatic components.

This interactional shift of the weights of the factors involved varies from case to case. Let's stay with the delusion with an example of the folie à deux. The partner (B), in whom the delusion was induced, has the same symptoms and the same firm hold on delusion as the partner (A) with a real schizophrenia. If, however, the two partners are separated, the delusion in partner (B) proves to be accessible to reason and disappears. However, if the two partners have lived together for a long time and the folie à deux has also existed for several years, and after separation of the two, it is not always possible to influence the delusion of partner B with arguments. We see that the bottom-up and thc top-down interaction depend on the weighting of the involved components.

Therefore, therapists should always *respect* the neuropsychodynamic paraconstructions as a counter-regulation, which is essential for the patient's stability to survive. This explains the patient's firm hold on the symptoms. Our therapeutic aim is to go beyond the paraconstructions, which means to use methods which have conclusive effects and can stabilize the spatiotemporal construction of the self and may "glue" together the self-fragments.

## 10.7 Case Examples

At first we will describe an example of a female schizophrenic patient suffering from a pregnancy paraconstruction.

### 10.7.1 Example of Pregnancy Paraconstruction and Treatment

The 24-year-old female student has fallen ill for the second time.

Three years ago, she had traveled from Germany to England for a student exchange. Already 3 weeks after her arrival in London, she felt observed and persecuted, became uncertain, and had severe sleep disorders. She had some ideas of delusions, which were not jet connected to a system, and she felt poisoned and suffered from visual and acoustic hallucinations. She was admitted to a psychiatric hospital in England. As the symptoms became worse and worse, the psychiatrists favored ECT treatment. The mother, who lived in Germany, was informed and categorically rejected this treatment. Therefore, the patient was quickly put into a plane and taken to our psychiatric hospital in Germany. When she was admitted to the ward, she was a girl-looking young woman. She suffered from serious cognitive disturbances, thought disruption, disintegration of her speech, schizophrenic ambivalence, and ambitendence. One of the results of the stress of the flight and transfer was that the disease became much worse and she was slipping into catatonic states. In the following, she had to be hospitalized for 1 whole year because of a schizophrenic psychosis with persistent severe catatonic symptoms. After 1 year of inpatient treatment, the outpatient care was continued for 2 years.

#### 10.7.1.1 Relapse

Now, she was to the clinic with a sudden paranoid hallucinatory symptom: people reproached her, she was "easy to have," and if she was not watchful at night, she would get pregnant, and she could smell that when it had happened. She was restless and troubled by many different delusional ideas, which were directed toward herself.

The patient had grown up as a single child with her constantly anxious and controlling mother. Her father, who had had also a psychosis, had committed suicide by hanging when the patient was only 5 years old.

After that, the daughter and mother lived together closely in a symbiotic relationship. Attempts of autonomy from the girl's side were answered from the mother with anxious helplessness and sometimes also strict reactions. A suitable triangulation figure in the sense of a replacement father had not been offered.

Now, at the age of 24, the patient had her first boyfriend. One time her mother was away on business, the boyfriend came to her house and lived with her for 2 weeks.

When the mother came back, she was horrified at the violation of her common privacy and threw the boyfriend out of the house. The 24-year-old girl could not bear the conflict between her own autonomous development and the symbiotic relation with the mother; because of her ego structural weakness, other compensatory mechanisms failed. The genetic disposition, which had been co-formed by psychotraumatic factors, was overstretched:

1. Psychosis of the father (genetic disposition)
2. Suicide of the father as the patient was 5 years old
3. Symbiotic relation with the mother
4. "Double binds" by the self-insecure mother
5. Previous mental development and delayed maturation with current autonomy conflict

The psychosis broke out for the second time at the age of 24 years, and the patient was admitted to the psychiatric hospital with the above-described symptoms. Then, after about 3 weeks of treatment, which included highly dosed antipsychotic medication, the florid and severe psychopathological symptoms decreased.

But then a telephone call of the boyfriend came. He told her that he had to part from her and that he was not able to tolerate her changed behavior which was caused by her illness. As a result of this relatively short information, which she did not tell us, the patient committed a suicide attempt. She went to an area of the general hospital where she could jump down from the third floor to the ground floor made of marble slaps. The falling down was accompanied by multiple physical injuries of her body. Her life could be saved only because she could be immediately treated in an intensive care unit of the same hospital building, and the necessary surgery could be done. After some weeks, she could be retransferred to our psychiatric ward. During the postoperative period, she had to stay in her bed for another several weeks. Now she developed a delusion of pregnancy, she noticed changes in the body and experienced child movements, and after a while, she could speak with the unborn child. (All tests for pregnancy had been negative.) The coenesthetic symptoms persisted firmly. There was no reaction to variously varied and highly dosed antipsychotic medication and other therapeutic methods. The mother was particularly horrified, asking again and again whether a pregnancy would be present. She blamed the ex-boyfriend that her daughter was in such a strange mental state. The extent to which the patient felt confirmed in her delusion of pregnancy can only be assumed.

#### 10.7.1.2 Explanation

We have seen the persistent firm symptom of pregnancy delusion in the schizophrenic psychosis as a *neuropsychodynamic paraconstruction*, which was important for the patient to stop further fragmentation of her spatiotemporal self. Now she experienced something new in her body. Altered body perceptions and voices could be related to a figure, i.e., child in the mother's body. In this way with the creative symptoms, she could hold on to the connection with the boyfriend; in some way, she could have got something from him. He was her first great love but he had abandoned her after the outbreak of her psychosis.

Any psychodynamic interpretation of her object loss at that time would have been still far too early for her psychotic disease condition. At this time, she was not able to work through her mourning; this was only possible after several months of intensive in- and outpatient treatment which included antipsychotic medication. Only then, under psychotherapeutic protection, she could take up contact with the ex-boyfriend once again by telephone and express her disappointment against him. When this had happened, very slowly it could be one of the reasons that she was able to let go of the symptoms of the pregnancy paraconstruction, supported by intensive psychotherapy.

What is the significance of the concept of neuropsychodynamic paraconstruction for therapy for the psychotic patient?

The conception of the symptom as a paraconstruction, which is currently necessary for the patient's stabilization of the spatiotemporal structure of her self, leads to a decisive paradigm shift of the therapist and the therapeutic team. The treating persons respect the current necessity of the symptom as protection from further self-fragmentation and also understand the meaning of the "tenacity" and the resistance of the symptom. They do not fight any more against the symptom and thus do not fight against the patient. Then the therapists sit with the patient "in the same boat." She feels no longer rejected but is better accepted and perhaps even partly understood.

Because of the paradigm change that the conception of the neuropsychodynamic paraconstruction offers, the basis for a reorganization of the therapeutic togetherness and a new common level of the therapeutic relationship is created, which has always a positive effect on the therapy.

Important is that the therapist, if he interprets a psychotic symptom as a neuropsychodynamic paraconstruction, recognizes and respects this as a currently unconscious restoration effort of the pathologically changed spatiotemporal structure of the self of the patient. Since the patient, in her psychotic disintegration, has brought about a creative act in the sense of a paraconstruction and thus has counteracted further self-dissolution, it is understandable that she must at first maintain and hold on intensively to this kind of paraconstruction. If, from a purely theoretical point of view, one could take away her paraconstruction, the disintegration would increase again, and the dissolution would have to be experienced as a worse version. As a result of the understanding interpretation, it is logical that patients remain so immovable and intense in such creative paraconstructions. If the therapist understands this mechanism, which runs in the patient's body and psyche, and is able to respect the current need for paraconstruction as protection against further decay, his attitude and behavior change. This change is usually perceived by the patients, they feel themselves to be taken more seriously and the relationship relaxes, which has a positive effect overall.

We have here a patient with an immature personality structure which was traumatized early in her development, she still lives in an intensive dependence on the mother, in which she grew up as a single child. Mother and daughter have a symbiotic relation. The steps of the daughter to autonomy, which would be adequate in her age development, are advised from outside and performed by female friends of the same age. She herself is not yet so far and is overwhelmed with regard to the intensity with which she has entered the relationship with the boyfriend. In this dilemma, a conflict tends to arise which promotes the outbreak of the psychosis and thus contributes to the sudden disintegration. As a result, we must work psychotherapeutically at the conditions which lead to a risk of disintegration: one factor is the entanglement in dependency.

This can be addressed, edited, and partially solved, by means of the transference to the substitute triangulation figure of the therapist, which promotes the development of autonomy. Adequate post-maturation thus becomes an opportunity to be able to bear similar conflicts better in the future. Then confrontations of conflict of this kind are no longer a cause for a jump into decompensation and thus to a psychotic relapse. We see further possibilities for therapy, among other methods, e.g., in the improvement of her creativity in connection with structure giving material (see Chap. 24).

After the hospital stay the patient is in outpatient psychotherapy. She is able to pursue her education as a foreign language correspondent. The psychosis is well compensated except for some residual symptoms such as ease of fatigue, loss of concentration, insecurity, and occasionally sensitive paranoid experience. Dark-skinned men who look Mediterranean would know all about her. They would talk so loudly about each other that she would have to listen to it in anticipation. She is confronted with it in Italian or Spanish language, when she is shopping, in the restaurants, or on the way to the training center. Then they make amusing remarks about her.

#### 10.7.1.3 In the Psychotherapeutic Setting

A drive-dynamic interpretation, which would be available from the patient's developmental history, could not yet be brought forward fruitfully at this time—1 year after discharge. Instead, a common view of a narcissistic component is rather possible.

It is a little fun to comment, therapist: "Everyone knows of you, you must be an important person. This is like a star on TV, everyone knows her."

After a little moment the patient and the therapist can laugh together, so the interpretation can be continued, therapist: "Thousands of people are interested in you and are also talking about that to each other." Laughing, the patient says, "I know this cannot be so, it is not possible, but I experience it when I have to pass the men and then hear certain words."

The therapist makes the remark that she should know that young men usually are making remarks about girls who are smart. But despite the fact that the therapist and the patient took a piece of reality seriously, she remained skeptical. The therapist is not quite sure whether it is a real distance step or identification with him. Some sessions later she declares that she is the ideal mobbing victim. She means the handicap of their legs, which are disfigured because of the multiple fractures after suicidal jump and sometimes caused their co-students to mock. Now she had become active and defended herself against one of the comrades. For the psychodynamic process, it is also crucial that she now perceives her own disability more consciously and thus can leave the prior delusional projections that had been necessary for her. After this, she is increasingly learning to ignore the "dark-haired southlanders and their remarks," as she says, "and I tell myself, 'Do not listen!'."

The therapy example shows the gradual transition from psychotic symptoms, which are understood as paraconstruction, to a more mature personality structure with defense mechanisms like they are to be seen with neurotic patients. Such an interpretative psychotherapy, as shown here, is only possible when corresponding ego strength (Federn 1978) has been achieved and the therapeutic relationship gives sufficient support.

## 10.7.2 Further Case Studies for Paraconstructions

### 10.7.2.1 Mutism-Paraconstruction

A 22-year-old young man who is treated in our psychiatric hospital is suffering from catatonic schizophrenia. When his parents visit him, we observe that he reacts more and more mutistically and turns his head to the wall in order *not* to see his father or his mother. We consider this situational mutism and negativism occurring in such situations as a present paraconstruction; it serves a purpose of protecting his disturbed order of his spatiotemporal self from a too much emotional state, since this could not be endured and would lead to further fragmentation (we interpret his behavior as a protection against emotional flooding and self-dissolution), and we call this a mutism-paraconstruction. Regarding this symptom, the neuronal aberrations have been described above.

Since we know that the young man used to spend his time with computer games, we managed to motivate him to paint on a laptop with the help of a painting program. Despite his catatonic restrictions, he can get involved. He experiences the computer with the offered functions as a "neutral counterpart," without the human emotional proximity that would be currently dangerous for him. He takes the opportunity to use the computer painting program more and more. After a while, he succeeded in expressing his psychopathological experience and thus communicating what was not possible to him before. In the picture, he fixed his body with horseshoe nails on the ground. In this way, he managed to communicate his inner experience.

### 10.7.2.2 Paraconstruction of Delusional Love

A 60-year-old female patient lives in delusional love to a conductor of an orchestra. When she turns on the radio, she hears him, he gives her messages, and he plays for her and goes to her wishes. In this paraconstruction, the patient experiences more meaning in her life, it becomes rich, otherwise emptiness and loneliness would occur.

Some paraconstructions have so much creativity that they are filled with rich inner experience. This explains additionally the strength of holding on to some conviction and the certainty which provides the obstinacy. In this love-paranoia paraconstruction, the question must be raised as to whether it is appropriate to go further therapeutically against it. It can be regarded as a respect for the patient, if she is left in her creative paraconstruction, accompanied, and kept away as far as possible from influences which may weaken her condition.

### 10.7.2.3 Othello Paraconstruction: Example of a Psychopathological Interaction of a Couple

In the German language, there is a saying:

> Eifersucht ist eine Leidenschaft, die eifrig sucht, was Leiden schafft.
>
> Jealousy is a passion, which is searching eagerly that what creates suffering.

Mrs. S., a pastor's wife, has been suffering for the last 10 years from the fact that her husband, the head of an evangelical church institution, seems to be unfaithful.

The two have been married for more than 30 years and have two grown-up children who live their own life. Mrs. S. has an apprenticeship as a medical technician, but has not been employed since the births of the children.

Her early childhood was overshadowed by events of the Second World War; the exhaustive escape of the mother as refugees with the 1-year-old child was an additional burden. The father, who had suffered from a bipolar disorder, had only met the girl when she was 6 years old. When he came back from the captivity of war to the family, she rejected him as long as she could think. The mother was an energetic, dominant woman, in whose family several schizophrenic diseases have occurred (see also Stierlin 1972). The marriage of Mrs. S. with her husband, whom she had met at the age of 16, was harmonious; the interaction was unconsciously asymmetric as she played the role of the weak and the needy, who could hardly make any decisions. And he took the role of the stronger and the more active part. She had always complained of weakness, lack of drive, sleep disturbances, multiple pain, and episodes of depressive moods, which were treated with antidepressants. For this reason, a job was no longer considered. For decades, everything went well, the children were raised, and the family built a house with a garden. The children went out of the house to study. Almost imperceptibly, she suspected more and more evidence that the husband had an intimate relationship with a female colleague in his office. Mrs. S. noticed color patches, on the upper shirts of her husband which she held for lipstick, found lint, woolen fibers, and fabric naps on his clothes, which she picked up with the tweezers and placed them in little bags with a date. She found blood spots and "pubic hair" in his underpants and refused to wash his laundry. That she also had to endure the smell of perfume and hairspray on his clothes, which she held for the summit of torture. Instead of confronting him directly with her findings, she wrote to him pieces of paper that she stuck or put into his clothes, where her "convictions" was noted.

She thus combined the demand that he should brush himself thoroughly before he entered the house and bought him a brush. The husband tried more and more to behave in such a way that he avoided anticipating possible suspicious events, which ultimately proved to be impossible, because everything he did was reinterpreted to his detriment. He was caught up in the paranoid system; he sat in the trap of a paradoxical construction. He told himself about his high workload, which distracted him, that he had had a lot of luck in life and now had the task to take the disease of his wife and to endure the situation.

In Mrs. S., however, there were also fluctuations in her subjective certainty; she sought help from a psychoanalyst, who seemed to understand her at first, and he suggested writing a letter to the husband's "beloved female colleague." When this had happened, the colleague looked up her boss (the husband) in his office and very annoyingly pushed Mrs. S.'s letter on the table and said she did not want anything like that. This was a very embarrassing situation for him, but he succeeded to explain that his wife suffered from a delusion of jealousy.

In the further course of the therapy of Mrs. S., the psychoanalyst slowly realized that it was a monothematic delusion. As he carefully tried to reassess her ideas, Mrs. S. broke off the therapy by simply not appearing anymore. The husband also experienced again and again that it was futile to talk with friends of the family about his situation because they did not know whether to believe him or her. The suffering and conviction of his wife were visible to all; even the children were unsure what to believe. There was, however, one friend who supported him. He was the teaching analyst of the said female colleague, who knew the reality from the other side. After 10 years of "prison life," it became so intolerable and the paradoxical life situation so unbearable that he decided to leave his beloved wife to save himself; in particular, he thought of the time of his retirement. This actually took place after several more years.

**Explanation**

With regard to the border between "normal" and pathological jealousy, there are certainly transitions, but in the course of time, the conviction becomes stronger and stronger and becomes the quality of a real delusion. There is always new evidence of the partner's unfaithfulness. Any explanation of the partner is immediately turned into the opposite. These are contrast ideas and contrast associations which Bleuler described as contrasting associations in Bleuler (1904); we (Hartwich 2007) described this process neuropsychodynamically as anticohesion tendencies of the spatiotemporal structure of the self which seems to be in danger to lose coherence. The anticohesion seems to be some kind of counter-regulation. Psychodynamically speaking, she strengthens the structure of her self with destructing the partner. She does not have to suffer on her own, but she can torment and possess him by dealing with his failures day and night. One of the main questions is whether people with a delusion of jealousy are able to love. The husband reported that he had never really felt loved by her, but he always thought that his love was great enough for both of them. Wurmser (2006) remarks that Jones (1967), Fenichel (1935), Pao (1969), and Coen (1987) also observed that the inability to love in pathological jealousy requires central attention.

#### 10.7.2.4 Coenesthetic Paraconstruction

A 31-year-old female patient, who is admitted to the orthopedic clinic 3 months after the birth of her first child, expresses the urgent desire that her cervical spine be surgically stiffened and made a firm corset. She is convinced that all her vertebrae have been mixed up, and she tries to rearrange the spine in her imagination. When she reaches the cervical spine from below in her phantasy, the head threatens to fall down. At that moment, she screams, gets very upset, loses contact with the environment, and lies down on the ground. After a while the contact relation to the environment comes back. She then asks for the abovementioned orthopedic procedures, including surgical intervention. We interpret this as a coenesthetic paraconstruction, which concretizes the inner dissolution experience. Her concretistic symbolic formation demands an equally concrete but here whimsical attempt at stabilization.

To the treatment technique, the psychiatric conciliar doctor goes to the paraconstruction. If it is currently respected as necessary, he declares to the patient that he does not consider the operation to be an appropriate measure but recommend a special medication to stabilize her spine. She agrees to go to the psychiatric ward, which is not far away in the same building. Here she agrees to take the medication to strengthen her backbone. A psychodynamic interpretation of the symptom would not have been appropriate in the present situation.

#### 10.7.2.5 Anticohesion

A patient who is a physicist experiences his two brain hemispheres as moving in opposite directions. He cannot stop them, so they move like a piston engine in the opposite direction, causing pain.

Another patient, a civil engineer, experiences the functions of his heart and his breathing as opposed to each other. When he watches the breath, the heart stops, and when he watches the heartbeat, the breathing stops. The basal functions of his vitality can only be used in alternative opposites.

By considering the brain, heart, and respiratory representations in the body self (Körperschema, Schilder 1925), we see in the symptoms of the two patients as variants in the dealing of the loss of coherence, which we would call *anticoherence*. This is not non-cohesion of the spatiotemporal self but a special constellation of the fragments in a paradoxical relation, namely, anticohesion, which counteracts the loss of cohesion, a connection that is at least a dialectical one (see Sect. 10.6.1.4 in this chapter).

All the examples clearly show how many different possibilities of new counter-regulation constructions (neuronal and psychodynamic) are provoked by the pathological change of the spatiotemporal structure of the self in the psychosis.

## 10.8 Postpartum Psychotic Disorders

Postpartum disorders are a group of diverse illnesses; affective disorders, psychotic states, or organic-like mental disorders that occur in the postpartum period. Here, we will describe the postpartum *psychosis*; this is a serious illness, which usually should be treated in a hospital because of the endangering of the newborn child and mother. The special feature of etiopathogenesis is that in addition to the risk factors described so far, the fact of birth with its physical and psychological changes can play a decisive role. The sudden reduction in hormone and transmitter levels often interferes with psychological stress during pregnancy and childbirth, which can lead to more difficult social conditions. We now report a prevalence rate of 1–2 per thousand of all childbirths (Kumpf-Tonsch et al. 2001; Riecher-Rössler 1997). The schizoaffective symptoms (Rohde and Marneros 1993a, b, c) are most frequently described in psychopathology. Hartwich and Grube (2015) report on the use of medication, the stop of breastfeeding, and the psychodynamic treatment in mother-child units.

### 10.8.1 Psychodynamic Characteristics of Postpartum Psychoses

In some postpartum psychotic disorders, one encounters a particularly conspicuous core symptom that hardly occurs in other psychotic disorders. It is a *serious disruption in the relationship of the mother to the newborn child*. The normally strong effects of joy and love for the child are not experienced for the mother. The natural and self-evident very high cathexis (investment energy), which is instinctively anchored everywhere in animals, is completely absent in these patients. This phenomenon could be called decathexis. The reaction to this experience is associated with incomprehension in oneself and with enormous feelings of guilt. Deprivation of care and suicidal behavior can be the consequences. Even the closest family members react with helplessness and incomprehension. In some cases, there is even a nihilistic delusion, which is dangerous for the newborn child and the mother if an extended suicide can occur.

### 10.8.2 Somatic and Neuronal Findings

As regards the neuronal experimental findings and the psychopathology of postpartum psychosis with depressive symptoms, there is an important investigation by Deligiannidis et al. (2013). They performed rs-fc analysis with seeds placed in the anterior cingulate cortex, and bilateral amygdala, hippocampi, and dorsolateral prefrontal cortices postpartum patients showed attenuation of connectivity for each of the tested regions and between corticocortical and corticolimbic regions vs. healthy comparison subjects. The perinatal concentrations of pregnanolone were not found to be different between groups. They emphasize:

> This is the first report of a disruption in the rs-fc patterns in medication-free subjects with PPD. This disruption may contribute to the development of PPD, at a time of falling neuroactive steroid concentrations.

Chase et al. (2014) reported disrupted posterior cingulate-amygdala connectivity in postpartum depressed women as measured with resting BOLD fMRI. They found that the "PCC–right amygdala connectivity was significantly disrupted in depressed compared to healthy mothers for low-frequency neuronal activity, showing a negative (inverse) coupling in the depressed group but not in the controls. PCC–right amygdala connectivity was positively correlated with PCC–parahippocampus connectivity. Resting connectivity patterns of positive co-activations in postpartum women mirrored the canonical DMN." They come to the conclusion that "these findings of reduced PCC–amygdala coupling raise the possibility that PPD might involve the disruption of outward, preventative aspects of Self-relevant thought and theory of mind/empathy processes." They hypothesize that the *severe disruption of the mother-child relationship* is addressed and assumed that the *disturbed amygdala PCC connectivity* can be related to how the mother is oriented to others in her social environment and especially to the newborn in her motherhood.

### 10.8.3 Therapeutic Remark

For postpartum disorders, especially those with psychotic symptoms, psychodynamic treatments should be used, beside pharmacotherapy. This means the inclusion of the child's father and the improvement of mother-child interaction. This can be done by baby massage, video mirroring, mother-child bath, mothers' art therapy in the group and individual, and mother's group therapy.

## 10.9 Prevention of Risk Groups for Later Psychotic Diseases

The offspring of mothers suffering from a postpartum psychosis are among the risk groups that may develop schizophrenic or schizoaffective psychoses as adolescents or in the early adulthood. Here, not only the psychotraumatizing events in the postnatal life but also the genetic dispositions are important. The problems in early diagnosis have been dealt with intensively over many years, in particular by Klosterkoetter (2002, 2013), Klosterkoetter et al. (2001), and Ruhrmann et al. (2014). It shows the use of a new four-step prediction index (European Prediction of Psychosis Study, EPOS-PI), which allows a hazard rate of 3.5 in the first till 85.1 in the fourth stage. This high probability of detection created for the first time a risk assessment, which is also important for the individual case, which should be a decisive advance for preventive treatment programs.

The most significant meta-analysis for psychosis prediction has also been based on the assessment of 27 studies worldwide involving a total of 2502 risk persons based on UHR and BS criteria (Fusar-Poli et al. 2012). The mean transition risk to a psychosis was 18% after 6 months, 22% after 1 year, 29% after 2 years, and 36% after 3 years and a further increase for the following years (Klosterkoetter 2013, p. 1304).

It is clear that psychotherapeutic methods, instead of antipsychotic medication, should be used for the adolescents or people in the early adulthood, who belong to the early high-risk group (HRS), with a straight line distance to a possible psychosis. However, these are not patients that are classifiable, but they are usually adolescents who should be cautiously addressed to preventive offers. It is very important to ensure that they are not stigmatized. For this purpose, Mueller et al. (2014, p. 22) remark for the treatment offerings: "… these services should provide a non-stigmatizing, low threshold setting for young help-seeking patients and their families." Individual and group therapeutic offers are basically related to dealing with cognitive disturbances well as the gradual training of stress situations, with the emphasis on addressing personal, emotionally intense difficulties and raising the threshold of abilities. Individual development problems and sensitivities require individual strategies. With regard to the evidence, Mueller et al. (2014) indicated that psychosocial interventions are safe and effective for patients with CHR (clinical high risk); they are well accepted by families and patients themselves. In individuals with FEP (first-episode psychosis), early detection and early intervention programs can reduce relapse when compared to commonly used therapies.

Hartwich (2006b) has investigated the subjective experience site on the question of more sophisticated psychotherapeutic approaches and then proposed the division into four types:

*Type I* Larger Group
: Prevailing is a feeling of recurring insufficiency, a failure in requirements and efficiency. There are cognitive disorders, especially attention deficit, as well as lack of drive and déjà vu experiences. Here, the often encountered autotherapeutic counter-regulation attempts should be intensified, e.g., the creativity, whose unfolding effects strengthen affect and cohesion.

*Type II* Small Group
: Here is predominantly the loss of the ego strength, the loss of experience of ego energy, and amotivation, which can be called hypocathexis whose neurobiological findings are described by Northoff (2011). Patients recall: "I was no longer right in me." Only later on, characteristic and persistent derealization and depersonalization experiences occur. Here too, therapeutic efforts, which can have a positive effect on the structural strength, should be given priority.

*Type III* Larger Group
: In addition, there is often a drug abuse; in patients who have frequently taken cannabis or other drugs, sometimes alcohol, in the course of the prodrome, in order to strengthen themselves and to relieve suffering, their report on psychopathology cannot be precisely assigned. Their experience is too much covered by drug experience. The young people experience an overcoming of their weakness by the substances mentioned; but it is a "game with the fire," which too much self-fragmentation promotes. As a result, abstinence efforts are the decisive first step in prevention.

*Type IV* Smallest Group
: The psychosis suddenly breaks out without clearly recognizable signs. The patients were not very noticeable, were often also very successful in school, or had completed an academic education. At any time—without external events being reported—the sudden outbreak of a schizophrenic psychosis, usually with paranoid hallucinatory symptoms, occurs. As a rule, these young people are not included in the preventive examinations.

## 10.10 Schizoaffective Psychoses

Angst (1980, 1986, 1987) from Zurich has scientifically and clinically worked out the definition and importance of the schizoaffective psychoses in such a way that they were adopted into all diagnosis systems (DSM, ICD). With regard to the subgroup differentiation of the heterogeneous syndromes, which are summarized under the term schizoaffective psychosis, psychodynamically two main aspects stand in a mutually changing mixture ratio. On the one hand, the manic depressive, bipolar component with the changing affectivity, the drive, and the dynamics is in the foreground. On the other hand, it is the schizophrenic component in which psychic structures can be disintegrated, self-fragmentation occurs, and counter-regulation can be observed. Family studies suggest a clear genetic link between bipolar I disorder, schizoaffective disorder, and schizophrenia.

### 10.10.1 Neuroimaging Studies in Schizoaffective Psychoses

Most of the neuroimaging studies are to be found on schizophrenic, affective, and bipolar patients. There are not many studies investigating the schizoaffective group separately. Cosgrove and Suppes (2013) report: "For DSM-5, existing nosological boundaries between bipolar I disorder and schizophrenia were retained and schizoaffective disorder preserved as an independent diagnosis since the biological data are not yet compelling enough to justify a move to a more continuously neurodevelopmental model of psychosis."

Madre et al. (2016) who reported a systematic review of neuropsychological and neuroimaging underpinnings for schizoaffective disorder found that neurocognitive and neuroimaging abnormalities in schizoaffective disorder resemble more schizophrenia than bipolar disorder. This is suggestive for schizoaffective disorder being a subtype of schizophrenia or being part of the continuum spectrum model of psychosis, with schizoaffective disorder being more skewed toward schizophrenia than bipolar disorder.

Amann et al. (2016) who investigated brain structural changes in schizoaffective disorder compared to schizophrenia and bipolar disorder come to the conclusion that comparing each patient group with the healthy control subjects that the patients with schizoaffective disorder and the patients with schizophrenia showed widespread and overlapping areas of significant volume reduction, but the patients with bipolar disorder did not.

### 10.10.2 Clinical Phenomenology

The intermediate-type "schizoaffective psychosis" has a number of advantages over schizophrenic diseases. Lithium, carbamazepine, valproate, and lamotrigine can be used as prophylactic agents. As Tsuang et al. (2000), Marneros et al (1992), Angst (1986), Harrow and Grossman (1984), and Samson et al. (1988) have investigated, their long-term prognosis is better than in the group of schizophrenics and worse than the group of affective psychoses. This depends presumably on the increased neuropsychodynamic structural level of the patients. In clinical observation in group psychotherapy, in which schizophrenic and schizoaffective patients are treated together, it can be observed that the latter have more solid ego strength than the schizophrenics (Hartwich and Grube 2000).

The long-term care of psychiatric patients also shows that some patients who were initially schizodominant are later transformed into more affective-dominant forms, and others later experience only maniform phases. From a psychodynamic point of view, it seems to be a good idea to distinguish between three different subtypes, based on Levitt and Tsuang (1988):

- Schizodominant
- Affective-dominant depressive
- Affective-dominant maniform

In all three subtypes, it is crucial that both the schizophrenic and affective components are perceived psychodynamically in the respective different weightings at the same time or at different times. The interdependence of schizophrenic and affective components causes complex patterns that manifest themselves in the symptoms and the patient's condition.

#### 10.10.2.1 Psychodynamic Interaction of the Schizo-Dominant Type

A 49-year-old patient has been suffering from recurrent schizoaffective episodes for more than 20 years, which are a mixture of schizophrenic symptoms and manic or also severely depressive symptoms. Now the family is confronted with the fact that he is withdrawing, talking increasingly incoherent, and committing a series of mistakes in the household. From the family, he has to be pushed to the simplest tasks like body care and eating. There are cognitive impairments in the form of severe attentional disorders.

The autistic withdrawal is the first attempt to escape from further fragmentation of his self. However, this is not enough. He seeks in Albert Einstein's writings an overall formula for a harmonious and peaceful social life all over the world. Now he succeeds in provoking an affective component (Kohut and Wolf et al. 1980). He "beams" himself to other continents to spread his ideal formula in a maniform behavior.

In a later psychodynamic working through of his psychotic experience, he tells us that he actively allows himself somehow to be overwhelmed by maniform toned affection and thereby shields himself from the external world. He cultivates his grandiose emotions and seems to be able to protect himself from further disintegration. He tries to stay at home and postpone admission to the psychiatric hospital as long as possible. He uses antipsychotic medication and lithium for treatment and prevention. In his numerous psychopathological fluctuations, he always prefers the same pattern as described above. Occasionally he must end up in the hospital when his counter-regulation attempts in the form of a maniform expansive paraconstruction, in which he experiences himself as the savior of the threatened world, is not sufficient.

##### Neuropsychodynamic Understanding

From the perspective of self-help, the patient has learned to develop his own protective mechanisms. The counter-regulation by enhancing grandiose emotions to protect his spatiotemporal self from disintegration on the phenomenal level seems to correspond with some new findings on the neuronal level. Martino et al. (2016) investigated patients who suffered from depressive and manic phases. The result of their study was that the contrasting symptoms of depression and mania may be related to opposite spatiotemporal patters in the resting state structure. The topographical balance between the default mode network (DMN) and the sensorimotor network (SMN) was increased in depression and decreased in mania. The subjectively intended shift of the balance toward DMN could be a stabilizing factor of the self.

However, the patient runs the risk that he cannot always sufficiently modulate his affective forces to avoid disintegration. Hering (2004, p. 98) presents a remarkable psychodynamic hypothesis (we should better say image or picture) concerning the interaction and symptom oscillation of the two dimensions: schizophrenic and affective. He starts from Kohut's concept of the vertical splitting, interpreting the two areas of the self, the healthy and the psychotic one, as separated by a "wall." The eruptive forces of the psychotic catastrophe are the panic experience of decay and dissolution. The eruptive forces push against the vertical barrier, tear it down, and flood the entire psyche.

On the neuronal level, that would happen in case there is not only a shift of the balance in the direction of the DMN but also a disbalance which leads to hyperconnectivity in the cortical midline structure of the brain which may correspond with schizophrenic self-fragmentation on the phenomenological level.

The idea behind this is that the neuronal virtual spatiotemporal structure does not correspond properly with the phenomenal spatiotemporal structure of the self-experience, so that psychoanalytically speaking, schizophrenics suffer from self-fragmentation (Kohut 1973; Kohut and Wolf 1980; Kernberg 1978) which may be seen as a trans-phenomenal concept. The "trans-phenomenal level targets the implicit yet operative matrix that underlies these anomalous subjective experiences" (Northoff 2016).

#### 10.10.2.2 Psychodynamic Interaction of Affective-Dominant Depressive Type

**Case Study**

Schizoaffective patient affective-dominant depressive type history and course: A 40-year-old female patient is currently being treated as an inpatient. Her disease has existed for 11 years. Because of schizoaffective episodes, she was so far eight times hospitalized. Now she is emotionally frozen in her depression and without any motivation. Before the admission, she lost her 8-year-old child to a caring family. In her current psychopathological condition, she does not recover from her emotional paralysis for weeks, despite careful low dosage of antidepressant medication. We see in her symptom of solidification a schizo-depressive paraconstruction with the function and the sense to protect the patient from disappointment, anger, and grief to the loss of her child. A stimulation of these emotions would probably become so strong that she would contribute to fragmentation and the formation of schizophrenic symptoms, which we have seen in prior relapses. The paraconstruction of the solidification, which can be compared with a "play dead reflex," is to be understood and taken into consideration in the therapy. This means, for example, that too high and too fast antidepressant medication would mobilize previously solidified emotions, which would expose the current structural level of the patient to a psychotic fragmentation and the danger of self-dissolution. In the psychotherapeutic support, it is necessary to contribute to the condition and slowly to catalyze the approaches to mourning.

#### Neuropsychodynamic Understanding

From a psychodynamic point of view, the affective force is a protective function in schizoaffective patients in the symptom oscillation between more object-related affective and more self-referenced schizophrenic constellation, which can manifest itself in deep depressive moods and solidification of the patient's feelings. This cannot be referred only to the pole of the maniform elevated states of mood. Not infrequently, in the process of negating, there is an enormous force which can provide protection against the disruption of the spatiotemporal construct of the self. A schizoaffective disease described by Hering (2004, p. 31), which showed a form of self-protective negation, was also mentioned: He had experienced himself as *as insensitive as a stone* and considered psychotherapy as useless because one should not deal with a worthless stone. It was only after the therapeutic intervention that something precious in the stone could be found that was slowly thawed and the rigidity was more flexible in its movements. Thus, the petrified feeling had previously a protective function.

### 10.10.2.3 Neuropsychodynamic Interaction of the Affective-Dominant Maniform Type

#### Case Study

A 40-year-old patient has been suffering from recurrent schizoaffective episodes for 12 years, which, despite phase prophylactic medication, occasionally lead to relapses which are so severe that he has to be hospitalized. In the meantime, he has more stable periods in which he is able to hold out and work in his profession. Whenever a depressive suction threatens and begins to take him down, he tries to tackle it. For this, he has developed various practices to "recharge," as he says.

For him, books with religious content, as well as "spiritual persons" that he visits, have a "trigger effect." In these times, he has set up his alarm clock 1–2 h before the usual time to wake up. Then he fills himself up with energy by reading these writings and taking religious rituals. Thereby he is gaining the strength for the day and can do his professional work. Nevertheless sometimes psychotic episodes occur, which he tolerates. Sometimes he loses the relation to the reality entirely and increases himself into his own overvaluation, and then he says, "I am God." In the sense of Kohut (1973), he achieves a mobilization of his narcissistic grandiose fantasies whose archaic forces provoke such a dangerous overstimulation of the ego that self-fragmentation takes place.

In his effort to escape the depressive suction, he creates a maniform-spiritual fantasy world, which elevates his self-esteem and stabilizes him. At the same time, however, the danger grows so that his structure is overloaded and schizophrenic fragmentations with the corresponding symptoms occur. In the longitudinal course, one could observe that the mixing of the schizophrenic and the affective dimension changed again and again. Sometimes his psychodynamic "charges" were successful or sometimes the dispositional weakness of the structure could not endure this kind of stimulation.

### 10.10.3 Therapy of Schizoaffective Psychoses in a Neuropsychodynamic Perspective

It is often worthwhile to look closely at the possibilities that psychotic patients find in order to protect themselves from disintegration, self-fragmentation, and deep depressive mood. Thus, in the special case of the schizoaffective psychoses, it is also necessary to understand how their affective dynamics are able to exert a binding strength (bonding strength, Bindungsstärke) that can compensate for their structural weakness. For neuropsychodynamic therapy approaches, it is therefore of crucial importance to develop the affective bonding energy and its positive binding power, which is capable of preventing self-fragmentation, and to use the protective quality in psychotherapy.

One of the ways we see is the *power* of *creativity* that is accessible to many people and can be systematically and well directedly enhanced.

People who work creatively or artistically and who play music, write poetry, paint, or sculpture all know the intense mobilization of their inner drive dynamics. "There is a state of passion and emotion, sometimes a state of ecstasy, which carries you away and let you forget the environment" (Hartwich 2012, p. 59).

In addition, we can learn from schizoaffective artists how to successfully protect themselves against psychotic relapses with their creative forces (Fryrear 2002). According to our hypothesis, in addition to antipsychotics, antidepressants, prophylactic medication, and psychotherapy, we also use creative methods in the therapy of schizoaffective psychoses to additionally use their own therapeutic potential. However, the therapeutic use of creative methods is not limited to specific diagnoses (see Hartwich and Fryrear 2002a, b and also Chap. 24 in this book).

The psychodynamics of the schizoaffective psychoses with their subtypes differs from the schizophrenia with regard to the affective dimension. Consequently, the treatments are to be set differently. The positive bonding force, which lies in affective energy with its protective quality against disintegration and fragmentation of the spatiotemporal structure of the self, can be used in neuropsychodynamic psychotherapy. However, it is necessary to slow down an excess of energy, which can be destructively manifested in the manic mood, at an early stage. The counter-regulations and the self-healing shown in the examples (see above) also help us to recognize the therapeutic potential and to use them in the treatment of schizoaffective psychoses. This is particularly true of procedures that encourage the creativity. In Chap. 16, we will describe the neuronal system in different brain areas, which seems to be responsible for that what we experience creativity and how this may be taken into a neuropsychodynamic context.

## 10.11 The Neuropsychodynamic Approach to Psychotherapy on Schizophrenia and Other Psychoses

In the past, classical psychoanalytical understanding and treatment of the illness by Freud's followers was found *not* to be successful enough, especially not in severe psychotic inpatients, their classical psychoanalysis failed. Our focus is to explain that many results of neuroimaging studies can be seen in relation to psychopathological symptoms and can even lead to new aspects of a neuropsychodynamic/neuropsychoanalytic approach which we call neuropsychodynamic psychotherapy.

The psychopathological symptoms are understandable in a psychological way and also from a somatic perspective. Therefore, any treatment should consider both aspects, e.g., psychotherapeutic methods and antipsychotic medication. It is by no means a matter of opting for one or the other therapy strategy. The key point is to combine bottom-up and top-down methods effectively in their interaction. The weighting of the respective components involved in the disease must be taken into account. Weighting is so important that it guides us with regard to the emphasis on therapeutic methods. This is true for the severity of the symptoms and the correlated neuronal aberrations on the one hand. On the other hand, psychological environmental factors can also cause neuronal aberrations; a good example is the folie à deux, which is described above.

Our emphasis is to find treatment methods which are able to connect self-fragments together or in other words which strengthen or even restore the spatiotemporal structure of the self. The bottom-up version would be to influence the spatiotemporal neuronal prerequisite of the possibility of the self-experience, e.g., with antipsychotic medication and other somatic therapies, while the top-down version would be to use psychotherapeutic and psychodynamic methods, including, e.g., creative therapies and self-mirroring, which may have an additional influence on neuronal activity in CMS (cortical midline structures).

### 10.11.1 Some Principles of Neuropsychodynamic Therapy

Not only for the group of those who belong to the early high-risk adolescents but also for acute and chronically schizophrenic patients, some basic neuropsychodynamic considerations and procedures will be described. Here, the basic principles of both individual and group psychotherapy will be presented. An important consideration in the treatment of the symptoms should be considered. Usually, it is about wanting to eliminate the symptoms like delusion, hallucinations, and other ego disorders quickly. Antipsychotics succeeds in many cases rapidly, in other cases usually only imperfectly. Although the neuropsychodynamic treatment of the psychoses is on the basis of psychopharmaceuticals, the treatment with antipsychotically effective drugs should be cautious from today's point of view, as in recent years, as Aderhold et al. (2015) have found in a meta-analysis evaluation of about 30 studies in this area which have shown evidence of a correlation between antipsychotic medication and their dosages and an additional reduction in the volume of the brain, which affects white and gray matter primarily in the frontal region of the brain. Additional means: there is a reduction to be found on schizophrenics, but those who did take antipsychotic medication over long time, the reduction was additionally worse.

For those affected, this can be associated with an increase in cognitive disorders.

From a neuropsychodynamic point of view many symptoms, not each, have a *meaning* and somehow a function in the present psychotic state. The description of the "functionality of dysfunctionality" in the sense of Mentzos (2011) is to be recalled here. That means, for example, a delusion or a hallucination to be interpreted as a paraconstruction (Hartwich 1997) which the patient is currently required in order *not* to get an even stronger inner dissolution in the sense of a more severe and dangerous self-fragmentation. Also the delusion and the hallucination are experienced in their subjective certainty as *part of their own self* and usually not as something foreign. Consequently, it is not surprising that the patient holds on firmly to his symptom. This is intended to make it clear that it is not in the foreground to directly "dispose" the symptoms psychotherapeutically. Rather, it is a matter of tackling what lies beyond or below the symptoms: the structural weakness that accompanies the risk of fragmentation and loss of cohesion, as well as the conflicts and "dilemmas" (Mentzos 2011), which have led to the aforementioned symptoms. This is the subject of neuropsychodynamic psychotherapy, which should influence positively the expressions of trans-phenomenal matrix. In the search for such influences, it is important to first orient one to the *autotherapeutic* attempts of the patients themselves. In doing so, we come across the patients' own efforts to increase their social contacts, how they want to intensify their body experience and how they use many kinds of creativity. Such autotherapeutic strategies should be included and, as far as possible, enhanced by the therapist.

In therapy sessions, as a rule, not an abstinent attitude, like traditional psychoanalytic setting, is advanced, but often an active structural strengthening and maturation promotion of the personality are striving. Much slower, more cautious, and more structured than in psychoreactive diseases, conflicts and reproductions are worked through in the transference, if it is already possible. Such interpretations are fruitful only to the extent that the persons concerned are capable of symbolizing; *concretistic* ideas cannot be interpreted, since the ability to abstract and to transmit meaning are usually too strongly restricted.

In people with risk factors who are not or not yet psychotic, interpretations can be more possible and appropriate than in the case of a manifest schizophrenia. When working with dreams, it must also be taken into account that the therapist is sometimes offered "dreamed" by the patient in order to be able to report indirectly about psychotic experiences, since the person concerned often has a fear of directly communicating them. This can be because the patient tests the therapist to what extent he can enter the psychotic sphere of experience and shares it with the patient without the usual reflex of increasing the antipsychotic medication.

However, what has not yet been sufficiently substantiated and empirically tested are the differently pronounced weightings of the many influencing factors in the *individual* disease fate. However, as therapists, we are always treating individual patients, and we are working here on an incompletely secured terrain as far as empirical science is concerned. Therefore, we depend on a certain degree of creative skillfulness in our experience and intuition. The severity of the weights of the respective components involved, whether more somatic-genetic or more psychological, must be detected in each individual case, the complexity of which is to be assessed in its various interactions, and thus the strength or fragility of the different structural levels of the individual patient or at least subgroups should be appraised. Since it has not yet been possible to measure the respective weightings of the individual components in their formative interactions scientifically for the *individual* case in a current situation or even in the course of time, we should assume that there are an infinite number of variations with subsequent psychotic symptoms, which are summarized as "schizophrenia or group of schizophrenia," more or less including schizoaffective psychoses. Since we, as therapists, usually work with individual patients, we should need a separate analysis of the intraindividual weighting patterns and their interactions with the environment in order to be helpful in the professional and personal encounter with the patient.

#### 10.11.1.1 Modification of Psychoanalytical Treatment Strategies

The term "modification" has been used to characterize the difference from classical traditional analytical approaches with regard to the special needs of people suffering from psychoses. This is about the setting: the patient and the therapist sit opposite; the frequency of the therapy setting is more variable. The therapist is more active with regard to promoting a positive transference, at least initially. He helps with the restructuring process, is reluctant to interpret unconscious material, observes the action dialogue, and is constantly alert, regarding the "threatening self-fragmentation" (Hering 2006) or "organismic panic" (Volkan 1994) and the "loss of the continuity of the self" (Pao 1979). Alanen (1997, 2001) describes in detail the "need-specific" treatment of schizophrenic psychoses.

**Modification of Psychoanalytical Treatment Strategies**

- Patient and the therapist sit opposite.
- In the setting, the therapist is more active.
- Time and setting frequency are more variable.
- Help with the restructuring of the self.
- Interpretations only restrained and rare.
- To be aware of the countertransference: the self-fragmentation risk of the patient.

The severity of the respective components involved, whether more somatic or more psychological, must be ascertained and detected in an individual case, and their complex interactions should be assessed, thus indicating the severity of the different structural levels of the individual psychosis. This is crucial for indication, non-indication, and contraindication, as well as the extent of the modification of psychoanalytic methods in therapy (Fig. 10.3).

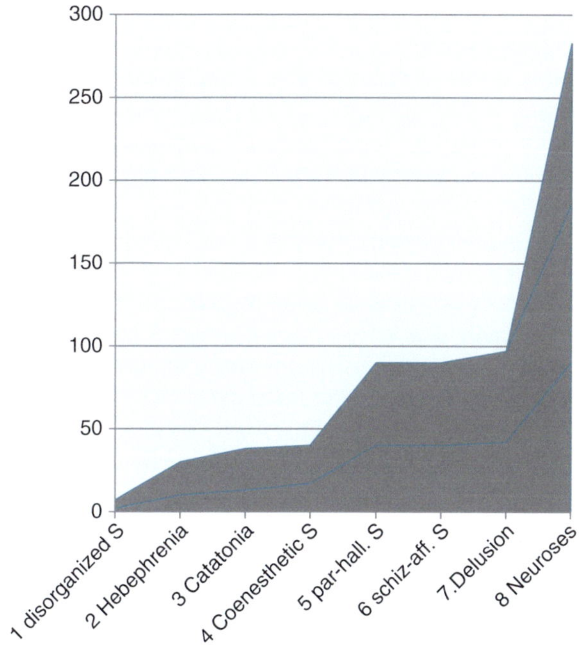


**Fig. 10.3** Assessment of structural strength/weakness in psychoses compared with neuroses from low (0) to high (300) strength (Hartwich 2013)

#### 10.11.1.2 Countertransference and Supervision

There are some important aspects to consider in the countertransference: the therapist cannot feel completely empathic to the experienced psychopathological phenomena of cognitive disturbances such as thought interferences, thought-break, attention deficit, and loss of the spatiotemporal self-experience. Here, he should always realize that his empathy is limited. If the therapist forgets that there are no complete common "experience accordance," he runs the risk of unconsciously fooling a "pseudo-understanding." Because our patients are usually quick to see if the therapist is truly authentic, they experience distance and isolation in such a therapeutic situation. An early relationship disturbance which may go back to the childhood can then be like remake in the transference. The therapist who observes and reflects these processes can perceive the opportunity and try to work with the patient at his early relationship constellations. If he does not reflect this, he runs the risk of his own impatient and occasionally annoying countertransference. Consequently, supervision is often helpful for the positive outcome of such therapies.

#### 10.11.1.3 Group Psychotherapeutic Procedures in Day Clinic and In- and Outpatient Treatment

In the treatment of neuropsychodynamically oriented group psychotherapy, the following basic characteristics can be seen as therapeutic effective and empirically supported (e.g., Hartwich and Schumacher 1985; Kanas 1986; Schwarz and

Matussek 1990; Schwarz 2001) for risk persons, acute and post-acute, as well as chronic schizophrenic patients:

In group therapy, social contact and stress training will always take place. If we assume that stressful experiences can have a strong relevance for the outbreak of a psychosis and for relapses, then it is a matter of stepping up the handling of comparable emotional events in the group therapy and exercising under the control so that the patient slowly learns adapt stress more and more.

In a group session, individual traumatizations can be processed psychodynamically from the biography of the individual members, and the person learns additionally from the experience of the other group participants. The question is still to be clarified whether an individual need-oriented approach is better suited to the individual patient than a program (manual) written prior for group therapy. The approach presumably depends on the level of training experience of the group leader. The beginner will rather prefer a concept in the form of a manual; the experienced will have a variety of therapeutic possibilities of variation in mind and can use them in the appropriate situation.

A further important aspect is the *group cohesion*, also referred to as the "we-feeling," which counteracts the isolation of the persons. Group cohesion can have an indirect positive effect on individual cohesion of the spatiotemporal structure of the self. In some group sessions, however, there is also a contradictory group dynamic. If more than one participant is very autistic, the forming of cohesion is prevented. In such a constellation, the therapist has the chance to recognize the discoherent fragments that are manifested in group dynamics as images of the self-fragmentation of the individual. Then he should act more structuring and protective. He only succeeds in doing so if he himself has learned to endure the self-fragmentation, which offers him as "chaos" in the group session, and he can learn "chaos ability."

In some situations, the group dynamic may develop a high degree of emotionality, which may be dangerous for the patients; but in a controlled way, it can contribute to the exercise, to be able to cope with emotional overloads and to generalize this to their everyday life.

In advanced and longer-running psychodynamic group therapies, there is the chance to work through the personal experiences, which may go back to the childhood of the patient with reference persons and life events. As in the case of individual therapy, transference and projections could be worked through referring to current situations and persons. The group situation then presents a favorable prerequisite for making conscious of the relationship between current conflicts and previous constellations. In our group therapies with schizophrenic and schizoaffective psychoses, we have made the same observations as Gabbard (2014), who writes: "For the patient who is stabilized on medication, weekly sessions of 60–90 min can serve to build trust and can provide a support group where patients can freely discuss concerns such as how to manage auditory hallucinations and how to deal with stigma of mental illness." However, this only applies if homogenous groups are formed; not included are affective disorders; personality disorders, in particular borderline patients; and organically caused disorders.

When group procedures are concerned in the broader sense, it is necessary to point out the promotion of creative processes (Hartwich and Fryrear 2002b), for example, painting in the group or painting with the help of a computer program (Hartwich and Brandecker 1997), sculpting with a hard stone for structuring (Hartwich and Weigand-Tomiuk 2002), playing music in the group actively and passively, and writing poems and stories and presenting them in the group. All in all, creative therapy methods have a wide field for creative ideas.

#### 10.11.1.4 Neuropsychodynamic Approach: One Possible Strategy: Compensation

Compensation would mean enhancing and increasing neuronal mechanisms which may be able to compensate aberrant connectivity, e.g., with creative therapeutic methods (Arieti 1974, 1976; Hartwich and Fryrear 2002a, b). When a person is in a state of creativity, enhanced neuronal activity in the temporal and parietal regions can be found, sometimes with a sudden increase of high frequency (summarized by Kandel 2012). Therapists may be able to strengthen these neuronal activities using therapeutic methods which stimulate and unfold creativity with the help of expressive media such as painting, sculpture, music, movement, and poetry. We postulate that such creative therapies could modify the hyperconnectivity in a positive way. In the future, more specific nuances in creative therapies could be investigated on the basis of neuronal compensation, which refers to the interaction and relation between the phenomenal and neuronal level.

Another compensatory mechanism could involve self-related processing (Northoff 2011, 2014a, b; Northoff et al. 2009). Neuroimaging studies have demonstrated that words and pictures that are highly related to the individual's self are considered to be more emotional than those that show a rather low degree of self-relatedness. On the neuronal level, a meta-analysis (Northoff et al. 2006) underlines that there is a concentration of neuronal activation in the cortical midline structures, the premotor, and the bilateral parietal cortex. Huang et al. (2016) showed that the MPFC region (medial prefrontal cortex) is closely linked to self-related processing.

In psychopathology, the mirror phenomenon is well known as "signe du miroir." This was first described by the French psychiatrists Delmas (1929) and Abély (1930). Some schizophrenic patients look in the mirror for hours in order to "find themselves"; this can be interpreted as an "autotherapeutic habit" (Scharfetter 1986) to stabilize themselves when there is a danger of deterioration. On the phenomenal level and even on the neuronal (prephenomenal) level, they may unconsciously use the activation arising from self-related processing. This corresponds with empirical studies about video mirroring in schizophrenics which showed an increase of ego strength in experimental investigations (Hartwich, Lehmkuhl 1979). The neuronal activation coming from the self-related processing could be enhanced using audio-visual mirroring systematically. There may be a positive influence on the abnormal resting state activity and its imbalance in interaction by improving the spatiotemporal structure of the self also on the phenomenal level.

#### 10.11.1.5 Neuronal Aspects of Countertransference

Countertransference, especially in the treatment of psychotic patients, should be given a broader definition. "This definition serves to attenuate the pejorative connotation of countertransference—unresolved problems in the treater that require treatment—and to replace it with a conceptualization that views countertransference as a major diagnostic and therapeutic tool that tells the treater a good deal about the patient's internal world"(Gabbard 2014). For the therapist working with schizophrenics, the usual countertransference is impatience, fear, anger, and feelings of distance. Often there is a countertransference obstacle, because the therapist has to bear the psychotic disintegration of his patient. In this case, he should learn what we call "chaos ability" (Hartwich 2007). He can learn this when he has worked with many serious ill psychotic inpatients. It may be considered normal that the therapist wants to protect himself from his own self-fragmentation; therefore, he builds countertransference resistance. When he becomes aware of this in his psychodynamic setting, he is more likely to realize the patient's danger of self-fragmentation at an early stage.

Neuropsychodynamic therapists can benefit by bearing *in mind the imagination* of normal and abnormal neuronal events in the brains of their psychotic patients, e.g., hyperconnectivity, DMN-CEN imbalance, and other dysfunctions in the neuronal network. The danger of their own self-fragmentation will then be reduced. By considering not only the phenomenal but also the neuronal (prephenomenal) perspective, the therapist's paradigm can be changed, e.g., resulting in a reduction of his countertransference resistance. This being the case, more therapists could attempt to treat schizophrenics and other psychoses neuropsychodynamically.

## References

Abély P. Le signe du miroir dans les psychoses et plus spécialement dans la démence précoce. Ann Med Psychol. 1930;88:28–36.

Aderhold V, Weinmann S, Hägele C, Heinz A. Frontale Hirnvolumenminderung durch Antipsychotika? Nervenarzt. 2015;86:302–23.

Alanen YO. Schizophrenia. Its origins and need-adapted treatment. London: Karnac books Ltd.; 1997.

Alanen YO. Schizophrenie. Stuttgart: Klett-Cotta; 2001.

Amann BL, Canales-Rodriges EJ, Madre M, et al. Brain structural changes in schizoaffective disorder compared to schizophrenia and bipolar disorder. Acta Psychiatr Scand. 2016;133(1):23–33.

Angst J. Verlauf unipolar depressiver, bipolar manisch-depressiver und schizoaffektiver Erkrankungen und Psychosen. Ergebnisse einer prospektiven Studie. Fortschr Neurol Psychiat. 1980;48:3–30.

Angst J. The course of schizoaffective disorders. In: Marneros A, Tsuang MT, editors. Schizoaffective psychoses. Berlin: Springer; 1986.

Angst J. Epidemiologie der affektiven Psychosen. In: Kisker KP, et al., editors. Psychiatrie der Gegenwart, vol. 5. Berlin: Springer; 1987.

Arieti S. Interpretation of schizophrenia. 2nd ed. New York: Basic Books; 1974.

Arieti S. Creativity. New York: Basic Books, Inc.; 1976.

Bateson G, Jackson DD, Weakland JH. Towards a theory of schizophrenia. Behav Sci. 1956;1:251–64.

Bateson G, Jackson DD, Haley J, et al. Schizophrenie und Familie. Frankfurt/M: Suhrkamp; 1978.
Benedetti G. Psychodynamik als Grundlagenforschung der Psychiatrie. In: Kisker KP, et al., editors. Psychiatrie der Gegenwart, Grundlagen und Methoden der Psychiatrie Bd I, vol. 1. Berlin: Springer; 1979. p. S 43–90.
Benedetti G. Psychotherapeutische Behandlungsmethoden. In: Kisker KP, et al., editors. Psychiatrie der Gegenwart, vol. 4. Heidelberg: Springer; 1987.
Bleuler E. Die negative Suggestibilität, ein physiologisscher Prototyp des Negativismus, der conträren Autosuggestion und gewisser Zwangsideen. Psychiat Neurol Wochenschr. 1904;6(249–253):261–3.
Bleuler E. Dementia praecox oder Gruppe der Schizophrenien. Leipzig: Deuticke; 1911.
Böker H, Northoff G. Die Entkoppelung des Selbst in der Depression: Empirische Befunde und neuropsychodynamische Hypothesen. Psyche Z Psychoanal. 2010;64:934–76.
Broadbent DE. Decision and stress. New York: Academic Press; 1971.
Chase HW, Moses-Kolko EL, Zevallos C, et al. Disrupted posterior cingulate-amygdala connectivity in postpartum depressed women as measured with resting BOLD fMRI. Soc Cogn Affect Neurosci. 2014;9:1069–75.
Coen SJ. Pathological jealousy. Int J Psychoanal. 1987;68:99–108.
Conrad K. Die Gestaltanalyse in der Psychiatrie. Stud Gen. 1952;5:503–14.
Conrad K. Die beginnende Schizophrenie. Stuttgart: Thieme; 1958.
Cosgrove VE, Suppes T. Informing DSM-5: biological boundaries between bipolar I disorder, schizoaffective disorder, and schizophrenia. BMC Med. 2013;11:127. https://doi.org/10.1186/1741-7015-11-127.
Deligiannidis KM, Sikoglu EM, Scott AS, et al. GABAergic neuroactive steroids and resting-state functional connectivity in postpartum depression: a preliminary study. J Psychiatr Res. 2013;47:816–28.
Delmas FA. Le signe du miroir dans la démence précoce. Ann Med Psychol. 1929;87:227–33.
Doege K, Kumar M, Bates AT, et al. Time and frequency domain event-related electrical activity associated with response control in schizophrenia. Clin Neurophysiol. 2010;121(10):1760–71.
Federn P. Ichpsychologie und die Psychosen. Frankfur/M: Suhrkamp; 1978. (Erstveröff. 1956)
Fenichel O. Beitrag zur Psychologie der Eifersucht. In: Fenichel O, editor. Aufsätze, vol. 1. Freiburg: Walter; 1935. p. 345–60.
Fleck S. The development of schizophrenia: a psychosocial and biological approach. In: Werbart A, Cullberg J, editors. Psychotherapy of schizophrenia: facilitating and obstructive factors. Oslo: Scandinavian Univ Press; 1992. p. S 179–92.
Freud S. Psychoanalytische Bemerkungen über einen autobiographisch beschriebenen Fall von Paranoia (Dementia paranoides) mit Nachtrag. In: GW, vol. 8. 6th ed. Frankfurt/M: Fischer; 1909–1913. p. S 239–320.
Freud S. Psychoanalytic notes on an autobiographical account of a case of paranoia, vol. 12. Standard ed. London: Hogarth Press; 1911.
Freud S. Neurose und Psychose. In: GW, vol. 13. 8th ed. Frankfurt/M: Fischer; 1920–1924. p. S 385–91.
Freud S. GW, vol. 14. Frankfurt/M: Fischer; 1925–1931. p. S 299.
Freud A. The ego and the mechanisms of defense (1936), in the writings of Anna Freud, vol. 2. revised ed. New York: International Universities Press; 1966.
Fromm-Reichmann F. Notes on the mother role in the family group. Bull Menn Clin. 1940;4:132–45.
Fryrear JL. The psychotic life of artist Dot Gori as told through words and art to Jerry Fryrear. In: Hartwich P, Fryrear JL, editors. Creativity: the third therapeutic principle in psychiatry. Sternenfels: Wissenschaft & Praxis; 2002.
Fusar-Poli P, Deste G, Smieskova R, et al. Cognitive functioning in prodromal psychosis: meta-analysis of cognitive functioning in prodromal psychosis. Arch Gen Psychiatry. 2012;69:562–71.
Gabbard GO. Psychodynamic psychiatry in clinical practice. 5th ed. Washington DC: American Psychiatric Publishing; 2014. p. 21.
Harrow M, Grossman L. Outcome in schizoaffective disorders: a critical review and reevaluation of the literature. Schizophr Bull. 1984;10:87–108.

Hartwich P. Schizophrenie und Aufmerksamkeitsstörungen. In: Zur Psychopathologie der kognitiven Verarbeitung von Aufmerksamkeitsleistungen. Berlin: Springer; 1980.

Hartwich P. Schizophrenien, kognitive Gesichtspunkte. In: Kisker KP, et al., editors. Psychiatrie der Gegenwart. 3rd ed. Berlin: Springer; 1987.

Hartwich P. Die Parakonstruktion: eine Verstehensmöglichkeit schizophrener Symptome. Vortrag Frankfurter Symposion: Schizophrenien—Wege der Behandlung. In: Hartwich P, Pflug B, Herausgeber. Schizophrenien—Wege der Behandlung. Sternenfels: Wissenschaft & Praxis; 1997.

Hartwich P. Wahn—Sinn und Antikohäsion. In: Hartwich P, Barocka A, editors. Wahn: definition, Psychodynamik, Therapie. Sternenfels: Wissenschaft & Praxis; 2004.

Hartwich P. Schizophrenie. Zur Defekt- und Konfliktinteraktion. In: Böker H, editor. Psychoanalyse und Psychiatrie. Heidelberg: Springer; 2006a. p. S 159–79.

Hartwich P. Schizophrene Prodromalzustände: Gibt es unterschiedliche Typen? Wie sind sie psychodynamisch zu verstehen und zu behandeln. In: Juckel G, Lempa G, Troje E, editors. Psychodynamische Therapie von Patienten im schizophrenen Prodromalzustand, Forum der psychoanalytischen Psychosentherapie, vol. 13. Göttingen: Vandenhoeck & Ruprecht; 2006b.

Hartwich P. Psychodynamisch orientierte Therapieverfahren bei Schizophrenien. In: Hartwich P, et al., editors. Schizophrene Erkrankungen. Sternenfels: Wissenschaft & Praxis; 2007. p. 33–98.

Hartwich P. Bildhauerei mit psychotisch Kranken. Die Bedeutung von Kreativität und Parakonstruktion. In: Mentzos S, Münch A, editors. Das Schöpferische in der Psychose, Forum der psychoanalytischen Psychosentherapie, vol. 28. Göttingen: Vandenhoeck & Ruprecht; 2012. p. S 56–70.

Hartwich P. Zu Indikation und Kontraindikation für psychoanalytisch modifizierte Behandlung der verschiedenen Schizophrenieformen, Forum der psychoanalytischen Psychosentherapie, vol. 29. Göttingen: Vandenhoek & Ruprecht; 2013. p. S 111–52.

Hartwich P, Brandecker R. Computer-based art therapy with inpatients: acute and chronic schizophrenics and borderline cases. The Arts in Psychotherapy. 1997;24(4):367–73.

Hartwich P, Fryrear JL. Einführung. In: Hartwich P, Fryrear JL, editors. Kreativität—Das dritte therapeutische Prinzip in der Psychiatrie. Sternenfels: Wissenschaft & Praxis; 2002a.

Hartwich P, Fryrear JL. Creativity, the third therapeutic principle in psychiatry. Sternenfels: Wissenschaft & Praxis; 2002b.

Hartwich P, Grube M. Psychodynamische Aspekte bei der Behandlung schizoaffektiver Psychosen. In: Böker H, editor. Depression, Manie und schizoaffektive Psychosen. Gießen: Psychosozial-Verlag; 2000.

Hartwich P, Grube M. Psychotherapie bei Psychosen. Neuropsychodynamisches Handeln in Klinik und Praxis. 3rd ed. Berlin: Springer; 2015.

Hartwich P, Lehmkuhl G. Audiovisual self-confrontation in schizophrenia. Arch Psychiat Nervenkr. 1979;227:341–51.

Hartwich P, Schumacher E. Zum Stellenwert der Gruppenpsychotherapie in der Nachsorge Schizophrener. Eine 5-Jahres-Verlaufsstudie. Nervenarzt. 1985;56:365–72.

Hartwich P, Weigand-Tomiuk H. Bildhauerei mit Marmor in der Psychiatrischen Klinik. In: Hartwich P, Fryrear JL, editors. Kreativität—Das dritte therapeutische Prinzip in der Psychiatrie. Sternenfels: Wissenschaft & Praxis; 2002.

Heimann H. Karl Wilhelm Idelers "Versuch einer Theorie des religiösen Wahnsinns"—nach 100 Jahren. Bibl Psychiat Neurol. 1957;100:68–78.

Hering W. Schizoaffektive Psychose. In: Psychodynamik und Behandlungstechnik. Göttingen: Vandenhoeck & Ruprecht; 2004.

Hering W. Psychodynamische Aspekte der schizoaffektiven Psychosen. In: Böker H, editor. Psychoanalyse und Psychiatrie. Heidelberg: Springer; 2006. p. S 181–91.

Holt DJ, et al. An anterior-to-posterior shift in midline cortical activity in schizophrenia during self-reflection. Biol Psychiatry. 2011;69(5):415–23.

Huang Z, Obara N, Davis H, Pokorny J, Northoff G. The temporal structure of resting-state brain activity in the medial prefrontal cortex predicts self-consciousness. Neuropsychologia. 2016;82(2016):161–170162.
Huber G, Penin H. Elektroenzephalogramme. Korrelationsuntersuchungen bei Schizophrenen. Fortschr Neurol Psychiat. 1968;36:641–59.
Ideler KW. Der religiöse Wahnsinn. Halle: Schwetschke; 1847. p. 11.
Jaspers K. Allgemeine Psychopathologie. 6th ed. Berlin: Springer; 1953.
Javitt DC. When doors of perception close: bottom-up models of disrupted cognition in psychosis. Annu Rev Clin Psychol. 2009;5:249–75.
Jones E. Jealousy. In: Jones E, editor. Papers of psychoanalysis. Boston: Beacon; 1967.
Jung R. Neurophysiologie und Psychiatrie. In: Gruhle HW, et al., editors. Psychiatrie der Gegenwart. I,1, Teil A. Berlin: Springer; 1967. p. S 325–928.
Jung CG. Über die Psychogenese der Schizophrenie. In: GW, Psychogenese der Geisteskrankheiten, vol. 3. Walter: Olten; 1979.
Jung R. Neurophysiologie und Psychiatrie. In: Kisker KP, et al., editors. Psychiatrie der Gegenwart. I, Teil 2. 2nd ed. Berlin: Springer; 1980. p. S 753–1103.
Kanas N. Group therapy with schizophrenics: a review of controlled studies. Int J Group Psychotherapy. 1986;36:339–51.
Kandel E. Das Zeitalter der Erkenntnis. Die Erforschung des Unbewussten in der Kunst, Geist und Gehirn von der Wiener Moderne bis heute. München: Siedler; 2012.
Kernberg OF. Borderline-Störungen und pathologischer Narzissmus. 2nd ed. Frankfurt am Main: Suhrkamp; 1978.
Klages W. Über Störungen des Raumgefühls bei Schizophrenen. Gedanken zu einer Paraordnung. In: Broekmann JM, Hofer G, editors. Die Wirklichkeit des Unverständlichen. Den Haag: Martinus Nijhoff; 1974. p. S. 241–7.
Klein M. New directions in psychoanalysis. New York: Basic Books; 1956.
Klosterkoetter J. Predicting the onset of schizophrenia. In: Haefner H, editor. Risk and protective factors in schizophrenia. Darmstadt: Steinkopff; 2002. p. S 193–206.
Klosterkoetter J. Prävention psychotischer Störungen. Nervenarzt. 2013;84(11):1299–309.
Klosterkoetter J, Hellmich M, Steinmeyer EM, et al. Diagnosing schizophrenia in the initial prodromal phase. Arch Gen Psychiatry. 2001;58:158–64.
Kohut H. Narzißmus. Frankfurt/M: Suhrkamp; 1973.
Kohut H, Wolf ES. Die Störungen des Selbst und ihre Behandlung. In: Peters UH, editor. Die Psychologie des 20. Jahrhunderts, vol. 10. Zürich: Kindler; 1980. p. S 667–82.
Kraepelin E. Psychiatrie. Ein Lehrbuch für Studierende und Ärzte. 3rd ed. Leipzig: Barth; 1889.
Kuhn TS. Die Struktur wissenschaftlicher Revolutionen. 2nd ed. Frankfurt/M: Suhrkamp; 1976.
Kumpf-Tonsch A, Schmid-Siegel B, Klier CM, et al. Versorgungsstrukturen für Frauen mit postpartalen psychischen Störungen—Eine Bestandsaufnahme für Österreich. Wien Klin Wochenschr. 2001;113:641–6.
Levitt JJ, Tsuang MT. The heterogeneity of schizoaffective disorder: implications for treatment. Am J Psychiat. 1988;145:926–36.
Lidz T, Cornelison A, Fleck S. Schizophrenia and the family. London: Tavistock; 1965.
Madre M, Canales-Rodriges EJ, Ortiz-Gil J, et al. Neuropsychological and neuroimaging underpinnings of schizoaffective disorder: a systematic review. Acta Psychiatr Scand. 2016;134(1):16–30.
Mahler MS. Symbiose und individuation. Stuttgart: Klett; 1972.
Maier W, Hawallek B. Neuentwicklung in der Erforschung der Genetik der Schizophrenen. In: Möller HJ, Müller N, editors. Schizophrenie. Wien: Springer; 2004. p. S 63–72.
Marneros A, Deister A, Rohde A. Comparison of long-term outcome of schizophrenic, affective and schizoaffective disorders. Br J Psychiatry. 1992;161:44–51.
Martino M, Magioncalda P, Huang Z, et al. Contrasting variability patterns in the default mode and sensorimotor networks balance in bipolar depression and mania. Proc Natl Acad Sci U S A. 2016;113(17):4824–9.

Mc Ghie A. Attention and perception in schizophrenia. p. 57–96. In: Maher BA, editor. Contributions to the psychopathology of schizophrenia. New York: Academic Press; 1977. p. 1–56.
Mentzos S. Psychodynamische Modelle in der Psychiatrie. Göttingen: Vandenhoeck & Ruprecht; 1991.
Mentzos S. Psychodynamische und psychotherapeutische Aspekte "endogener" Psychosen. In: Hartwich P, Haas S, editors. Pharmakotherapie und Psychotherapie bei Psychosen. Sternenfels: Wissenschaft & Praxis; 1996. p. S 17–29.
Mentzos S. Die "endogenen" Psychosen als die Psychosomatosen des Gehirns. In: Müller T, Matejek N, editors. Ätiopathogenese psychotischer Erkrankungen. Göttingen: Vandenhoeck & Ruprecht; 2000. p. S 13–33.
Mentzos S. Psychodynamik des Wahns. In: Schwarz F, Maier C, editors. Psychotherapie der Psychosen. Stuttgart: Thieme; 2001.
Mentzos S. Lehrbuch der Psychodynamik. Göttingen: Vandenhoeck & Ruprecht; 2011.
Mueller H, Laier S, Bechdolf A. Evidence-based psychotherapy for the prevention and treatment of first-episode psychosis. Eur Arch Psychiatry Clin Neurosci. 2014;264(Suppl 1):17–25.
Mulert C, Leicht G, Hepp P, et al. Single-trial coupling of the gamma-band response and the corresponding BOLD signal. NeuroImage. 2010;49(3):2238–47.
Northoff G. Neuropsychoanalysis in practice. Oxford: Oxford Univ Press; 2011.
Northoff G. Das disziplinlose Gehirn—Was nun Herr Kant? Auf den Spuren unseres Bewusstseins mit der Neurophilsophie. München: Irisiana; 2012.
Northoff G. Unlocking the brain, Coding, vol. I. Oxford: Oxford Univ Press; 2014a.
Northoff G. Unlocking the brain, Consciousness, vol. II. Oxford: Oxford Univ Press; 2014b.
Northoff G. Is schizophrenia a spatiotemporal disorder of the brain's resting state? World Psychiatry. 2015;14(1):34–5. https://doi.org/10.1002/wps.20177.
Northoff G. Neuro-philosophy and the healthy mind. Learning from the unwell brain. New York: W.W. Norton & Company; 2016.
Northoff G, Boeker H. Orbitofrontal cortical dysfunction and "sensomotor regression", a combined study of fMRI and personal constructs in catatonia. Neuropsychoanalysis. 2003;5:149–75.
Northoff G, Qin P. How can the brain's resting state activity generate hallucinations? A 'resting state hypothesis' of auditory verbal hallucinations. Schizophr Res. 2011;127:202–14.
Northoff G, Stanghellini G. How to link brain and experience? Spatiotemporal psychopathology of the lived body. Front Hum Neurosci. 2016;10:172. https://doi.org/10.3389/fnhum.2016.00172.
Northoff G, Heinzel A, Bermpohl F, et al. Reciprocal modulation and attenuation in the prefrontal cortex: an fMRI study on emotional-cognitive interaction. Hum Brain Mapp. 2004;21:202–12.
Northoff G, Heinzel A, de Greck M, et al. Self-referential processing in our brain—a meta-nalysis of imaging studies on the self. NeuroImage. 2006;31(1):440–57.
Northoff G, Schneider F, Rotte M, et al. Differential parametric modulation of self-relatedness and emotions in different brain regions. Hum Brain Mapp. 2009;30(2):369–82.
Pao PN. Pathological jealousy. Psychoanal Q. 1969;38:616–38.
Pao PN. Schizoprenic disorders. Theory and treatment from a psychodynamic point of view. New York: International Univ Press; 1979.
Petrilowitsch N. Beiträge zur Strukturpsychopathologie. Basel: Karger; 1958.
Quin P, Northoff G. How is our self related to midline regions and the default-mode network? Neuroimage. 2011;57:1221–33.
Riecher-Rössler A. Psychiatrische Störungen und Erkrankungen nach der Geburt. Fortschr Neurol Psychiatr. 1997;65:97–107.
Robinson JJD, Nils-Frederic Wagner NF, Northoff G. Is the sense of agency in schizophrenia influenced by resting-state variation in self-referential regions of the brain? Schizophr Bull. 2015;42(2):270–6. https://doi.org/10.1093/schbul/sbv/102.
Rohde A, Marneros A. Psychosen im Wochenbett: Symptomatik, Verlauf und Langzeitprognose. Geburtshilfe Frauenheilkd. 1993a;53:800–10.
Rohde A, Marneros A. Postpartum psychoses: onset and long-term course. Psychopathology. 1993b;26:203–9.

Rohde A, Marneros A. Zur Prognose der Wochenbettpsychosen: Verlauf und Ausgang nach durchschnittlich 26 Jahren. Nervenarzt. 1993c;64:175–80.
Ruhrmann S, Schultze-Lutter F, Schmidt S, et al. Prediction and prevention of psychosis: current progress and future tasks. Eur Arch Psychiatry Clin Neurosci. 2014;264(Suppl 1): 9–16.
Sacks O. Foreword. In: Solms M, Turnbull O, editors. The brain and the inner world. An introduction to the neuroscience of subjective experience. London: Karnac Books Ltd; 2002. p. 12.
Samson JA, Simpson JC, Tsuang MT. Outcome studies of schizoaffective disorders. Schizophr Bull. 1988;14:543–54.
Sartre JP. Being and nothingness. An essay on phemonemological ontology. Translated by Barnes HE, New York: Philosophical library, Inc, L'é'tre et le néant. Essai d'ontologie phénoménologique (1943). Paris: Librairie Gallimard; 1956.
Scharfetter C. General psychopathology. Cambridge: University Press; 1980.
Scharfetter C. Schizophrene Menschen. 2nd ed. München: Urban & Schwarzenberg; 1986.
Scharfetter C. Das weite Spektrum bedürfnisangepasster Therapien bei Schizophrenien. In: Hartwich P, Pflug B, editors. Schizophrenien—Wege der Behandlung. Sternenfels: Wissenschaft & Praxis; 1999.
Scharfetter C. Wahn im Spektrum der Selbst- und Weltbilder. Sternenfels: Wissenschaft & Praxis; 2003.
Scharfetter C. Scheitern in der Sicht auf Psychopathologie und Therapie. Sternenfels: Wissenschaft & Praxis; 2012.
Schilder P. Das Körperschema. Berlin: Springer; 1925.
Schmitt A, Malchow B, Keeser D, et al. Neurobiologie der Schizophrenie. Aktuelle Befunde von der Struktur zu den Molekülen. Nervenarzt. 2015;86:324–31.
Schneider K. Klinische Psychopathologie. 6th ed. Stuttgart: Thieme; 1962.
Schwarz F. Gruppenprozess und Gruppenpsychotherapie. In: Schwarz F, Maier C, editors. Psychotherapie der Psychosen. Stuttgart: Thieme; 2001. p. 102–9.
Schwarz F, Matussek P. Die Beurteilung der Psychosen-Psychotherapie aus der Sicht des Patienten. In: Matussek P, editor. Beiträge zur Psychodynamik endogener Psychosen. Heidelberg: Springer; 1990. p. 190–237.
Solms M, Turnbull O. The brain and the inner world. An introduction to the neuroscience of subjective experience. London: Karnac Books Ltd; 2002.
Stemich-Huber M. Heraklit. Der Werdegang des Weisen. Amsterdam: Grüner; 1996.
Stierlin H. Family dynamics and separation patterns of potential schizophrenia. In: Rubinstein D, Alanen YO, editors. Psychotherapy of schizophrenia. Amsterdam: Excerpta Medica; 1972.
Stransky. Schizophrenie und intrapsychische Ataxie. Jb Psychnat. 1914;36:485.
Tienari P. Interaction between genetic vulnerability and family environment. Acta Psychiatr Scand. 1991;84:460–5.
Tienari P, Wynne LC, Morning J, et al. The Finnish adoptive family study of schizophrenia. Implications for family research. Br J Psychiatry Suppl. 1994;23:20–6.
Tsuang MT, Simpson JC, Fleming JA. Schizoaffektive Erkrankungen. In: Helmchen H, Henn F, Lauter H, Sartorius N, editors. Psychiatrie der Gegenwart, Schizophrene und affektive Störungen, vol. 5. 4th ed. Berlin: Springer; 2000.
Vanhaudenhuyse A, Demertzi A, Schabus M, et al. Two distinct neuronal networks mediate the awareness of environment and of self. J Cogn Neurosci. 2011;23:570–8.
Venables PH. Input dysfunction in schizophrenia. In: Maher BA, editor. Contributions to the psychopathology of schizophrenia. New York: Academic press; 1977. p. 1–56.
Volkan VD. Identification with the therapist's function and ego-building in the treatment of schizophrenia. Br J Psychiatry Suppl. 1994;23:77–82.
Wellek A. Das Problem des seelischen Seins. Meisenheim: West-Kultur; 1953.
Whitfield-Gabrieli S, et al. Hyperactivity and hyperconnectivity of the default network in schizophrenia and in first-degree relatives of persons with schizophrenia. Proc Natl Acad Sci U S A. 2009;106(4):1279–84.

Wiebking C, Duncan NW, Tiret B, et al. GABA in the insula–a predictor of the neural response to interoceptive awareness. NeuroImage. 2014a;86:10–8.
Wiebking C, Duncan NW, Qin P, et al. External awareness and GABA—a multimodal imaging study combining fMRI and [18F]- flumazenil-PET. Hum Brain Mapp. 2014b;35:173–84.
Wurmser L. Pathologische Eifersucht. Dilemma von Liebe und Macht. Forum Psychoanal. 2006;22:3–22.
Ziehen T. Psychiatrie. 2nd ed. Leipzig: Hirzel; 1902.

# Depressive Syndromes 11

Heinz Boeker and Georg Northoff

**Abstract**

The *self* is a core dimension in depression. It is attributed to negative emotions (e.g. failure, guilt). The *increased inward focus* in depression is connected with a *decreased environmental focus*.

The development of *neuropsychodynamic hypotheses* of the altered self-reference is based on the investigation of the emotional-cognitive interaction in depressed patients. It may be hypothesized that the increased negative self-attributions—as typical characteristics of an increased self-focus in depression—may result from *altered neuronal activity in subcortical-cortical midline structures in the brain* (especially from *hyperactivity in the cortical-subcortical midline regions and hypoactivity in the lateral regions*).

A mechanism-based approach was developed focussing on the psychodynamic, psychological and neuronal mechanisms in healthy and depressed persons.

The increased resting state activity in depression is especially associated with an *increased resting state activity in the default mode network (DMN)*. By means of this, changes in the complete spatiotemporal structure of the intrinsic activity of the brain and the dysbalance between default mode network and executive network (EN) are induced.

It is neither lesions nor disturbances of adaptive neuronal mechanisms which generate depressive symptoms, but rather *increasingly dysfunctional mechanisms of compensation on the basis of the increased resting state activity*.

H. Boeker (✉)
Department of Psychiatry, Psychotherapy, and Psychosomatics, Psychiatric University Hospital Zurich, University of Zurich, Zurich, Switzerland
e-mail: heinz.boeker@bli.uzh.ch

G. Northoff
Mind, Brain Imaging, and Neuroethics,
Institute of Mental Health Research, University of Ottawa, Ottawa, ON, Canada
e-mail: georg.northoff@theroyal.ca; www.georgnorthoff.com



H. Boeker et al. (eds.), *Neuropsychodynamic Psychiatry*,
https://doi.org/10.1007/978-3-319-75112-2_11

Possible *therapeutic consequences* of the neuropsychodynamic approach to depression involve the necessary emotional attunement in psychotherapy of depressed patients and the adequate timing of therapeutic interventions The hypotheses which have been developed in the context of the neuropsychodynamic model of depression may be used for more specific psychotherapeutic interventions, aiming at specific mechanisms of compensation and defence, which are related to the increased resting state activity and the disturbed resting state-stimulus interaction.

## 11.1 Introduction

Contrary to mourning, in which the loss of the significant other is elevated by a symbolizing process (Segal 1956), the depressive syndrome does not have a symbolizing function but is the consequence of the inhibition or the somatopsychic-psychosomatic dead end of depression (Gut 1989; Boeker 2002, 2003a). At an early stage, Freud differentiated depression, paraphrased as "melancholia", from mourning (Trauerarbeit): "In mourning it is the world which has become poor and empty; in melancholia it is the ego itself" (Freud 1917, p. 246).

This citation underlines the functional perspective on depressive symptoms: a primary useful reaction to loss (withdrawal of the mourning person to stabilize one's self) develops into becoming dysfunctional in the further cause (lack of emotional resonance; predominance of negative, dysfunctional cognitions; and extreme withdrawal behaviour).

Freud stressed the ego-regression from object cathexis to narcissism and pointed to the loss of reality function connected with this:

> ...; then, owing to a real slight or disappointment coming from this loved person, the object-relationship was shattered. The result was not the normal one of a withdrawal of the libido from this object and a displacement of it on to a new one, but something different, for whose coming-about various conditions seem to be necessary. The object-cathexis proved to have little power of resistance and was brought to an end. But the free libido was not displaced on to another object; it was withdrawn into the ego. There, however, it was not employed in any unspecified way, but served to establish an *identification* of the ego with the abandoned object. Thus the shadow of the object fell upon the ego, and the latter could henceforth be judged by a special agency, as though it were an object, the forsaken object. In this way an object-loss was transformed into an ego-loss and the conflict between the ego and the loved person into a cleavage between the critical activity and the ego as altered by identification. (Freud 1917, p. 249)

The important relation between loss of the object and loss of the self will be focussed on in a neuro-scientific perspective. By this we hope that one of the most challenging psychoanalytical theories of depression can be explained, especially the inadequate understanding of the special characteristics of the actual depressive psychopathology (depressive affect, anhedonia, disturbance of drive, cognitive dysfunction).

It was Edith Jacobson (1971, p. 315) who criticized the psychoanalytical concepts of depression, especially referring to patients with recurrent episodes or chronic depression ".... as scarcely plausible or contradictory ... ". She pointed to a neglect of Freud's "Ergänzungsreihen" of causal factors, by which the constitutional and hereditary factors of the development of a transference were considered significant. Jacobson recommended a "multifactorial, psychosomatic approach" as the basis for a complex theory of affective disorders (and of schizophrenia).

Inhibition thus became a problem in the established psychoanalytical theory of depression (Boeker 2001, 2003b, 2005; Boeker and Northoff 2010).

## 11.2 Inhibition, Stupor and Cognitive Dysfunction

The phenomenon of inhibition in depression is connected with a disturbance of initiation, with psychomotor stupor and with cognitive dysfunctions. Focussing on a multidimensional theory of depression as the foundation for the understanding and treatment of depressed patients, a neuropsychodynamic model of the disturbance of the emotional self-reference in depression will be presented based on empirical findings.

The limitations of the psychoanalytical theory of depression are related to the circularity of different etiological factors and the consequences of the somatopsychic final loop of the multidimensional dynamics in depression. Besides the usual vegetative symptoms (resulting from the involved disturbance of the autonomic nervous system), the prefrontal cortical dysfunctions in patients with severe depression should be underlined. Neuronal dysfunction in depression is connected with severe neuropsychological deficits (e.g. mnestic deficits, disturbances in attention and executive function, cf. Boeker and Grimm 2012; Boeker et al. 2012). The neuronal dysfunction should be taken into consideration in the course of the therapeutic relationship and when choosing what should be focussed on therapeutically, especially since it cannot be changed on a symbolic level by direct interventions (cf. Boeker 2003b).

A neuropsychodynamic concept of depression has a complex somatopsychic focus especially on the condition of the self (cf. Boeker 1999; Boeker et al. 2000c; Kratzsch 2001), the subjective experiences and the way in which these experiences are encoded and symbolized. The mechanisms of defence and compensation and the processes of emotion regulation are central in the neuropsychodynamic model of the disturbed emotional self in depression.

The principles of neuronal integration (reciprocal modulation, modulation by means of functional unit, top-down modulation and modulation through reversal) are presented a neurophysiological correlates of the mechanisms of defence and compensation (especially of somatization, introjection and sensorimotor regression). It may be hypothesized that these mechanisms represent inhibition in depression in a paradigmatic way. Furthermore, the self-experience in depressed patients, which correlates with the disturbances in self-reference, will be focussed on.

The development of such a psychodynamic theory of depression should take into account critical comments on the correlation between neuroscience and psychoanalysis and the complexity of subjective experience in the first-person perspective and the impossibility to localize this subjective experience in specific brain regions (cf. the discussion in the first part of this book). The starting point for the neuropsychodynamic theory of depression is a psychic-orientated term of localization, which considers the biological, psychological and social dimension of depression, with the differentiation between higher and lower regions of functions being of only secondary importance. In this neuropsychodynamic concept, the different modes of correlation between biological, psychological and social dimensions are central. These correlations depend on a horizontal localization, which is virtual and functional dynamic. Contrary to a merely biologically oriented perspective in which the brain is generally looked upon as isolated from the psychosocial context, it is not a question of an "isolated brain" but an "embedded brain" (Northoff 2000, 2004, 2013; Northoff and Boeker 2006; Northoff et al. 2004). If the critical remarks on the impossibility of localizing psychodynamic mechanisms in the brain are taken seriously, first-person neuroscience should indeed search for different ways of finding correlations between subjective experience and neuronal conditions. Instead of the neuronal localization in one or more regions, the neuronal integration over different regions will be focussed on here.

## 11.3 Neuronal Integration

Neuronal integration characterizes the coordination and modulation of neuronal activity over multiple brain regions. The interaction between brain regions far apart from each other is considered necessary for the development of specific functions, e.g. emotions or cognition (cf. Friston 2003; Price and Friston 2002).

It may be hypothesized that defence mechanisms as complex emotional-cognitive interactions are not localized in specific or isolated brain regions but depend on the interaction between different brain regions, i.e. neuronal integration. Therefore isolated brain regions that are far apart from each other should be connected with one another by means of *connectivity*, which represents the relation between the neuronal activities in different brain regions. Besides this, there is anatomical connectivity, which describes the anatomical substrates of the nervous system. Furthermore, a differentiation should be made between functional and effective connectivity (cf. Friston and Price 2001, p. 277):

*Functional connectivity* represents the "correlation between distant neurophysiological events", which may depend either on direct interaction between these events or on different factors which modulate different events. In the first case, the correlation results from the interaction itself; in the second case, the correlation may result by means of different factors, e.g. stimulus related procedures with common input or stimulus-induced oscillations.

In contrast to functional connectivity, *effective connectivity* describes the direct interaction between brain regions. Effective connectivity relates to the direct influence of one neuronal system to another, either on the synaptic level or on the level of neuronal interaction in different regions (macro-level, cf. Friston and Price 2001). In the further course, connectivity of the macro-level is assumed.

On the basis of connectivity, the neuronal activity between brain regions far away from each other can be adapted, coordinated and harmonized. This coordination and adaptation relies on certain principles of *neuronal integration* (Northoff 2004; Northoff and Boeker 2006). These principles represent functional mechanisms, which play an important role in the organization and coordination of neuronal activity in different brain regions. Four of these principles have been investigated in our study of emotional-cognitive interaction by means of neuroimaging:

- Top-down modulation
- Reciprocal modulation
- Modulation through functional unit
- Modulation through reversal

It may be hypothesized that each of these four principles of neuronal integration may be associated with specific mechanisms of defence and compensation.

## 11.4 Top-Down Modulation

Top-down modulation can be described as modulation of regions in a lower hierarchy by those in a higher hierarchy. A common example, therefore, is the modulation of neuronal activity in subcortical regions by cortical regions. Premotoric/motoric cortical regions can modulate the neuronal activity in the subcortical basal ganglia (e.g. the nucleus caudatus and the striatum, cf. Masterman and Cummings 1997; Northoff 2002a, b). A further example is the top-down modulation of the primary visual cortex by prefrontal cortical regions, which is necessary for visual processing (Lamme 2004). The top-down modulation resembles the concept of the "re-entrant circularity" (Tononi and Edelman 2000) and feedback modulation (Lamme 2001). These concepts focus on the exchange of information and the adaptation of neuronal activity in a certain brain region in accordance with another remote region. As a result, the neuronal activity in a lower region can be adapted, filtered and attuned in accordance with the neuronal activity in the higher region.

In the context of cognitive-emotional interaction in depressed patients, the focus is on the medial prefrontal cortex (MPFC): It could be shown in multiple studies that the neuronal activity in the medial prefrontal cortex as well as in the amygdala is part of emotional processing (Phan et al. 2002; Murphy et al. 2003). It may be assumed that their functional relation is characterized by means of the top-down modulation of the amygdala by the medial prefrontal cortex (Shin et al. 2005; Pessoa et al. 2002; Pessoa and Ungerleider 2004; Davidson 2002). Furthermore, it may be hypothesized that medial prefrontal cortical regions enact top-down control over the neuronal activity in the insula (Nagai et al. 2004). The insula has a tense and reciprocal relation to the subcortical medial regions, for instance, the hypothalamus and the periaqueductal grey (PAG) (Panksepp 1998a, b).

Both the amygdala and the subcortical medial regions play an important role in the regulation of the internal somatic functions, whereas the medial prefrontal cortical regions are more associated with emotional processing (Phan et al. 2002; Murphy et al. 2003; Northoff and Bermpohl 2004).

All these three regions—the medial prefrontal cortex, the amygdala and subcortical medial regions—have tense and reciprocal connections. Therefore, *circular modulation* between all these regions may be assumed.

Possibly there is not only a top-down modulation but also a bottom-up modulation (cf. Fig. 11.1).

A hierarchically lower region modulates the activity in a hierarchically higher region by means of the bottom-up modulation. In this way subcortical midline regions may, for instance, modulate the neuronal activity in the medial prefrontal cortex by means of the insula. Accordingly, it may be assumed that the bottom-up and top-down modulation co-exist in the same region at the same time (cf. Fig. 11.1). The simultaneous activity of the bottom-up and the top-down modulation may cause reciprocal adaptation between emotional processing and processing of internal body functions in a functional perspective.

Recent studies underline that the processing of the somatic and therefore interoceptive stimuli represents a decisive component of *empathy*. The assumption of a close relationship between interoceptive awareness and empathy is based on results underlining that the same brain regions are involved in both processes, for instance, the insula and the anterior cingulate cortex. A functional MRI study underlined the direct circularity between interoception and empathy. The neuronal activity is increased during interoception in the bilateral insula and in different medial cortical regions (subgenual anterior cingulate cortex, SACC; dorsomedial prefrontal cortex; DMPFC; posterior cingulate cortex, PCC; and precuneus) during empathy. These

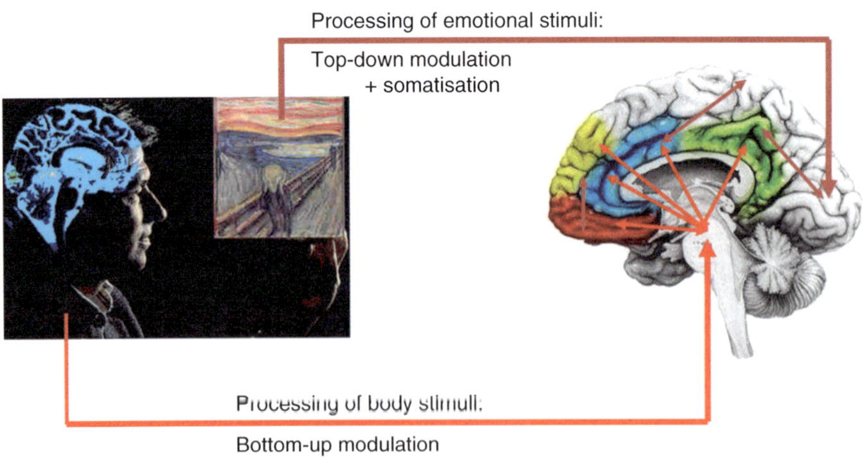


**Fig. 11.1** Top-down and bottom-up modulation

results point to a specific interaction between empathy and interoception in contrast to exteroception (Ernst et al. 2013). Interestingly, dysfunctional activity patterns in the insula and the anterior cingulate cortex are found in alexithymic patients.

The personality pattern of *alexithymia* is connected with a disturbed identification and characterization of emotions, especially in the discrimination of somatic feelings and emotional sensation. Ernst et al. (2014) showed in a functional imaging study investigating the connection between aspects of alexithymia (by means of the Toronto Alexithymia Scale, TAS-20) the interoceptive awareness (investigated by means of the Body Perception Questionnaire, BPQ) and the concentrations of glutamate and GABA (GABA concentration in the left insula and the ACC) by means of three tesla-magnet resonance spectroscopy, a close relationship between alexithymia and interoceptive awareness on the behavioural level. A concentration of glutamate in the left insula was associated positively with alexithymia as well as with the subscore "reactivity of the autonomous nervous system" in the BPQ, whereas GABA concentrations in the ACC were selectively associated with alexithymia. These results underline for the first time the close relationship between alexithymia, interoceptive awareness and GABA and glutamate concentrations in the anterior cingulate cortex and the insula. The increased excitatory transmission, mediated by glutamate, and the correlated increased activity in the insula can be interpreted as an expression of the increased interoceptive awareness in alexithymia (Ernst et al. 2014).

The simultaneous occurrence of the top-down and bottom-up modulation corresponds in a psychological perspective to the predominance of emotional awareness, in contrast to somatic awareness. The experience and awareness of internally or externally generated emotions come to the fore, whereas the awareness of the body remains in the background. In this way, a predominating external focus may be explained, in which attention is directed to other persons and events in the outside world. In contrast, the internal focus—that is, the attention to one's own body—remains in the background.

## 11.5 Somatization

The functional equilibrium between the bottom-up and top-down modulation may be disturbed in the course of regressive processes and defence mechanisms connected with these. Regression may be understood as a re-actualization of earlier functional levels and the correlated dominance of somatic reactions instead of emotional and cognitive patterns of reaction. "Somatization" may be characterized as a specific form of mental processing and intensive perception of somatic disturbances and will be referred to in the following. The mechanisms of somatization can be observed paradigmatically in depressive patients who often experience somatic symptoms—especially autonomous-vegetative symptoms—in a subjective way. It may be hypothesized that somatization in depression is correlated with disturbed equilibrium of bottom-up and top-down modulation between emotional and internal processing of the body.

On the functional level, somatization may point to predominance of internal processing of the body in contrast to emotional processing. Primary signals of the internal controlling centres of the body are processed, whereas processing of the emotional stimuli originated either internally or externally remains in the background. The balance between internal processing of the body and emotional processing is adapted to a new functional level in the following course. Accordingly, it can be assumed that depressed patients with strong somatization react much more intensively to internal stimuli from the body than to internal or external emotional stimuli. Furthermore, depressed patients show among other things extremely strong autonomous-vegetative reactions (for instance, heart rate variability; Bar et al. 2004; Guinjoan et al. 1995). Finally, depressed patients show reduced reactions to externally induced emotions, for instance, in the context of social interaction. This was already underlined by results from earlier studies on communication in depression (cf. Coyne 1976a, b, 1985; Coyne et al. 1987).

From the psychological perspective, somatization is reflected by an increased awareness of one's own body and the internal functions of the body. Depressed patients shift their attention away from their own or other's emotions to their own body functions. Depressed patients do not observe their own emotions, but rather their own body functions. They do not observe others' emotions, but rather their own body.

In summary, empirical results underline that depressed patients with strong somatization show increased attention to the body and decreased emotional attention. Thus, depressed patients have deficits in theory of mind tasks focussing on social interaction capabilities (subjects are confronted with social interactions in the form of drawings and have to empathize with the role of the various persons they see and predict their reactions). The deficits in social interaction found in depressed patients may be based on the shift in attention focus from the outside to the inside. On the neuropsychological level, this increased internal focus is reflected in changes in attention and theory of mind, which was shown empirically both in theory of mind tasks (Kerr et al. 2003; Inoue et al. 2004) and in findings of attention deficits (Sheppard 2004; Murphy et al. 1999; Paradiso et al. 1997).

From a physiological point of view, the altered functional balance between bottom-up and top-down modulation may correspond to the altered neuronal activity in the medial prefrontal cortex, the amygdala and subcortical medial regions (cf. Mayberg 2003a; Liotti et al. 2002; Elliott et al. 2002).

## 11.6 Reciprocal Modulation and Introjection

Neuroimaging studies showed a pattern of opposite signal changes in the medial and lateral prefrontal cortex during emotional-cognitive interaction (Goel and Dolan 2003a, b; Northoff and Bermpohl 2004; Northoff et al. 2004). These results correspond to the assumption of functional mechanisms of reciprocal modulation and reciprocal reduction during emotional-cognitive interaction. Reciprocal modulation

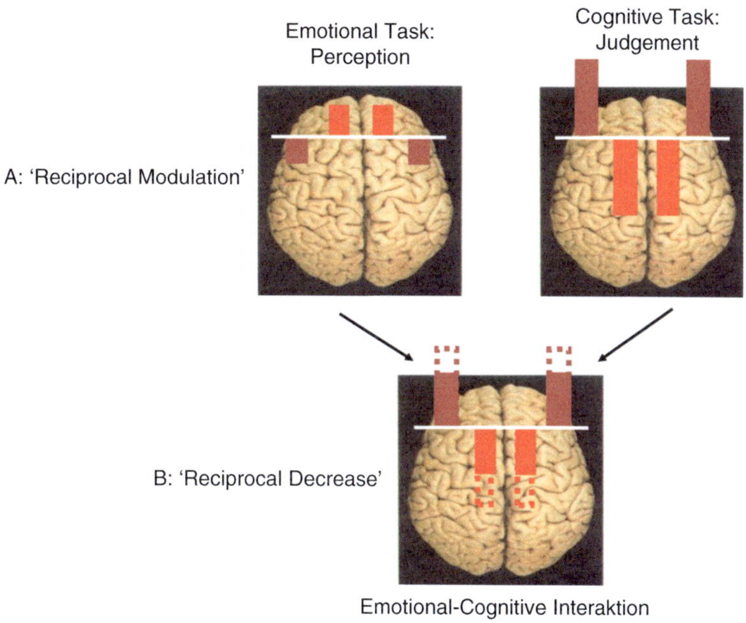


**Fig. 11.2** Reciprocal modulation and introjection. Introjection: disturbances of the emotional-cognitive modulation—the experience of the outside world is changed to an experience of the inner self (internal focus)

is defined by signal changes in opposite directions (i.e. increases and decreases of signals). It is already known that emotional processing (when an emotional picture is perceived) leads to an increase of signals in the medial prefrontal cortical regions as well as to decreases of signals in the lateral prefrontal cortex (Phan et al. 2002; Murphy et al. 2003; Northoff et al. 2004). In contrast, cognitive tasks (e.g. judgement or assessment) produce an opposite pattern of signal increases in the lateral prefrontal cortex and signal decreases in the medial prefrontal cortex. This corresponds to the functional mechanism of reciprocal modulation (see Fig. 11.2; cf. Northoff et al. 2004).

Interestingly, analogous patterns of reciprocal modulation were also observed in other cortical regions, for instance, in the medial and lateral orbitofrontal cortex, in the right and left motor cortex, in the striatal and extra-striatal visual cortex, in the subgenual anterior cingulum and in the right prefrontal cortex, in the sub-/pre- and supergenual anterior cingulum as well as in the visual and auditory cortex (overview in Northoff et al. 2004). On the basis of the mentioned empirical results, it may be assumed that emotional-cognitive interaction is association with the functional mechanism of reciprocal reduction: if a cognitive task comprises an emotional component (e.g. when assessing an emotional picture), fewer signal decreases result in the medial prefrontal cortical regions and simultaneously fewer signal increases in the lateral prefrontal cortical regions. This process was characterized as *attenuation* (synonym: decrease; cf. Northoff et al. 2004). A reciprocal attenuation can be referred to because this process occurs in the medial as well as in the lateral prefrontal cortical regions in opposite directions (i.e. fewer signal decreases and increases).

On the basis of these empirical results, Phillips et al. (2003) and Mayberg (2003a, b) developed a model of the altered reciprocal functional interaction between ventromedial and dorsolateral prefrontal cortex (DLPFC) in MDD (model of ventrodorsal dissociation) where reciprocal modulation of neural activity in depression is reduced because of reduced deactivation in the medial regions and reduced activation of the DLPFC (Grimm et al. 2008; Carhart-Harris et al. 2008). The disturbed modulation of the lateral prefrontal cortical resting state most probably contributes to a reduced stimulus-induced activity triggered by cognitive stimuli. From a neuropsychodynamic perspective, the reduced resting state-stimulus activity and the reduced exteroceptive-neuronal interaction contribute to a reduced constitution of objects which also results from the reduced emotional valence and reward. Current object relationship experiences lose their emotional significance.

Reduced reciprocal modulation—as the third part of the reduced resting state-stimulus interaction—contributes to a reduced activation of cognitive processes by exteroceptive stimuli. In this way, the "object cathexis" is reduced (Carhart-Harris et al. 2008). Furthermore, it may be assumed that the disturbed reciprocal modulation may be an adaptive mechanism by which the depressive self is finally disconnected from the significance of current object relationship experiences. From a psycho-energetic perspective, the result is a further reduction of object cathexis. The depressed patient attempts to constitute and cathect compensational objects from his/her inner world. The inner world encompasses interoceptive stimuli of the body and cognitive stimuli instead of external objects.

It may be assumed that the reciprocal modulation and attenuation can be altered in introjective mechanisms on a neuropsychodynamic level. *Introjection* is—in an operational definition—characterized by the shift of object focus from the outside to the inside in subjective experience. The subject-object relationship is no longer directed to the outside, but to the inside. The experience of the outside is changed into an experience of the inner self.

The defence mechanism of introjection can be observed paradigmatically in depressed patients. These patients tend to internalize their conflicts with others and shift the aggression, which was primarily directed towards others, against their own self. It may be hypothesized that this introjection in depression may be associated with the abnormal reciprocal modulation during emotional-cognitive interaction (Malancharuvil 2004; Adroer 1998; Deci et al. 1994; Berman and McCann 1995).

From a functional and psychological perspective, introjective processes in depressed patients may be connected to disturbances of emotional-cognitive readaptation. Depressed patients are no longer able to adequately assess their own emotional and body experience. The assessment of one's own conditions is "subjectively" distorted and decoupled from "objective" reality. This subjective distortion is seen in a marked negativity of the assessment of one's own emotions and the body and also of the assessment of others' emotions and the events in the outside world. This extreme negativity corresponds to the "negative or attentional bias" (cf. Elliott et al. 2002; Gotlib et al. 2004).

The results of a study on the perception of faces (Gotlib et al. 2004), for instance, pointed out that the "negative or attentional bias" may be connected to interpersonal dysfunction in depressed patients. Further studies are necessary to explain how and why the "negative or attentional bias" apparently induces a decoupling of the "subjective" assessment from the "objective" reality.

On the basis of the generated empirical results, it may be assumed that the disturbed reciprocal modulation and attenuation during emotional-cognitive interaction in depression are of outstanding importance for the development of introjective defence and compensation mechanisms. In the context of the disturbed reciprocal modulation, the depressed patient is no longer able to adequately perceive, assess and evaluate his/her own experience. The abnormal reciprocal attenuation contributes to the depressed patient being no longer able to connect his/her own emotional experience with his/her own cognition, ultimately resulting in a decoupling of experience and cognition. In the course of this decoupling, the reciprocal modulation of emotion and cognition is prevented. The depressed patient himself/herself no longer has access to his/her own cognition, thus preventing adequate assessment. Similarly, the perceived emotions of others have no access to their own cognition, which prevents a realistic emotional evaluation. Emotional conflicts with others are internalized and the resulting aggression is directed against the own self instead of others. From this perspective, introjection is the result of a strong inner focus and a weak focus on the outside world.

From a physiological point of view, it may be assumed that the altered reciprocal modulation and attenuation in depressed patients—with a strong inner focus and introjective defence—are correlated with changes in neuronal activity patterns in the medial and lateral prefrontal cortex. The assumption of a pathologically increased reciprocal modulation between medial and lateral prefrontal cortex is in line with numerous studies underlining the hyperactivity in the medial prefrontal cortex and the hypoactivity in the lateral prefrontal cortex during emotional stimulation (Mayberg 2003a; Liotti et al. 2002; Elliott et al. 2002).

## 11.7 Modulation by Means of Functional Unity and Sensorimotor Regression

A further example of functional mechanisms of emotional-cognitive interaction is the development of functional units in different brain regions over time. Such transient functional units could be identified on the basis of psychophysiological features or functional connectivity in the involved regions (Friston 1998, 2003; Friston et al. 1998a, b, 2003; Friston and Penny 2003; Friston and Price 2003). The cortical midline structures (CMS, Northoff and Bermpohl 2004), for instance, show a continuously high level of neuronal activity also under resting state conditions (e.g.

when passively regarding a fixation cross; Gusnard et al. 2001; Gusnard and Raichle 2001; Mazoyer et al. 2001; Raichle 2001; Raichle et al. 2001). Furthermore, the regions within the cortical midline structures are characterized by dense anatomical connections. In addition, investigations on the functional activity in the CMS found an increase in functional activity in the CMS between anterior and posterior CMS regions in the resting state, whereas this connectivity decreased during active cognitive tasks.

The participation of both anterior and posterior midline structures is in line with results from further studies on cognitive, emotional and social processing (overview in Northoff and Bermpohl 2004). Furthermore, signal decreases were found both in the orbitomedial prefrontal cortex (OMPFC) and the parietal cortex (PC) during cognitive tasks requiring attention. In addition, these regions show an increased circular connectivity (Greicius et al. 2003; Raichle 2003). In summary, these empirical results deliver convincing evidence for the existence of the CMS and a functional unit, which is particularly active and cohesive in the resting state (Greicius et al. 2003; Wicker et al. 2003a, b, c).

With this in mind, it may be assumed that modulation through functional unit is altered in regressive processes involving sensory and motor functions (cf. Fig. 11.3). Sensorimotor regression can be defined as a defence and compensation mechanism, which is activated when conflicts and anxieties can no longer be solved by means of cognitive and emotional functions and when somatic and particularly sensorimotor functions are involved. The defence mechanism or the sensorimotor regression can be paradigmatically observed in patients with catatonia ("scared stiff"; Boeker et al. 2000c). It may be hypothesized that the sensorimotor regression in catatonia is connected with altered modulation by functional unit in the CMS (overview in Northoff et al. 2003, 2004; Boeker et al. 2000a, b).

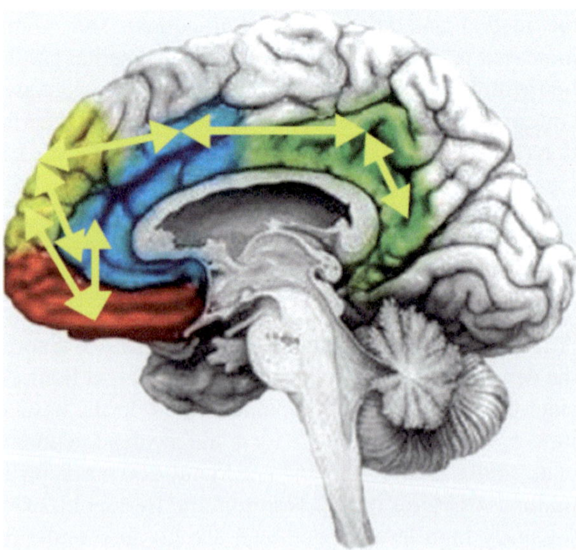

**Fig. 11.3** Modulation through functional unit and sensorimotor regression. *Arrows:* alterations in the functional connectivity between orbitofrontal cortex, medial prefrontal and premotor/motor cortex in patients suffering from stupor—transformation of emotional into motor symptoms—symptomatic overlap of stupor and conversion

## 11.8 Self-Referential Processing and Ego-Inhibition

How are the sensory stimuli, which reach our brain, processed further and finally transformed into actions?
What happens during this processing phase?
How are the original sensory stimuli transformed?

Certain sensory stimuli are related to one's own person, contrary to other stimuli which are rather related to other persons and the outside world. Accordingly, self-referential stimuli can be differentiated from non-self-referential stimuli (Northoff and Boeker 2006; Boeker 2004). This differentiation is valid not only for sensory stimuli but also for emotional and cognitive stimuli.

From a functional perspective, self-referential processing points to a simple distinguishing process, the distinction between self and non-self. This process is decisive for the distinction between one's own stimuli and those from others and therefore for the distinction between self and the outside world. Moreover, self-referential processing may be a prerequisite condition for the development of a concept of one's own self, the so-called mental or phenomenal self as a subject of experience (Damasio 1999; Panksepp 1998a, b; Northoff 2004; Northoff et al. 2005).

From a psychological perspective, self-referential processing may be manifested in the possibility of a subjective experience of one's own *self* or *ego* (both terms are used synonymously in the following; Freud 1914; Dennecker 1989; Kohut 1977). By marking certain stimuli as self-referential, they can be experienced subjectively, i.e. from the individual subject's or ego perspective (Northoff and Boeker 2006). Owing to the fact that, besides the internal and sensory stimuli of one's own body, emotions and cognitions are also examined with regard to their self-reference, emotions and cognitions can be attributed to one's own self. The subject constituted primarily by one's own body is "filled", so to speak, with certain related emotions and cognitions.

In depressed patients there is possibly an increased processing of the internal body stimuli—as described above—whereas emotional stimuli are reduced. This dysbalance contributes to the relation of the self-referential processing, especially to the stimuli of one's own body. In this way, there may be an increased experience of one's own body with the experience of one's own ego as body.

In contrast, the decrease of emotional processing leads to a greatly reduced relation of the self-referential processing to emotional stimuli: emotions are increasingly less related to one's own self. The self can no longer be "filled" with emotions and cognition but remains "empty" and emotionally inhibited: this emotional inhibition changes into an *ego-emptiness* and an *ego-inhibition*, respectively.

Self-referential processing can be related to the medial prefrontal cortical regions, the above-mentioned cortical midline structures (CMS), independent of the different functional domains. The cortical midline structures include the medial

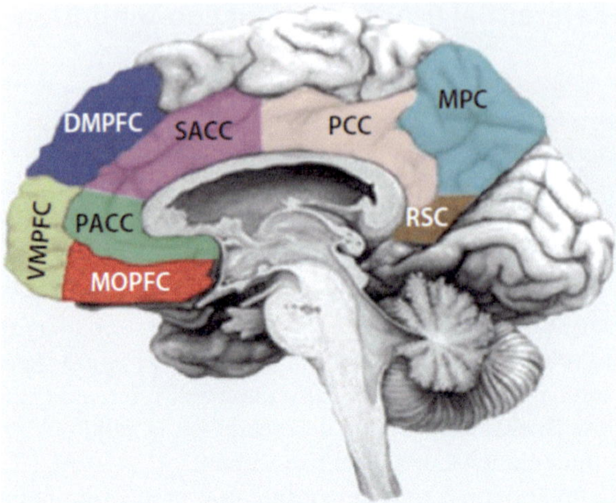


**Fig. 11.4** Cortical midline structures (CMS). *DMPFC* dorsomedial prefrontal cortex, *MOPFC* medial orbitofrontal cortex, *MPC* medial prefrontal cortex, *PACC* perigenual anterior cingulate cortex, *PCC* posterior cingulate cortex, *SACC* subgenual anterior cingulate cortex, *RSC* retrosplenial cortex, *VMPFC* ventromedial prefrontal cortex (from Northoff and Bermpohl 2004, pp. 102–107; with kind permission from Elsevier)

orbitofrontal cortex (MOPFC), the ventro- and dorsomedial prefrontal cortex (VMPFC, DMPFC), the medial parietal cortex (MPF) and the anterior and posterior cingulate cortex (ACC, PCC, cf. Fig. 11.4).

It may be assumed that the ego-inhibition and ego-emptiness in depressed patients are connected with a dysfunction of the CMS. Most probably the interaction of the CMS as a functional unit is disturbed in depression. The reduced emotional self-referential processing in depressed patients may result from this.

## 11.9 Depression as a Psychosomatic Disorder of Emotional Regulation

The self and the specific alterations in the experience of the self are a central dimension in depressed patients. The depressive mood signalizes to the person concerned how much he/she depends on relations and emotional ties, particularly after stressful life events, such as the loss of significant others or significant psychosocial roles. The depressive experience acts as an emotional signal which strengthens emotional ties. At the same time, it is the expression of a "dead end" (Gut 1989), in which a person gets stuck after a process which has been going on a long time involving the development of different defence and compensation mechanisms. Therefore the dangers and challenges which the self is confronted with during its development and through interaction with significant others are reactivated when depressive mood is experienced. Accordingly, when depressive mood is experienced, it may be interpreted as an expression of the experiences of a self which is focussed on social resonance.

Depressive mood should be distinguished from mourning and is difficult to define linguistically. In the early days of psychiatry, the psychopathological term *anhedonia* was used to describe feelings of emptiness and lack of joy. Experiencing depressive mood is often connected with distorted cognition and particularly with extreme pessimism and excessive, unfounded self-reproach (overview of the cognitive-affective structure in depressed patients in Boeker and Northoff 2010). This process often goes hand in hand with distorted perception of the body and its vegetative functions.

During their depressive illness and the period afterwards, it is only with great difficulty that many depressed patients are able to talk about what they have experienced and find a way to approach it. A rare example of such an attempt is the description of depression as the "pain at the loss of the self which can no longer be suppressed" by Miller (1987). This experience of losing the self can become manifest in an extreme and sometimes bizarre way in the course of psychotic depression, particularly as nihilistic delusion.

### 11.9.1 Cotard Syndrome

"Cotard syndrome" was described by the psychiatrist, in Paris, as a special type of agitated melancholia/depression. What is special about this syndrome is a "délire des negations", which is understood to be a hypochondriac psychotic syndrome as part of an anxious-agitated depressive disorder, as a specific example of "delusional depression".

Unlike in "hypochondriac delusion", in "nihilistic delusion" it is not so much the dysfunction of existing organs which is important but rather the experience of "not being". Cotard syndrome can develop in different disorders (not only in depression but also in bipolar affective disorders, schizophrenia, brain tumours and personality disorders). Cotard syndrome may be considered an extreme form of the experience of self-loss, especially in severe depression.

Nihilistic ideas in Cotard syndrome include, for instance, the depressed patient's conviction that he/she has lost his/her strength, mind, emotions or certain organs (usually the stomach, intestines and brain). The most serious form of the syndrome is when the subject's own existence and the existence of the world are denied. These patients are absolutely convinced that they have died. In another group of depressed patients suffering from depressive delusion, the experience of self-loss is correlated with the psychotic conviction of having been burdened with irreparable guilt (delusion of guilt), of becoming impoverished without any hope of change or relief (delusion of impoverishment) or of suffering from an incurable tumour or from Alzheimer disease (hypochondriac delusion).

The strain these patients feel and their considerable suicide risk should at all times be kept at the forefront of the therapeutic focus. Nevertheless, although these patients stubbornly cling to their convictions or seek some kind of punishment for their supposedly bad behaviour, it can be felt in the therapeutic relationship that the sometimes bizarre symptoms of the depressive delusion have a protective mental function in the light of the experienced loss of the self. This protective function is seen when the self attains a certain degree of stabilization when being confronted with the possible risk of complete disintegration and elimination. Even though the paranoid symptoms can be extremely painful, the delusion still remains related to feelings. The depressive patient tries to experience these feelings in view of the risk of completely losing his/her self.

**Case Study: Mr. O.: "I am Dead"**

Mr. O. was referred to the Psychiatric University Hospital at the age of 61 with the diagnosis of hypochondriac delusion and with suspected nihilistic delusion. The depressive disorder was first seen several months previously. He had increasingly withdrawn and finally tried twice to commit suicide (the first time sustaining a stab wound near the carotid artery and the second time driving his car into a tree). Previous treatment with antidepressants and neuroleptics had proven to be inadequate.

At the beginning of the hospital treatment, Mr. O. did not want any contact. He was retarded in his movements and his drive was considerably reduced. His formal thinking was blocked with intense rumination. Mr. O. was convinced that he could no longer swallow or eat or drink anything and that his bones were disintegrating. He had passive suicidal ideas connected with the hope of being put out of his misery.

In the years before going into hospital, Mr. O. had had to face a number of stressful life events: his wife was suffering from a chronic neurological illness, finally being confined to a wheelchair and increasingly needing intensive care. Mr. O. lost his job when his company went bankrupt. He started withdrawing from social contact, keeping to himself, and focussed on his somatic complaints. After various gastrointestinal examinations, a dysplasia of the oesophagus was suspected.

A recurrent depressive disorder, currently severe, with psychotic symptoms (ICD-10: F33.3) was diagnosed at the Psychiatric University Hospital. Mr. O. was suffering from a depressive-psychotic syndrome with hypochondriac and nihilistic delusions. He was absolutely convinced that he was suffering from a fatal disease of the gullet and bones and that he was a hopeless case whom nobody could help; he was convinced that he was already dead.

Because of his therapy-resistant depressive-psychotic symptoms, Mr. O. was given electroconvulsive therapy (ECT). Soon after the fourth ECT session, his mood started to improve and his facial expression started to relax. After the seventh treatment session, he became increasingly able to perceive the outside world, made contact with other patients and started to get his appetite back. The treatment was continued after the eighth ECT session (out-patient maintenance ECT). The patient's mood and drive remained stable over the following years, and he was able to pursue his hobbies and to tackle appropriately the problems relating to his wife's illness.

### 11.9.2 The Neuropsychodynamic Perspective of Depression

The circular interactions between the disturbed regulation of the self and the connected neurobiological processes, especially the regulation and adaptation of neuronal activity patterns and neurotransmission, will be considered from a neuropsychodynamic perspective. To avoid any misunderstandings, we do not intend to describe the neuronal correlates of psychodynamic processes or to reduce psychodynamic processes to neuronal mechanisms and thus define the neuronal correlates of the former. Rather, the approach we are presenting here is based on so-called neuronal predispositions, which may contribute to the development of a depressive disorder depending on biological, psychological and social conditions. Accordingly, we will not focus on the causal relationship between neuronal mechanisms and psychodynamic processes, but the transition between the neuronal and psychodynamic levels of consideration. The hypotheses developed on the basis on empirical results must necessarily remain hypothetical and sketchy. This is because our current knowledge is limited or because of principle reasons, such as the limitations of our brain-based epistemological abilities.

When developing neuropsychodynamic hypotheses of depression, the multifaceted scientific results on the biology and neurobiology of depression should be considered (for an overview see Holsboer-Trachsler and Vanoni 2007). The different biological, psychological and social dimensions of depression may be related to one another in a circular way using a system-theoretical perspective (cf. Boeker 2002, 2003a). This model of depression as a psychosomatosis of emotion regulation (see Fig. 11.5) conceptualizes depression as a biological condition, which

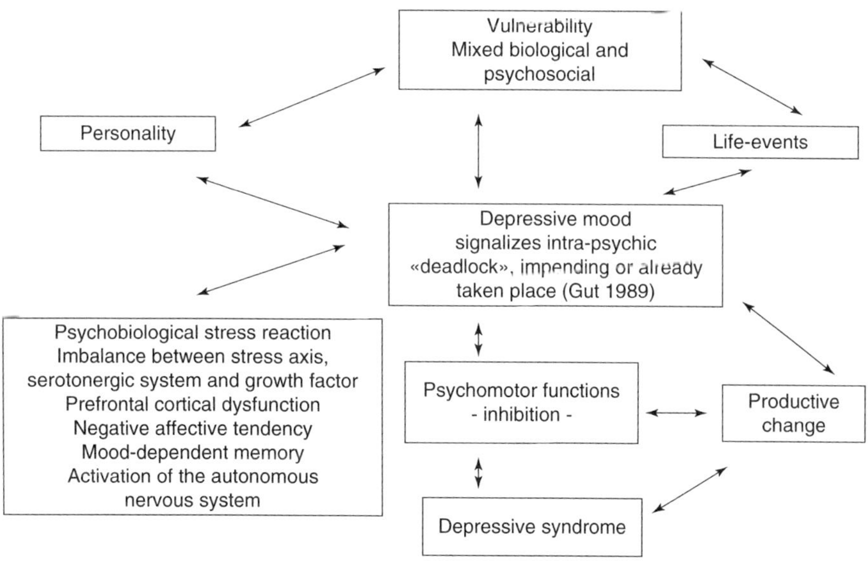


**Fig. 11.5** Depression as psychosomatosis of emotion regulation (from Boeker 2009; with kind permission from Wissenschaft & Praxis publishers)

develops in different stages involving circular interrelations between psychological and neurobiological processes. Besides the mixed biological and psychosocial vulnerability, this model considers the influence of personality, cognitive structure, current and chronically stressful life events and the psychobiological stress reaction often induced by experiences of separation and involving neurophysiological and cognitive disturbances as well as dysfunctional compensation patterns (cf. Boeker 2009).

It may be assumed that the initial neuronal network patterns are adapted, modified and transformed by means of the sequential expression of specific genes and genetic combinations, depending on the way they are used throughout life (Wiesel 1994). Experience-based changes of monoaminergic afferences themselves may be triggered for further reaching use-dependent changes of adapted synaptic relations. This sequence of complex adaptive reorganization processes may be manifested as a mental disorder in particularly vulnerable or genetically predisposed individuals. Post (1992) hypothesized that psychosocial stressors under certain conditions lead to long-term changes in gene expression. This process leads to changes in neuropeptides and the neuronal microstructure, and a spatial-temporal cascade of adaptation processes develops, which underlies the cognitive changes seen in the course of many depressive disorders. Whereas depressive episodes are triggered by life events at the beginning of the illness, in the later course, the life events or stressors inducing the manifest symptoms get smaller and smaller and can hardly be identified externally. The experience of the depressive episode and the connected neurotransmitter and peptide changes may leave behind memory traces which may cause a predisposition for further episodes. These processes, also known as *kindling effect*, contribute to the lowering of the vulnerability threshold. From this system-theoretical perspective, under specific environmental conditions and a specific psychosocial constellation, relatively minor disturbances may contribute to the development of an altered personality structure. This is accompanied by intrapsychic conflicts and tensions and a largely unconscious cognitive structure. In a longer adaptation and compensation process, secondary and partly additional somatic changes and disturbances may occur (cf. Mentzos 2009).

## 11.10 Psychodynamics and Psychopathology of the Self in Depression

In view of the development of the variety of psychoanalytic theories of depression (overview in Boeker 1999, 2005; Gabbard 2014; Mentzos 1995, 2009), we will focus on the essential dimensions of depression, which are related to Freud's outstanding contribution *Mourning and Melancholia* (Freud 1917). According to this, the reactivation of earlier experiences of loss in childhood; the introjection of the lost objects in childhood, which are connected with negative emotions; and the current loss of object relations, which are connected to the loss of the self, may be described as psychodynamic essentials of depression.

### 11.10.1 Early Object Loss

The early loss of objects in childhood is part of the biopsychosocial vulnerability of persons likely to develop a depressive disorder in later life. The fixation on the mental representation of the lost objects involves great psycho-energetic exertion, which Freud compared to an "open wound" taking psychic energy from the self and object representations. This regressive process leads to both the outer world and the self being completely depleted of representations.

### 11.10.2 Introjection of the Lost Object

Depressed persons develop an increased inner focus and are no longer able to focus on the outer world. This self-focussed attention is the focus of perception of interpersonal relationships. This information is derived from sensory perceptions which react to changes in somatic activity (Ingram 1990). The attention shifts from the exteroceptive sensory system to the interoceptive sensory system, which processes somatic stimuli.

### 11.10.3 Loss of Actual Object Relations

The loss of actual object relations represents the third psychodynamic key feature of depression. The mental representation of the lost object includes a disregarding encountering other objects in the actual environment of the adult. Others are exclusively or predominantly perceived in the perspective of the lost object. The melancholic inhibition differs from mourning decisively:

> In mourning we found that the inhibition and loss of interest are fully accounted for by the work of mourning in which the ego is absorbed. In melancholia, the unknown loss will result in a similar internal work and will therefore be responsible for the melancholic inhibition. The difference is that the inhibition of the melancholic seems puzzling to us because we cannot see what it is that is absorbing him so entirely. The melancholic displays something else besides which is lacking in mourning—an extraordinary diminution in his self-regard, an impoverishment of his ego on a grand scale. (Freud 1917, pp. 245–246)

Freud underlined that melancholia is characterized by "…a lowering of the self-regarding feelings to a degree that finds utterance in self-reproaches and self-revilings, and culminates in a delusional expectation of punishment" (Freud 1917, p. 244). Some lines later Freud amplified: "We should regard it as an appropriate comparison, too, to call the mood of mourning a 'painful' one. We shall probably see the justification for this when we are in a position to give a characterization of the economics of pain" (Freud 1917, p. 244).

Object relationships are replaced by means of a regressive cathexis of the self which contributes to an increase of the splitting in the self. This was phenomenologically described as increased self-focus.

### 11.10.4 Phenomenological and Psychopathological Aspects

On the basis of the three psychodynamic key features of depression and the connected alteration of the self-experience of the depressed (increased self-focus, attribution of the self with negative emotions and increase of the cognitive processing of the self), a phenomenological and psychopathological approach to the self of depressed patients is developed.

Depressed patients are characterized by an increased self-focus; they focus on their own self, whereas they are no longer able to change their focus on others. A social-psychological perspective of depression starts from a self-focussed attention as the focus of the perception of interpersonal relationships; this includes informations from sensoric perceptions, which react on changes of the activity of the body (Ingram 1990). The perception of the depressed person is no longer focussed on the relationship to his/her social environment but on his/her own self as a primary focus ("decreased environmental focus"). With regard to the sensoric perception, it can be assumed that the attention shifts from the exteroceptive sensory system (signalizes environmental events and the relationship between the subject and her/his environment) to the interoceptive system, which processes the stimuli of the body.

A great number of studies with different measures and methodologies underline the self-focussed perception in depression. The results converge in an increased and possibly altered level of self-focussed attention in depression (Ingram 1990). It is unclear up to now whether this increased self-focus exclusively exists on an explicit, conscious level—according to our definition—or already on an implicit, unconscious level.

### 11.10.5 Attribution of Negative Emotions to One's Own Self

The self in depression is attributed to negative emotions (failure, guilt, hypochondriac fear of illness and death, in some cases connected with depressive delusion). Positive emotions are no longer connected with one's own self. These scrupulous tendencies correlate with the number of former suicide attempts and are risk factors for suicidal behaviour.

### 11.10.6 Increase of Cognitive Processing of One's Own Self ("Self-Focus")

The cognitive processing of the own self (rumination) increases the depressive mood and develops into an increasingly dysfunctional mechanism of compensation which is connected with the self focussed attention and a repetitive focus of one's own negative emotions (Ingram 1990; Treynor et al. 2003; Rimes and Watkins 2005). This increasingly so-called *analytical self-focus* contrasts with the reduced experience-based self-focus (*experiential self-focus*).

The analytical self-focus corresponds to the term *self-awareness* or *self-consciousness*, whereas the experiential self-focus relates to the features of *self-related processing*.

? What do these phenomenologically oriented descriptions mean in relation to the above-mentioned pschodynamic core mechanisms (see Sect. 11.10)?

It may be hypothesized that the increased inward focus in depression corresponds with psychodynamic processes, especially introjections and identification of the self with a lost object. The connection between introject and self with negative affects which is postulated in a psychodynamic perspective corresponds to the phenomenological/psychopathological description of the association of the self with negative emotions. Finally, the significance of the actual loss of objects corresponds with the increased cognitive processing of the self because the cognitive processing no longer focusses on the actual loss but on the loss of objects with which the self is identified ("decreased environmental focus").

### 11.10.7 The Decoupling of the Self in Depression

The development of neuropsychodynamic hypotheses of the altered self-reference in depression is based on the investigation of the emotional-cognitive interaction in depressed patients, which focussed on the neurophysiological correlates of depressive inhibition and the neurophysiological substrates of negative cognitive schemes and the neuropsychological deficits (overview in Boeker and Northoff 2005, 2010; Northoff et al. 2002, 2005, 2007; Grimm et al. 2008, 2009; Boeker et al. 2012; Walter et al. 2009). To sum up, depression may be characterized by reduced neuronal activity in the left dorsolateral prefrontal cortex and increased activity in the right dorsolateral prefrontal cortex. The neuronal activity in the left dorsolateral prefrontal cortex cannot be modified by emotional valence. The severity of depression correlates with the activity in the right dorsolateral prefrontal cortex. Connected with the reduced deactivation in the pregenual ACC (default mode network), depressed persons cannot shift their attention from themselves to the outside world (Grimm et al. 2008). The degree of helplessness and the severity of depressive symptoms correlate with the reduced deactivation in the PACC and PCC. The signal intensities in different subcortical and cortical midline regions (DMPFC, SACC, precuneus, ventral striatum, DMT) were reduced significantly. On the basis of these empirical results, it may be concluded that the increased negative self-attributions—as typical characteristics of an increased self-focus in depression—may result from altered neuronal activity in subcortical-cortical midline structures in the brain (especially from hyperactivity in the cortical-subcortical midline regions and hypoactivity in the lateral regions) (Fig. 11.6).

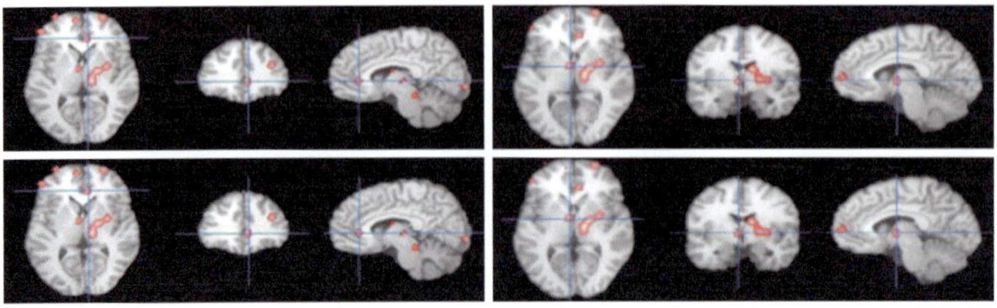

**Fig. 11.6** Neurophysiology of depression: hyperactivity in the cortical-subcortical midline regions and hypoactivity in the lateral regions (from Alcaro et al. 2010 with the kind permission of Elsevier)

## 11.11 Neuropsychodynamic Hypotheses on the Disturbed Self-Reference in Depression

On the basis of neuropsychological, neurophysiological and neurochemical findings and, as we mentioned, psychodynamic dimensions of depression, neuropsychodynamic hypotheses on the disturbed self-reference in depression were developed and related to psychodynamic and specific neuronal mechanisms of depression.

### 11.11.1 Increased Resting State Activity and Reactivation of Early Object Loss

The resting state activity in the brain represents the intrinsic activity of the brain and should be differentiated from the brain's activity induced by somatic stimuli or outside stimuli. It may be hypothesized that the empirically validated induced resting state activity in depression is a predisposition for reactivation of early object loss experiences in the subject.

PET studies in major depression underline the decreased resting state activity especially in the lateral anterior cortical midline regions (PACC, VMPFC, compare Mayberg 2002, 2003a, b; Phillips et al. 2003). Alterations in neural activity were shown in ventral regions of the so-called default mode network (DML) in depressed patients (reduced deactivation, i.e. negative BOLD reactions, compare Greicius et al. 2007; Grimm et al. 2009; Sheline et al. 2009). Furthermore a translational meta-analysis of resting state studies in depressed patients and in animal models confirms resting state hyperactivity in ventral, cortical midline regions (PACC, VMPFC, Alcaro et al. 2010; Fitzgerald et al. 2007; Price and Drevets 2010; Drevets et al. 2008).

In contrast to the anterior midline regions, posterior midline regions (PCC, precuneus/cuneus) and the superior temporal gyrus (STG) show hypoactivity in the resting state (Alcaro et al. 2010; Heinzel et al. 2009). The hyperactivity in the anterior midline regions and the hypoactivity in the posterior midline regions result from a disturbed balance between anterior and posterior midline regions in acute depression and a disturbance in the default mode network in depression (Raichle et al. 2001; Buckner et al. 2008).

Relation to psychodynamic mechanisms
? How can these results be related to psychodynamic mechanisms, especially the reactivation of early loss in childhood?

A possible scenario in childhood shall be outlined at first; afterwards the situation in adulthood shall be outlined in a second step. A great number of results underline that the VMPFC (especially on the right side) is an area of convergence inside the central anterior subcortical-cortical midline regions, which is of special importance during the early development. Early traumatization causes alterations in developing processes which may contribute to the development of early immature defence mechanisms (Feinberg 2011).

Traumatization in early childhood was found in a large subgroup of depressed patients (experiences of loss, divorce of parents, physical or sexual abuse, compare Boeker 2000; Gabbard 2005; Nemeroff et al. 2003). Traumatic life events may cause biological alterations on the genetic, hormonal or anatomic-structural level (Feder et al. 2009). It may be hypothesized that traumatic life experiences interfere with the development of the VMPC and especially the ventral anterior subcortical-cortical midline regions as an essential part of the default mode network (DMN). Anterior midline regions are especially involved in the processing of the degree of self-reference of different stimuli, whereas the posterior regions are likely to be involved in the processing of social and non-self-related stimuli (Qin and Northoff 2011).

Furthermore it may be assumed that early traumatic experience of object loss is associated with the desperate attempt to relate the self to the lost object in order to develop a self-object relationship and to experience the lost object as a self-object. In connection with this, a possible hyperactivation, especially in the anterior midline regions, may be induced, which finally contributes to a dysbalance in the posterior midline regions with a consecutive hypoactivation. Finally a specific neuronal pattern may be assumed in a psycho-energetic perspective, which is mediated by hyperactivity in the anterior midline regions with consecutive dysbalance in the posterior midline regions and the complete default mode network when attempting to generate an object cathexis.

? How can this neuronal pattern related to the occurrence of depression and the reactivation of early object loss?

The reactivation of early object loss presupposes autobiographical memories and their retrieval. The event which triggers the depressive episode in the adult resembles the event in early childhood (as a subjectively experienced disappointment or loss). It was shown that the retrieval of autobiographical memories is mediated by the anterior and posterior cortical midline regions (Buckner et al. 2008). In healthy persons a connection between the retrieval of autobiographical memories and the self-focus in the VMPFC, PCC and precuneus could be demonstrated (see Sajonz et al. 2010).

The retrieval of early traumatic experiences in the context of current experiences of object loss triggers a reactivation of the same neuronal patterns used in the early development of a relationship to the object with a consecutive hyperactivity in the anterior midline regions and a dysbalance in the posterior regions and in the default mode network.

In view of the early object loss, depressed patients develop a psychological predisposition to attribute a high degree of self-reference to lost or disappointing objects. Therefore, a neuronal predisposition for the development of hyperactivity in the resting state in anterior midline regions when object loss is experienced in adulthood is induced. Accordingly it can be hypothesized that the psychological and neuronal predisposition corresponds with the psychodynamic predisposition, the reactivation of early object loss. When actual stimuli induce less stimulus-induced activity in the anterior midline regions because of the abnormally increased resting state activity, the brain keeps predominantly occupied with itself, i.e. that the earlier experienced contents and the objects are reactivated by the increased resting state activity accordingly or that the representations of the earlier objects are rerepresented in the actual context (Fig. 11.7).

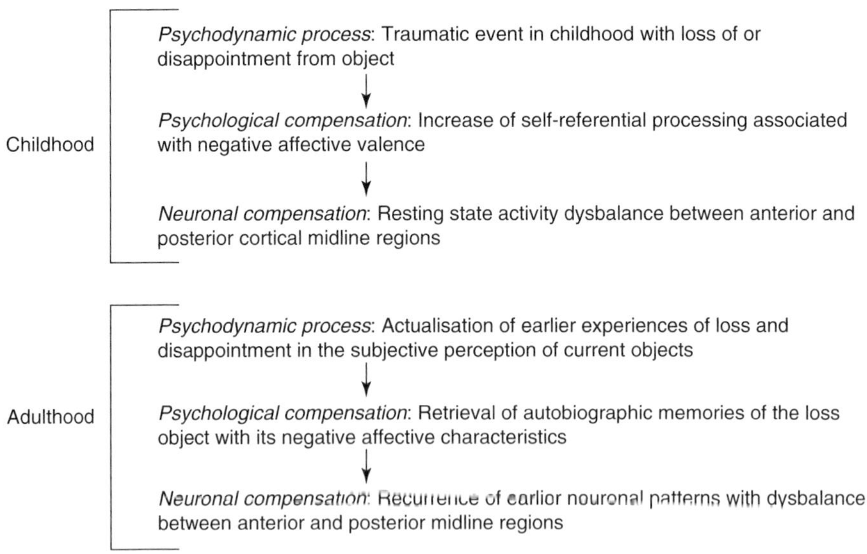


**Fig. 11.7** Resting state activity: dysbalance and reactivation of lost objects

### 11.11.2 Reduced Resting State-Stimulus Interaction and Introjection in Correlation with Negative Emotions

Resting state-stimulus interaction includes the interaction between two different modes of neural activity: the intrinsic activity of the brain and the stimulus-induced activity (induced by stimuli from the outside world and/or the own body). It may be hypothesized that resting state-stimulus interaction in depression is reduced because of increased resting state activity and that the resting state-stimulus interaction is associated with introjective processes in the interaction between self and the object world connected with negative emotions (Fig. 11.8).

#### 11.11.2.1 Results of Neuroimaging Studies

Functional activation paradigms (emotional, cognitive) showed dysfunctional activation patterns, especially hyperactivity, in the ventral cortical midline regions in patients with major depression during resting state and emotional stimulation (Elliott et al. 1998, 2002; Mayberg et al. 1999; Canli et al. 2004; Fu et al. 2004; Davidson et al. 2003). In other brain regions, especially in the reward system (VS/N. accumbens, right and left amygdala), dysfunctional activation patterns were found during positive and/or negative emotional stimulation in MDD (Lawrence et al. 2004; Surguladze et al. 2005; Canli et al. 2004; Kumari et al. 2003). These dysfunctional activation patterns may be interpreted as the neurophysiological basis of the

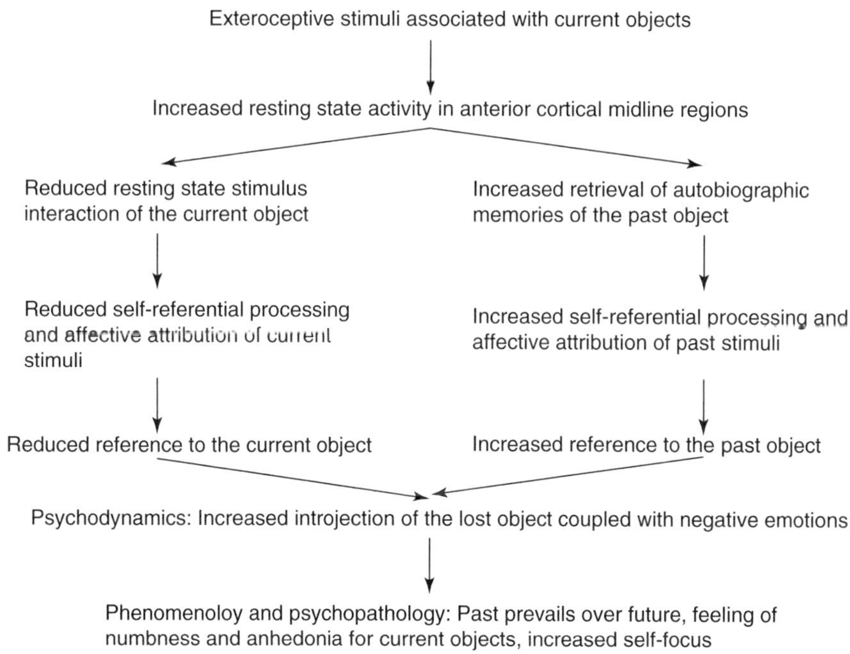


**Fig. 11.8** Reduced resting state-stimulus interaction and negative affect

“negative affective bias”, that is, the focussing of negative emotions which is related to the inability to process positive emotions (Phillips et al. 2003; Mayberg 2003a, b; Heller et al. 2009; Heinzel et al. 2009; Grimm et al. 2009). The reduced deactivation in cortical-subcortical regions was correlated with the severity of depression and the degree of hopelessness. A direct association between reduced resting state-stimulus interaction and the severity of depression may be assumed (Grimm et al. 2009).

Further studies focussed on the biochemical basis of the reduced resting state-stimulus interaction. It could be shown that increased resting state activity in the PACC is associated with the concentration of the neurotransmitter glutamate (Walter et al. 2009; Northoff et al. 2007; Alcaro et al. 2010; Sanacora 2010). A dysbalance between neuronal inhibition and excitation in the anterior midline regions in depressed patients may be assumed.

? The question is to what extent does the increased resting state activity and the reduced resting state-stimulus interaction in these brain regions depend on a disturbed glutamatergic excitation?

The induced resting state activity in the brain inhibits the neuronal processing of stimuli from the outside world. Stimuli from the outside world cannot be related to one’s own self or connected with emotional valence. Nevertheless, these psychological mechanisms are still active and are mediated by the induced resting state activity. From a psycho-energetic perspective, the energy which is usually used for the development of self-relatedness and the connection with emotions (emotional valence) is related to early stimuli which are related to early object loss and the induced resting state activity.

The above-mentioned meta-analysis (Alcaro et al. 2010) did not only include resting state data but also results from studies on glutamate and GABA in humans as well as in animals. If we assume that the resting state hyperactivity in the PACC in humans depends on the disturbed glutamatergic metabolism, a disturbed expression/sensitivity of glutamatergic receptors can be expected (e.g. NMDA and AMPA receptors). This assumption is consistent with results from animal studies which predominantly found an increase of NMDA and a decrease of AMPA receptor sensibility/expression and moreover a reduction of NMDA receptors during antidepressant treatment in the above-mentioned resting state regions. These observations are consistent with the effects of ketamine on functional PACC activity in the therapy of depressed patients (Northoff et al. 1997; Salvadore et al. 2009; Zarate Jr et al. 2006).

Our hypothesis of a glutamatergic modulation of the increased resting state activity in the PACC is supported by another study (Walter et al. 2009). It could be shown that a fMRI-marker of the possible resting state hyperactivity in the PACC (i.e. a reduced negative BOLD answer; NBA) correlated with the concentration of glutamate in the same brain region in depressed patients. This correlation was not found in healthy persons which underlines that the resting state activity was not associated with glutamate but more with GABA as an inhibitory transmitter (cf.

Northoff et al. 2007). These results support our assumption of a glutamatergic neuronal excitation of the resting state activity in the PACC with a simultaneous decoupling of the GABAergic neuronal inhibition (cf. Alcaro et al. 2010; Sanacora 2010). On the background of these results, a dysbalance between neuronal inhibition and excitation in the anterior cerebral midline regions in depressed can be concluded. This dysbalance leads to an abnormal increase of the resting state activity in the brain by means of which stimuli from the outside induce smaller or no stimuli-induced activity in turn. In other words the high resting state activity in the brain thus blocks the neuronal processing of stimuli of the outside.

Relation to psychodynamic mechanisms
? How can these results be related to psychodynamic dimensions or mechanisms of defence and compensation (especially introjection) and furthermore correlated with negative affects?

The anterior midline regions (e.g. PACC, VMPFC, DMPFC) contribute in a specific way to the development of self-related processing and the emotional valence of stimuli; together with the subcortical midline regions, they are included in one network. All these regions show an increased activity which has an effect on the resting state-stimulus interaction. The consequence is that the stimuli which have to be processed are no longer modulated by the resting state activity of the brain. In a psychological view, the actual stimulus can no longer relate to one's own self and connect with emotional valence accordingly. Nevertheless these psychological mechanisms which enable self-reference and emotional reference are still active and are mediated by the increased resting state activity at the same time. But in case that both psychological mechanisms (and their energy) are not used for the processing of the actual stimulus and the actual encounter with objects, so they are related to the earlier stimuli which may be connected with the former loss of objects. In an energetic perspective, the energy which is usually used to generate self-reference and the connection with affects (emotional valence) during actual stimuli is related to earlier stimuli which were related to the former loss of the object together with the increased resting state activity. This change of the self-reference and the emotional reference away from the actual stimuli to the former objects gets along with significant phenomenologic and psychopathologic alterations and psychodynamic processes. In a phenomenologic perspective, depressed patients predominantly focus on the past and not on the present and on the future (Grimm et al. 2009). In a psychopathologic perspective, it can be expected that depressed patients no longer perceive and experience the actual stimuli or objects in a subjective way. It may be assumed that this process is expressed in the "feeling of feelingness". This focus on the earlier experienced stimuli manifests in the increase of the self or inner focus.

The change from the actual to former stimuli and with this related object may contribute in a psychodynamic perspective—as a defence mechanism—to an increase of introjective processes. In this context stimuli which are related to the early loss of objects are reactualized in the actual context and get along with an increased resting state activity in the anterior midline regions. Because these regions mediate self-reference and emotional valence (independent of their origin and the nature of the stimulus), this processing continues also in this case. The reactualized stimuli of the past which are connected with early experiences of the loss of objects are related to one's self and connected with negative emotional valence. The dysbalance between the earlier and the actual stimuli also gets along with a dysbalance of the emotional valence and a change from positive to negative affects (because the earlier stimuli are associated with loss and disappointment and are reactivated during adulthood).

Stimuli which are connected with the early loss of objects are introjected and connected with negative affects in the emotional context. In an energetic context, these stimuli in adulthood are hypermetabolized which manifests in an excessive amount of self-reference and negative affects. This *hypermetabolizing* occurs on the expense of the actual stimuli (*hypometabolizing*), in which in extreme cases, any kind of self-reference and emotional valence may be completely absent.

The change from current to early stimuli may contribute to an increase of introjective processes in a neuropsychodynamic perspective. The reactivated stimuli from the past are related to the self and negative emotional valence. Stimuli which are connected with early object loss are introjected and related with negative emotions. It may be assumed that this process corresponds with an introjective type of depression (Blatt 1998) which is characterized by an increased self-reference, a decreased interpersonal relatedness and an increased intellectualization, reaction building and rationalization as well as by loneliness, loss and dysphoria.

? Which relation exists between the dysbalance of earlier and actual stimuli or objects and the observation of an altered balance between neuronal inhibition and excitation, i.e. GABA and glutamate?

*The resting state-stimulus interaction affords neuronal inhibition, which is mediated by GABA. By means of neuronal inhibition, the resting state activity can be suppressed. By this energy is available for the processing of incoming stimuli in an inverse proportion. If the suppression of the resting state activity does not happen any longer, the incoming stimuli can no more be associated, which results in a decreased resting state-stimulus interaction.*

### 11.11.3 Reduced Resting State-Stimulus Interaction and the Loss of Current Object Relations

It may be hypothesized that the modulation of the resting state activity by means of stimulus-induced activity and the resting state-stimulus interaction in depression is reduced because of the increased resting state activity. Reduced resting state-stimulus interaction probably leads to dysfunctional development of the neuronal

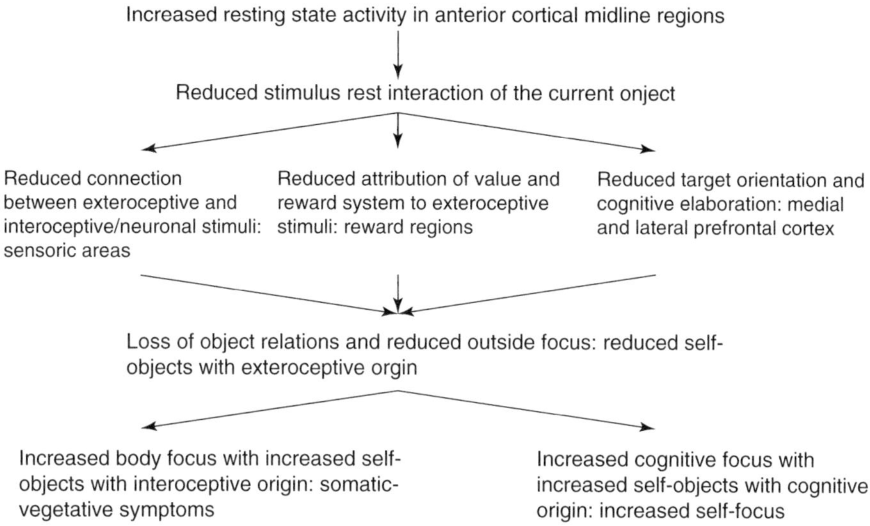


**Fig. 11.9** Reduced resting state-stimulus interaction, reduced outside focus and increased self-focus

structure and organization which is associated with dysfunctional processing of current experiences of loss. The reduced resting state-stimulus interaction can be seen in three different patterns:

- First, interoceptive stimuli no longer modify the resting state or baseline activity of the brain (Wiebking et al. 2010).
- Second, exteroceptive stimuli are no longer associated with value and reward (Pizzagalli et al. 2009; Smoski et al. 2009; Kumar et al. 2008; Dichter et al. 2009).
- Third, exteroceptive stimuli no longer induce or constitute cognitive processing (Goel and Dolan 2003a, b; Northoff et al. 2004; Grimm et al. 2006; cf. Fig. 11.9).

#### 11.11.3.1 Reduced Interaction of Exteroceptive Stimuli

Wiebking et al. (2010) studied the neuronal activity during exteroceptive and interoceptive perception (heartbeat and counting of the cardiac frequence) in relation to the resting state activity of the brain. Exteroceptive stimuli (heartbeat) contributed to a neuronal activation in the bilateral anterior insula in depressed patients, if it was related to the previous resting state activity of the brain. But if the exteroceptive-induced activity was viewed independently from the previous resting state activity, significant differences were found between healthy persons and depressed patients. The resting state activity in depressed was increased compared to healthy persons, as the reduced deactivation during stimulus-induced activity had shown already. Thus the increased resting state activity in the bilateral anterior insula contributed to a reduced interaction of the exteroceptive stimulus with the previously increased resting state activity. Analogous results were not found during the interoceptive stimuli (counting of the heartbeat), by means of which the specificity of the reduced resting state-stimulus interaction during exteroceptive stimuli was underlined.

The importance of the increased resting state activity in the bilateral anterior insula was further supported through its correlation with the severity of depression: the higher the resting state activity in these regions was, the more severe the depression was experienced in the self-assessment (BDI). Furthermore depressed patients showed increased scores for body perception and stress (Body Perception Questionnaire; BPQ). This result is consistent with the increased perception and attention to one's own body and the unspecific somatic symptoms, which are characteristic of depression. Whereas the BPQ scores in healthy persons correlated with the resting state activity in the insula significantly, this correlation was not found in depressed patients: Therefore it could be concluded that the increased resting state activity in depressed was decoupled from the perception and experience of one's own body. The increased resting state activity and the consecutively reduced resting state-stimulus interaction are psychologically and psychopathologically of great importance.

#### 11.11.3.2 Processing of Value and Reward

The second component of the reduced resting state-stimulus interaction relates to the processing of value and reward. The reward network includes central regions the ventral tegmental area (VTA), the ventral striatum (VS) and the ventromedial prefrontal cortex (VMPFC). Actual studies underline a reduced activity during the processing of reward tasks through exteroceptive stimuli just in these regions of depressed patients (Pizzagalli et al. 2009; Smoski et al. 2009; Kumar et al. 2008; Dichter et al. 2009).

#### 11.11.3.3 Cognitive Processing

The third component of the reduced resting state-stimulus interaction relates to the reciprocal pattern of neuronal activity between the medial and lateral prefrontal cortex which could be shown in different studies (Goel and Dolan 2003b; Northoff et al. 2004; Grimm et al. 2006) and which were described as reciprocal modulation (Northoff et al. 2004). In healthy persons exteroceptive stimuli lead to a deactivation, i.e. to negative BOLD answers, in the anterior medial cortical regions, whereas they lead to an increased neuronal activity in the lateral prefrontal cortical regions (e.g. in the dorsolateral prefrontal cortex; DLPFC) during cognitive and emotional stimulation. Controversary in depressed patients—as already mentioned—not only a hyperactivity in the ventral cortical midline regions was found but also a hypoactivity in the left DLPFC as well as during emotional and during cognitive processing either (Lawrence et al. 2004; Davidson et al. 2003; Keedwell et al. 2005). On the basis of these results, Phillips et al. (2003) and Mayberg (2003b) proposed a model of altered reciprocal functional connections between ventromedial and dorsolateral prefrontal cortex in major depression (model of the ventrodorsal dissociation in major depression). The reciprocal modulation is reduced in depression because of the decreased deactivation in the medial regions and the decreased activation of the dorsolateral prefrontal cortex (Grimm et al. 2008; Carhart-Harris and Friston 2010), i.e. there is an opposite activity in the medial and lateral prefrontal cortex. When the medial regions are strongly deactivated, a strong activation in the lateral regions results; when the lateral regions are less activated, a less deactivation results in the medial regions.

Therefore exteroceptive stimuli do not lead to a decreased deactivation in the medial cortical regions but also to a decreased activation in the lateral regions

(DLPFC), which are especially associated with cognitive processing (Grimm et al. 2006, 2008, 2009). The disturbed modulation of the lateral prefrontal cortical rest activity probably leads to a decreased stimulus-induced activity, which is triggered by cognitive stimuli.

Psychodynamic context
? Of what significance are these three different components of the reduced resting state-stimulus interaction in a psychodynamic context?

In depression exteroceptive stimuli do not induce the existing resting state activity of the brain any longer. The reduced resting state-stimulus activity contributes to a reduced exteroceptive-neuronal interaction. It may be assumed, accordingly, that the exteroceptive stimuli and the objects which are related to them are introjected with less probability. The consecutive reduced constituting of the objects also results from the second component of the reduced resting state-stimulus interaction, the reduced processing of value and reward. The neuronal activity of the reward system—as the basis of processing of value induced by stimuli—is extremely reduced in depression. In a psychodynamic context, it can be assumed that the selection of objects and the possibility to identify with them are more and more reduced because of the reduced ability of processing of value of stimuli. Actual object relation experiences get increasingly unimportant.

The reduced reciprocal modulation—as third component of the reduced resting state-stimulus interaction—implies that exteroceptive stimuli no longer induce the activation of cognitive processes (e.g. goal-oriented cognitions). This interferes with the *object cathexis*, which Carhart-Harris and Friston (2010) associate in a neuropsychological perspective with goal-oriented cognition and activity within the dorsolateral prefrontal cortex. The important thing here is that the reciprocal modulation itself is not disturbed in depression, but rather the causes of the reduced medial deactivation have to be explored. These may consist in the reduced rest activity and the reduction of exteroceptive stimuli, in which the latter is connected with the reduction of the previous exteroceptive-neuronal processing of values. Neuronal activity is not induced in the medial cortical regions by the incoming exteroceptive stimuli. In this perspective the apparent abnormal reciprocal modulation in depression represents a process of compensation aiming at to adapt the brain with regard to the processing of exteroceptive stimuli.

Accordingly the neurophysiologic problem in depression is not a lesion or a deficit of the reciprocal modulation but the maintenance and the adaptive function of this mechanism. The depressive self is encoupled from the experienced significance of actual object relation experiences finally. In a psycho-energetic perspective, a reduction of the object cathexis results.

? Which are the mechanisms of compensation which are developed by the depressed patient in this situation?

To answer this question, Freud's metaphor of an open wound may be picked up: The depressed tries to constitute compensatory objects and to cathect them. In view of the inability to constitute objects of exteroceptive origin, i.e. from the outside, and to introject them, the depressed patient constitutes and introjects objects from the inside. The internal environment includes interoceptive stimuli from the body and cognitive stimuli. Instead of exteroceptive-interoceptive and exteroceptive-cognitive connections, neuronal-interoceptive and interoceptive-cognitive connections are developed from now on. By this the depressed attempts to compensate the loss of exteroceptive stimuli. In a psychodynamic perspective, external objects, i.e. objects of exteroceptive origin from the outside, are replaced by internal objects, i.e. objects of interoceptive and cognitive origin. The outstanding aim is to constitute objects and to develop object relations. Therefore in a self-psychological perspective, this can be understood as a compensation of the loss of external or sensory self-objects by internal or somatic and cognitive self-objects.

Regarding psychopathological symptoms, the constitution of internal or somatic and cognitive *self-objects* contributes to the development of somatic symptoms. The body is perceived in an altered way (cf. Wiebking et al. 2010). If the constitution of cognitive self-objects dominates, the depressed suffers from distinctly distorted negative cognitive schemes and ruminations, which are experienced as painful and tormenting. Because of the loss of exteroceptive objects, the focus of attention is directed to one's own self and the self-objects of cognitive origin. In a phenomenological perspective, this was described by us as increased cognitive processing of one's own self with ruminations and a dysbalance between the experiential and the analytical self-focus (see Sect. 11.10.6).

## 11.12 Neuronal Substrates of Depression and Implications for Psychotherapy

### 11.12.1 Prefrontal Cortical Dysfunction in Depression

What are the neuronal substrates of depression focussed on in psychotherapeutic interventions on a neuropsychodynamic basis in the framework of multidimensional treatment of depression?

The dysfunction of the limbic-cortical networks (among others increased resting state activity in the medial prefrontal cortex (MPFC) is of central importance. An overview of the results of prefrontal cortical functions will be presented in the following sections. On this empirical basis, a hypothesis of the physiology of prefrontal cortical functions can be developed. This hypothesis is mainly based on the interaction between:

- The dorsolateral prefrontal cortex (DLPFC; W9, 46)
- The ventrolateral prefrontal cortex (VLPFC; W44, 47)
- The orbitomedial cortex (OMPFC; W10, 14, 25)
- The anterior cingulum (AC 24, 32)

### 11.12.2 Neuropsychological Functions in the Prefrontal Cortex

Prefrontal cortical functions were related to both cognitive and emotional functions. Different functions were associated specifically and exclusively with different prefrontal cortical areas, so that from a neuropsychological perspective, a "double dissociation" may be assumed. The results referring to such a "double dissociation" between OMFC, AC, VLPFC and DLPFC will be presented in the following, focussing on neuropsychological functions.

#### 11.12.2.1 Dissociation Between Orbitomedial Frontal Cortex (OMFC) and Dorsal and Ventrolateral Prefrontal Cortex (DLPFC/VLPFC)

The medial orbitofrontal cortex can be associated with emotional processing, whereas the lateral prefrontal cortex is more or less associated with cognitive processing (Sarazin et al. 1998). The lateral prefrontal cortex is specifically associated with both working memory and executive functions (Sarazin et al. 1998), whereas the medial orbitofrontal cortex is more predominantly associated with "behavioural functions", "response inhibition" and "decision-making" (Bechara et al. 2000; Rogers et al. 1999).

#### 11.12.2.2 Dissociation Between Orbitomedial Frontal Cortex (OMFC) and Ventrolateral Prefrontal Cortex (VLPFC)

The OMFC is associated with emotional processing, especially with negative emotional processing (cf. Northoff et al. 2000), whereas the VLPFC is involved more with emotional monitoring and "inhibition of associations in emotions" (Drevets and Raichle 1998). Furthermore, the observed reciprocal patterns of activation and deactivation in both regions during negative and positive emotions underline such a differentiated participation of the OMFC and the VLPFC in emotional processing (Drevets and Raichle 1998; Northoff et al. 2000). The OMPF most likely processes emotional contents, whereas the VLPFC monitors, represents and perhaps also inhibits and manipulates these contents, if they do not fit in the relevant context of behaviour (Elliott et al. 2000).

#### 11.12.2.3 Dissociation Between Ventrolateral Prefrontal Cortex (VLPFC) and Dorsolateral Prefrontal Cortex (DLPFC)

The VLPFC selects behavioural strategies on the basis on active retrieval and storage and therefore can compare and organize different and adequate behavioural responses. In contrast, the DLPFC more likely participates in online observation and potential manipulation of these contents, which are retrieved and stored in the VLPFC (Stern et al. 2000; Rogers et al. 1999). The relation between VLPFC and DLPFC is expressed paradigmatically during "decision-making". The VLPFC selects advantages where there are concrete or external "cues" in a course of a specific situation, whereas the DLPFC tends more to generate a decision in an accidentally generated context without concrete or external cues (Paulus et al. 2001). Therefore it can be assumed that the VLPFC for the most part participates in "intuitive decisions" characterized by fast processing, unconsciousness and a strong relation with emotional processes (Bechara et al. 2000). In contrast, the DLPFC seems to participate in "logical decisions" characterized by slow processing, consciousness and manipulation of the respective behavioural strategy.

In summary, the relationship between OMFC, VLPFC and DLPFC may be characterized by different ways of participating in emotional processing, emotional monitoring, emotional observation, "decision-making" and "working memory".

### 11.12.3 Connectivity in the Prefrontal Cortex

The analysis of the connectivity in the prefrontal cortex is mostly based on studies with Makaken (primates); there are only a few results generated in humans.

#### 11.12.3.1 Neuronal Networks in the Prefrontal Cortex

Three neuronal networks in the prefrontal cortex can be differentiated:

- Orbitomedial (W10-14, 25)
- Lateral network (W45, 46, 8a, 8b)
- Mediating network (W9, 24)

Because the mediating network with areas 9 and 24 includes the greatest variety of connections of nearly all prefrontal cortical areas, it may be regarded as an integrating network, connecting the orbitomedial and lateral network. Furthermore, it should be taken into account that the different areas of the orbitomedial and lateral prefrontal network are not homogenous in themselves. Within the orbitomedial network, areas 10, 11 and 15 may be differentiated from areas 12, 13 and 14. Within the lateral network, area 45 in particular shows a highly individual pattern of connectivity, characterized by a high amount of efferencies and an extremely low number of afferencies. Therefore, area 45 may modulate the lateral prefrontal cortical function.

#### 11.12.3.2 Afferencies and Efferencies

- The OMFC is connected with the mediotemporal lobe (hippocampus, entorhinal cortex, parahippocampus), hypothalamus and brain stem.
- The VLPFC is strongly connected with different sensory areas, the amygdala, the parietal cortex (posterior) and premotor/motor cortex.
- The DLPFC is connected with the mediotemporal lobe, the parietal cortex and the premotor/motor cortex.

#### 11.12.3.3 Connectivity Between Different Areas in the Prefrontal Cortex

There is bilateral connectivity between different areas within the orbitomedial network (WC10-14), as well as between the different connections within the lateral network (W45, 46, 8a, 8b, 9). Furthermore, bilateral connectivity also exists between the anterior cingulated (W24) and areas within the lateral network (W8, 9, 45, 46). There is also unilateral connectivity from the ventrolateral cortical area 45 to orbitomedial areas (W10-14), but not vice versa. Unilateral connectivity also exists from the dorsolateral prefrontal area 46 to the orbitomedial area W10 and 11. Finally unilateral connectivity was found from the orbitomedial areas W10-13 to the mediating areas W9 and 24. There is no connectivity between inferior orbitomedial areas (W13, 14, 25) and lateral prefrontal areas (8a, 45, 46).

To sum up, there is connectivity between OMFC, VLPFC and DLPFC, and this may be characterized by the contrast between OMFC on the one hand and VLPFC and DLPFC on the other. The orbitomedial and lateral prefrontal networks are connected by a group of regions, which can be subsumed as medial prefrontal cortex or anterior cingulated (W9 and W24). Furthermore, it should be taken into consideration that the connectivity between OMFC and VLPFC is unilateral, whereas the connectivity between OMFC and DLPFC and between VLPFC and DLPFC is bilateral.

#### 11.12.3.4 Time Courses

The time course of activation in the prefrontal cortex can be characterized by the following steps (cf. Northoff 2004):

- Early activation in the OMFC after 200 ms and in the DLPFC after 300 ms for a period of 20–50 ms at a time
- Late activation in both OMFC and DLPFC with the same temporal sequence (300 and 400 ms) as during the early activation but with a longer duration of activity (100–300 ms or even longer)
- Dissociation between the right and left DLPFC in the time course; the right DLPFC is activated earlier (150–300 ms) than the left DLPFC (400 ms)
- Faster processing for non-aversive (100–200 ms) than for aversive (200–400 ms) stimuli; furthermore, also faster processing for the discrimination of negative stimuli (100–200 ms) than for positive stimuli (150–300 ms) in the neurons of the OMFC
- Tonic activity in AC between 1000 and 4000 ms, contrasting with the phasic activity in the DLPFC

#### 11.12.3.5 Hypothesis: Different Modes of Processing in the Prefrontal Cortex

Different modes of processing may be assumed in the prefrontal cortex:

- *Early feedforward processing* from OMFC to DLPFC via anterior cingulate in the temporal range between 200 and 300 ms after presentation of the emotional stimulus. This early processing may be related to emotional experience or specific emotions.
- *Mediating feedback processing* from DLPFC to OMFC via VLPFC in the timeframe between 300 and 400 ms, which may be seen neuropsychologically in emotional judgement concerning the presence or absence of emotions.
- *Late re-entrant processing* from OMFC to DLPFC via anterior cingulated in the timeframe between 400 and 600 ms, which may be seen neuropsychologically in

the ability to judge emotionally concerning the classification of feelings as positive or negative.
- *Output processing* between DLPFC and premotor/motor cortex and/or parietal cortex between 600 and 1000 ms, which may be seen neuropsychologically in the motor reaction and/or the imaginative association concerning the respective emotion.

In conclusion, it may assumed that there is *resonant processing* between OMFC, anterior cingulated (AC), DLPFC and VLPC, which is seen in early feedforward processing (200–300 ms), mediating feedback processing (300–400 ms) and late re-entrant processing (400–600 ms) and also in output processing (600–1000 ms). Furthermore, the following neuropsychological hypothesis may be assumed: such resonant processing in the prefrontal cortex is responsible for the transformation of emotional experience into behaviour which is generated by the interaction between emotion and cognition.

## 11.13 Prefrontal Cortical Function and Pathophysiology and Psychotherapy of Depression

### 11.13.1 Pathophysiological Results in Depression

Depression may be characterized by a great variety of findings in both the brain stem and the prefrontal cortex. Particular consideration will be given here to findings from the prefrontal cortex, which includes the OMFC, VLPFC and DLPFC (cf. Table 11.1).

#### 11.13.1.1 Post-mortem Findings in Depressive Disorders

Rajkowska (1999, 2000) found significant changes in the neuronal size and density in the OMFC, particularly in layer II, whereas in the VLPFC and DLPFC, the larger neurons in layers III–VI were significantly reduced in size and density. Because of these results, it may be concluded that particularly supragranular layers are involved in the OMFC, whereas in the VLPFC and DLPFC, mainly infragranular layers are modified. Moreover, in the anterior cingulate, a reduction of GABAergic neurons, chiefly in layer II, was found in depressed patients.

#### 11.13.1.2 Functional Imaging Findings in Depression

Hyperactivity in the OMPFC, in the subgenual region and in the VLFC was found in acute depression, whereas a reduced activity was found in the right DLPFC, anterior cingulated and caudatum (Mayberg et al. 1999; Liotti and Mayberg 2001; Brody et al. 1999). Interestingly, normalization of the hyperactivity and also of the hypoactivity in the corresponding regions with a reduction of the activation in the OMFC and the VLPFC and furthermore an increase of the activation in the right DLPFC and the anterior cingulate was found after remission (Mayberg et al. 1999; Brody et al. 1999). An alteration in the right DLPFC in depression can be further supported by findings showing a reduction of P300 (300–600 ms) in the right frontal

**Table 11.1** Prefrontal cortical processing, affective, cognitive and psychomotor symptoms and therapy of depression

| Prefrontal cortical processing | Pathophysiology and dysfunction | Cognitive phenomenology | Cognitive neuropsychology | Cognitive-behavioural therapy |
|---|---|---|---|---|
| *Feedforward* Medial orbitofrontal to dorsolateral prefrontal cortex over anterior cingulate: 200–300 ms | – Abnormal synchronization<br>– Hyperfunction with no possibility of modulation | – Intensity of feelings without cognitive control<br>– Negative emotional bias | – Attention: change reduced and focus maintained<br>– Set: initiation maintained and change reduced | – Imagination<br>– Change from attention and emotion influenced chronometrically and by TMS |
| *Feedback* Dorsolateral to medial orbitofrontal cortex over ventrolateral cortex: 300–400 ms | – Reduced output from dorsolateral prefrontal<br>– Hypofunction | – "Thematic over congruency"<br>– "Emotions without content" | – Working memory: reduced manipulation<br>– Reduced "intuitive decisions" | – Cognitive restructuring<br>– Training of emotional working memory |
| *Re-entrant* Medial orbitofrontal to dorsolateral prefrontal cortex over anterior cingulate: 400–600 ms | – Delayed processing<br>– Hypofunction with reduced processing | – Lack of flexible reasoning<br>– Focus on the present | – Working memory: reduced manipulation and online availability of information Altered response to feedback | – Monitoring and awareness of temporal dimensions<br>– Self-monitoring with emotional and cognitive restructuring |
| *Output* Dorsolateral prefrontal cortex to premotor/ motor cortex: 600–1000 ms | – Reduced output from dorsolateral prefrontal cortex<br>– Hypofunction | – Psychomotor slowing<br>– Behavioural deactivation | – Extended reaction time<br>– Deficits in planning and guessing tasks | – Imagining motion sequences<br>– Behavioural physiological activation |

electrodes (F4) in the EEG of depressed patients during an attention task; P300 showed a reduced amplitude and a late latency (Liotti and Mayberg 2001).

The involvement of the DLPFC is also supported by findings on the therapeutic effect of transcranial magnetic stimulation (TMS) in the left or right DLPFC in depressed patients (Schutter et al. 2001; Tormos et al. 1997; Pascual-Leone et al. 1996).

To sum up, the OMPFC, VLPFC and DLPFC showed the following alterations in depression:

– Differential involvement of supra- and infragranular layers in the OMFC (supragranular) and VLPFC/DLPFC (infragranular)

- Reciprocal hyperactivation and hypoactivation in OMPFC/VLPFC and DLPFC
- Specific involvement of the right DLPFC
- Abnormal interaction between emotion and cognition with an abnormal top-down or bottom-up modulation between OMFC, VLPFC and DLPFC

### 11.13.2 Feedforward Processing: Abnormal Synchronization and Chronometric-Orientated Emotional Imagination

#### 11.13.2.1 Pathophysiology

On the basis of the pathophysiological findings in depression, we may hypothesize abnormally strong feedforward processing from the OMFC to the DLPFC via the anterior cingulated between 200 and 300 ms. This hypothesis is supported by the following findings in depression:

- Hyperactivity in both the medioorbitofrontal cortex and the sub- and pregenual regions of the anterior cingulate (Drevets 2001)
- Post-mortem findings concerning changes in the supragranular neurons in the OMFC and anterior cingulated, which may lead to alterations in the inhibitory interneurons with a consecutive neuronal disinhibition
- Normalization of the hyperactivity in the OMFC and anterior cingulate after remission of depression
- Early alterations in working memory between 157 and 210 ms in depressed patients, which may reflect depressed patients' ability to hyperfocus their attention (Pelosi et al. 2000)

Such abnormal feedforward processing may lead to abnormal synchronization of the neuronal activity in the OMFC and interior cingulate with consecutive hypoactivity in the DLPFC. Both the OMFC and the AC may be partly involved in generating tonic activity (Raichle et al. 2001; Gevins et al. 1996, 1997), which is in contrast to the phasic activity, for instance, in the DLPFC. This abnormal synchronization may be characterized by tonic hyperactivity which makes phasic activity impossible. This hypothesis is supported by the following findings:

- No possibility of modulation by a different area (e.g. the amygdale; cf. Garcia et al. 1999; Rauch et al. 2000), which may be altered in depression (Drevets 2000) and the DLPFC. This results therefore in insufficient modulating ability by neuronal input.
- Fixed time sequence of the time course of neuronal activity between OMFC, anterior cingulate and DLPFC: the neuronal activity in these networks is "time-locked".
- No variation of intensity and degree of neuronal activity in the spatiotemporal domains in the time interval between 200 and 300 ms: absence of variability.

- Fixed sequence of the spatial extension of neuronal activity between OMFC, anterior cingulate and DLPFC. The neuronal is "space-locked".
- Lack of neuronal output with decreased processing of neuronal activity to other regions such as the DLPFC and VLPFC. No forward processing with a lack of neuronal output.

In contrast to depression, the feedforward processing from the medioorbitofrontal cortex (MOFC) to the DLPFC via the anterior cingulate may not be disturbed in schizophrenia because no alterations were found in the MOFC in schizophrenic patients. Although the neuronal output from the MOFC and the anterior cingulate in depression do not seem to reach the DLPFC, the DLPFC shows considerable changes in schizophrenic patients.

#### 11.13.2.2 Phenomenology and Neuropsychology

The hyperactivity in MOFC and anterior cingulate may be related to the increased intensity of emotions or emotions experienced by depressed patients. Because the MOFC and amygdala are especially involved in generating negative emotions, hyperactivity with abnormal synchronization of the MOFC may result in a predominance of negative emotions in depressed patients, which makes positive emotional processing play only a secondary role. Because of hyperactivity in the MOFC, modulation of the MOFC which would be necessary for processing positive emotions (cf. Northoff et al. 2000) may no longer be possible, so that this inability to modulate contributes considerably to the fixation of negative emotions. Therefore, depressed patients may be tied to negative emotional processing on the neuronal level, which may lead in these patients to the negative affective bias described above (Murphy et al. 1999).

Because of the abnormal synchronization of the anterior cingulate with the MOFC, the activity in the former is both time- and space-locked, so that variability with adaptation to the corresponding neuronal conditions is not possible. The hyperactivity in the anterior cingulate in depression may account for the sustained capability to focus attention (cf. Paulus et al. 2001; Elliott et al. 2000). As predominant negative emotional processing and the sustained capability to focus attention occur simultaneously, the negative emotional experience is abnormally increased. The abnormal synchronization may be manifested in the inability to change the attention so as to have different foci of attention or to change between different emotional states.

The abnormal synchronization of the neuronal activity between the MOFC and anterior cingulate leads to an alteration in the output with consecutive changes in forward processing to the DLPFC. The intrinsic activity of the DLFPC possibly stays the same, which may account for the sustained capability of set initiation. In contrast to the sustained basic activity in the DLPFC in depression, this basic activity or the DLPFC itself is fundamentally disturbed in schizophrenia, so that the set initiation is also disturbed in schizophrenic patients. In contrast to the intrinsic activity, the extrinsic activity which describes the input to the DLPFC is altered in depression. This may be explained by the lack of forward processing with no input from the anterior cingulate. Therefore, the set-shifting or mental flexibility, which may be induced by external input, is disturbed, so that initiation in depressed patients is still preserved, but flexible adaptation is disturbed.

#### 11.13.2.3 Psychotherapy

From a neuronal perspective, the *main aim* should be to *overcome abnormal synchronization*.

Compensation for not being able to modulate and for the lack of variability, and also for the time- and space-locked neuronal activity, and the reduced output from the anterior cingulate is of central importance. This compensation may possibly induce the specific constellation of negative affective bias, hyperintensity of emotions, sustained attention focus with reduced change of attention and the dissociation between set initiation and shifting.

The ability to modulate may be supported in psychotherapy by positive emotional stimulation using *relaxation techniques* in combination with *imagination exercises*, e.g. imagining positive emotions. It should, of course, be taken into account that patients in an acute depressive state may no longer be able to imagine emotions. In such cases, abnormal synchronization may be counteracted by *transcranial magnetic stimulation (TMS)* which acts chronometrically and is applied in the time range between 200 and 300 ms over the DLPFC. Such chronometrically oriented TMS may be able to form a link with the emotional imagination. Moreover, it could open the way to "unlock" time- and space-locked neuronal activity and thereafter bring about stabilization by positive emotional imagination.

Furthermore, the emotional imagination should be combined with attention tasks, especially attention shift. Thus, the shift between experiencing negative emotions could take place while imagining positive emotions. Patients should also try and focus their attention on something other than emotions and thus train their *set-shifting* ability. For instance, patients may focus their attention on their own bodies and their movements which could be combined with physiotherapy. Focussing their attention on non-emotional contents may increase neuronal output and thus compensate the deficits in forward processing from the anterior cingulate to the DLPFC.

In summary, the primary therapeutic aim in the context of abnormal feedforward processing is to overcome abnormal synchronization with time- and space-locked neuronal activity. In patients with mild or moderate depression, abnormal synchronization may be overcome in psychotherapy by focussing on the positive emotional imagination and set-shifting. In contrast, chronometrical transcranial magnetic stimulation with an application of single stimuli (about 300 ms) over the DLPFC may be necessary in patients with severe depression; this application may be connected directly to positive emotional imagination and set-shifting. It may be assumed that overcoming abnormal synchronization in feedforward processing may account for the sudden improvement or "sudden gain" which can be observed, for instance, during cognitive-behavioural therapy in depressed patients (Tang and DeRubeis 1999).

### 11.13.3 Feedback Processing: Reduced Feedback and Cognitive Restructuring

#### 11.13.3.1 Pathophysiology

The input to the DLPFC is reduced due to abnormal synchronization, resulting in a reduced output from the DLPFC to the VLPFC and MOPFC in the time interval between 300 and 400 ms. The assumption of reduced dorsolateral and ventrolateral prefrontal cortical output in depression can be supported by the following findings:

- Hypofunction in the DLPFC during acute depression
- Normalization of the DLPFC's function in remission
- Inverse correlation between the hypofunction of the DLPFC and the MOFC/AC and hyperfunction during acute depression (Liotti and Mayberg 2001)
- Reduced amplitude and late latency of the P300 over the DLPFC in depressed patients during an attention task
- Post-mortem findings of a reaction in infragranular layers of the pyramidal neurons in the DLPFC and VLPFC in depressed patients, which is compatible with the assumed disturbed output from the DLPFC and VLPFC

On the basis of these findings, reduced feedback from the DLPFC to the MOFC via the VLPFC can be assumed in the time interval between 300 and 400 ms in depression. This reduced feedback processing has the following implications:

- Delayed neuronal processing from the DLPFC to the VLPFC and from the VLPFC to the MOFC: lack of correct timing between DLPFC and VLPFC
- Alteration of spatial distribution of neuronal activity from the DLPFC to the VLPFC
- Reduced output processing from the DLPFC and VLPFC
- Reduced input to the MOFC from the VLPFC
- Reduced ability to modulate the MOFC via the VLPFC

In contrast to depression, in schizophrenia it should be assumed that feedback from the DLPFC to the MOFC via the VLPFC is not only reduced (as in depression) but also disorganized and fundamentally altered because the DLPFC appears to be altered primarily in schizophrenic patients.

#### 11.13.3.2 Phenomenology and Neuropsychology

Depressed patients show a congruence between mood and emotion, which has been described as a "thematic over-congruence". As shown above, modulation in the DLPFC was determined by the abnormal synchronization in the "feedforward processing" from the MOFC to the DLPFC via the anterior cingulated, resulting in a focus on and restriction to negative emotions. Abnormal coupling of the anterior cingulate and the DLPFC may lead to the initiation of predominantly negative cognitions, which can no longer be connected with positive emotions because of the lack of set-shifting. Feedback processing from the DLPFC to the VLPFC, which are connected to each other through reciprocal connectivity, may lead to alterations in working memory (Pelosi et al. 2000) so that negative thoughts can no longer be modified. The inability to modulate negative thoughts consequently contributes to the predominance of negative emotions, making other forms of cognition, for instance, positive thoughts, impossible. The predominance of negative cognitions is then completely congruent with the negative emotions, so that a "thematic over-congruence" develops due to this abnormal, rigid and unilateral coupling of emotion and cognition.

Furthermore, the reduced function of working memory, which is connected with reduced and delayed neuronal activity in the DLPFC and VLPFC, makes it impossible to keep thoughts on-line, so that all other forms of content and thoughts are excluded. When working memory deficits and therefore the depression in general become so severe that it becomes absolutely impossible to stay on-line, not even negative thoughts can be represented any longer, so that the patient is conscious of merely the emotion but not the thoughts. This is how "emotions lacking content" (Brenner 1991) develops. In this case, the patients experience only feeling and emotions, but not thoughts.

Depressive patients show severe deficits in "decision-making" (cf. Murphy et al. 2001), which are closely related to the functions of the ventromedial and lateral prefrontal cortex. Decision-making is mostly based on "intuitive decisions" which remain largely subconscious and should therefore be distinguished from "logical decisions". The latter are mainly characterized by consciousness (Northoff et al. 2000). In the event of reduced feedback from the DLPFC and the VLPFC to the MOFC, decisions requiring a certain cognitive component, even if they are made unconsciously, can no longer be carried out correctly. On a symptomatological level, these deficits in "intuitive decision-making" become manifest in the inability to shift emotions, so that depressed patients can no longer connect their cognitions or unconscious cognitions with the corresponding emotions.

#### 11.13.3.3 Psychotherapy

The main aim of therapeutic intervention concerning reduced feedback processing is overcoming the "thematic over-congruence" between emotions and cognitions.

Even negative emotions should be connected with positive cognitions and vice versa. A connection between negative emotions and positive cognitions may be made possible by imagining cognitions or, in other words, cognitional imagination of positive thought content. In contrast to this, connecting negative cognitions with positive emotions may be possible by means of emotional imagination. This reciprocal connection between emotional and cognitional content, with dissociation of both contents, may lead to a loosening of the "thematic over-congruence" between emotion and cognition, which cognitive-behavioural therapy of depression tries to achieve through cognitive restructuring (Beck 1962; McGinn 2000; Deckersbach et al. 2000).

In addition to cognitive restructuring, "thematic over-congruence" may be overcome by training the working memory since it supports both the manipulation and representation of thoughts. Training the working memory may facilitate cognitive restructuring, so that both cognitive training and cognitive restructuring should be used complementarily.

The cognitive imagination can be connected with the current emotional experience of patients with emotions lacking content, so that are again able to dissociate emotional content from cognitional content. Therefore, although insufficient, training the working memory seems necessary, since otherwise imaginary cognitions cannot be represented on-line. The complementary connecting of cognitive imagination and working memory may lead to the reestablishment of flexible coupling, with the possibility of dissociating cognition from emotion, which is necessary for adequate "decision-making" in depressed patients, so that they can shift again between cognition and emotion.

### 11.13.4 Re-entrant Processing: Reduced "Re-entry" and Awareness of Thoughts

#### 11.13.4.1 Pathophysiology

Because of abnormal synchronization in the MOFC/AC and reduced feedback from the VLPFC, reduced re-entrant processing (i.e. re-entry from the MOFC to the DLPFC via the anterior cingulate) in the time interval between 400 and 600 ms may be assumed. This assumption can be supported by the following findings in depression:

- Reduced amplitude in P300 in the time interval between 300 and 600 ms during an attention task (Liotti and Mayberg 2001)
- Alterations in negative and positive potentials between 400 and 800 ms during a working memory task (Pelosi et al. 2000)
- Hypofunction in the DLPFC in acute depression
- Therapeutic response to transcranial magnetic stimulation over the DLPFC in depressed patients

Re-entrant processing should be distinguished from feedback processing, because the former, unlike the latter, is based on the same anatomical structures that were used during previous neuronal processing. Alterations in re-entrant processing have the following implications:

- Delayed late processing in the MOFC, AC and DLPFC, which could be blocked by tonic hyperactivity with abnormal synchronization in early time intervals
- Reduced activation in the MOFC, AC and DLPFC in late time intervals (400–600 ms)
- Abnormal relational interference between early and late processing over the same anatomical regions
- Alterations in the spatial and temporal relation between the MOFC/AC and DLPFC with potential spatial and/or temporal desynchronization in late processing

Unlike in depression, re-entrant processing in schizophrenia appears to be not only reduced but also chaotic, because one of the main nerve centres (the DLPFC) has both functional and structural lesions, making functional reorganization necessary, which is not the case in depression.

#### 11.13.4.2 Phenomenology and Neuropsychology

Depressed patients may be characterized by a reduction in flexible thinking, which could be closely connected with working memory deficits and thus with the late function of the DLPFC. The DLPFC becomes active mainly in late time intervals between 300 and 600 ms during working memory tasks (Gevins et al. 1996, 1997); this corresponds temporally with the late re-entrant processing. The reduction of inflexible thinking, therefore, may be closely connected with working memory deficits, making it impossible for conscious thoughts to be kept on-line or manipulated. Moreover, the inability to keep thoughts on-line contributes to the loss of the time dimension in subjective experience, so that patients are forced to focus on the present, having lost the past and the future in subjective experience. Depressed patients are no longer able to integrate past and future events into present experiences. Instead, subjective experience is determined solely by the prevailing negative emotions and cognitions in the present, blocking access to other emotions either from the past or still yet to happen.

Further, reduced re-entrant processing to the DLPFC is possibly also connected with the inability to control thoughts or cognitions. Thoughts are initiated in early processing, but they cannot be controlled and thus attributed to the self because of the reduced re-entry in late processing, which means that depressed patients subjectively experience the inability to control the cognition they themselves initiated.

Finally, depressed patients give altered responses to feedback in planning and guessing tasks, which may be due to a neuronal delay in the activity in the medial orbitofrontal cortex (Elliott et al. 1998). Reduced and delayed activity in the MOFC is in line with the assumption of reduced re-entrant processing in the late time intervals, because reacting and replying to feedback in planning tasks require both feedback processing from the DLPFC to the MOFC and late re-entrant processing from the MOFC to the DLPFC.

#### 11.13.4.3 Psychotherapy

The aim of therapeutic intervention in late processing is to restore the subjective experience of different time dimensions, in order to free the way for flexible thinking and cognition as well as self-monitoring.

It may be possible, for instance, to restore the subjective experience of time dimensions by the patient imagining possible events and then validating them based on their realistic character in the present. Furthermore, an attempt could be made to work with the contents of the patient's memory in such a way as to enable a comparison of past and present events and thoughts and thus to modulate them both ways.

Self-monitoring could be supported by training the awareness of thoughts and cognitions, e.g. by *meditation*. Flexible thinking and cognition can also be achieved through cognitive restructuring in cognitive-behavioural therapy (McGinn 2000; Beck 1962; Deckersbach et al. 2000). These therapeutic goals may be more easily reached on the basis of the restored subjective experience of time dimensions.

### 11.13.5 Output Processing: Reduced Output and Psychomotor Activation

#### 11.13.5.1 Pathophysiology

It has been shown that the dorsolateral prefrontal cortex is hypoactive in depression. Because of the strong connections from the DLPFC to the premotor/motor cortex, that is, the main output regions, reduced activity in the DLPFC may also lead to reduced input to the premotor/motor cortex in the time interval between 600 and 1000 ms. This contributes to reduced internal initiation and execution of actions into movements. Assumed reduced output from the DLPFC with consecutively reduced input to the premotor/motor cortex can be supported by the following findings:

- Alterations in the latency of the readiness potential which correspond to the deficits in the internal initiation of movements in depressed patients (Northoff et al. 2000)
- Alterations, but not major deficits, in cortical motor activation
- Severe psychomotor retardation in acutely depressed patients

Owing to the hypofunction of the DLPFC with reduced input to the premotor/motor cortex in late time intervals (600–1000 ms), functional decoupling of the prefrontal cortex, on the one hand, and the premotor/motor cortex, on the other hand, may be assumed.

#### 11.13.5.2 Phenomenology and Neuropsychology

The reduced internal initiation of action and movement leads to psychomotor slowing, retardation and inactivation, as can often be observed in depressed patients and—as an extreme—in catatonic patients. These deficits are manifest neuropsychologically in longer reaction times (Liotti and Mayberg 2001) and in greater deficits in planning of actions and executive functions (e.g. in the "Tower of London" test). It is exactly these longer reactions times as well as the executive deficits that may possibly be accounted for by the functional decoupling of the prefrontal cortex, on the one hand, and the premotor/motor cortex, on the other hand.

#### 11.13.5.3 Psychotherapy

The main aim of therapeutic intervention concerning altered output processing focusses on psychomotor and physiological activation as well as sensorimotor imagination, leading to reactivation of the DLPFC and the premotor/motor cortex and also to recoupling of the DLPFC and the premotor/motor cortex.

## 11.14 A Neuropsychodynamic Approach to Depression: Conclusion and Therapeutic Consequences

A neuropsychodynamic approach to depression may be summarized as follows: The self and the changes in self-experience are core dimensions in depression and of psychoanalytical theories of depression. The experience of self-related depression can be characterized as the experience of the loss of the self. A mechanism-based approach was developed, focussing on the psychodynamic, psychological and neuronal mechanisms in healthy and depressed persons. On the basis of empirical results concerning emotional-cognitive interaction in depression, neuropsychodynamic hypotheses of the self in depression were developed:

First, it may be assumed that the empirically validated increased resting state activity in depression is a predisposition for the reactivation of experiences of early loss. The term "experiences of object loss" focusses not only on traumatic relationship experiences but also encompasses the loss of the self in a significant relationship structure.

Second, it may be hypothesized that the resting state-stimulus interaction in depression is reduced because of the increased resting state activity and that it corresponds with introjective processes of the self in the relationship with objects (correlated with negative emotions).

Third, it may be hypothesized that the modulation of the resting state activity by means of stimulus-induced activity and the resting state-stimulus interaction in depression is reduced because of the increased resting state activity. Dysfunctional development of the neuronal structure and organization results from the reduced resting state-stimulus interaction, as displayed in the processing of current experiences of loss.

The increased resting state activity in depression is especially associated with an increased resting state activity in the default mode network (DMN). By means of this, changes in the complete spatial-temporal structure of the intrinsic activity of the brain and the dysbalance between default mode network and executive network (EN) are induced. The reciprocal or negative interaction between DMN and EN is shifted in the direction of the DMN. This dysbalance causes an abnormal increase in the internal mental contents, whereas externally oriented actions are decreased. The increased inward focus (with strong ruminations) and a reduced outward focus (with a reduced relationship to the outside world) are core symptoms in depression. The depressed patient is no longer able to differentiate between external stimuli and his/her own self (caused by the increased resting state activity in the DMN which cannot be modified by external stimuli).

The question of why adaptive mechanisms are activated in the disturbed context of the increased resting state activity may be answered by mentioning the central aim of these neuropsychodynamic mechanisms: to maintain at all costs the subjective existence of the self in view of the experienced threat of loss of the self. It is neither lesions nor disturbances of adaptive neuronal mechanisms which generate depressive symptoms, but rather increasingly dysfunctional mechanisms of compensation on the basis of the increased resting state activity.

### 11.14.1 A Spatiotemporal Approach to Depression

As has been shown, different kinds of cognitive symptoms in depression may be connected with different neuropsychological functions, which in turn can be attributed to different forms of processing in the prefrontal cortex. These different forms of processing in the prefrontal cortex can be characterized by spatiotemporal patterns of neuronal activity, which are altered in a specific way during depression. In depression, there is hypofunction in the medial orbitofrontal cortex with consecutive changes in feedforward, feedback and re-entrant processing in the prefrontal cortex, anterior cingulate, DLPFC and VLPFC, respectively. The changes in prefrontal processing in depression are mostly related to an abnormal increase and synchronization in feedforward processing from the ventromedial prefrontal cortex, as well as to the resulting abnormal reduction of feedback and re-entrant processing.

Psychotherapeutic interventions in depression should focus on restoring these various forms of processing in the prefrontal cortex. In this context, emotional, cognitional and motor imagination and working memory training, chronometrically oriented cognitive therapy and awareness of the time dimension play an important role in the psychotherapy of depression. These various inputs, based on neurophysiological mechanisms, could well complement existing psychotherapeutic approaches such as psychodynamic psychotherapy, cognitive-behavioural therapy, interpersonal therapy (IPT), mindfulness-based cognitive therapy (MBCT) and CBASP. Considering the cognitive phenomenology described above and the underlying physiological mechanisms, new psychotherapeutic approaches may be developed in the future on the basis of the specific change in processing in the prefrontal cortex, in the sense of "phenomenologically and physiologically based neuropsychodynamic psychotherapy".

Some limitations of this neuropsychodynamic approach to depression concern the problem of investigating psychodynamic dimensions of depression by means of operationalized studies (compare Boeker and Northoff 2010; Boeker et al. 2013). Further studies are necessary to validate the increased resting state activity in depression.

### 11.14.2 Therapeutic Consequences

Possible therapeutic consequences of the neuropsychodynamic approach to depression involve the necessary emotional attunement in psychoanalytic psychotherapy of depressed patients and the adequate timing of therapeutic interventions (confer Stern 1985; Boeker 2003b; Boeker et al. 2012). The hypotheses which have been developed in the context of the neuropsychodynamic model of depression may be used for more specific psychotherapeutic interventions, aiming at specific mechanisms of compensation and defence, which are related to the increased resting state activity and the disturbed resting state-stimulus interaction. Moreover, in a future "brain-based psychoanalytical psychotherapy" of depression, the enabled processes of development and separation will be based on new experiences in the context of the therapeutic relationship.

## References

Adroer S. Some considerations in the structure of the self and its pathology. Int J Psychoanal. 1998;79:681–96.
Alcaro A, Panksepp J, Witczak J, et al. Is subcortical-cortical midline activity in depression mediated by glutamate and GABA? A cross-species translational approach. Neurosci Biobehav Rev. 2010;34(4):592–605.
Bar KJ, Greiner W, Jochum T, et al. The influence of major depression and its treatment on heart rate variability and pupillary light reflex parameters. J Affect Disord. 2004;82(2):245–52.
Bechara A, Damasio H, Damasio AR. Emotion, decision making and the orbitofrontal cortex. Cereb Cortex. 2000;10:295–307.
Beck A. Thinking and depression. Arch Gen Psychiatry. 1962;9:36–45.
Berman SM, McCann JT. Defense mechanisms and personality disorders: an empirical test of Millon's theory. J Pers Assess. 1995;64(1):132–44.
Blatt SJ. Contributions of psychoanalysis to the understanding and treatment of depression. Am J Psychoanal Assoc. 1998;46:723–52.
Boeker H. Selbstbild und Objektbeziehungen bei Depressionen: Untersuchungen mit der Repertory Grid-Technik und dem Gießen-Test an 139 PatientInnen mit depressiven Erkrankungen. Monographien aus dem Gesamtgebiete der Psychiatrie. Darmstadt: Steinkopff-Springer; 1999.
Boeker H. Depression, Manie und schizoaffektive Psychosen. Psychodynamische Theorien, einzelfallorientierte Forschung und Psychotherapie. Giessen: Psychosozial Verlag; 2000.
Boeker H. Depression, Manie und schizoaffektive Psychosen: psychodynamische Theorien, einzelfallorientierte Forschung und Psychotherapie. Giessen: Psychosozial-Verlag; 2001.
Boeker H. Depressionen: psychosomatische Erkrankungen des Gehirns? In: Boeker H, Hell D, editors. Therapie der affektiven Störungen. Psychosoziale und neurobiologische Perspektiven. Stuttgart: Schattauer; 2002. p. 183–205.
Boeker H. Sind Depressionen psychosomatische Erkrankungen? Vierteljahresschr Naturforsch Ges Zürich. 2003a;148:1–16.
Boeker H. Symbolisierungsstörungen bei schweren Depressionen: Zur Bedeutung psychosomatischer Circuli vitiosi bei depressiv Erkrankten. In: Lahme-Gronostaj H, editor. Symbolisierung und ihre Störungen. Frankfurt: Deutsche Psychoanalytische Vereinigung; 2003b. p. 149–64.
Boeker H. Persons with depression, mania and schizoaffective Psychosis – investigations of cognitive complexity, self-esteem, social perception and object relations by means of the Repertory Grid-Technique. In: Klapp BF, et al., editors. Role repertory grid and body grid – Construct psychological approaches in psychosomatic research. Frankfurt: VAS – Verlag für Akademische Schriften, Reihe Klinische Psycholinguistik; 2004. p. S3–20.
Boeker H. Melancholie, Depression und affektive Störungen: Zur Entwicklung der psychoanalytischen Depressionsmodelle und deren Rezeption in der Klinischen Psychiatrie. In: Boeker H, editor. Psychoanalyse und Psychiatrie – Historische Entwicklung, Krankheitsmodelle und therapeutische Praxis. Berlin: Springer; 2005. p. 115–58.
Boeker H. Selbst und Körper in der Depression: Herausforderungen an die Therapie. Schweiz Arch Neurol Psychiatr. 2009;160(5):188–99.
Boeker H, Grimm S. Emotion und Kognition bei depressiv Erkrankten. In: Boeker H, Seifritz E, editors. Psychotherapie und Neurowissenschaften. Integration – Kritik – Zukunftsaussichten. Bern: Verlag Hans Huber; 2012. p. 309–51.
Boeker H, Northoff G. Desymbolisierung in der schweren Depression und das Problem der Hemmung: Ein neuropsychoanalytisches Modell der Störung des emotionalen Selbstbezuges Depressiver. Psyche – Z Psychoanal. 2005;59:964–89.
Boeker H, Northoff G. Die Entkopplung des Selbst in der Depression: empirische Befunde und neuropsychodynamische Hypothesen. Psyche – Z Psychoanal. 2010;64:934–76.
Boeker H, Northoff G, Lenz C, et al. Die Rekonstruktion der Sprachlosigkeit: Untersuchungen des subjektiven Erlebens ehemals katatoner PatientInnen mittels modifizierter Landfield-Kategorien. Psychiatr Prax. 2000a;27:389–96.

Boeker H, von Schmeling C, Lenz C, et al. Subjective experience of catatonia: construct-analytical findings by means of modified Landfield categories. In: Scheer JW, editor. The person in society: challenges to a constructivist theory. Giessen: Psychosozial-Verlag; 2000b. p. S303–16.
Boeker H, Hell D, Budischewski K, et al. Personality and object relations in patients with affective disorders: idiographic research by means of the repertory grid-technique. J Affect Disord. 2000c;60:53–60.
Boeker H, Schulze J, Richter A, Nikisch G, Schuepbach D, Grimm S. Sustained cognitive impairments after clinical recovery of severe depression. J Nerv Ment Dis. 2012;200(9):773–6.
Boeker H, Richter A, Himmighoffen H, et al. Essentials of psychoanalytic process and change: how can we investigate the neural effects of psychodynamic psychotherapy in individualised neuro-imaging? Front Hum Neurosci. 2013;7:355.
Brenner C. A psychoanalytic perspective on depression. J Am Psychoanal Assoc. 1991;39(1):25–43.
Brody A, Saxena S, Silverman D, et al. Brain metabolic changes in major depressive disorder from pre- to posttreatment with paroxetine. Psychiatry Res Neuroimaging. 1999;91:127–39.
Buckner RL, Andrews-Hanna JR, Schacter DL. The brain's default network: anatomy, function, and relevance to disease. Ann N Y Acad Sci. 2008;1124:1–38.
Canli T, Sivers H, Thomason ME, et al. Brain activation to emotional words in depressed vs healthy subjects. Neuroreport. 2004;15:2585–8.
Carhart-Harris RL, Mayberg HS, Malizia AL, et al. Mourning and melancholia revisited: correspondences between principles of Freudian metapsychology and empirical findings in neuropsychiatry. Ann General Psychiatry. 2008;24:7–9.
Carhart-Harris RL, Friston KJ. The default-mode, ego-function and free energy: a neurobiological account of Freudian ideas. Brain. 2010;133:1265–83.
Coyne JC. Depression and response of others. J Abnorm Psychol. 1976a;85:186–93.
Coyne JC. Toward an interactional description of depression. Psychiatry. 1976b;39:28–40.
Coyne JC. Studying depressed persons' interactions with strangers and spouses. J Abnorm Psychol. 1985;94(2):231–2.
Coyne JC, Kessler RC, Tal M, et al. Living with a depressed person. J Consult Clin Psychol. 1987;55(3):347–52.
Damasio AR. How the brain creates the mind. Sci Am. 1999;281(6):112–7.
Davidson RJ. Anxiety and affective style: role of prefrontal cortex and amygdala. Biol Psychiatry. 2002;51(1):68–80.
Davidson RJ, Irwin W, Anderle MJ, Kalin NH. The neural substrates of affective processing in depressed patients treated with venlafaxine. Am J Psychiatry. 2003;160:64–75.
Deci EL, Eghrari H, Patrick BC, et al. Facilitating internalization: the self-determination theory perspective. J Pers. 1994;62(1):119–42.
Deckersbach T, Gershumy B, Otto M. Cognitive-behavioral therapy for depression. Psychiatr Clin North Am. 2000;23(4):795–809.
Dennecker FW. Das Selbst-system. Psyche. 1989;43:577–608.
Dichter GS, Felder JN, Petty C, et al. The effects of psychotherapy on neural responses to rewards in major depression. Biol Psychiatry. 2009;66:886–97.
Drevets W. Neuroimaging studies of mood disorders. Biol Psychiatry. 2000;48:813–29.
Drevets W. Neuroimaging and neuropathological studies of depression. Curr Opin Neurobiol. 2001;11:240–9.
Drevets W, Raichle M. Reciprocal suppression of rCBF during emotional versus higher cognitive processes. Cognit Emot. 1998;12(3):353–835.
Drevets WC, Price JL, Furey ML. Brain structural and functional abnormalities in mood disorders: implications for neurocircuitry models of depression. Brain Struct Funct. 2008;213(1-2):93–118.
Elliott R, Sahakian BJ, Michael A, et al. Abnormal neural response to feedback on planning and guessing tasks in patients with unipolar depression. Psychol Med. 1998;28:559–71.
Elliott R, Rubinzstein JS, Sahakian BJ, et al. Selective attention to emotional stimuli in a verbal go/no-go task: an fMRI study. NeuroReport. 2000;11:1739–44.
Elliott R, Rubinsztein JS, Sahakian BJ, et al. The neural basis of mood-congruent processing biases in depression. Arch Gen Psychiatry. 2002;59(7):597–604.

Ernst J, Northoff G, Boeker H, et al. Interoceptive awareness enhances neural activity during empathy. Hum Brain Mapp. 2013;34:1615–24.

Ernst J, Boeker H, Hättenschwiler J, et al. The association of interoceptive awareness and alexithymia with neurotransmitter concentrations in insula and anterior cingulate. Soc Cogn Affect Neurosci. 2014;9:857–63.

Feder A, Nestler EJ, Charney DS. Psychobiology and molecular genetics of resilience. Net Rev Neurosci. 2009;10(6):446–57.

Feinberg TE. Neuropathologies of the self: clinical and anatomical features. Conscious Cogn. 2011;20:75–81.

Fitzgerald PB, Sritharan A, Daskalakis ZJ, et al. A functional magnetic resonance imaging study of the effects of low frequency right prefrontal transcranial magnetic stimulation in depression. J Clin Psychopharmacol. 2007;27:488–92.

Freud S. Zur Einführung des Narzissmus. GW 10. 1914.

Freud S. Mourning and melancholia. SE XIV; 1917. pp. 239–60.

Friston KJ. Imaging neuroscience: principles or maps? Proc Natl Acad Sci U S A. 1998;95(3):796–802.

Friston K. Learning and inference in the brain. Neural Netw. 2003;16(9):1325–52.

Friston KJ, Penny W. Posterior probability maps and SPMs. NeuroImage. 2003;19(3):1240–9.

Friston KJ, Price CJ. Dynamic representations and generative models of brain function. Brain Res Bull. 2001;54(3):275–85.

Friston KJ, Price CJ. Degeneracy and redundancy in cognitive anatomy. Trends Cogn Sci. 2003;7(4):151–2.

Friston KJ, Fletcher P, Josephs O, et al. Event-related fMRI: characterizing differential responses. NeuroImage. 1998a;7(1):30–40.

Friston KJ, Josephs O, Rees G, et al. Nonlinear event-related responses in fMRI. Magn Reson Med. 1998b;39(1):41–52.

Friston KJ, Harrison L, Penny W. Dynamic causal modelling. NeuroImage. 2003;19(4):1273–302.

Fu CHY, Williams SCR, Cleare AJ, et al. Attenuation of the neural response to sad faces in major depression by antidepressant treatment: a prospective, event-related functional magnetic resonance imaging study. Arch Gen Psychiatry. 2004;61:877–89.

Gabbard G. Psychodynamic psychiatry in clinical practice. Arlington: American Psychiatric Press; 2005.

Gabbard GO. Psychodynamic psychiatry in clinical practice. 5th ed. Washington DC: American Psychiatric Publishing; 2014.

Garcia R, Voulmba R, Baudry M, et al. The amygdala modulates prefrontal cortex activity relative to conditioned fear. Nature. 1999;402:294–6.

Gevins A, Smith M, Le J, et al. High resolution evoked potential imaging of the cortical dynamics of human working memory. Electroencephalogr Clin Neurophysiol. 1996;98:327–48.

Gevins A, Smith M, McEvoy L, et al. High resolution EEG mapping of cortical activation related to working memory. Cereb Cortex. 1997;7:374–85.

Goel V, Dolan RJ. Explaining modulation of reasoning by belief. Cognition. 2003a;87(1):11–22.

Goel V, Dolan RJ. Reciprocal neural response within lateral and ventral medial prefrontal cortex during hot and cold reasoning. NeuroImage. 2003b;20(4):2314–21.

Gotlib IH, Krasnoperova E, Yue DN, et al. Attentional biases for negative interpersonal stimuli in clinical depression. J Abnorm Psychol. 2004;113(1):121–35.

Greicius MD, Krasnow B, Reiss AL, et al. Functional connectivity in the resting brain: a network analysis of the default mode hypothesis. Proc Natl Acad Sci U S A. 2003;100(1):253–8.

Greicius MD, Flores BH, Menon V, et al. Resting-state functional connectivity in major depression: abnormally increased contributions from subgenual cingulate cortex and thalamus. Biol Psychiatry. 2007;62:429–37.

Grimm S, Schmidt CF, Bermpohl F, et al. Segregated neural representation of distinct emotion dimensions in the prefrontal cortex – an fMRI study. NeuroImage. 2006;30(1):325–40.

Grimm S, Beck J, Schüpbach D, et al. Imbalance between left and right dorsolateral prefrontal cortex in major depression is linked to negative emotional judgment. An fMRI study in severe major depressive disorder. Biol Psychiatry. 2008;63:369–76.

Grimm S, Ernst J, Boesiger P, et al. Increased self-focus in major depressive disorder is related to neural abnormalities in subcortical midline structures. Hum Brain Mapp. 2009;30(8):2617–27.
Guinjoan SM, Bernabo JL, Cardinali DP. Cardiovascular tests of autonomic function and sympathetic skin responses in patients with major depression. J Neurol Neurosurg Psychiatry. 1995;59(3):299–302.
Gusnard DA, Raichle ME. Searching for a baseline: functional imaging and the resting human brain. Nat Rev Neurosci. 2001;2(10):685–94.
Gusnard DA, Akbudak E, Shulman GL, et al. Medial prefrontal cortex and self-referential mental activity: relation to a default mode of brain function. Proc Natl Acad Sci U S A. 2001;98(7):4259–64.
Gut E. Productive and unproductive depression. London: Tavistock and Routledge; 1989.
Heinzel A, Grimm S, Beck J, et al. Segregated neural representation of psychological and somatic-vegetative symptoms in severe major depression. Neurosci Lett. 2009;456(2):49–53.
Heller AS, Johnstone T, Shackman AJ, et al. Reduced capacity to sustain positive emotion in major depression reflects diminished maintenance of fronto-striatal brain activation. Proc Natl Acad Sci U S A. 2009;106(52):22445–50.
Holsboer-Trachsler E, Vanoni C. Depression in der Praxis. 3rd ed. Wessobrunn: Sozio-medico Verlag; 2007.
Ingram RE. Self-focused attention in clinical disorders: review and a conceptual model. Psychol Bull. 1990;107(2):156–76.
Inoue Y, Tonooka Y, Yamada K, et al. Deficiency of theory of mind in patients with remitted mood disorder. J Affect Disord. 2004;82(3):403–9.
Jacobson E. Depression. Comparative studies of normal, neurotic and psychotic conditions. New York: New Intern Universities Press; 1971.
Keedwell PA, Andrew C, Williams SCR, et al. The neural correlates of anhedonia in major depressive disorder. Biol Psychiatry. 2005;58:843–53.
Kerr N, Dunbar RI, Bentall RP. Theory of mind deficits in bipolar affective disorder. J Affect Disord. 2003;73(3):253–9.
Kohut H. The restoration of the self. International Universities Press, New York; 1977. Deutsch (1979) Die Heilung des Selbst. Frankfurt: Suhrkamp.
Kratzsch S. Depressionen: Erleben und Selbst in der depressiven Erkrankung. In: Milch W, editor. Lehrbuch der Selbstpsychologie. Stuttgart: Kohlhammer; 2001. p. 191–213.
Kumar P, Waiter G, Ahearn T, et al. Abnormal temporal difference reward-learning signals in major depression. Brain. 2008;131:2084–93.
Kumari V, Mitterschiffthaler MT, Teasdale JD, et al. Neural abnormalities during cognitive generation of affect in treatment-resistant depression. Biol Psychiatry. 2003;54:777–91.
Lamme VA. Blindsight: the role of feedforward and feedback corticocortical connections. Acta Psychol. 2001;107(1-3):209–28.
Lamme VA. Separate neural definitions of visual consciousness and visual attention; a case for phenomenal awareness. Neural Netw. 2004;17(5-6):861–72.
Lawrence NS, Williams AM, Surguladze S, et al. Subcortical and ventral prefrontal cortical neural responses to facial expressions distinguish patients with bipolar disorder and major depression. Biol Psychiatry. 2004;55:578–87.
Lichtenberg JD. Psychoanalysis and infant research. Hillsdale: The Analytic Press; 1983.
Liotti M, Mayberg H. The role of functional neuroimaging in the neuropsychology of depression. J Clin Exp Neuropsychol. 2001;23:121–36.
Liotti M, Mayberg HS, McGinnis S, et al. Unmasking disease-specific cerebral blood flow abnormalities: mood challenge in patients with remitted unipolar depression. Am J Psychiatry. 2002;159(11):1830–40.
Malancharuvil JM. Projection, introjection, and projective identification: a reformulation. Am J Psychoanal. 2004;64(4):375–82.
Masterman DL, Cummings JL. Frontal-subcortical circuits: the anatomic basis of executive, social and motivated behaviors. J Psychopharmacol. 1997;11(2):107–14.
Mayberg H. Depression, II: localization of pathophysiology. Am J Psychiatry. 2002;159:1979.

Mayberg HS. Modulating dysfunctional limbic-cortical circuits in depression: towards development of brain-based algorithms for diagnosis and optimised treatment. Br Med Bull. 2003a;65:193–207.

Mayberg HS. Positron emission tomography imaging in depression: a neural systems perspective. Neuroimaging Clin N Am. 2003b;13:805–15.

Mayberg HS, Liotti M, Brannan SK, et al. Reciprocal limbic-cortical function and negative mood: converging PET findings in depression and normal sadness. Am J Psychiatry. 1999;156:675–82.

Mazoyer B, Zago L, Mellet E, et al. Cortical networks for working memory and executive functions sustain the conscious resting state in man. Brain Res Bull. 2001;54(3):287–98.

McGinn L. Cognitive-behavioral therapy for depression. Am J Psychother. 2000;54:257–62.

Mentzos S. Depression und Manie; Psychodynamik und Psychotherapie affektiver Störungen. Zürich: Vandenhoeck und Ruprecht; 1995.

Mentzos S. Lehrbuch der Psychodynamik. 5th ed. Göttingen: Vandenhoeck & Ruprecht; 2009.

Miller A. Zeitkurven. Ein Leben. Frankfurt: Fischer; 1987. Englisch: Timebends. New York: Grove Press.

Murphy FC, Sahakian JS, Rubinsztein A, et al. Emotional bias and inhibitory control processes in mania and depression. Psychol Med. 1999;29:1307–21.

Murphy FC, Rubinzstein JS, Michael A, et al. Decision-making cognition in mania and depression. Psychol Med. 2001;31:679–93.

Murphy FC, Nimmo-Smith I, Lawrence AD. Functional neuroanatomy of emotions: a meta-analysis. Cogn Affect Behav Neurosci. 2003;3(3):207–33.

Nagai Y, Critchley HD, Featherstone E, et al. Activity in ventromedial prefrontal cortex covaries with sympathetic skin conductance level: a physiological account of a "default mode" of brain function. NeuroImage. 2004;22(1):243–51.

Nemeroff CB, Heim CM, Thase ME, et al. Differential responses to psychotherapy versus pharmacotherapy in patients with chronic forms of major depression and childhood trauma. PNAS. 2003;100:14293–6.

Northoff G. Das Gehirn: Eine neurophilosophische Bestandesaufnahme. Paderborn: Mentes-Verlag; 2000.

Northoff G. Catatonia and neuroleptic malignant syndrome: psychopathology and pathophysiology. J Neural Transm. 2002a;109(12):1453–67.

Northoff G. What catatonia can tell us about "top-down" modulation: a neuropsychiatric hypothesis. Behav Brain Sci. 2002b;25(5):555–77. discussion 578–604

Northoff G. Philosophy of the brain. The brain problem. Amsterdam: John Benjamin Publishing; 2004.

Northoff G. Unlocking the brain. Volume 2: consciousness. Oxford: Oxford University Press; 2013.

Northoff G, Bermpohl F. Cortical midline structures and the self. Trends Cogn Sci. 2004;8(3):102–7.

Northoff G, Boeker H. Principles of neuronal integration and defense mechanisms: neuropsychoanalytic hypothesis. Neuropsychoanalysis. 2006;8(1):69–84.

Northoff G, Eckert J, Fritze J. Glutamatergic dysfunction in catatonia? Successful treatment of three acute akinetic catatonic patients with the NMDA antagonist amantadine. J Neurol Neurosurg Psychiatry. 1997;62:404–6.

Northoff G, Richter A, Gessner M, et al. Functional dissociation between medial and lateral spatiotemporal activation in negative and positive emotions: a combined FMRI/MEG study. Cereb Cortex. 2000;10:93–107.

Northoff G, Bogerts B, Baumgart F, et al. Orbitofrontal cortical dysfunction and "sensorimotor regression": a combined study of fMRI and personal constructs in catatonia. Neuropsychoanalysis. 2002;4:149–75.

Northoff G, et al. Emotional-behavioral disturbances in catatonia: a combined study of psychological self-evaluation and fMRI. Neuropsychoanalysis. 2003;3:151–67.

Northoff G, Heinzel A, Bermpohl F, et al. Reciprocal modulation and attenuation in the prefrontal cortex: an fMRI study on emotional-cognitive interaction. Hum Brain Mapp. 2004;21(3):202–12.

Northoff G, Richter A, Bermpohl F, et al. NMDA hypofunction in the posterior cingulate as a model for schizophrenia: an exploratory ketamine administration study in fMRI. Schizophr Res. 2005;72(2-3):235–48.
Northoff G, Walter M, Schulte RF, et al. Gaba concentrations in the human anterior cingulate cortex predict negative BOLD responses in fMRI. Nat Neurosci. 2007;10(12):1515–7.
Panksepp J. Affective neuroscience: the foundations of human and animal emotions. New York: Oxford University Press; 1998a.
Panksepp J. The periconscious substrates of consciousness: affective states and the evolutionary origins of the self. J Conscious Stud. 1998b;5(5-6):566–82.
Paradiso S, Lamberty G, Garvey M, et al. Cognitive impairment in the euthymic phase of chronic unipolar depression. J Nerv Ment Dis. 1997;185(12):748–54.
Pascual-Leone A, Catala M, Pascual-Leone Pascual A. Lateralized effect of rapid rate transcranial magnetic stimulation of the prefrontal cortex on mood. Neurology. 1996;46:499–502.
Paulus M, Hozack N, Zauschner B, et al. Prefrontal, parietal, and temporal cortex networks underlie decision-making in the presence of uncertainty. NeuroImage. 2001;13:91–100.
Pelosi L, Slade T, Blumhardt LD, et al. Working memory dysfunction in depression: an event-related potential study. Clin Neurophysiol. 2000;111:1531–43.
Pessoa L, Ungerleider LG. Neuroimaging studies of attention and the processing of emotion-laden stimuli. Prog Brain Res. 2004;144:171–82.
Pessoa L, McKenna M, Gutierrez E, et al. Neural processing of emotional faces requires attention. Proc Natl Acad Sci U S A. 2002;99(17):11458–163.
Phan KL, Wager T, Taylor SF, et al. Functional neuroanatomy of emotion: a meta-analysis of emotion activation studies in PET and fMRI. NeuroImage. 2002;16(2):331–48.
Phillips ML, Drevets WC, Rauch SL, et al. Neurobiology of emotion perception II: implications for major psychiatric disorders. Biol Psychiatry. 2003;54:515–28.
Pizzagalli DA, Holmes AJ, Dillon DG, et al. Reduced caudate and nucleus accumbens response to rewards in unmedicated individuals with major depressive disorder. Am J Psychiatry. 2009;166(6):702–10.
Post RM. Transduction of psychosocial stress into the neural biology of recurrent affective disorder. Am J Psychiatry. 1992;149:99–1010.
Price JL, Drevets WC. Neurocircuitry of mood disorders. Neuropsychopharmacology. 2010;35(1):192–216.
Price CJ, Friston KJ. Degeneracy and cognitive anatomy. Trends Cogn Sci. 2002;6(10):416–21.
Qin P, Northoff G. How is our self related to midline regions and the default-mode network? NeuroImage. 2011;57:1221–33.
Raichle ME. Cognitive neuroscience. Bold insights. Nature. 2001;412(6843):128–30.
Raichle ME. Functional brain imaging and human brain function. J Neurosci. 2003;23(10):3959–62.
Raichle ME, MacLeod AM, Snyder AZ, et al. A default mode of brain function. Proc Natl Acad Sci U S A. 2001;98(2):676–82.
Rajkowska G. Morphometric evidence for neuronal and glial prefrontal cell pathology in major depression. Biol Psychiatry. 1999;45:1085–98.
Rajkowska G. Postmortem studies in mood disorders indicate altered numbers of neurons and glial cells. Biol Psychiatry. 2000;48:766–77.
Rauch S, Whalen P, Sin L, et al. Exaggerated amygdala response to masked facial stimuli in post-traumatic stress disorder: a functional MRI study. Biol Psychiatry. 2000;47:769–76.
Rimes KA, Watkins E. The effects of self-focused rumination on global negative self-judgments in depression. Behav Res Ther. 2005;43:1673–81.
Rogers R, Oweb A, Williams E, et al. Choosing between small, likely rewards and large, unlikely rewards activates inferior and orbital prefrontal cortex. J Neurosci. 1999;20(19):9029–38.
Sajonz B, Kahnt T, Margulies DS, et al. Delineating self-referential processing from episodic memory retrieval: common and dissociable networks. NeuroImage. 2010;50(4):1606–17.
Salvadore G, Cornwell BR, Colon-Rosario V, et al. Increased anterior cingulate cortical activity in response to fearful faces: a neurophysiological biomarker that predicts rapid antidepressant response to ketamine. Biol Psychiatry. 2009;65:289–95.

Sanacora G. Cortical inhibition, gamma-aminobutyric acid, and major depression: there is plenty of smoke but is there fire? Biol Psychiatry. 2010;67:397–8.
Sarazin M, Pillon B, Giannakopoulos P, et al. Clinicometabolic dissociation of cognitive functions and social behavior in frontal lobe lesions. Neurology. 1998;51(1):142–8.
Schutter D, Honk J, Postma A, et al. Effects of slow rTMS at the right DLPFC on EEG asymmetry and mood. Neuroreport. 2001;12:445–7.
Segal H. Bemerkungen zur Symbolbildung. In: Both-Spillius E, editor. Melanie Klein heute (Bd 1). München: Verlag Internat Psychoanalyse; 1956.
Sheline YI, Barch DM, Price JL, et al. The default mode network and self-referential processes in depression. Proc Natl Acad Sci U S A. 2009;106(6):1942–7.
Sheppard LC. How does dysfunctional thinking decrease during recovery from major depression? J Abnorm Psychol. 2004;113(1):64–71.
Shin LM, et al. A functional magnetic resonance imaging study of amygdala and medial prefrontal cortex responses to overtly presented fearful faces in posttraumatic stress disorder. Arch Gen Psychiatry. 2005;62(3):273–81.
Smoski MJ, Felder J, Bizzell J, et al. fMRI of alterations in reward selection, anticipation, and feedback in major depressive disorder. J Affect Disord. 2009;118(1-3):69–78.
Stern DN. The interpersonal world of the infant. New York: Basic Books; 1985. Deutsch. Die Lebenserfahrung des Säuglings. Stuttgart: Klett-Cotta; 1992.
Stern C, Owen A, Tracey I, et al. Activity in ventrolateral and mid-dorsolateral prefrontal cortex during nonspatial working memory. NeuroImage. 2000;11:392–9.
Surguladze SA, Young AW, Senior C, et al. Recognition accuracy and response bias to happy and sad facial expressions in patients with major depression. Neuropsychology. 2005;18:212–8.
Tang T, DeRubeis R. Sudden gains and critical sessions in cognitive-behavioral therapy for depression. J Consult Clin Psychol. 1999;67:894–904.
Tononi G, Edelman GM. Schizophrenia and the mechanisms of conscious integration. Brain Res Rev. 2000;31(2-3):391–400.
Tormos J, Canete C, Tarazona F, et al. Lateralized effects of self-induced sadness and happiness on corticospinal excitability. Neurology. 1997;49:487–91.
Treynor W, Gonzalez R, Nolen-Hoeksema S. Rumination reconsidered: a psychometric analysis. Cogn Ther Res. 2003;27:247–59.
Walter M, Henning A, Grimm S, et al. The relationship between aberrant neuronal activation in the pregenual anterior cingulate, altered glutamatergic metabolism and anhedonia in major depression. Arch Gen Psychiatry. 2009;66(5):478–86.
Wicker B, Keysers C, Plailly J, et al. Both of us disgusted in My insula: the common neural basis of seeing and feeling disgust. Neuron. 2003a;40(3):655–64.
Wicker B, Perrett DI, Baron-Cohen S, et al. Being the target of another's emotion: a PET study. Neuropsychologia. 2003b;41(2):139–46.
Wicker B, Ruby P, Royet J-P, et al. A relation between rest and the self in the brain? Brain Res Rev. 2003c;43(2):224–30.
Wiebking C, Bauer A, de Greck M, et al. Abnormal body perception and neural activity in the insula in depression: an fMRI study of the depressed material me. World J Biol Psychiatry. 2010;11(3):538–49.
Wiesel TN. Genetics and behaviour. Science. 1994;264:16–47.
Zarate CA Jr, Singh JB, Carlson PJ, et al. A randomized trial of an N-methyl-D-aspartate antagonist in treatment-resistant major depression. Arch Gen Psychiatry. 2006;63:856–64.

# Manic and Bipolar Syndromes

# 12

Heinz Boeker, Simone Grimm, Peter Hartwich,
and Georg Northoff

**Abstract**

Patients with bipolar disorders show numerous neuropsychological impairments, not only during depressive and manic episodes but also after remission of symptoms. These dysfunctions are associated with structural and functional changes in cortical and limbic brain regions and have a profound impact on patients' psychosocial functioning. This chapter gives an overview on relevant neuroscientific and neuropsychological findings in bipolar disorders. Furthermore, the personality structure and the interpersonal relationships of bipolar patients are taken into account.

In a neuropsychodynamic perspective, the hypernomic und ambiguity-intolerant behaviour of bipolar patients may be looked upon as an attempt to cope with an impending threat of a collapsing self-worth regulation and the shame which results from the experience of mania. In case of an instability of the self-image, a narcissistic slight or a disappointment (with an impairment of the ideal

H. Boeker (✉)
Department of Psychiatry, Psychotherapy, and Psychosomatics, Psychiatric University Hospital Zurich, University of Zurich, Zurich, Switzerland
e-mail: heinz.boeker@bli.uzh.ch

S. Grimm
Charité, Universitätsmedizin Berlin, Klinik für Psychiatrie und Psychotherapie, Berlin, Germany
e-mail: simone.grimm@charite.de

P. Hartwich
Department of Psychiatry, Psychotherapy, Psychosomatics, General Hospital of Frankfurt am Main, Teaching Hospital University Frankfurt, Frankfurt, Germany
e-mail: Prof.Hartwich@casytec.de

G. Northoff
Mind, Brain Imaging, and Neuroethics, Institute of Mental Health Research, University of Ottawa, Ottawa, ON, Canada
e-mail: georg.northoff@theroyal.ca; www.georgnorthoff.com



H. Boeker et al. (eds.), *Neuropsychodynamic Psychiatry*,
https://doi.org/10.1007/978-3-319-75112-2_12

self) may induce a regression to the precursor stages of the ideal self (regressive activation of the grandiose self) or an actualization of other intrapsychic elements of the self-worth regulation and their precursors (e.g. complete submission to the claims of a rigid, rigorous superego). This may be interpreted as an attempt to cope with the narcissistic dysbalance and the danger of a depressive reaction.

The conceptualization of manic symptoms as mood modulators has important implications for the understanding and treatment of mania. The confrontation with the "temptation of mania" is of fundamental importance in the treatment of mania. The effectiveness of the drug treatment and the compliance of the patients depend on their attitude to mania essentially. In this context mania may not only be comprehended as the experience of a fatal disorder but also as a condition which is intended and desired. The neuropsychodynamic view of manic symptoms as mood modulators constitutes an important basis for the dialogue with the patients and provides them with a better understanding of their painful renunciation of the manic sense of euphoria.

"Mania is not a luxary that one can easily afford."

Kay Redfield Jamison (1995)

## 12.1 Introduction

"When I am high I couldn't worry about money if I tried so. The money will come from somewhere; I am entitled; God will provide. Credit cards are disastrous, personal checks worse. Unfortunately for manics anyway, mania is a natural extension of economy. What with credit cards and bank accounts there is little beyond reach...." (Jamison 1995).

This self-description of Kay Redfield Jamison corresponds to the descriptions which every psychiatrist gets to know from his manic patients. The manic symptom constellation may be easily diagnosed. Therefore it can also be understood that the different diagnostic systems (DSM-IV/DSM-5; ICD-10) show a greater symptomatologic uniformity of mania compared to depression. But this symptomatologic uniformity should not be equated with a syndromatologic or phenomenologic uniformity. There is rather a very great variety of different symptoms in manic episodes comparing different patients and in the course of the disorder in each single case, respectively (Goodwin and Jamison 1990).

The most frequent symptoms in mania concern affect, drive, behaviour, formal thinking and attention. The core symptoms of mania are based on the increase of affect and vitality. The most frequent affective symptoms in mania are characterized by irritability, euphoria, oscillations of mood and expansivity. Nevertheless it should be kept in mind that bipolar patients suffer most of the time (more than 70%) from depression (Denicoff et al. 2000)! Furthermore patients with bipolar disorder often present initially with a major depressive episode. This may be one reason for the late diagnosis of bipolar disorder in many cases.

The criteria for mania are summarized in Table 12.1 according to DSM-IV/DSM-5:

The new *Diagnostic and Statistical Manual of Mental Disorders, Fifth Edition* (DSM-5) has a number of changes to bipolar disorders (also known to some by their old label, manic depression). According to the American Psychiatric Association (APA) and the publisher of the DSM-5, only minor changes were made to this category of diagnoses.

In order to enhance the accuracy of diagnosis and facilitate earlier detection in clinical settings, the primary criteria for manic and hypomanic episodes (criterion A) now include an emphasis on changes in activity and energy—not just mood.

**Table 12.1** DSM-IV/DSM-5 criteria for manic episodes

| |
|---|
| A. A distinct period of abnormally and persistently elevated, expansive, or irritable mood, lasting at least 1 week (or any duration if hospitalization is necessary) |
| B. During the period of mood disturbance, three (or more) of the following symptoms have persisted (four if the mood is only irritable) and have been present to a significant degree<br>1. Inflated self-esteem or grandiosity<br>2. Decreased need for sleep (e.g. feels rested after only 3 h of sleep)<br>3. More talkative than usual or pressure to keep talking<br>4. Flight of ideas or subjective experience that thoughts are racing<br>5. Distractibility (i.e. attention too easily drawn to unimportant or irrelevant external stimuli)<br>6. Increase in goal-directed activity (either socially, at work or school, or sexually) or psychomotor agitation<br>7. Excessive involvement in pleasurable activities that have a high potential for painful consequences (e.g. engaging in unrestrained buying sprees, sexual indiscretions, or foolish business investments) |
| C. The symptoms do not meet criteria for a mixed episode |
| D. The mood disturbance is sufficiently severe to cause marked impairment in occupational functioning or in usual social activities or relationships with others or to necessitate hospitalization to prevent harm to self or other, or there are psychotic features |
| E. The symptoms are not due to the direct physiological effects of a substance (e.g. a drug of abuse, a medication, or other treatment) or a general medical condition (e.g. hyperthyroidism)<br>*Note*: Manic-like episodes that are clearly caused by somatic antidepressant treatment (e.g. medication, electroconvulsive therapy, light therapy) should not count towards a diagnosis of Bipolar I disorder<br>What differentiates manic and hypomanic episodes or states from each other is the severity, duration, and from a psychological point of view the experience of each patient. The DSM-IV sets 4 days as the minimum duration for a hypomanic episode and states that the mood has to be "clearly different from the usual non-depressed mood". But the two most important criteria (D and E) that refer to functional impairment essentially summarize the major difference between a hypomanic and a manic episode. The hypomanic episode is associated with an "unequivocal change in functioning that is uncharacteristic of the person when not symptomatic" (criterion C of hypomanic episode) and "the episode is not severe enough to cause marked impairment in social or occupational functioning, or to necessitate hospitalization, and there are no psychotic features" (criterion D)<br>Two more episodes that are often present in bipolar disorder are major depressive and mixed episodes. Their symptoms and diagnostic criteria are given below. According to the DSM-IV, the diagnostic criteria and symptoms of a major depressive episode are the same both in bipolar and unipolar disorders. Mixed episodes by definition can only be present in bipolar disorder as they require the presence of both a depressive and a manic episode |

### 12.1.1 Mixed Episode

The DSM-IV diagnosis of bipolar I disorder, mixed episode, requiring that the individual simultaneously meet full criteria for both mania and major depressive episode, has been removed from the DSM-5.

Instead, a new specifier, "with mixed features," has been added, according to the APA, which can be applied to episodes of mania or hypomania when depressive features are present. It can also be applied to episodes of depression—such as in the context of major depressive disorder or bipolar disorder—when features of mania/hypomania are present.

According to the APA, the "DSM-5 allows the specification of particular conditions for other specified bipolar and related disorder, including categorization for individuals with a past history of a major depressive disorder who meet all criteria for hypomania except the duration criterion (i.e., at least 4 consecutive days). A second condition constituting an other specified bipolar and related disorder is that too few symptoms of hypomania are present to meet criteria for the full bipolar II syndrome, although the duration is sufficient at 4 or more days."

The manic feelings the person experiences should be severe enough to cause difficulty or impairment in their ability to function at work, with friends or family, at school or other important areas in their life. Symptoms also cannot be the result of substance use or abuse (e.g. alcohol, drugs, medications) or caused by a general medical condition.

Inflated self-esteem is typically present, ranging from uncritical self-confidence to marked grandiosity, and may reach delusional proportions. Individuals may give advice on matters about which they have no special knowledge. Despite lack of any particular experience or talent, the individual may embark on writing a novel or composing a symphony or seek publicity for some impractical invention. Grandiose delusions are common (e.g. having a special relationship to God or to some public figure from the political, religious, or entertainment world).

Almost invariably, there is a decreased need for sleep. The person usually awakens several hours earlier than usual, feeling full of energy. When the sleep disturbance is severe, the person may go for days without sleep and yet not feel tired.

Manic speech is typically pressured, loud, rapid and difficult to interrupt. Individuals may talk nonstop, sometimes for hours on end, and without regard for others' wishes to communicate. Speech is sometimes characterized by joking, punning and amusing irrelevancies. The individual may become theatrical, with dramatic mannerisms and singing. Sounds rather than meaningful conceptual relationships may govern word choice (i.e. clanging). If the person's mood is more irritable than expansive, speech may be marked by complaints, hostile comments, or angry tirades.

The individual's thoughts may race, often at a rate faster than can be articulated. Frequently there is flight of ideas evidenced by a nearly continuous flow of accelerated speech, with abrupt changes from one topic to another. For example, while talking about a potential business deal to sell computers, a salesperson may shift to discussing in minute detail the history of the computer chip, the industrial revolution, or applied mathematics. When flight of ideas is severe, speech may become disorganized and incoherent.

A person in a manic episode may easily lose attention. Distractibility is evidenced by an inability to screen out irrelevant external stimuli (e.g. the interviewer's tie, background noises or conversations or furnishings in the room). There may be a reduced ability to differentiate between thoughts that are germane to the topic and thoughts that are only slightly relevant or clearly irrelevant.

The increase in goal-directed activity often involves excessive planning of, and excessive participation in, multiple activities (e.g. sexual, occupational, political, religious). Increased sexual drive, fantasies and behaviour are often present. The person may simultaneously take on multiple new business ventures without regard for the apparent risks or the need to complete each venture satisfactorily. Almost invariably, there is increased sociability (e.g. renewing old acquaintances or calling friends or even strangers at all hours of the day or night), without regard to the intrusive, domineering and demanding nature of these interactions. Individuals may also display psychomotor agitation or restlessness by pacing or by holding multiple conversations simultaneously (e.g. by telephone and in person at the same time). Some individuals write a torrent of letters on many different topics to friends, public figures, or the media.

Expansiveness, unwarranted optimism, grandiosity and poor judgement often lead to an imprudent involvement in pleasurable activities such as buying sprees, reckless driving, foolish business investments and sexual behaviour unusual for the person, even though these activities are likely to have painful consequences. The individual may purchase many unneeded items without the money to pay for them. Unusual sexual behaviour may include infidelity or indiscriminate sexual encounters with strangers.

Very often conflictuous relationship patterns and negative reactions of partners, family members, friends, or neighbours are induced by the manic behaviour. Others are often extremely spoiled and disturbed; many relationships get lost. Mania leaves broken crockery in the social environment of the patient "and brings no luck".

The prognosis of mania and bipolar disorder is the worst the earlier manic symptoms get manifest in the course of the disorder. The manic episode with psychotic features may get along with a suicidal risk (Simpson and Jamison 1999).

## 12.2 Neuroscientific and Neuropsychological Findings in Bipolar Disorder

Patients with bipolar disorders show numerous neuropsychological impairments—not only during depressive and manic episodes but also after remission of symptoms. These dysfunctions are associated with structural and functional changes in cortical and limbic brain regions and have a profound impact on patients' psychosocial functioning. The following chapter will give a brief overview on relevant neuroscientific and neuropsychological findings in bipolar disorders.

### 12.2.1 Structural and Functional Brain Imaging

Several studies showed structural changes in cortical and subcortical brain regions. Especially, enlarged ventricles and sulci as well as cortical hyperintensities have been reported (Kempton et al. 2008). Furthermore, there is evidence for larger amygdala volume (Strakowski et al. 2012) and demyelinization of fibre tracts from the amygdala to hippocampus and prefrontal cortex (Benedetti et al. 2011) as well as to frontotemporal areas of emotion regulation (Versace et al. 2010; Phillips and Kupfer, 2013). Increases in brain volume as a possible result of lithium administration have also been described (Lyoo et al. 2010), even though recent findings strongly suggest that this might rather be an artefact due to an effect of lithium on BOLD response (Cousins et al. 2013). Functional studies using BOLD fMRI show altered activation patterns in ventrolateral prefrontal cortex in bipolar patients during emotional and cognitive tasks (Lim et al. 2013). The amygdala is a key region within a frontal-limbic network of emotion regulation and displays increased resting state activity as well as hyperreactivity in response to emotional stimuli even in euthymic periods in bipolar patients (Strakowski et al. 2012; Chen et al. 2011). A recently introduced model attempts to integrate findings of morphological and functional studies and postulates two ventral prefrontal networks of emotion regulation (Strakowski et al. 2012). These consist of repetitive circuits for information processing and modulate activity of the amygdala and other limbic regions. One of these networks is predominantly modulated by the ventrolateral prefrontal cortex and is crucial for the processing of external emotional signals such as facial expressions showing anger, happiness, fear, etc. (Townsend and Altshuler 2012). The other network is closely associated with ventromedial prefrontal cortex and modulation of internal emotional reactions. This model therefore explains bipolar disorders with a dysfunction in circumscribed key regions as well as with aberrant connectivity between brain regions that have comparable impact on networks of affect and emotion regulation (Phillips et al. 2008). Bipolar disorders are characterized by dysfunctions within as well as between these regions, which accordingly result in dysfunctional regulation of affects.

### 12.2.2 Neuropsychological Impairments

Severe cognitive impairments are a frequent symptom of bipolar disorders and occur in at least 30% of patients (Gualtieri and Morgan 2008). A study by Osher et al. (2011) compared cognitive performance in euthymic bipolar patients, healthy controls and subjects with minimal cognitive impairment (MCI; which is often interpreted as an early indicator for dementia) and showed that performance of bipolar patients was comparable to that of MCI controls and was even below that in the case of attention. The profile and severity of cognitive dysfunctions is influenced by several clinical and demographical variables. Accordingly, patients with bipolar I disorder are more severely impaired than those with bipolar II (Hsiao et al. 2009). While symptom severity per se seems to be not very relevant, duration of illness, duration of hospitalization, age and age at onset are considered negative predictors for the severity of cognitive deficits (Van Gorp et al. 1998; Christensen et al. 1997; Beblo et al. 2011). Cognitive impairments also occur in euthymic patients and mainly concern the domains of attention/concentration, processing speed, inhibitory control, executive functions as well as learning and memory. Furthermore, some studies also reported these impairments in first-degree relatives of patients, which furthermore underline that cognitive dysfunctions might be considered traits and improve our understanding of the dissociation between clinical and cognitive symptoms (Torres et al. 2007; Bonnin et al. 2010; Bora et al. 2009). Improvements in cognitive function, particularly in the domains memory and executive functions, are a strong predictor for functional remission (Bearden et al. 2011). Martinez-Àran et al. (2007) report that psychosocial functioning of bipolar patients is more strongly associated with neuropsychological than with clinical parameters. While they could not find an association between illness duration, number and type of episodes, number of hospitalizations and suicide attempts with psychosocial functioning, there was a strong link with executive functions and memory performance. Based on such findings, it has been discussed whether cognitive dysfunctions might be the crucial component of a particular endophenotype of bipolar disorder (Bora et al. 2009).

Sustained impairments in euthymic periods have been described for executive functions, verbal learning and memory and sustained attention. Six extensive meta-analyses describe pronounced deficits particularly in executive functions (Arts et al. 2009; Bora et al. 2009; Kurtz and Gerraty 2009; Mann-Wrobel et al. 2011; Robinson et al. 2006; Torres et al. 2007). The term executive functions comprises a number of distinct abilities. According to Karnath and Sturm (2002), it describes cognitive processes of planning and execution crucial for information processing and action control. Executive functions are necessary to plan actions using several intermediate steps in order to reach a higher goal, to focus attention on relevant information and to inhibit unfit actions. Executive functions also include verbal fluency, attentional shifts and working memory (Lezak 1995). Executive dysfunctions in bipolar patients concern selectively the ability to change concepts, response inhibition and planning (Robinson et al. 2006; Arts et al. 2009). These dysfunctions have been associated with aberrations in the prefrontal cortex. Post- mortem studies showed reductions in glial and neuronal density (Cotter et al. 2002; Ongur et al. 1998; Rajkowska et al. 2001) and altered activation during executive tasks in the ventrolateral prefrontal cortex (Blumberg et al. 2003). It has been suggested that executive dysfunctions might be causal for memory deficits in bipolar patients, since memory tests also require structuring of information to be learned as well as attentional shifts (Arts et al. 2009; Robinson et al. 2006).

Further studies are needed to conclude whether these neuropsychological domains and deficits should be considered separate entities. Bonnin et al. (2010) showed that verbal memory performance and executive functions were the only cognitive parameters predicting the functional level of patients 4 years later. However, there is increasing evidence that deficits in sustained attention are also present in euthymic periods (Bora et al. 2005; Clark and Goodwin 2004; Liu et al. 2002; Torres et al. 2007). Sustained attention describes the ability to focus attention

**Table 12.2** Cognitive function and structural and functional deficits in bipolar disorder

| Cognitive function | Structural deficits | Functional deficits |
|---|---|---|
| Memory | • Volume decreases in<br>– DLPFC<br>– Hippocampus<br>– Entorhinal cortex | • DLPFC hyperactivity<br>• Hypoactivity in<br>– Anterior cingulate<br>– VLPFC |
| Attention/working memory | • Volume decreases in DLPFC | • DLPFC hyperactivity<br>• Anterior cingulate hypoactivity |
| Executive functions | • Volume decreases in DLPFC<br>• Enlarged ventricles | • DLPFC hyperactivity<br>• Hypoactivity in<br>– Anterior cingulate<br>– VLPFC<br>– Orbitofrontal cortex<br>– Temporal cortex |

over longer periods of high-stimulus frequency. Clark and Goodwin (2004) as well as Bora et al. (2009) describe deficits in sustained attention in bipolar patients as an indicator for a chronic disease or as a trait marker occurring independent of clinical symptoms.

To summarize, bipolar patients show numerous cognitive deficits that are associated with structural and functional changes in cortical and limbic brain regions (Table 12.2). These cognitive deficits greatly influence patients' psychosocial functioning. A better understanding of the underlying pathophysiology might therefore well be the key for more effective neuropsychological therapies.

## 12.3 Premorbid Personality in Bipolar Disorder

The study of the association between the premorbid personality and affective disorders has a long clinical tradition and was focussed on a great number of empirical studies (for an overview, see Himmighoffen 2000). These studies focus on the question whether the premorbid personality structure contributes to the development of an affective disorder and how it influences the course of the disorder.

The nosologic discrimination of the bipolar manic-depressive psychoses, unipolar depression and unipolar mania, which was hold by Leonhard (1963), was validated by the investigations of Angst (1966) and Perris (1966). Besides statistically significant differences in the familial predisposition, gender ratio, the age at primary manifestation, the duration and frequency of the episodes, both authors also found differences of the premorbid personality. Thus they verified the syntonic temperament (with personality traits such as mental balance and living in harmony with the environment), which had been described by Bleuler (1922), more often in patients with bipolar affective psychoses than in patients with unipolar depression. In the latter they found in contrast frequently occurring psychoasthenic, pedantic, orderly and scrupulous personality traits.

The results of the personality research in bipolar affective disorders proved to be incoherent and contradictory in part. Whereas bipolar patients showed far less irregularities in their premorbid personality structure than unipolar depressed patients and were largely similar to psychologically healthy persons in some studies (cf. Möller and von Zerssen 1987; Möller 1992), a greater number of older and more actual studies pointed to the specificities of the premorbid personality and the family and partnership relationships as well of bipolar patients (Kröber 1993; for an overview cf. Himmighoffen 2000). There is a lot of evidence that the premorbid or also the interval personality of patients with bipolar affective disorders is characterized by an unstable self-esteem, which is compensated by means of an extreme adaptation and striving after performance. Personality traits as trustfulness in order, unquestioning acceptance of authority and compulsiveness further more vitality, inclination to aggression and striving for autonomy and independence. The partnerships of bipolar patients in case were often characterized by the dependence on the other and the strong wishes for emotional nearness and contact. There was a strong tendency to idealize partners, to avoid conflicts and to struggle for harmony (Matussek et al. 1986). These interpersonal conditions were conceptualized in the role-dynamic approach of Kraus (1991). Kraus described the so-called "hypernomic" behaviour (nomos = law) and "ambiguity intolerance" of affect psychotic patients. According to Kraus "hypernomic" behaviour is characterized by a strong orientation towards following normative rules, and therefore it is less individualized and predominantly other-directed.

"Ambiguity intolerance" indicates the incapability to bear contradictory feelings towards another person and to perceive contradictory features of the other person. According to Kraus "hypernomic" and "ambiguity-intolerant" behaviour serve the maintenance of an identity which is based on external elements, especially on role relationships.

## 12.4 Neuropsychodynamics of Bipolar Disorder and of Mania

In a neuropsychodynamic perspective, the hypernomic and ambiguity-intolerant behaviour of bipolar patients may be looked upon as an attempt to cope with an impending threat of a collapsing self-worth regulation. In case of an instability of the self-image, a narcissistic slight or a disappointment (with an impairment of the ideal self) may induce a regression to the precursor stages of the ideal self (regressive activation of the grandiose self) or an actualization of other intrapsychic elements of the self-worth regulation and their precursors (e.g. complete submission to the claims of a rigid, rigorous superego; cf. Chap. 4). This may be interpreted as an attempt to cope with the narcissistic dysbalance and the danger of a depressive reaction.

A further attempt to come to terms with the threatened narcissistic homeostasis and to cope with the narcissistic dysbalance concerns the forced focus on others who may provide recognition of one's achievements (corresponding to Kraus' hypernomia concept or the interpersonal defence and the so-called psychosocial arrangements; cf. Mentzos 1984; Willi 1975). However, the narcissistic balance which can be achieved by this means stays labil. Intrapsychic conflicts are suspended by interpersonal mechanisms of defence and compensation only temporarily; by means of these interpersonal mechanisms, the concrete features and behaviours of one partner enable, support and stabilize the neurotic (or psychotic) defence of conflicts and the satisfaction of wishes of the other partner.

The results of the study of Himmighoffen (2000) are remarkable in this context: The results of the self-image of persons with bipolar disorders who were investigated during the so-called symptom-free interval (applying the Giessen-Test) showed that bipolar patients differed from the standard sample or persons without mental disorders in essential features of their self-image and corresponded more with the self-image of unipolar depressed patients.

In bipolar and in unipolar manic patients, an unstable self-worth image was also found beyond the episodes, in the so-called symptom-free interval (low self-esteem, depressive mood, strong emphasis on the importance of the social role and acceptance corresponding to a "role-orientated identity"). In this context Kröber (1993) characterizes mania as a "revocation of the compromise" (with the demands of others and the social environment and the social norms). The compromise under these conditions may be compared with the "secured dominance of the other" and as the consequence of the repetitive failure when trying to realize self-oriented needs and desires.

Mentzos (2009) hypothesized the existence of a superego in bipolar patients which had developed by means of the identification with the father contrary to the superego in unipolar depressive patients which had developed by means of identification with the mother.

These different intrapsychic and interpersonal patterns used for the regulation of the narcissistic homeostasis enable bipolar patients—so we may assume—to suspend the dominance of their "paternally determined superego" at least temporarily. In unipolar depressed patients, this kind of suspension of the "motherly determined superego" may not be possible, because it would result in a much greater existential danger of the self (cf. Böker 1999).

The relationships between personality structure, conflict, mechanisms of defence and compensation and interpersonal dynamics are summarized in Table 12.3 in an ideal-typical manner:

With regard to the increased emotional arousal in bipolar affective disorders, the suggestions of Kipp and Stolzenburg (2009) are helpful, particularly in view of adequate therapeutic interventions. The authors understand manic symptoms as mood modulators and describe different modes of tuning in everyday situations and their psychodynamic meaning:

Modes of tuning which are connected with the approximation of the ego ideal (superego) and the ego:

- Use of psychotropic substances, ecstasy
- Narcissistic activities
- Relief of superego by means of regression in groups

**Table 12.3** Personality, conflict, mechanisms of defence and compensation and interpersonal dynamics in bipolar affective disorders (cf. Böker 1999)

| Depressive syndrome | Unipolar depression | Bipolar affective disorder |
|---|---|---|
| Personality | Typus melancholicus (Tellenbach 1961) | "Cyclothym" |
| Structure, conflict and mechanisms of defence and compensation | – Actualization of the basic conflict (self-worth conflict)<br>– Regression to the archaic superego/grandiose self<br>– Suppression of aggression<br>– Insufficient separation of the ambivalent object (superego/self)<br>– Structural impairment of ego functions<br>– In toto introjection of the ambivalent object | – Relative narcissistic equilibrium by means of alternating mobilization of archaic superego and grandiose self<br>– Internalization of incompatible self-objects or object representations Söldner and Matussek 1990)<br>– "Paternally determined superego": Superego contains incompatible proportions of motherly and paternally determined superego |
| Depressive syndrome | Unipolar mania | Schizoaffective disorder |
| Personality | Typus manicus (Von Zerssen 1977) | "Mixture" of cycloid and schizoid traits (Bleuler 1922; Kretschmer 1977) |
| Structure, conflict and mechanisms of defence and compensation | – Regression to the grandiose self on the background of paternally determined superego (following the reduction of self-esteem)<br>– Suspension of the paternally-determined superego<br>– Defence of defensive emptiness<br>– Aggression to secure basic self-existence | – Combination von schizophrenic identity conflict (self-identity vs. symbiotic relationship with the object) and depressive self-value conflict (self-value vs. object value)<br>– Regressive actualization of the archaic superego (schizo-depressive disorder) and of the grandiose self (schizo-manic disorder and mixed schizoaffective disorder) in combination with an actualization of the primary autonomy-dependence conflict (loss of the self-object differentiation, productive-psychotic defence of fragmentation anxieties) |

Modes of tuning which are connected with the desire for symbiotic relationships:

- Narcissistic mirroring and symbiotic experiences in relationship:
  - *Modes of tuning which are connected with symbiotic experiences with nature*
  - *Modes of tuning which are connected with denial of reality*
- Forgetting of conflicts
- "Positive thinking"

Modes of tuning which are connected with distance to objects

- Feelings of strength and independence in active separations

Modes of tuning which are connected with the phallic-narcissistic defence of schizoid anxieties:

- Hyperactivity, among others sexual hyperactivity

In a neuropsychodynamic perspective, the conceptualization of manic symptoms as mood modulators has important implications for the understanding and treatment of mania. A therapeutic perspective opens up which considers the "function of mania". We could also speak of a "positivism of mania", i.e. the need to understand the psychic function of the manic symptoms when the manic patient is confronted with the dilemma of mania and the increasing dysfunctionality of the coping mechanisms: The above-described modes of tuning are not only connected with manic symptoms, but they are also the reason for the continuation of the disorder.

The confrontation with the "temptation of mania" is of fundamental importance in the treatment of mania. The effectiveness of the drug treatment and the compliance of the patients depend on their attitude to mania essentially. In this context mania may not only be comprehended as the experience of a fatal disorder but also as a condition which is intended and desired. The neuropsychodynamic view of manic symptoms as mood modulators constitutes an important basis for the dialogue with the patients and provides them with a better understanding of their painful renunciation of the manic sense of euphoria. Elia (1983) had already pointed to the striking paradox of mania "…, that a relatively asthenic ego tries by means of a deep regression to restore life and its relationships, to restart, even if it occurs in the pathological way we know (translated by H.B.)". Mania represents in this perspective a last resort to exist and to be accepted; for the patients themselves, it is often the only pattern of personal development.

In such a perspective, the defence reaction and the defended anxiety and weakness of the manic patient may be perceived in the therapeutic encounter by means of counteridentification. Kipp and Stolzenburg (2000) underline that the manic patient may experience—so to say with a "tentative finger"—that the therapist is fully interested in the patient, not only during the time span of the therapeutic session. According to the respective requirements, the therapeutic setting has to be modified.

In the therapeutic relationship, a reduction of the exhausting dichotomy is enabled, "...either to be god or nothing, as a therapist as well as a patient...(translated by H.B.)" (Elia 1983). A change in the therapy is marked by means of the experience that certain events or encounters induce the arousal of the manic patients and that the patients decide to avoid or limit such situations.

Based on the knowledge that the triggers for mania often are not joyful events and experiences which would increase self-confidence but rather painful losses, separations and experiences which undermine self-confidence (as in case of depression), Mentzos (1996) described mania as an alternate solution of the depressive dilemma (the conflict between being dependant on the object and autonomous self-value): as an extreme revocation of obedience, connected with a denial of binding wishes and the development of expansive, ego-orientated activities (without showing any consideration towards social norms and realities). The therapeutical relevance of such a conceptualization of mania as an alternate solution of the depressive conflict is the possible experience of the patients that the therapists understand their feelings and behaviour not only as defence (denial) but also as something that may tend to be positive: the justified, although "dangerous" termination of a long-standing submission (cf. Mentzos 1995). In the same sense Schwarz (2014) described mania as a creative attempt at a solution of hitherto unresolved developmental steps.

The consideration of neuropsychodynamic correlations is already important during the acute state of the disorder: Contrary to the harmonizing conflict avoidance and the submission under the "secured dominance of others" (Matussek et al. 1986), some patients try to realize their autonomy wishes and delegate control functions and lose anxieties to their relatives or to the members of the therapeutical team. Kröber (1992) pointed to the similarity of the psychiatric hospital and the bipolar family: Both may be characterized as "custodial, well-established team, inflexible, having a high sense of responsibility for others". Under these circumstances, there may be the danger that the existing antagonistic intrapsychic conflicts and tensions may be increased by institutional mechanisms additionally. Furthermore, Kröber recommends to consider an outpatient treatment (as far as this may be possible). In any case the understanding of the neuropsychodynamics of mania and bipolar disorder may avoid a further complication of the developing "action dialogues", not least because of the reflexion of the countertransference-related reaction of the therapeutical teams.

Clinically, there is often a continuous transition from the good mood to the hypomanic moods, to the manic high spirits, to megalomania and then to aggressively irritated mania. The dynamics of the development can be rapid or slow as well as stand still at any stage. Psychotherapeutic interventions are often only possible after the control and deceleration of the progressive process by means of psychopharmaceuticals. Most psychotherapeutic treatments, whether individual or group psychotherapy, are focused on the prevention of relapses. Since the treatments then take place in the interval time, in this period it is usually to be noticed a peculiar and *constant denial* of the events with regard to the manic phases which had taken place. In this context, Gabbard (2014) points to the results of the study by Ghaemi et al. (1995), in which 28 manic patients were examined:

"The investigators found that even when all other symptoms of mania had improved or remitted, insight remained notably absent." (p. 234)

The personal psychiatric interview from patients who are euthymic and who had suffered from prior manic episodes underscores this tendency to deny. Very often maniform moods, up to manic diseases, are not reported spontaneously. There is to be seen a noncompliance and an ongoing lack of insight.

What may be the reason for this serious obstacle to our psychotherapeutic efforts?

On the one hand, periods of good mood, high spirits and hypomanic episodes and even mania are not experienced by the person as abnormal or as an illness. In this case the family members, who are usually suffering from the patient's behaviour, are to be questioned.

On the other hand, the events and actions in the manic episode, in which the patient feels irresistible and exalted above all other men, in which money is squandered and sexual debaucheries and insults are made, are in retrospect often very embarrassing and extremely shameful. Consequently it is understandable that the memory about this will be avoided. In the psychiatric interview or in a psychotherapeutic session, there is usually unconsciously and sometimes consciously a resistance against a revival of these events. (For resistance, see Chap. 8.)Why is the sense of shame so bad that its defence mechanisms must be so strong?

Jean Paul Sartre (1943/1956) has shown for the experience of shame that, unlike the feeling of guilt, the look of the other as well as the assessment by other people is the decisive factor.

Sartre (1943) writes:

"Now shame, …, is shame of self; it is the recognition of the fact that I am indeed that object which the Other is looking at and judging" (p. 261).

"But in order to me to be what I am, it suffices merely that the other look at me" (p. 262).

"Through the Other's look I live myself as fixed in the midst of the world, as in danger, as irremediable" (p. 268).

The fixing eyes of the family members on him because of the events of the prior mania as well as the eyes of the psychiatrist have a *spatial* and a *temporal* dimension. The actual necessary *spatial distance* from the other person, who judges or convicts him, is suddenly abolished by being looked at him. Also the otherwise existing temporal course melts into a momentary look. Thus the other's look, figuratively speaking, becomes an arrow shot, penetrating the naturally protecting defence mechanisms and causing destruction inside. The look that can uncover the sense of shame goes deep into the body and, in addition to the psychological pain, triggers strong physical reactions, including blushing, increasing heart rate and powerlessness.

Sartre: "But the Other's look is not only apprehended as spatializing, it is also temporalizing" (p. 266).

"As temporal-spatial object in the world, as an essential structure of a temporal-spatial situation in the world, I offer myself to the Other's appraisal" (p. 267).

## 12.5 Neuropsychodynamic-Orientated Psychotherapy in Bipolar Disorder

The above-mentioned problem of the defence of shame while denying the bitterly embarrassing behaviour in a prior manic episode, therefore, is understandably opposed to a necessary treatment. As a result, Hartwich and Pfeffer (2007, 2015) emphasized this aspect in the treatment of relapse prophylaxis in bipolar disorders. The experience teaches that in group psychotherapy "experts are among themselves", and when they share embarrassing experiences, little is taken amiss. In a 4-year psychodynamic group therapy with bipolar outpatients after inpatient stay, the aim was to investigate the effect on relapse prevention compared to a control group treated with the usual means of a psychiatric ambulance. It turned out that the patients in the group therapy knew well not only the value of the prophylactics but also the danger of hypomanic states, which can announce a coming up full mania. The participants' comments are typical: "Hey, you are over the top today." Or "You need your lithium, do not throw it into the trash can, the error I have made and later bitterly regret". Sometimes even the warning comes: "You have to recognize the great mood as a danger!" It is impressive how openly the patients deal with each other. It is more like that you can finally talk about the experiences without being condemned. One finds allies, and the embarrassment of the feeling of shame is mitigated by the understanding looks and remarks of the other group members instead of hurting.

The collectivization of the sense of shame produces a common and mutual space in the group, and the common experience leads to a relaxation of the formerly denied and partly frozen feelings in the course of many group sessions. The fact that shame can be felt again among the group members can be seen as a therapeutic improvement. This gradually relieves the need for protection of the repression, the trivialization ceases, and the splitting off or even the maniform defence is hardly needed any more. However, this takes time, often a lot of months.

In the empirical study mentioned above, the therapy group had significantly better values for the variables of the integrative capacity of the disease and less hospitalization days than the control group, which could be related to the treatment of the shame. Thus, the significance of the psychodynamic treatment of the shame aspect for the history of manic diseases was pointed out, which should concern not only group but also individual psychotherapies.

Special consideration should be given to the countertransference at the beginning of maniform mood change. Frequently the therapist is "infected" by the good mood of his patient and then acts together with him. Instead, it is necessary to recognize that the raised mood can initially be a maniform defence, which is supposed to conceal the shame about the events that have taken place in a previously experienced manic episode. It is about recognizing such an alteration of the mood early enough as a danger of an incipient manic dynamic, which can "gallop" uncontrollably to the full image of a mania. Crucial is that both the therapist and the patient become aware of this problem and act accordingly.

Psychotherapeutic and sociotherapeutic interventions in bipolar affective disorders are of great importance. This results not least because of the following knowledge:

- Contrary to previous assumptions, bipolar patients show a significant reduction of their functionality in the long-term course.
- Compared with unipolar depression, there is a greater probability in bipolar affective disorders to suffer from frequent episodes.
- The length of the symptom-free interval is a predictor of the probability to relapse during the next 4 years.
- There is a high probability of relapse also with lithium prophylaxis.
- A deterioration of the social adaptation was found in ca. 60% of bipolar patients, whereas a gradual improvement of the coping strategies developed in 40% of the investigated patients during the first 7 years after hospitalization.

- Psychosocial variables as "expressed emotions" and "life events" represent a significant role in the relapses of the disorder.
- More than half of the inpatient bipolar patients are not completely symptom-free upon completion of the treatment.
- Psychosocial impairments in the long-term course and high relapse rates are also found in patients who are treated with mood stabilizers.

The above-mentioned results underline the importance of psychotherapeutic interventions in patients with bipolar disorder. A Cochrane study showed that all of the included psychotherapeutic methods contributed to an improvement of the compliance of bipolar patients and had a positive impact on the coping mechanisms of the patients (for further details and references cf. Hollon and Ponniah 2010; see Table 12.4). In the process of neuropsychodynamic-orientated psychotherapy, the focus is on personality, the dysfunctionality of interpersonal coping mechanisms, and especially on the increased emotional arousal.

**Table 12.4** Psychotherapy in bipolar affective disorders

| | |
|---|---|
| *All Psychotherapeutic methods:*<br>– Improvement of the compliance of the patients | Cochrane (1984, 1987) |
| *CBT* | |
| – Case studies | Paykel (1995) |
| – Improvement of the adherence to drug treatment, reduction of hospitalization rates<br>– No difference concerning frequency of episodes | Cochrane (1984) |
| – Meta-analysis (19 controlled studies, heterogeneity of the variables): Insufficient empirical basis | Meyer and Hautzinger (2000) |
| – Reduction of the anxiety symptomatology, but time-stable negative cognitions (in comparison with unipolar depression) | Zaretsky et al. (1999) |
| – Potentially effective prevention of bipolar episodes during 12 months, no reduction of relapse rates for manic episodes after 18 months | Lam et al. (2003, 2005) |
| – No difference in comparison with TAS, 40% discontinuation of therapy | Scott et al. (2006) |
| *Interpersonal and social Rhythm therapy (IPSRT)* | Frank et al. (1999) |
| – No difference between IPSRT and intensive clinical management (ICM) during the actual episode | |
| – Reduction of the remission rate during 2 years of maintenance treatment | |
| – "Possibly efficacious" as supplementary therapy during the acute treatment of depression in bipolar disorder | Hollon and Ponniah (2010) |
| *Sleep management* | Wehr et al. (1998) and Wirz-Justice et al. (1999) |
| *Inpatients-family interventions* | Clarkin et al. (1999) |
| *Family-focused treatment (FFT)*<br>– Reduction of relapse rates of depressive episodes | Miklowitz and Goldstein (1997), Johnson et al. (2000), and Miklowitz et al. (2007) |

**Table 12.4** (continued)

| | |
|---|---|
| *Psychoeducation*<br>– No advantages in comparison with TAU after 18 months, but after 5 years<br>– Efficacious (in combination with pharmacotherapy): Reduction of relapse rates of hypomanic/manic episodes | Cochran (1984), Van Gent et al. (1988, 1993), and Hollon and Ponniah (2011) |
| *Programme for the identification of early warning signs*<br>– Advantages in comparison with TAU (after 18 months; duration of remission, social functioning level, level of employment)<br>– Reduction of "expressed emotions"<br>– Structured group psychotherapy ("life goals programme"): Improvement of quality of life (symptom control, realization of significant life goals) | Honig et al. (1997), Baur and McBride (1996), and Baur et al. (1998) |
| *Family-focused psychoeducation (Fountoulakis 2010)*<br>– Identification of early warning signs and triggering factors<br>– Improvement of the compliance | Fountoulakis (2010) |
| *Psychodynamic psychotherapy*<br>– Couple therapy: Psychosocial stabilization<br>– Group psychotherapy: Psychosocial stabilization | Davenport et al. (1997) and Kanas and Cox (1998) |

There is only a relatively few number of psychotherapy studies in bipolar patients until now. Hollon and Ponniah (2010) tried to categorize the effectiveness of psychotherapy in bipolar disorders and therefore especially analysed studies on psychoeducation, CBT, "family-focused therapy" and "interpersonal and social rhythm therapy". They concluded that a stable therapeutic relationship across several phases contributes to a successful treatment essentially.

An efficacious psychotherapy in bipolar disorder at least comprises the following essentials:

- Outpatient group therapy
- Psychoeducation
- Self-observation of mood swings, life events, behaviour and thinking
- Reflexion of expectations and moral standards
- Support of competencies for self-management of mood swings and early warning signs
- Normalization and stabilization of sleep-wake and social life rhythms
- Stress management
- Activity management
- Increase of self-efficacy beliefs
- Participation of partners and family members
- Preparatory confrontation with possible crises and relapses

In summary the therapeutic interventions for the improvement of the course of bipolar disorders should start as early as possible, because the coping of the disorder is all the better the less episodes bipolar patients had experienced so far (Carlson et al. 2000). In any case pharmacotherapy has to be part of an integrated therapeutic approach. Thereby the therapists must get away from the idea "to adapt the patient to a medical-pharmacological rational therapy" (Kröber 1993). In this process the experience of the disorder and the coping of bipolar patients have to be considered. The memory of the acute symptoms is for many patients connected with shame. The last manic episode is characterized as mainly positive at the same time. Patients with only few episodes until now prefer social and interactional explanations of the causes of the illness. The medical model of the disorder usually is only accepted in the longer course of the disorder. The acceptance of the medical model of the disorder often correlates with resignation and the feeling of subjective helplessness (cf. Kröber 1993).

A fundamental information about the course and treating possibilities of the disorder contributes to the increase of the compliance of the patient and to the development of more adequate mechanisms of compensation. Therefore the establishment of a stable therapeutical relationship, in which the problems resulting from countertransference also have to be constantly reflected by the therapists, is a precondition. The outpatient group therapy contributes to the promotion of adequate coping mechanisms and social competence and enables the identification of early warning signs and the experience of intrapsychic and interpersonal vitious circles and their solution in the actual group situation (Böker 2002). The support of the communication in the family of the patient by means of couple therapy and family-oriented treatment should be taken into account. Psychotherapeutic interventions not only contribute to a reduction of partnership conflicts (as possible triggers of relapses) but also to an improvement of the adherence to pharmacological treatment.

## References

Angst J. Zur Ätiologie und Nosologie endogener depressiver Psychosen. Berlin: Springer; 1966.
Arts B, et al. Meta-analyses of cognitive functioning in euthymic bipolar patients and their first-degree relatives. Psychol Med. 2009;39(3):525.
Bearden CE, et al. The impact of neurocognitive impairment on occupational recovery of clinically stable patients with bipolar disorder: a prospective study. Bipolar Disord. 2011;13(4):323–33.
Beblo T, et al. Specifying the neuropsychology of affective disorders: clinical, demographic and neurobiological factors. Neuropsychol Rev. 2011;21(4):337–59.
Benedetti FP, et al. Disruption of white matter integrity in bipolar depression as a possible structural marker of illness. Biol Psychiatry. 2011;69(4):309–17.
Bleuler E. Die Probleme der Schizoidie und der Syntonie. Zeitschrift für die gesamte Neurologie und Psychiatrie. 1922;78(1):373–99.
Blumberg HP, et al. A functional magnetic resonance imaging study of bipolar disorder: state- and trait-related dysfunction in ventral prefrontal cortices. Arch Gen Psychiatry. 2003;60(6):601–9.
Böker H. Selbstbild und Objektbeziehungen bei Depressionen: Untersuchungen mit der Repertory Grid-Technik und dem Giessen-Test an 139 PatientInnen mit depressiven Erkrankungen. Monographien aus dem Gesamtgebiete der Psychiatrie. Darmstadt: Steinkopff-Springer; 1999.
Böker H. Psychotherapie bei bipolaren affektiven Störungen. In: Böker H, Hell D, editors. Therapie der affektiven Störungen. Psychosoziale und neurobiologische Perspektiven. Stuttgart: Schattauer; 2002. p. 230–45.

Bonnin CM, et al. Clinical and neurocognitive predictors of functional outcome in bipolar euthymic patients: a long-term, follow-up study. J Affect Disord. 2010;121(1–2):156–60.
Bora E, et al. Cognitive endophenotypes of bipolar disorder: a meta-analysis of neuropsychological deficits in euthymic patients and their first-degree relatives. J Affect Disord. 2009;113(1–2):1–20.
Carlson GA, Bromert EJ, Sievers S. Phenomenology and outcome of subjects with early- and adult-onset psychotic mania. Am J Psychiatry. 2000;127:213–9.
Chen CH, et al. A quantitative meta-analysis of fMRI studies in bipolar disorder. Bipolar Disord. 2011;13(1):1–15.
Christensen H, et al. A quantitative review of cognitive deficits in depression and Alzheimer-type dementia. J Int Neuropsychol Soc. 1997;3(6):631–51.
Clark L, Goodwin GM. State- and trait-related deficits in sustained attention in bipolar disorder. Eur Arch Psychiatry Clin Neurosci. 2004;254(2):61–8.
Cotter D, et al. The density and spatial distribution of GABAergic neurons, labelled using calcium binding proteins, in the anterior cingulate cortex in major depressive disorder, bipolar disorder, and schizophrenia. Biol Psychiatry. 2002;51(5):377–86.
Cousins DA, et al. Lithium, gray matter, and magnetic resonance imaging signal. Biol Psychiatry. 2013;73(7):652–7.
Denicoff KD, Leverich GS, Nolen WA, et al. Validation of the prospective NIMH-life chart-method (NIMH-LCM-p) for longitudinal assessment of bipolar illness. Psychol Med. 2000;30:1391–7.
Elia C. Der psychodynamische Zugang zum manischen Patienten. In: Benedetti C, et al., editors. Psychosentherapie. Stuttgart: Hippokrates; 1983. p. 263–317.
Gabbard GO. Psychodynamic psychiatry in clinical practice. 5th ed. Washington DC: Amer Psychiatric Publ; 2014.
Ghaemi SN, Stoll SL, Pope HG. Lack of insight in bipolar disorder: the acute manic episode. J Nerv Ment Dis. 1995;183:464–7.
Goodwin FK, Jamison KR. Manic-depressive illness. Oxford, NY: Oxford University Press; 1990.
van Gorp WG, et al. Cognitive impairment in euthymic bipolar patients with and without prior alcohol dependence - a preliminary study. Arch Gen Psychiatry. 1998;55(1):41–6.
Gualtieri CT, Morgan DW. The frequency of cognitive impairment in patients with anxiety, depression, and bipolar disorder: an unaccounted source of variance in clinical trials. J Clin Psychiatry. 2008;69(7):1122–30.
Hartwich P, Pfeffer F Die bipolare Psychodynamik in der Gruppenpsychotherapie. Pilotstudie zur Wirksamkeit. Vortrag auf dem Kongress "Bipolare Störungen" am 21.09.2007 an der Ruhruniversität Bochum, published in Hartwich P, Grube M (2015) Psychotherapie bei Psychosen, Kap Größenwahn und Scham. Heidelberg: Springer; 2007. p. 117–26.
Himmighoffen H. Selbstwertproblematik und psychosoziale Bewältigungsstrategien bei PatientInnen mit bipolaren affektiven Psychosen. In: Böker H, editor. Depression, Manie und schizoaffektive Psychosen. Psychodynamische Theorien, einzelfallorientierte Forschung und Psychotherapie. 3rd ed. Giessen: Psychosozial-Verlag; 2000. p. 227–43.
Hollon SD, Ponniah K. A review of empirically supported psychological therapies for mood disorders in adults. Depress Anxiety. 2010;27:891–932.
Hsiao YL, et al. Neuropsychological functions in patients with bipolar I and bipolar II disorder. Bipolar Disord. 2009;11(5):547–54.
Jamison KR. An unquiet mind. New York: Vintage Books; 1995.
Karnath HO, Sturm W. Störungen von Planungs- und Kontrollfunktionen. Klinische Neuropsychologie. H. W. P. K. Stuttgart: Thieme; 2002.
Kempton MJ, et al. Meta-analysis, database, and meta-regression of 98 structural imaging studies in bipolar disorder. Arch Gen Psychiatry. 2008;65(9):1017–32.
Kipp J, Stolzenburg H-J. Stimmungsmodulation und die Psychodynamik der Manie. Psyche – Zeitschrift für Psychoanalyse. 2000;54:544–66.
Kraus A. Neuere psychopathologische Konzepte zur Persönlichkeit manisch-depressiver. In: Mundt C, Fiedler P, Lang H, Kraus A, editors. Depressionskonzepte heute. Berlin: Springer; 1991.

Kretschmer E. Körperbau und Charakter. 26th edition, 1921. Berlin, Heidelberg, New York; 1977.
Kröber H-L. Akute Krisen bei Manien. Nervenheilkunde. 1992;11:1–3.
Kröber H-L. Bipolare Patienten im Intervall: Persönlichkeitsstörungen und Persönlichkeitswandel. Nervenarzt. 1993;64:318–23.
Kurtz MM, Gerraty RT. A meta-analytic investigation of neurocognitive deficits in bipolar illness: profile and effects of clinical state. Neuropsychology. 2009;23(5):551–62.
Leonhard K. Die präpsychotische Temperament bei den monopolaren und bipolaren phasischen Psychosen. Psychiatr Neurol. 1963;146:105–15.
Lezak M. Neuropsychological assessment. Oxford, NY: Oxford University Press; 1995.
Lim CS, et al. Longitudinal neuroimaging and neuropsychological changes in bipolar disorder patients: review of the evidence. Neurosci Biobehav Rev. 2013;37(3):418–35.
Liu SK, et al. Deficits in sustained attention in schizophrenia and affective disorders: stable versus state-dependent markers. Am J Psychiatry. 2002;159(6):975–82.
Lyoo IK, et al. Lithium-induced gray matter volume increase as a neural correlate of treatment response in bipolar disorder: a longitudinal brain imaging study. Neuropsychopharmacology. 2010;35(8):1743–50.
Mann-Wrobel M, et al. Meta-analysis of neuropsychological functioning in euthymic bipolar disorder: an update and investigation of moderator variables. Bipolar Disord. 2011;13(4):334–42.
Martinez-Aran A, et al. Functional outcome in bipolar disorder: the role of clinical and cognitive factors. Bipolar Disord. 2007;9(1–2):103–13.
Matussek P, Luks O, Seibt G. Partner relationships of depressives. Psychopathology. 1986;19:143–56.
Mentzos S. Neurotische Konfliktverarbeitung. Frankfurt/M: Fischer; 1984.
Mentzos S. Depression und Manie; Psychodynamik und Psychotherapie affektiver Störungen. Göttingen: Vandenhoeck und Ruprecht; 1995.
Mentzos S. Lehrbuch der Psychodynamik. Die Funktion der Dysfunktionalität psychischer Störungen. Göttingen: Vandenhoeck & Ruprecht; 2009.
Möller HJ. Zur Bedeutung und methodischen Problematik der psychiatrischen Persönlichkeitsforschung: Der "Typus melancholicus" und andere Konzepte zur prämorbiden Persönlichkeit von Patienten mit affektiven Psychosen. In: Marneros A, Phillip M, editors. Persönlichkeit und psychische Erkrankung. Berlin: Springer; 1992. p. 45–65.
Möller HJ, von Zerssen D. Prämorbide Persönlichkeit von Patienten mit affektiven Psychosen. In: Kiska KP, Lauter H, Meyer J-E, Müller C, Strungren E, editors. Psychiatrie der Gegenwart Band V. Berlin: Springer; 1987. p. 165–79.
Ongur D, et al. Glial reduction in the subgenual prefrontal cortex in mood disorders. Proc Natl Acad Sci U S A. 1998;95(22):13290–5.
Osher Y, et al. Computerized testing of neurocognitive function in euthymic bipolar patients compared to those with mild cognitive impairment and cognitively healthy controls. Psychother Psychosom. 2011;80(5):298–303.
Paykel ES. Handbook of affective disorders. New York: Guilford; 1995.
Perris C. A study of bipolar (manic-depressive) and unipolar recurrent depressive psychosis. Acta Psychiat Scand Suppl. 1966;194:1–89.
Phillips ML, Kupfer DJ. Bipolar disorder diagnosis: challenges and future directions. Lancet. 2013;381(9878):1663–71.
Phillips ML, et al. A neural model of voluntary and automatic emotion regulation: implications for understanding the pathophysiology and neurodevelopment of bipolar disorder. Mol Psychiatry. 2008;13(9):833–57.
Rajkowska G, et al. Reductions in neuronal and glial density characterize the dorsolateral prefrontal cortex in bipolar disorder. Biol Psychiatry. 2001;49(9):741–52.
Robinson LJ, et al. A meta-analysis of cognitive deficits in euthymic patients with bipolar disorder. J Affect Disord. 2006;93(1–3):105–15.
Sartre JP. Being and nothingness. An essay on phemonemological ontology. Translated by Barnes H E Philosophical Library, Inc, New York, L'é'tre et le néant. Essai d'ontologie phénoménologique (1943). Paris: Librairie Gallimard; 1956.

Schwarz F. Psychodynamische Psychotherapie bei bipolaren Störungen. Z Psychiatr Psychol Psychother. 2014;62(4):273–81.

Simpson SG, Jamison KR. The risk of suicide in patients with bipolar disorders. J Clin Psychiatry. 1999;60(Suppl 2):53–6.

Söldner M, Matussek P. Kindheitspersönlichkeit und Kindheitserlebnisse bei Depressiven. In: Matussek P, editor. Beiträge zur Psychodynamik endogener Psychosen. Berlin, Heidelberg, New York: Springer; 1990. p. 134–62.

Strakowski SM, et al. The functional neuroanatomy of bipolar disorder: a consensus model. Bipolar Disord. 2012;14(4):313–25.

Tellenbach H. Melancholie. Zur Problemgeschichte, Typologie, Pathogenese und Klinik. Berlin, Göttingen, Heidelberg: Springer; 1961.

Torres IJ, et al. Neuropsychological functioning in euthymic bipolar disorder: a meta-analysis. Acta Psychiatr Scand Suppl. 2007;434:17–26.

Townsend J, Altshuler LL. Emotion processing and regulation in bipolar disorder: a review. Bipolar Disord. 2012;14(4):326–39.

Versace A, et al. Abnormal left and right amygdala-orbitofrontal cortical functional connectivity to emotional faces: state versus trait vulnerability markers of depression in bipolar disorder. Biol Psychiatry. 2010;67(5):422–31.

Willi J. Die zweier Beziehung. Spannungsursachen – Störungsmuster – Klärungsprozesse – Lösungsmodelle. Hamburg: Rowohlt; 1975.

Wirz-Justice A, Quinto C, Cajochen C, Werth E, Hock C. A rapid-cycling bipolar patient treated with long nights, bed rest, and light. Biol Psychiatry. 1999;45(8):1075–7.

Zerssen von D. Premorbid personality and affective psychoses. In: Burrows GD, editor. Handbook of studies on depression. Amsterdam: Excerpta Medica; 1977. p. 79–103.

# Anxiety Disorders

# 13

Peter Hartwich, Heinz Boeker, and Georg Northoff

**Abstract**

This chapter is about pathological anxiety in psychiatric disorders: phobic disorders, panic disorders, generalized anxiety disorder, and separation anxiety. We describe two typical case examples: agoraphobia and social phobia with all their complexity.

Neurobiological research emphasizes that the perception of new, threatening stimulus constellations generates an unspecific activity pattern in memory-storing associative cortical and subcortical structures. The activity of the HPA axis presents a protecting function in the sense of an emergency reaction. This has sustaining consequences for the functions of neurons and glial cells: gene expression of neurons, production of nervous growth factors, dendritic and axonal growth, development of dendritic spines and synaptic contacts, and reorganization of neural and synaptic connections in the brain.

Neuroimaging studies of the *effects of psychotherapy* in patients with anxiety disorders show that psychotherapy leads to a decrease of the differences between patients and healthy persons in the sense of normalization. This is not only interesting as a demonstration of the neurobiological effects of psycho-

P. Hartwich (✉)
Department of Psychiatry, Psychotherapy, Psychosomatics, General Hospital of Frankfurt am Main, Teaching Hospital University Frankfurt, Frankfurt, Germany
e-mail: Prof.Hartwich@casytec.de

H. Boeker
Department of Psychiatry, Psychotherapy, and Psychosomatics, Psychiatric University Hospital Zurich, University of Zurich, Zurich, Switzerland
e-mail: heinz.boeker@bli.uzh.ch

G. Northoff
Mind, Brain Imaging, and Neuroethics,
Institute of Mental Health Research, University of Ottawa, Ottawa, ON, Canada
e-mail: georg.northoff@theroyal.ca; www.georgnorthoff.com



H. Boeker et al. (eds.), *Neuropsychodynamic Psychiatry*,
https://doi.org/10.1007/978-3-319-75112-2_13

therapy but also enables a better understanding of the effects and mechanisms of psychotherapy. Accordingly, in neuropsychodynamic psychotherapy not new or compensatory networks are developed, but instead functional networks are trained, regulated, and normalized, which also can be found in healthy persons. Especially a reduction of the amygdala activity is found in anxiety disorders. In some anxiety disorders, the PFC activity is reduced and on this way "normalized"; in others rather the control of the PFC on the amygdala is improved, without changes of the PFC activity. On this background the neurobiological models of anxiety especially start from a disturbed balance between hyperactive limbic emotional regions (amygdala and insula) and the dysfunctional cingulate control.

Anxiety is a central feature of most psychiatric illnesses, but it is also a universal human experience.

It was 1844 that Kiekegaard (1981) described anxiety as a fundamental element of our lives, in which the individual becomes conscious of his insecurity in the world and realizes the uncertainty associated with freedom. Sartre takes up this point of view that in experiencing fear, the individual gains consciousness from his freedom. "My fear is free and manifests my freedom into my fear, and I have chosen myself as fearful in this or that circumstance" (Sartre 1956, 1943, p. 445).

This chapter is not intended to address the fact that anxiety, as a human phenomenon, always has the purpose of avoiding dangers and preserving health and life. It is about pathological anxiety in psychiatric disorders.

## 13.1 Anxiety Syndromes as Psychiatric Disorders

The patient feels restricted, uncertain, driven into a corner, and feels frightened with a reduced capacity of reflection and narrowed field of perception. This can be accompanied by palpitations, constricted throat, cardiac pains, wide pupils, raised pulse rate and blood pressure, sweating and increased muscular tonus.

Scharfetter (1985) emphasized that on the one hand there are transitions between the defined syndromes and on the other hand the individual's ability to react to danger, and to deal with fear is very different not only intraindividually but also interindividually.

When investigating anxiety phenomena scientifically, it is additionally complicated by the following: the experienced anxiety can only be specified subjectively and cannot be measured objectively, unless the somatic accompaniments are recorded. Consequently, the subjectively experienced anxiety is usually assessed with questionnaire tests, and the physiological anxiety correlates can be registered (pulse, blood pressure, EEG, EMG, neuroendocrine substances, etc.).

## 13.2 Classification of Anxiety Syndromes (DSM-5)

Gabbard (2014) emphasized that anxiety disorders underwent major changes in the development of diagnostic systems, so in DSM-5 the core anxiety disorders are now phobic disorders, generalized anxiety disorder, panic disorder, and separation anxiety.

- Phobic disorders:
  - Agoraphobia
  - Social phobia
  - Specific (isolated) phobia
- Panic disorder (episodic paroxysmal anxiety)
- Generalized anxiety disorder
- Separation anxiety

## 13.3 Neurobiological Research of Anxiety Syndromes

In a neuroscientific, systemtheoretical, and evolutionary perspective, anxiety and stress may be looked upon as stressors for the adaptive self-organization of the structure and function of living systems. Anxiety contributes to open inherited behaviour programs and to modify, adapt, and reorganize the respective neural networks. The perception of new, threatening stimulus constellations generates an unspecific activity pattern in memory-storing associative cortical and subcortical structures. The activity of the HPA axis presents a primarily protecting function in the sense of an emergency reaction. This has sustaining consequences for the functions of neurons and glial cells: gene expression of neurons, production of nervous growth factors, dendritic and axonal growth, development of dendritic spines and synaptic contacts, and reorganization of neural and synaptic connections in the brain.

Psychosocial factors are often the most important triggers of the neural stress reaction; these include conflicts, loss of competence, psychosocial roles, and psychosocial support.

In a *neuropsychodynamic* perspective, anxiety and coping with anxiety is an integrative aspect of the development of the subject as a psychic entity through the encounter of the organism with the outside world and the processing of the respective somatic reactions. Unconsciousness and consciousness are gradients of a continuum, which includes the outside world, the subject, and the interaction between subject and outside world.

Neuroscientific models of anxiety disorders which start from the induction of an anxiety reaction in the respective person (on the background of objectively undangerous situations and stimuli, e.g. social situations or narrow rooms), within certain brain regions, e.g. the amygdala, the insula, and the anterior cingulate cortex (ACC), are strongly activated (Etkin 2012; Gross and Canteras 2012; Brühl et al. 2014).

Cognitive models of anxiety disorders start from the increased arousal and a stronger reactivity to potentially dangerous stimuli (cf. Bishop 2007). On the basis of this increased arousal, a specific anxiety may develop by means of conditioning processes (connection between an actually undangerous stimuli with an unpleasant stimulus). The anxiety disorder then is kept up by means of avoidance: avoidance reduces the anticipatory anxiety for the short term but increases the experienced "danger" of the feared stimulus and therefore increases the expected anxiety and the avoiding behaviour on the long term. The anxiety bias of selective perception is in the centre of the neurocognitive models of anxiety: the actual state anxiety contributes to a modulation of the amygdala responsiveness. In contrast to prefrontal down-regulation of amygdala activity, the current state anxiety modulates amygdala responsiveness (cf. Fig. 13.1), trait anxiety modulates prefrontal activation (cf. Fig. 13.2).

Consequently the modification of cognitive patterns is one of the main aims of cognitive behavioural therapy (CBT). Within the guiding idea is "Change the mind and you change the brain" (Paquette et al. 2003; cf. Fig. 13.3a, b). The study investigated the biological effect of a 4-week exposition-based CBT in 12 patients with spider phobia by means of fMRI. It could be shown that the neural activity in the occipital, inferior frontal regions and in the amygdala when watching pictures of

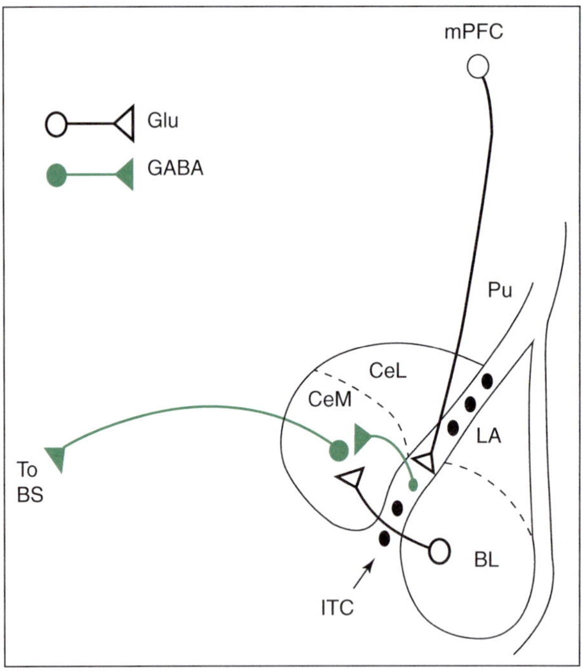


**Fig. 13.1** Neurocognitive mechanisms of anxiety. Prefrontal down regulation of amygdala activity. LA Nucleus lateralis amygdalae, Cel Nucleus centralis lateralis amygdalae, CeM Nucleus medialis centralis, BS Brain stem, BL Nucleus basolateralis amygdalae, PU Putamen, Glu Excitatoric glutamatergic projections, GABA Inhibitoric GABAergic projections, ITC Intermediate GABAergic neurons, mPFC medial prefrontal cortex (cf. Bishop 2007; with friendly permission from Elsevier)

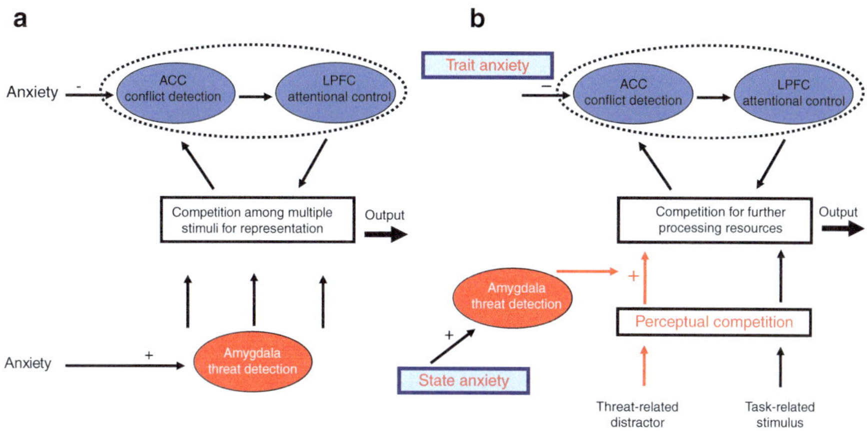


**Fig. 13.2** Neurocognitive mechanisms of anxiety. Anxiety bias of selective perception. (**a**) Modulation of amygdala responsivity through state-anxiety, (**b**) Modulation of prefrontal activation through trait-anxiety (cf Bishop 2007; with friendly permission from Elsevier

spiders was increased in the beginning, but after the therapy, the neural activity was reduced in these regions. Multifaceted studies showed that cognitive strategies (e.g. new or modified cognitive valuation) activate the prefrontal cortex (PFC) and within regulate the activity of limbic structures (Kohn et al. 2014; Diekhof et al. 2011; Brühl et al. 2015).

Neurobiological models of the effectiveness of exposition therapy start from the assumption that exposition does not "extinguish" the existing connection between stimulus and reaction "physically" but moreover that a new learning of the not existing danger is involved. This new association then suppresses the existing anxiety association but which also may be activated again. Summarizing, the neuroscientific studies underline that the medial prefrontal cortex contributes to a downregulation of the amygdala activity.

A possible neuropsychodynamic perspective with regard to the development of anxiety becomes apparent because of following reasons:

- Each change of interpretation in the neural context, which is associated with danger, provokes motoric answers, which react against the danger.
- Unconscious processing: amygdala.
- Conscious processing: cortex.
- Sources of dangers: outer and inner objects.
- Flight: danger from the outside.
- Defence mechanisms: inner dangers.
- Differentiation on the neural level: lack of empirical data (concerning the transition) until now.
- Access to different spatio-temporal "distributed patterns" of memory by means of psychoanalysis (free association), which are important for the development of anxiety. The anxiety comprehends the whole personality and exists in different psychic formations.

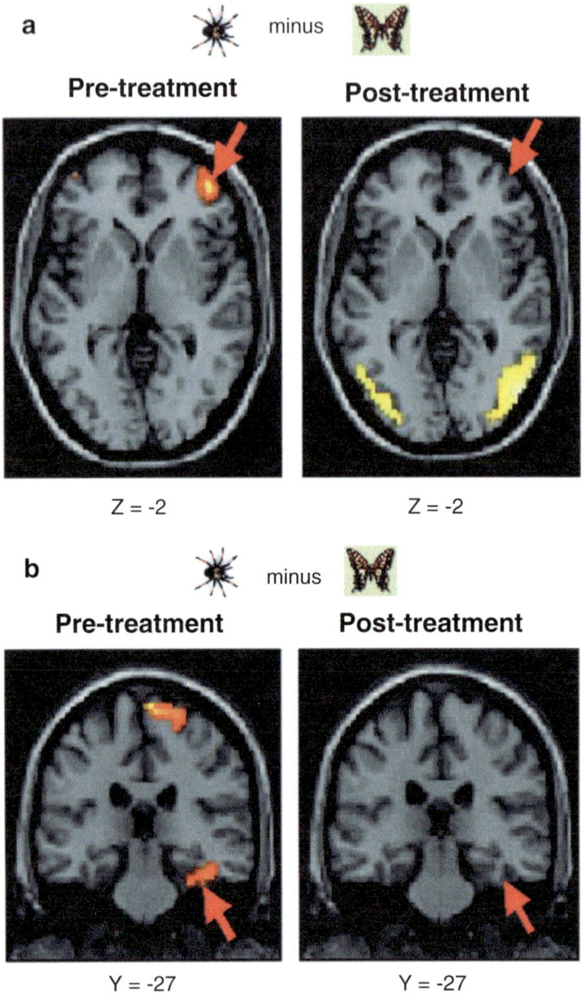


**Fig. 13.3** Effects of CBT (Cognitive Behavioural Therapy) on spider phobia. (**a**) Activation of right DLPFC (dorsolateral prefrontal cortex) before CBT in patients with spider phobia, missing activation of right DLPFC after CBT, (**b**) Activation of right parahippocampal gyrus (BA 36) before CBT, missing activation of right parahippocampal gyrus after CBT (cf. Paquette et al. 2003; with friendly permission from Elsevier)

In a neuropsychodynamic perspective, anxiety may be induced, when inner and outer conflicts are defended deficiently (panic model of Shear et al. 1993, neurotic experience of unknown self-components). A possible loss of relationships also may induce a strong experience of threat and anxiety (cf. Bowlby 1977). Accordingly neuropsychodynamic psychotherapy aims at changing the ego-structural weaknesses, the unconscious conflicts, the suppressed aggressive feelings, and/or the relationship conflicts. Because the avoidance of anxiety is often maintained, exposition strategies may be integrated in the therapy (Beutel et al. 2010).

As a summary, neuroimaging studies of the effects of psychotherapy in patients with anxiety disorders show that psychotherapy leads to a decrease of the differences between patients and healthy persons in the sense of a normalization. This is not only interesting as a demonstration of the neurobiological effects of psychotherapy but also enables a better understanding of the effects and mechanisms of psychotherapy. Accordingly, in psychotherapy not new or compensatory networks are developed, but instead functional networks are trained, regulated, and normalized, which also can be found in healthy persons. Especially a reduction of the amygdala activity is found in anxiety disorders. In some anxiety disorders, the PFC activity is reduced and on this way "normalized"; in others rather the control of the PFC on the amygdala is improved, without changes of the PFC activity. On this background the neurobiological models of anxiety especially start from a disturbed balance between hyperactive limbic emotional regions (amygdala and insula) and the dysfunctional cingulate control.

## 13.4 Concepts of Etiopathogenesis

With regard to theories of anxiety, the learning of theoretical and the psychodynamic approaches has to be mentioned. Classical and operant conditional models were the starting point in learning theoretical dimension. In the attribution theory, it is assumed that anxiety is generated by making the own person responsible for the anxiety-inducing events.

Rüger (1986) pointed out that a comprehensive general theory of anxiety should be able to explain the biological and psychological mechanisms of anxiety starting from the perception of the threat, the respective psychophysiologic processing of the stimuli, and the anxiety reaction as well; he thinks that the "arousal-concept therefore is suitable" (pp. 46–47):

> Internal (e.g. drive impulses) or external (e.g. objective threat) stimuli are perceived and according the genetic equipment and the former life experiences validated in regard of the strength of the threat and the possible threat for the organism or the integrity of the individual; they induce an "arousal-reaction" in the central nervous system with the above mentioned physiological procedures, accompanied by the subjective feeling of anxiety. This "preparedness attitude" induces mechanisms of anxiety compensation, in the case of an consciously perceived threat as coping mechanism, in the case of unconscious irrational conflicts rather as classical defence mechanism. (translated by H.B.)

Psychodynamic concepts assume that anxiety endangers the ego experience in its cohesion and continuity and therefore triggers psychic defence mechanisms. These include regression as well as mature and so-called immature defence mechanisms. They are able to reduce the experience of anxiety. If these mechanisms are not sufficient because of the existing structure weakness of the personality or unbearable intensity of the threatening situation, anxiety disorders develop. It was already in the beginning of psychoanalysis and psychodynamics that the discussion about anxiety and its illness consequences took a central place. Freud (1892–1899, p. 255) proposed the term "anxiety neurosis". Anxiety is an emotion, which gets conscious in the ego. There the anxiety impulses from the id, and the psychic representations are controlled. The repressed affects may find an expression as symptoms according to the functions of the defence mechanisms. Because the origin of the anxiety stimulating situation is often unconscious, the psychoanalytic or psychodynamic treatment has the aim that the patient gets aware of these unconscious reasons in order to work them through. @Less the intellectual recognition of the origin is decisive, but more the conscious re-experience of the accompanying affects, which are often derived from a former, less mature developmental period of the human, then to be worked through in the therapy of the adult with her/his more mature structure.

Gabbard (2014) pointed to the existence of the original anxiety affects in each person and underlined that these may be triggered by stress or traumatic events. Therefore the therapist has to be creative in order to perceive the individual and specific anxiety components of the patient.

## 13.5 The Neuropsychodynamic Approach

Mentzos (2011) describes the origin of symptoms of phobia as follows: it consists in a repression of the original anxiety-producing content preceding the origin of the phobia and a subsequent displacement of the associated danger or the fear of a relatively insignificant externality. Greenson (1959) has formulated the creation of the symptoms in phobia: one form of anxiety is used as a defence against another anxiety.

Mentzos points out that it does not remain with the displacement, a central component of this mode is the preventive avoidance. In addition to the actual anxiety symptoms in the patients, the avoidance behaviour often leads to considerable restrictions on their normal life.

*Agoraphobia* is an unpleasant phenomenon that affects patients if they have to go alone across an empty square, a wide street, or a plain. They then have the feeling of having no hold and develop serious fears with additional vegetative symptoms. As a rule, they develop an avoidance behaviour in that they, e.g. do not leave their home any more.

*Claustrophobia*, often confused in the people's mouth, however, refers to the fear of closed spaces, such as lifts, trams, trains, prisons, etc.

*Specific phobia* is the abnormal fear of certain objects or situations, such as thunderstorms, dark cellar, and to have a certain disease (AIDS, melanoma, and syphilis).

Exaggerated fears that someone has when he is in a group of people and believes to be critically observed are called *social phobias*. This includes, for example, speaking in front of the group and exam in group situations. Behind this is the fear of shame, which can be accompanied by hand tingling, excessive sweating, intense blushing, heart palpitations, and so-called blackouts.

**Case Study: Agoraphobia**

A 30-year-old teacher has mostly been staying at home with his wife for 2 years and is learning Portuguese language. If he has to go to the city alone on an important date, he goes the way he knows well and where he knows above all that there are doctors' offices. Thus, he laboriously moves from doctor's plate to doctor's plate and has to overcome himself again and again until he reaches the goal. Sometimes there are panic situations with heart rashes and other vegetative symptoms, but mostly it is with the fear of this fear.

In psychotherapy, the death of the mother, which happend 2 years ago, is treated as a triggering situation. In many drawings, in which he portrays his dreams and the psychotherapeutic process, it is about clinging scenes and separation situations. After some months of intensive psychodynamic treatment, he succeeds in coming to the therapy session without being accompanied. Gradually, he also succeeded in transferring his transference wishes to the early childhood constellation as a single child of a mother who had lost her husband early and had bound the boy strongly to herself.

One year after the end of the therapy, I get a postcard from Portugal where it can be read that he can make his dream come true, namely, a backpacking tour through Portugal alone.

### 13.5.1 Social Phobia and Complexity

One should take into account that there are many phobic disorders that have a complex substructure. Behind what is called social phobia, for example, exam nerves, many problems can be hidden: temporary or permanent impairments, work disturbances, profound problems, identity crises, addiction, neurotic conflicts, or psychotic dispositions and even psychotic episodes.

Tabbert-Haugg (2003) describes the different aspects of the depth psychology and manifestation of diagnoses, which are first classified as exam nerves or job problems as symptoms of social phobia in late adolescents and other patients who are looking for a psychotherapist. There are, on the one hand, forms that are easy and temporary, which are curable by immediate crises intervention and, on the other hand, forms which represent psychic dysfunction in a threshold situation. These are typical phase-specific conflicts with the reactivation of pre-Oedipal constellations causing narcissistic disorders and those on a psychotic level. The psychodynamic treatment methods consist of crisis interventions, focal therapy, and deep psychological and analytical psychotherapy, depending on the kind and severity of the disorder.

**Case Study: An Initial Symptom of Social Phobia (Exam Nerves)**
We summarize a case from Tabbert-Haugg (2003, pp. 150–152): A young Arab student suffers from severe work problems and exam nerves after his father's death.

At first the sudden death and the burial ceremony were not communicated by the family to him in order not to disturb his preparations for the exam. However, when he got to know about the death of his father, the unconscious aggressive conflicts with the father came up violently, and his mental abilities were considerably impaired. Unconscious feelings of revenge and the magically increased fear of destroying the father in overly increasing aggression could produce a profound regression with unbearable feelings of guilt. In case he would pass the exam, the long-planned takeover of the paternal position in the home country by the family clan would be dealt. He would become patriarch of the family over sisters and mother. His mother always adored him as a partner, since she consoled herself as the wife number two of the son's father (Islam). However, he was dethroned in puberty by the father's strict authoritarian patriarchy, having, contrary to his own will, to go to an international school and later to begin his studies prescribed by the father. The therapist's attempt to enable him to continue his preparation for exam by means of medical certificates, so that he could still take his exam, was boycotted by the patient consciously or unconsciously. The treating psychoanalyst writes: "He has given up and expects his expulsion. In my countertransference, I feel sad, angry, and powerless. The 'nightmare exam' of the student became for me a lost struggle for a psychotherapeutic success".

## 13.6 Discussion

This example shows that – as it is often the case – even before the occurrence of "social phobia", deeper conflicts are beyond, for example, with the father, who determines his way of life against the patient's will, the situation of migration with all its conflicts of adaption and insults, as well as the identity crisis as a result of very different cultural traditions. The reactivation of depressive and narcissistic features as well as loss of compensation on psychotic structure level is not uncommon.

## 13.7 Psychotherapeutic Outcome

In many cases today, phobic disorders are treated with behavioural therapeutic methods. Whether psychodynamic or learning theoretically based therapies (CBT) are more effective or not, Leichsenring et al. (2013) found that both therapies are efficient and do not differ referring to the result. The response rate was a little better in behavioural therapy (60%) than in psychodynamic therapy (52%); but this did not differ statistically significantly. Overall, the study shows that psychodynamic therapy is no better or worse than behavioural therapy, based on the positive efficacy with regard to the stronger response, which is also related to the long-term effect (Leichsenring et al. 2014). From the results of the multicentre study, it should be emphasized that 40–48% of the patients were neither significantly improved by the behavioural therapy nor by psychodynamic psychotherapy. The authors point out that these numbers correspond too many other studies regarding non-responsiveness to therapies.

## 13.8 Panic Disorder (Episodic Paroxysmal Anxiety)

A meta-analysis of 20 studies (Kossowsky et al. 2013) indicated that children with the diagnosis "separation anxiety disorder" significantly increases the risk to develop panic disorders later on and also any anxiety disorders.

Panic attacks as episodic paroxysmal anxiety can occur all of a sudden and unexpectedly with heart palpitations, chest pain, chocking fit, dizziness, and alienation feelings. For minutes, the fear of death and loosening control occur. The anxiety disorders in the DSM-5 version are worked out by Gabbard (2014) under psychodynamic aspects. One of the possible causes is the loss experience which is not adequately coped with. In a study with more than 1000 pairs of twins (Kendler et al. 1992), panic disorder was associated with both parental divorce and death. The meta-analysis of Kossowsky et al. (2013) argues for the separation hypothesis.

A further etiological factor in female patients is physical and sexual abuse in childhood (Stein et al. 1996). It is also discussed that a predisposing neurophysiologic vulnerability exists in panic disorders; this is discussed by Gabbard (2014) under a neuropsychodynamic concept: De Masi (2004) assumes that severe traumatic anxiety can be triggered by conditioned stimuli, which are connected with prior danger situations. This model integrates neuroscientific findings with a psychodynamic understanding. In this model, he refers to the work of LeDoux (1996), who pointed out that unconscious memories of fear situations anchored in the amygdala may leave indelible traces in the brain. The amygdala is the first area of the brain, which is activated in the case of a fear signal. "This activation may be entirely unconscious, and the fight-flight response may take over before the thalamus has time to relay information to the cortex." (Gabbard 2014, p. 265). As a result, rational thinking cannot be applied to the situation.

## 13.9 Generalized Anxiety Disorder

The generalized anxiety disorder is then diagnosed when generalized, persistent, and free-floating anxiety exists, according to ICD-10 F40.1 over a period of at least several weeks and according to DSM-5 up to 6 months. It is the most common disease, with the highest rate of comorbidity. Goisman et al. found in a multicentre study already in 1995 that almost 90% of the patients thus diagnosed had another anxiety disorder in their lives (Goisman et al. 1995).

Leichsenring et al. (2009) found in a randomized controlled study that the treatment results of short-term psychodynamic psychotherapies and CBT (cognitive behavioural therapy) differed not statistically significantly. From medical practitioners, anxiety-relieving drugs are often prescribed. However, it is worthwhile to listen to the patient's history of the illnesses, in order to select the appropriate therapeutic method for him. Some will be referred to expert behavioural therapists, and others will require psychodynamic therapy; in both cases a temporary psychopharmacological co-treatment may be necessary, but this must be strictly limited in time. The psychodynamic therapies depend on the kind of disease that is so diverse that the one can be managed with a short therapy of several weeks, the other needs a longer psychodynamic therapy of several months, and some patients need a psychoanalytic treatment of some years.

## References

Beutel ME, Stark R, Pan H, et al. Changes of brain activation in pre-post-short-term psychodynamic in-patient psychotherapy: an fMRI study of panic disorder patients. Psychiatry Res Neuroimaging. 2010;184(2):96–104.
Bishop SJ. Neurocognitive mechanisms of anxiety: an integrative account. Trends Cogn Sci. 2007;11(7):307–16.
Bowlby J. The making and breaking of affectional bonds. I. Aetiology and psychopathology in the light of attachment theory. An expanded version of the fiftieth Maudsley lecture, delivered before the Royal College of Psychiatrists, 19 November 1976. Br J Psychiatry. 1977;130(3):201–10.
Brühl AB, Delsignore A, Komossa K, Weidt S. Neuroimaging in social anxiety disorder – a meta-analytic review resulting in a new neurofunctional model. Neurosci Biobehav Rev. 2014;47:260–80.
Brühl AB, Herwig U, Rufer M, Weidt S. Neurowissenschaftliche Befunde zur Psychotherapie von Angststörungen. Z Psychiatr Psychol Psychother. 2015;63(2):109–16.
De Masi F. The psychodynamic of panic attacks: a useful integration of psychoanalysis and neuroscience. Int J Psychoanal. 2004;85:311–36.
Diekhof EK, Geier K, Falkai P, Gruber O. Fear is only as deep as the mind allows: a coordinate-based meta-analysis of neuroimaging studies on the regulation of negative affect. NeuroImage. 2011;58:275–85.
Etkin A. Neurobiology of anxiety disorders: from neural circuits to novel solutions? Depress Anxiety. 2012;29(5):355–8.
Freud S. Studien über Hysterie. Gesammelte Werke Bd 1, vol. 1977. Frankfurt/M: Fischer; 1892–1899. p. 75–312.
Gabbard GO. Psychodynamic psychiatry in clinical practice. 5th ed. Washington, DC: American Psychiatric; 2014.
Goisman RM, Goldenberg I, Vasile RG, et al. Comorbidity of anxiety disorders in a multicenter anxiety study. Compr Psychiatry. 1995;36:303–11.
Greenson RR. Phobia, anxiety and depression. J Am Psychoanal Assoc. 1959;7:663–74.
Gross CT, Canteras NS. The many paths to fear. Nat Rev Neurosci. 2012;13(9):651–8.
Kendler KS, Neale MC, Kessler RC, et al. Childhood parental loss and adult psychopathology in women: a twin study perspective. Arch Gen Psychiatry. 1992;49:109–16.
Kiekegaard S. Der Begriff Angst. GW Abt 11/12. Gütersloher Taschenbücher. Gütersloh: Siebenstern; 1981. (Erstveröff. 1844) [893].
Kohn N, Eickhoff SB, Scheller M, et al. Neural network of cognitive emotion regulation–an ALE meta-analysis and MACM analysis. NeuroImage. 2014;87:345–55.
Kossowsky J, Pfaltz MC, Schneider S, Taeymans J, Locher C, Gaab J. The separation anxiety hypothesis of panic disorder revisited: a meta-analysis. Am J Psychiatry. 2013;170:768–81.
LeDoux J. The emotional brain: the mysterious underpinnings of emotional life. London: Weidenfeld & Nicolson; 1996.
Leichsenring F, Salzer S, Jäger U, et al. Short-term psychodynamic psychotherapy and cognitive-behavioral psychotherapy in generalized anxiety disorder: a randomized, controlled study. Am J Psychiatry. 2009;166:875–81.
Leichsenring F, Salzer S, Beutel ME, et al. Psychodynamic therapy and cognitive-behavioral therapy in social anxiety disorder: a multicenter randomized controlled trial. Am J Psychiatry. 2013;170:759–67.
Leichsenring F, Salzer S, Beutel ME, et al. Long-term outcome of psychodynamic therapy and cognitive-behavioral therapy in social anxiety disorder. Am J Psychiatry. 2014;171:1074–82.
Mentzos S. Lehrbuch der Psychodynamik. Die Funktion der Dysfunktionalität psychischer Störungen. 5th ed. Göttingen: Vandenhoeck & Ruprecht; 2011.
Paquette V, Levesque J, Mensour B, et al. "Change the mind and you change the brain": effects of cognitive behavioral therapy on the neural correlates of spider phobia. NeuroImage. 2003;18(2):401–9.

Rüger U. Angst. In: Müller C, editor. Lexikon der Psychiatrie. 2nd ed. Heidelberg: Springer; 1986. p. S 43–8.

Sartre JP. Being and nothingness. An essay on phenomenological ontology. Translated by Barnes HE, New York: Philosophical library, Inc., L'é'tre et le néant. Essai d'ontologie phénoménologique (1943), Paris: Librairie Gallimard; 1956.

Scharfetter C. Allgemeine psychopathologie. 2nd ed. Stuttgart: Thieme; 1985.

Shear MK, Cooper AM, Klerman GL, et al. A psychodynamic model of panic disorder. Am J Psychiatry. 1993;150:859–79.

Stein MB, Walker JR, Anderson G, et al. Childhood physical and sexual abuse in patients with anxiety disorders and in a community sample. Am J Psychiatry. 1996;153:275–7.

Tabbert-Haugg C. Alptraum Prüfung, Gestörtes Prüfungsverhalten als Ausdruck von Schwellenängsten und Identitätskrisen. Stuttgart: Pfeiffer, Klett-Cotta; 2003.

# Obsessive-Compulsive Disorders

# 14

Peter Hartwich, Heinz Boeker, and Georg Northoff

**Abstract**

In two case reports, we describe typical obsessive-compulsive disorders with overlapping diagnoses.

The characteristic defensive operations of the ego consist in isolation of affect, intellectualization, reaction formation, undoing, and displacement. The great variety of these defense mechanisms results in the diverse symptoms in obsessive-compulsive disorder. Different psychodynamic interpretations (Freud, Gabbard, Mentzos) are described. One of the most fundamental problems and threatening feelings of obsessive-compulsive persons is to get out of control.

New access to the syndrome shows neurobiological investigations. There seems to be a relevant dysfunction of cortico-striato-thalamo-cortical loops. This is connected with a dysbalance of different neuromodulators and neurotransmitters. It can be characterized by hyperactivity. Biochemically, hyperactivity in this loop is related to dopamine and glutamate.

We put these neurobiological findings into a neuropsychodynamic context. One of the main defense mechanisms is displacement, the shift from thought to action—the latter replaces the former. The compulsive action is experienced as dystonic to the own self.

P. Hartwich (✉)
Department of Psychiatry, Psychotherapy, Psychosomatics, General Hospital of Frankfurt am Main, Teaching Hospital University Frankfurt, Frankfurt, Germany
e-mail: Prof.Hartwich@casytec.de

H. Boeker
Department of Psychiatry, Psychotherapy, and Psychosomatics, Psychiatric University Hospital Zurich, University of Zurich, Zurich, Switzerland

G. Northoff
Mind, Brain Imaging, and Neuroethics,
Institute of Mental Health Research, University of Ottawa, Ottawa, ON, Canada
e-mail: georg.northoff@theroyal.ca; www.georgnorthoff.com



H. Boeker et al. (eds.), *Neuropsychodynamic Psychiatry*,
https://doi.org/10.1007/978-3-319-75112-2_14

If the direct loop is strong, excitation takes over inhibition with the subsequent abnormally strong initiation of behavior and movement. This leaves no room for the spontaneous thoughts as they are generated on the cortical level—they simply displaced by the movements. As the subcortical generation of movements takes over and *is no longer in balance* with the cortical generation of internally oriented cognition including self-relatedness, the initiated movement is experienced as dystonic rather than syntonic to the self. Hence, the neuropsychodynamic conceptualization is well compatible with the neurobiological findings and may even further specify it on a functional level.

For neuropsychodynamic therapy we see here a window which may allow new therapeutic strategies which are able to influence the abovementioned balance in a positive way.

In normal life obsessive-compulsive features are to be found when there is a threatening uncertainty of the individual, a group, or a whole society. Usually mankind builds rituals, as we find them in magical actions and in the religions as religious practices or even everyday rituals. We can thus see that rituals serve to stabilize the self. This form of self-stabilization can be, however, abnormally exaggerated when the individual suffers from the rituals, and society ascribes the behavior as abnormal or morbid.

## 14.1 Case Report (1)

For more than 20 years, Mr. M. has been afflicted by inner restlessness. So he has become accustomed to wander several miles a day, which gives him some relief, but there is often an obsessional counting of steps, which goes back to his childhood. For 2 years, the depressive mood of anxiety and agitation, which was stressed especially in the morning, has become stronger. In the evenings, it is usually better. Its primary personality has anancastic (obsessional) features. He is plagued by obsessional thinking, obsessional impulses, and obsessional acts. He must, for example, go back to his car several times to check whether he really locked it. Even to the door of his apartment, he has to go back several times to check whether he has really closed and locked it. In his profession as an external auditor at the government financial office, he is right because his perfectionism is positive for his work. However, he needs a long time for everything, since he has to control the papers over and over again. Often he feels tormented by obsessional thinking and compulsions. He knows that they have no sense, but he is trapped to have the compulsive rituals. If he does not follow them, he becomes psychomotorically restless, gets fear, and feels desperate. From time to time, he is convinced that he has a serious illness, e.g., rectal cancer, cardiac infarction, brain tumor, etc. This leads him to be examined by medical specialists. The medical statement that no pathological findings exist only calms him briefly. Then he goes to another doctor and is examined again. He is desperate that he himself is not able to believe that the statements of the experts are correct.

The father died in the Second World War. The inner father is anchored in him as upright, overcorrect, and strict with clear morality. Even in the regular therapies, to which he always appears punctual and correct, he avoids giving details of the father (splitting).Since the bombing of her hometown during the war, the mother had often periods of weeping. He had a good relationship with her. She also died early, shortly after the war. The inner mother is a kind, but uncertain woman, giving little support. He has no siblings.

His wife used to be a competitive sportswoman; with her he has an intimate symbiotic relationship, and they have no children.

## 14.2 Psychodynamics

After a few sessions, a series of dreams come: memory pictures of his war experiences, tanks, machine guns, dead men. He associates that the war is the uncertainty par excellence. Something worse than what he had experienced could not exist for him. He reports that he was forced to be a soldier at the age of 16 and experienced the most terrible of the Eastern war front. What he has trapped in his memory can be emotionally recaptured. In the psychotherapeutic sessions, he reports anxiety and nightmares after a few more weeks, e.g., he cannot run away when a truck drives into his room and pushes him away. He associates an overwhelming power that sweeps him away. Then he talks about the 3 years when he was a prisoner of war in Belgium, where he had to work underground in a coal mine and had been camping in the open air. He had been hungry all the time and many comrades died.

Since he had already been afraid as a teenager in small rooms, the hard work in the coal mines was an extreme burden, since he had to enter the pit every day and work in the narrow tunnels. It was very dusty; one could see little and was not protected. He often suffered from panic, but as a prisoner of war he had no choice. After the dismissal, and even to this day, it was not customary to talk about his experiences.

In therapy, it became the subject that this particular personal suffering has never been accepted by society. The chance to speak now about the encapsulated trauma, to bring back some of the experience of suffering and anxiety in the sessions, could only be perceived to a small extent. There was a pronounced isolation of affect and rationalization, which could not be further questioned. It was important to respect and accept these defense mechanisms as protective means of his spatiotemporal self.

In his transference there are idealizing features with self-evident obedience and keeping distance. In the countertransference he is experienced as a well-behaved patient while the therapist has the feeling of running against a wall again and again.

## 14.3 Psychopathology

When we consider the DSM-5 criteria for obsessive-compulsive personality disorder, we find orderliness, perfectionism, and mental and interpersonal control at the expense of flexibility and openness (American Psychiatric Association 2013). The person is "preoccupied with details, rules, lists, order, and organization." Gabbard (2014) points out that there is some controversy regarding the extent of the overlap between OCD (obsessive-compulsive disorder) and OCPD (obsessive-compulsive personality disorder) in DSM 5. But OCPD and OCD should be discussed separately when treatment indications are considered.

As Kaplan and Sadock (1984, p. 438–439) define:

> Obsessive refers to an idea or thought, compulsive refers to an urge or impulse to action that, when put into operation, leads to a compulsive act. Obsession and compulsion have certain features in common: (1) An idea or an impulse obtrudes itself insistently, persistently, and impelling into the person's conscious awareness. (2) A feeling of anxious dread accompanies the central manifestation and frequently leads the person to make countermeasures against the initial idea or impulse. (3) The obsession or compulsion is ego-alien – that is, it is experienced as being foreign to and not a usual part of one's experience of oneself as a psychological being; it is undesired, unacceptable, and uncontrollable.

In contrast to alien influences that occur in schizophrenia, the patient knows that the obsession and compulsion are something which he generates himself. Also obsessions are not necessarily absurd in content, but their persistence and monotonous repetition is experienced as absurd. As Scharfetter (1980) points out, obsessional impulses of a dangerous and disturbing kind, e.g., to knife one's own child, are hardly ever translated into action, even if the patient suffers a lot from the fear to follow the disturbing impulse. Usually he uses all his energy to resist the impulse.

In the case report, we also see typical obsessive acts: The patient's obsessional controlling of the doors of his car or his condominium and obsessional counting. Other examples are obsessional cleaning and washing, because of the fear of bacteria. That is often accompanied by rituals or avoidances.

## 14.4 Case Report (2) with Overlapping Diagnoses

### 14.4.1 Depressive Obsessive-Compulsive Disorder and Anorexia Nervosa

Today the female patient is 54 years old. An anorexia nervosa existed when she was 18 years old. At the age of 30, depressive episodes dominated for several years. Now she is dominated by obsessive-compulsive symptoms. If she leaves the house, she becomes restless and fearful: *Did I close the apartment door, have I switched off the electric stove*? She runs back and controls everything, and then she goes back to the street. Once again, she is concerned whether she has actually locked the front door this time. She always has to check her wardrobe, she fears that clothes could be stolen. She also buys new clothes by Internet shopping, which she does not unpack. There is also a quarrel with the husband, who hardly finds space in the apartment, since the rooms are furnished with clothes boxes.

She has problems to sleep in the night, because she has to control many things in the house again and again.

Psychodynamically, obsessive-compulsive disorder and anorexia have something in common: a *pathologically excessive need for control* and a need for making the life safe, which ultimately can never be fully satisfied.

This common ground shows that in the course of a patient's life, different manifestations of symptoms occur: in the time of puberty and adolescence, anorexia nervosa. After recovery from anorectic symptoms, later as an adult she suffers from depression with nihilistic ideas. When these symptoms calmed down during a long treatment period, she then has predominantly obsessive symptoms in advanced life.

If psychodynamic psychiatrists treat these patients at intervals for more than two to four decades, it becomes evident how symptoms can change. It is the structural ego-strength with its defense mechanisms that change. Crucial is the profound and deep anxiety, which in this case was caused by war events and family poverty in the patient's childhood and a genetic disposition on the part of the father. Based on the deeply anchored experience from childhood, the forming structure of a spatiotemporal self with a temporally enduring security of the existence of a coherent self has not been able to develop. This deep insecurity runs through all the developmental periods in later life and can lead to different syndromes; many symptoms can have a function that serves to secure the self.

It would be important to get to know the neurobiological conditions and changes of these children's brain and later to investigate which neurobiological aberrations are compensated for further development and which aberrations cannot be compensated for but can be detected in a different neural pattern. Such findings could be important to help to find the adequate strategies in therapy.

## 14.5 Epidemiology

Lifetime prevalence is estimated from 1% to 3% (Kessler et al. 2005a, b). A lot of comorbidity with other severe psychiatric illnesses is to be observed, e.g., depression, anxiety disorder, trauma- and stressor-related disorders, personality disorders, anorexia nervosa, etc. Klein et al. (2016) point out that the mentioned comorbidity is not an exception, but it is *the rule* of obsessive-compulsive disorders. Therefore, there are considerable differences in the frequency of occurrence of obsessive-compulsive disorders in the scientific literature.

## 14.6 About Etiopathogenesis

The etiology and pathogenesis of obsessive-compulsive disorder is still unknown today. There is still no generally accepted hypothesis which covers all clinical obsessive-compulsive syndromes. In the field of psychodynamics, there are some working hypotheses referring to the old concept of obsessive-compulsive neurosis. Freud (1908) sees the symptoms of compulsion as an expression, namely, of unconscious motifs and ideas. These cannot be reconciled with their own punitive superego, which is why feelings of guilt occur. Thus the symptoms are to be understood as defense mechanisms in order to repress drive.

In Freud's view the obsessive-compulsive disorder is contrary to hysteria and phobia, not the expression of a rather balanced compromise between id, ego, and superego. The ego in obsessive-compulsive disorder is no longer able to defend the neurotic conflict by one main form of symptom creation alone. Due to the extremely forced conflict between id and superego, the ego has to develop a growing number of defense mechanisms which enable to bear the neurotic conflict. The characteristic defensive operations of the ego consist in isolation of affect, intellectualization, reaction formation, undoing, and displacement. The great variety of these defense mechanisms results in the diverse symptoms in obsessive-compulsive disorder.

Freud assumed that the forced neurotic conflict in OCD is the result of regression to the anal-sadistic phase of the organization of drives. This regressive movement gets along with a more destructive component of aggression and a growing ambivalence toward the objects. Nevertheless, the genital organization of the libido is not completely lost. Freud (1908) underlined that the symptoms in OCD protect the ego against the complete loss of the object. In this context ego ideal and superego in OCD represent the anxiety against object loss.

In OCD there is not only a regression of the libido but also a regression of the ego: In Freud's view the regression of the ego reaches to the magical-animistic organizational level of the drives and gets along with a failure of reality testing. Additionally the borderlines between subject and outer world get lost, reality testing is out of order in favor of the lust-unlust principle, and thinking and acting are equated resulting in a thought disorder, respectively. Narcissistic cathexis substitutes object cathexis. The psychic energy of the drives is no longer neutralized and thus enables the ego to develop the feeling of omnipotence and of magical powers of its organs.

Freud's hypothesis stressed that the ego in OCD patients is not affected as a whole in the regressive process: The regression only concerns those parts of the ego which react on the suppressed drives. This partial ego regression may explain that the actual conscious ego is confronted with its own neurotic obsessive-compulsive symptoms without understanding them.

More modern concepts are described by Gabbard (2014, p. 580):

> Because obsessive-compulsive patient find both anger and dependency consciously unacceptable, they defend against those feelings with defenses such as reaction formation and isolation of affect. In a counterdependent effort to deny any dependency on anyone, many obsessive-compulsive persons go to great length to demonstrate their independence and their 'rugged individualism'. Similarly they strive for complete control over all anger, and they may even appear deferential and obsequious to avoid any impression of harboring angry feelings.

Mentzos (2011) places the mode of processing in the foreground. While the nature of the conflicts and the level of organization of the structure of the personality can vary widely, the modus forms a common constant. This modus can serve to stop the psychotic disintegration or the psychotic self-fragmentation before the outbreak of a schizophrenic psychosis. Likewise, in severe depression, there are obsessive-compulsive symptoms which are to be understood as attempts to ward off depression.

What does Mentzos mean by the term modus? He explains that, in the modus, both the intrapsychic defense constellation and the dominant nature of the individual's relationship to himself and to the object (other human beings) are expressed in a characteristic manner. Thus, for example, a hysterical modus, a depressive modus, and a modus of the individual with obsessive-compulsive disorder can be differentiated.

Dümpelmann and Northoff (2016) point out that the strengthening of autonomy is abnormally pronounced by compulsive thinking and compulsive action.

One of the most important fundamental problems and threatening feelings of obsessive-compulsive persons is *to get out of control*. This is annoying for themselves and also for partners, especially for intimacy. The fear of the patient to get out of control is also an obstacle in any treatment, not only the psychotherapeutic or psychodynamic one. The countertransference is usually characterized by feelings of anger and even aggression, which the therapist experiences very strongly. The psychotherapy is not effective as long as the therapist becomes aware that his countertransference feelings belong to the patient and are the patient's aggressions.

According to Dümpelmann and Northoff (2016), the compulsive disorder can be described by two key features: (1) the disturbed and distorted form of autonomy by the compulsions and (2) the disturbed self-perception with shifts of unconscious thoughts and actions.

## 14.7 Neurobiological Investigation

Mavrogiorgou (2014) points out that during the last 20 years, the knowledge about the obsessive-compulsive disorder has been expanded by many neurobiological studies. There seems to be a relevant dysfunction of cortico-striato-thalamo-cortical loops. This is connected with a dysbalance of different neuromodulators and neurotransmitters.

Specifically, the loop between orbitofrontal cortex (OFC), anterior cingulate cortex (ACC), and caudate nucleus (CN) is altered. The findings show anatomo-structural as well as functional changes which are quite consistent, well replicated, and specific for OCD. These regions and the loop as whole can be characterized by hyperactivity: there is too much neural activity, i.e., excitement in the regions and the loop as whole (Pauls et al. 2014; Nakao et al. 2014). Biochemically, hyperactivity in this loop is related to dopamine and glutamate (Pauls et al. 2014).

The loop connects cortical and subcortical regions in a feedback loop. The OFC and ACC serve as cortical nodes that relay information to the subcortical CN and the striatum. From there on the information can be relayed back to the cortex in a direct, i.e., excitatory, or indirect, i.e., inhibitory, way. The direct pathway leads from CN over globus pallidus internal/substantia nigra over the thalamus back to the cortex where it leads to strengthening of neural excitation. In contrast, the indirect loop is inhibitory as it includes yet another subcortical relay station, the subthalamic nucleus and external globus pallidus (Pauls et al. 2014). OCD is now characterized by a dysbalance between these loops: the direct excitatory cortico-subcortical-cortical loop is abnormally strong, while the indirect inhibitory loop is abnormally weak – behavioral impulses and goal orientation as generated on the cortical level can no longer be inhibited and subsequently lead to the kind of repetitive and stereotypical behavior of OCD. Yet another region that may be central is the insula that is strongly activated during disgust and has therefore also been associated with OCD.

How can we put these neurobiological findings into a neuropsychodynamic context? One of the main defense mechanisms in OCD is displacement, the shift from thought to action – the latter replaces the former. Moreover, the compulsive action is experienced as dystonic to the own self – they are estranged action which is experienced as dystonic rather than syntonic relative to the own self of the patient.

How can we explain both displacement and ego-dystonia? The OFC and ACC are part of the cortical midline structure where internally oriented cognition like self-related processing and spontaneous thoughts are generated (see chapter on self). In contrast, the subcortical component of the cortical-subcortical loop is strongly involved in generating movements as the main function of the basal ganglia like substantia nigra, CN, globus pallidus (internal and external), and subthalamic nucleus.

If now the direct loop is strong, excitation takes over inhibition with the subsequent abnormally strong initiation of behavior and movement. This leaves no room for the spontaneous thoughts as they are generated on the cortical level—they simply displaced by the movements. As the subcortical generation of movements takes over and is no longer in balance with the cortical generation of internally oriented cognition including self-relatedness, the initiated movement is experienced as dystonic rather than syntonic to the self. Hence, the neuropsychodynamic conceptualization is well compatible with the neurobiological findings and may even further specify it on a functional level.

Finally, the OFC has also been associated with early traumatic life experience by Nakao et al. (2013), and Duncan et al. (2015) showed that early traumatic life events in childhood are manifest in neuronal changes in OFC and ACC neuronal activity changes as for instance in increased entropy, i.e., disorder in neuronal activity patterns. That is well compatible with the often observed early traumatic life events in OCD patients as, for instance, in our first case. Early traumatic life events may thus lead to neuronal instability with decreased neuronal continuity in OFC and ACC which, psychodynamically, may be manifest on an unstable temporally less continuous self. As the self is temporally less stable and continuous, it is more prone to displacement by movements and actions as related to the over-excitation of the direct cortical-subcortical-cortical loop. Hence, again, neurobiological findings and psychodynamic descriptions converge with the latter being apparently constituted on the basis of the former.

## References

American Psychiatric Association. Diagnostic and statistical manual of mental disorders. 5th ed. Washington, DC: American Psychiatric Association; 2013.

Dümpelmann M, Northoff G. Zwangssyndrome. In: Böker H, et al., editors. Neuropsychodynamische Psychiatrie. Berlin: Springer; 2016. p. 297–308.

Duncan NW, Hayes DJ, Wiebking C, et al. Negative childhood experiences alter a prefrontal-insular-motor cortical network in healthy adults: a preliminary multimodal rsfMRI-fMRI-dMRI study. Hum Brain Mapp. 2015;36:4622–37.

Freud S. Character and anal erotism. In: The standard edition of the complete psychological works of Sigmund Freud, vol. 9. Translated and edited by Strachey J. London, Hogarth Press, 1959; 1908. pp 167–175.

Gabbard GO. Psychodynamic psychiatry. Washington DC: American Psychiatric; 2014.

Kaplan HI, Sadock BJ. Modern synopsis of comprehensive textbook of psychiatry/III. Baltimore: William & Wilkins; 1984.

Kessler RC, Berglund P, Demler O, et al. Life-time prevalence and age-of-onset distributions of DSM-IV disorders in the national comorbidity survey replication. Arch Gen Psychiatry. 2005a;62:593–602.

Kessler RC, Chiu WT, Demler O, et al. Prevalence, severity and comorbidity of 12-month DSM-IV disorders in the National Comorbidity Survey Replication. Arch Gen Psychiatry. 2005b;62:617–27.

Klein JP, Zurowski B, Wolff J, et al. Leitliniengerechte stationäre psychiatrisch-psychotherapeutische Behandlung der Zwangsstörung. Nervenarzt. 2016;87(7):731–8.

Mavrogiorgou P. Zwangsstörungen. In: Juckel G, Edel MA, editors. Neurobiologie und Psychotherapie. Stuttgart: Schattauer; 2014. p. 128–34.

Mentzos S. Lehrbuch der Psychodynamik. Die Funktion der Dysfunktionalität psychischer Störungen. Göttingen: Vandenhoeck & Ruprecht; 2011.

Nakao T, Matsumoto T, Morita M, et al. The degree of early life stress predicts decreased medial prefrontal activations and the shift from internally to externally guided decision making: an exploratory NIRS study during resting state and self-oriented task. Front Hum Neurosci. 2013;7:339. https://doi.org/10.3389/fnhum.2013.00339.

Nakao T, Okada K, Kanba S. Neurobiological model of obsessive-compulsive disorder: evidence from recent neuropsychological and neuroimaging findings. Psychiatry Clin Neurosci. 2014;68(8):587–605. https://doi.org/10.1111/pcn.12195.

Pauls DL, Abramovitch A, Rauch SL, Geller DA. Obsessive-compulsive disorder: an integrative genetic and neurobiological perspective. Nat Rev Neurosci. 2014;15(6):410–24. https://doi.org/10.1038/nrn3746.

Scharfetter C. General psychopathology. Cambridge: Cambridge University Press; 1980.

# Somatization and Bodily Distress Disorder

# 15

Moritz de Greck

**Abstract**

Somatization refers to the formation of bodily symptoms—including pain, vegetative symptoms, and others—that do not have an appropriate medical explanation.

From a psychodynamic perspective, somatization is caused by the repression of subjectively unbearable emotions, conflicts, and memories. A neurobiological correlate of this mechanism is the altered function of the brain regions that are involved in the processing of autobiographical memory.

If intensified pain perception is the leading symptom, subjectively experienced social separation can very often be identified as a trigger mechanism. From a neurobiological point of view, this phenomenon is caused by shared neural presentations for the processing of pain and social isolation.

## 15.1 Introduction

Somatization, in other words, the appearance of somatic symptoms that do not have a medical explanation, has puzzled patients and physicians for more than 2000 years, and it continues to puzzle patients, physicians, and psychotherapists.

During the quest for an understanding of the mechanisms that underlie somatization, many important achievements have been made, such as the establishment of psychoanalysis by Josef Breuer and Sigmund Freud.

As indicated by the profound changes made in the categorization of somatization-related disorders in the current or coming versions of diagnostic manuals (Diagnostic and Statistical Manual of Mental Disorders, version 5

M. de Greck
Department of Psychiatry, Psychosomatic Medicine and Psychotherapy, Johann Wolfgang Goethe University Hospital, Frankfurt am Main, Germany



H. Boeker et al. (eds.), *Neuropsychodynamic Psychiatry*,
https://doi.org/10.1007/978-3-319-75112-2_15

[DSM-V] and the International Statistical Classification of Diseases and Related Health Problems version 11 [ICD-11]), the concept of somatization is still subject to substantial changes.

This chapter aims to explore important psychodynamic and neurobiological mechanisms of the fascinating phenomena of somatization, which form at the interface between mind and body.

## 15.2 Case Report: Mrs. B.—Part 1

### 15.2.1 Presentation

When Mrs. B. presented for outpatient therapy, she was suffering from pain at different body sites, and she had been experiencing aching in her neck, back, elbows, and wrists for 4 years. The pain was constant, but it could vary in intensity on a day-to-day basis, with different locations being dominant at different times. Mrs. B. suffered enormously from this pain and had been seeking treatment from a number of different doctors. Radiological diagnostics showed small ossifications of the spine, which might have accounted, in part, for the back and neck pain, but there were no significant findings that could explain the elbow and wrist complaints. Mrs. B. also reported repeated incidences of sudden transitory paralysis of her arms and hands, for which neurological diagnostics showed no relevant findings. While suffering from this paralysis, Mrs. B. was unable to perform any activities with her hands. Mrs. B. believed her pain was related to fibromyalgia, a disease she could identify with and frequently informed herself about by using the internet and other media. However, she could not explain her transitory paralysis.

At the start of her treatment, Mrs. B. was 39 years old. She had been married for 10 years and had three children, aged 7, 4, and 2 years. She had been working successfully as a lawyer until the birth of her second child. She then gave up her career to concentrate on caring for her children. The family was financially secure with her husband's income.

## 15.3 Somatization: The Evolution of a Psychosomatic Concept

### 15.3.1 From the Ancient to the Medieval: Hysteria, Hypochondria, and Witchery

The first concepts of somatization; in other words, the development of somatic symptoms without an adequate medical cause, can be traced back to the ancient Greeks and Romans (Morschitzky 2007). For instance, the terms "hysteria" and "hypochondria" reach back to these times. The term "hysteria" (derived from the Greek word for uterus, "hystera") refers to the ancient belief that unexplainable somatic symptoms in female patients were caused by a wandering uterus.

While the diagnosis of hysteria was made almost only in women, men suffering from somatic symptoms without an adequate underlying cause were diagnosed with hypochondria.

The term "hypochondria" is also derived from the Greek, from the terms "hypo" and "chondros", the first meaning "under" and the latter, "cartilage". The expression "hypochondria" refers to complaints that were felt under the rib cartilage, which, according to the Greek physician Galen, were frequently accompanied by fear and sadness (Morschitzky 2007). In Europe, the concepts of hysteria and hypochondria were regarded as valid until the thirteenth century, when the belief in witches and witchcraft developed. Typical psychosomatic symptoms, such as convulsions, trances, or loss of sensibility, were now interpreted as a sign that a patient suffering from these symptoms was possessed by evil spirits or had had sexual contact with the devil (Morschitzky 2007). Hundreds of years before there were humane approaches to these disturbing phenomena, these patients were threatened by unbearable torture and death.

### 15.3.2 Eighteenth and Nineteenth Centuries: Briquet, Charcot, Janet, and the French Era

In the late eighteenth century and throughout the nineteenth century, as Paris became the world center of medical development, major contributions regarding unexplained somatic symptoms were made by Parisian physicians. For instance, Paul Briquet (1796–1881) put an end to the association of unexplained symptoms and the uterus (Mai 1982). Briquet, the chief internist at the Paris Hôpital de la Charité from 1846 onward, was confronted with a great number of patients with unexplainable symptoms, which were in those times diagnosed as "hysteric". Through his collection of 430 case studies, Briquet detected the influence of stress and unhappy life situations in the development of "hysteria". Later, owing to his contributions in the field, unexplained somatic symptoms were known as "Briquet's syndrome".

Jean-Martin Charcot (1825–1893) was the chief doctor at the Paris Salpêtrière Hospital from 1862 onward. The Salpêtrière had the largest psychiatric department in Europe at that time, and Charcot was responsible for more than 6000 inpatients (Schuchart 2017). Charcot used hypnosis to provoke and eradicate somatic ("hysteric") symptoms, and he linked such symptoms to traumatic events (Morschitzky 2007).

Pierre Janet (1859–1947) was a psychologist and student of Charcot's at the Salpêtrière. He also investigated "hysteric" patients and described the role of the "subconscious" and "dissociation" in the development of hysteric symptoms. His work can be seen as a basis on which Breuer and Freud later developed psychoanalysis.

### 15.3.3 Breuer and Freud: The Origin of Psychoanalysis

In 1883, Josef Breuer (1842–1925), a famous general practitioner in Vienna, treated his patient Anna O., who suffered from a variety of medically unexplained symptoms, such as aphasia, paralysis, and disorders of visual perception. Most of Anna

O.'s symptoms would now be categorized as conversion disorder; however, her neuralgic pain might have fulfilled the criteria of a pain disorder, i.e., a form of somatization-related disorder. In order to help Anna O., Breuer experimented with different techniques, which included systematically talking about those situations in which a specific symptom had occurred. This technique of talking about ("Aberzählen") led (at least temporarily) to a reduction in the strength of her symptoms. Sigmund Freud (1856–1939), who had attended to Charcot's lectures in Paris in 1885, published a case report on Anna O. in 1894, together with Breuer. This is now seen as the beginning of psychoanalysis. It is thus fair to say that the riddle of unexplained somatic symptoms played an important role in the development of psychoanalysis (and thus in the development of psychotherapy).

### 15.3.4 Schur: Desomatization and Resomatization

Max Schur (1897–1969) was a psychoanalyst; he was a friend of Sigmund Freud and his personal physician. Schur developed the influential concepts of desomatization and resomatization (Schur 1955) as mechanisms underlying unexplained somatic symptoms. During normal psychomotor development, the development of affect leads to fewer and fewer somatic reactions (for example, fewer instances of heart rate being increased during a period of anxiety). This occurs owing to the person's increasing ability to control affect through mental processes ("secondary process thinking"). Schur termed this process "desomatization". If however, secondary process thinking deteriorates (for instance, because of the activation of repressed traumatic memories), somatic reactions might reappear and lead to somatic symptoms–a process called "resomatization". Since the traumatic memories, or in other words, the actual causes of the somatic symptoms, remain repressed, the appearance of the somatic symptoms cannot be explained by the patient. Schur described these symptoms as "concomitants" of repressed affect. He distinguished them from "equivalents" of affect (in particular, anxiety), which he considered as memories of emotional states from a preverbal developmental stage (Schur 1955).

Schur's concept resembles that of alexithymia, which has been investigated by a number of studies in recent years. Alexithymia refers to a reduced capacity to verbalize affect and a tendency to develop emotional fantasies (Taylor 1984). As numerous studies have shown, somatizing patients show alexithymic features (Bach and Bach 1996; Bailey and Henry 2007; Bankier et al. 2001; Burba et al. 2006; Duddu et al. 2003; Grabe et al. 2004; Mattila et al. 2008). These features are congruent with Schur's concept of repressed emotions being the underlying cause of somatization-related disorders.

### 15.3.5 Alexander: Vegetative Neuroses

The Hungarian-American psychoanalyst Franz Alexander (1891–1964) emphasized the difference between conversion and "vegetative neuroses" (Alexander

1977), which would today be categorized as somatoform symptoms. Alexander was able to explain that conversion referred to unconsciously induced dysfunctions of perception or motor systems that helped the patient to solve an emotional conflict. Vegetative neuroses (i.e., somatoform disorders), in contrast, developed as a consequence of longstanding emotional states that could not be resolved or discharged. In more detail (see Alexander 1977 and Chap. 8), In more detail (see Alexander 1977, Chap. 8 ), explained that vegetative neuroses develop as a consequence of the paramount need of the nervous system to maintain homoeostasis.

The voluntary, somatic, section of the nervous system is engaged with maintaining successful contact with the external world, while the vegetative section of the nervous system is responsible for "internal affairs". In emergency situations, the sympathetic nervous system prepares the organism with "fight or flight" responses (for instance, by increasing heart rate or breathing rate). In rest situations, however, the parasympathetic nervous system triggers repair and assimilation processes (such as digestion).

In terms of Alexander's concept, vegetative neuroses can develop in two different ways:

1. If an organism prepares itself for an emergency situation by activating the sympathetic nervous system, and then inhibits the relevant action (i.e., "fight" or "flight"), this may lead to the development of conditions such as arterial hypertension. The organism remains in a constant state of preparation, with long-lasting activation of the sympathetic nervous system.
2. If the organism reacts to stress not by activating the sympathetic nervous system, but rather the parasympathetic nervous system, diseases such as stress-induced gastric ulcers may develop. While this may sound counterintuitive at first, it is further explained by Alexander as follows: Instead of actively confronting a stress-inducing situation (as in the case of increased sympathetic activation), the organism reacts paradoxically and relies on infantile behavioral patterns of retreat. The organism (unconsciously) seeks a protective mother-object, and anticipates feeding (hence the increased parasympathetic activation). Here again, the organism remains in a long-lasting state of preparation, since it does not actually receive the (unconsciously) anticipated feeding.

Alexander continues by comparing these "vegetative neuroses" to "psychoneuroses". Every neurosis implies the retreat from an action and the escape to an autoplastic process. In psychoneuroses motoric activity is replaced by psychological activity (acting in phantasy). Vegetative neuroses, in contrast, develop, since prolonged emotional tensions lead to vegetative changes. In addition, Alexander separates vegetative neuroses from conversion disorders: whilst conversion disorders (alike psychoneuroses) are mediated by the voluntary somatic nervous system, vegetative neuroses develop on the basis of a dysfunctional vegetative nervous system.

Owing to his profound contributions to the field of psychosomatic medicine, Alexander became known as the "father" of psychosomatic medicine (New York Times 1964).

### 15.3.6 Somatization as a Defense Mechanism

A number of authors have conceptualized somatization in terms of a defense mechanism that is implemented by a person in order to deal with an unbearable conflict. Aversion from the external world and concentration on the inner world play important roles in regard to somatization as a defense mechanism. In other words, somatizing patients retreat from processes in their external reality (mostly because these processes are conflictive and apparently unbearable) and focus instead on bodily processes and phenomena (Hansell and Mechanic 1985; Witthöft and Hiller 2010). This makes it easier for the patient to deal with emotionally burdensome situations and conflicts in the short-term, but these conflicts stay unresolved. At the same time, the patient acquires increased sensitivity to bodily phenomena (Nakao and Barsky 2007). With regard to this overall less-than-optimal outcome (i.e., the patient retreats from important conflicts and, in turn, has an increased awareness of bodily disturbances, which leads to further problems, such as more visits to doctors), somatization was categorized as a rather immature defense mechanism (Bond et al. 1983).

The American psychoanalyst George Vaillant (1977) defined hypochondria as a defense mechanism; this is helpful for understanding somatization. According to Vaillant, the hypochondriac patient has introjected the belief of a disappointing and frustrating mother-object. Instead of being angry with the disappointing object, which might lead to unbearable anxieties, the patient is angry about and afraid of his/her bodily complaints. Not infrequently, during the course of treatment, the therapist also becomes the target of the patient's anger, since the therapist is also experienced as a disappointing object who is unable to help the patient.

### 15.3.7 Post-Psychodynamic Concepts: Today's Classification of Somatization and Bodily Distress Disorders

Current classification systems, such as the DSM and the ICD, describe somatization-related disorders with regard to objectively observable behaviors, while etiological mechanisms are regarded as playing only a minor role. Somatization-related disorders were categorized as "functional disorders" in the DSM-II and ICD-8 and in the ICD-9 (Hiller and Rief 2005). The expression "functional" referred to somatic symptoms that were caused by functional somatic abnormalities, although no structural abnormalities could be observed.

In the DSM-III and DSM-IV, as well as the ICD-10 (which is the current diagnostic system used in most countries outside the United States), somatization-related disorders were categorized as "somatoform disorders". The expression "somatoform" refers to the feature that somatization led to symptoms that have the form, but not the identity, of a somatic disease (Hiller and Rief 2005). With the transition from the DSM-IV to the DSM-V, somatization-related disorders have once more been reconceptualized. The category "somatic symptom disorder" is now favored, rather than somatoform disorder. In addition, significant changes have been made concerning the diagnostic criteria. Most importantly, the absence of a medical explanation is no longer required. Instead, the diagnosis of somatic symptom disorder refers to patients who show significant distress and impairment because of somatic symptoms. It is argued that these new criteria are advantageous compared with the previous ones, for several reasons:

1. The new concept is supposed to be more useful in primary care. It emphasizes the role of positive symptoms (i.e., distressing somatic symptoms and abnormal thoughts, feelings, or behavior) rather than the absence of a medical explanation.
2. The new concept is an improvement over the old concept of somatoform disorders, which overlapped different sub-diagnoses (American Psychiatric Association 2013, p. 309).

The soon-to-appear ICD-11 (to be published in 2018) follows in the same manner; the diagnosis "Somatoform Disorder" will be omitted, and instead the diagnosis "Bodily Distress Disorder" will be introduced.

In the chapter Somatic Symptoms and Related Disorders, the DSM-V allows differentiation between the following diagnoses with regard to somatizing (i.e., medically unexplained) symptoms: Somatic Symptom Disorder (300.82), Illness Anxiety Disorder (300.7), Conversion Disorder, Psychological Factors Affecting Other Medical Conditions (316), Factitious Disorder (300.19), Other Specified Somatic Symptom and Related Disorders (300.89), and Unspecified Somatic Symptom and Related Disorders (300.82) (American Psychiatric Association 2013).

### 15.3.8 Difference Between Somatization and Conversion

Somatization and conversion disorders have much in common. For instance, they refer to medically unexplained symptoms and they not infrequently occur together in one and the same patient. In particular, patients formerly described as "hysteric" often show a combination of conversion and somatization.

There are, however, a number of criteria that differentiate these two disorders:

- While somatization-related symptoms appear in association with vegetatively innervated organs, conversion disorders affect muscular systems that are innervated by the voluntary somatic nervous system.
- Somatization-related symptoms develop slowly, while conversion-related symptoms often appear suddenly and can also disappear suddenly. In other words, the "leap from psyche to soma" (Alexander 1977) is much faster in conversion-related symptoms than in somatization-related symptoms. This characteristic is associated with their origins. While somatization is related to changes in the vegetative nervous system, conversion symptoms are caused by the voluntary somatic nervous system. In other words, behind a conversion symptom is always the unconscious decision to show this specific symptom.

- Somatization-related symptoms are often caused by a repressed affect, while conversion-related symptoms are often caused by unconscious conflicts, the conversion disorder being a mechanism to deal with these conflicts.
- Somatizing patients are often burdened by their symptoms. Patients with conversion disorder, however, often impress by an affective insensibleness ("belle indifference"). A reason for this characteristic is that patients with conversion disorder can solve agonizing conflicts with the help of their symptoms, which leads to feelings of relief.
- Conversion disorder symptoms often have a symbolic function (e.g., dissociative paralysis of a leg in a patient with a separation conflict: this patient wants to go, but is also afraid to go, and seemingly cannot go). Somatizing symptoms, however, seldom have a symbolic meaning (Alexander 1977; Ermann 2006; Mentzos 2009).

## 15.4 Case Report: Mrs. B.—Part 2

### 15.4.1 Diagnostic and Psychodynamic Assessment

Mrs. B. fulfilled the criteria for a somatic symptom disorder with predominant pain (300.82) according to the DSM-V (American Psychiatric Association 2013). She was also diagnosed with a conversion disorder with weakness and paralysis (F44.4).

With regard to significant relationship experiences, Mrs. B. reported that she was an unwanted child of her parents, and that, from early childhood, she had to take care of herself. Her mother was not able to react empathetically to Mrs. B.'s need for care and support, so that Mrs. B. eventually gave up and relied on herself. These experiences still played a role in her life at the time of her admission to the clinic. Mrs. B. explained that she had similar feelings with regard to her husband. She felt that he did not support her as much as she wished with managing the household and caring for their three children. It became clear, however, that Mrs. B. could not explain her need for stronger support to her husband, since she was too afraid of being rejected (again), which would have led to a conflict and intense feelings of sadness and anger. In addition, Mrs. B. expected that she alone should deal successfully with this situation.

## 15.5 The Neurobiological Basis of Somatization and Bodily Distress

### 15.5.1 Pain Perception

Pain is a complex subjective experience. In general, the perception of pain is preceded by impending or actual tissue damage. Nerve fibers (nociceptors) detect this tissue damage and react by increasing their neuronal activity (nociceptive pain). Nociceptors react to a range of different threats, such as high temperature, mechanical stimulation, or chemical impacts. The neurons of these nociceptors are located within the spinal ganglia. Neuronal activity is transferred via the dorsal root into the dorsal horn of the spinal cord. Here, activity is transposed from the first neuron to the second neuron. The axons of the second neurons lead, as the spinothalamic tract, to the thalamus. Within the thalamus, activity is transposed from the second neuron to a third neuron. The axons of these third neurons lead neuronal activity to the postcentral gyrus, to the anterior cingulate cortex, and to the insular cortex (Craig 2003). The somatosensory cortex, which is essential for the processing of external perceptions, is located within the postcentral gyrus. The anterior cingulate cortex plays an important role in the processing of affective and motivational components of pain (such as the initiation of an anger response after pain perception). The insular cortex is crucially involved in the processing of interoceptive perceptions from the internal world of the body. It processes not only pain, but also other qualities, such as temperature.

The transposition of neuronal activity from the first neuron to the second neuron can be inhibited via descending pathways originating in the reticular formation of the brain stem (Bushnell et al. 1999; Martin 2003; Trepel 2012). This mechanism was described for the first time in 1965 by Melzack and Wall as the "gate control theory" (Melzack and Wall 1965). This mechanism can explain why, for example, moods and affect have a strong impact on pain processing. In addition, the analgesic effects of some antidepressants can be explained by the gate control theory.

### 15.5.2 Emotional Pain Processing

Besides the mere somatic perception of pain, emotional components also play an important role in pain processing (Price 2000; Rainville 2002; Shackman et al. 2011). Pain is, in general, an unpleasant subjective experience that induces an anxiety-like feeling of threat or anger (depending on the extent of pain and the threat experienced in a specific situation). In the view of Craig (2003), pain is a homoeostatic emotion, similar to other homoeostatic emotions such as temperature, thirst, or hunger. In other words, similarly to hunger, which will motivate an organism to eat when energy levels drop below a homoeostatic level, the pain system will motivate the organism to get rid of the painful stimulus. Neuroimaging studies have shown that pain perception leads to activity within the somatosensory cortex, anterior cingulate cortex, insular cortex, thalamus, premotor cortex, and the cerebellar vermis (Casey 1999; Coghill et al. 1999; Svensson et al. 1997). The role of the somatosensory cortex is seen in the localization of pain and the evaluation of its intensity (Bushnell et al. 1999). The anterior cingulate cortex is essential for the affective evaluation of the pain-causing event as a threat; it directs attention and controls vegetative processes (Martin 2003; Rainville 2002). The insular cortex is also involved in the somatosensory processing of pain stimuli. It plays an important role in the affective evaluation of the pain-causing events and the control of vegetative processes. The activity of the premotor cortex and the cerebellum is related to motor preparation processes (Coghill et al. 1999).

### 15.5.3 Pain and Social Isolation

Expressions such as "heartache" or painful separation illustrate the close association of physical pain with unpleasant feelings of separateness from an attachment figure. The two entities have an impending or actual threat in common: while in physical pain there is a danger of physical damage, in the pain of separation there is a danger of being separated from or excluded by the peer group. Since humans live in social groups, and since the survival of a single human excluded by his or her group is significantly jeopardized, it is understandable that the two situations are evaluated as similarly dangerous, and that the situations feel subjectively similar and lead to neural activity in similar networks. MacDonald and Leary (2005) conceptualize that it makes good sense, from an evolutionary point of view, that social species have a neurobiological system that punishes those members who do not prevent social exclusion or who do not try to react to an impending social exclusion as fast and as well as they can. Concordantly, social exclusion leads to neural activity in the same brain areas as those that are involved in the emotional processing of physical pain. As shown in a brain imaging study by Eisenberger et al. (2003), the anterior cingulate cortex in healthy participants showed increased neural activity when the subjects felt socially excluded. Later, Landa et al. (2012) described a common neurobiological system for physical and social pain. They further suggested that disturbed development, including insecure attachment and problems in affect regulation, may lead to hypersensitivity to physical and social pain. This view was supported by the study of Nickel and Egle (2006), who found an association of somatization and sexual or physical abuse in patients suffering from low back pain. The assumption that a hypersensitive pain system is the neurobiological basis that underlies somatization-related disorders was further supported by the studies of Egloff et al. (2014) and Gupta et al. (2007). From an evolutionary point of view, an overlap of systems for physical pain and social pain seems plausible. Physical damage is dangerous, in particular for those members of a social group who are isolated and cannot rely on social support in the case of further deterioration of their physical health.

The connection between the physical and social pain system was further demonstrated in a fascinating animal study, showing that 30-min social interaction led to endogenous brain opioid release in rats (Panksepp and Bishop 1981).

### 15.5.4 Neurobiological Correlates of Altered Pain Processing in Somatization

In 2011, Browning et al. published a meta-analysis of 21 original research papers about neurobiological changes in patients suffering from medically unexplained complaints (Browning et al. 2011). In many of the included studies, the authors found increased neural activity of the anterior cingulate cortex and insular cortex, two key regions crucially involved in the processing of pain. In their investigation of healthy subjects, Coghill et al. (2003) were able to show that the anterior cingulate cortex showed stronger neural activity induced by pain stimuli if participants were more sensitive to pain. In sensitive persons, such as patients suffering from subjective electrosensitivity, the mere anticipation of being exposed to electromagnetic mobile phone radiation led to increased activity in the anterior cingulate cortex and insular cortex (Landgrebe et al. 2008).

Taken together, these studies suggest that a hypersensitive pain system may underlie somatization-related disorders.

### 15.5.5 Neurobiological Correlates of Altered Emotional Processing in Somatization

Psychodynamic models hypothesize that the suppression or repression of emotions plays an important role in the development of somatization (Alexander 1977; Schur 1955). In a functional magnetic resonance imaging (fMRI) study, neurobiological correlates for this psychological process could be identified. Somatizing patients were asked to share the emotional states of faces showing different emotional expressions. While doing this, somatizing patients showed reduced neural activity in the bilateral parahippocampal gyrus (and other regions), in comparison to healthy control subjects (de Greck et al. 2012). The parahippocampal gyrus is known to be involved in autobiographical memory processes (Gardini et al. 2006). It shows increased activity when conflicting autobiographical contents are processed (Loughead et al. 2010). Reduced activity of the parahippocampal gyrus was thus interpreted as a possible neurobiological correlate for repressed emotional memories. This is congruent with a study by Schmeing et al. (2013), who found reduced neural activity of the parahippocampal gyrus in healthy subjects when they were asked to freely associate about conflicting topics. Interestingly, patients who participated in an inpatient multimodal psychodynamic psychotherapy group (which included approaches that aimed to increase insight into unconscious psychodynamic conflicts) showed normalization of parahippocampal gyrus activity when they were trying to share the emotional states of emotional faces (de Greck et al. 2013).

### 15.5.6 Neurobiological Correlates of Altered Attention Shifting

Somatization implies the reduction of attention directed to the external world and a concentration of attention towards processes of the internal bodily world (Hansell and Mechanic 1985; Nakao and Barsky 2007; Witthöft and Hiller 2010). Investigating the neural mechanisms underlying this attention shift, de Greck et al. (2011) used fMRI to investigate brain activity induced by the processing of relevant external stimuli (in this study, visual cues, which led to a reward if the subject responded rapidly enough to them). The postcentral gyrus and the ventroposterior thalamus (as well as other regions) showed reduced activity in somatizing patients. Both the postcentral gyrus and the ventroposterior thalamus play an important role in the processing of external stimuli.

## 15.6 Neuropsychodynamic Integration

Neurobiological findings can signifcantly contribute to our knowledge about somatization and its underlying mechanisms.

With regard to pain disorders, the knowledge of an overlap of neural systems for physical and social pain is helpful. The anterior cingulate cortex and the insular cortex are not only activated by pending or actual physical tissue damage, but also by social exclusion. This explains why traumatic childhood experiences that lead to a reduced ability to deal with social exclusion experiences can also cause increased sensitivity to pain stimuli, which results in an increased risk of suffering from pain disorders.

With regard to those somatization-related disorders in which not pain, but vegetative symptoms, are the predominant complaint, knowledge about the underlying neurobiological basis might also be helpful. Owing to traumatic childhood experiences, many somatizing patients have emotional deficits; for instance, reduced emotional awareness caused by repressed emotions. As a consequence, these patients cannot rely on their emotions to develop motivations and concepts to deal with stressful situations in a specific way. Thus, significant problematic situations or conflicts remain unresolved. This may lead to chronic stress and the long-term increase of vegetative neural activity. In addition, these patients have only a limited capacity to understand emotion-related vegetative changes, such as an increased heart rate, as correlates of emotional states. These bodily changes are thus experienced as irritating and disturbing, and they induce anxieties that a not-yet-detected disease might be responsible. Somatizing patients are, on the one hand, confronted with burdensome conflictive situations for which they do not feel any motivation to resolve. On the other hand, these patients experience disturbing vegetative bodily phenomena for which they cannot find an explanation. Reduced activity of the parahippocampal gyrus might be a neurobiological correlate of reduced access to repressed emotional memories.

## 15.7 Implications of Therapy

Somatizing patients are famous for being a therapeutic challenge. They often reject psychodynamic models of symptom formation and rely on somatic models instead. In addition, many somatizing patients have had disappointing or traumatizing experiences with significant people in their lives; thus, the establishment of a positive therapeutic relationship is often difficult owing to negative transference. Furthermore, the therapist's patience is often challenged by the patient's skeptical attitude, on the one hand, and their profound difficulties, on the other hand. Hence, the formation of a positive therapeutic relationship is essential. A detailed history of somatic complaints, a flexible readiness to accept somatic disease models, patience, and calmness might be helpful in establishing such a relationship.

The models and the results presented here suggest that the following approaches might be important in psychotherapeutic work with somatizing patients:

- Patients suffering from chronic pain may have experienced episodes of social exclusion. And often they subjectively identify themselves with being socially isolated, which is not necessarily in accordance with an objective assessment of the patient's social situation. Identifying situations that are perceived as social isolation, and bringing these situations into the patient's consciousness, is thus essential. Only when patients are consciously aware of their perceptions are they are able to react voluntarily. Very often, once patients are able to overcome a situation of subjectively perceived isolation, the somatized symptoms lose significance.

- Patients reporting vegetative complaints will probably benefit from therapeutic approaches that focus on affective awareness and differentiation of affect. The therapeutic aims might also include working through emotions that were repressed because of traumatic childhood experiences. Once these emotions, with their vegatative bodily phenomena, are accessible and understandable for the patient, the patient can develop motivations to deal with the burdensome conflictive situations.
- Psychoeduction about the neurobiological and psychodynamic associations presented here might also be helpful for somatizing patients.

From a neuro psychodynamic point of view, the following questions might be helpful in the treatment of somatizing patients:

If pain symptoms are predominant, does the patient identify himself/herself with being socially excluded or isolated?
If a patient suffers from medically unexplained vegetative symptoms, what are the conflicts and emotions that cannot be experienced consciously?

## 15.8 Case Report: Mrs. B.—Part 3

### 15.8.1 Psychotherapy Results

At first, Mrs. B. was very ambivalent about whether she should engage in psychodynamic treatment. Eventually, however, she decided to commit to psychodynamic outpatient treatment, for which a total of 50 sessions was planned, with one 50-min session per week. Mrs. B. tended not to take medication; very seldom, she took very low doses of ibuprofen for her pain. She was also very skeptical about psychodynamic concepts concerning her pain and paralysis. However, because the therapeutic relationship was characterized by a positive, accepting tone, Mrs. B. tolerated psychodynamic explanations of the possible reasons underlying increases or decreases in her symptom load. During the course of the therapy, the connection between Mrs. B.'s inner conflict with her husband and her childhood experiences became more and more clear to her. Mrs. B. felt that she was left alone in managing the household and caring for the children; however, she did not dare to ask her husband for support, since she anticipated rejection, which would be followed by intense feelings of sadness and anger. At around the tenth month of psychotherapy, however, Mrs. B. eventually addressed her feelings toward her husband. And, since her husband did indeed start to support her more, she experienced a signficant reduction in the paralytic symptoms. Mrs. B.'s pain-related symptoms decreased more slowly after she started work again, in a part-time position as a lawyer. In addition, Mrs. B. reactivated old friendships and became more active in her free time, and this significantly increased her quality of life. Until the end of her course of psychotherapy, Mrs. B. continually reported a steady decrease in her pain symptoms. This was also related to increased time spent on pleasant activities, which did not leave her much time to focus on bodily complaints.

### 15.8.2 Neuro Psychodynamic Findings

Mrs. B.'s pain symptoms were probably related to increased activity in her anterior cingulate and insular cortices, triggered by her subjective view that she was separated from and isolated from important life sectors (career and free time activities). From a psychodynamic point of view, it is important that Mrs. B. had had to rely on herself from early childhood onward. At the beginning of the psychotherapy sessions, Mrs. B. was very ambivalent about her need for more support from her husband, which she repressed out of fear of being rejected again. Regarding this, she had transferred her relationship experiences from her mother onto her husband. After Mrs. B. told her husband about her wish for more support, however, and gained his help, she was able to reactivate important sectors of her life, such as work and free time activities, which led to significantly reduced feelings of exclusion. This might have led to the normalization of neural activity in her anterior cingulate cortex and insular cortex, thus leading to a desensitization of her pain system. In addition, Mrs. B. was able to focus more on pleasant events in her external reality, which led to decreased concentration on unpleasant bodily experiences.

## References

Alexander F. Psychosomatische Medizin. Grundlagen und Anwendungsgebiete. 3. unveränderte Auflage. Berlin: Walter de Gruyter; 1977.

American Psychiatric Association. Diagnostic and statistical manual of mental disorders. Arlington: American Psychiatric Association; 2013.

Bach M, Bach D. Alexithymia in somatoform disorder and somatic disease: a comparative study. Psychother Psychosom. 1996;65(3):150–2.

Bailey PE, Henry JD. Alexithymia, somatization and negative affect in a community sample. Psychiatry Res. 2007;150(1):13–20.

Bankier B, Aigner M, Bach M. Alexithymia in DSM-IV disorder: comparative evaluation of somatoform disorder, panic disorder, obsessive-compulsive disorder, and depression. Psychosomatics. 2001;42(3):235–40.

Bond M, Gardner ST, Christian J, Sigal JJ. Empirical study of self-rated defense styles. Arch Gen Psychiatry. 1983;40(3):333–8.

Browning M, Fletcher P, Sharpe M. Can neuroimaging help us to understand and classify somatoform disorders? A systematic and critical review. Psychosom Med. 2011;73(2):173–84.

Burba B, Oswald R, Grigaliunien V, Neverauskiene S, Jankuviene O, Chue P. A controlled study of alexithymia in adolescent patients with persistent somatoform pain disorder. Can J Psychiatr. 2006;51(7):468–71.

Bushnell MC, Duncan GH, Hofbauer RK, Ha B, Chen JI, Carrier B. Pain perception: is there a role for primary somatosensory cortex? Proc Natl Acad Sci U S A. 1999;96(14):7705–9.

Casey KL. Forebrain mechanisms of nociception and pain: analysis through imaging. Proc Natl Acad Sci U S A. 1999;96(14):7668–74.
Coghill RC, McHaffie JG, Yen YF. Neural correlates of interindividual differences in the subjective experience of pain. Proc Natl Acad Sci U S A. 2003;100(14):8538–42.
Coghill RC, Sang CN, Maisog JM, Iadarola MJ. Pain intensity processing within the human brain: a bilateral, distributed mechanism. J Neurophysiol. 1999;82(4):1934–43.
Craig AD. Pain mechanisms: labeled lines versus convergence in central processing. Annu Rev Neurosci. 2003;26:1–30.
de Greck M, Bölter AF, Lehmann L, Ulrich C, Stockum E, Enzi B, Hoffmann T, Tempelmann C, Beutel M, Frommer J, Northoff G. Changes in brain activity of somatoform disorder patients during emotional empathy after multimodal psychodynamic psychotherapy. Front Hum Neurosci. 2013;7:410.
de Greck M, Scheidt L, Bölter AF, Frommer J, Ulrich C, Stockum E, Enzi B, Tempelmann C, Hoffmann T, Han S, Northoff G. Altered brain activity during emotional empathy in somatoform disorder. Hum Brain Mapp. 2012;33(11):2666–85.
de Greck M, Scheidt L, Bölter AF, Frommer J, Ulrich C, Stockum E, Enzi B, Tempelmann C, Hoffmann T, Northoff G. Multimodal psychodynamic psychotherapy induces normalization of reward related activity in somatoform disorder. World J Biol Psychiat. 2011;12(4):296–308.
Duddu V, Isaac MK, Chaturvedi SK. Alexithymia in somatoform and depressive disorders. J Psychosom Res. 2003;54(5):435–8.
Egloff N, Cámara RJ, von Känel R, Klingler N, Marti E, Ferrari ML. Hypersensitivity and hyperalgesia in somatoform pain disorders. Gen Hosp Psychiatry. 2014;36(3):284–90.
Eisenberger NI, Lieberman MD, Williams KD. Does rejection hurt? An fMRI study of social exclusion. Science. 2003;302(5643):290–2.
Ermann M. Psychosomatische Medizin und Psychotherapie. 5. überarbeitete Auflage. Stuttgart: W. Kohlhammer GmbH; 2006.
Gardini S, Cornoldi C, De Beni R, Venneri A. Left mediotemporal structures mediate the retrieval of episodic autobiographical mental images. NeuroImage. 2006;30(2):645–55.
Grabe HJ, Spitzer C, Freyberger HJ. Alexithymia and personality in relation to dimensions of psychopathology. Am J Psychiatry. 2004;161(7):1299–301.
Gupta A, McBeth J, Macfarlane GJ, Morriss R, Dickens C, Ray D, Chiu YH, Silman AJ. Pressure pain thresholds and tender point counts as predictors of new chronic widespread pain in somatising subjects. Ann Rheum Dis. 2007;66(4):517–21.
Hansell S, Mechanic D. Introspectiveness and adolescent symptom reporting. J Hum Stress. 1985;11(4):165–76.
Hiller W, Rief W. Why DSM-III was right to introduce the concept of somatoform disorders. Psychosomatics. 2005;46(2):105–8.
Landa A, Peterson BS, Fallon BA. Somatoform pain: a developmental theory and translational research review. Psychosom Med. 2012;74(7):717–27.
Landgrebe M, Barta W, Rosengarth K, Frick U, Hauser S, Langguth B, Rutschmann R, Greenlee MW, Hajak G, Eichhammer P. Neuronal correlates of symptom formation in functional somatic syndromes: a fMRI study. NeuroImage. 2008;41(4):1336–44.
Loughead JW, Luborsky L, Weingarten CP, Krause ED, German RE, Kirk D, Gur RC. Brain activation during autobiographical relationship episode narratives: a core conflictual relationship theme approach. Psychother Res. 2010;20(3):321–36.
Macdonald G, Leary MR. Why does social exclusion hurt? The relationship between social and physical pain. Psychol Bull. 2005;131(2):202–23.
Mai FM. The forgotten avant-garde. Trends Neurol Sci. 1982;5:67–8.
Martin JH. Neuroanatomy: text and atlas. 3rd ed. New York: McGraw-Hill Medical Publishing Division; 2003.
Mattila AK, Kronholm E, Jula A, Salminen JK, Koivisto AM, Mielonen RL, Joukamaa M. Alexithymia and somatization in general population. Psychosom Med. 2008;70(6):716–22.
Melzack R, Wall PD. Pain mechanisms: a new theory. Science. 1965;150(3699):971–9.

Mentzos S. Lehrbuch der Psychodynamik. Die Funktion der Dysfunktionalität psychischer Störungen. 2. Auflage. Göttingen: Vandenhock und Ruprecht; 2009.
Morschitzky H. Somatoforme Störungen. Wien: Springer; 2007.
Nakao M, Barsky AJ. Clinical application of somatosensory amplification in psychosomatic medicine. Biopsychosoc Med. 2007;1:17.
New York Times. Dr. Franz Alexander, 73, dies. New York Times; 1964.
Nickel R, Egle UT. Psychological defense styles, childhood adversities and psychopathology in adulthood. Child Abuse Negl. 2006;30(2):157–70.
Panksepp J, Bishop P. An autoradiographic map of (3h)diprenorphine binding in rat brain: effects of social interaction. Brain Res Bull. 1981;7(4):405–10.
Price DD. Psychological and neural mechanisms of the affective dimension of pain. Science. 2000;288(5472):1769–72.
Rainville P. Brain mechanisms of pain affect and pain modulation. Curr Opin Neurobiol. 2002;12(2):195–204.
Schmeing JB, Kehyayan A, Kessler H, Do Lam AT, Fell J, Schmidt AC, Axmacher N. Can the neural basis of repression be studied in the MRI scanner? New insights from two free association paradigms. PLoS One. 2013;8(4):e62358.
Schuchart S. Berühmte Entdecker von Krankheiten. Jean-Martin Charcot begründete die moderne Neurologie. Dtsch Ärztebl Int. 2017;7:68.
Schur M. Comments on the metapsychology of somatization. Psychoanal Study Child. 1955;10:119–64.
Shackman AJ, Salomons TV, Slagter HA, Fox AS, Winter JJ, Davidson RJ. The integration of negative affect, pain and cognitive control in the cingulate cortex. Nat Rev Neurosci. 2011;12(3):154–67.
Svensson P, Minoshima S, Beydoun A, Morrow TJ, Casey KL. Cerebral processing of acute skin and muscle pain in humans. J Neurophysiol. 1997;78(1):450–60.
Taylor GJ. Alexithymia: concept, measurement, and implications for treatment. Am J Psychiatry. 1984;141(6):725–32.
Trepel M. Neuroanatomie. 5. Auflage. München: Elsevier Urban & Fischer; 2012.
Vaillant GE. Adaptation to life. Boston, MA: Harvard University Press; 1977.
Witthöft M, Hiller W. Psychological approaches to origins and treatments of somatoform disorders. Annu Rev Clin Psychol. 2010;6:257–83.

# Anorexia and Bulimia

# 16

Peter Hartwich and Georg Northoff

**Abstract**

The movement of the psychiatric history back and forth between somatic and psychiatric etiopathogenesis has been for anorexia nervosa for more than 300 years. Today we see the causes of the different disease types: early anorexia nervosa, anorexia nervosa in puberty, chronic, and late anorexia nervosa, as well as bulimia as somatic and psychodynamic. A neuropsychodynamic approach combines psychodynamic models and experiences with somatic aspects. In particular the neuronal *imbalance* between self-referential processing and sensory processing of one's own body experience leads to a *paraconstruction* of the self-mentalization and de-physicalization of the body. For the therapy of the various forms of disease, future-oriented approaches result.

Our neuropsychodynamic concept consists of the following aspects: mostly there is to be found a disposition in the family with psychiatric illnesses. In puberty the patients have autonomy conflicts; the result is a self-alteration which is followed by a disbalance between self-referential and sensorimotor processing, then the imbalance crosses the threshold, and the paraconstruction of self-mentalization of one's own body follows with the typical overly controlling behavior and severe eating problems.

We describe examples of a typical case in puberty and also a case of late anorexia.

P. Hartwich (✉)
Department of Psychiatry, Psychotherapy, Psychosomatics, General Hospital of Frankfurt am Main, Teaching Hospital University Frankfurt, Frankfurt, Germany
e-mail: Prof.Hartwich@casytec.de

G. Northoff
Mind, Brain Imaging, and Neuroethics,
Institute of Mental Health Research, University of Ottawa, Ottawa, ON, Canada
e-mail: georg.northoff@theroyal.ca; www.georgnorthoff.com



H. Boeker et al. (eds.), *Neuropsychodynamic Psychiatry*,
https://doi.org/10.1007/978-3-319-75112-2_16

In some rare cases, there is even a transition to schizophrenia regarding to the concept of the danger of losing the spatiotemporal structure of the self depending on the level of the personality structure.

All this shows that neuropsychodynamic psychotherapy focuses on strengthening the self.

## 16.1 Historical: Psychological or Organic

Anorexia nervosa is one of the oldest well-described diseases, in which the controversial discussion about somatic and psychodynamic etiological aspects has been discussed in the course of the psychiatric history. A satisfactory explanation of their etiopathogenesis and really effective healing methods are still not available today. Anorexia nervosa has a high mortality rate among all mental disorders (Arcelus et al. 2011) and even is higher than bulimia nervosa.

Eggers (1980) points to a first description of Simone Porta o Portio from the year 1500. It was, however, Morton (1689), in England under the designation "phthisis nervosa," who indicated the still valid cardinal symptoms: cachexia, anorexia, amenorrhea, and constipation. Gull (1874) coined the name anorexia nervosa. He gave an in-depth clinical description and regarded the stubborn and implausible behavior as psychologically caused. In France Lasègue (1873) and Axenfeld and Huchard (1883) coined the terms "anorexia hysterique" and later "anorexia mental." In the German-speaking world, Freud (1904-1905, S22) writes a critical remark on therapy:

"We shall not use psychoanalysis if we are dealing with the rapid elimination of threatening phenomena, for example with hysterical anorexia."

Charcot (1886), who initially thought of a form of hysteria, had to concede, however, on the basis of some deaths in his anorexia nervosa patients that this went beyond hysteria.

The influence of Simmonds (1916) led to the assumption of an organic hypothesis, describing pituitary cachexia, the cause of which is the destruction of the pituitary front lobe. On account of the external similarity of the patients, the cause of the anorexia nervosa was found to have been discovered. Hormonal disorders were still assumed by Zutt (1948). Sheehan (1937) succeeded in the clarification, followed by Decourt (1953), Oberdisse et al. (1965), and Fey and Hauser (1970), who were able to reject the hypothesis of a primary organic disease in the sense of pituitary partial insufficiency by exact hormonal analyzes.

Nowadays Amianto et al. (2016) understand anorexia nervosa and bulimia as a disorder of the self and its functions. They point out studies that patients who suffer from anorexia nervosa have functional and structural alterations in a wide network of brain areas, including the precuneus, cingulate cortex, insula, temporal poles, thalamus, hypothalamus, nucleus caudatus (Amianto et al. 2013; Cowdrey et al. 2014), and hippocampus-amygdala complex (Wittman et al. 2010). These findings seem not to be caused by starvation and malnutrition effects on the brain because they occur at an early stage of anorexia nervosa and may, as Riem et al. (2012) point out, overlap with the brain circuits related to attachment functions, which are associated with the onset of the disorder.

## 16.2 Psychodynamic Hypotheses of Bulimia

*Bulimia* was later differentiated from anorexia nervosa, e.g., by Russel (1979). The patients are characterized by binge eating and inappropriate compensatory methods to prevent weight gain. With regard to psychopathology and psychodynamics, depressive mood and higher sexual activity, ego weakness, and separation difficulties from their parents are mentioned. Many psychodynamic aspects mentioned in the literature overlap with those of the anorexia nervosa. Fichter (1985) notes critically that from the 1970s onward, "a true boom" of bulimia research came up. Krüger et al. (2010) point out that in bulimic women, the disease is often overlooked for many years, as they conceal their eating disorder from close relatives at an almost normal weight. As a psychodynamic basis for the disorder, Reich (2010) describes a disorder of identity that is displaced into the body in which the bulimic patients would experience their body or parts of it as defective and flawed. Consequently, the shame before narcissistic exposure is a central anxiety, which is controlled and regulated by the bulimic symptoms.

## 16.3 Psychodynamic Hypotheses of the Anorexia Nervosa

In the description of the psychodynamic hypotheses, the reader will ask why is so much "old" literature quoted in this chapter? The reason is that the 1960s–1980s were the heyday of psychodynamic understanding; this formed the basis for treatment strategies. In the following years, attention was focused more on behavioral therapy strategies, which advantaged to interrupt the vicious circle, caused by starvation, losing weight, and organic psychological effects. This already sufficed in a number of improving cases; although the criterion improvement was often measured only in kilograms of body weight.

Langdon-Brown (1937) and Speer (1958) considered anorexia nervosa as a form of schizophrenia because of the therapist's lack of ability to emphasize with the patients and the patient's strong defense against the therapy. Hartwich (1974) points out that the psychosis is not only a question of the diagnostic reference system, but also a number of cases have already become known in which the transition from the anorexia nervosa to schizophrenia has been clearly demonstrated. He quotes cases described by Aubert and Peigné (1965), Meyer (1961), Cermak and Ringel (1960), Stäubli-Fröhlich (1953), as well as his own case observations. In one of these rare cases, the following psychopathological transitions could be observed: after the recovery from the full and acute paranoid-hallucinatory schizophrenia, a 37-year-old female patient changed into an anorexia nervosa, which then lasted for several years. Psychodynamically, it was obvious that the patient, with the severe eating disorder, was able to hold onto a symptom that saved her from slipping back into a schizophrenic self-fragmentation.

A psychodynamic conception was formulated by Thomä (1961) in which the fear of physical urges, including the rejection of sexuality, causes a process of regression. Clauser (1964) emphasizes the defenses of femininity, and Richter (1965) puts the disturbance of the self-integration, due to an extreme dependence on the mother in the foreground. The image of the overly dominant mother, for whom Theander (1970) also used the term "overprotective," is one of the most frequently mentioned psychodynamic factors. Bruch (1973, 1978) also argues that the development of the anorexia nervosa goes together with a disturbed mother-child relationship, and it is important to note the fact that in the development of the child, only the needs of the mother are allowed instead of an autonomous development. This, according to Minuchin et al. (1978), leads to the inability of the child to separate itself from the mother and to develop a personal feeling of the own body.

Scharfetter (1980) formulates the hypothesis that it is the rejection of the girl's role as a sexually mature woman; the adult woman is the negative model to be avoided. In this endeavor, a regressive behavior takes place with a morbid defense of dominant oral impulses and immature sexuality, which the girl experiences as a threat. Fantasies of oral pregnancies and disgust of a sexually mature body and of food were in the foreground.

Mentzos (2011, p. 200) describes the body as an object relationship with constantly occupying externalizations and internalizations. The model that one's own body can be regarded as an object and the way in which it is treated as an object relation is useful for understanding the relationship to one's body in anorexic and bulimic patients. Furthermore, he emphasizes proximity to addiction in the sense of a pathological compromise solution as an addictive mode of conflict and trauma processing.

Gabbard (2014) summarizes the aspects of the psychodynamic understanding of anorexia nervosa (p. 361):

- A desperate attempt to be special and unique
- An attack on the false sense of self fostered by parental expectations
- An assertion of a nascent true self
- An attack on a hostile maternal introject viewed as equivalent to the body
- A defense against greed and desire
- An effort to make others—rather than the patient—feel greedy and helpless
- A defensive attempt to prevent unmetabolized projections from the parents from entering the patient
- An escalating cry for help to shake the parents out of their self-absorption and make them aware of the child's suffering
- In some cases a dissociative defense into separate self-states as a way of regulating intense affect

Gabbard adds cognitive features, including misperception of one's own body image and obsessive-compulsive thoughts and rituals.

Schauenburg et al. (2009) mention in the ANTOP (Anorexia Nervosa Treatment of Outpatients) study that daughters with their symptoms constantly rebel against their parents. Because of their illness, they have no feelings of guilt.

Hartwich and Steinmeyer pursued a different approach, namely, an empirically supported multidimensional approach by introducing a mathematical path analysis (causal analysis) into psychiatry in 1974. In doing so, it is possible to describe the influence of factors and complicated relationships as well as interactional weight shifts more clearly than with the mere description. In investigating which factors influence the severity of the disease, it has been shown that the occurrence of psychiatric disorders in the family is a direct negative-acting condition. With regard to disturbed family relations, a further criterion, namely, the disturbance in childhood development, is necessary in order to influence the disease severity. Competitor mothers and dominant mothers, mutually exclusive, strongly influence the disease severity, whereby the causal analysis additionally shows that the personality and behavior of the child can also provoke the dominant behavior of the mother.

Garfinkel and Garner (1982) and Garfinkel et al. (1986) provided a purely descriptive, multifactorial approach to the question of etiopathogenesis, by using all previously known explanatory approaches and their weightings, which were empirically determined or assumed. On the basis of numerous case studies, Mester (1981, p. 289) comes to the psychodynamic statement:

At the center of psychodynamics, which leads into an anorexia nervosa, therefore, is the inability of these young people to assume roles and positions according to mature biological laws…. The apparent coquetting with death is used to maintain mental stability: in anorexia the feeling of self-identity, self-availability and control is desperately restored…. The destruction of the body-self-unity creates a defect, which already brings the disease a little in the direction of psychosis.

Bruch (1982) proposed that at the root of anorexia nervosa was fundamentally a deficit of the self. She argued those individuals who suffer from anorexia nervosa function with a "false self," which implies that these individuals may not discriminate between their own and their caregivers' expectations and needs.

Fichter (1985) reported 12 patients with an anorexia nervosa as well as a schizophrenic or schizoaffective psychosis. In several cases it was observed that the eating disorder had temporarily completely disappeared with the onset and during the existence of psychotic symptoms.

## 16.4 Classification of the Disease Forms

### 16.4.1 Anorexia Nervosa in Prepuberty, Puberty, and Adolescence

It is the most common form with a distribution peak between the ages of 15 and 18. Diagnostically, the Quetelet's index is 17.5 and below, loss of weight through food refusal, self-induced vomiting, laxative abuse, exaggerated sporting activities, disorder of body-experience (Leiberleben), amenorrhea, and compulsive rituals.

### 16.4.2 Early Anorexia

The disease develops in small children in the 4–8 months of life; the eating disorders rarely persist beyond the time of puberty.

### 16.4.3 Chronic Anorexia and Late Anorexia

The disease begins in the childhood with persistent struggles in the family about the food intake. In the puberty an improvement of the symptoms and the difficulties in ingestion of food are usually observed. A severe cachexia is rather the exception, also rarely amenorrhea. As a rule, the affected persons remain underweight. In case of stressful life events, anorectic phases occur.

When patients show the pronounced symptoms of anorexia nervosa often with severe cachexia for the first time in the mature age approximately after the age of 35, one calls it late anorexia.

### 16.4.4 Male Anorexia Nervosa

The frequency of male anorexia nervosa is about 7% of the total disease. The nosological classification is not uncontroversial. According to Fichter (1985), the incidence of suicidal thoughts and trials, as well as identity crises, bulimia attacks, vomiting, and obsessiveness with regard to eating, is noticeable.

### 16.4.5 Bulimia

The bulimia is distinguished from the anorexia nervosa by the fact that a cachexia is not usually observed; the food behavior is characterized by "seizures" of excessive eating of large amounts of food with subsequent self-induced vomiting. Pronounced depressive moods are frequent and are seen as a result of the main symptoms. With regard to the description of psychodynamic pathogenesis, there are many overlaps but also differences (see Sect. 16.2) with the anorexia nervosa.

### 16.4.6 Case Study of an Anorexia Nervosa in Puberty

A 15-year-old female patient looked like a skeleton; Morton (1689) used already the expression "skin-covered skeleton". In her fur-skinned hair, she looked like a boy and reported precociously and with high energy on her symptoms. For 3 years her weight had fallen from 42 to 30 kg with a size of 157 cm.

She is always afraid of eating too much and thus overload the often aching stomach. The menstrual bleeding lasted only 6 months after she entered her 14th year. She is freezing all the time, her hands look blue, and she complains of constipation, insomnia, and insatiable hunger. She is convinced firmly to suffer from an organic disease. In school class she is laughed at, mocked, and called "the baby." Her situation has become so intolerable that she and her mother want help.

The continuously speaking mother, who pushes her daughter close to herself, tries to reach special conditions for her child referring the nutrition in the hospital.

From the family: the grandfather died of a nervous disease. The grandmother represents her vegetarianism with sense of conscience. The father, a teacher, must constantly eat diet because of chronic gastritis. The family lives out of the reform house, rejects sugar and preserves, and prefers wheat mash, rice mash, self-picked tea, and self-pounded sauerkraut. The 11-year elder sister is the favorite of the mother. The 4-year elder brother is a nurse and visits the family only sporadically.

As a third and last child, she developed quickly and without difficulty. She has played with puppets passionately and cultivated individual and close friendships, and sport and movement games were especially appreciated. At grammar school she has shone through ambition. Piano playing, ballet, and riding were her hobbies. At the age of 12, she had an unexpected fear of eating. At that time, she was annoyed by the close relationship between mother and elder sister. The girl wanted to be like the mother, so beautiful, witty, and artistically gifted. But the sister and mother thought and did almost the same. There was no way getting close to the mother. She was best understood by her brother, who massaged her stomach whenever she had a stomach ache.

Because of the discomfort and the weight loss, the mother moved with her from doctor to doctor and finally went to a nonmedical practitioner, who diagnosed a pancreatic disease and gave her laxatives for months. Finally, only in the bed of the mother could the child be prevented from freezing.

Since then, she has lost contact with the class community; she has not been interested in beat music, boys, and parties. She said, she would have liked to be a boy.

In the weeks before hospitalization, she had painted 30 pictures, in which she was portrayed, e.g., as in "mysterious chains," "in the sea," "sleepless," "many thoughts," "among many eyes of the others," and "my fear" (see Fig. 16.1).

One part of the therapy sessions consisted of projecting the slides on the screen to a large extent which we had photographed from her pictures. This method stimulated the patient to explain and to associate in order to be able to understand as much as possible which psychodynamic aspects should be treated. Here is one example (Fig. 16.1):

The picture "my fear":

An existential fear of losing the self is expressed in the creative drawing.

In this picture, figurative figures of treelike structures are traversed. In particular, a thin, naked figure is affected, the head of which is juxtaposed in several sizes, both in en face and profile. The body is divided; the one half is drawn over from the arms and claws, to the other side. The patient expresses herself to this picture:

> Inside I am divided; my two halves cannot connect with each other. The left side is dominated by commandments and prohibitions; here is order, security and clarity. I always need someone to tell me what to do. The leftmost person is my mother. On the right side are represented the unknown powers, of which I am greatly afraid. I will be drawn from long tentacles to the unknown to become a small, helpless child (far right in the picture). The great, life-giving sun is shining, but cannot intervene.

**Fig. 16.1** The patient called the drawing "my fear." An existential fear of losing the self is expressed in this creative drawing. In this picture, treelike structures traverse the human body. In particular, a thin, naked figure whose head can be seen in several sizes, both in en face and profile, is affected. The body is split; half of it is pulled over to the other side by the arms and claws

The improvement of strengthening the self, autonomy and self-assurance, demarcation from the family, ability to eat, gaining of weight, etc. were only gradually possible after many months (see Hartwich 1974).

#### 16.4.6.1 Discussion

Although the cachexia was evident, we initially in the inpatient treatment refrained from enhancing the therapeutic accentuation of the eating problem. Otherwise the pathological interaction from the family would act in the context of a countertransference, and thus the "struggle" for the eating behavior and the psychotherapeutic treatment of the behind existing psychodynamic problems would not have been possible. Except those cases in which cachexia is life-threatening and/or the accompanying cerebral impairment causes a brain-organically conditioned psychopathology which is so severe that an initial artificial nutrition (e.g., stomach tube) is necessary in order to prevent health damage and even death. If furtive self-induced vomiting is occurring, it is especially important to obtain serum electrolytes regularly and to watch for the development of hypokalemia. Only after a period of somatic treatment, there is a much better starting point for psychotherapy.

From autopsies of deaths of this disease, findings of cerebral atrophy have been published repeatedly in the context of severe cachexia. We (Zeumer et al. 1982) used a systematic study of cachectic anorexia nervosa cases at the time of inpatient admission compared with the treatment after a period of 2 months and found that in CT, the so-called atrophy signs were declined. Consequently, because of the *reversible* change of the brain caused by cachexia, we speak of a cerebral dystrophy.

### 16.4.7 Case Study of Late Anorexia

The 47-year-old, intellectually high-gifted and perfectionist woman suffered from anorexia nervosa so severely that a hospital stay of 17 months was necessary because of life-threatening symptoms and suicidal behavior. In her life she had suffered from varying symptoms, which had been described as psychosomatic (abdominal pain, migraine, backache, dizziness, neurodermatitis). Now, after the sudden death of two close relatives, a weight loss of more than 15 kg had occurred in the last few months, so that the patient with a body size of 172 cm, the body weight was below 38 kg during the first weeks of clinical treatment. During the time of the treatment, of which she spent about 7 months mostly lying in bed, she painted more than 2000 pencil drawings. She expressed her inner images, consisting of memories from childhood and later life, her dreams, and her psychodynamic process. The pictures that had just been drawn became the main object of the almost daily psychotherapeutic sessions. After successful treatment and discharge, she wrote an extensive book (508 pages) about her experiences, from which the following is literally quoted:

"In the initial phase, anorexia for me meant mainly tortuous non-eating, while at the same time I could eat a horse. Before that was a time of inscrutable appetite deficit during which I had continually weight loss. The anorexia itself had taken hold of me very slowly, almost creepily, but finally she held me relentlessly in her catches and let me feel my powerlessness. It was bitter, very bitter. I suffered from loss of self-esteem, felt hopelessly abandoned and degraded. What had hitherto been my strength, my mind, it failed miserably, and at this stage of the disease I realized the extent of the lack of freedom. In relation to this slavery, the sense of humiliation grew gigantically as a direct correspondence to the yearning for death." (Matern-Scherner 1994)

## 16.5 Anorexia Nervosa and Schizophrenia

There are rare cases with a particular psychodynamic aspect where the transition from anorexia nervosa to schizophrenia and vice versa is confirmed. Already Mester (1981) pointed out that the symptoms of anorexia nervosa could serve to maintain a mental stability from a psychodynamic perspective; in a desperate way, the feeling of one's own identity, self-availability, and self-control is restored.

There is a special experience that drives the parents and therapists sometimes crazy when dealing with anorexia nervosa patients, namely, the particular obstinacy with which those defend the not eating. This feature, which was sometimes referred as delusion or compulsion, should be understood with regard to its function. It is possible that in some cases, or even many cases, the stubborn clinging on the symptoms mentioned have even a protective function. The coherence of their self, which goes into the experience of their body self, which is so important in this developmental stage, is threatened. The threat can be of different origin in the individual case, from the family constellation, the social environment, or development conflicts. If the coherence of the self is in danger to be destroyed, what was described in some rare anorexia nervosa cases, the slipping into a schizophrenic psychosis would occur. In none of the 12 cases described by Fichter (1985) had the psychosis been before the eating disorder, in several cases the eating disorder temporarily disappeared with the onset and during the existence of psychotic symptoms.

In this sense, the psyche-brain-unit unconsciously protects itself from further decay with the aid of the anorectic symptoms. The stubborn defense of non-eating in these rare cases can be regarded as a laborious attempt in the neuropsychodynamic sense of anorectic paraconstruction (Hartwich 1997, 2006) to avoid self-fragmentation. This emphasizes that it is not a purely psychodynamic defense mechanism but that neurobiological aspects are added.

Schizophrenic diseases in families of affected persons have not been counted as frequently, as first published by Theander (1970), and a direct inheritance of the genetic disposition cannot be assumed. However, it has also been reported by a path analysis (Hartwich and Steinmeyer 1974) that the severity of the disease is increased when psychiatric disorders (depression, other psychoses, anorexia nervosa, people without an exact diagnosis, who have spent several months in psychiatric hospitals) occur in the family.

What Is the Difference Between Schizophrenia and Anorexia Nervosa?

Regarding to the concept of weakness of the self and the danger of losing the spatiotemporal structure of the self (see Chap. 3 Schizophrenia and Other Psychoses), we see two different levels of personality structures. In the danger of losing the connection of the self is so severe that the patient needs paraconstructions in order to help the self to stabilize in some way even with a lot of symptoms, e.g., delusions. In anorexia nervosa there is also a danger of losing the spatiotemporal stability of the self but in a higher and more mature level of the personality structure, so we can interpret the symptoms and the typical personality features as a firm hold to stabilize the self, even with psychopathological reactions.

## 16.6 Neuropsychodynamic Understanding

There are many studies on brain-organic and other somatic findings in anorexia nervosa patients, but the results correspond to secondary symptoms. If the normal development of the brain in this period of life is in itself normal, then the characteristic years in which anorexia and bulimia of girls and young women fall are characterized by a developmental period in which the maturity of the ego and other aspects of personality development are not yet brought to an end. As Northoff says (2011, p. 314)

> Maturation of cortical synchronization and neuronal networks in early adulthood goes through a period of transient destabilization in late adolescence before being organized in the most stable and mature way. Late adulthood may thus be considered a critical period for constituting stable and more precise cortico-cortical synchronization. Some preliminary evidence suggests that the process of transient destabilization and mature stabilization in late adolescence and early adulthood may be related to changes in cortico-cortical myelination and GABA which both undergo changes in exactly these time periods and may thereby significantly impact neural wiring and synchronization in early adulthood (Di Cristo 2007; Uhlhaas et al. 2009a, b).

Either an individual passes through this time with their typical conflicts without psychological disturbances, which is usually the case, or in some cases, it comes to psychiatric disorders. These may manifest themselves as psychoses or many other diseases or in some cases as anorexia nervosa or bulimia.

With regard to the etiopathogenesis of the anorexia nervosa and bulimia, "monothematic" explanations, such as rejection of femininity, sexual hostility, victims of the dominant mother, etc., fall short. Thus, a mono-causal explanation of the mentioned illnesses cannot be assumed today. If we try to weigh the factors emphasized in the literature, we have a network that is made up of many conditions, which can each intensify and weaken each other. At present, it is not possible to formulate the dynamics of the network or the conditional constellations in such a way that a scientifically justified pathogenetic pattern can be described for anorexia nervosa and also bulimia. Thus, in a neuropsychodynamic understanding, we depend on each individual case of illness to be assessed in its individual conditional structure in order to derive the corresponding psychodynamic treatment from it. For this purpose it seems to be helpful to follow the new approach of Amianto et al. (2016) who interpret anorexia nervosa as a disorder of the self.

The idea is—as mentioned above—what Tasca and Balfour (2014) and Williams et al. (2015) also emphasize that anorexia nervosa and bulimia or binge eating disorder consist of symptoms which try unconsciously to manage painful internal experiences related to a deficit of the self.

The concept from Amianto et al. (2016) is based on the hypothesis of a deficiency in the development of the self. The authors consider the development of the self in terms of its synchronic (i.e., experienced in the moment) and diachronic (i.e., experienced as continuous over time) aspects. "Both synchronic and diachronic aspects of the self are relevant to the expression of eating disorder symptoms. Further, the maturation of the self is interwoven with the development of attachment functioning from infancy to adolescence. This interplay between these developmental processes of the self and of attachment could be crucial in understanding the pathogenesis of eating disorders" (Amianto et al. 2016, p. 1). They also suggest a neurobiological link between the theory of the self in anorexia nervosa and even bulimia and the disturbances in spatiotemporal functioning of the brain. The authors emphasize a neurobiological pathway by which deficiencies in the self is related to attachment functions in individuals with eating disorders.

Neurobiologically, this may be evidence of what Northoff (2015a) called temporal dysfunction (see Northoff 2015b for details) in the resting state of the brain, thus disrupting the perception of time and the integration of the self across past and present experiences. The model suggests the disturbances in the brain's resting state may be specifically associated with spatial functioning possibly related to experiencing one's own body as an integrated aspect of the self.

In a neuropsychodynamic understanding, the fear of losing the continuity of the self which manifests in losing the control of one's own actions and feelings is very high among patients who suffer from anorexia nervosa. Such a loss of control would mean that experience and behavior must be given to "uncontrolled powers" (from the outside and/or from the inside). This feared loss of independence, no longer being able to determine itself, is a crucial basic problem. So the patients develop a *counterregulation* in form of an excessive overly control of their actions and especially of their bodies.

From the neuronal point of view, there are present abnormal changes in the intrinsic activity in the brain of these patients and that these may specifically affect sensory areas (such as the sensorimotor cortex) and the midline regions. Sensory areas and the insula (near the secondary sensory cortex) show abnormal connections, i.e., functional connectivity with the middle line regions; this means that the connection between self, i.e., self-referential processing (middle lines) and sensory processing of one's own body (insula, sensory cortex), is disbalanced (Lavagnino et al. 2014). The latter enslave the former; the self is somehow enslaved by his own body. So there is a *paraconstruction of the self*—it is no longer mentally constructed by one's own thoughts but by one's own body.

But since the self is primarily perceived as mental and not physical, such an experience of a mental self can only be achieved by the abnormal reduction of the physical contents of the body—the body is abnormally emaciated, which is then neuropsychodynamically quasi an attempt or paraconstruction of a "self-mentalization of the body." Only when the body becomes nonphysical can it be experienced as a substitute for the original mental self. This is accompanied by compulsive behavior, which can be explained by neuronal overlaps between anorexia and compulsive disorder (see Chap. 14 in this book).

**Neuropsychodynamics of the Anorexia nervosa**

•**Symptoms:** typical overly controlling behavior and severe eating problems

↑

•**Paraconstruction:** of the self-mentalization of one's own body

↑

•**Threshold:** imbalance crosses the threshold, sensomotric processing with body experience dominates self-referential processing with the experience of self

↑

•**Dysbalance:** between self-referential and sensorimotor processing

↑

•**Self-alteration:** fear to lose the spatio-temporal continuity of the self represented in the spatio-temporal characteristics of the resting state in the midline structures of the brain

↑

•**Conflicts:** defense and denial of feminitniy and sexuallity, obstruction of autonomy development, overly dominant mother, dysgust of fat body

↑

•**Disposition:** psychiatric diseases in the family, depression, psychoses, personality disorder, eating disorder

## 16.7 Neuropsychodynamic Therapy Concepts

Schauenburg et al. (2009) emphasize that many months to years of psychotherapy are necessary to cure, depending on the extent of chronification. In the ANTOP (Anorexia Nervosa Treatment of Outpatients) study, OPD (Operationalized

Psychodynamic Diagnostics) were used; the severity of the disease, extent of denial, central interactions with parents, the importance of central conflict issues, and structural abilities are recorded. Zipfel et al. (2014) report the results of the ANTOP study about two manual-based outpatient treatments for anorexia nervosa—focal psychodynamic therapy and enhanced cognitive behavior therapy—versus optimized treatment as usual. In the multicenter, randomized controlled study, there was no difference in weight gain recorded by the end of treatment among the three groups. However patients allocated focal psychodynamic therapy had higher recovery rates compared with those with optimized treatment as usual at 12-month follow-up. Compared with cognitive behavior therapy, psychodynamic treatments are less directive, and the interpersonal relationships are more important to induce insight.

Mentzos (2011, p. 273) describes the phenomenon of self-injury: "One is faced with the paradoxical situation that a disorder that actually causes dysfunction is at the same time a function." This perspective we transfer also to the symptoms of the anorexia nervosa. For the neuropsychodynamic understanding and the psychotherapeutic approach, this means a change in the attitudes of the therapist, as described by Hartwich (2006) for dealing with paraconstruction. For some psychotherapists, this could mean a change in their paradigm.

Especially the attitude of countertransference (see Chap. 9) has a chance to change to better understanding. Because one often hears about the pathologically distorted variant of self-determination statements such as "illness-unreasonable," "lying, untruthful," "denying," "stubborn, obstinate" and "wangling every therapist out." After a while, the treating team is completely exhausted and annoyed as the patients maintaining power of the pathological symptoms seem to be inexhaustible. The countertransference leads to the statement (Freyberger 1980, p. S418): "We are of the opinion that in principle any anorexia nervosa treatment should be initiated by means of a tube feeding (with or without the use of psycho pharmaceuticals)."

Because of the understanding of the functionality of the symptoms and the "paradigm shift" of the therapist, we see the emphasis on the *respect of the protective function* of the paraconstruction consisting of typical symptoms. This should be signaled to the patient. This can build the basis for a psychotherapeutic process, which makes it possible to work through those factors which previously triggered the loss of self-control. This can be the case of intrusive behavior of close reference persons, whether psychically or physically and sexually. In other cases, fixations on magical ideas or development-related hormonal and other physical changes that enter into a pathological interaction with social factors, such as being a scapegoat, to be laughed at, the fear to be deformed, and much more. If the pathological constellation, which is important for the individual case, can be effectively treated until the individual treated patient regains the true self-control, she can let go of the pathological form of the body control and achieve a lasting improvement up to healing. An additional neuropsychodynamic treatment concept of the future would involve the above-described threshold situation in an overall treatment plan and, in addition to the psychodynamic approaches mentioned, try to reconcile the imbalance of the dominant sensory processing of the own body with the self-referential processing which corresponds to self-experience and self-strengthening.

## References


Amianto F, Northoff G, Daga GA, Fassino S, Tasca GA. Is anorexia nervosa a disorder of the self? A psychological approach. Front Psychol. 2016;7:849. https://doi.org/10.3389/fpsyg.2016.00849.

Amianto F, Caroppo P, D'Agata F, Spalatro A, Lavagnino L, Caglio M, et al. Brain volumetric abnormalities in patients with anorexia and bulimia nervosa: a voxel-based morphometry study. Psychiatry Res. 2013;213:210–6. https://doi.org/10.1016/j.pscychresns.2013.03.010.

Arcelus J, Mitchell AJ, Wales J, Nielsen S. Mortality rates in patients with anorexia nervosa and other eating disorders: a meta-analysis of 36 studies. Arch Gen Psychiatry. 2011;68:724–31.

Aubert P, Peigné F. Deux observations d'anorexie mentale masculine. Ref Neuropsychiat Infant. 1965;12:515–21.

Axenfeld A, Huchard H. Traité des Névroses. Paris: Baillière; 1883.

Bruch H. Eating disorders: obesity, anorexia nervosa, and the person, within. New York: Basic Books; 1973.

Bruch H. The golden cage: the enigma of anorexia nervosa. Cambridge: Harvard University Press; 1978.

Bruch H. Anorexia nervosa: therapy and theory. Am J Psychiatry. 1982;139:1531–8. https://doi.org/10.1176/ajp.139.12.1531.

Cermak I, Ringel E. Zum Problem der Anorexia nervosa. Z Nervenheilk Wien. 1960;17:152–82.

Charcot JM. Neue Vorlesungen über die Krankheiten des Nervensystems, insbesondere über Hysterie. Deutsch. Übersetzug von S. Freud, Leipzig/Wien; 1886.

Clauser G. Das Anorexia-nervosa-Problem unter besonderer Berücksichtigung der Pubertätsmagersucht und ihrer klinischen Bedeutung. Ergebn Inn Med Kinderheilk. 1964;21:97–164.

Cowdrey FA, Filippini N, Park RJ, Smith SM, McCabe C. Increased resting state functional connectivity in the default mode network in recovered anorexia nervosa. Hum Brain Mapp. 2014;35:483–91. https://doi.org/10.1002/hbm.22202.

Decourt J. Die Anorexia nervosa. Deut Med Wschr. 1953;47:1619–1622 u. 1661-1664.

Di Cristo G. Development of cortical GABAergig circuits and its implications for neurodevelopmental disorders. Clin Genet. 2007;72(1):1–8.

Eggers C. Anorexia nervosa und adipositas. In: Spiel W, Hrsg. Die Psychologie des 20. Jh. Bd XII. Zürich: Kindler; 1980, pp. 576–622.

Fey M, Hauser GA. Die Postpubertätsmagersucht. Bern Stuttgart Wien: Huber; 1970.

Fichter MM. Magersucht und Bulimia. Berlin-New York- Heidelberg: Springer; 1985.

Freud S. Über Psychotherapie, GW 5 Die Freudsche psychoanalytische Methode. Fischer Frankfurt am Main. 1904-1905;1981:S11–26.

Freyberger H. Gastroenterologische Erkrankungen. In: Hahn P, Hrsg. Die Psychologie des 20. Jh Bd IX,1. Zürich: Kindler; 1980. pp. 410–41.

Gabbard GO. Psychodynamic psychiatry in clinical practice. 5th ed. Washington, DC: American Psychiatric Press; 2014.

Garfinkel PE, Garner DM. Anorexia nervosa. A multidimensional perspective. New York: Brunner-Mazel Publishers; 1982.

Garfinkel PE, Garner DM, Rodin G. Anorexia nervosa und Bulimie. In: Kisker KP et al., Hrsg. Psychiatrie der Gegenwart. Bd.1 Neurosen, Psychosomatische Erkrankungen, Psychotherapie. Berlin Heidelberg: Springer; 1986. pp. 103–24.

Gull W. Anorexia nervosa (Apepsia hysterica, anorexia hysterica). Transact Clin Soc. 1874;7:22.

Hartwich P. Die Psychopathologie der Anorexia nervosa in Federzeichnungen einer Kranken. In: Broekmann JMU, Hofer G, Hrsg. Die Wirklichkeit des Unverständlichen. Den Haag: Martinus Nijhoff; 1974. pp. 248–72.

Hartwich P. Die Parakonstruktion: eine Verstehensmöglichkeit schizophrener Symptome. Vortrag Frankfurter Symposion: Schizophrenien – Wege der Behandlung. In: Hartwich P, Pflug B, Hrsg. Schizophrenien – Wege der Behandlung. Sternenfels: Wissenschaft & Praxis; 1997. pp. 19–28.

Hartwich P. Schizophrenie. Zur Defekt- und Konfliktinteraktion. In: Böker H, Hrsg. Psychoanalyse und Psychiatrie. Heidelberg: Springer; 2006. pp. 159–79.
Hartwich P, Steinmeyer E. Strukturmodell zur Darstellung krankheitserschwerender Faktoren der Anorexia nervosa mittels Pfadanalyse. Arch Psychiat Nervenkr. 1974;219:297–312.
Krüger C, Reich G, Buchheim P, Cierpka M. Essstörungen und Adipositas: Epidemiologie – Diagnostik – Verläufe. In: Reich G, Cierpka M, Hrsg. Psychotherapie der Essstörungen. 3. Aufl. Stuttgart: Thieme; 2010, pp. 27–61.
Langdon-Brown W. Anorexia nervosa. Lancet. 1937;1:473–4.
Lasègue EC. De l'anorexie hysterique. Arch Gén Méd. 1873;21:385–411.
Lavagnino L, Amianto F, D'Agata F, Huang Z, Mortara P, Abbate-Daga G, Marzola E, Spalatro A, Fassino S, Northoff G. Reduced resting-state functional connectivity of the somatosensory cortex predicts psychopathological symptoms in women with bulimia nervosa. Front Behav Neurosci. 2014;8:270. https://doi.org/10.3389/fnbeh.2014.00270. eCollection 2014.
Matern-Scherner E. Kein Buch zum Verschlingen, Mager-Sucht-Maskeraden. Frankfurt am Main: R.G. Fischer; 1994.
Mentzos S. Lehrbuch der Psychodynamik. Göttingen: Vandenhoeck & Ruprecht; 2011.
Mester H. Die anorexia nervosa. Berlin: Springer; 1981.
Meyer JE. Das Syndrom der Anorexia nervosa. Katamnestische Untersuchungen. Arch Psychiat Nervenkr. 1961;202:31–59.
Minuchin S, Rosman BL, Blaker L. Psychosomatic families: anorexia nervosa in context. Cambridge: Harvard University Press; 1978.
Morton R. Phthisiologia or a treatise of consumption. London: Smith; 1689.
Northoff G. Neuropsychoanalysis in practice. Oxford: University Press; 2011.
Northoff G. Is schizophrenia a spatiotemporal disorder of the brain's restingstate? World Psychiatry. 2015a;14:34–5. https://doi.org/10.1002/wps.20177.
Northoff G. Resting state activity and the "stream of consciousness" inschizophrenia—neurophenomenal hypotheses. Schizophr Bull. 2015b;41:280–90. https://doi.org/10.1093/schbul/sbu116.
Oberdisse K, Solbach HG, Zimmermann H. Die endokrinologischen Aspekte der Anorexia nervosa. In: Meyer JEU, Feldmann H, Hrsg. Anorexia nervosa. Stuttgart: Thieme; 1965. pp. 21–33.
Reich G. Psychodynamischen Aspekte der Bulimie und Anorexie. In: Reich G, Cierpka M, Hrsg. Psychotherapie der Essstörungen. 3. Aufl. Stuttgart: Thieme; 2010. pp. 72–92.
Richter HE. Die dialogische Funktion der Magersucht. In: Meyer JE, Feldmann H, Hrsg. Anorexia nervosa; 1965. pp. 108–12.
Riem MM, Bakermans-Kranenburg MJ, van IJzendoorn MH, Out D, Rombouts SA. Attachment in the brain: adult attachment representations predict amygdala and behavioral responses to infant crying. Attach Hum Dev. 2012;14:533–51. https://doi.org/10.1080/14616734.2012.727252.
Russel GFM. Bulimia nervosa. An ominous variant of anorexia nervosa. Psychol Med. 1979;9:429–48.
Scharfetter C. Entwicklungskrisen. In: Peter UH, Hrsg. Psychologie des 20. Jahrhunderts. Zürich: Kindler; 1980. pp. 315–20.
Schauenburg H, Friedrich H-C, Wild B, Zipfel S, Herzog W. Fokale psychodynamische Psychotherapie der anorexia nervosa. Ein Behandlungsmanual. Z Psychotherapeut. 2009;54:270–80.
Sheehan HL. Post-partum necrosis of anterior pituitary. J Path Bact. 1937;45:42.
Simmonds M. Über Kachexie hypophysären Ursprungs. Dtsch Med Wschr. 1916;42:190–8.
Speer E. Magersucht und Schizophrenie. Stuttgart: Thieme; 1958.
Stäubli-Fröhlich M. Probleme der Anorexia nervosa. Schweiz Med Wschr. 1953;35:811–7. 837-841
Tasca GA, Balfour L. Attachment and eating disorders: are view of current research. Int J Eat Disord. 2014;47:710–7. https://doi.org/10.1002/eat.22302.
Theander S. Anorexia nervosa. A psychiatric investigation of 94 female patients. Acta Psych Scand Suppl. 1970;214:1–194.
Thomä H. Anorexia nervosa. Bern/Stuttgart: Huber/Klett; 1961.

Uhlhaas PJ, et al. The development of neural synchrony reflects late maturation and restructuring of functional networks in humans. Proc Natl Acad Sci U S A. 2009a;106(24):9866–71.

Uhlhaas PJ, et al. Neural synchrony in cortical networks: history, concept and current status. Front Integr Neurosci. 2009b;3:17.

Williams K, King J, Fox JR. Sense of self and anorexia nervosa: a grounded theory. Psychol Psychother. 2015;89:211–28. https://doi.org/10.1111/papt.12068.

Wittman ML, Lovero KL, Lane SD, Paulus MP. Now or later? Striatum and insula activation to immediate versus delayed rewards. J Neurosci Psychol Econ. 2010;3:15–26. https://doi.org/10.1037/a0017252.

Zeumer H, Hacke W, Hartwich P. A quantitative approach to measuring the cerebrospinal fluid space with CT. Neuroradiology. 1982;22:193–7.

Zipfel S, Wild B, Groß G, Friedrich HC, et al. Focal psychodynamic therapy, cognitive behaviour therapy, and optimised treatment as usual in outpatients with anorexia nervosa (ANTOP study): randomised controlled trial. Lancet. 2014;383(9912):127–37.

Zutt J. Das psychiatrische Krankheitsbild der Pubertätsmagersucht. Arch Psychiat Nervenkr. 1948;180:776–849.

# Traumatogenic Disturbances: PTSD, Complex PTSD and Trauma-Related Disorders

# 17

Clara Mucci, Andrea Scalabrini, and Georg Northoff

**Abstract**

Regarding traumatogenic disturbances, there were many changes and new developments during last years both in the psychodynamic understanding and in neuroscience. DSM-5 made major changes on PTSD but didn't recognise the diagnosis of complex PTSD, which is now established in the PDM-2, and it is accepted by the task force for the expected ICD-11.

In this work our aim is to shed a novel light on the neuropsychodynamic understanding of trauma and its manifestations both at a psychological and at a neuroscientific level.

The first distinction to be made is that between the so-called man-made trauma, i.e. trauma due to the violence of another human being, or even within a relationship, and traumatisation resulting from a natural catastrophe (earthquake, typhoon, etc.).

We need also to distinguish different levels of interpersonal traumatisations: (1) severe lack of attunement between child and caregiver (early relational trauma) from severe neglect, maltreatment, abuse and incest, resulting in emotional dysregulation, distortion of reality and destructiveness of the self or aggressiveness against the other; (2) maltreatment and abuse and identification with the aggressor, where the two sides, victim and aggressor, remain embedded within the psyche and intertwined within the personality, repeating a chain of violence; and (3) massive trauma and the consequence for traumatic generations. The effects of trauma of human agency on the brain will be discussed.

C. Mucci · A. Scalabrini
Department of Psychological, Health and Territorial Sciences (DiSPuTer), G. d'Annunzio University of Chieti-Pescara, Chieti, CH, Italy

G. Northoff (✉)
Mind, Brain Imaging, and Neuroethics, Institute of Mental Health Research, University of Ottawa, Ottawa, ON, Canada
e-mail: georg.northoff@theroyal.ca; www.georgnorthoff.com



H. Boeker et al. (eds.), *Neuropsychodynamic Psychiatry*,
https://doi.org/10.1007/978-3-319-75112-2_17

Adverse early experiences (early life stress, ELS) have a profound impact on individual differences in stress responsiveness and are associated with an increased vulnerability for psychiatric disorders, such as depression, in later life. Furthermore, there is increasing evidence that ELS also modulates the development of the oxytonergic system. In adults with childhood maltreatment, it was shown that the suppressing effect of OXT on cortisol levels is significantly reduced. A recent study rather supports the hypothesis of inverse or even detrimental oxytocin effects in subjects with ELS experiences. The presented findings demonstrate how crucial it is to consider environmental factors and particularly early social experiences prior to a therapeutic administration of oxytocin.

## 17.1 The Concept of PTSD and the Previous Understanding of Trauma

> I do not want to take drugs for my nightmares, because I must remain a memorial to my dead friends.
>
> Vietnam Vet (quoted in Caruth, ed., *Explorations in Memory*, p. vii, Preface)

The diagnostic category of PTSD was introduced in 1980 in DSM-III by the American Psychiatric Association mainly as a consequence of research and clinical observations about the symptoms that veterans from the Vietnam War reported. In the political climate of those years, different syndromes, such as the battered woman syndrome, the Vietnam veterans syndrome and the abused child syndrome, were subsumed under the new diagnoses; at that time it was mostly a compilation of symptoms without serious regard to the aetiology of the disorder, a problem which in our view partly still remains nowadays.

The new political climate brought about by movements for racial equality, feminist movements and by the general social awareness that came about after the civil movements of the 1970s contributed to the recognition of the traumatic effects of extreme traumatisation. In addition, it was not until the 1970s that the full consequences of the genocides of the twentieth century, mainly the Shoah, were being acknowledged in their full psychological impact for the first time.

In parallel, also contributing to the DSM-III in those same years, another group of researchers and psychiatrists were creating a diagnostic system for dissociative disorders arriving at a new classification. Initially, there was no awareness that the two diagnostic categories had overlaps. But a broader diagnostic system came to be recognised by DSM-III-R and by DSM-IV.

In the same years, the pioneer work of survivor psychoanalysts such as Judith Kestenberg, Milton Jukovy, Martin Bergmann, Henry Krystal and John Lifton and Dori Laub, among the others, were establishing a clear clinical connections between their persecutions and the survivors' symptoms and in some case in the further generations, showing also signs of intergenerational transmission.

Interestingly for the theoretical and historical field we are reconstructing, it had been Freud himself who had first pointed out at the real effects of trauma on the nervous system and on the body and the psyche, through the recognition of the combat war neuroses. But what Freud had recognised as psychic traumatisation impacting onto the body and the psyche for combat war was by Freud himself disavowed in the theory of trauma he had developed when treating hysterical women in the bourgeois Vienna of the end of the nineteenth century, women suffering mostly from the effects of incest (Breuer and Freud 1957).

Notoriously he had moved from acknowledging the full real impact of the abuse (which at the time he called "seduction", so that his trauma theory in those years went after the name of "seduction theory") onto the body and the mind of the hysterics to a theory of trauma that, while recognising the pathological effects of that kind of interpersonal traumatisations, put the stress on the psychic conflicts in the mind of the suffering subject to account for the pathological consequences. Traumatisation came mostly after the event, with a "nachtraglichkeit" effect. In this way he was disclaiming the full impact of the real traumatisation (see for this discussion also Mucci 2008), to privilege the fantasmatic or intrapsychic core of the pathology on the interpersonal/external causes.

Paradoxically, psychoanalysis, created to make sense of the traumatisations of those women as a unique technique that traced back physical symptoms (paralysis, aphasia, blindness) to a traumatic, psychological reason, was at risk of forgetting its origin, to the point that Werner Bohleber felt the need, in the 2007 IPA Congress dedicated precisely to "Memory, Repetition and Process and psychoanalysis and Today's Culture", to state authoratively: "Psychoanalysis, originally undertaken in order to discover repressed childhood memories, is now in danger of becoming a treatment technique that actually fades out history" (Bohleber 2007, p. 109).

The connection between external, environmental and interpersonal cause of the pathological effects of traumatisation has therefore come fully to the fore only in the last 30 years.

### 17.1.1 Dissociation, Not Repression, as a Defence from Trauma

Another major consequence of Freud's prevailing approach on trauma and the psychopathological effects of it was the undervaluation of one of the main consequences of traumatisation: dissociation itself. He maintained that hysterical women had "repressed" the stressful event and its emotional impact so that the cause had been removed from consciousness The main pathologies his patients suffered from at the end of the nineteenth century, which he called "psychoneurosis", i.e. obsessive neuroses and hysterical neuroses, were caused by a mechanism of defence he called repression. Freud's view of the psychopathological mechanism of repression as the major cause/mechanism of neurotic illness prevailed for several decades, even if two of his contemporaries, at the time totally disregarded, Pierre Janet and Sandor Ferenczi, had already pointed out that overbearing and overwhelming traumatisations of the kind of war and severe abuse created powerful dissociative effects as a response and defence to trauma (see also Lingiardi and Mucci 2014).

The introduction of the diagnosis of PTSD as a specific syndrome caused by trauma created a debate in the mental health practitioners and experts that is in many ways still open nowadays, as to how and what to define trauma and its destructive consequences. On the one hand, finally the traumatic effects of "extreme events" or what in DSM-III-R had been defined "outside the range of human experience" had found a recognition, and yet the very idea of considering events such as rape, war and genocide "outside the range of human experience" was controversial enough and even at risk of creating paradoxically a sort of normalisation (see Laura Brown's famous reply to the controversy in Caruth, ed. 1995 trauma explorations in memory, "not outside the range) and even a medicalisation of a form of suffering that was brought about by real-life and interpersonal experiences.

### 17.1.2 PTSD in DSM-5

In the last version of DSM-5, appeared in 2013, the diagnosis regarding trauma-related disorders had major changes: (a) it established a new DSM-5 diagnostic category, "Trauma and Stressor-Related Disorders" for PTSD (and acute stress disorder, adjustment disorders and others so that PTSD is no longer classified as an anxiety disorder), (b) PTSD was reconceptualised broadly to include post-traumatic anhedonic/dysphoric externalising and dissociative clinical presentations along with the original fear-based anxiety disorder, and (c) it established preschool and dissociative subtypes.

Two dissociative specifications were included, one based on depersonalisation (experience of being an outside observer of or detached from oneself, e.g. feeling as if "this is not happening to me" or one were in a dream) and one about derealisation (experience of unreality, distance or distortion (e.g. "things are not real")). The elements of dissociation have been included on the basis of more clinical work (Scalabrini et al. 2017; Lyssenko et al. 2017) and also neuroscientific findings, as demonstrated, for instance, by Lanius et al. (2010a, b, 2012) and Schore (2002).

Three lines of research converged towards the recognition of this dissociative element: (1) symptom assessments, (2) treatment outcomes and (3) psychobiological studies. From the taxometric and epidemiological studies, it was evident that the 15–30% suffering from the dissociative subtype were mostly men who had experienced repeated traumatisation and early adverse experiences; they had comorbid psychiatric disorders and evidenced greater suicidality and functional impairment. The subtype also replicated cross-culturally.

Studies by Lanius and colleagues indicate that depersonalisation and derealisation responses in PTSD are distinct from re-experiencing/hyperarousal reactivity. Those individuals who re-experience their traumatic memories and showed concomitant psychophysiological hyperarousal exhibited *reduced* activation in the medial prefrontal and the rostral anterior cingulate cortex and increased amygdala reactivity. These reactivation responses are thought to be mediated by failure of prefrontal inhibition or top-down control of limbic regions, in coherence with Allan Schore's interdisciplinary research (Schore 1994, 2003a, b, 2012).

In contrast, the group who exhibited symptoms of depersonalisation and derealisation showed increased activation in the rostral anterior cingulate cortex and the medial prefrontal cortex. Depersonalisation/derealisation responses are suggested to be mediated by midline prefrontal inhibition of the limbic regions. The prefrontal inhibition of the limbic regions is also investigated by Nakao et al. (2013) with similar findings.

As a consequence, the treatment for these individuals needs to be different (Lanius et al. 2012; Mucci 2017, 2018; Schore 2012). Individuals with those dissociative traits respond better to treatment with affective and interpersonal regulation in addition to exposure-based therapies. For the assessment, together with the Clinical Administered PTSD Scale (CAPS), which includes items assessing depersonalisation ("Have there been times when you felt as if you were outside of your body, watching yourself as if you were another person?") and derealisation ("Have there been times when things going on around you seemed unreal or very strange and unfamiliar?"), there are several self-report rating scales that assess dissociative symptomatology. These include the Dissociative Experiences Scale (DES), the Somatoform Dissociation Scale (SDQ), the Multiscale Dissociation Inventory (MDI), the Traumatic Dissociation Scale (TDS) and the Stanford Acute Stress Reaction Questionnaire (SASRQ).

The reason for this dissociative response to trauma needs to be further explained. We will provide a developmental etiological explanation linked to levels of interpersonal traumatisation, from early relational trauma (first level, Schore 1994) to second level of severity (incest, abuse, active maltreatment) (Ferenczi 1932a, b; Perry 1999; Cloitre et al. 2009; Courtois 1996; Mucci 2013, 2017) to a third level, namely, massive social trauma of the kind experienced in extermination camps, in war and in genocides, as evidenced, for instance, in the Shoah survivors (Lifton 2012; Laub and Lee 2002; Laub and Auerhahn 2017; Krystal 1968; Kestenberg 1980; Mucci 2013, 2017, 2018).

## 17.2 Towards the Necessity of a Recognition of Complex PTSD (CPTSD)

DSM-5, besides dropping the subjective element of criterion A, added sexual violence to actual or threatened death or serious injury and further specified the kinds of exposure as (1) direct experiencing, (2) witnessing in person, (3) learning of events happening to close family members or friends and (4) repeated work-related exposure (as the case of police officers repeatedly exposed to details of child abuse or first responders collecting human remains). But so far DSM-5 does not recognise that the most severe symptomatology depends on interpersonal violence such as abuse, maltreatment and/or incest occurring repeatedly in the home and family on the hands of people whom the child tends to trust and depend on for daily care and protection, what PDM-2 (Psychodynamic Diagnostic Manual-2) and ICD-11 (International Classification of Diseases-11) has recognised under the rubric of complex PTSD (see also van der Van der Kolk 2002; Herman 1992; Van der Kolk and D'Andrea 2010; De Bellis 2001; Schore 2013; Lanius et al. 2010b).

The most severe interpersonal trauma is in fact due to attachment trauma, or early relational trauma, with lack of a proper affective regulation and attunement between mother and child (Schore 2000), to the rupture of the trust with a caregiver (betrayal trauma, Freyd 1996) or to the break of the "empathic dyad" in the human connection (Laub and Auerhahn 1989). This often unseen or unrecognised and widely spread "hidden epidemic", that is, the impact of early life trauma on health and disease (as defined by Lanius et al. 2010b), leads to lack of emotional regulation, impulsivity and lack of effortful control, with extremely low self-esteem in the victim and the introjection, in case of maltreatment and violence, of the negative affects of the persecutor. In this way the traumatisation from a human and relational external source, from the outside, becomes internalised in the form of an internal persecutor as exemplified by Ferenczi in Freud's time, as early as Ferenczi (1932a), and as explained with the mechanism of the *Alien Self* (Bateman and Fonagy 2006) and an internalised victim-persecutor dyad (Mucci 2013, 2014) with further revictimisations leading to destructive behaviour and externalisation onto others or onto one's body, as happens in borderline patients (Jurist et al. 2008), who cut and damage themselves in various ways.

Even if infant research, psychoanalysis and neuroscience have repeatedly pointed out at the extremely severe consequences of interpersonal trauma within the family and in relation with other human beings (since they are potential figures of attachment) for the self and as a potential cause for mental pathologies, trauma resulting from long-term abusive maltreatment and incest is still unrecognised by DSM-5, and the requests by several clinician and researcher to include complex PTSD in DSM-5 have been disattended. The American Psychiatric Association in rejecting van der Kolk's request to introduce the diagnoses of developmental trauma disorder gave the following response: "The notion that early childhood adverse experiences lead to substantial developmental is more clinical intuition that a research-based fact. There is no known evidence of developmental disruption that were preceded in time in a causal fashion by any type of trauma syndrome" (reported by van der Kolk 2014, p. 149).

The ACE extensive epidemiological research has also shown how traumatisations stemming from psychological, physical and sexual abuse in at-risk or dysfunctional families, namely, families presenting intermarital violence, antisocial behaviour, mental illness and drug abuse (Felitti et al. 1998), correlate with both psychological pathologies (depression, suicidality, addiction to alcohol and drugs, eating disorders, personality disorders) and physical illness, such as metabolic disease, heart and circulation disease, liver and kidneys disease and disorders of the immune system. This epidemiological research has been carried out on 18,000 subjects.

Cumulative trauma or complex PTSD results in psychopathologies and psychosomatic diseases.

PDM-2 (Lingiardi and McWilliams 2017) acknowledges complex PTSD in the second edition (CPTSD) in the subsection S41.3 (pp. 190–193). The forthcoming ICD-11 (expected 2018) is probably including complex PTSD.

The mechanism of formation of psychological disorders based on affect dysregulation and of diseases of the kind mentioned above is easily explainable with the neurobiology of stress (with affect dysregulation producing excess cortisol with activation of HPA with the disrupture of all serotoninergic, adrenergic, dopaminergic and immune system and subsequent disruption of the various biological systems, as shown also in the research carried out by Renè Spitz in the 1940s in the USA in which deprived children in American orphanages fell ill because of infections due to the poor defences of the immune system (caused by excess cortisol in reaction to extreme stressful conditions)). Interestingly, the same reactions were noticed in rodent families (see also Hofer 2005, 2006), with the same disruption of neurobiological systems under severe.

### 17.2.1 Dissociation Is Only in Trauma of Human Agency

One of the problems of a definition of PTSD even in DSM-5, with the most recent definition as "Trauma and Stress-Related Disorders", is that it still does not differentiate between trauma due to natural disaster and trauma due to human agency and parental care and abuse.

The disorganisation of attachment that follows early relational trauma and abuse is what causes the vulnerability for dissociation and the pathologies with a dissociative structure. In fact, it has been authoratively pointed out (Cassidy and Shaver 2002; Liotti 1992a, b) that the child does not dissociate as a consequence of a natural catastrophe such as earthquakes and typhoons or even in accidents like car accidents and other non-intentional human violence (Liotti 1992a, b). The dissociative element is a defence from trauma of human agency (which has a consequence for the treatment: what has been disrupted neurobiologically in a lack of relationship needs to be repaired from therapies that lead to affect regulation and implicit reparation of distorted or dysfunctional images of self and other).

In fact 80% of maltreated children (Carlson et al. 1989) or children with pre- or post-natal maternal alcohol and cocaine use have been found to be insecure-disorganised (type D) in the AAI. These are children who have reduced stress tolerance, and instead of finding comfort and safety in their parents, they are frightened and alarmed by them. In Main and Hesse's description of the disruptive behaviour of these disorganised children, we notice at the same time the activation of both energy-expending sympathetic and energy-saving parasympathetic components of the ANS.

As Laub and Auerhahn argue in *Failed empathy*: "since the traumatic state cannot be represented it cannot be modified by interpretation … and what is initially requested by the therapy is not the elucidation of the conflict but the restructuring of a new relationship and a new connection between self and other" (Laub and Auerhahn 1989, p. 392).

## 17.3 Three Levels of Interpersonal Traumatisations and Impact onto the Psyche and the Body

### 17.3.1 Level 1: Early Relational Trauma

Linking interdisciplinary interpersonal neurobiology, developmental psychology, neuroscience, attachment and trauma studies, we distinguish three levels of interpersonal traumatisation (and of severity of abuse), resulting with disorganised attachment in the child and dissociation both in the child and in the adult.

Level 1 is early relational trauma, i.e. traumatisation due to disattunement between the needs of the child and the capacity to provide care and containment and reassurance and safety from the caregiver, (Schore 2003a; Solomon and George 1999; Cassidy and Shaver 2002); it creates disorganised attachment, which leaves a vulnerability towards a dissociative response (if there is no repair in future relationships).

Neurobiologically, since the caregiver works as the hidden regulator of all systems, the child is left in a state of affect dysregulation with hypoarousal, and the further step is a vulnerability for dissociation, even towards future traumatisations not of human agency (Schore 1994; Mucci 2013).

In this regard, it should be noted that DSM-5 has introduced the attachment reactive disorder, which stems from this lack of attunement and containment on the part of the caregiver, but it has been included within the depressive disorders. This might create the risk or the misunderstanding of considering what is a dysfunction in the early relationship between caregiver and child, which could be and should be repaired in a relationship through adequate emotional care and support, to a mood disorder to be treated pharmacologically in the child.

Nowadays in experimental research it is well established that the resting state indicates that self-related processing may be considered as an ongoing process/activity (Northoff et al. 2006; Northoff and Panksepp 2008; Bai et al. 2016). Such ongoing intrinsic brain activity might allow us to maintain a continuous and temporally extended sense of self and of relatedness. In this way, the normal high resting-state activity can be seen as a "physiological baseline" that is indicative of mental health (Northoff and Bermpohl 2004; Northoff et al. 2006). In a recent study (Nakao et al. 2013), it is shown how early life stress (ELS) is associated neuronally with decreased functional connectivity within the default mode network (DMN) in the resting state. Moreover, it is linked with greater deactivation of medial prefrontal cortex (MPFC) during a self-oriented task. Although DMN shows large amplitudes of very low-frequency oscillations and strong involvement during self-oriented tasks, these features' relation to ELS remains unclear.

Another important recent research on negative childhood experiences by Duncan and colleagues (Duncan et al. 2015) has shown how negative childhood experiences (NCE) can have long-term effects on the structure and function of the brain. Alterations have been noted in grey and white matter, in the brain's resting state, on the glutamatergic system and on neural and behavioural responses to aversive stimuli. These effects can be linked to psychiatric and psychological disorders such as depression, anxiety and personality disorders that are influenced by excessive exposure to early life stressors and who experienced early relational trauma. These findings highlight the impact of NCEs on multiple inter-related brain systems. In particular, they highlight the role of a prefrontal-insular-motor cortical network in the processing and responsivity to aversive stimuli and its potential adaptability by NCEs.

In case of optimal care, the caregiver functions as the socio-emotional regulator during the first stages of prenatal and post-natal development. From the attunement or lack of attunement between caregiver and child and their constant interaction stems the basic affective regulation that is critical for the activation and maintenance and development of the limbic circuits. Since the limbic system mielinates during the first year and a half of the child and the right hemisphere is deeply connected to the limbic system, this first year and a half of development of human life during which the attachment bond with a caregiver is created permanently affects future development, especially in the right hemisphere, which develops first. From the development of the limbic areas (amygdala, thalamus, hypothalamus, hippocampus), determinant for affective response and future empathic connection with the other, and which responds to arousal-seeking and aversion-avoiding reactions, the relationship with a non-traumatised caregiver, capable of insuring this constant regulation of the child through a pattern of attunement and capacity for prompt repair in the normal moments of dysregulation, insures the proper development of cortical areas, especially the orbitofrontal and ventral-medial areas that guarantee, when adequately developed, the control of emotional input and determine the capacity for future self-regulation, ideation and control of higher capacities, starting from the first hemisphere going to the proper development of the left hemisphere, which starts a critical growth after the second year of life (Cozolino 2002; Henry 1993; Schore 1994; Siegel 1999; Sasso 2007).

The orbitofrontal areas undergo a fundamental growth between 10 and 12 months of the child (Diamond and Doar 1989; Schore 1994).

As Schore writes cogently, "At the end of the first year right lateralised cortical-subcortical circuits imprint, in implicit procedural memory, an internal working model of attachment which encodes strategies of affect regulation that nonconsciously guide the individual through interpersonal contexts" (Foreword to Bromberg 2011, p. xii). Orbitofrontal areas, between cortical and subcortical areas, control ANS sympathetic and parasympathetic reactions, therefore the regulation of impulses and the control/inhibition of them. At birth, no areas of the orbitofrontal regions are functioning. It is the progressive maturation of limbic circuits, thanks to the attachment relation that connects (with a special input at the end of the first year), amygdala, cingulate cortex, insula and orbitofrontal areas (Schore 1994, 2000, 2002). Through the connection with hypothalamus and brain stem, regulation controls instinctual and bodily drives, through a representational system that is unique for the human being, what in psychoanalysis goes under the object relation theory, a relationship connected through an affect (see Pandya and Yeterian 1991; see also Kernberg 1977).

When the caregiver has unresolved traumatisations (what in AAI is called U, unresolved trauma), the dissociated parts of the personality and of the behaviour can be transferred to the limbic system of the child in the attachment connection, which connects mostly the right limbic brain of the infant to the right limbic brain of the (still traumatised) mother, and the attachment relationship itself becomes a vehicle for intergenerational transmission traumatisation and therefore for possible dissociation in the child even in lack of actual maltreatment and abuse (one more reason to treat traumatisation in the caregiver as soon as possible).

We can speak therefore of an intersubjective and psychogenic mechanism of intergenerational traumatisation and of the dissociative defences towards overwhelming and disturbing negative effects in the child (Schore 2010). Research shows that severe maternal dysfunctions (not paternal, unless the main caregiver is the father), such as mental disorder, unresolved mourning especially in the first 2 years of the child, addiction and traumatisations, correlate with dissociation in the child (Liotti 1992a, b; Draijer and Langeland 1999; Roelofs et al. 2002; Carlson and Sroufe 1995; Main and Hesse 1990; Ainsworth and Eichberg 1991).

To summarise it should be remembered that:

1. Secure attachment is the best protection for prevention of future PTSD reaction to trauma of both natural and human agency.
2. Disorganised attachment in the mother can cause disorganisation in the child and therefore may create the vulnerability for future dissociation and pathological responses even in the lack of maltreatment, abuse, violence and incest and even through a sort of silent traumatisation (Kogan 1995; Laub 1995; Abraham and Torok 1994).

The attachment bond is now recognised as the main organising principle for optimal growth and neurobiological, cognitive, emotional, affective and social capacities in the child (Schore 1994, 2010; Cozolino 2014; Siegel 1999), well beyond those fundamental qualities of assuring security and protection as recognised by Bowlby's studies (Bowlby 2005). It affects the proper development of the right hemisphere which develops first and is therefore the neurobiological substratum of attachment bonds and what we now consider the neurobiological substratum of the implicit memory influencing deeply memories of self and other in relation (Schore 2003a, b; Diamond et al. 1964).

### 17.3.2 Level 2: Maltreatment and Abuse and Identification with the Aggressor: Ferenczi and the Traumatic Cycle

In addition to this first level of traumatisation, we define a second level of traumatisation that not only fosters affect dysregulation but the internal mechanism of identification with both the victim and the persecutor (according to Ferenczi as defined by Mucci 2013, 2016, 2017, 2018).

In the psychoanalytic field, in opposition to Freud, the traumatisation of the second level (abuse, maltreatment, severe emotional and physical deprivation and incest) had been highlighted by Ferenczi (1932a, b) when he develops his theory of interpersonal trauma. As early as 1932 in his *Clinical Diary* (on March 25, 1932, "Psychic Bandage"), Ferenczi describes how the overwhelming experience of repeated maltreatment and abuse especially in the family leaves a permanent mark and results precisely in a "permanent split in the personality" with "lack of integration of emotional unit" and ultimately in a change in the personality of the victim (Ferenczi 1932a, p. 69).

The child treats with dissociation both as psychic response (i.e. material split form conscious, negated or disavowed, not repressed) and physical response as a sort of "feigned death" of the kind that present time neuroscientific explanation has described as parasympathetic mechanisms of vertical disconnection between limbic system and higher cortical areas (Shore, p. 117, in Dell and O'Neil 2010), resulting a freezing response:

This is Ferenczi in the famous passage in on "Fragmentation" (*Clinical Diary*, 21 February 1932):

> A child is the victim of overwhelming aggression, which results in "giving up the ghost", … with the firm conviction that this self-abandonment (fainting) means death. However, it is precisely this complete relaxation induced by self-abandonment that may create more favorable conditions for him to endure the violence …. Therefore someone who has "given up the ghost" survives this death physically and with a part of his energy begins to live again; he even succeeds in reestablishing unity with the pretraumatic personality, although this is usually accompanied by memory lapses and retroactive amnesia of varying duration. But this amnesic piece is actually a part of the person, who is still "dead", or exists permanently in the agony of anxiety. The task of the analysis is to remove this split. (*Clinical Diary*, p. 39)

The extraordinary accuracy of this description of the dissociative traumatic reaction resulting even in a fainting of the body, a freezing response, has been confirmed by neurophysiological findings, as in the research by Porges (2011), as the polyvagal response leading to blunting and analgesia (compatible with the "shrinking of conscious experience" as described by Janet). More than a defence, and certainly not an intentional or even partially intentional defence, the neurophysiology of trauma describes a collapse of mental and psychical resources as a response to the overwhelming external experience more than to the intrapsychic defence at work.

Dissociation is to be understood neurobiologically and neurodevelopmentally as a "primitive coping strategy of affect regulation, … a loss of vertical connectivity between cortical and subcortical limbic areas within the right hemisphere" (Schore 2009; Dell and O'Neil 2010, p. 117).

In exact correspondence with Ferenczi's description, this is how Schore has recently described the process of extreme traumatisation and the parasympathetic mechanism:

> The dissociative metabolic shutdown state is a primary regulatory process, used throughout the life-span, in which the stressed individual passively disengages in order to conserve energies, foster survival by the risky posture of 'feigning death', and allow the restitution of depleted resources by immobility. In this passive hypometabolic state heart rate, blood pressure, and respiration are decreased, while pain numbing and blunting endogenous opiates are elevated. It is this energy-conserving parasympathetic (vagal) mechanism that mediates the "profound detachment" of dissociation. ("Forward", to Bromberg 2011, xvii)

The neurobiological effects of trauma inform the development of images of self and other in those critical moments for physical and psychological development. In terms of introjection of interpersonal dynamics, identification and emotions, the child introjects both the dissociated guilt and the aggressiveness of the persecutor (Ferenczi 1932a, b):

The children feel physically and morally helpless, their personality is still too insufficiently consolidated for them to be able to protest even if only in thought. The overwhelming power and authority of the adults render them silent; often they are deprived of their senses. *Yet that very fear, when it reaches its zenith, forces them automatically to surrender to the will of the aggressor, to anticipate each of the wishes and to submit to them; forgetting themselves entirely*, to identify totally with the aggressor. As a result of the identification with the aggressor, let us call it introjection, the aggressor disappears as external reality and becomes intrapsychic instead of extra -psychic. Yet the most important transformation with the adult partner, an identification based on fear, calls forth, *is the introjection of the guilt feeling of the adult.* (Ferenczi 1932b, pp. 297–298, emphasis in the text mine)

The identification with the aggressor, with the internalisation of the rage against one's body or against the other, explains much of the destructiveness against self and other in personality disorders and other severe mental pathology (Mucci 2016, 2017, 2018).

### 17.3.3 Level 3: Massive Trauma

Massive social trauma as genocide, war and extermination of the kind of the Shoah can cause severe traumatisation with PTSD symptoms, dissociation, depression, insomnia, chronic pain and suicidal attempts. Lifton, Krystal, Kestenberg, Kogan and Laub among the other clinicians have thoroughly discussed the effects of massive trauma as in the Shoah, distinguishing effects on the first generations and possible effects on the following generations (we want to stress, through the mediation of attachment) (see the work of clinicians such as Grubrich-Simitis 1981; Kogan 1995a, b; Kestenberg 1980; Laub 1995; Mucci 2013) and the extensive research of Yehuda and McFarlane (1995), Yehuda et al. (1996, 2001), Yehuda et al. (2005, 2007) and Sagi-Schwartz et al. (2008).

Dori Laub, a Shoah survivor himself, has described the traumatogenic mechanism as the rupture of the empathic dyad and the impossibility of going back to an internalised good object that gets destroyed in the presence of traumata of human agency; trauma is therefore characterised by a specific non-figurability and a symbolic unrepresentable status, precisely because an internal witness has been destroyed, so that the possible retrieval of the original (dissociated) traumatisation is possible only in the presence of a totally present and fully committed other (Laub and Finchelstein 2010; Mucci 2013, 2017, 2018).

Laub also has pointed out at the peculiar epistemologic status of traumatisation as a state of "knowing-not knowing". Neuropsychologically this state could be described with the peculiar implicit knowledge of the traumatic effects embodied by the body through implicit memory that disavows the actual explicit and declarative knowledge of the traumatic events, erases them but leaves an implicit, non-conscious trace in the body.

Trauma also follows an internal process of intergenerational transmittance through the identification of the alive with the dead and through the children's stepping into the wishes of the parents and of other relatives, through dynamics of projective identification, dissociative displacements and reversals.

### 17.3.4 Traumatic Generations

According to Laub there certainly is an unconscious principle of traumatic transmission between and among generations that we have explained through attachment mechanism of transposition and disorganised attachment in the parent and as a consequence in the child. These transmissions might happen through silent dissociative splits and identification with the split parts of the parents and also through the destructive identification with the persecutor. In some case, the child (second generation or third) is viewed as the persecutor. Or it is the second generation that feels the burden and the psychological necessity in a reversal of structure, to care about the psychological needs of the traumatised parent to reduce his/her/the parent's suffering and to try to "repair" in the parent the effects of the traumatisation, in this way disrupting one's life or reducing it to a false self-kind of existence.

In addition, it is likely that the second generation might inherit symbolically a fantasmatic wish to elaborate the traumatisation that the first generations have actually undergone in reality (Mucci 2013). This explanation is not far from the symbolic and fantasmatic kernel of the trauma inherited by the second generation through the mysterious "cript" encysted intersubjectively (see Abraham and Torok 1994). Children psychologically "inherit" the problematic themes and issues of the parents, through the porous and labile boundaries of the traumatised parents (which might have dissociated, denied the trauma or might have alexythimically displaced and removed it) (Fig. 17.1). Boundary problems are always in the traumatised and as a consequence in their generations. Through this intermingling of roles and role reversals, second generations might have the need to care for the parents and repair for them their trauma (a pattern often easily traceable in second-generation AAI of parents who are survivors of massive traumata) (Mucci 2015).

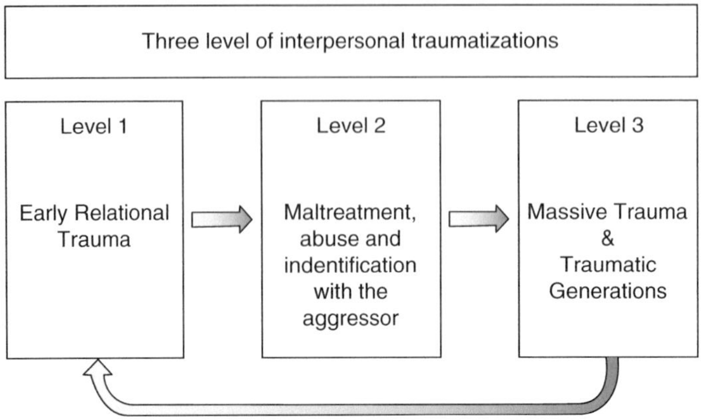


**Fig. 17.1** Model of interpersonal traumatisations

## 17.4 Traumatic Consequences

Although the DSM-5 still does not recognise the link between present self-destructive behaviour and past trauma as a cause of present mental illness, there is nonetheless clinical evidence of this link: at least 50% of people seeking help for various disorders, from depression to personality disorders and dissociative disorders (according to US data) (Schwartz and Perry 1994; Perry and Pollard 1997; Felitti and Anda), have been victims of maltreatment and abuse, so that early relational trauma (with insecure attachment), in combination with deprivation, abuse and maltreatment, in the presence of vulnerability, seems to be the prerequisite (van der Kolk et al. 1991; Fonagy; Liotti; Schore) for the establishment of a borderline disorder proper (as in DSM-5) and in personality disorders in general, what Kernberg terms borderline organisation, a structure that includes hysteric-histrionic, borderline, narcissistic, schizoid and paranoid disorders (Gabbard 2002; Baker et al. 1992; Gunderson 2008; Gunderson and Sabo 1993; Zanarini et al. 1989, 1997; Schore 1997).

As a consequence of repeated maltreatment and abuse, (e.g. complex PTSD, which has been recognised in PDM-2, as CPTSD), the child adapts her/his behaviour to the environment, and in this way a permanent cognitive and affective distortion and a twist in personality is initiated.

### 17.4.1 Effects of Trauma of Human Agency on the Brain

Trauma allows a unique perspective to how psychological mechanism impacts onto the brain or how the stimuli presented in the environment might contribute to foster health and well-being and optimal growth or on the contrary to create malfunctioning, distortion, dissociation, negative affects and lack of empathy, reducing even areas of the brain and causing cellular death, especially in the right hemisphere.

Several studies of neuroimaging point at a volumetric reduction of parietal and hippocampal areas in severely traumatised groups such as veterans and abused and depressed women (Gurvits et al. 2002; Stein et al. 1997; De Bellis et al. 1999). This probably happens because of the reduction of connections among areas because of deficit in neurotransmitters of serotoninergic, adrenergic and glutamate (GABA) type.

Patients with borderline personality disorders with a past of abuse and losses also show hippocampal reduction (Driessen et al. 2000; Schmahl et al. 2003; Brambilla et al. 2004).

Interestingly, the state of arousal which impacts the HPA axis in individual with a long story of deprivation, maltreatment and abuse is easily activated within a subthreshold level mechanism that is called kindling (and this is even in subjects with borderline personality disorders; see Rinne et al. 2002). The stress-related dysregulation that activates the HPA axis with excess level of glucocorticoids damages the neural tissue through mechanisms of neural destruction, inhibition neurogenesis and alterations in the processes of myelination (McEwen 1998).

Hippocampus seems to be particularly receptive to this neurotoxic activation due to excess cortisol and glucocorticoids. We should remember that hippocampus is not only involved in declarative memory but together with hypothalamus, amygdala and prefrontal areas is particularly implied in the circuits that are activated as a response to traumatic situations. The reduction in volume of the hippocampus therefore is not only determined by but interferes with the subsequent response to the proper response to the hyperactivation of the system in response to stressors. In combination with the fact that PET has evidenced that the blood flux is reduced in the anterior cingulate in response to traumatic activation, the two elements (hippocampus deficit and empathy deficit in the cingulate area) might explain some of the symptoms (emotional dysregulation, amnesia and dissociation, difficulties in verbalising the experience and mentalisation).

On the contrary, PTSD seems to be related to hyperactivation of the amygdala and subsequent volumetric increase (Shin et al. 2004) (hippocampal reduction though has been evidence also for schizophrenia and depressive disorder). Teicher sees also a reduction in the corpus callosum as a result of severe and protracted traumatisation (Teicher 2000).

Therefore, trauma resulting from early deprivation proves how the development of the brain depends on optimal care during childhood.

Interestingly, and in connection with Schore's findings, Van der Kolk (2014) has observed a strong lateralisation of the activity of the right brain when traumatic memories are reactivated, in addition with a striking reduction of activity of Broca's area in the left hemisphere, the area notoriously implicated in language and verbalisation of meaningful experiences. The possibility of attributing new meaning to the traumatic events restructures the brain activity making it more balanced and deactivates the excessive response of the limbic system while augmenting the possibility of cortical awareness and explicit verbalisation.

Moreover, Lanius et al. (2010a, b, 2012) have shown in functional magnetic resonance a predominant right-hemispheric activation in post-traumatic stress disorder (PTSD) patients, while they are dissociating: patients dissociate in order to escape the overwhelming emotions for which they have no words.

Implicit traumatic memories thus encoded and encysted in the right brain, in connection with amygdala activation more than cortical awareness, have to find a path towards explicit consciousness and verbal expression, which in turn means that a self with wider awareness has been restored to reality.

This is the route of psychotherapy. A psychotherapy that is fine-tuned between the patient and therapist through the right brain, in any case, is particularly useful for this kind of (traumatised, right brain lateralised) patient. Narration processes seem to be able to perform the neuronal integration that is missing or poor in traumatised minds.

Psychotherapy will work on the implicit, bodily imprints, the traces of internalised representations of past relationship starting from attachment traces. As Siegel confirms (Siegel 1999), when implicit memory is retrieved, the neural net profiles that are reactivated involve circuits on the brain that are a fundamental part of our everyday experience of life: behaviours, emotions and images that are encoded in non-verbal operational models of the mind are there in the here and now of the session.

Destructive patterns are there repeated and enacted in moments of one-to-one exchanges in which both participants in the dialogue rehearse a right-brain-implicit model that has been interiorised (but the therapist can make use of left strategies as well through language and interpretation, too). Especially severe patients, like the borderline ones with severe history of early relational trauma or the complex PTSD, prone as they are to a mass of unregulated emotions and massive primitive defences, will enact a flood of unstable and violent affects that is mostly unconscious in the sense that they cannot be cortically controlled and balanced. Any occasion in the limits of the setting of the therapy will re-enhance and reactivate the emotional *Sturm und Drang*; these patients are subject to the continuous storm of enactments that will on one hand enable the repetition of what has not been dominated and understood consciously and on the other hand will be the object of the exploration and the exchange with the ongoing dialogic mental and bodily process of the therapeutic encounter: it is the process that guides the content; this is why content per se and interpretation per se will not affect the process. In other words, the content of the interaction needs to be "embedded" in relational experience that embodies what they call "implicit relational knowing"—an ongoing process that is itself part of the content (Bromberg 2012).

The present PTSD diagnosis is not adequate to describe the consistent and long-term suffering and personality distortion that follow repeated and continual abuse by a caregiver. These kinds of cumulative traumatisations are often the reason for adult psychopathology and are difficult to trace back immediately to childhood abuse, since memory blocks are connected to disorganised attachments and dismissive attitudes. These kinds of cumulative traumata result in psychopathologies and psychosomatic diseases. Specifically, –given the direct effect of the affective bond on the development of areas and circuits linked to empathy, to relationship with others and to implicit models of self-other, self-esteem, including the capacity for self-care and altruism, together with cognitive, social, affective and emotional development, –people who have been severely deprived or abused are at risk of developing depression, anxiety, dissociation, personality disorders, compulsive disorders, obesity and eating disorders and addictions.

Moreover, recent evidence from brain imaging indicates that subcortical signals of neuronal arousal are positively related to the degree of experienced affect, while higher brain regions are more typically negatively correlated with cognitive emotional experiences (Liotti and Panksepp 2004; Northoff et al. 2009).

This strongly suggests, in line with our point of view, that the lower regions actively generate primal affective states, while the higher regions may be regulating, reprocessing and dampening them. Likewise, such interactions help explain how lower brain arousals may disrupt cognitive processing and can results in the inability in regulating these affective states.

Thus it can be concluded that clinicians and psychotherapist need to take into account that to work with traumatised people, they need to address all the relational verbal and non-verbal skills such as enactments that "potentially allow for the reorganization of cortical (orbitofrontal), subcortical (amygdala) connectivity" (Schore 2012) and are particularly necessary to work at the "core self" level and re-establish from there the "physiological secure base" to maintain a continuous and temporally extended sense of self and sense of relatedness that in our view can be related in fMRI research, not only with the task evoked activity of the brain but also with the intrinsic ongoing activity of the so-called resting-state condition.

## 17.5 Adverse Early Experiences as a Risk Factor for Increased Stress Vulnerability and Psychiatric Disorders: Implications for the Therapeutic Use of Oxytocin

Simone Grimm

Adequate neuronal and physiological reactions to transitory changes in homeostasis during the experience of stress are crucial for human survival and well-being (McEwen 1998). Disturbances of homeostasis increase vulnerability for the effects of cumulative stress experiences in the way of adverse social experiences such as sexual, physical and emotional abuse, but also physical and emotional neglect occurring before puberty (Pechtel and Pizzagalli 2011). These adverse early experiences (early life stress, ELS) have a profound impact on individual differences in stress responsiveness (Davidson and McEwen 2012; McEwen 2012) and are associated with an increased vulnerability for psychiatric disorders, such as depression, in later life (Kendler et al. 2004; Burke et al. 2005). Converging evidence from animal models and human studies indicates that ELS causes persisting changes to hypothalamic-pituitary-adrenal axis (HPA) reactivity with altered cortisol responses to psychosocial stress (Sanchez et al. 2001; Pryce et al. 2005; Heim et al. 2001, 2008; Elzinga et al. 2008; Carpenter et al. 2007, 2009; Klaassens et al. 2009). It has been suggested that the inconsistent findings of either increased or decreased stress responsiveness after ELS might be explained by a trajectory of initial hyperactivation of the HPA system progressing to a state of chronic stress hyporeactivity (Pryce et al. 2005; Fries et al. 2005) as a type of counterregulatory adaptation after sustained exposure to stress during development (Miller et al. 2007). This proposed mechanism mirrors findings of blunted stress responsiveness in animals exposed to chronic stress (Saltzman et al. 2006; Sterlemann et al. 2008). Over time, experimentally stressed animals persistently show lower levels of hippocampal mineralocorticoid receptor expression, suggesting long-term alterations in gene expression regulation. Along this line, ELS is associated with structural and functional changes in brain regions implicated in neuroendocrine control and emotional regulation (Kaffman and Meaney 2007; Lupien et al. 2009; Plotsky et al. 2005; Heim and Binder 2012; van Harmelen et al. 2010; Vythilingam et al. 2002; Buss et al. 2007; Dannlowski et al. 2012; Edmiston et al. 2011; Burghy et al. 2012).

Individual stress reactivity in central nervous system is regulated via the coordinated activity of a limbic and prefrontal network (e.g. amygdala, hippocampus, medial prefrontal cortex; Ulrich-Lai and Herman 2009). As a key region within this limbic-prefrontal network, amygdala is reciprocally connected with regions implicated in autonomic and neuroendocrine control, memory and salience processing (Freese and Amaral 2009). Activity and connectivity of amygdala signal emotional salience of stimuli and initiate adequate reactions in brain and body (LeDoux 2000). Changes in functional connectivity between amygdala and other limbic-prefrontal regions have been demonstrated before and after stress induction (Clewett et al. 2013; Hermans et al. 2011; Vaisvaser et al. 2013; van Marle et al. 2010; Veer et al. 2011). Furthermore, connectivity between amygdala and hippocampus is increased as an immediate as well as a delayed reaction to stress (Ghosh et al. 2013; Vaisvaser et al. 2013). Hippocampus inhibits the HPA axis and gets negative feedback by its hormonal end product cortisol (Herman et al. 2005). Connectivity between amygdala and hippocampus is inversely associated with cortisol concentration (Henckens et al. 2012; Vaisvaser et al. 2013) and predicts capacity of the HPA axis to restore homeostasis after perturbation (Kiem et al. 2013). While these findings show that functional connectivity between amygdala and hippocampus plays a crucial role for reactivity to transient stress—particularly for maintenance and restoration of homeostasis via neuroendocrine control—a recent study demonstrated that not only transient but also cumulative stress early in life results in altered connectivity between these regions (Fan et al. 2014).

Furthermore, there is increasing evidence that ELS also modulates the development of the oxytocinergic system. Oxytocin (OXT) is a neuropeptide that modulates HPA axis activity, and the intranasal administration of OXT attenuates physiological as well as behavioural reactivity to psychosocial stress (Heinrichs et al. 2003; Quirin et al. 2011; Linnen et al. 2012; Ditzen et al. 2009). Accordingly, OXT is increasingly used for the treatment of psychiatric disorders (MacDonald et al. 2011; Guastella et al. 2008; Labuschagne et al. 2011). Results so far show an inconsistent or even detrimental effect of OXT (MacDonald et al. 2011; Olff et al. 2013; Pincus and Lukowitsky 2010), which might be related to not sufficiently taking into account early social experiences of patients (Simeon et al. 2011). Decreased urinary concentrations of OXT have been measured in maltreated children (Fries et al. 2005), and Heim et al. (2009) demonstrated decreased cerebrospinal fluid (CSF) OXT concentrations in women with ELS. These findings are in accordance with animal studies showing that early nurturing experiences induce persistent alterations in OXT receptor levels in rats and decreased CSF OXT concentrations in nursery-reared rhesus monkeys compared with mother-reared controls (Winslow et al. 2003). In recent years there have been several studies that not only showed altered OXT sensitivity but rather a detrimental OXT effect on physiological and neuronal markers of stress reactivity in subjects with adverse childhood experiences (Meinlschmidt and Heim 2007; Grimm et al. 2014; Feeser et al. 2014; Fan et al. 2014). In adults with childhood maltreatment, it was shown that the suppressing effect of OXT on cortisol levels is significantly reduced (Meinlschmidt and Heim 2007). Depending on intensity and duration of early adverse experiences, effects of OXT seem to not only be reduced but rather inverse. While subjects with early life stress experiences show blunted stress reactivity (attenuated increase in cortisol, reduced hippocampal deactivation) during psychosocial stress, administration of OXT increases their hormonal reactivity and hippocampal deactivation, thereby actually increasing stress reactivity (Grimm et al. 2014).

If reduced limbic deactivation and HPA axis responsivity during psychosocial stress are indeed markers for biological resilience in ELS subjects, the OXT-induced changes might be considered a detrimental effect of the treatment. On the other hand, one might also interpret these changes as a "normalisation" of neural activity patterns and hormonal reactivity since they correspond to those seen in control subjects during the placebo condition.

A recent study rather supports the hypothesis of inverse or even detrimental oxytocin effects in subjects with ELS experiences. Feeser et al. (2014) demonstrated better emotion recognition in subjects with ELS experiences. While in subjects without ELS experiences previous findings of improved emotion recognition after OXT administration were confirmed, this was not the case in subjects with ELS experience.

However, all these studies were conducted in healthy subjects with early adverse experiences, which increased their vulnerability for psychiatric disorders, but who had nevertheless not developed any kind of psychopathology. Therefore, the investigated cohorts might be considered particularly resilient and not representative for subjects with early adverse experiences. Nevertheless, the presented findings demonstrate how crucial it is to consider environmental factors and particularly early social experiences prior to a therapeutic administration of oxytocin.

## References

Abraham N, Torok M. The shell and the kernel: renewals of psychoanalysis, vol. 1. Chicago: University of Chicago Press; 1994.

Ainsworth MDS, Eichberg C. Effects on infant-mother attachment of mother's unresolved loss of an attachment figure, or other traumatic experience. In: Parkes CM, Stevenson-Hinde J, Marris P, editors. Attachment across the life cycle, vol. 3. New York: Tavistock; 1991. p. 160–83.

Bai Y, Nakao T, Xu J, Qin P, Chaves P, Heinzel A, et al. Resting state glutamate predicts elevated pre-stimulus alpha during self-relatedness: a combined EEG-MRS study on "rest-self overlap". Soc Neurosci. 2016;11(3):249–63.

Baker L, Silk KR, Westen D, Nigg JT, Lohr NE. Malevolence, splitting, and parental ratings by borderlines. J Nerv Ment Dis. 1992;180:258–64.

Bateman A, Fonagy P. Mentalization-based treatment for borderline personality disorder: a practical guide. Oxford: Oxford University Press; 2006.
Bohleber W. Remembrance, trauma and collective memory: the battle for memory in psychoanalysis. Int J Psychoanal. 2007;88(2):329–52.
Bowlby J. A secure base: clinical applications of attachment theory, vol. 393. London: Taylor & Francis; 2005.
Brambilla P, Soloff PH, Sala M, Nicoletti MA, Keshavan MS, Soares JC. Anatomical MRI study of borderline personality disorder patients. Psychiatry Res Neuroimaging. 2004;131(2):125–33.
Breuer J, Freud S. Studies on hysteria. MANCA; 1957.
Bromberg PM. The shadow of the tsunami and the growth of the relational mind. New York: Routledge; 2011.
Bromberg PM. The shadow of the tsunami: and the growth of the relational mind. New York, NY: Routledge; 2012.
Burghy CA, et al. Developmental pathways to amygdala-prefrontal function and internalizing symptoms in adolescence. Nat Neurosci. 2012;15(12):1736–41.
Burke HM, Davis MC, Otte C, Mohr DC. Depression and cortisol responses to psychological stress: a meta-analysis. Psychoneuroendocrinology. 2005;30(9):846–56.
Buss C, et al. Maternal care modulates the relationship between prenatal risk and hippocampal volume in women but not in men. J Neurosci. 2007;27(10):2592–5.
Carlson EA, Sroufe LA. Contribution of attachment theory to developmental psychopathology. In: Cicchetti D, Cohen DJ, editors. Developmental psychopathology, theory and methods, vol. 1. Oxford: John Wiley & Sons; 1995. p. 581–617.
Carlson V, Cicchetti D, Barnett D, Braunwald KG. Finding order in disorganization: lessons from research on maltreated infants' attachments to their caregivers. In: Cicchetti D, Carlson V, editors. Child maltreatment: theory and research on the causes and consequences of child abuse and neglect. Cambridge: Cambridge University Press; 1989.
Carpenter LL, et al. Decreased adrenocorticotropic hormone and cortisol responses to stress in healthy adults reporting significant childhood maltreatment. Biol Psychiatry. 2007;62(10):1080–7.
Carpenter LL, et al. Effect of childhood emotional abuse and age on cortisol responsivity in adulthood. Biol Psychiatry. 2009;66(1):69–75.
Cassidy J, Shaver PR, editors. Handbook of attachment: theory, research, and clinical applications. New York: Rough Guides; 2002.
Clewett D, Schoeke A, Mather M. Amygdala functional connectivity is reduced after the cold pressor task. Cogn Affect Behav Neurosci. 2013;13:501–18.
Cloitre M, Stolbach BC, Herman JL, Kolk BVD, Pynoos R, Wang J, Petkova E. A developmental approach to complex PTSD: childhood and adult cumulative trauma as predictors of symptom complexity. J Trauma Stress. 2009;22(5):399–408.
Courtois CA. Healing the incest wound: adult survivors in therapy. New York: Norton; 1996.
Cozolino L. The neuroscience of psychotherapy: building and rebuilding the human brain, Norton series on interpersonal neurobiology. New York: Norton; 2002.
Cozolino L. The neuroscience of human relationships: attachment and the developing social brain (Norton series on interpersonal neurobiology). New York: Norton; 2014.
Dannlowski U, et al. Limbic scars: long-term consequences of childhood maltreatment revealed by functional and structural magnetic resonance imaging. Biol Psychiatry. 2012;71(4):286–93.
Davidson RJ, McEwen BS. Social influences on neuroplasticity: stress and interventions to promote well-being. Nat Neurosci. 2012;15:689–95.
De Bellis MD. Developmental traumatology: the psychobiological development of maltreated children and its implications for research, treatment, and policy. Dev Psychopathol. 2001;13(3):539–64.
De Bellis MD, Baum AS, Birmaher B, Keshavan MS, Eccard CH, Boring AM, et al. Developmental traumatology, part I: biological stress systems. Biol Psychiatry. 1999;45:1259–70.
Dell PF, O'Neil JA, editors. Dissociation and the dissociative disorders: DSM-V and beyond. New York: Routledge; 2010.

Diamond A, Doar B. The performance of human infants on a measure of frontal cortex function, the delayed response task. Dev Psychobiol. 1989;22(3):271–94.
Diamond MC, Krech D, Rosenzweig MR. The effects of an enriched environment on the histology of the rat cerebral cortex. J Comp Neurol. 1964;123(1):111–9.
Ditzen B, et al. Intranasal oxytocin increases positive communication and reduces cortisol levels during couple conflict. Biol Psychiatry. 2009;65(9):728–31.
Draijer N, Langeland W. Childhood trauma and perceived parental dysfunction in the etiology of dissociative symptoms in psychiatric inpatients. Am J Psychiatr. 1999;156(3):379–85.
Driessen M, Herrmann J, Stahl K, Zwaan M, Meier S, Hill A, Osterheider M, Peterson D. Magnetic resonance imaging volumes of the hippocampus and the amygdala in women with borderline disorder and early traumatization. Arch Gen Psychiatry. 2000;57(12):1115–22.
Duncan NW, Hayes DJ, Wiebking C, Tiret B, Pietruska K, Chen DQ, Rainville P, Marjańska M, Ayad O, Doyon J, Hodaie M, Northoff G. Negative childhood experiences alter a prefrontal-insular-motor cortical network in healthy adults: a preliminary multimodal rsfMRI-fMRI-MRS-dMRI study. Hum Brain Mapp. 2015;36(11):4622–37.
Edmiston EE, et al. Corticostriatal-limbic gray matter morphology in adolescents with self-reported exposure to childhood maltreatment. Arch Pediatr Adolesc Med. 2011;165(12):1069–77.
Elzinga BM, et al. Diminished cortisol responses to psychosocial stress associated with lifetime adverse events a study among healthy young subjects. Psychoneuroendocrinology. 2008;33(2):227–37.
Fan Y, Herrera-Melendez AL, Pestke K, Feeser M, Aust S, Otte C, Pruessner JC, Böker H, Bajbouj M, Grimm S. Early life stress modulates amygdala-prefrontal functional connectivity: implications for oxytocin effects. Hum Brain Mapp. 2014;35(10):5328–39.
Feeser M, Fan Y, Weigand A, Gärtner M, Aust S, Böker H, Bajbouj M, Grimm S. The beneficial effect of oxytocin on avoidance-related facial emotion recognition depends on early life stress experience. Psychopharmacology. 2014;231(24):4735–44.
Felitti VJ, Anda RF, Nordenberg D, Williamson DF, Spitz AM, Edwards V, Koss MP, Marks JS. Relationship of childhood abuse and household dysfunction to many of the leading causes of death in adults. Am J Prev Med. 1998;14(4):245–58.
Ferenczi S. In: Dupont E, editor. The clinical diary of Sándor Ferenczi. Cambridge: Harvard University Press; 1932a. 1988.
Ferenczi S. Confusion of tongues between adults and the child (the language of tenderness and the language of [sexual] passion) (Trans. J. M. Masson & I. Loring), Appendix C. In: Masson JM, editor. The assault on truth: Freud's suppression of the seduction theory. New York: Random House; 1932b. 1984.
Freese JL, Amaral DG. Neuroanatomy of the primate amygdala. In: Whalen PJ, Phelps EA, editors. The human amygdala. New York: The Guilford Press; 2009. p. 3–42.
Freyd JJ. Betrayal trauma: the logic of forgetting childhood abuse. Cambridge: Harvard University Press; 1996.
Fries E, Hesse J, Hellhammer J, Hellhammer DH. A new view on hypocortisolism. Psychoneuroendocrinology. 2005;30(10):1010–6.
Gabbard G. Psychodynamic psychiatry. Arlington, VA: American Psychiatric Publications; 2002.
Ghosh S, Laxmi TR, Chattarji S. Functional connectivity from the amygdala to the hippocampus grows stronger after stress. J Neurosci. 2013;33:7234–44.
Grimm S, Pestke K, Feeser M, Aust S, Weigand A, Wang J, Wingenfeld K, Pruessner JC, Böker H, Bajbouj M. Early life stress modulates oxytocin effects on limbic system during acute psychosocial stress. Soc Cogn Affect Neurosci. 2014;9(11):1828–35.
Grubrich-Simitis I. Extreme traumatization as cumulative trauma: psychoanalytic investigations of the effects of concentration camp experiences on survivors and their children. Psychoanal Study Child. 1981;36(1):415–50.
Guastella AJ, Mitchell PB, Dadds MR. Oxytocin increases gaze to the eye region of human faces. Biol Psychiatry. 2008;63(1):3–5.
Gunderson JG. Borderline personality disorder: a clinical guide. Washington, DC: American Psychiatric Press; 2008.

Gunderson JG, Sabo AN. The phenomenological and conceptual interface between personality disorder and PTSD. Am J Psychiatry. 1993;150:19–27.
Gurvits TV, Shenton ME, Hokama H, Ohta H, Lasko NB, Gilbertson MW, Orr SP, Kikinis SP, Jolesz FA, McCarley RW, Pitman RK. Magnetic resonance imaging study of hippocampal volume in chronic, combat-related post-traumatic stress disorder. Biol Psychiatry. 2002;40(11):1091–9.
van Harmelen AL, et al. Reduced medial prefrontal cortex volume in adults reporting childhood emotional maltreatment. Biol Psychiatry. 2010;68(9):832–8.
Heim C, Binder EB. Current research trends in early life stress and depression: review of human studies on sensitive periods, gene-environment interactions, and epigenetics. Exp Neurol. 2012;233(1):102–11.
Heim C, Newport DJ, Bonsall R, Miller AH, Nemeroff CB. Altered pituitary-adrenal axis responses to provocative challenge tests in adult survivors of childhood abuse. Am J Psychiatry. 2001;158:575–81.
Heim C, Newport DJ, Mletzko T, Miller AH, Nemeroff CB. The link between childhood trauma and depression: insights from HPA axis studies in humans. Psychoneuroendocrinology. 2008;33(6):693–710.
Heim C, et al. Lower CSF oxytocin concentrations in women with a history of childhood abuse. Mol Psychiatry. 2009;14(10):954–8.
Heinrichs M, Baumgartner T, Kirschbaum C, Ehlert U. Social support and oxytocin interact to suppress cortisol and subjective responses to psychosocial stress. Biol Psychiatry. 2003;54(12):1389–98.
Henckens MJ, van Wingen GA, Joels M, Fernandez G. Corticosteroid induced decoupling of the amygdala in men. Cereb Cortex. 2012;22:2336–45.
Henry JP. Psychological and physiological responses to stress: the right hemisphere and the hypothalamo-pituitary-adrenal axis, an inquiry into problems of human bonding. Integr Physiol Behav Sci. 1993;28(4):369–87.
Herman JL. Complex PTSD: a syndrome in survivors of prolonged and repeated trauma. J Trauma Stress. 1992;5(3):377–91.
Herman JP, Ostrander MM, Mueller NK, Figueiredo H. Limbic system mechanisms of stress regulation: hypothalamo-pituitary-adrenocortical axis. Prog Neuro-Psychopharmacol Biol Psychiatry. 2005;29:1201–13.
Hermans EJ, van Marle HJ, Ossewaarde L, Henckens MJ, Qin S, van Kesteren MT, et al. Stress-related noradrenergic activity prompts large-scale neural network reconfiguration. Science. 2011;334(6059):1151–3.
Hofer MA. The psychobiology of early attachment. Clin Neurosci Res. 2005;4(5):291–300.
Hofer MA. Psychobiological roots of early attachment. Curr Dir Psychol Sci. 2006;15(2):84–8.
Jurist EL, Slade AE, Bergner SE. Mind to mind: infant research, neuroscience, and psychoanalysis. New York: Other Press; 2008.
Kaffman A, Meaney MJ. Neurodevelopmental sequelae of postnatal maternal care in rodents: clinical and research implications of molecular insights. J Child Psychol Psychiatry. 2007;48(3-4):224–44.
Kendler KS, Kuhn JW, Prescott CA. Childhood sexual abuse, stressful life events and risk for major depression in women. Psychol Med. 2004;34(8):1475–82.
Kernberg O. The structural diagnosis of borderline personality organization. In: Borderline personality disorders: the concept, the syndrome, the patient. New York: International Universities Press; 1977. p. 87–121.
Kestenberg JS. Psychoanalyses of children of survivors from the Holocaust: case presentations and assessment. J Am Psychoanal Assoc. 1980;28(4):775–804.
Kiem SA, Andrade KC, Spoormaker VI, Holsboer F, Czisch M, Samann PG. Resting state functional MRI connectivity predicts hypothalamus-pituitary-axis status in healthy males. Psychoneuroendocrinology. 2013;38:1338–48.
Klaassens ER, et al. Effects of childhood trauma on HPA-axis reactivity in women free of lifetime psychopathology. Prog Neuro-Psychopharmacol Biol Psychiatry. 2009;33(5):889–94.

Kogan I. The cry of mute children: a psychoanalytic perspective of the second generation of the Holocaust. London: Free Association Books; 1995.
Krystal H, editor. Massive psychic trauma. New York: International Universities Press; 1968.
Labuschagne I, Phan KL, Wood A, Angstadt M, Chua P, Heinrichs M, Stout JC, Nathan PJ. Medial frontal hyperactivity to sad faces in generalized social anxiety disorder and modulation by oxytocin. Int J Neuropsychopharmacol. 2011;14:1–14.
Lanius RA, Brand B, Vermetten E, Frewen PA, Spiegel D. The dissociative subtype of posttraumatic stress disorder: rationale, clinical and neurobiological evidence, and implications. Depress Anxiety. 2012;29(8):701–8.
Lanius RA, Vermetten E, Loewenstein RJ, Brand B, Schmahl C, Bremner JD, Spiegel D. Emotion modulation in PTSD: clinical and neurobiological evidence for a dissociative subtype. Am J Psychiatr. 2010a;167(6):640–7.
Lanius RA, Vermetten E, Pain C, editors. The impact of early life trauma on health and disease: the hidden epidemic. Cambridge: Cambridge University Press; 2010b.
Laub D. Truth and testimony: the process and the struggle. In: Trauma: explorations in memory. Baltimore: John Hopkins Press; 1995. p. 63.
Laub D, Auerhahn N. Knowing and not knowing. In: Psychoanalysis and holocaust testimony: unwanted memories of social trauma. Baltimore: John Hopkins Press; 2017. p. 32.
Laub D, Auerhahn NC. Failed empathyDOUBLEHYPHENa central theme in the survivor's holocaust experience. Psychoanal Psychol. 1989;6(4):377.
Laub D, Finchelstein F. Memory and history from past to future: a dialogue with Dori Laub on trauma and testimony. In: Memory and the future. London: Palgrave Macmillan; 2010. p. 50–65.
Laub D, Lee S. Thanatos and massive psychic trauma: the impact of the death instinct on knowing, remembering, and forgetting. J Am Psychoanal Assoc. 2002;51(2):433–64.
LeDoux JE. Emotion circuits in the brain. Ann Rev Neurosci. 2000;23(1):155–84.
Lifton RJ. Death in life: survivors of Hiroshima. Chapel Hill: University of North Carolina Press; 2012.
Lingiardi V, McWilliams N, editors. Psychodynamic diagnostic manual: PDM-2. New York: Guilford Publications; 2017.
Lingiardi V, Mucci C. Da Janet a Bromberg, Passando per Ferenczi. Psichiatr Psicoterapia. 2014;33(1):41–62.
Linnen AM, Ellenbogen MA, Cardoso C, Joober R. Intranasal oxytocin and salivary cortisol concentrations during social rejection in university students. Stress. 2012;15(4):393–402.
Liotti G. Disorganized/disoriented attachment in the etiology of the dissociative disorders. Dissociation. 1992a;5:196–204.
Liotti G. (1992b) Disorganizzazione dell'attaccamento e predisposizione allo sviluppo di disturbi funzionali della coscienza. In: Ammaniti M, Stern DN (a cura di) Attaccamento e psicoanalisi. Laterza: Bari, 219-232.
Liotti M, Panksepp J. Imaging human emotions and affective feelings: implications for biological psychiatry. In: Textbook of Biological Psychiatry. Hoboken: Wiley; 2004. p. 33–74.
Lupien SJ, McEwen BS, Gunnar MR, Heim C. Effects of stress throughout the lifespan on the brain, behaviour and cognition. Nat Rev Neurosci. 2009;10(6):434–45.
Lyssenko L, Schmahl C, Bockhacker L, Vonderlin R, Bohus M, Kleindienst N. Dissociation in psychiatric disorders: a meta-analysis of studies using the dissociative experiences scale. Am J Psychiatr. 2017;175:37–46.
MacDonald E, Dadds MR, Brennan JL, Williams K, Levy F, Cauchi AJ. A review of safety, side-effects and subjective reactions to intranasal oxytocin in human research. Psychoneuroendocrinology. 2011;36:1114–26.
Main M, Hesse E. Parents' unresolved traumatic experiences are related to infant disorganized attachment status: is frightened and/or frightening parental behavior the linking mechanism? In: Greenberg MT, Cicchetti D, Cummings M, editors. Attachment in the preschool years: theory, research and intervention. Chicago: University of Chicago Press; 1990. p. 161–82.

van Marle HJ, Hermans EJ, Qin S, Fernandez G. Enhanced resting-state connectivity of amygdala in the immediate aftermath of acute psychological stress. NeuroImage. 2010;53:348–54.
McEwen BS. Stress, adaptation, and disease - allostasis and allostatic load. Ann N Y Acad Sci. 1998;840:33–44.
McEwen BS. Brain on stress: how the social environment gets under the skin. Proc Natl Acad Sci U S A. 2012;109(Suppl 2):17180–5.
Meinlschmidt G, Heim C. Sensitivity to intranasal oxytocin in adult men with early parental separation. Biol Psychiatry. 2007;61(9):1109–11.
Miller GE, Chen E, Zhou ES. If it goes up, must it come down? Chronic stress and the hypothalamic-pituitary-adrenocortical axis in humans. Psychol Bull. 2007;133(1):25–45.
Mucci C. Il dolore estremo: il trauma da Freud alla Shoah. Borla. 2008.
Mucci C. Beyond individual and collective trauma: intergenerational transmission, psychoanalytic treatment, and the dynamics of forgiveness. London: Karnac; 2013.
Mucci C. Trauma, healing and the reconstruction of truth. Am J Psychoanal. 2014;74(1):31–47.
Mucci C. Attachment trauma, intergenerational transmission and resilience: What an AAI on a Second generation Shoah survivor may teach us. Poster presented at the 7th International Attachment Conference, (IAC), The New School, New York, August 6–8, 2015.
Mucci C. Implicit memory, unrepressed unconscious, and trauma theory: the turn of the screw between contemporary psychoanalysis and neuroscience. In: Craparo G, Mucci C, editors. Unrepressed unconscious, implicit memory and clinical work. London: Karnac; 2016.
Mucci C. Ferenczi's revolutionary therapeutic approach. Am J Psychoanal. 2017;77(3):239–54.
Mucci C. Borderline bodies. Affect regulation therapy for personality disorders. New York, NY: W. W. Norton & Company; 2018.
Nakao T, Matsumoto T, Morita M, Shimizu D, Yoshimura S, Northoff G, Yamawaki S. The degree of early life stress predicts decreased medial prefrontal activations and the shift from internally to externally guided decision making: an exploratory NIRS study during resting state and self-oriented task. Front Hum Neurosci. 2013;7:339.
Northoff G, Heinzel A, De Greck M, Bermpohl F, Dobrowolny H, Panksepp J. Self-referential processing in our brain—a meta-analysis of imaging studies on the self. NeuroImage. 2006;31(1):440–57.
Northoff G, Panksepp J. The trans-species concept of self and the subcortical–cortical midline system. Trends Cogn Sci. 2008;12(7):259–64.
Northoff G, Bermpohl F. Cortical midline structures and the self. Trends Cogn Sci. 2004;8(3):102–7.
Northoff G, Schneider F, Rotte M, Matthiae C, Tempelmann C, Wiebking C, Bermpohl F, Heinzel A, Danos P, Heinze H-J, Bogerts B, Walter M, Panksepp J. Differential parametric modulation of self-relatedness and emotions in different brain regions. Hum Brain Mapp. 2009;30(2):369–82.
Olff M, Frijling JL, Kubzansky LD, Bradley B, Ellenbogen MA, Cardoso C, Bartz JA, Yee JR, van Zuiden M. The role of oxytocin in social bonding, stress regulation and mental health: an update on the moderating effects of context and interindividual differences. Psychoneuroendocrinology. 2013;38:1883–94.
Pandya DN, Yeterian EH. Prefrontal cortex in relation to other cortical areas in rhesus monkey: architecture and connections. Prog Brain Res. 1991;85:63–94.
Pechtel P, Pizzagalli DA. Effects of early life stress on cognitive and affective function: an integrated review of human literature. Psychopharmacology. 2011;214:55–70.
Perry BD. Maltreated children: experience, brain development and the next generation. New York: Norton; 1999.
Perry BD, Pollard R. Altered brain development following global neglect in early childhood. Proceedings from the Annual Meeting of the Society for Neuroscience, New Orleans; 1997.
Pincus AL, Lukowitsky MR. Pathological narcissism and narcissistic personality disorder. Ann Rev Clin Psychol. 2010;6:421–46.
Plotsky PM, et al. Long-term consequences of neonatal rearing on central corticotropin-releasing factor systems in adult male rat offspring. Neuropsychopharmacology. 2005;30(12):2192–204.
Porges SW. The polyvagal theory: neurophysiological foundations of emotions, attachment, communication, and self-regulation, Norton series on interpersonal neurobiology. New York: Norton; 2011.

Pryce CR, et al. Long-term effects of early-life environmental manipulations in rodents and primates: Potential animal models in depression research. Neurosci Biobehav Rev. 2005;29(4-5):649–74.
Quirin M, Kuhl J, Dusing R. Oxytocin buffers cortisol responses to stress in individuals with impaired emotion regulation abilities. Psychoneuroendocrinology. 2011;36(6):898–904.
Rinne T, De Kloet ER, Wouters L, Goekoop JG, DeRijk RH, van den Brink W. Hyperresponsiveness of hypothalamic-pituitaryadrenal axis to combined dexamethasone/corticotropin-releasing hormone challenge in female borderline personality disorder subjects with a history of sustained childhood abuse. Biol Psychiatry. 2002;52(11):1102–12.
Roelofs K, Keijsers GP, Hoogduin KA, Näring GW, Moene FC. Childhood abuse in patients with conversion disorder. Am J Psychiatr. 2002;159(11):1908–13.
Sagi-Schwartz A, van IJzendoorn MH, Bakermans-Kranenburg MJ. Does intergenerational transmission of trauma skip a generation? No meta-analytic evidence for tertiary traumatization with third generation of Holocaust survivors. Attach Hum Dev. 2008;10(2):105–21.
Saltzman KM, Weems CF, Carrion VG. IQ and posttraumatic stress symptoms in children exposed to interpersonal violence. Child Psychiatry Hum Dev. 2006;36(3):261–72.
Sanchez MM, Ladd CO, Plotsky PM. Early adverse experience as a developmental risk factor for later psychopathology: evidence from rodent and primate models. Dev Psychopathol. 2001;13(3):419–49.
Sasso G. The development of consciousness: an integrative model of child development, neuroscience and psychoanalysis. London: Karnac Books; 2007.
Scalabrini A, Cavicchioli M, Fossati A, Maffei C. The extent of dissociation in borderline personality disorder: a meta-analytic review. J Trauma Dissociation. 2017;18(4):522–43.
Schmahl CG, Vermetten E, Elzinga BM, Bremner JD. Magnetic resonance imaging of hippocampal and amygdala volume in women with childhood abuse and borderline personality disorder. Psychiatry Res Neuroimaging. 2003;122(3):193–8.
Schore AN. Affect regulation and the origin of the self: the neurobiology of emotional development. London: Psychology Press; 1994.
Schore AN. Early organization of the non-linear right brain and development of a predisposition to psychiatric disorders. Dev Psychopathol. 1997;9:595–631.
Schore AN. Attachment and the regulation of the right brain. Attach Hum Dev. 2000;2(1):23–47.
Schore AN. Dysregulation of the right brain: a fundamental mechanism of traumatic attachment and the psychopathogenesis of posttraumatic stress disorder. Aust N Z J Psychiatry. 2002;36(1):9–30.
Schore AN. Affect dysregulation and disorders of the self, Norton series on interpersonal neurobiology. New York: Norton; 2003a.
Schore AN. Affect regulation and the repair of the self, Norton series on interpersonal neurobiology. New York: Norton; 2003b.
Schore AN. Relational trauma and the developing right brain. Ann N Y Acad Sci. 2009; 1159(1):189–203.
Schore AN. Relational trauma and the developing right brain: the neurobiology of broken attachment bonds. In: Baradon T, editor. Relational trauma in infancy: psychoanalytic, attachment and neurocontributions to parent–infant psychotherapy. London: Routledge; 2010. p. 19–47.
Schore AN. The science of the art of psychotherapy, Norton series on interpersonal neurobiology. New York: Norton; 2012.
Schore AN. Relational trauma, brain development, and dissociation. In: Treating complex traumatic stress disorders in children and adolescents: scientific foundations and therapeutic models. New York: The Guilford Press; 2013. p. 3–23.
Schwartz ED, Perry BD. The post-traumatic response in children and adolescents. Psychiatr Clin North Am. 1994;17:311–26.
Shin LM, Orr SP, Carson MA, Rauch SL, Macklin ML, Lasko NB, et al. Regional cerebral blood flow in the amygdala and medial prefrontalcortex during traumatic imagery in male and female vietnam veterans with ptsd. Arch Gen Psychiatry. 2004;61(2):168–76.
Siegel DJ. The developing mind, vol. 296. New York: Guilford Press; 1999.
Simeon D, Bartz J, Hamilton H, Crystal S, Braun A, Ketay S, Hollander E. Oxytocin administration attenuates stress reactivity in borderline personality disorder: a pilot study. Psychoneuroendocrinology. 2011;36:1418–21.

Solomon J, George C. The place of disorganization in attachment theory: linking classic observations with contemporary findings. In: Attachment disorganization. New York: Guilford Press; 1999.

Stein MB, Koverola C, Hanna C, Torchia MG, McClarity B. Hippocampal volume in women victimised by childhood sexual abuse. Psychol Med. 1997;27(4):635–56.

Sterlemann V, Ganea K, Liebl C, Harbich D, Alam S, Holsboer F, et al. Long-term behavioral and neuroendocrine alterations following chronic social stress in mice: implications for stress-related disorders. Horm Behav. 2008;53(2):386–94.

Teicher M. Wounds that time won't heal: the neurobiology of child abuse. Cerebrum. 2000;2(4):50–67.

Ulrich-Lai YM, Herman JP. Neural regulation of endocrine and autonomic stress responses. Nat Rev Neurosci. 2009;10:397–409.

Vaisvaser S, Lin T, Admon R, Podlipsky I, Greenman Y, Stern N, et al. Neural traces of stress: cortisol related sustained enhancement of amygdala-hippocampal functional connectivity. Front Hum Neurosci. 2013;7:313.

Van der Kolk BA. The assessment and treatment of complex PTSD. In: Treating trauma survivors with PTSD. Washington, D.C.: American Psychiatric Publishing; 2002. p. 127–56.

Van der Kolk BA. The body keeps the score. Brain, mind, and body in the healing of trauma. New York: Penguin Books; 2014.

Van der Kolk BA, d'Andrea W. Towards a developmental trauma disorder diagnosis for childhood interpersonal trauma. In: The impact of early life trauma on health and disease: the hidden epidemic. Cambridge: Cambridge University Press; 2010. p. 57–68.

Van der Kolk BA, Perry JC, Herman JL. Childhood origins of self-destructive behavior. Am J Psychiatry. 1991;148:1665–71.

Veer IM, Oei NY, Spinhoven P, van Buchem MA, Elzinga BM, Rombouts SA. Beyond acute social stress: increased functional connectivity between amygdala and cortical midline structures. NeuroImage. 2011;57:1534–41.

Vythilingam M, et al. Childhood trauma associated with smaller hippocampal volume in women with major depression. Am J Psychiatry. 2002;159(12):2072–80.

Winslow JT, Noble PL, Lyons CK, Sterk SM, Insel TR. Rearing effects on cerebrospinal fluid oxytocin concentration and social buffering in rhesus monkeys. Neuropsychopharmacology. 2003;28(5):910–8.

Yehuda R, Engel SM, Brand SR, Seckl J, Marcus SM, Berkowitz GS. Transgenerational effects of posttraumatic stress disorder in babies of mothers exposed to the World Trade Center attacks during pregnancy. J Clin Endocrinol Metabol. 2005;90(7):4115–8.

Yehuda R, Golier JA, Tischler L, Harvey PD, Newmark R, Yang RK, Buchsbaum MS. Hippocampal volume in aging combat veterans with and without post-traumatic stress disorder: relation to risk and resilience factors. J Psychiatr Res. 2007;41(5):435–45.

Yehuda R, Halligan SL, Grossman R. Childhood trauma and risk for PTSD: relationship to intergenerational effects of trauma, parental PTSD, and cortisol excretion. Dev Psychopathol. 2001;13(3):733–53.

Yehuda R, McFarlane AC. Conflict between current knowledge about posttraumatic stress disorder and its original conceptual basis. Am J Psychiatr. 1995;152(12):1705–13.

Yehuda R, Teicher MH, Trestman RL, Levengood RA, Siever LJ. Cortisol regulation in posttraumatic stress disorder and major depression: a chronobiological analysis. Biol Psychiatry. 1996;40(2):79–88.

Zanarini MC, Gunderson JG, Marino MF, Schwartz EO, Frankenburg FR. Childhood experiences of borderline patients. Compr Psychiatry. 1989;30:18–25.

Zanarini MC, Williams AA, Lewis RE, et al. Reported pathological childhood experiences associated with the development of borderline personality disorder. Am J Psychiatry. 1997;154:1101–6.

# Personality and Personality Disorders 18

Andrea Scalabrini, Clara Mucci, and Georg Northoff

**Abstract**

Personality is a complex construct that is closely related to the subjective experiences between internal world and external reality, a sense of subjectivity that is referred to the concept of self in psychiatry and psychodynamic literature.

The self-referential activity of the brain is also central in neuroscientific studies; indeed the intertwining of scientific advances in various areas from neuroscience to infant research studying the interactive regulatory processes between the caregiver and the baby is facilitating the integration of various forms of knowledge about the complex construct of personality and self.

Moreover during last years the DSM-5 (2013) in the attempt to ameliorate the assessment of personality disorders (PDs) maintained the categorical personality model contained in DSM-4 but also added an alternative model for PDs, which emphasizes specifically the construct of self and interpersonal relatedness, also considered core aspects for the psychodynamic assessment of personality.

In the first section of this chapter, we attempt to highlight the psychodynamic and neural understanding of different categorical PDs (as listed in DSM-5), while in the second section, we propose a neuropsychodynamic model based on the rest-self overlap (or containment) and rest simulus interaction that can shed a novel light on the multidisciplinary study of the *self* and *interpersonal relatedness* considering the altered brain activity in regard to different attachment

A. Scalabrini · C. Mucci
Department of Psychological, Health and Territorial Sciences (DiSPuTer), G. d'Annunzio University of Chieti-Pescara, Chieti, CH, Italy

G. Northoff (✉)
Mind, Brain Imaging, and Neuroethics,
Institute of Mental Health Research, University of Ottawa, Ottawa, ON, Canada
e-mail: georg.northoff@theroyal.ca; www.georgnorthoff.com



H. Boeker et al. (eds.), *Neuropsychodynamic Psychiatry*,
https://doi.org/10.1007/978-3-319-75112-2_18

histories and personality organizations. The aim of this project is to create a bridge between the different organizations and structures of personality and their neurobiological underpinnings.

## 18.1 Introduction

Personality is a broad and complex construct that is difficult to conceptualize, and it is considerably more than what it is possible to observe if we consider only behaviors and symptoms. How can we approach personality though?

Personality is more related to what one is and to what one experiences consciously and unconsciously in relation to the internal world and to the external reality, a sense of subjectivity that we can call self and/or identity.

This emphasis on the subjective experience and on the balance between the internal and external experiences caused many discussions during the last years regarding how to scientifically propose a model for personality disorders (PDs) in DSM-5.

Indeed at the beginning of the DSM-5 revision process, it was widely expected that personality disorders would have a dimensional component, but during the closing weeks, it was decided that the categorical personality model contained in DSM-4 would be reprinted in DSM-5 with no essential changes in the criteria (Zachar et al. 2016).

Thus the DSM-5 contains a hybrid model of categories and dimensions, which is extended in Section III "Emerging Measures and Models" and proposed as the alternative model for PDs (AMPD).

Here in this chapter our attempt is to propose a neuropsychodynamic model of personality and personality disorders considering the centrality of the self for DSM-oriented assessment, for psychodynamic understanding, and for neuroscience. We will first focus on the categorigal classification of personality disorders of the DSM-5, emphasizing the psychodynamic aspects that are characteristic of each disorder. In the subsequent section, we will underline the necessity of a neuropsychodynamic model for personality disorders starting from the novel emphasis given to the construct of self and relatedness in clinical psychology, in the long-standing psychodynamic organization of personality, and last but not least in neuroscience and research on the brain structure and functionality.

## 18.2 Categorical Classification of Personality Disorder

DSM-5 conceptualized a personality disorder as an enduring pattern of inner experience and behavior that deviates markedly from the expectations of the individual's culture, which is pervasive and inflexible, which has an onset in adolescence or early adulthood, and which is stable over time, and leads to distress or impairment.

Personality disorders are grouped into three clusters based on descriptive similarities. Cluster A includes paranoid, schizoid, and schizotypal personality disorders. Individuals with these disorders often appear odd or eccentric. Cluster B includes antisocial, borderline, histrionic, and narcissistic personality disorders. Individuals with these disorders often appear dramatic, emotional, or erratic. Cluster C includes avoidant, dependent, and obsessive-compulsive personality disorders. Individuals with these disorders often appear anxious or fearful. It should be noted that this clustering system, although useful in some research and educational situations, has serious limitations and has not been consistently validated (American Psychiatric Association 2013).

## 18.3 Cluster A

### 18.3.1 Paranoid Personality Disorder (PPD)

**Definition** Paranoid personality disorder represents a pattern of distrust and suspiciousness such that others' motives are interpreted as malevolent.

#### 18.3.1.1 Psychodynamic Aspects

The lack of trust in others seems to be the core feature of paranoid personality disorder (PPD), and, following Kernberg's classification, this disorder takes place in the low-level borderline personality organization. They occupy the introjective, self-definition end of Blatt's continuum from relatedness to self-definition (Blatt and Blass 1992).

Primitive and persecutory introjects determine the presence of an accusatory and violent moral instance (primitive and harsh superego). The individual cannot tolerate the bad aspects of the self; thus the bad characteristics are split and projected onto others. The PPD individual works to make his or her projection "fit" the projective target.

Along with an image of self and of other devalued and damaged, there is another one of the opposite polarity, grandiose and violent. For the subject it is impossible to integrate the dyads and ends up in attributing the negative aspects to external figures, perceiving himself as a victim of a persecutor.

The key defense mechanisms are splitting and projective identification; representations of self and others are "totally good" and not integrated with "totally bad" characteristic or vice versa.

For the paranoid individual, anyone is a potential enemy, and everyone is watched with distrust, waiting for a false step to confirm their representation of reality. It gives rise to circumspection, coldness, tension, and control; reality testing is maintained even though there may be psychotic episodes.

From a psychodynamic point of view, the central problem of the paranoid subject is therefore the inability to deal with depressive feelings caused by the awareness of the coexistence, both in them and in others of good parts and bad parts. This scenario can be either caused by an excessive degree of aggressiveness as well as an empathically deficient maternal environment that does not provide a sufficient management of depressive anxiety and the ability to worry about the others.

From an affective point of view, paranoid personality is characterized by the feeling of anger and desire for revenge but also suffers from overwhelming fears as a combination of shame and fear. McWilliams (2011) underlines that shame is considered a threat for both paranoid personality disorder and narcissistic personality disorder. The narcissistic personality is ashamed if certain aspects of the self (fragile and devaluated) are exposed to others, so that all the efforts are in the direction to impress the others so that they cannot find out. PPD individuals instead spend their own energies to overturn the intention of the other persons (according to their mental representation) to humiliate them. An obstacle in the therapy is that the therapist as well can be considered as a bad and a persecutory object.

To our knowledge there are no published laboratory studies that investigated specifically at PPD. Instead, such studies have been done in samples that include patients with Schizophrenia spectrum disorder and/or other personality disorders, which have included patients with PPD.

It has been highlighted that cognitive impairment similar to, but usually less severe than, the deficits found in schizophrenia is one of the hallmarks of the prototypical cluster A disorder, SZPD (Siever et al. 2002). No studies have examined these parameters in PPD itself, but degradations in information acquisition and processing appear to be a risk factor for dimensional paranoia.

### 18.3.2 Schizoid Personality Disorder (SPD)

**Definition** Schizoid personality disorder indicates a pattern of detachment from social relationships and a restricted range of emotional expression.

#### 18.3.2.1 Psychodynamic Aspects

The main problem in schizoid individuals is a deficit in the ability to relate with the others. Bowlby (1980) and Winnicott (1960) propose that the etiology of the disorder is given by a severe deprivation in a context of early attachment, particularly in regard to the avoidance of the attachment and maternal care severely inadequate.

From this early relational scenario comes the choice of retreat (from emotion, from affects, from feelings) and of the *false self* as protection for the *true self*. This results in a split of the self in unintegrated images: internally the subject is sensitive and emotionally dependent, while externally he/she appears self-sufficient and distant. According to Winnicott (1960) it is possible to hypothesize a sense of inauthenticity of schizoid life: "if the mother/caregiver failed in giving what he/she needed no one else would succeed." Their isolated superiority could result from the defense of possible relational delusions but also by their insatiability, moreover may have its origin in rejecting the incursions of an overcontrolling or overintrusive other (caregiver).

On the other side Klein (1946) believes that the schizoid defense is used to avoid the persecutory anxiety that derives from the projection of the own aggression onto objects. The schizoid defends him/herself from the anxiety of destroying the loved object by withdrawal. The characteristic defense mechanisms of the schizoid individual are splitting retreat, autistic fantasy. Once they have constructed their own fantasy world, they do not seem to be shaken to get gratification or even commit themselves to avoid punishment. In this regard Clarkin et al. (1997) argue that there is a deficit in the capacity to discriminate pleasure and pain.

They may appear notably detached, or they may behave in a socially appropriate way while privately attending more to their inner world than to the surrounding world of human beings (Fairbairn 1952).

As suggested by McWilliams (2011), working with schizoid patients implies that the therapist is open enough to a certain degree of authenticity and a level of awareness of emotions and imagery. It also requires to leave a certain degree of emotional space in order not to repeat the role of the overintrusive or abandoning object.

### 18.3.3 Schizotypal Personality Disorder (SZDP)

**Definition** Schizotypal personality disorder represents a pattern of acute discomfort in close relationships, cognitive or perceptual distortions, and eccentricities of behavior.

#### 18.3.3.1 Psychodynamic Aspects

The schizotypal personality disorder (SZDP) individual appears strange and eccentric and, in addition, psychotic features and symptoms are emphasized in comparison with schizoid personality. There are problems in the ideation and in communication, detachment from the consensually accepted representation of reality, ideas of reference, and unusual perceptual experiences.

Schizotypal individual uses a very particular language, often incomprehensible to the listener, characterized by digressions, dizzying, loosening associative links, and inconsistency. As with schizoid disorder, schizotypal disorder involves a temperamental genesis and early relational deficits. Choice of retreat in an imaginary world can be considered as a solution to the conflict.

This disorder belongs to the dimension of schizotypy and shares with schizophrenia genetic-phenomenological characteristics in response to treatment.

Psychodynamically, in line with Psychodynamic Diagnostic Manual 2 (PDM-2), schizotypy is considered as a trait common in schizoid psychologies (the most severe trait in a continuum) and not as a type of personality. Indeed, Westen and colleagues (2012) empirically identified a grouping of patients in a clinical sample that they labeled "schizoid-schizotypal," characterized by "pervasive impoverishments, and peculiarities in, interpersonal relationships, emotional experience, and thought processes" (p. 280).

Cognitive deficits, together with social/interpersonal and affective processes impairments, represent functionally, clinically, and neurobiologically significant manifestations of SZDP and schizotypy in general.

Prefrontal-dependent cognitive processes, specifically working memory and context processing, have been studied in SZDP. The intense focus on cognitive dysfunction in SZDP has been largely driven by the critical role of working memory impairments on functional outcomes in schizophrenia.

The caudate, which receives input primarily from dorsolateral prefrontal regions and has been implicated in higher-order cognitive processes, appears to be involved in the cognitive abnormalities of SZDP. Specifically, caudate volumes have been shown to be smaller in patients with SZDP, and greater volume reduction was related to poorer cognitive performance (Levitt et al. 2002); in addition it has been found that greater aberrant morphology of the right caudate in patients with SPD was also related to cognitive impairment (Levitt et al. 2004, 2009).

Functional imaging studies have also begun to reveal the neural substrates associated with working memory impairments: patients with SPD, compared to healthy control participants, showed attenuated working memory-associated activation of the left ventral prefrontal cortex, superior frontal gyrus, intraparietal cortex, and posterior inferior gyrus (Koenigsberg et al. 2005). In another study, activation of the left posterior cingulate gyrus and deactivation of the superior temporal gyrus, insula, and middle frontal gyrus were both attenuated during a working memory task in patients with SPD compared to healthy controls (Vu et al. 2013).

In regard to the neural underpinnings of social/interpersonal and affective processes, brain imaging studies have recently showed that schizotypal patients exhibit exaggerated habituation of amygdala response to affectively valenced social visual stimuli compared to healthy control participants (Hazlett et al. 2012). Premkumar and colleagues have characterized differences in neural activity patterns in response to social rejection in low- vs. high-schizotypy, nonclinical participants. Specifically, they found a significant difference in response to social rejection in the dorsal anterior cingulate cortex (dACC), right superior frontal gyrus, and left ventral prefrontal cortex in the low- vs. high-schizotypy group (Premkumar et al. 2012).

## 18.4 Cluster B

### 18.4.1 Antisocial Personality Disorder (ASPD)

**Definition** Antisocial personality disorder indicates a pattern of disregard for, and violation of, the rights of others.

#### 18.4.1.1 Psychodynamic Aspects

Antisocial individuals seem to live in a pre-social affective world where emotions and feelings are experiences exclusively in relation to oneself but not to others. Antisocial subjects do not experience gratitude and remorse and often feel anger and contempt; as narcissistic subjects they can assume devaluing and denigrating attitudes toward others, but in addition they inflict consciously psychic or physical pain.

Many authors emphasize the disinhibition features as a key component of this disorder (e.g., Siever and Davis 1991). In addition we may find early relation and familiar histories of very severe abuse and deprivation. Often the attachment is totally compromised, and the child perceives the parent as a stranger, bad, and unworthy of trust and builds up a grandiose self that can give him some sense of self-sufficiency; it can be considered a precursor of the omnipotent control. Relationally they can refuse any affective bond, or they can attempt to bind with others through manipulation of power and destruction

These individuals are characterized by lack of empathy and humanity, inability to see others as individuals with their own feelings and needs. There is no remorse for what their injurious actions produce on other people, and there is no possibility to have feeling about that. It needs to be taken into account the superego, whose compromise is at the basis of the moral deficit.

The antisocial subject disclaims all responsibility and attributes to the outside world all the problems and responsibility. According to Winnicott, antisocial personality develops around serious maternal mirroring failures and can be paradoxically interpreted as a hope of obtaining what has not happened and what has not offered to them. Kernberg (1984) places ASPD at the low level of borderline personality organization at the extreme pole of a narcissistic continuum, before the so-called malignant narcissism, constituted by individuals with a narcissistic personality disorder characterized in addition by paranoid and antisocial features and ego-syntonic aggressiveness.

The prognosis for these individuals is very scarce, and, as proposed by PDM-2, any therapeutic influence are better if antisocial or psychopath individual has reached midlife or later and thus felt a decline in physical power and encountered limits to omnipotent strivings.

The experimental literature defined two subgroups, one constituted by antisocial reactive individuals and the other constituted by antisocial psychopathic individual. The review written by Blair (2010) reveals that individuals who present with an increased risk for reactive, but not instrumental, aggression show increased amygdala responses to emotionally evocative stimuli. This suggests that such individuals are primed to respond strongly to an inappropriate extent to threatening or frustrating events. In contrast, individuals with psychopathic tendencies show decreased amygdala and orbitofrontal cortex responses to emotionally provocative stimuli or during emotional learning paradigms. This suggests that such individuals face difficulties with basic forms of emotional learning and decision-making.

fMRI research also strongly support the hypothesis that amygdala and orbitofrontal cortex (OFC) functioning is disrupted in individuals with psychopathic tendencies and that neural circuit is strongly relevant for the moral system (Blair, 2007). Other systems may also be affected, but there is a paucity of studies.

For example a recent resting-state fMRI study (Tang et al. 2013) showed abnormal resting-state connectivity in ASPD patient in default mode network (DMN) and the attention network.

Interestingly decreased functioning of the DMN may manifest as difficulties in self-relatedness functions such as adaptively regulating emotions, future planning, or self-inspection.

Individuals with ASPD demonstrated decreased functional connectivity between regions of the default mode and attention networks. Moreover decreased functioning between the attention network, together with the DMN, with cerebellar network may result in deficits of transmission in the implementation of cognitive control and self-regulation. These results could be interpreted as inefficient transmission between DMN, which detects conflict, and the attention network, which implements increased cognitive control to resolve conflict in future trials.

A person with ASPD, characterized by these resting-state features may act impulsively and inappropriately, resulting in antisocial behaviors.

### 18.4.2 Borderline Personality Disorder (BPD)

**Definition** Borderline personality disorder is a pattern of instability in interpersonal relationships, self-image, and affects and marked impulsivity.

#### 18.4.2.1 Psychodynamic Aspects

There are several psychodynamic theorizations of the borderline personality disorder. Kernberg (1975) focuses on the rapprochement phase (18–24 months) as theorized by Mahler (1971), during which the child, after moving away from his/her mother and starting to acquire self-consciousness as a separate entity and after exploring the environment and enjoying the skills concerning the new acquired autonomy, returns to the caregiver driven by the fear of losing her but doesn't find the response he/she expected because of the ambivalence of the mother-caregiver. It results in confusion at the identity level, extremely vulnerable to separation events, unable to tolerate loneliness and to live without the anxiety connected to the relational distance or to the relational intimacy.

According to Kernberg, these individuals are unable, due to the innate quote of aggressiveness, to integrate into a single representation the "totally good" and the "totally bad" images of the self and of the object, to modulate their affects, or sublimate their impulses, to plan and to finalize their behavior. At a conflict level, the child would not be able to cope with the anxiety and guilt associated with such integration that endangers the self and object representations.

There is a severe oscillation between the times when the subject feels good and surrounded by good people and others where he considers himself worthless and frustrating (identity diffusion, feeling of emptiness, suicidal impulses) and tends to see who is close to him as wicked, enemy, and abandoning (rage and hostility).

This way of functioning is supported by defensive mechanisms such as splitting and projective identification. Theories focusing primarily on relational deficit, trauma, and attachment (Fonagy and Target, 1997) show that most of BPD patients have insecure attachment styles associated with traumatic experiences; specifically scholars have identified a relevant disorganized/disoriented or "type D" insecure attachment style (e.g., Liotti 2004; Holmes 2014; Main and Solomon 1986).

This attachment pattern results in chronic and long-term difficulties in tolerance and affects regulation (Fonagy et al. 1995) with compromised mentalization ability (to recognize internal states in self and others that underlie behaviors) and inability to experience continuity of the self and others (Bromberg 2000; Chefetz 2015; Meares 2012). These relational features are particularly active with people who trigger the attachment system (therapist included), causing them to be treated with confusing state of minds and dissociative states in case of BPD individuals with history of human trauma (Mucci 2013, 2017; Scalabrini et al. 2017a).

In regard to etiology, there is evidence for a genetic vulnerability (Kernberg and Caligor 2005; Paris 1993; Siever and Davis 1991; Stone 1980; Torgersen 2000), for origins in an early attachment disorder (Guidano and Liotti 1983), for developmental arrest (Bateman and Fonagy 2004; Fonagy and Target 2002; Masterson 2013), and for the effects of severe relational trauma (Meares 2012; Mucci 2013).

The relative weight of each of these factors varies from person to person. It's quiet clear that within the same BPD levels we have a large variability of different types of BPD patients. These individuals are notoriously difficult patients, partly because they may challenge ordinary therapeutic limits and evoke intense countertransference reactions and partly because they require specific treatment models (e.g., as mentalization-based treatment; transference-focused psychotherapy, or dialectical behavior therapy).

Supporting the psychodynamic understanding of BPD, fMRI research reported consistent results over years regarding the affective system and the processing of anxiety in BPD; particularly it has been showed how BPD, compared with healthy control, is characterized by increased amygdala activation when viewing aversive emotion-inducing slides (Koenigsberg et al. 2009a, b) or when viewing pictures of human emotional facial expressions (Minzenberg et al. 2007). Moreover during the recall of an unresolved life event, BPD patients showed bilateral activations of the amygdala (Beblo et al. 2006).

Other studies reported the involvement of the prefrontal cortex in BPD in the attempt to control intensive emotions elicited by negative stimuli (Herpertz et al. 2001) or the minor involvement of the anterior cingulate cortex (ACC) in cognitive coping strategies (Koenigsberg et al. 2009a, b). Thus, patients with borderline personality disorder do not seem to engage the cognitive control regions to the extent that healthy individuals do, which might contribute to the affective instability of this disorder. Other studies indicate the existence of different neural networks in BPD with or without comorbid PTSD syndrome (Driessen et al. 2004; Kraus et al. 2009). Altogether this research indicates a dysfunctional frontolimbic network, which involves the anterior cingulate cortex, the orbitofrontal cortex, the dorsolateral prefrontal cortex, the hippocampus, and the amygdala, in BPD individuals.

### 18.4.3 Histrionic Personality Disorder (HDP)

**Definition** Histrionic personality disorder is a pattern of excessive emotionality and attention seeking.

#### 18.4.3.1 Psychodynamic Aspects

Kernberg (1984) differentiates the histrionic personality disorder (borderline personality organization) from the hysteric personality disorder (neurotic personality organization): the unintegration of the representation of the self and of the object and related affects takes into account the affective instability, the precariousness of relationships, and the impulse control associated with HPD. Usually hysteric-histrionic individuals are preoccupied with gender, sexuality, and their relation to power. As pointed out by McWillaims (2011), HPD patients unconsciously consider their own sex as weak, defective, or inferior and the opposite sex as powerful, exciting, frightening, and enviable, and as a consequence their behavior is primarily focused on seductiveness and attention seeking.

While hysteric disorder reveals a prevalence of Oedipal conflicts (seductiveness that results in the impossibility of satisfying sexual relations) expressed in a very regressive way, in the histrionic disorder sexual behavior is a means to satisfy the drive desires and the most primitive self and ego needs.

Often histrionic patients cannot keep in mind of being in a person's important thought, and the only way to maintain a closeness to these people is to use their seductiveness in search of a sensory contact as a substitute for the mental and psychic proximity they need.

In their story it is possible to find serious deficit in maternal care along a continuum constituted by real inadequacy of the caregiver and excessive pressing child needs. In addition it is possible to find a kind of cultural/familiar devaluation of their gender. The defensive structure is organized around the mechanisms of splitting, projective identification, dissociation, denial, acting out with a cognitive style impressionistic, poor in details, and unspecific.

To our knowledge there are no studies that investigated solely the neurobiological underpinnings of the HPD; however there are several studies on the affective instability that can be considered one of the core features of HPD individuals. Indeed, while BPD is the personality disorder most closely associated with affective instability, the DSM-4 and the DSM-5 criteria for histrionic personality disorder also include two criteria related to affective instability: "rapidly shifting and shallow expressions of emotions" and " …exaggerated expression of emotion" (APA 1994, p. 658).

Affective instability is not a unitary construct and encompasses shifts in affective valence, rapidity of mood shifts, short risetime, delayed time to return to baseline, and reactivity to internal and external stimuli (Koenigsberg 2010). It could arise from an increased sensitivity of neural systems involved in the generation of an emotional state or an impairment in emotional control mechanisms. The former includes the amygdala, the insula, the pregenual and subgenual anterior cingulate cortices, the orbitofrontal cortex, and the ventromedial and ventrolateral prefrontal cortices (Phillips et al. 2003).

It can be hypothesized that HPD patients have an impairment in top-down control that could also contribute to affective instability (Dillon and Pizzagalli 2007). Specifically, the dorsal anterior cingulate, dorsomedial, dorsolateral prefrontal, orbitomedial prefrontal cortices and hippocampus have been implicated in the conscious control of emotion (Dillon and Pizzagalli 2007; Phillips et al. 2003), and these regions downregulate amygdala activity.

### 18.4.4 Narcissistic Personality Disorder (NPD)

**Definition** Narcissistic personality disorder is a pattern of grandiosity, need for admiration, and lack of empathy.

#### 18.4.4.1 Psychodynamic Aspects

Narcissistic individuals show difficulties in regulating self-esteem, ambitions, and ideals that are the fundamental core of narcissistic personality disorder.

According to Kohut (1971), the basis of the subjective experience of these subjects is the empathic failure of parents who have failed to respond adequately to the natural demands and needs of the child's self. Personality remains locked in that phase, and the infant self becomes an omnipotent grandiose self.

Inversely, Kernberg (1984) emphasizes the development of a pathological grandiose self that is characterized by the condensation of ideal aspects of self and others with the real self. These constellations inevitably lead to the inflation of the ego and the impoverishment of the superego which remain somewhat primitively organized, leading to mild to moderate antisocial features (Kernberg 1984). The individual believes of being self-sufficient, does not recognize the other's needs, consider him/herself with special abilities and special powers, and devalue constantly the external reality and the others. Kernberg (1984) characterizes the most problematic type of narcissistic individual long a continuum with "malignant narcissism" (i.e., narcissism blended with sadistic aggression, antisocial and paranoid features, and ego-syntonic aggressivity), a condition that he places on a continuum with the antisocial personality disorder (Rosenfeld 1964, 1987).

The difference in the psychodynamic theory between the two authors can be explained by the different patients they treated in their career: Kohut worked principally with outpatient higher-level personality organization individuals, while Kernberg was more exposed to inpatient low-level personality organization subjects, more aggressive and more primitive.

The DSMs' narcissistic personality disorder describes the more grandiose or arrogant version of narcissistic personality (first described by Reich 1933, as the "phallic narcissistic character") but doesn't consider the other face of the coin: there are many individuals who look for psychological help feeling ashamed, avoiding relationships with others, with difficulty in engaging in activities in a long run, looking a bit suspicious and diffident, but, internally, they are preoccupied with grandiose fantasies. Rosenfeld (1987) distinguished between the "thick-skinned" and "thin-skinned" narcissist; Akhtar (1989) between the "overt" and "covert" (shy) patient; Gabbard (1989) between the "oblivious" and the "hypervigilant" types; Masterson (1993) between the "exhibitionistic" and "closet" types; and Pincus et al. (Pincus et al. 2014; Pincus and Roche 2011) between the "grandiose" and "vulnerable" ones.

In all theories there is an agreement in considering narcissism as a problem in self-regulation, where the individuals are disengaged from the social/empathic processing toward the other and are constantly worried about their selves, trying to get the admiration from the other to not show their inner aspects of the self. Thus, contemporary clinicians adopt an integrated approach working with NPD individuals, being more confrontative with defense mechanisms when they are salient and being more empathically attuned when the vulnerability is more accessible.

In neuroscience little research has been done with narcissism to directly measure the neural mechanisms behind the trait or the disorder.

A recent structural diffusion tensor imaging study found that narcissism goes along with weakened frontostriatal connectivity of white matter tracts (Chester et al. 2015). The authors interpret their findings in terms of a neural disconnect between brain regions responsible for self-representation (medial frontal cortex) and reward (ventral striatum) suggesting that narcissistic individuals lack an intrinsic system for self-rewarding activity and at the same time strive from external reward.

Cascio et al. (2014) reported that narcissistic individuals display increased activation in the "social pain network" (dorsal ACC, subgenual ACC, and anterior insula) following social exclusion in a cyberball paradigm. It has to be noted that highly narcissistic individuals did not report elevated feelings of social exclusion in self-report measure, which lead the authors to conclude that "narcissists' social pain [is] seen only in the brain" (p. 335).

In a recent fMRI study pictures of emotional faces were presented and participants were asked to empathize with the person in the picture. Participants high on narcissistic traits displayed decreased deactivation of right anterior insula during processing of emotional faces (Fan et al. 2011) suggesting a lack in empathizing process that the authors interpreted as indicative of an increased self-focus among narcissistic individuals. These results are confirmed and extended by another study that directly investigated the relationship between the resting-state spontaneous activity and task-evoked activity in social processing (Scalabrini et al. 2017b). Narcissism has been found to be associated with right anterior insula positively in resting state while negatively associated during social processing. These results suggest an increased preoccupation for the self, more specifically the bodily and interoceptive self (Craig 2009; Gu et al. 2013) during a rest/mind-wandering period (Smallwood and Schooler 2006) while a disengagement in regard to activity that involves the "other."

## 18.5 Cluster C

### 18.5.1 Avoidant Personality Disorder (APD)

**Definition** Avoidant personality disorder is a pattern of social inhibition, feelings of inadequacy, and hypersensitivity to negative evaluation.

#### 18.5.1.1 Psychodynamic Aspects

The main feature of subjects with avoidant personality disorder is the extreme sensitivity to rejection, which determines their pathological behavior. Shame is the main emotion in this disorder and is expressed in several ways such as feeling of being unable to compete with others, feeling of being defeated physically or mentally, feeling of being disgusting, etc.

The subject is afraid of all those situations in which he/she's forced to be revealed or where he/she can be noticed. From this phenomenology depends all the typical symptoms characterizing these patients: social anxiety, rejection of relationship, fear of criticism, and extreme shyness. They live in a state of mind believing that "if I retire nothing will hurt me." Extremely high ideal aspects accompanied by inflexible and severe stances characterize the superego. Internal saboteurs of these patients are constituted by fear of the judgment of the others and sense of self-defeating.

APD can be reserved and introverted, with a tendency to feel inferior and inadequate, indecisive, and inhibited; moreover they seem to have difficulties in recognizing and describing their feelings identifying their anxiety-charged thoughts, connecting them to their environmental triggers, and mastering them.

There are different kinds of anxiety such as separation anxiety (fear of losing the attachment object), castration anxiety (fear of damage of the body, especially about sexual mutilation), moral anxiety (dread of violating one's core values), and annihilation anxiety (as conceptualized by Kohut in 1977 describing the concept of disintegration anxiety) that is the terror of fragmentation with consequent loss of the sense of self or the terror of destruction based on prior traumatic events (see chapter on traumatogenic disorders).

The etiology of ASD may lie in affective dysregulation (Schore 2003a) and consequent failure to have developed coping strategies or defenses that mitigate normal developmental fears. These individuals usually are characterized having had experiences with a caregiver who, because of the caregiver's own anxiety and attachment style, could not function as an affective regulator and covey a sense of support and as secure base to permit the developments of a sense of agency for the self.

In fMRI research little has been studied on ASD; however studies of generalized social anxiety disorder, a disorder thought to be associated with ASD, have identified decreased insula-dorsal anterior cingulate functional connectivity relative to healthy subjects during viewing of fearful faces (Klumpp et al. 2012). In another study based on a habituation task, it has been reported that the thalamus, parahippocampal gyrus, ventrolateral prefrontal cortex, and dorsal anterior cingulate in ASD patients were not activated as strongly as in the healthy subjects during viewing of repeated versus novel negative pictures. In addition, insula connectivity to extensive cortical regions, including the rostral anterior cingulate, the medial and dorsolateral prefrontal cortex, and the posterior cingulate, did not increase to the extent that it did in the healthy subjects. These results suggest that the failure of this connectivity to increase adequately may contribute to impaired behavioral habituation and consequent faulty emotion regulation in ASD patients as well in BPD (Koenigsberg et al. 2014).

## 18.5.2 Dependent Personality Disorder (DPD)

**Definition** Dependent personality disorder is a pattern of submissive and clinging behavior related to an excessive need to be taken care of.

### 18.5.2.1 Psychodynamic Aspects

People with a dependent personality disorder are characterized by the total inability to live autonomously. They define themselves in relation to others, feeling secured and satisfied in interpersonal context, particularly with the primary attachment relationship. DPD individuals become dysregulated when expected to depend on their own resources and use defense mechanism as somatization and acting out to elicit the care in others.

An important variant of dependency is the passive-aggressive pattern, in which the dependent relationship is characterized by hostility and negative valence. In this variant individuals defined themselves in opposition to others; this makes difficult to pursue their goals and their directedness and in addition triggers aggression and mistreatment from the others.

Another variant is constituted by counterdependency (Bornstein 1993), which is characterized by individuals that may define themselves as the source of other's dependency, keeping out of awareness their needs and vulnerability via denial and reaction formation defense mechanisms.

The etiology of DPD has been associated with failure in the child's dependence on the nourishment received from the caregiver. The internalized message is that independence is a source of dangers and often is associated with hyper-involved and intrusive mothers. Usually parents refused children whenever there was an attempt toward separation-autonomy. There are similarities with the anaclitic aspects of borderline patients, but dependent subjects differ in the quality of relationships, characterized by submission and angry appetites, which are not manifested as in borderline patients but systematically denied defensively.

## 18.5.3 Obsessive-Compulsive Personality Disorder (OCPD)

**Definition** Obsessive-compulsive personality disorder is a pattern of preoccupation with orderliness, perfectionism, and control.

### 18.5.3.1 Psychodynamic Aspects

The main difference between the obsessive-compulsive disorder (OCD) and obsessive-compulsive personality disorder (OCPD) is that the first is experienced as ego-dystonic, while the OCPD is experienced as ego-syntonic.

Perfectionism and inflexibility are the main features of people with OCPD. The core of this disorder is related with the great uncertainty that these people perceive about themselves and about their value; indeed there are common psychodynamic elements with narcissistic personality disorder and problems in self-regulation and maintenance of self-esteem.

Many aspects of the OCPD are probably due to the childish need to show perfectionism and deserving of love in front of their own parents who are perceived as both demanding and distant. The subject feels a desire of love and attachment but in parallel feels that his/her wishes are not destined to be satisfied; hence a deep and strong rage emerges from all past experiences with people which were felt as no more emotionally available.

Both the desire for dependency and anger are perceived to be wrong because they are destructive for the object, so that the subject tries to overcome them with defense mechanism such as isolation of affects, intellectualization, reaction formation, and retroactive cancelation. Unconsciously remains the anger, the unfulfilled need of affection and attention, and the feeling of being not loved and not appreciated that continue to threat the individual's self.

Psychoanalytic scholars (Fisher and Greenberg 1985; Salzman 1980) suggest that the core affect of people with obsessive-compulsive personality is the fear of losing control. Most obsessive thoughts and compulsive actions involve efforts to undo or counteract impulses toward destructiveness, greed, and messiness.

The superego is severe and extremely demanding: these individuals are highly self-critical (as well critical toward others).

These subjects are trapped in an ambivalence that forces them to suffocate desires and emotions to adopt the values of others as true. Despite all the efforts of perfection, they rarely experience a sense of satisfaction.

Beneath an ordered and rigid external appearance, OCPD individuals are internally preoccupied with underlying issues of control and caught in an unconscious conflict between feeling that they must submit to others' demands (which elicits rage and shame) or rebel and defy them (which elicits anxiety and fear of retaliation).

Rigidity, order, and intellectualization defend against awareness of the underlying conflict and the emotions that accompany it.

To our knowledge there are no fMRI studies on the OCPD while there are several on obsessive-compulsive disorder (OCD) that have greatly increased our understanding of the neural mechanisms behind the OCD symptomatology. Although the replicability among these studies has been imperfect, they strongly link obsessive-compulsive symptoms with activation of the orbitofrontal cortex, with less consistent involvement of the anterior cingulate gyrus, the striatum, the thalamus, the lateral frontal and temporal cortices, the amygdala, and the insula (Saxena et al. 2001).

## 18.6 Toward a Neuropsychodynamic Understanding of Personality Disorders and Personality Organizations

### 18.6.1 A Psychodynamic View on Personality

Psychodynamically, the concept of personality refers to the dynamic and individual way concern to the organization of enduring patterns of behavior, cognitions, emotions, motivations, and way of relating to oneself and others.

The development of personality can be traced as a part of the experience of oneself in relation to the world, and it develops from the early relational experiences with the animate world (Trevarthen and Aitken 2001. Psychological and psychiatric problems are always intertwined with personality and need to be appreciated within the relational and cultural context of the individual.

Kernberg and Caligor (2005) defined personality as to refer to the dynamic integration of the totality of a person's subjective experience and behavior patterns, including:

(a) *Conscious concrete and habitual behaviors, experiences of self and of the surrounding world, conscious, explicit psychic thinking, and habitual desires and fears*
(b) *Unconscious behavior patterns, experiences and views, and intentional states*

Personality derives from the human organism's capacity to experience subjective states that reflect the internal condition of the body as well as of the perception of external environment. Thus, personality refers to the self and to the continuous exchange between internal and external stimuli and the way of processing and dealing with them.

It includes discrete psychic functions, such as affects, perception, cognition, and instrumental as declarative memory, and various levels of self-reflective functions, from relatively simple mirroring of perceived and intended motor movements and perceived sensory experiences to complex self-reflective evaluation of cognitive and affective states (Kernberg 2016).

#### 18.6.1.1 Personality Organization: Integration Between Descriptive and Structural Understanding of PDs

First it needs to be noted that the term "borderline" when used by psychodynamic clinicians and researcher, as introduced by Kernberg (1967), denotes a level of personality organization and has a different meaning from the "borderline" term used to describe the personality disorder in DSMs, conceptualized as a list of traits to describe a categorical diagnosis, in which only one specific prototype of personality organization is labeled "borderline personality disorder."

From a psychodynamic point of view, a description of personality pathology needs to include (1) the descriptive features of the disorder, (2) the structural organization underlying the descriptive features, and (3) a framework theory about the patient's psychodynamics.

Descriptive features inform about maladaptive personality features and relationship with significant others and constitute a useful approach to formulate a descriptive diagnosis.

Complementarlly, the structural assessment informs about the severity of the personality pathology through the assessment of the individual's experiences of the relationship with the self and with others (identity), defense mechanisms, and reality testing.

Kernberg (1984) developed a theoretical classification of personality disorders that combines categorical (i.e., DSM oriented) and dimensional (i.e., relative degree of infusion of mental life with aggression, and introversion vs. extroversion) constructs for understanding PDs.

Personality can be expressed in several ways: at the behavioral level, personality pathology is characterized by the inhibition, the exaggeration of certain behaviors or contradictory behaviors together with symptomatic expressions. Behind this wall we can look at what is defined as the structural level of personality which can be organized by the sense of identity, intrinsically constituted by the sense of self and others (coherent and integrated in normal personality while incoherent, not integrated, or "diffuse" in borderline conditions).

The borderline personality organization (BPO) contains both specific personality disorders as described in DSM-5 Sect. 18.6 (American Psychiatric Association 2013) and other PDs not mentioned in the DSM (e.g., hypomanic, sadomasochistic, hypochondriasis, malignant narcissism) (Kernberg and Caligor 2005).

Focusing on the "neurotic level of personality organization," it is possible to explain maladaptive personality features within the context of (1) a normal identity, (2) the predominance of higher-level, repression-based defense mechanism, and (3) intact reality testing.

Considering the "borderline level of personality organizations," the maladaptive personality rigidity is characterized by (1) identity diffusion, (2) lower-level, splitting-based, defense mechanism, and (3) variable reality testing.

Finally the "psychotic level of personality organization" is characterized by (1) identity diffusion, (2) lower-level, splitting-based, defense mechanism, and (3) lack of reality testing.

#### 18.6.1.2 Intrapsychic and Neurobiological Organization of Personality

The psychodynamic approach in assessing the level of organization implies the basic assumption that the psychological structure is connected with the underlying neurobiological correlates in the development of self in relation with other. In more general terms, the intrapsychic structures represented by object relations theory reflect a second, intrapsychic level of organismic organization, based on a primary, neurobiological one (Kernberg 2015). It is proposed that primitive mental mechanism associated with splitting and their derivatives would be based on biological, subcortical limbic developments of separate positive and negative affective systems, and their potential integration would be based on a cortical level of processing of emotional experience originally sharply dissociated (Roth 2009; Kernberg 2015; Schore 2015, Schore 2003a). The intrapsychic structure, or personality organization, could reflect a second level of organization based upon a primary neurobiological one; however what remains unclear is how the development of the self and personality features influence the neuronal processing in different contexts.

The intertwining of scientific advances in various areas from neuroscience (e.g., Northoff et al. 2006; Northoff and Panksepp 2008; Panksepp and Biven 2012) to infant research studying the interactive regulatory processes between the caregiver and the baby (e.g., Schore 2000, 2001; Lyons-Ruth 2008; Beebe and Lachmann 2014) is facilitating the integration of various forms of knowledge about the complex construct of personality and self.

#### 18.6.1.3 Self-Definition and Interpersonal Relatedness in Personality Disorders

Personality disorders (PDs) have always been a core object of interest in clinical psychology and psychiatry, and they are considered as associated with fundamental disturbances of self and interpersonal relations, problems that extend in severity within and across disorders.

In diagnosing PDs we should take in account that there is an evidence that the person's psychology causes significant distress to self or others, is of long duration, and is so much a part of the person's experience that he or she cannot remember, or easily imagine, being different (PDM-2, Lingiardi and McWilliams 2017).

Personality can be considered as a dimension or a continuum from healthy features (coherent sense of self and personal identity, engagements in satisfying relationships, relatively flexible functioning when stressed by external events or internal conflict, expression of impulses in a manner appropriate to the situation, internalized moral values in accordance with behavior) to maladaptive-disturbed features (identity diffusion and incoherent sense of self in time, problems in self-other differentiation and relatedness, lack or transient loss in reality testing, problems in affect regulation, attention and learning, inflexibility and rigidity in several domains).

Historically only in 1938, Stern proposed the borderline concept to describe a group of patients who were apparently "unanalyzable" and did not reach the criteria to be placed either in neurotic or in psychotic diagnosis.

Several authors tried to classify and give a name to this group of patients (e.g., Knight 1953 "Borderline States of the Ego"), but only in the 1960s, as previously specified, Kernberg (1967) proposed a broader concept of borderline personality organization, which included primarily the evaluation of the identity, and the evaluation of defense mechanisms and reality testing, as closely associated with the continuity and coherence of the sense of self.

Few years later Kohut (1971) postulated a failure of the development of a cohesive sense of self during its own development and in relation with the environment, leading to fragmentation of the body, mind, self, and self-object.

Lately Fonagy et al. (2007) proposed that the construction of the sense of a subjective self is a fundamental aspect of acquiring knowledge about the world through the caregiver's pedagogical communicative displays which in this context focuses on the child's thoughts and feelings. The parent-infant dyad can be considered as the first intersubjective encounter that predispose to the development of the self and its agency in healthy individuals, while it predisposes to the development of the *alien self* (Fonagy et al. 2010) in borderline conditions.

Hence it can be considered that the concept of self seems to constitute a predisposition for individual differences in behavior, affects, cognition, and sensorimotor expression, i.e., one's personological profile.

#### 18.6.1.4 The Alternative Model for Personality Disorder in DSM-5

In line with psychodynamic tradition and taking into account the increasing interest for the concept of self, the DSM-5 and the proposed ICD-11 are moving toward a self and other classification of personality, an empirically based dimensional model for maladaptive personality traits.

In the AMPD personality disorders are not following categorical criteria while they are characterized by impairments in personality *functioning* and pathological personality traits. First the clinician needs to evaluate the impairments in the personality functioning (concerning specifically the self and interpersonal impairments), and only if this criterion is considered clinically relevant it is possible to proceed with the trait-based diagnosis.

It proposes that different degrees of impairments in levels of self-definition and interpersonal functioning are central to defining personality disorders, which range from no impairment to extreme impairments as expressed in a profound inability to reflect on the self together with severe impairments in self-other boundaries (self-impairments) and in significant impairments in the awareness and understanding of the thoughts, feelings, and motivations of others (interpersonal impairments; Skodol 2012).

The notion that interpersonal relatedness and self-definition issues are central in personality disturbances has been strongly influenced by attachment theory (Fonagy and Luyten 2009; Fonagy et al. 2010; Levy 2005) and contemporary interpersonal formulations (Pincus 2005). Theory and research in this field have addressed the role of early caregiving relationships in the development of representations of self and others in both normal and disrupted developments (Blatt et al. 1997).

This conceptualization is very much coherent with several psychoanalytic formulations that point out how relatively satisfactory caring experiences are potentially facilitating the development of a differentiated and cohesive sense of self and a capacity for increasingly mature interpersonal relatedness and capacity for intimacy (Blatt and Blass 1990, 1992; Blass and Blatt 1996; Kernberg 1984; Kohut 1971).

This emphasis on the concept of self and its development through the relation with the environment and significant others also emphasize the relational quality of the self.

Indeed, several authors are emphasizing how the dual exchange between caregiver and the infant, continuously modulated, influences epigenetically the structure and the formation of the growing subject, organizing the mind-body-brain interceptive and exteroceptive connections in relation to the other (Mucci 2018; Schore 2012).

#### 18.6.1.5 Neuroscientific Correlates of the Self

In particular authors as Northoff and Bermpohl (2004), Northoff et al. (2006), Damasio (2010, 2012), and Panksepp and Biven (2012) emphasized the existence of a complex, distributed, and functionally based system of the self.

The core self is described as a trans-species functional entity based in medial midbrain structures and extending to deep, subcortical forebrain regions. This system is implicated in an unconscious perceived of "felt" sense of an embodied self, arising from basic proprioceptive and sensorimotor processes.

Panksepp and Biven (2012) have also suggested that the core self as mediated by these subcortical midline structures allows the operation of core automatic "self-referential processing." This allows the linking of external events to the motivational and emotional impulses of the organism. The core self functions have been conceptualized as the basis of an ultimate, more complex, reflective, idiographic self that permits awareness, as opposed to raw experiences, of phenomenal-affective contents. These two different aspects of the self are mutually regulating.

Northoff (2015) emphasized how the self has been operationalized in many experimental studies in term of self-relatedness (SR) and how SR influences behavioral performance. SR has been associated with functions as basic as perception (Sui et al. 2012, 2013), action (Frings and Wentura 2014), reward (de Greck et al. 2008), and emotions (Northoff et al. 2009; Phan et al. 2004). Interestingly the author (Northoff 2016a, b, c) also emphasized how the brain activity for SR is linked with the resting state suggesting how it is possible to conceptualize a rest-self overlap (Bai et al. 2016). Following his view it is possible to make the assumption that the self-reflecting basic aspect of the brain shows a certain functionality and connectivity related to the self when our brain is in a task-free state.

Indeed functional aspects of the self-system, involving self-processing, perception of the self, and perception of self in relation to others, are distributed through higher cortical midline structures (CMS), notably the medial prefrontal cortex, anterior cingulate gyrus, precuneus, and posterior cingulate cortex (for a review, see Qin and Northoff 2011). Interestingly these structures form a part of the default mode network (DMN).

The DMN also comprises the posterior cingulate, anterior cingulate, and medial prefrontal cortices, the precuneus, and temporoparietal junctions (McKiernan et al. 2006). This network is activated when subjects are at "rest" and deactivated during performance of cognitively demanding tasks (Andrews-Hanna et al. 2010). Uddin et al. (2009) and Northoff et al. (2006) argue that the only kind of stimuli available to the brain during "rest" is internal and includes memory formation and retrieval, introspection, and ongoing monitoring of the self and its social relationships.

As such, this suggests that a disruption of resting-state activity might underlie the abnormal self-concept that appears to characterize PDs.

Further frontal neural systems involved in self-processes include a right (lateralized) frontoparietal network involved in self-recognition and self-awareness and social understanding. These include more lateral structures overlapping with the distribution of mirror neuron areas, which might also serve as a partial basis for recognition of intentional mental states in others and symbolic mental activity, as well as for bodily imitation (Panksepp and Biven 2012; Knox 2010; Schore 2012; Siegel 2015; Cozolino 2014; Iacoboni 2009; Rizzolatti and Sinigaglia 2008). Further, right hemispheric areas, notably the right orbitofrontal cortex, are implicated in aspects of the self, such as SR, awareness of the self in relation to others, subjective sense of continuity and coherence, and a sense of an embodied self. The early growth and maturation of these regions is experience dependent and requires nurturing self-other interactions. When an individual is denied these positive experiences, serious failures of development occur. As such, processes of the self can be adversely affected (Schore 2004; Mucci 2017).

It appears that much of the formation of the self occurs through the internalization of benign or adverse interpersonal and sociocultural experiences (Roth and David Sweatt 2011). This process of relational internalization is enabled by the human capacity for intersubjectivity, attunement, and empathy and the predisposition to joint meaning-making and companionship, which are present from birth (Trevarthen and Aitken 2001; Stern 2000; Tronick 2007; Lyons-Ruth 2008).

#### 18.6.1.6 Self and Attachment: New Heading

Self-related processing or relational processing is closely linked to the primary sense of attachment, namely, the primary outward orientation to attach to objects (Brockman 2002). Similarly both attachment and self-related processing (Northoff 2011) enable the constitution and differentiation between self and others. Brockman (2002): "Attachment begins before any sense of self and before any sense of object to attach to" (p. 90).

Attachment plays a fundamental role in shaping personality, the self of individuals and the sense of relatedness; more specifically disorganized attachment stemming from parental traumatization and early relational trauma creates not only the vertical disconnection that is called dissociation in the mind-brain-body system but also the impulsivity and lack of effortful control that are characteristics of borderline pathologies or personality disorders in general (Mucci 2017). This is due to the fact that early traumatization does not allow the connection between limbic areas and amygdala (the limbic system, mostly connected with emotions, especially in the right hemisphere) and superior orbitofrontal areas, which are the areas of adult control and agency, and also of intentionality, decisionmaking, and so on. In few words all the "superior qualities" include imagination creativity and activation of planning. This creates the dysfunctions typical of borderline pathologies. They are pathologies created in long-term abuse and in dysfunctional families (Mucci 2013; Felitti et al. 1998).

A recent multimodal resting-state fMRI (rsfMRI), aversion task fMRI, glutamate magnetic resonance spectroscopy (MRS), and diffusion magnetic resonance imaging (dMRI) combined with the Childhood Trauma Questionnaire (CTQ) in healthy subjects aimed to examine the impact of negative childhood experiences on the brain (Duncan et al. 2015).

Interestingly and in line with our proposal, the research showed how increased measures of individual negative childhood experiences (NCEs) were related to lower levels of medial prefrontal cortex (mPFC) glutamate levels and how the degree of NCEs may impact resting-state activity properties in the mPFC—a key region within the default mode and affective processing networks (Daniels et al. 2011; Roy et al. 2012). Specifically increased instances of reported NCEs were related to increased entropy measures of mPFC.

The link between NCEs, glutamate, and entropy complements and extends previous studies showing an effect of early life stressors on EEG frequency bands which predict signal entropy (Bruce et al. 2009; McFarlane et al. 2005), and prior results demonstrating an effect of stress on mPFC resting-state activity CTQ scores and mPFC glutamate and entropy correlated with neural BOLD responses to the anticipation of aversive stimuli in regions throughout the aversion-related network, with strong correlations between all measures in the motor cortex and left insula.

Structural connectivity strength, measured using mean fractional anisotropy, between the mPFC and left insula correlated to aversion-related signal changes in the motor cortex. These findings highlight the impact of NCEs on multiple interrelated brain systems. In particular, they highlight the role of a prefrontal-insular-motor cortical network in the processing and responsivity to aversive stimuli and its potential adaptability by NCEs.

#### 18.6.1.7 Neuropsychodynamic Model of Self and Personality Functioning

Although the neural mechanisms underpinning and determining normal and disordered mental function are clearly highly important, it is not possible to fully conceive of or describe a range of psychological phenomena, including some disturbed in BPD, such as subjective awareness and sense of coherence, identity, or more sophisticated executive functions, in these terms.

This leads us to hypothesize that self-specific information of individuals may be present in the resting state and that the study of neural architecture of the resting state may represent the predisposition of every individual to act in the world (Fig. 18.1).

Moreover we may suggest how the study of the interaction between the ongoing activity of the brain and the role of individual personality differences in interaction with task-induced activity may be considered as a further investigation to better understand the human brain.

We are moving toward a new way in classifying and studying personality that can develop a bridge between neuroscience and clinical psychodynamic

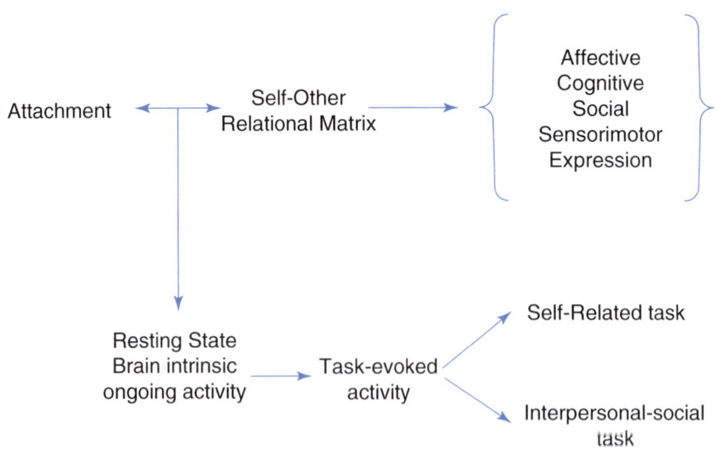


**Fig. 18.1** Neuropsychodynamic model of self, personality, and brain: rest-self-containment and rest-stimulus interaction

psychology: (1) self-based and (2) brain-based predisposition for individual differences, regarding the dimensional continuum from adaptive to maladaptive personality features, which characterize the individual differences to experience the relation between internal world and external reality given that our relational experiences with the world shape our self and in parallel our brain.

The study of personality has to take into account the so-called rest-stimulus interaction (Northoff et al. 2010) as a way to conceive, from a neuronal point of view, how the brain's intrinsic activity with the own particular spatiotemporal schemata encodes self-specific information (rest-self overlap, e.g., Bai et al. 2016) of past and (possible) future input-output relationship (Northoff 2016a, b, c).

In this context we propose a novel conceptualization of the link between self-brain states and personality organization. In our view different and interconnected states of the self (i.e., *self-relational alignment, self-constitution, self-manifestation, self-expansion)* are embedded in the intrinsic activity of the brain which, as we already pointed out, predispose the expression of personality in the world:

First we propose that *self-relational alignment* is a prerequisite that gives the framework for the other states of the self. It is essentially linked with the relational continuity, which can also be conceptualized in a neuro-ecological continuum between brain and the external world. It is given by the first relational encounter with a caregiver and his/her capacity to attune with the mind-brain of the infant to give the possibility of a secure environment where the infant's brain can start to exchange mutual informations with the world. The infant's brain start to become part of the time and the space of the world by the relational alignment with the spatiotemporal structure of the animate and inanimate reality. This attunement capacity predisposes the constitution of the self.

*Self-constitution* represents the building blocks of the consciousness processing: it includes the perception of time and the perception of space. It is linked with the ownership of own body, location of self in space, authorship and control of own actions, and difference between fantasy and reality. It is linked with the capacity to distinguish self from nonself and internal from the external. Thus it's strongly connected with reality testing and it is the self-state that distinguish psychotic organization of personality from the others. In the case of psychotic individuals, at the neuronal level the relationship from cortical midline structure (CMS) and somatosensory network is altered, resulting in lack of differentiation in processing intrinsic and extrinsic stimuli.

*Self-manifestation* represents actual consciousness in the present moment: it includes the experience of time and space with the perception of environment and identification with social reality; the cognitive functions as thinking, imaging, metalizing; and the affective and motivational system. It is particularly linked with the degree integration of self and significant others and, in parallel, with the quality of defense mechanism (higher order, lower order). In several personality disturbances, there is an altered relationship between the subcortical-cortical limbic circuits with orbitofrontal regions that are deputed to regulate emotions and mentalize affects.

*Self-expansion* is linked with stable and integrated aspects of the self in time and space: (a) autobiographical self, (b) social self, (c) linguistic self, and (d) mental self. There is the capacity to inhibit behaviors and to tolerate the ambivalence of the affects considering past, present, and future. The self and the consciousness can be expanded in time and space without losing the capacity to differentiate internal from external stimuli. Higher is the capacity to self-expand, higher will be the integrations of various aspects of self and others and more mature the quality of defense mechanism. In normal expansion there is an increased connectivity between DMN and other networks in learning from new experiences with the world, while in some neurotic disturbances, the coherence between certain networks can be lost according to the impossibility to elaborate the experience given the repression-based psychological defense mechanism.

How these different states of the self are linked to the intrapsychic structure of individuals?

When we consider the personality organization or intrapsychic structure (characterized by identity integration, defense mechanism, and reality testing) considered as a second level of organization, we may take into account a primary neurobiological one which, in our hypothesis, may be characterized by the brain's intrinsic activity in term of form or structure characterized by space and time and hence spatiotemporal structure (Northoff 2016a, b).

Since the brain's intrinsic activity can be characterized by particular and merely individualized spatiotemporal structure, we suppose that any contents (whether affective, cognitive, social or sensorimotor) and their underlying extrinsic activity must first and foremost be integrated within the brain's intrinsic (internal) activity and in its spatiotemporal structure. The degree and the way the contents and their extrinsic activity are integrated into the brain's intrinsic activity determine how we perceive them into consciousness and hence how we make experience of them, i.e., our subjective or personal approach to the external reality.

In this view our personality can be directly related to our subjective experience of temporal flow, which is related to our temporal structure of the brain's intrinsic activity, thus the temporal balance between infraslow/slow and fast oscillations (Northoff 2017).

In *psychotic personality organization*, on a psychological level it is possible to find diffusion of identity (not integrated and fragmented), primitive defense mechanism, and loss of reality testing; on a neurobiological level we find disturbance at a level of self-constitution, a disruption in the global organization of the brain's intrinsic activity: whole topography over all networks and frequency range are disrupted, and, for instance, the usual negative correlations between the default mode network (DMN) and the control executive network (CEN) that are usually characterized by an anticorrelation are in psychosis transformed into a positive correlation which in turn may lead to the breakdown of the rest-self overlap where there is a self-assignment to either internal or external stimuli (Carhart-Harris 2013; Carhart-Harris et al. 2014). In this case we may hypothesize that there is no possibility to differentiate the internal world from the external reality; there is a fragmentation of the perception of time and space, which results in the identity diffusion or fragmentation of self-constitution. We may hypothesize that this psychotic organization shows severe impairments at a pre-phenomenal level of experience on the spatiotemporal structure of brain's intrinsic activity.

In *borderline personality organization*, characterized by intact reality testing, primitive defense mechanism based on splitting and projective identification, and identity diffusion, at a neurobiological level, we can see how the whole brain topography and organization between networks are preserved, but the balance between them can show abnormalities, as in the case of bipolar disorder (Magioncalda et al. 2015; Martino et al. 2016), and can be hypothesized how the relation with the external stimuli can be impaired. We may find abnormalities in the rest-task interaction (e.g., on narcissistic personality features, Scalabrini et al., in press) and lack of integration in brain's networks; for example, in borderline personality disorders, we can observe alterations in orbitofrontal cortex and connected subcortical regions (amygdala and hippocampus) (Koenigsberg et al. 2009a, b; Minzenberg et al. 2007; Enzi et al. 2013). Moreover other studies have revealed the functional neuroanatomy of borderline disorders that are associated with the hypersensitivity, intolerance for aloneness, and attachment fears typical of patients in this broad diagnostic group (Buchheim et al. 2008; Fertuck et al. 2009; King-Casas et al. 2008).

We may hypothesize that individuals with borderline organization of personality show severe impairments at a pre-reflective level of experience, at a self-manifestation state, where it is possible to have information about implicit experience of self, body, others, time, and space as related to the spatiotemporal structure of brain's intrinsic activity and consciousness.

Regarding the *neurotic personality organization*, characterized by intact reality testing, mature defense mechanism based on repression, and integrated sense of self and identity, we may find some compromission at the self-expansion state where at a neurobiological level the whole brain's topography, organization between networks, and balance between them are preserved, but their coherence is not given, so we may expect that in neurotic organization we may find a decreased coherence between networks and decreased cross-frequency coupling while the spatiotemporal structure by itself is well integrated. Thus these individual have difficulty to expand themselves in time and to finalize their motives because of the internal conflict which are shaping their personality. We may hypothesize that the neurotic organization doesn't show impairments at pre-phenomenal and at a pre-reflective level of experience, but it's possible that some contents at the reflective level of the experience (explicit experience of cognitive, affective, social, sensorimotor functions) are not accessible to the consciousness so that they are repressed and they will manifest themselves in some incoherence in brain's functioning to process either internal and external stimuli.

In sum we propose that different states of self intrinsically connected to the spontaneous brain spatiotemporal organization represent a primary neurobiological organization that is connected with different impairments in the second level of organization represented by the intrapsychic structure or personality organization of individuals.

The advance in fMRI studies that comprise the study of the intrinsic brain activity (during resting state), the task-evoked brain's activity, and the interaction of the two conditions may shed a novel light in the understanding of personality pathology.

### 18.6.2 Clinical Case Example: The Case of F

F. was a patient of 24 years old who asked for clinical help because of her self-defeating and self-destructive behaviors such as cutting her wrist (not too severely), alcohol abuse, and incapacity to pursue her goals in study and work and with severe difficulty in maintaining a relationship.

F. was very smart, with an IQ approximately >120, but she never succeeded in school after she was 16 years old, and she decided to quit college after first 2 years because she was not constant in studying and she was convinced that the study of languages was not in line with her artistic talent. Moreover she wanted to work to save some money and travel in South America to learn Argentinean tango and to find the love of her life.

At the time of the first consultation she didn't have any job, she was fired because of her irresponsibility and her "temper" in treating with clients. She was not economically autonomous, and she was receiving money from her parents.

About relationship and sexuality, F. described her relational life as confused and *quiet messy*. Her contact with men was always connected to alcohol abuse and promiscuous situations: "I've been drinking too much so I don't really remember where I spent last night and with who… when I woke up I found some bites on my body and I was frightened…so I called my friend who told me what happened last night."

She never had a relationship longer than few months saying that at the beginning she is always very passionate, but after a while she get bored of people who are not good anymore for her. At the same time if she felt of being abandoned, she would act desperately to get attention.

Her self-esteem was very low at that time of the first consultation, and she presented herself as a shy person and ashamed of her problems; however internally she had grandiose features such as: "I am different from anybody and I don't want to reach any compromise…it is better to be the last of a mass of people then to be in the middle."

She was impulsive in the area of alcohol and drugs, in spending money without thinking at any consequences. Moreover she had outburst of anger with parents, and she would easily get angry for nothing throwing things: "I'm always fighting with my roommate and last time I threw away a mug."

She also reported that she couldn't think about herself because she is *nothing* and feels empty all the time. When she felt to be under stress, she reported to cut her wrist and her arms: "I feel the pain and in these moments…it is strange but I feel kind of being alive."

F. resulted to have problems mainly in the area of the self-manifestation, since she cannot experience a sense of self-continuity. What she experiences is continuously distorted or acted impulsively, resulting in a lack of effective expansion in time. Moreover it can be noticed how there is a disconnection between the affective experiences and the capacity to regulate her emotions, wishes, and self-defeating drives. It can be hypothesized that there is a disconnection between the orbitofrontal regions and cortico-limbic circuits suggesting the difficulty in integrating various aspects of self and others in a whole and multifaceted sense of self.

F. reached the extremes to be diagnosed with a borderline personality disorder at a categorical level (*unstable or diffuse sense of self/identity, a pattern of unstable interpersonal relationship, frantic efforts to avoid abandonment, impulsivity in spending, sex and substance abuse, self-mutilating behaviors, difficulty controlling anger, chronic feeling of emptiness, and dissociative attitude*); in addition she presents narcissistic features (*sense of entitlement, believe of being special—in a covert manner, often preoccupied with fantasies of ideal love*).

#### 18.6.2.1 Familiar History of F

F. is the second child, and she has a brother 6 years older than her. Parents immigrated from the south of Italy to Germany when the brother was born, and then they came back when the mother was pregnant with F. The pregnant mother was depressed (Green 1993), and she had a history of abuse; in addition she was mistreated repetitively from the family of the husband when they come back to Italy. In detail the mother would lose weight instead of gaining it when was pregnant with F. From F. narrative the mother was severely depressed and under medication during her childhood, so the main caregiver was the father. He was a simple hardworking man who was taking care of her child. F. used to sleep between their parents until she was 15 years old. She was very close to her father until she became adolescent, and he started to develop a sexual interest for the daughter.

The father can be defined as incestual (Racamier 2010): he was obsessed with her body and used to spy on her in the restroom, to open her bedroom unexpectedly without knocking, and to gaze on her sexual parts inappropriately. She started to become angrier and angrier with him, while she did not find any support from the mother. Once she finally disclosed the sexual attention of the father to the mother, she reacted denying and telling her she was crazy. Subsequently they kicked her off the house when she was 23 years old (1 year before she asked for psychological and psychiatric help)

It can be hypothesized that the incapacity of the mother, because of her mental illness and her history of abuse, to take care of the child as an infant and as a young girl has resulted in a disorganized attachment (Liotti 2004; Fonagy et al. 2010; Schore 2003a, b) characterized by affective dysregulation, uncoherent sense of self and others and incapacity to tolerate and manage impulsivity, and lack in mentalization. In terms of object relation theory, the internalized good object is not consistent and thus is not felt as soothing and capable of mood and affective regulation. Moreover the atmosphere at home can be considered traumatic in the sense of a continual distortion of what happens at home and a negation of the truth of incestual attention from the third part (Van Der Kolk 1987, 1988). The attachment disorganized patterns together with the traumatic experiences explained the development of a dissociative attitude of F. especially connected with drinking. Through the act of drinking, she could disconnect from her reality, and she could depersonalize and forget what happens as a way to detach from herself and from reality. The proneness to disconnect from herself and from reality is closely related to dissociation phenomena which resulted from traumatic experiences. This brings us back to the self-relational alignment that we consider as a framework where the subject can start to exchange mutual informations with the world. Given the familiar and attachment history of F., here we can see how the traumatic events impaired her ability to attune and align with her own sense of self and with the environment. From a neuropsychodynamic point of view, it can be hypothesized a higher degree of entropy in the regions connected with self-referential activity and attachment, the anterior cingulate cortex (ACC) and the ventromedial prefrontal cortex (VmPFC).

Together with descriptive features mentioned before, F. is framed in the low-level borderline personality organization as characterized by identity (unintegrated and discontinued sense of self and of significant others) and sexual diffusion (sexual promiscuity and object inconsistency); use of massive primitive defenses as splitting, projective identification, dissociation; and maintenance of reality testing but with severe cognitive distortions (a certain proneness to dissociation).

She was referred to a psychodynamic psychotherapist for twice a week treatment tailored for borderline personality disorder.

## References


Akhtar S. Narcissistic personality disorder: descriptive features and differential diagnosis. Psychiatr Clin North Am. 1989;12(3):505–29.

American Psychiatric Association. Diagnostic and statistical manual of mental disorders (DSM-5®). Washington, DC: American Psychiatric Publication; 2013.

American Psychiatric Association (APA). Diagnostic and statistical manual of mental disorders (DSM-IV). 1994.

Andrews-Hanna JR, Reidler JS, Sepulcre J, Poulin R, Buckner RL. Functional-anatomic fractionation of the brain's default network. Neuron. 2010;65(4):550–62.

Bai Y, Nakao T, Xu J, Qin P, Chaves P, Heinzel A, et al. Resting state glutamate predicts elevated pre-stimulus alpha during self-relatedness: a combined EEG-MRS study on rest-self overlap. Soc Neurosci. 2016;11(3):249–63.

Ball JS, Links PS. Borderline personality disorder and childhood trauma: evidence for a causal relationship. Curr Psychiatry Rep. 2009;11(1):63–8.

Bateman AW, Fonagy P. Mentalization-based treatment of BPD. J Personal Disord. 2004;18(1):36–51.

Beblo T, Driessen M, Mertens M, et al. Functional MRI correlates of the recall of unresolved life events in borderline personality disorder. Psychol Med. 2006;36:845–56.

Beebe B, Lachmann FM. The origins of attachment. New York: Routledge; 2014.

Blair RJR. The amygdala and ventromedial prefrontal cortex in morality and psychopathy. Trends Cogn Sci. 2007;11(9):387–92.

Blair RJR. Neuroimaging of psychopathy and antisocial behavior: a targeted review. Curr Psychiatry Rep. 2010;12(1):76–82.

Blass RB, Blatt SJ. Attachment and separateness in the experience of symbiotic relatedness. Psychoanal Q. 1996;65(4):711–46.

Blatt SJ, Blass RB. Attachment and separateness: a dialectic model of the products and processes of development throughout the life cycle. Psychoanal Study Child. 1990;45:107–27.

Blatt SJ, Blass RB. Relatedness and self-definition: two primary dimensions in personality development, psychopathology, and psychotherapy. In: Barron JW, Eagle MN, Wolitzky DL, editors. Interface of psychoanalysis and psychology. Washington, DC: American Psychological Association; 1992. p. 399–428.

Blatt SJ, Auerbach JS, Levy KN. Mental representations in personality development, psychopathology, and the therapeutic process. Rev Gen Psychol. 1997;1(4):351.

Bornstein RF. The dependent personality. New York: Guilford Press; 1993.

Bowlby J. Attachment and loss, vol. 3. New York: Basic; 1980.

Brockman R. Self, object, neurobiology. Neuropsychoanalysis. 2002;4(1):89–101.

Bromberg P. Review of standing in the spaces: essays on clinical process, trauma, and dissociation. Hillsdale: Analytic Press; 2000.

Bruce ENEN, Bruce MCMC, Vennelaganti S. Sample entropy tracks changes in EEG power spectrum with sleep state and aging. J Clin Neurophysiol. 2009;26:257.

Buchheim A, Erk S, George C, Kächele H, Kircher T, Martius P, Walter H. Neural correlates of attachment trauma in borderline personality disorder: a functional magnetic resonance imaging study. Psychiatry Res Neuroimaging. 2008;163(3):223–35.

Carhart-Harris R. Psychedelic drugs, magical thinking and psychosis. J Neurol Neurosurg Psychiatry. 2013;84(9):e1.

Carhart-Harris RL, Leech R, Hellyer PJ, Shanahan M, Feilding A, Tagliazucchi E, Nutt D. The entropic brain: a theory of conscious states informed by neuroimaging research with psychedelic drugs. Front Hum Neurosci. 2014;8:20.

Cascio CN, Konrath SH, Falk EB. Narcissists' social pain seen only in the brain. Soc Cogn Affect Neurosci. 2014;10(3):335–41.

Chefetz RA. Intensive psychotherapy for persistent dissociative processes: the fear of feeling real (Norton series on interpersonal neurobiology). New York: Norton; 2015.

Chester DS, Lynam DR, Powell DK, DeWall CN. Narcissism is associated with weakened frontostriatal connectivity: a DTI study. Soc Cogn Affect Neurosci. 2015;11(7):1036–40.

Clarkin JF, Lenzenweger MF, Livesley WJ. Major theories of personality disorder. Arch Gen Psychiatry. 1997;54(10):967–8.

Clarkin JF, Yeomans FE, Kernberg OF. Psychotherapy for borderline personality: focusing on object relations. Washington, DC: American Psychiatric Publication; 2007a.

Clarkin JF, Levy KN, Lenzenweger MF, Kernberg OF. Evaluating three treatments for borderline personality disorder: a multiwave study. Am J Psychiatry. 2007b;164(6):922–8.

Cozolino L. The neuroscience of human relationships: attachment and the developing social brain (Norton series on interpersonal neurobiology). New York: Norton; 2014.

Craig AD. How do you feel—now? The anterior insula and human awareness. Nat Rev Neurosci. 2009;10(1):59.

Damasio A. Neuroscience and psychoanalysis: a natural alliance. Psychoanal Rev. 2012;99(4):591–4.

Damasio AR. Self comes to mind: constructing the conscious brain. New York: Random House; 2010.

Daniels JK, Frewen P, McKinnon MC, Lanius RA. Default mode alterations in posttraumatic stress disorder related to early-life trauma: a developmental perspective. J Psychiatry Neurosci. 2011;36:56–9.

de Greck M, Rotte M, Paus R, Moritz D, Thiemann R, Proesch U, Northoff G. Is our self based on reward? Self-relatedness recruits neural activity in the reward system. NeuroImage. 2008;39(4):2066–75.

Diamond D, Blatt SJ, Stayner DA, Kaslow N, Auerbach J, Lowyck B. Manual for the differentiation-relatedness scale. New Haven: Yale University; 2014a.

Diamond D, Clarkin JF, Levy KN, Meehan KB, Cain NM, Yeomans FE, et al. Change in attachment and reflective function in borderline patients with and without comorbid narcissistic personality disorder in transference focused psychotherapy. Contemp Psychoanal. 2014b;50(1–2):175–210.

Dillon DG, Pizzagalli DA. Inhibition of action, thought, and emotion: a selective neurobiological review. Appl Prev Psychol. 2007;12(3):99–114.
Driessen M, Beblo T, Mertens M, et al. Posttraumatic stress disorder and fMRI activation patterns of traumatic memory in patients with borderline personality disorder. Biol Psychiatry. 2004;55:603–11.
Duncan NW, Hayes DJ, Wiebking C, Tiret B, Pietruska K, Chen DQ, Hodaie M. Negative childhood experiences alter a prefrontal-insular-motor cortical network in healthy adults: a preliminary multimodal rsfMRI-fMRI-MRS-dMRI study. Hum Brain Mapp. 2015;36(11):4622–37.
Enzi B, Doering S, Faber C, Hinrichs J, Bahmer J, Northoff G. Reduced deactivation in reward circuitry and midline structures during emotion processing in borderline personality disorder. World J Biol Psychiatry. 2013;14(1):45–56.
Fairbairn WR. Psychological studies of the personality. London: Routledge/Kegan Paul; 1952.
Fals-Stewart M, Birchler GR, Kelley ML. Learning sobriety together. A randomized clinical trial examining behavioral couples therapy with alcoholic female patients. J Consult Clin Psychol. 2006;74:579–91.
Felitti VJ, Anda RF, Nordenberg D, Williamson DF, Spitz AM, Edwards V, Marks JS. Relationship of childhood abuse and household dysfunction to many of the leading causes of death in adults: The Adverse Childhood Experiences (ACE) study. Am J Prev Med. 1998;14(4):245–58.
Fertuck EA, Jekal A, Song I, Wyman B, Morris MC, Wilson ST, Stanley B. Enhanced 'Reading the Mind in the Eyes' in borderline personality disorder compared to healthy controls. Psychol Med. 2009;39(12):1979–88.
Fisher S, Greenberg RP. The scientific credibility of Freud's theories and therapy. New York: Columbia University Press; 1985.
Fonagy P, Luyten P. A developmental, mentalization-based approach to the understanding and treatment of borderline personality disorder. Dev Psychopathol. 2009;21(4):1355–81.
Fonagy P, Target M. Attachment and reflective function: their role in self-organization. Dev Psychopathol. 1997;9(4):679–700.
Fonagy P, Target M. Early intervention and the development of self-regulation. Psychoanal Inq. 2002;22(3):307–35.
Fonagy P, Target M, Gergely G. Psychoanalytic perspectives on developmental psychopathology. Dev Psychopathol. 1995;1:504–54.
Fonagy P, Gergely G, Target M. The parent–infant dyad and the construction of the subjective self. J Child Psychol Psychiatry. 2007;48(3-4):288–328.
Fonagy P, Luyten P, Bateman A, Gergely G, Strathearn L, Target M, Allison E. Attachment and personality pathology. In: Psychodynamic psychotherapy for personality disorders: a clinical handbook, vol. 1. Washington, D.C.: American Psychiatric Publication; 2010. p. 37–88.
Frings C, Wentura D. Self-priorization processes in action and perception. J Exp Psychol. 2014;40(5):1737.
Gabbard GO. Two subtypes of narcissistic personality disorder. Bull Menn Clin. 1989;53:527–39.
Green A. The dead mother. Psyche. 1993;47(3):205–40.
Gu X, Hof PR, Friston KJ, Fan J. Anterior insular cortex and emotional awareness. J Comp Neurol. 2013;521(15):3371–88.
Guidano VF, Liotti G. Cognitive processes and emotional disorders: a structural approach to psychotherapy. New York: Guilford Press; 1983.
Hazlett EA, Zhang J, New AS, et al. Potentiated amygdala response to repeated emotional pictures in borderline personality disorder. Biol Psychiatry. 2012;72:448–56.
Herpertz SC, Dietrich TM, Wenning B, et al. Evidence of abnormal amygdala functioning in borderline personality disorder: a functional MRI study. Biol Psychiatry. 2001;50:292–8.
Holmes J. The search for the secure base: attachment theory and psychotherapy. New York: Routledge; 2014.
Hopwood CJ, Malone JC, Ansell EB, Sanislow CA, Grilo CM, McGlashan TH, Gunderson JG. Personality assessment in DSM-5: empirical support for rating severity, style, and traits. J Personal Disord. 2011;25(3):305–20.

Iacoboni M. Mirroring people: the new science of how we connect with others. New York: Macmillan; 2009.
Kernberg O. Borderline personality organization. J Am Psychoanal Assoc. 1967;15(3):641–85.
Kernberg OF. A systems approach to priority setting of interventions in groups. Int J Group Psychother. 1975;25(3):251–75.
Kernberg OF. Severe personality disorders: psychotherapeutic strategies. New Haven: Yale University Press; 1984.
Kernberg OF. Neurobiological correlates of object relations theory: the relationship between neurobiological and psychodynamic development. In: International forum of psychoanalysis, vol. 24, No. 1. New York: Routledge; 2015. pp. 38–46.
Kernberg OF. What is personality? J Personal Disord. 2016;30(2):145–56.
Kernberg OF, Caligor E. A psychoanalytic theory of personality disorders. In: Lenzenweger MF, Clarkin JF, editors. Major theories of personality disorder. 2nd ed. New York: Guilford Press; 2005. p. 114–56.
Kerr IB, Finlayson-Short L, McCutcheon LK, Beard H, Chanen AM. The 'self' and borderline personality disorder: conceptual and clinical considerations. Psychopathology. 2015;48(5):339–48.
King-Casas B, Sharp C, Lomax-Bream L, Lohrenz T, Fonagy P, Montague PR. The rupture and repair of cooperation in borderline personality disorder. Science. 2008;321(5890):806–10.
Klein M. Notes on some schizoid mechanisms. Int J Psychoanal. 1946;27:99.
Klumpp H, Angstadt M, Phan KL. Insula reactivity and connectivity to anterior cingulate cortex when processing threat in generalized social anxiety disorder. Biol Psychol. 2012;89:273–6.
Knight RP. Borderline states. Bull Menn Clin. 1953;17(1):1.
Knox J. Self-agency in psychotherapy: attachment, autonomy, and intimacy. New York: Norton; 2010.
Koenigsberg HW. Affective instability: toward an integration of neuroscience and psychological perspectives. J Personal Disord. 2010;24(1):60–82.
Koenigsberg HW, Buchsbaum MS, Buchsbaum BR, et al. Functional MRI of visuospatial working memory in schizotypal personality disorder: a region-of-interest analysis. Psychol Med. 2005;35:1019–30.
Koenigsberg HW, Fan J, Ochsner KN, et al. Neural correlates of the use of psychological distancing to regulate responses to negative social cues: a study of patients with borderline personality disorder. Biol Psychiatry. 2009a;66:854–63.
Koenigsberg HW, Siever LJ, Lee H, et al. Neural correlates of emotion processing in borderline personality disorder. Psychiatry Res. 2009b;172:192–9.
Koenigsberg HW, Denny BT, Fan J, Liu X, Guerreri S, Mayson SJ, Siever LJ. The neural correlates of anomalous habituation to negative emotional pictures in borderline and avoidant personality disorder patients. Am J Psychiatr. 2014;171(1):82–90.
Kohut H. The analysis of the self a systematic psychoanalytic approach to the treatment of narcissistic personality disorders. New York: International Press; 1971.
Kraus A, Esposito F, Seifritz E, et al. Amygdala deactivation as a neural correlate of pain processing in patients with borderline personality disorder and co-occurrent posttraumatic stress disorder. Biol Psychiatry. 2009;65:819–22.
Lenzenweger MF, Clarkin JF, Kernberg OF, Foelsch PA. The Inventory of Personality Organization: psychometric properties, factorial composition, and criterion relations with affect, aggressive dyscontrol, psychosis proneness, and self-domains in a nonclinical sample. Psychol Assess. 2001;13(4):577.
Levitt JJ, McCarley RW, Dickey CC, et al. MRI study of caudate nucleus volume and its cognitive correlates in neuroleptic-naive patients with schizotypal personality disorder. Am J Psychiatry. 2002;159:1190–7.
Levitt JJ, Westin CF, Nestor PG, et al. Shape of caudate nucleus and its cognitive correlates in neuroleptic-naive schizotypal personality disorder. Biol Psychiatry. 2004;55:177–84.
Levitt JJ, Styner M, Niethammer M, et al. Shape abnormalities of caudate nucleus in schizotypal personality disorder. Schizophr Res. 2009;110:127–39.

Levy KN. The implications of attachment theory and research for understanding borderline personality disorder. Dev Psychopathol. 2005;17(4):959.

Lingiardi V, McWilliams N. Psychodynamic diagnostic manual second edition (PDM-2). New York: Guilford Press; 2017.

Liotti G. Trauma, dissociation, and disorganized attachment: three strands of a single braid. Psychotherapy. 2004;41(4):472.

Lyons-Ruth K. Contributions of the mother–infant relationship to dissociative, borderline, and conduct symptoms in young adulthood. Inf Ment Health J. 2008;29(3):203–18.

Magioncalda P, Martino M, Conio B, Escelsior A, Piaggio N, Presta A, Ferri F. Functional connectivity and neuronal variability of resting state activity in bipolar disorder—reduction and decoupling in anterior cortical midline structures. Hum Brain Mapp. 2015;36(2):666–82.

Mahler MS. A study of the separation-individuation process: and its possible application to borderline phenomena in the psychoanalytic situation. Psychoanal Study Child. 1971;26(1): 403–24.

Main M, Solomon J. Discovery of an insecure-disorganized/disoriented attachment pattern: procedures, findings and implications for the classification of behavior. In: Brazelton TB, Yogman M, editors. Affective development in infancy. Norwood: Ablex; 1986.

Martino M, Magioncalda P, Huang Z, Conio B, Piaggio N, Duncan NW, Inglese M. Contrasting variability patterns in the default mode and sensorimotor networks balance in bipolar depression and mania. Proc Natl Acad Sci. 2016;113(17):4824–9.

Masterson JF. The emerging self: a developmental, self and object relations approach to the treatment of closet narcissistic disorder of the self. New York: Brunner/Mazel; 1993.

Masterson JF. Psychotherapy of the borderline adult: a developmental approach. New York: Routledge; 2013.

McFarlane A, Clark CR, Bryant RA, Williams LM, Niaura R, Paul RH, Hitsman BL, Stroud L, Alexander DM, Gordon E. r brain correlates of negative childhood experience r r 4635 r the impact of early life stress on psychophysiological, personality and behavioral measures in 740 non-clinical subjects. J Integr Neurosci. 2005;4:27–40.

McKiernan KA, D'angelo BR, Kaufman JN, Binder JR. Interrupting the "stream of consciousness": an fMRI investigation. NeuroImage. 2006;29(4):1185–91.

McWilliams N. Psychoanalytic diagnosis: understanding personality structure in the clinical process. New York: Guilford Press; 2011.

Meares R. A dissociation model of borderline personality disorder. New York: Norton; 2012.

Minzenberg MJ, Fan J, New AS, Tang CY, Siever LJ. Fronto-limbic dysfunction in response to facial emotion in borderline personality disorder: an event-related fMRI study. Psychiatry Res. 2007;155:231–43.

Morey LC, Skodol AE, Grilo CM, Sanislow CA, Zanarini MC, Shea MT, McGlashan TH. Temporal coherence of criteria for four personality disorders. J Personal Disord. 2004;18(4):394–8.

Mucci C. Beyond individual and collective trauma: intergenerational transmission, psychoanalytic treatment, and the dynamics of forgiveness. London: Karnac; 2013.

Mucci C. Ferenczi's revolutionary therapeutic approach. Am J Psychoanal. 2017;77(3):239–54.

Mucci C. Borderline bodies. Affect regulation therapy for personality disorders. New York: W.W. Norton; 2018.

Northoff G. Neuropsychoanalysis in practice: brain, self and objects. New York: Oxford University Press; 2011.

Northoff G. Is schizophrenia a spatiotemporal disorder of the brain's resting state? World Psychiatry. 2015;14(1):34–5.

Northoff G. Is the self a higher-order or fundamental function of the brain? The "basis model of self-specificity" and its encoding by the brain's spontaneous activity. Cognit Neurosci. 2016a;7(1-4):203–22.

Northoff G. Spatiotemporal psychopathology I. no rest for the brain's resting state activity in depression? Spatiotemporal psychopathology of depressive symptoms. J Affect Disord. 2016b;190:854–66.

Northoff G. Spatiotemporal psychopathology II: how does a psychopathology of the brain's resting state look like? Spatiotemporal approach and the history of psychopathology. J Affect Disord. 2016c;190:867–79.
Northoff G. The brain's spontaneous activity and its psychopathological symptoms–Spatiotemporal binding and integration. Prog Neuro-Psychopharmacol Biol Psychiatry. 2017;80(Pt B):81–90.
Northoff G, Bermpohl F. Cortical midline structures and the self. Trends Cogn Sci. 2004;8(3):102–7.
Northoff G, Panksepp J. The trans-species concept of self and the subcortical–cortical midline system. Trends Cogn Sci. 2008;12(7):259–64.
Northoff G, Heinzel A, De Greck M, Bermpohl F, Dobrowolny H, Panksepp J. Self-referential processing in our brain—a meta-analysis of imaging studies on the self. NeuroImage. 2006;31(1):440–57.
Northoff G, Schneider F, Rotte M, Matthiae C, Tempelmann C, Wiebking C, Bogerts B. Differential parametric modulation of self-relatedness and emotions in different brain regions. Hum Brain Mapp. 2009;30(2):369–82.
Northoff G, Qin P, Nakao T. Rest-stimulus interaction in the brain: a review. Trends Neurosci. 2010;33(6):277–84.
Panksepp J. The periconscious substrates of consciousness: Affective states and the evolutionary origins of the SELF. J Conscious Stud. 1998;5(5-6):566–82.
Panksepp J, Biven L. The archaeology of mind: neuroevolutionary origins of human emotions. New York: Norton; 2012.
Paris J. The treatment of borderline personality disorder in light of the research on its long term outcome. Can J Psychiatry. 1993;38(Suppl 1):28–34.
Phan KL, Taylor SF, Welsh RC, Ho SH, Britton JC, Liberzon I. Neural correlates of individual ratings of emotional salience: a trial-related fMRI study. NeuroImage. 2004;21(2):768–80.
Phillips ML, Drevets WC, Rauch SL, Lane R. Neurobiology of the emotion perception I: the neural basis of normal emotion perception. Biol Psychiatry. 2003;54:504–14.
Pincus AL. A contemporary integrative interpersonal theory of personality disorders. Major Theor Pers Disord. 2005;2:282–331.
Pincus AL, Roche MJ. Narcissistic grandiosity and narcissistic vulnerability. In: Campbell WK, Miller JD, editors. Handbook of narcissism and narcissistic personality disorder. Hoboken: Wiley; 2011. p. 31–40.
Pincus AL, Cain NM, Wright AGC. Narcissistic grandiosity and narcissistic vulnerability in psychotherapy. Personal Disord. 2014;5:439–43.
Premkumar P, Ettinger U, Inchley-Mort S, et al. Neural processing of social rejection: the role of schizotypal personality traits. Hum Brain Mapp. 2012;33:695–706.
Qin P, Northoff G. How is our self related to midline regions and the default-mode network? NeuroImage. 2011;57(3):1221–33.
Racamier PC. L'inceste et l'incestuel. Malakoff: Dunod; 2010.
Reich W. On character analysis. Psychoanal Rev. 1933;20:89.
Rizzolatti G, Sinigaglia C. Mirrors in the brain: how our minds share actions and emotions. Malakoff: Oxford University Press; 2008.
Ronningstam E. Narcissistic personality disorder: Facing DSM-V. Psychiatr Ann. 2009;39(3):11–121.
Rosenfeld H. On the psychopathology of narcissism: a clinical approach. Int J Psychoanal. 1964;45(2-3):332–7.
Rosenfeld H. Impasse and interpretation. London: Routledge; 1987.
Roth G. Aus Sicht des Gehirns [from the perspective of the brain]. Frankfurt am Mein: Suhrkump; 2009.
Roth TL, David Sweatt J. Annual Research Review: Epigenetic mechanisms and environmental shaping of the brain during sensitive periods of development. J Child Psychol Psychiatry. 2011;52(4):398–408.
Roy M, Shohamy D, Wager TD. Ventromedial prefrontal-subcortical systems and the generation of affective meaning. Trends Cogn Sci. 2012;16:147–56.

Salzman L. Treatment of the obsessive personality. New York: Aronson; 1980.
Saxena S, Bota RG, Brody AL. Brain-behavior relationships in obsessive-compulsive disorder. Semin Clin Neuropsychiatry. 2001;6(2):82.
Scalabrini A, Cavicchioli M, Fossati A, Maffei C. The extent of dissociation in borderline personality disorder: a meta-analytic review. J Trauma Dissociation. 2017a;18(4):522–43.
Scalabrini A, Huang Z, Mucci C, Perrucci MG, Ferretti A, Fossati A, Romani GL, Northoff G, Ebisch SJ. How spontaneous brain activity and narcissistic features shape social interaction. Sci Rep. 2017b;7(1):9986.
Schore AN. Attachment and the regulation of the right brain. Attach Hum Dev. 2000;2(1):23–47.
Schore AN. Effects of a secure attachment relationship on right brain development, affect regulation, and infant mental health. Inf Ment Health J. 2001;22(1-2):7–66.
Schore AN. Affect dysregulation and disorders of the self (Norton series on interpersonal neurobiology). New York: Norton; 2003a.
Schore AN. Affect regulation and the repair of the Self (Norton series on interpersonal neurobiology). New York: Norton; 2003b.
Schore AN. The human unconscious: the development of the right brain and its role in early emotional life. In: Emotional development in psychoanalysis, attachment theory and neuroscience. London: Routledge; 2004. p. 33–62.
Schore AN. The science of the art of psychotherapy (Norton series on interpersonal neurobiology). New York: Norton; 2012.
Schore AN. Affect regulation and the origin of the self: the neurobiology of emotional development. Routledge; 2015.
Siegel DJ. The developing mind: how relationships and the brain interact to shape who we are. New York: Guilford; 2015.
Siever LJ, Davis KL. A psychobiological perspective on the personality disorders. Am J Psychiatry. 1991;148(12):1647.
Siever LJ, Koenigsberg HW, Harvey P, Mitropoulou V, Laruelle M, Abi-Dargham A, Buchsbaum M. Cognitive and brain function in schizotypal personality disorder. Schizophr Res. 2002;54(1):157–67.
Skodol AE. Personality disorders in DSM-5. Annu Rev Clin Psychol. 2012;8:317–44.
Smallwood J, Schooler JW. The restless mind. Psychol Bull. 2006;132(6):946.
Sroufe LA. An organizational perspective on the self. In: The self in transition: infancy to childhood; 1990. p. 281–307.
Stern A. Psychoanalytic investigation of and therapy in the border line group of neuroses. Psychoanal Q. 1938;7(4):467–89.
Stern DN. The interpersonal world of the infant: a view from psychoanalysis and developmental psychology. Karnac Books; 1985.
Stern DN. The interpersonal world of the infant a view from psychoanalysis and developmental psychology: A view from psychoanalysis and developmental psychology. New York: Basic Books; 2000.
Stone MH. The borderline syndromes: constitution, personality, and adaptation. New York: McGraw-Hill; 1980.
Sui J, Humphreys GW. The integrative self: how self-reference integrates perception and memory. Trends Cogn Sci. 2015;19(12):719–28.
Sui J, Chechlacz M, Humphreys GW. Dividing the self: distinct neural substrates of task-based and automatic self-prioritization after brain damage. Cognition. 2012;122(2):150–62.
Sui J, Liu M, Mevorach C, Humphreys GW. The salient self: the left intraparietal sulcus responds to social as well as perceptualsalience after self-association. Cereb Cortex. 2013;25(4):1060–8.
Sui J, Liu M, Mevorach C, Humphreys GW. The salient self: the left intraparietal sulcus responds to social as well as perceptual-salience after self-association. Cereb Cortex. 2015;25(4):1060–8.
Tang Y, Jiang W, Liao J, Wang W, Luo A. Identifying individuals with antisocial personality disorder using resting-state FMRI. PLoS One. 2013;8(4):e60652.
Torgersen S. Genetics of patients with borderline personality disorder. Psychiatr Clin N Am. 2000;23(1):1–9.

Trevarthen C, Aitken KJ. Infant intersubjectivity: research, theory, and clinical applications. J Child Psychol Psychiatry. 2001;42(1):3–48.
Tronick E. The neurobehavioral and social-emotional development of infants and children. New York: WW Norton & Company; 2007.
Tyrer P. The problem of severity in the classification of personality disorder. J Personal Disord. 2005;19(3):309–14.
Uddin LQ, Clare Kelly AM, Biswal BB, Xavier Castellanos F, Milham MP. Functional connectivity of default mode network components: correlation, anticorrelation, and causality. Hum Brain Mapp. 2009;30(2):625–37.
Van Der Kolk BA. The psychological consequences of overwhelming life experiences. In: Psychological trauma. Washington, DC: American Psychiatric Press; 1987. p. 1–30.
Van der Kolk BA. The trauma spectrum: the interaction of biological and social events in the genesis of the trauma response. J Trauma Stress. 1988;1(3):273–90.
Vu MA, Thermenos HW, Terry DP, et al. Working memory in schizotypal personality disorder: fMRI activation and deactivation differences. Schizophr Res. 2013;115:113–23.
Westen D, Shedler J, Bradley B, DeFife JA. An empirically derived taxonomy for personality diagnosis: bridging science and practice in conceptualizing personality. Am J Psychiatr. 2012;169:273–84.
Winnicott DW. The theory of the parent-infant relationship. Int J Psychoanal. 1960;41:585.
Zachar P, Krueger RF, Kendler KS. Personality disorder in DSM-5: an oral history. Psychol Med. 2016;46(1):1–10.

# Dissociative Syndromes

# 19

Carsten Spitzer

**Abstract**

Dissociation can likewise be conceived as state (of consciousness), as personality trait in terms of a disposition to dissociate, as collective name for a heterogeneous group of psychopathological features as well as psychophysiological response to traumatic experiences or as defence mechanism. A neuropsychodynamic model of dissociation could be conceptualized as follows: An external or internal stimulus leads to neuronal activation patterns that manifest on the level of subjective experience as highly adverse and intense emotions. Due to the negative valence and the (hyper-)arousal, these affective states impede upon an adequate integration with cognitive and self-referential information by a 'bottom-up' hyperactivation of prefrontal areas which, in turn, inhibit top-down cortical-subcortical networks which are essential for memory functioning, (self-)consciousness and agency as well as body control. Dissociation results from a cortical-subcortical inhibition and serves to cope with both external traumatic stress as well as interpersonal and intrapsychic conflicts by means of a subjective decontextualization.

## 19.1 Introduction

> This girl, who was bubbling over with intellectual vitality … embellished her life in a manner which probably influenced her decisively in the direction of her illness, by indulging in systematic day-dreaming, which she described as her 'private theatre'. … There developed in rapid succession a series of severe disturbances which were apparently quite new: … convergent squint (diplopia), markedly increased by excitement; complaints that the walls of the room seemed to be falling over … contracture and anaesthesia of the right upper, and,

C. Spitzer
Asklepios Fachklinikum Tiefenbrunn, Rosdorf, Germany
e-mail: c.spitzer@asklepios.com



H. Boeker et al. (eds.), *Neuropsychodynamic Psychiatry*,
https://doi.org/10.1007/978-3-319-75112-2_19

> after a time, of the right lower extremity. … It was while the patient was in this condition that I undertook her treatment, and I at once recognized the seriousness of the psychical disturbance with which I had to deal. Two entirely distinct states of consciousness were present which alternated very frequently and without warning … she would complain of having 'lost' some time and would remark upon the gap in her train of conscious thoughts …. (Breuer and Freud 1955)

This very condensed extract of one of the most famous case reports of psychoanalysis specifies several phenomena, which are currently subsumed under the category of dissociation and which refer to core mental functions such as self-awareness, identity knowledge or self-consciousness, memory for personal relevant material, perception of self and environment as well as sensory and motor functions. In our everyday practical life contexts, these mental operations seem to 'happen automatically and by themselves' and are usually implicit or preconscious. Only in cases of marked and clinically meaningful disturbances of these functions, the relevance of their coordinated interplay and their successful integration for our first-person perspective and phenomenal self becomes evident. Their successful integration seems to be crucial for our experience of ourselves as subject, of our feeling of agency and 'ownership' of our mental as well as bodily processes. Not surprisingly, there has always been a great fascination surrounding the dissociative syndromes.

## 19.2 Definition and Phenomenology

Considering the intriguing and iridescent conceptual history, which has evolved from and been influenced by various forces (Ellenberger 1970; Van der Hart and Dorahy 2009), it is not surprising that dissociation lacks a single and coherent definition. It can likewise be conceived as state (of consciousness), as personality trait in terms of a disposition to dissociate, as collective name for a heterogeneous group of psychopathological features as well as psychophysiological response to traumatic experiences or as intrapsychic defence mechanism (Cardena 1994; Dell and O'Neil 2009). Even a phenomenological approach remains problematic, because there is an ongoing controversy about the question which symptoms should be classified as dissociative (Cardena 1994; Dell and O'Neil 2009; Holmes et al. 2005; Nijenhuis and van der Hart 2011). For example, this applies to certain alterations in consciousness, depersonalization and derealization, flashbacks as a form of intentionally uncontrollable hypermnesia or to distinct bodily symptoms. Additionally, on closer and more detailed inspection, it becomes evident that the classification of clinical phenomena as dissociative is implicitly or explicitly grounded on theoretical models of dissociation. As a possible solution for this predicament, it may be helpful—quasi as suboptimal compromise—to consult the current classification systems.

Both ICD and DSM coincide that the key feature of dissociation refers to a disintegration or disruption of the normally integrative functions of consciousness, memory, personal identity and perception of the environment. The ICD-10 extends this loss of integration to the neurophysiological functions of the sensory, sensibility and motor systems. This is adopted by the DSM-5, which even expands this idea by assuming that virtually every area of psychological functioning can be affected by dissociation: 'Dissociative disorders are characterized by a disruption of and/or discontinuity in the normal integration of consciousness, memory, identity, emotion, perception, body representation, motor control, and behaviour. Dissociative symptoms can potentially disrupt every area of psychological functioning' (American Psychiatric Association 2013, p. 291).

From a phenomenological point of view, dissociative symptoms and syndromes are very heterogeneous and manifest in different ways:

- Amnesia for personally relevant material, i.e. memory deficit that relates to the autobiographical memory
- Alterations in (the qualitative state) of consciousness, e.g. narrowing of the field of consciousness like in trance, usually combined with feelings of alienation, an imminent ego fragmentation and emotional numbing
- Depersonalization and derealization, particularly with an as if experience
- Disturbances in the way that space and time are experienced, e.g. the tunnel vision often described by trauma survivors or the impression that time passes like in slow-motion
- Altered sense of identity as well as
- Bodily phenomena such as analgesia, anaesthesia or movement and sensory disorders (e.g. deafness or blindness)

On the level of categorical diagnoses, it might be helpful to differentiate between *dissociative disorders of consciousness* (dissociation affects mental functions) and *conversion disorders* (dissociation affects the body); however, the overlap ad comorbidity of these categories is high (Brown et al. 2007; Spitzer et al. 1999). Taking a wider perspective than ICD-10 and DSM-5, the dissociative disorders of consciousness comprise dissociative amnesia, fugue, stupor, trance and possession states, depersonalization/derealization disorder, Ganser's syndrome and multiple personality disorder which should better be termed dissociative identity disorders (DID) as in DSM-5. The conversion disorders comprise dissociative convulsions, dissociative disturbances of motor functions (e.g. paralyses) and disorders with dissociative anaesthesia and sensory loss.

## 19.3 Neurobiological Findings and Neuropsychodynamic Considerations

To date, it is impossible to determine the one and only, genuine and unique neurobiology of dissociation considering the heterogeneity and complexity of the affected mental functions. Moreover, there is an ongoing controversy about the assumption of a single, basic and uniform psychobiological mechanism underlying all dissociative phenomena. For heuristic reasons, it might be reasonable to virtually deconstruct the concept of dissociation and to differentiate between the following aspects. Such a differentiation is certainly necessary for meaningful neuropsychodynamic considerations:

- A general predisposition to dissociate in terms of a personality trait, i.e. a dissociative tendency
- Mnestic disorders within the framework of dissociation
- Dissociation as altered states of consciousness and detachment from the self and the world
- Conversion disorders

### 19.3.1 Dissociative Tendency or Dissociation Predisposition

Twin studies have indicated that 40–50% of variance in dissociative tendencies and dissociative phenomena, respectively (as assessed by the Dissociative Experiences Scale[1]), in both children and adolescents as well as adults can be attributed to genetic factors (Jang et al. 1998; Becker-Blease et al. 2004; Pieper et al. 2011). Preliminary molecular genetic findings suggest that alterations in the serotonergic system may play a role in the aetiology of dissociation: Participants with the SS genotype of the functionally relevant polymorphism of the promotor region of the serotonin transporter gene (5-HTTLPR) reported more dissociative symptoms compared to participants with the other genotypes (Pieper et al. 2011). Polymorphisms in a gene involved in the glucocorticoid receptor-regulating cochaperone of stress proteins (the FKBP5 gene) might explain up to 14% of the variance in peritraumatic dissociation in medically injured children (Koenen et al. 2005). Possibly, gene-environment interactions might be crucial, too. For example, in patients with obsessive-compulsive disorder (OCD), the interplay between physical neglect and the 5-HTT genotype predicted the extent of dissociative symptoms (Lochner et al. 2007). Likewise, in patients with bipolar disorders, there is evidence that the association between dissociation and childhood trauma is mediated by polymorphisms of the brain-derived neurotrophic factor (BDNF) gene and the catechol-*O*-methyltransferase (COMT) gene (Savitz et al. 2008). Interestingly, genetic variations in the COMT gene have been associated with pain, and individuals with a COMT pain vulnerable genotype experienced more dissociative symptoms in the emergency department after motor vehicle collisions than those without this genotype (McLean et al. 2011). From a neuropsychodynamic stance, it must be underscored that these genetic variations are involved in the physiological regulation of stress and anxiety.

If the dissociation predisposition is rooted in the individual genetic make-up, it is obvious to consider the selective or evolutionary advantage. Of note, already the German psychiatrist Ernst Kretschmer stated that dissociative reactions represent

[1] The Dissociative Experiences Scale (DES) is an internationally wide-spread and well-established self-report measure with good psychometric properties.

both onto- and phylogenetically acquired and individually modulated stereotypes of behaviour which can be considered as derivatives of the flight or play-dead reflex, respectively (1923/1960). Similarities between animal and human behavioural patterns in the face of threat and assault have recently been rediscovered and interpreted within the framework of dissociation (Nijenhuis et al. 1998).

Regarding ontogenetic influences it has to be underlined that the brain's plasticity is shaped by environmental factors, especially during sensitive developmental periods. Early-life stress and unfavourable attachment experiences provoke the release and modify the expression of several stress mediators and neurotransmitters within specific brain regions. The interaction of these mediators with developing neurons and neuronal networks may lead to long-lasting structural and functional 'scars' associated with cognitive and emotional consequences (Chen and Baram 2016), which may become clinically evident as dissociative tendencies. This line of reasoning is supported by converging evidence from psychoanalysis, cognitive developmental psychology, attachment research and neurobiology. Of particular interest are the findings of prospective studies investigating developmental pathways to dissociation (Ogawa et al. 1997; Dutra et al. 2009). In a prospective longitudinal study of children considered at high risk for poor developmental outcomes at birth due to poverty, participants were followed-up four times across 19 years. In addition to physical abuse and neglect in their first 2 years, abusive experiences of their mothers, avoidant attachment style and a psychologically unavailable caretaker were significant predictors of dissociation in young adulthood (Ogawa et al. 1997). Another longitudinal study with an even more detailed analysis of the mother-infant interaction indicated that dissociation in young adulthood was significantly predicted by disrupted maternal communication and lack of parental responsiveness in infancy, while traumatic experiences (apart from childhood verbal abuse) did not add to the prediction of dissociation (Dutra et al. 2009). In contrast to many cross-sectional studies retrospectively assessing childhood trauma and reporting an association with dissociative phenomena in adulthood, prospective longitudinal approaches clearly support the notion that the quality of the early mother-infant interaction is at least as important for the development of the self as traumatic stress early in life, possibly even more important.

### 19.3.2 Mnestic Disorders within the Framework of Dissociation

Dissociative amnesia (DA) is characterized by an inability to recall important personal information, usually of a traumatic or stressful nature, that is too extensive to be explained by ordinary forgetfulness. Thus, DA affects the autobiographical memory. Usually, this kind of amnesia is retrograde, incomplete, localized, selective or systematized and can vary in its intensity; generalized and anterograde variants have been described, too (Staniloiu and Markowitsch 2012, 2014).

Memory, its relevance for the self-system and its environmental adaptation as well as its neurobiological foundations are highly complex, and their delineation is far beyond the scope of this chapter. In the context of DA, it is important to distinguish between implicit and explicit memory systems: While implicit mental processes are not bound to consciousness (also called anoetic), explicit mental operations require consciousness (i.e. noetic). They can be influenced by will and are experienced as 'my' mental actions, i.e. the person is aware that he himself is the agent of these operations. The episodic autobiographical memory is part of the explicit systems and develops ontogenetically latest. Within the hierarchy of long-term memory systems, it is considered highest and as conjunction of subjective time, autonoetic consciousness and the experiencing self (Staniloiu and Markowitsch 2012). To qualify as autobiographical experience, it is a prerequisite that the information is affectively coloured and being processed with reference to the self, space, time and context. For the encoding of autobiographical information, limbic structures and the prefrontal cortex play a major role, while consolidation and storage of memory initially require hippocampal, later cortical association networks. The retrieval involves right-hemispheric frontotemporal structures and limbic regions (Kopelman 2002; Staniloiu and Markowitsch 2012).

In DA, the intentional recall of autobiographical material is usually compromised. This retrieval deficit has mainly been explained by two different mechanisms. It was proposed that emotional hyperarousal resulting from severe psychosocial stress activates (right) prefrontal control and executive systems which in turn inhibit medial temporal and diencephalon memory retrieval mechanisms leading to the clinical condition of DA (Kopelman 2002; Bell et al. 2011). This assumption is in line with similar theories of dissociation where top-down executive inhibition is postulated (Sierra and Berrios 1998; Lanius et al. 2010). An alternative model posits that traumatic experiences cause the release of stress-related hormones resulting in a functional disconnectivity of frontotemporal networks, particularly of the right hemisphere (Markowitsch et al. 1997; Staniloiu and Markowitsch 2012, 2014). Of note, this so-called mnestic block syndrome is assumed to be caused by a failure to activate or engage the relevant structures and networks which are important for the retrieval of autobiographical memories, while the other model's hypothesis refers to an active inhibition (Bell et al. 2011). Possibly, these two theories do not exclude each other but can be integrated by considering that some studies have focused on the resting state and others have applied an activation paradigm. While resting state studies are generally consistent with the model of the mnestic block syndrome, activation studies support the theory of an executive inhibition (Bell et al. 2011).

In addition to these 'classical' forms, DA is often found in traumatized individuals.

> Mr. R. having survived the accident of an airplane that crashed shortly after take-off was interviewed only a few hours after the tragedy. He remembered having carried an unconscious passenger right next to him to the emergency exit where the rescue team took over; he was not able to recall any further details of the disaster. He just remembered to fasten his seatbelt while riding across the maneuvering area; his next memory referred to a paramedic offering him tea in the medical tent. Other survivors of the airplane crash reported that Mr. R. actively engaged in the recovery of many other passengers.

Neurobiological as well as neuropsychological studies suggest that traumatic experiences are encoded and stored in fragments and in different memory systems. Some elements of the traumatic experience are overly well represented in the implicit memory (e.g. sensory-based encoding resulting in a perceptual memory trace), but there is no sufficient elaboration and integration into episodic memory. The information in the so-called trauma memory is being perceived as fragmented with regard to sequence, coherence, content and affective significance as related to oneself and the surrounding world; thus traumatic memories cannot sufficiently be symbolized in a narrative (Sartory et al. 2013). Of note, the hallmarks of these peritraumatic or anterograde DAs are abnormalities in the encoding, elaboration and further processing, but not the retrieval of already stored information. This essential difference lies at the heart of a recent refinement of the dissociation construct that distinguishes between two qualitatively distinct phenomena: 'detachment' and 'compartmentalization' (Holmes et al. 2005).

The constitutive feature of detachment is the subjective experience of an altered state of consciousness characterized by an 'alienation' of oneself or the external world (i.e. detachment). During these altered states, there is often an absence or flattening of emotional 'colouring' and the impression that it is not oneself who is the experiencing agent. In contrast, compartmentalization is characterized by a disrupted interplay of normally integrated mental (sub)systems leading to a partial or even complete failure to deliberately control processes and actions that can normally be influenced by an act of volition. Applying these considerations to DA, it can be conceptualized within both forms of dissociation. Detachment—often caused by traumatic experiences—relates to encoding abnormalities, while compartmentalization refers to a retrieval deficit and the inability to deliberately gain access to the episodic autobiographical memory (Holmes et al. 2005). On a descriptive and clinical level, detachment becomes prototypically evident in depersonalization and derealization; conversely, conversion phenomena are a clear-cut manifestation of compartmentalization. In posttraumatic stress disorder, both forms of dissociation can be observed (Holmes et al. 2005).

### 19.3.3 Altered States of Consciousness and Detachment from the Self and the World

Paul Schilder, one of the psychoanalytical pioneers investigating depersonalization, gave a concise account of this condition:

> To the depersonalized individual the world appears strange, peculiar, foreign, dream like. Objects appear at times strangely diminished in size, at times flat. Sounds appear to come from a distance. The tactile characteristics of objects likewise seem strangely altered, but the patients complain not only of the changes in their perceptivity but their imagery appears to be altered. Patients characterise their imagery as pale, colourless and some complain that they have altogether lost the power of imagination. The emotions likewise undergo marked alteration. Patients complain that that they are capable of experiencing neither pain or pleasure; love and hate have perished with them. They experience a fundamental change in their

> personality, and the climax is reached with their complaints that they have become strangers to themselves. It is as though they were dead, lifeless, mere automatons. The objective examination of such patients reveals not only an intact sensory apparatus, but also an intact emotional apparatus. All these patients exhibit natural affective reactions in their facial expressions, attitudes, etc.; so that it is impossible to assume that they are incapable of emotional response (Schilder 1928).

In addition to the peculiar alterations in feeling described as emotional numbness, a split in the ego in an experiencing self and an observing self has repeatedly been underlined as the central feature of depersonalization by many psychoanalytical authors (e.g. Guralnik and Simeon 2010). These psychodynamic considerations correspond well with recent results from psychophysiological and neuroimaging studies (Sierra and David 2011). Dysfunctional interactions between phylogenetically older and younger brain structures and neural networks leading to a disruption of somatosensory processing have been identified to be involved in depersonalization (Sierra and David 2011). Consistently, activation of the multimodal association cortex (BA 39 as a part of the parietal-temporal-occipital area) was found to be closely associated with the degree of depersonalization (Simeon et al. 2000). This area lies near to the junction of temporal, occipital and parietal secondary association cortices involved in the processing of visual, auditory and somatosensory (body-knowledge) information. It plays a major role in the cross-modal association of this information as well as in the comparison of currently processed information with prior experiences; thus, it contributes to the constitution of subjectivity and emotion regulation. The multimodal association cortex is structurally and functionally connected with the anterior and posterior parts of the insula, which serves important functions in interoceptive awareness of body states, in the experience of bodily self-awareness, sense of body ownership and agency as well as the generation of emotional feeling. In patients with depersonalization disorder, the insular cortex was found to be hypoactive, possibly resulting from an inhibition by a hyperactive right ventrolateral prefrontal cortex (BA 47) involved in emotion recognition, particularly with respect to negative emotions. Further evidence indicates lower activity in the amygdala and the hypothalamus in depersonalization. Taken together, fronto-limbic and parieto-limbic disconnections are assumed as core neurobiological mechanisms underlying depersonalization (Sierra and David 2011). In line with these hypotheses, studies on patients with the dissociative subtype of posttraumatic disorder (PTSD) reveal an excessive activation of prefrontal cortical areas and the anterior part of the cingulate gyrus, involved in arousal modulation and emotion regulation, in combination with a reduced activity in the amygdala and the right anterior insula (Lanius et al. 2010).

These neurobiological findings are quite consistent with the psychodynamic conceptualization of depersonalization as defence mechanism against negative emotions such as anxiety, shame or disgust that result from immediate experiences or are associated with memories. If the intensity of these affective states becomes a threat to the self, they get shut down, are experienced from a distance and are detached from the self. However, the psychophysiological studies do not indicate a total lack of emotional responses but suggest that the peripheral physiological correlates of the emotional response are blunted due to a selective attenuation of sympathetic autonomic arousal (Sierra and David 2011). One might speculate that DPD patients experience negative emotions without the corresponding bodily sensations, and this mismatch evokes the feeling of alienation.

### 19.3.4 Pseudoneurological Conversion Disorders

The DSM-5 subsumes the pseudoneurological conversion disorders under the category of somatic symptom disorders and terms them functional neurological symptom disorder. In contrast, the ICD-10 considers them as dissociative disorders. There are good arguments for both positions, and from a neuropsychodynamic perspective, there is considerable overlap and similarity.

> Ms. N., 19 years old, single and without any vocational qualification, was referred to the emergency department of a university hospital by her general practitioner due to the acute onset of a slack and complete paralysis of her right arm that had occurred the same morning. Her neurological and general physical examinations, as well as additional assessments including magnetic resonance imaging of the head, an electroencephalogram (EEG), extensive laboratory studies, cerebrospinal fluid diagnostics as well as neuro-physiological tests (evoked potential, electroneurographs) were unremarkable suggesting that organic causes could be discarded. Suspecting a conversion disorder, a psychiatric-psychotherapeutic consultation was initiated. At the examination, Ms. N. was oriented to all qualities; there was no evidence for an altered state of consciousness, disorientation, attention or memory deficits. Interestingly, she did not appear to be worried or concerned by her paralysis at all. Asked about potential psychosocial stressors, she reported to have given birth to a healthy boy just a week ago. The intense and stormy relationship to the father of her son was characterized by ambivalent feelings on both sides; he left her shortly after she had told him of expecting his child. After this depressing end of the affair, Ms. N. initially wanted to have an abortion, but her mother convinced her to continue the pregnancy. Immediately after the normal and uncomplicated delivery and seeing her son for the first time, she cried out: 'He looks like his father!'

In addition to the above-mentioned conversion phenomena with motor symptoms manifesting as paralysis, gait abnormalities, tremor or astasia, there is a multitude of other conversion symptoms including anaesthesia, sensory loss (e.g. deafness or blindness) as well as psychogenic non-epileptic seizures (PNES). This diversity may explain why neurobiological, particularly functional neuroimaging, studies have not yet revealed any strikingly consistent finding as indicated by systematic and critical reviews (Browning et al. 2011). However, there is emerging evidence for neuropsychodynamically plausible alterations corresponding to neurobiological correlates of other dissociative syndromes as depicted above. Convergent neuroimaging findings implicate dysfunctions in extensive brain circuits comprising the dorsolateral prefrontal, ventromedial prefrontal and inferior frontal cortices, components of the hippocampal formation, the temporoparietal junction, cingulum as well as those cortical areas pivotal for the respective neurophysiological system (van Beilen et al. 2010; Aybek et al. 2014; Burke et al. 2014; Perez et al. 2015). In sensory and motor conversion disorder, hypoactivity in contralateral somatosensory and primary motor cortices accompanied by concurrent hyperactivity in prefrontal regions, cingulum and angular gyrus has been reported. Taken together, it is suggested that a top-down or cortical-subcortical inhibition or decoupling of networks and pathways mediating emotion regulation, self-referential processes and perceptual awareness as well as motor planning and coordination might be the central neurobiological mechanisms in conversion. Alterations in self-referential processing might be interpreted as neural correlate of inter- and intra-systemic conflicts in a psychodynamic sense. In other words, conversion disorders can be conceptualized as the phenomenological result of a failure in the integration of contradictory affective, cognitive and self-referential information leading to dysfunctions in the motor and sensory systems.

## 19.4 Unresolved Issues

Dissociative identity disorder (DID) is characterized by at least two distinct identities or dissociated personality states each with its own affective, cognitive and behavioural pattern (American Psychiatric Association 2013; Dorahy et al. 2014). The different labels referring to these identities such as identity or personality states, dissociative subsystems, self- or ego states as well as alters do not only indicate terminological confusion but also conceptual ambiguity. From both a psychodynamic and a neurobiological perspective, the constitutive essence of constructs such as ego, self, identity, consciousness and personality remains to be determined. One essential question relates to the problem if dissociative parts of the personality are actually self-aware and self-conscious in the sense of subjectivity with a phenomenal, first-person perspective (Nijenhuis and van der Hart 2011). Translating these considerations into neuropsychodynamic terms, one might ask if multiple selves are possible in one brain (Reinders et al. 2003). This would be the premise for the clinically seminal symptom of switching, i.e. the observable change between different alter self-states. Taking this into account, inconsistent neurobiological findings with regard to DID do not come as a surprise (Dorahy et al. 2014). While some studies do not report differences in activation or perfusion patterns between distinct dissociative parts, other investigations indicate divergent activation and perfusion patterns related to different identity states. In case series of DID patients, the switching process seems to be accompanied by the activation and inhibition of a varying array of neural networks and structures (reviewed by Dorahy et al. 2014). However, it is far from certain that these alterations in neuronal activity are specific to these so-called switches, and it may well be that they can also be found in related phenomena, e.g. regression. In short, neurobiological approaches to DID do not provide definite evidence to clarify its ambiguous nosological status. From a clinical point of view, there is no doubt that patients having experienced severe and chronic early life stress and interpersonal traumatization exhibit different personality configurations making it feasible to conceive DID as the most complex, childhood-onset form of posttraumatic stress disorder (Dalenberg et al. 2012). Correspondingly, a recent neuroimaging study found different and opposing activation patterns in neural networks involved in emotion modulation in hypo- and hyperaroused identity states (Reinders et al. 2014). These results are very similar to the findings reported for the neurobiologically and clinically distinct subtypes of PTSD (Lanius et al. 2010).

To date, a neuropsychodynamic interpretation of phenomena and processes such as the unconscious choice of symptom, the high affinity for symbolic symptoms and important unconscious identification mechanisms remains to be determined. From a psychodynamic point of view, these aspects are essential for approaching and understanding the patients and make it possible to draw conclusions on their basic intrapsychic or interpersonal conflicts, structural deficits and traumatic experiences.

The same applies to communicative features: Dissociative symptoms and disorders become manifest in an interpersonal field, and the reactions of relevant others and of the environment play a key role not only in the pathogenesis but also in the maintenance and treatment. The importance of the secondary gain was already underscored by Freud himself. Thus, dissociative phenomena serve as a good example to illustrate that the brain is a socially embedded organ (cf. Chap. 4 this volume). The 'hysterical communication' and the unconscious dissociative enactments (Bromberg 2008) can be conceived as alterations in the self-brain-environment relation.

## 19.5 Preliminary Attempt of an Integration and Therapeutic Implications

Despite the tentativeness of the above-mentioned findings and considerations and the necessity of replication studies, it may be possible to draw some converging lines of neurobiological evidence that correspond well to clinical models. Taking into account the vast phenomenological heterogeneity of dissociative symptoms, syndromes and disorder, the involvement of a multitude of neural networks does not come as a surprise including cortical and subcortical structures such as the prefrontal cortex, the multimodal association cortex as a part of the parietal-temporal-occipital area, the limbic system and cortical midline structures. In line with the phenomenology of dissociation, these networks play a crucial role in the processing and integration of affective signals, stimuli from the external environment, the body and the inner world of the person as well as the encoding, storage and retrieval of these information. In particular, cortical midline structures are considered the most relevant functional circuit with respect to the processing of self-referential affective and cognitive information resulting in the experience of an 'I' or a phenomenal self (Northoff and Bermpohl 2004; Musholt 2013).

A neuropsychodynamic model of dissociation could be conceptualized as follows: an external stimulus (e.g. traumatic stress) or an internal signal (e.g. a conflict between sexual desires and the interdiction of the superego) lead to neuronal activation patterns that manifest on the level of subjective experience as highly adverse and intense emotions such as anxiety, shame, disgust and anger, i.e. the unpleasant and incompatible ideas according to Freud. Both the highly negative valence and the (hyper-)arousal of these affective states lead to a failure of their adequate integration with cognitive and self-referential information by a 'bottom-up' hyperactivation of (right) prefrontal areas which, in turn, inhibit top-down cortical-subcortical networks which are essential for memory functioning, self-consciousness and agency as well as body control. In short, the dysfunctional interaction between cortical and subcortical circuits results in dissociative phenomena. The relevance of the indicated lateralization to right hemispherical processes remains speculative, but there is converging evidence both from the fields of neurobiology and developmental psychology for this notion (Schore 2002). Our proposed concept can even be extended to a stress-vulnerability model: a neuropsychodynamically conceived predisposition for dissociation results from unfavourable developmental conditions characterized by traumatic stress and disrupted child-caretaker interactions (e.g. failures in affect-mirroring), which—on a neurobiological level—interfere with the formation of stable and functional neural networks pivotal for the development of a healthy self-system and ego functions. While dissociative symptoms emerge from little psychosocial stress in subjects with a high tendency to dissociate, severe traumatic experiences can evoke dissociation even in mentally healthy people without a respective predisposition. Dissociation becomes clinically manifest whenever the psychosocial strain disrupts the individual protection shield or is too excessive for the brain's information processing capacities, respectively. The intentionally wide term 'psychosocial strain' encompasses both external traumatic experiences as well as interpersonal and intrapsychic conflicts. Initially, dissociation enables the individual to experience the negative emotions associated with the psychosocial burden less intensely and helps to cope with it by a subjective decontextualization. In case this kind of coping procedure is successful, the person feels a short-term relief, and by means of operant conditioning, the threshold for dissociation-inducing signals becomes lower and lower, i.e. there is a generalization of the dissociative response to psychosocial stress.

From a clinical and therapeutic perspective, it is vital to identify dissociative phenomena as early as possible. Psychometrically sound diagnostic tools such as self-report measures and extensive interviews can aid the assessment process and help to establish the diagnosis of the most prominent dissociative disorders (Frankel 2009). However, dissociative symptoms are also frequent in other mental disorders (Putnam et al. 1996). Because dissociation mainly represents a specific subjective experience that can hardly be observed from the outside, the respective clinical phenomena are often missed. Their early identification is particularly important, because dissociation predicts poor treatment outcome (Spitzer et al. 2007). There is good evidence that emotional, amygdala-based learning processes are inhibited during dissociative states (Ebner-Priemer et al. 2009), and even a most intense working through will miss the mark if the patient's dissociation goes unrecognized. The different therapeutic approaches for the treatment of dissociative symptoms and disorders converge in their focus on control and thus agency, self-efficacy, improvement of ego-functioning and strengthening the self (for a review see the respective chapters in Dell and O'Neil 2009).

## References

American Psychiatric Association. Diagnostic and statistical manual of mental disorders. 5th ed. Arlington: American Psychiatric Publishing; 2013.
Aybek S, Nicholson TR, Zelaya F, et al. Neural correlates of recall of life events in conversion disorder. JAMA Psychiat. 2014;71:52–60.
Becker-Blease KA, Deater-Deckard K, Eley T, et al. A genetic analysis of individual differences in dissociative behaviors in childhood and adolescence. J Child Psychol Psychiatry. 2004;45:522–32.
van Beilen M, Vogt BA, Leenders KL. Increased activation in cingulate cortex in conversion disorder: what does it mean? J Neurol Sci. 2010;289:155–8.
Bell V, Oakley DA, Halligan PW, et al. Dissociation in hysteria and hypnosis: evidence from cognitive neuroscience. J Neurol Neurosurg Psychiatry. 2011;82:332–9.
Breuer J, Freud S. Studies on hysteria. In: Strachey J, editor. The standard edition of the complete psychological works of Sigmund Freud, vol. 2. London: Hogarth; 1955. p. 1–319.
Bromberg PM. Mentalize THIS!: dissociation, enactment, and clinical process. In: Jurist E, Slade A, Bergner S, editors. Mind to mind: infant research, neuroscience, and psychoanalysis. New York: Other Press; 2008. p. 414–34.
Brown RJ, Cardena E, Nijenhuis ERS, et al. Should conversion disorder be reclassified as dissociative disorder in DSM-V? Psychosomatics. 2007;48:369–78.
Browning M, Fletcher P, Sharpe M. Can neuroimaging help us to understand and classify somatoform disorders? A systematic and critical review. Psychosom Med. 2011;73:173–84.
Burke MJ, Ghaffar O, Staines WR, et al. Functional neuroimaging of conversion disorder: the role of ancillary activation. Neuroimage Clin. 2014;30:333–9.
Cardena E. The domain of dissociation. In: Lynn SJ, Rhue RW, editors. Dissociation: theoretical, clinical, and research perspectives. New York: Guilford Press; 1994. p. 365–94.
Chen Y, Baram TZ. Toward understanding how early-life stress reprograms cognitive and emotional brain networks. Neuropsychopharmacology. 2016;41:197–206.
Dalenberg CJ, Brand BL, Gleaves DH, et al. Evaluation of the evidence for the trauma and fantasy models of dissociation. Psychol Bull. 2012;138:550–88.
Dell PF, O'Neil JA, editors. Dissociation and dissociative disorders: DSM-V and beyond. New York: Routledge; 2009.
Dorahy MJ, Brand BL, Sar V, et al. Dissociative identity disorder: an empirical overview. Aust N Z J Psychiatry. 2014;48:402–17.
Dutra L, Bureau JF, Holmes B, et al. Quality of early care and childhood trauma: a prospective study of developmental pathways to dissociation. J Nerv Ment Dis. 2009;197:383–90.
Ebner-Priemer UW, Mauchnik J, Kleindienst N, et al. Emotional learning during dissociative states in borderline personality disorder. J Psychiatry Neurosci. 2009;34:214–22.
Ellenberger HF. The discovery of the unconscious. The history and evolution of dynamic psychiatry. New York: Basic Books; 1970.
Frankel AS. Dissociation and dissociative disorders: clinical and forensic assessment with adults. In: Dell PF, O'Neil JA, editors. Dissociation and dissociative disorders: DSM-V and beyond. New York: Routledge; 2009. p. 571–83.
Guralnik O, Simeon D. Depersonalization: standing in the spaces between recognition and interpellation. Psychoanal Dialogues. 2010;20:400–16.
Holmes EA, Brown RJ, Mansell W, et al. Are there two qualitatively distinct forms of dissociation? A review and some clinical implications. Clin Psychol Rev. 2005;25:1–23.
Jang KL, Paris J, Zweig-Frank H, et al. Twin study of dissociative experience. J Nerv Ment Dis. 1998;186:345–51.
Koenen KC, Saxe G, Purcell S, et al. Polymorphisms in FKBP5 are associated with peritraumatic dissociation in medically injured children. Mol Psychiatry. 2005;10:1058–9.
Kopelman MD. Disorders of memory. Brain. 2002;125:2152–90.

Lanius RA, Vermetten E, Loewenstein RJ, et al. Emotion modulation in PTSD: clinical and neurobiological evidence for a dissociative subtype. Am J Psychiatry. 2010;167:640–7.

Lochner C, Seedat S, Hemmings SM, et al. Investigating the possible effects of trauma experiences and 5-HTT on the dissociative experiences of patients with OCD using path analysis and multiple regression. Neuropsychobiology. 2007;56:6–13.

Markowitsch HJ, Calabrese P, Fink GR, et al. Impaired episodic memory retrieval in a case of probable psychogenic amnesia. Psychiatry Res. 1997;74:119.e26.

McLean SA, Diatchenko L, Lee YM, et al. Catechol O-methyltransferase haplotype predicts immediate musculoskeletal neck pain and psychological symptoms after motor vehicle collision. J Pain. 2011;12:101–7.

Musholt K. A philosophical perspective on the relation between cortical midline structures and the self. Front Hum Neurosci. 2013;7:536. https://doi.org/10.3389/fnhum.2013.00536.

Nijenhuis ER, van der Hart O. Dissociation in trauma: a new definition and comparison with previous formulations. J Trauma Dissociation. 2011;12:416–45.

Nijenhuis ER, Spinhoven P, Vanderlinden J, et al. Somatoform dissociative symptoms as related to animal defensive reactions to predatory imminence and injury. J Abnorm Psychol. 1998;107:63–73.

Northoff G, Bermpohl F. Cortical midline structures and the self. Trends Cogn Sci. 2004;8:102–7.

Ogawa JR, Sroufe LA, Weinfield NS, et al. Development and the fragmented self: longitudinal study of dissociative symptomatology in a nonclinical sample. Dev Psychopathol. 1997;9:855–79.

Perez DL, Dworetzky BA, Dickerson BC, et al. An integrative neurocircuit perspective on psychogenic nonepileptic seizures and functional movement disorders: neural functional unawareness. Clin EEG Neurosci. 2015;46:4–15.

Pieper S, Out D, Bakermans-Kranenburg MJ, et al. Behavioral and molecular genetics of dissociation: the role of the serotonin transporter gene promoter polymorphism (5-HTTLPR). J Trauma Stress. 2011;24:373–80.

Putnam FW, Carlson EB, Ross CA, et al. Patterns of dissociation in clinical and nonclinical samples. J Nerv Ment Dis. 1996;184:673–9.

Reinders AA, Nijenhuis ER, Paans AM, et al. One brain, two selves. NeuroImage. 2003;20:2119–25.

Reinders AA, Willemsen A, den Boer JA, et al. Opposite brain emotion-regulation patterns in identity states of dissociative identity disorder: a PET study and neurobiological model. Psychiatry Res. 2014;223:236–43.

Sartory G, Cwik J, Knuppertz H, et al. In search of the trauma memory: a meta-analysis of functional neuroimaging studies of symptom provocation in posttraumatic stress disorder (PTSD). PLoS ONE. 2013;8(3):e58150.

Savitz JB, van der Merwe L, Newman TK, et al. The relationship between childhood abuse and dissociation. Is it influenced by catechol-O-methyltransferase (COMT) activity? Int J Neuropsychopharmacol. 2008;11:149–61.

Schilder P. Depersonalization. Introduction to psychoanalytic psychiatry. Nerv Ment Dis. 1928;50:120.

Schore AN. Advances in neuropsychoanalysis, attachment theory, and trauma research: implications for self psychology. Psychoanal Inq. 2002;22:433–84.

Sierra M, Berrios GE. Depersonalization: neurobiological perspectives. Biol Psychiatry. 1998;44:898–908.

Sierra M, David AS. Depersonalization: a selective impairment of self-awareness. Conscious Cogn. 2011;20:99–108.

Simeon D, Guralnik O, Hazlett EA, et al. Feeling unreal: a PET study of depersonalization disorder. Am J Psychiatry. 2000;157:1782–8.

Spitzer C, Spelsberg B, Grabe HJ, Mundt B, Freyberger HJ. Dissociative experiences and psychopathology in conversion disorders. J Psychosom Res. 1999;46:291–4.

Spitzer C, Barnow S, Freyberger HJ, Grabe HJ. Dissociation predicts symptom-related treatment outcome in short-term inpatient psychotherapy. Aust N Z J. 2007;41:682–7.

Staniloiu A, Markowitsch HJ. Towards solving the riddle of forgetting in functional amnesia: recent advances and current opinions. Front Psychol. 2012;3:403. https://doi.org/10.3389/fpsyg.2012.00403.

Staniloiu A, Markowitsch HJ. Dissociative amnesia. Lancet Psychiatry. 2014;1:226–41.

Van der Hart O, Dorahy MJ (2009) History of the concept of dissociation. In: Dell PF, O'Neil JA (eds) Dissociation and dissociative disorders: DSM-V and beyond. Routledge, New York, pp 3-26

# Addictive Disorders

# 20

Moritz de Greck and Georg Northoff

**Abstract**

Addictive disorders cause profound changes in behavior and subjective experience. The addictive behavior will preoccupy the patient's thinking and dominate his motivations. It will become essential for the patient's well-being.

From a neurobiological point of view, the reward system (including nucleus accumbens, amygdala, and the prefrontal cortex) plays a key role in the formation of addictive disorders.

From a psychodynamic perspective, the four psychologies of psychoanalysis (including drive theory, ego psychology, object relations theory, and self psychology) developed complementary concepts which are helpful in our understanding of the function of addictive disorders.

The knowledge of the neurobiological underpinnings and psychodynamic concepts of addictive disorders leads to new implications for their treatment.

## 20.1 Introduction

Humans can develop a wide spectrum of addictive disorders. It is useful to differentiate between substance-related and non-substance-related addictive disorders. Substance-related disorders include alcohol addiction, benzodiazepines

M. de Greck (✉)
Department for Psychiatry, Psychosomatic and Psychotherapy,
Johann Wolfgang Goethe University Hospital, Frankfurt am Main, Germany

G. Northoff
Mind, Brain Imaging, and Neuroethics,
Institute of Mental Health Research, University of Ottawa, Ottawa, ON, Canada
e-mail: georg.northoff@theroyal.ca; www.georgnorthoff.com



H. Boeker et al. (eds.), *Neuropsychodynamic Psychiatry*,
https://doi.org/10.1007/978-3-319-75112-2_20

addiction, opioid addiction, and others. On the other hand, pathological gambling, computer game addiction, Internet addiction, and sexual addiction are included in non-substance-related addictive disorders. For many substances and activities, there is a recreational, non-pathological form of consumption. If a person loses his ability to control consumption, however, and is instead subjected by inner obsession and if feelings, thoughts, and activities are dominated by consumption, a recreational use transforms into an addiction. In addition, behavioral patterns, similar to addictive activities, occur in several other psychiatric disorders such as, for instance, eating disorders or obsessive compulsive disorders.

## 20.2 Case Report: Mr. J. (26 Years): Part 1

### 20.2.1 Admission

Mr. J., a 26-year-old student with engineering major, was admitted to day-clinic treatment. The treatment was recommended by his outpatient psychotherapist. The duration of the treatment was planned to last for several weeks. Mr. J. was the only child of his parents. His relationship with his parents was rather oriented on achievements than keenly empathic. At the time of admission, Mr. J. was living alone. He was receiving financial support by his parents. He was also earning money by a part-time job as a software engineer. He separated from his girlfriend (a teaching student of the same age) about a year before admission; since then, he was single. During the admission interview, Mr. J. explained that he spent more and more time in the Internet, in particular browsing through news and information sites. His Internet time increased to about 6 h/day at the time of admission. His Internet activities had troubled him since about 2 years. For instance, he often argued with his girlfriend about his Internet activities, which finally led to the separation after 3 years of relationship. He intended to reduce his Internet time for several times; however, he could not control his urges and gave up each time. During the last months, he should have spent more efforts on his study. Due to poor preparation, he recently failed in an important test. As a consequence, he can now only finish his study with a delay of 1 year. He knew that he should have concentrated more on his study; however, the urge to surf the Internet was stronger. He took the averted graduation as a reason to seek for therapeutic support. Having spent few sessions with his outpatient therapist, though, his therapist recommended to switch to a day-clinic treatment, since the outpatient treatment seemed not sufficient to deal with his problems. During the admission interview, Mr. J. also mentioned that he was unsatisfied with his lack of social contacts and activities, as well as his few sporting activities. He described himself as a socially cautious person; however, this got more severe during the last years.

### 20.2.2 Diagnostic Criteria

The German psychiatrist and psychoanalyst Mentzos (2009) described a restrained inner freedom, restricted flexibility, and repeated loss of control as key features occurring in all addictive diseases. Driving forces behind addictive behaviors are the urges to achieve pleasant affective states of relaxation and/or stimulation.

The psychologist M. Griffiths defined six core criteria of addictive disorders (Griffiths 2005; Andreassen et al. 2012): salience (i.e., the addictive activity dominates thinking and behavior), mood modification (i.e., the activity modifies or improves mood), tolerance (i.e., increasing amounts of the activity are required to achieve previous effects), withdrawal (i.e., unpleasant feelings occur when the activity is discontinued or suddenly reduced), conflict (i.e., the activity causes conflicts in relationships, work, education, or other activities), and relapse (i.e., addictive disorders imply a tendency to revert to earlier patterns of the activity after abstinence or control).

The DSM-5 (American Psychiatric Association 2013) lists 11 criteria, as common diagnostic characteristics independent from the specific substance of abuse:

1. The substance is taken in larger amounts or more often than intended.
2. There is a desire or there have been attempts to cut down substance use.
3. A great deal of time is spent in order to obtain the substance, use it, and recover from its effects.
4. There is craving for the substance.
5. Substance use results in failures to fulfill obligations at work, at school, or at home.
6. Substance use is continued despite social or interpersonal problems caused by the substance use.
7. Important social, occupational, or recreational activities are given up or reduced to use the substance.
8. The substance is used, although it is physically hazardous.
9. The substance is used, despite knowledge of physical or psychological problems caused by the substance.
10. Tolerance develops, i.e., increasing amounts of the substance are needed to obtain the same effect.
11. Withdrawal develops in substance-free episodes.

## 20.3 Case Report: Mr. J. (26 Years): Part 2

### 20.3.1 Diagnostic Criteria

Mr. J. fulfilled several criteria for addiction. For instance, Mr. J. was unsatisfied with his Internet behavior and the amounts of time spent online. In the past, he had made several vain attempts to cut down his online time. But instead he steadily increased his Internet time. Due to his Internet activities, he got problems with his girlfriend, finally leading to their separation. In addition, he wasn't able to fulfill important obligations concerning his study.

### 20.3.2 Psychodynamic Concepts

### 20.3.3 The Formation of Addictive Disorders: Perspectives of Different

#### 20.3.3.1 Psychoanalytic Schools

The psychoanalyst Pine (1988) differentiated four conceptually separable psychological perspectives, which developed within the psychoanalytic framework: drive theory, ego psychology, object relations theory, and self psychology. Each of these psychoanalytic schools has improved our understanding of psychodynamic mechanisms underlying addictive disorders enormously.

### 20.3.4 Drive Theory

Drive theory was developed by S. Freud. Until the invention of drive theory, symptoms of addictive disorders were seen as a consequence of toxic effects of the drug of abuse. The motives underlying drug intake were not object of much consideration. S. Freud was the first to emphasize the question: what led addicted patients to consume a noxious drug (Radó 1933)? In view of drive theory, addictive disorders form on the basis of disturbed libido development, and addictive behaviors serve as a surrogate for prohibited gratifications of the sexual drive. Freud viewed masturbation as the origin of all addictive disorders (which might follow, Subkowski 2008; Freud 1962). Apart from this, Freud wrote rather scarcely about addictive disorders (Subkowski 2008).

According to drive theory, the ecstasy induced by drug consumption serves as a proxy for the genital orgasm (Mentzos 2009), or, in other words, as described by the Hungarian psychoanalyst S. Radó, drug intakes induce a "pharmacogen orgasm" (Chessick 1960; Radó 1926).

Other authors, for instance, D. Hartmann, described a regression to or in some cases a fixation on the oral phase of psychic development. Hence, the drug serves as a symbolic representation of mother's milk (Hartmann 1969). This is in accordance with the observation that in some ecstatic states, the individual totally withdraws from external objects, as described by the (object relationship theory oriented) psychoanalyst Rosenfeld (1960). In addition, Hartmann (1969) observed that sexual activities of adolescent drug addicts were predominantly restricted to autoerotic and masturbatic behaviors. There are, however, also opposite phenomena, as, for example, sexual disinhibition after alcohol intake (Rosenfeld 1960) (Fig. 20.1).

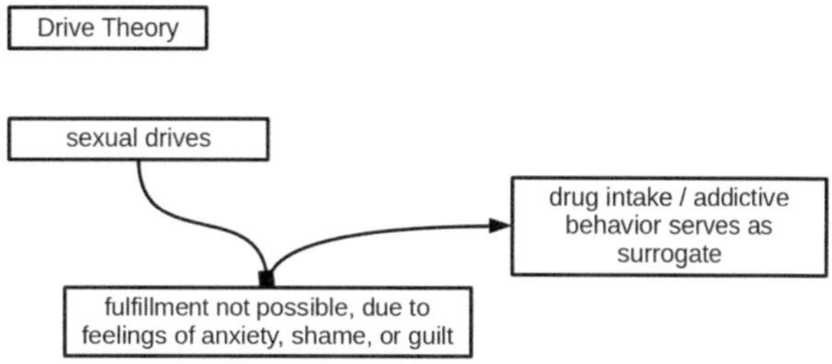


**Fig. 20.1** Formation of addictive disorders from a drive theory perspective

### 20.3.5 Ego Psychology

From an ego-psychological point of view, addictive behavior is the only way possible for a structurally dysfunctional ego to achieve a minimum level of psychic stability. The Swiss-American psychoanalyst Wurmser (1997) described this as a dysfunctional symbolization ("hyposymbolization"): addictive patients often lack the ability to meaningfully articulate significant feelings. These are then instead expressed in the form of somatic complaints, bodily unease, or social accuses.

Different ego-psychological authors stress that it is not the desire for pleasant affective states but rather the desire to mitigate unbearable tensions that drives addictive behavior (Hartmann 1969; Rost 2001; Savitt 1963). Due to a structural weakness of the ego, these patients can mitigate these tension states only by referring to drugs (Hildebrandt 2007; Rosenfeld 1960).

In view of Radó (1933), addicted patients suffer from deficits in their self-care development and are incapable to autonomously stabilize the self-esteem. They are thus sensitive to frustrations which can induce agonizing states of "anxiety depressions." According to the German psychoanalyst Rost (2001), these anxiety-depression states resemble a primitive affect ("Ur-Affekt"), which is comparable with infantine death anxiety. Since these affective states are experienced as unbearable, patients will try everything to get rid of them. If—in this situation—a patient experiences (often by accident) that a certain behavior (e.g., shopping, browsing the Internet, drinking alcohol) leads to an alleviation of these unpleasant affective states, there is a risk that this person will turn to this behavior more frequently in the future (which is homologous to instrumental learning). The addictive behavior thus takes the place of self-care. What follows is a retreat from reality: psychic well-being is not achieved by dealing with demands of the "real" external world but instead by substance intake or addictive behaviors. Since these are much more simple and faster than arranging with issues of the real world, they often feel "magical."

Similarly, Wurmser (1997) views addictive behavior as a possibility to artificially defend against overwhelmingly unpleasant effects of loneliness, void, disillusionment, and anger. Wurmser then turns to defects of the superego and conceptualizes addictive disorders as a "turning syndrome"

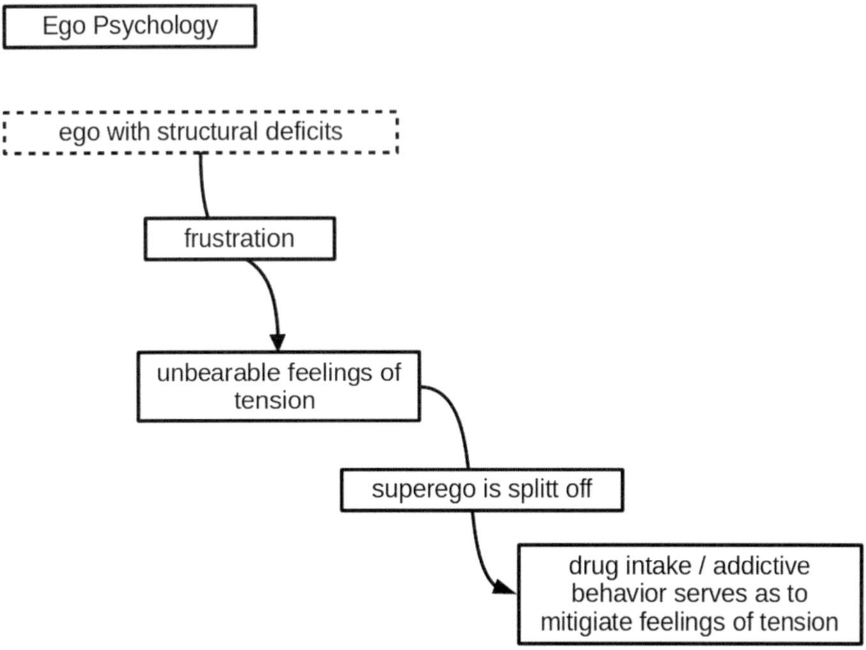


**Fig. 20.2** Formation of addictive disorders from an ego-psychological perspective

("Kehrtwende-Syndrom"): addicted patients predominantly tend to use the defense mechanism splitting in order to achieve emotional stability. According to Wurmser, splitting describes a complex but unstable defense construct which combines the defense mechanisms repression, abnegation, regression, projection, externalization, and dissociation. Within one moment, mature superego functions (including a functioning conscience, ideals, and responsibility) deteriorate. The superego collapses and cannot control the addictive behavior any longer. The mature superego is replaced by a more primitive one, which tolerates the self-destructiveness of the addictive behavior and sometimes even appreciates it as a means of self-punishment.

The important role of superego defects in the development of addictive disorders is further emphasized by several other authors. Freud wrote about a pleasant disinhibition due to the shutdown of superego commandments caused by drug intake (Freud 1905). In addition, Rost (2001) stresses the role of superego defects in addicted patients and quotes the famous statement: "The super-ego can be dissolved in alcohol."

A very different point of view takes Wernado et al. (2006) by focusing on the effect of psychotropic substances on ego functions such as affect tolerance, affect control, tolerance toward frustrations, impulse control, the capability to anticipate, or the capability to judge. In the course of an addictive development, it does not seldomly happen that these ego functions do not serve the ego anymore but are rather subjected to assure substance intake (e.g., acquisition, consumption, Fig. 20.2).

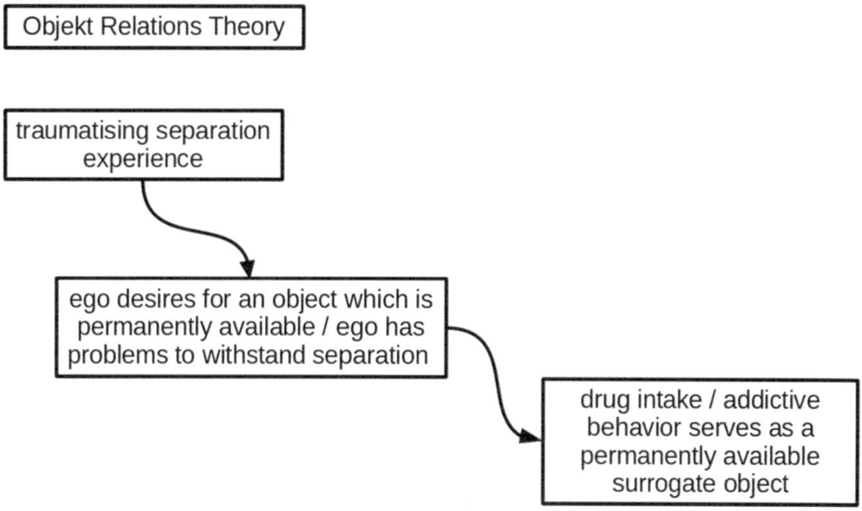


**Fig. 20.3** Formation of addictive disorders from an object relations theoretical perspective

### 20.3.6 Object Relations Theory

Within the school of object relations theory, addictive behavior is seen as a surrogate object, which is permanently available. At this, addictive behavior is often installed to avoid painful feelings of separateness. Rosenfeld (1960) hypothesizes that drug intake represents on an unconscious level the attempt to incorporate a deceased object. D. Hartmann, when working with drug-addicted adolescents, was able to show that one of them indeed had lost a parent (Hartmann 1969). The drug thus serves as a substitute for a human object. While a real human object has limited abilities and follows own interests, however, the drug is permanently present and freely available (Wernado et al. 2006). Wurmser (1997) describes that sedating drugs (such as opioids or alcohol) act as a "motherly claustrum" (i.e., as the fantasized mother's womb) which is on the one hand desired (in terms of a "claustrophilia") and on the other hand feared (in the sense of a "claustrophobia", Fig. 20.3).

### 20.3.7 Self Psychology

From a self-psychological view, unempathic relationships between a child and his caregiver can lead to deficits in the cohesion, strength, and harmony of the child's self (Kohut 1975; Wolf 1998). Drug consumption or addictive activities can then serve as a means to stabilize a chronic instable self (Wurmser 1997). This stabilization by drug intake or addictive behaviors, however, is short termed. It is followed by feelings of shame and guilt, which lead to a renewed destabilization of the self. To emphasize this twofold effect, the German psychoanalyst W. Milch introduced the expression "pathological self-object" in order to differentiate it from original self-object (e.g., empathic relationships) (Milch 2001).

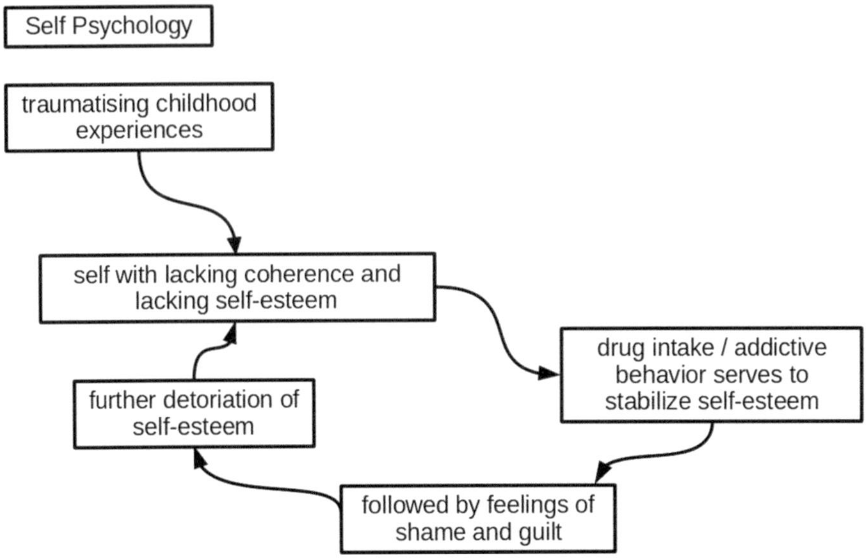


**Fig. 20.4** Formation of addictive disorders from a self-psychological perspective

Wurmser (1997) conceptualizes a vicious circle of a narcissistic breakdown, which underlies addictive behavior: a disappointment or frustration leads to intense, uncontrollable feelings of anger, shame, and desperation. These feelings turn into a vague, unbearable tension, which induces a "desire for relief." An irrepressible urge to act is the consequence, which leads to aggressions against one's self and others. This is followed by a split-off of the superego (a "global denial of all superego contents"), which renders the superego temporarily meaningless. Eventually, the drug is lustfully consumed. The consumption is followed, however, by renewed feelings of disappointment, shame, and guilt (Fig. 20.4).

## 20.4 Case Report: Mr. J. (26 Years): Part 3

### 20.4.1 Biographical Background

When interviewed about significant biographical events, Mr. J. came to speak about an episode which he experienced in his childhood. Mr. J.'s parents separated when he was 9 years due to his father's extramarital affair. Mr. J. remembered to have taken site for his mother and stayed with her after the separation, while he felt strong feelings of anger and hate toward his father. To Mr. J.'s surprise, however, his parents reunited 2 years later. Besides being surprised, Mr. J. wasn't able to bring back his emotions connected to this incident. In the course of his psychotherapy, however, Mr. J. was able to elaborate that he experienced not only the separation of his parents but also their reunion as painful rejections.

### 20.4.2 Neurobiological Underpinnings

The dopaminergic reward system—key system of all addictive disorders—leads to profound alterations of subjective experience and behavior. Addictive activities dominate the patient's motivations and cognitions nearly completely. Besides the addictive behavior, there is nothing left to be enjoyed. In healthy individuals, the dopaminergic reward system induces behavioral patterns which serve two purposes: to preserve the organism itself and to preserve its species. For instance, the search for water and nutritious food, relationships, and sexual contacts are initiated and rewarded by the dopaminergic reward system (Frank et al. 2010; Goldstone et al. 2009; Hollmann et al. 2013; Kringelbach et al. 2003; Simmons et al. 2005; Stoléru et al. 2012; Wang et al. 2004). The dopaminergic reward system, however, also plays a key role in the development of addictive disorders. All substances which can lead to addiction have a key feature in common, which is their ability to activate the reward system without having any evolutionary or self-preservatory advantages (Di Chiara and Bassareo 2007; Heinz et al. 2004; Robbins and Everitt 1999; Wise 1996; Wise and Bozarth 1987). Drug consumption thus induces a feeling of desire for renewed intake as if the organism had eaten a nutritious food in a state of hunger or as if he had something similar essential for the preservation of his self or his species.

### 20.4.3 Important Components of the Reward System

Important components of the reward system include the ventral tegmental area, the ventral striatum which contains the nucleus accumbens, the ventral pallidum, the amygdala, the prefrontal cortex, the orbitofrontal cortex, and the insular cortex (Berridge and Robinson 2003). The ventral tegmental area is the origin of the dopaminergic reward system. Here is the location of dopaminergic neurons that send their axons to the ventral striatum and thus can activate this region. The main purpose of the ventral tegmental area is to influence the organism's behavior, so that actions, which had led to a result which was better than expected (positive prediction error), will be repeated more frequently in the future (Bayer and Glimcher 2005). The ventral striatum and the nucleus accumbens contain two distinguishable sets of neurons: neurons which are activated by dopaminergic stimulation from the ventral tegmental area lead to motivation for specific actions (which corresponds to the feeling of "wanting"), while neurons which are under the influence of opioid neurotransmission lead to affective pleasure (which corresponds to the feeling of "liking"; Berridge and Robinson 2003). The ventral pallidum is involved in these motivational and affective processes similar to the ventral striatum (Berridge and Robinson 2003; Cromwell and Berridge 1993). The amygdala plays an important role in the emotional evaluation of relevant stimuli, which are important for survival or species preservation. It is essential in attributing salience to important stimuli (Berridge and Robinson 2003; Grüsser and Thalemann 2006). Neurons of the prefrontal cortex, which send their

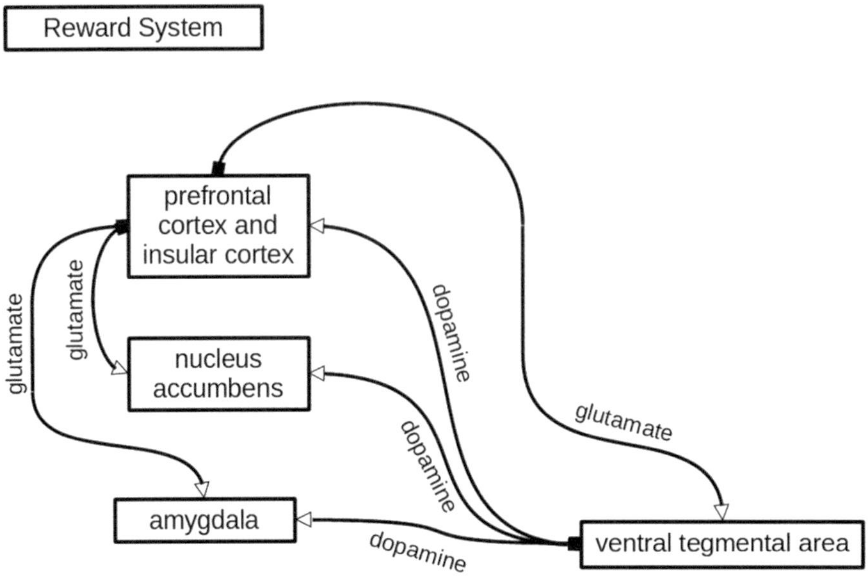


**Fig. 20.5** Key structures of the reward system (simplified after Berridge and Robinson 2003)

activating axons to the ventral tegmental area, also play an important role (Karreman and Moghaddam 1996). The prefrontal cortex is important for the development of long-term motivations and aims. By a positive evaluation of the actual situation with regard to long-term aims, it can induce a tonic baseline activity in the ventral tegmental area and ventral striatum (Davey et al. 2008). Artificial stimulation of the prefrontal cortex of alcohol-addicted patients with transcranial direct current stimulation leads to reduced craving for alcohol (Boggio et al. 2008). Similar results were obtained with transcranial magnetic stimulation in cocaine-addicted patients: again the stimulation of prefrontal cortex led to reduced craving for cocaine (Camprodon et al. 2007). The orbitofrontal cortex and insular cortex are important to estimate the reward value of a stimulus. For this stored similar experiences have to be evaluated (Berridge and Robinson 2003; Grüsser and Thalemann 2006) (Fig. 20.5).

### 20.4.4 The Function of the Reward System in a Nutshell

The following sketch may illustrate the functions of different reward structures:

A hungry girl decides to climb an apple tree to pick an apple. She clambers the tree and chooses an apple which shines particularly red. Since the apple tastes better than expected, the girl concludes to look out for red apples more often.

The decision to climb on the tree is made in the prefrontal and orbitofrontal cortices. Here are memories stored, regarding worthwhile previous ventures to pick an apple. The insular cortex has memories about the delicious taste of apples. Prefrontal and insular cortices activate the ventral tegmental area, which in turn activates the ventral striatum using its dopaminergic neurons. The dopaminergic activity in the ventral striatum induces the actual motivation to climb on the tree. It makes the girl wanting to eat the apple. The ventral striatum is also responsible to convert the decision into action, i.e., it induces and maintains goal-directed actions. The consumption of the apple, which is more delicious than expected, also leads to increased activity in the ventral striatum, this time however induced by opioid neurotransmission, which makes the girl to like the apple. Also the orbitofrontal cortex gets activated by the positive tasty experience, which will increase the probability that the girl will look for red apple in the future. In other words, what is liked will be more frequently wanted (Berridge 2003; Berridge and Robinson 2003; Cardinal et al. 2002; Grüsser and Thalemann 2006).

### 20.4.5 Neurobiological Alterations in Addiction

Inherited or acquired alterations of activity of the reward system lead to an increased risk for addictive disorders. For instance, Volkow et al. (2006) were able to show that a high dopamine D2 receptor availability protects against alcoholism in individuals who had a high inherited risk to develop alcohol addiction. This finding is confirmed by animal studies which were able to show that high dopamine D2 receptor availability leads to reduced alcohol consumption (McBride et al. 1993; Stefanini et al. 1992). In addition, patients suffering from alcohol addiction or Internet addiction had a reduced availability of dopamine D2 receptors in the ventral striatum (Heinz et al. 2004; Kim et al. 2011; Martinez et al. 2005).

Animal experiments show that organisms try to maintain a homeostatic baseline level of dopamine in the ventral striatum (Hoebel et al. 1983; Robbins and Everitt 1999). It is thus hypothesized that in healthy humans, this homeostatic baseline activity can be achieved by weak reinforcers found in everyday life (e.g., tasty foods or joyful relationships). If, however, an individual has inherited a decreased dopamine activity, there is a high risk that this individual may turn to stronger dopamine activators (such as drugs or addictive behaviors) to achieve a sufficient baseline activity (Reuter et al. 2005; Robbins and Everitt 1999).

If an addictive disorder has come into existence, the situation gets more complicated though, since the organism reacts to the high load of dopamine with a downregulation of its receptors. This is the reason why in the course of an addictive development, increasing doses are needed to reach the same effect. If the substance of abuse or the addictive behavior is omitted, the organism finds itself in a situation which is even worse compared to before the development of the addiction. Since the reward system has downregulated its responsiveness, it is now harder than before, to maintain a sufficient homeostatic baseline dopamine activity by relying on natural reinforcers (such as food or relationships; Robbins and Everitt 1999; Rossetti et al. 1992). This was shown in a study investigating detoxified alcoholics: when compared to healthy control subjects, the ventral striatum of detoxified alcoholics showed reduced activity when they expected or received a reward (Beck et al. 2009, Fig. 20.6).

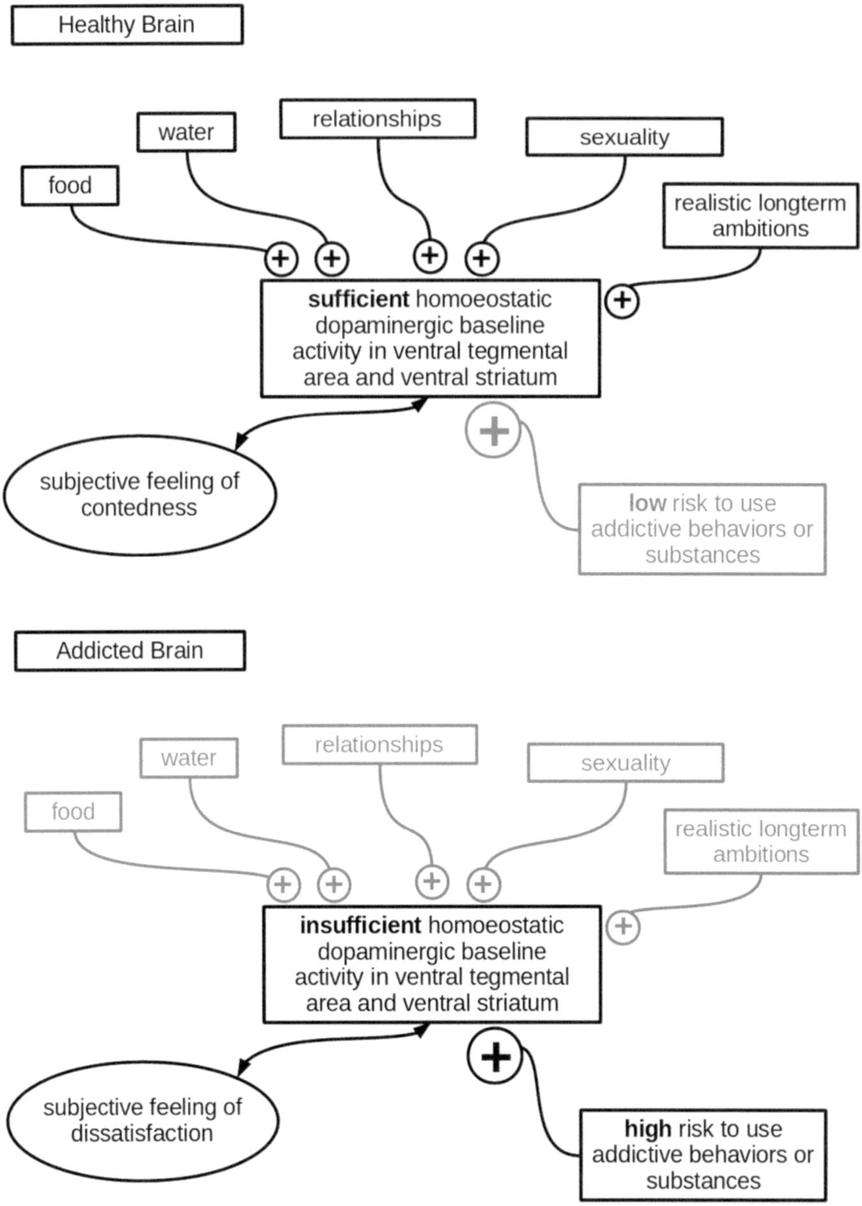


**Fig. 20.6** The healthy brain is able to maintain a sufficient baseline activity in its ventral tegmental area and ventral striatum by relying on weak dopamine activators such as food, water, relationships, sexuality, or realistic long-term ambitions. This corresponds to the subjective feeling of general contentedness. The healthy brain is comparatively safe to withstand strong dopamine activators such as addictive behaviors or drugs

The addicted brain, however, is not able to maintain a sufficient baseline activity in its reward system by weak reinforcers. This may be due to inherited alterations of the dopamine D2 receptors or the consequence of a downregulation of the reward system due to addictive dopaminergic overstimulation. This corresponds to the subjective feeling of general dissatisfaction. As a consequence, there is a high risk to rely on strong dopamine activators such as addictive behaviors or drugs to achieve a homeostatic baseline activity.

### 20.4.6 Psychopharmacological Approaches to Addiction

As recent psychopharmacological studies with addicted patients have shown, there are two promising medications available in the treatment of addictive disorders: naltrexone and acamprosate.

Naltrexone is an opioid antagonist which blocks the opioid-induced activation of the reward system. Thus, the hedonic components of drug abuse are reduced and the positive prediction error is getting smaller, which will decrease the frequency of drug consumption in the future. As studies have shown, naltrexone is able to reduce the number of alcohol relapses significantly (Rösner et al. 2008; Srisurapanont and Jarusuraisin 2005).

In addition, the putative glutamate antagonist acamprosate is able to reduce alcohol relapses significantly (Rösner et al. 2008).

### 20.4.7 Neuropsychodynamic of Addiction

Psychodynamic and neurobiological concepts of addiction show a manifold overlap.

With regard to drive theory concepts, the urge for food or sexuality can indeed be compared with addictive craving, since both are neurobiologically mediated by reward activity (Di Chiara and Bassareo 2007; Frank et al. 2010; Goldstone et al. 2009; Heinz et al. 2004; Hollmann et al. 2013; Kringelbach et al. 2003; Robbins and Everitt 1999; Simmons et al. 2005; Stoléru et al. 2012; Wang et al. 2004; Wise 1996)

In addition, ego-psychological theories correlate with neurobiological findings. For instance, many drugs are capable to mitigate states of anxiety. For example, alcohol and benzodiazepines can modulate the GABA-A receptor, which leads to reduced anxiety (Benkert and Hippius 2010).

With regard to object relations theory, a similar mechanism exists, since endogenous opioids play an important role in the hedonic interactions. As was shown in a study investigating rats, social interactions led to endogenous brain opioid release (Panksepp and Bishop 1981). A lack of rewarding social interactions may thus lead to a higher risk to consume opioids in order to maintain a homeostatic baseline level.

From a self-psychological view, there is an overlap between the concept of an incoherent self (i.e., a self-state dominated by feelings of insecurity, lack of self-esteem, lack of self-calming capabilities, and a lack of realistic aims) and a state of chronically reduced baseline dopamine activity. The prefrontal cortex may play an important role at this, since the prefrontal cortex can induce a tonic baseline activity in the reward system, when the individuum feels itself in congruence with long-term goals and interests (e.g., aims concerning relationships or one's career; Davey et al. 2008).

### 20.4.8 Neuropsychodynamically Informed Implications for the Treatment of Addictive Disorders

Elaborating on suggestions made by Volkow et al. (2004), a neuropsychodynamically informed therapeutic approach should be based on the following components:

- Psychotherapeutic interventions should focus on (pleasant) social interactions since these have a direct influence on baseline dopamine activity in reward regions. For instance, it might be helpful to encourage addicted patients to reactivate friendships or join social sports or other social activities.
- In addition, the psychotherapeutic work should focus on the patient's goals and ambitions. Often, addicted patients have difficulties to describe these or even have lost them in the course of their addictive history. Often it takes time to newly define them. Emphasis should be put on the development of realistic goals and ambitions, since these may lead to a higher baseline dopamine activity and can thus decrease the needs for additional (artificial) dopamine activators such as drugs or addictive behaviors.
- Drug consumption leads to reward activity (and the correlating feelings of pleasure) on the short term but increases unpleasant feelings of guilt and shame on the long term. Thus, psychotherapeutic interventions, which increase self-control in this view, would be beneficial.
- Psychodynamic biographic work may include approaches to identify traumatic experiences, in which the patient felt separated, alone, or lost. If these memories are unconscious, the patient may improve, when they are brought to consciousness, since the patient can then deal with these memories (and similar events which might occur to him in his everyday life) in a more adequate way. Relevant for the patient and his symptoms is the subjectively experienced extent of separation or social exclusion, which may differ from his objectively observable reality. For instance, Mr. J. had stable contact to his mother and father; however, subjectively he still felt painfully excluded and separated.
- In cases of more severe addictions with frequent relapses, one might also consider to prescribe drugs which modulate the dopamine system or can decrease craving (e.g., naltrexone and acamprosate).

## 20.5 Case Report: Mr. J. (26 Years): Part 4

### 20.5.1 Therapy

At the start of his psychotherapy, an arrangement was made that Mr. J. would take notes about his Internet activities (i.e., how much time per day he spent on which websites) and his urges to browse the Internet. His records were then evaluated once a week. While it was unclear to Mr. J. what was behind his urgent needs to surf the Internet at the beginning of his treatment, this got more obvious during the course of his therapy. When his weekly records were evaluated together with Mr. J., he learned that in particular situations, what Mr. J. experienced as social rejection led to increased needs to join the Internet. For instance, Mr. J. felt a strong desire to surf the Internet and eventually gave in to that need, after a fellow student canceled an appointment with him. At the beginning of his treatment, Mr. J. wasn't even aware that he was sensitive to situations like this. After Mr. J. had learned about this connection, however, he was able to deal with such situations in a different way. For example, Mr. J. assured himself if appointments would take place or not.

In the course of his treatment, psychodynamic connections could be made between his actual behavior (i.e., his uncontrollable urges to browse the Internet triggered by situations that were experienced as rejection) and traumatic childhood experiences: Mr. J.'s parents separated when he was 9 years due to his father's extramarital affair. They later reunited, however, which was very surprising for Mr. J. Subjectively, Mr. J. felt and experienced both situations, the separation of his parents and their surprising reunion, as painful exclusion. Since Mr. J. felt helpless in both situations, however, Mr. J. repressed the hurtful feelings of pain, sadness, and anger about the perceived rejection. As a consequence, Mr. J. had difficulties to deal with similar situations in his later life. His Internet addiction (i.e., the search for exciting content and interesting websites) helped him to avoid the painful feelings triggered by these situations.

During the course of his therapy, however, Mr. J. observed reduced needs to surf the Internet, and his records showed that his average Internet time decreased significantly. Reasons for this included that Mr. J. was more aware about his behavior, trigger situations, and the psychodynamic connections. In addition, Mr. J. began to change his free time activities; for instance, he took part in sporting activities again.

## References

American Psychiatric Association. Diagnostic and statistical manual of mental disorders. Arlington: American Psychiatric Association; 2013.

Andreassen CS, Torsheim T, Brunborg GS, Pallesen S. Development of a Facebook addiction scale. Psychol Rep. 2012;110(2):501–17.

Bayer HM, Glimcher PW. Midbrain dopamine neurons encode a quantitative reward prediction error signal. Neuron. 2005;47(1):129–41.

Beck A, Schlagenhauf F, Wüstenberg T, Hein J, Kienast T, Kahnt T, Schmack K, Hägele C, Knutson B, Heinz A, Wrase J. Ventral striatal activation during reward anticipation correlates with impulsivity in alcoholics. Biol Psychiatry. 2009;66(8):734–42.

Benkert O, Hippius H. Kompendium der Psychiatrischen Pharmakotherapie. 8th ed. Heidelberg: Springer; 2010.

Berridge KC. Pleasures of the brain. Brain Cogn. 2003;52(1):106–28.

Berridge KC, Robinson TE. Parsing reward. Trends Neurosci. 2003;26(9):507–13.

Boggio PS, Sultani N, Fecteau S, Merabet L, Mecca T, Pascual-Leone A, Basaglia A, Fregni F. Prefrontal cortex modulation using transcranial dc stimulation reduces alcohol craving: a double-blind, sham-controlled study. Drug Alcohol Depend. 2008;92(1-3):55–60.

Camprodon JA, Martínez-Raga J, Alonso-Alonso M, Shih MC, Pascual-Leone A. One session of high frequency repetitive transcranial magnetic stimulation (rtms) to the right prefrontal cortex transiently reduces cocaine craving. Drug Alcohol Depend. 2007;86(1):91–4.

Cardinal RN, Parkinson JA, Hall J, Everitt BJ. Emotion and motivation: the role of the amygdala, ventral striatum, and prefrontal cortex. Neurosci Biobehav Rev. 2002;26(3):321–52.

Chessick RD. The pharmacogenic orgasm ïn the drug addict. Arch Gen Psychiatry. 1960;3:545–56.

Cromwell HC, Berridge KC. Where does damage lead to enhanced food aversion: the ventral pallidum/substantia innominata or lateral hypothalamus? Brain Res. 1993;624(1-2):1–10.

Davey CG, Yücel M, Allen NB. The emergence of depression in adolescence: development of the prefrontal cortex and the representation of reward. Neurosci Biobehav Rev. 2008;32(1):1–19.

Di Chiara G, Bassareo V. Reward system and addiction: what dopamine does and doesn't do. Curr Opin Pharmacol. 2007;7(1):69–76.

Frank S, Laharnar N, Kullmann S, Veit R, Canova C, Hegner YL, Fritsche A, Preissl H. Processing of food pictures: influence of hunger, gender and calorie content. Brain Res. 2010;1350:159–66.

Freud S. Der Witz und seine Beziehung zum Unbewussten. Frankfurt: S. Fischer; 1905.

Freud S. Aus den Anfängen der Psychoanalyse - Briefe an Wilhelm Fließ. Abhandlungen und Notizen aus den Jahren 1887-1902. S. Fischer: Frankfurt; 1962.

Goldstone AP, Prechtl de Hernandez CG, Beaver JD, Muhammed K, Croese C, Bell G, Durighel G, Hughes E, Waldman AD, Frost G, Bell JD. Fasting biases brain reward systems towards high-calorie foods. Eur J Neurosci. 2009;30(8):1625–35.

Griffiths M. A 'components' model of addiction within a biopsychosocial framework. J Subst Use. 2005;10(4):191–7.

Grüsser SM, Thalemann CN. Verhaltenssucht. 1st ed. Bern: Verlag Hans Huber; 2006.

Hartmann D. A study of drug-taking adolescents. Psychoanal Study Child. 1969;24:384398.

Heinz A, Siessmeier T, Wrase J, Hermann D, Klein S, Grüsser SM, Grüsser-Sinopoli SM, Flor H, Braus DF, Buchholz HG, Gründer G, Schreckenberger M, Smolka MN, Rösch F, Mann K, Bartenstein P. Correlation between dopamine d(2) receptors in the ventral striatum and central processing of alcohol cues and craving. Am J Psychiatry. 2004;161(10):1783–9.

Hildebrandt HA. Psychoanalyse der sucht - eine kritische bilanz. Psychoanalyse Widerspruch 36; 2007.

Hoebel BG, Monaco AP, Hernandez L, Aulisi EF, Stanley BG, Lenard L. Self-injection of amphetamine directly into the brain. Psychopharmacology. 1983;81(2):158–63.

Hollmann M, Pleger B, Villringer A, Horstmann A. Brain imaging in the context of food perception and eating. Curr Opin Lipidol. 2013;24(1):18–24.

Karreman M, Moghaddam B. The prefrontal cortex regulates the basal release of dopamine in the limbic striatum: an effect mediated by ventral tegmental area. J Neurochem. 1996;66(2):589–98.

Kim SH, Baik SH, Park CS, Kim SJ, Choi SW, Kim SE. Reduced striatal dopamine d2 receptors in people with internet addiction. Neuroreport. 2011;22(8):407–11.

Kohut H. Narzissmus: eine Theorie der psychoanalytischen Behandlung narzisstischer Persönlichkeitsstörungen. Frankfurt am Main: Suhrkamp; 1975.

Kringelbach ML, O'Doherty J, Rolls ET, Andrews C. Activation of the human orbitofrontal cortex to a liquid food stimulus is correlated with its subjective pleasantness. Cereb Cortex. 2003;13(10):1064–71.

Martinez D, Gil R, Slifstein M, Hwang DR, Huang Y, Perez A, Kegeles L, Talbot P, Evans S, Krystal J, Laruelle M, Abi-Dargham A. Alcohol dependence is associated with blunted dopamine transmission in the ventral striatum. Biol Psychiatry. 2005;58(10):779–86.

McBride WJ, Chernet E, Dyr W, Lumeng L, Li TK. Densities of dopamine d2 receptors are reduced in cns regions of alcohol-preferring p rats. Alcohol. 1993;10(5):387–90.

Mentzos S. Textbook of psychodynamics. The function of dysfunctionality of psychic disorders. (German: Lehrbuch der Psychodynamik. Die Funktion der Dysfunktionalität psychischer Störungen.). 2nd ed. Göttingen: Vandenhock und Ruprecht; 2009.

Milch W. Lehrbuch der Selbstpsychologie. Stuttgart: Verlag W. Kohlhammer; 2001.

Panksepp J, Bishop P. An autoradiographic map of (3h)diprenorphine binding in rat brain: effects of social interaction. Brain Res Bull. 1981;7(4):405–10.

Pine F. The four psychologies of psychoanalysis and their place in clinical work. J Am Psychoanal Assn. 1988;36:571–96.

Radó S. The psychic effects of intoxicants: an attempt to evolve a psycho-analytical theory of morbid cravings. Int J Psychoanal. 1926;7:396–413.

Radó S. The psychoanalysis of pharmacothymia (drug addiction). Psychoanal Q. 1933;2:1–23.

Reuter J, Raedler T, Rose M, Hand I, Gläscher J, Büchel C. Pathological gambling is linked to reduced activation of the mesolimbic reward system. Nat Neurosci. 2005;8(2):147–8.

Robbins TW, Everitt BJ. Drug addiction: bad habits add up. Nature. 1999;398(6728):567–70.

Rosenfeld HA. On drug addiction. Int J Psychoanal. 1960;41:467–75.

Rösner S, Leucht S, Lehert P, Soyka M. Acamprosate supports abstinence, naltrexone prevents excessive drinking: evidence from a meta-analysis with unreported outcomes. J Psychopharmacol. 2008;22(1):11–23.

Rossetti ZL, Melis F, Carboni S, Gessa GL. Dramatic depletion of mesolimbic extracellular dopamine after withdrawal from morphine, alcohol or cocaine: a common neurochemical substrate for drug dependence. Ann N Y Acad Sci. 1992;654:513–6.

Rost WD. Psychoanalyse des Alkoholismus. In: Theorie, diagnostik, Behandlung. 6th ed. Stuttgart: Klett-Cotta; 2001.

Savitt RA. Psychoanalytic studies on addiction: Ego structure in narcotic addiction. Psychoanal Q. 1963;32:43–57.

Simmons WK, Martin A, Barsalou LW. Pictures of appetizing foods activate gustatory cortices for taste and reward. Cereb Cortex. 2005;15(10):1602–8.

Srisurapanont M, Jarusuraisin N. Opioid antagonists for alcohol dependence. Cochrane Database Syst Rev. 2005;12:CD001867.

Stefanini E, Frau M, Garau MG, Garau B, Fadda F, Gessa GL. Alcohol-preferring rats have fewer dopamine d2 receptors in the limbic system. Alcohol Alcohol. 1992;27(2):127–30.

Stoléru S, Fonteille V, Cornélis C, Joyal C, Moulier V. Functional neuroimaging studies of sexual arousal and orgasm in healthy men and women: a review and meta-analysis. Neurosci Biobehav Rev. 2012;36(6):1481–509.

Subkowski P. Störungen der Trieborganisation in Suchtentwicklungen. In: Psychodynamik der Sucht: Psychoanalytische Beiträge zur Theorie. Göttingen: Vandenhoeck & Rupprecht; 2008. p. 51–90.

Volkow ND, Fowler JS, Wang GJ. The addicted human brain viewed in the light of imaging studies: brain circuits and treatment strategies. Neuropharmacology. 2004;47(Suppl 1):3–13.

Volkow ND, Wang GJ, Begleiter H, Porjesz B, Fowler JS, Telang F, Wong C, Ma Y, Logan J, Goldstein R, Alexoff D, Thanos PK. High levels of dopamine d2 receptors in unaffected members of alcoholic families: possible protective factors. Arch Gen Psychiatry. 2006;63(9):999–1008.

Wang GJ, Volkow ND, Telang F, Jayne M, Ma J, Rao M, Zhu W, Wong CT, Pappas NR, Geliebter A, Fowler JS. Exposure to appetitive food stimuli markedly activates the human brain. NeuroImage. 2004;21(4):1790–7.

Wernado M, Blaufuß J, Jacob A, Kannenberg S. Spezifische Interventionen auf der Basis der analytischen/analytisch orientierten Therapie bei psychischen Störungen im Zusammenhang mit psychotropen Substanzen. In: Lehrbuch der Psychotherapie: 2 Psychoanalytische und tiefenpsychologisch fundierte Therapie. 3rd ed. München: CIP-Medien; 2006. p. 97–108.

Wise RA. Neurobiology of addiction. Curr Opin Neurobiol. 1996;6(2):243–51.

Wise RA, Bozarth MA. A psychomotor stimulant theory of addiction. Psychol Rev. 1987;94(4):469–92.

Wolf ES. Theorie und Praxis der psychoanalytischen Selbstpsychologie. Frankfurt: Suhrkamp; 1998.

Wurmser L, editor. Die verborgene Dimension. Göttingen: Vandenhoeck & Ruprecht; 1997.

# Part III

# Neuropsychodynamic Perspectives

# Principles of Neuropsychodynamic Therapy

# 21

Peter Hartwich, Heinz Boeker, and Georg Northoff

**Abstract**

*Neuropsychodynamics* refers to the *social embedded brain* and the *relation* between the individual brain, environment, and the subjective experience of the personal identity over time. We do not know how the brain transforms its neuronal activities into mental features, but we know that it is a fact.

In many psychiatric illnesses such as schizophrenia, bipolar disorder, anorexia nervosa, obsessive-compulsive disorder, and other psychiatric illnesses, we assume abnormalities of the spatiotemporal self. Many symptoms, not all, of these diseases can be seen as an attempt to reorganize the self and its brain-based spatiotemporal relation to the world.

This approach offers novel treatment options with the aim to develop strategies to modulate the spatiotemporal structure of the resting state's alignment to the world by applying spatially and temporally modulating stimuli. This could be done by, e.g., creative therapies and other neuropsychodynamic therapeutic methods in the sense of top-down modulation. In the sense of bottom-up modulation, this could also be done by psychopharmaceuticals, magnetic stimulation, etc.

P. Hartwich (✉)
Department of Psychiatry, Psychotherapy, Psychosomatics, General Hospital of Frankfurt am Main, Teaching Hospital University Frankfurt, Frankfurt, Germany
e-mail: Prof.Hartwich@casytec.de

H. Boeker
Department of Psychiatry, Psychotherapy, and Psychosomatics, Psychiatric University Hospital Zurich, University of Zurich, Zurich, Switzerland
e-mail: heinz.boeker@bli.uzh.ch

G. Northoff
Mind, Brain Imaging, and Neuroethics, Institute of Mental Health Research, University of Ottawa, Ottawa, ON, Canada
e-mail: georg.northoff@theroyal.ca; www.georgnorthoff.com



H. Boeker et al. (eds.), *Neuropsychodynamic Psychiatry*,
https://doi.org/10.1007/978-3-319-75112-2_21

The main question is how functional activity of the brain is influenced by therapeutic methods. The basis of neuropsychodynamic treatment is the patient-therapist relationship with their fitting, transference, and countertransference. Neurobiological and psychological changes go hand in hand and influence each other.

## 21.1 Introduction

Neuropsychodynamic psychiatry is a new approach to the understanding of psychiatric disorders and the treatment of these diseases. This leads to an extension of what Gabbard (2014, p 5f) writes: "Psychodynamic psychiatry is an approach to diagnosis and treatment characterized by a way of thinking about both patient and clinician that includes unconscious conflicts, deficits and distortions of intrapsychic structures, and internal object relations and that integrates these elements with contemporary findings from the neurosciences."

*Neuropsychodynamics* refers to the *social embedded brain*. "The mind therefore is intimately bound up with the first-person observational perspective. This is the only perspective from which everything we observe can be grounded in a background sense of self, which is ultimately generated by our inner awareness of living in a physical body" (Solms and Turnbull 2002, p. 77). Understanding the processes, mechanisms and structures at the border between psyche and soma, between brain (neural activities) and consciousness new paths will be opened up in the therapy of psychiatric disorders. The crucial point is the *relation* between the individual brain, environment, and the subjective experience of the personal identity over time. We do not know how the brain transforms its neuronal activities into mental features, but we know that it is a fact. And also the other way around: mental features transformed the neuronal activity and with it sometimes the structure of the brain. Northoff (2016, p. 19) assumes, "that the resting state of the brain and its particular spatial-temporal configuration provide the necessary input to bridge the gap between what we observe as neuronal activity and what we experience as mental features."

In many psychiatric illnesses such as schizophrenia, bipolar disturbance, anorexia nervosa, affective-compulsive and other disturbances, we assume abnormalities of the spatial-temporal self. Many symptoms, not all, of these diseases can be seen as an attempt to reorganize the self and its relation to the world. In this understanding, what we call symptoms are more or less compensatory strategies to rebuild the continuity of the spatial-temporal structure of the self. Mostly the psychopathology is characterized by spatial and temporal abnormalities in the resting state of the cortical midline and other structures which correspond with the spatial and temporal consciousness, as an "essential structure of a temporal-spatial situation in the world" (Sartre 1956, p. 267). It is based on the "statistically and spatial-temporally based unity between brain and environment, the environment-brain unity" (Northoff 2014,

p. 366). That means the spatial-temporal structure of the stimuli from the environment links the spatial-temporal resting state activity of the brain and vice versa.

Northoff (2016, p. 169) emphasizes: "Once we understand both of these areas better we may have the opportunity to view schizophrenic symptoms in a spatial-temporal context. We would then need to characterize the various symptoms in spatial-temporal terms and relate them to specific spatial-temporal features in the brain's resting state activity. This endeavor could be called a spatial-temporal approach to psychiatric disorders such as schizophrenia." Today we would not restrict this kind of understanding to schizophrenia but also to all the other psychiatric diseases, where the continuity of the spatial-temporal self is in danger to be disturbed.

This approach can offer novel treatment options with the aim to develop strategies to modulate the spatial-temporal structure of the resting state by applying spatially and temporally modulating stimuli. This could be done by, e.g., creative therapies and other neuropsychodynamic therapeutic methods in the sense of top-down modulation. In the sense of bottom-up modulation, this could also be done by psychopharmaceuticals, magnetic stimulation, etc.

We assume, the better the empathy of the therapist and the better the professional relation between patient and therapist is, the more effective can be the modulation of the abnormal resting state and the hyperconnectivity in the cortical midline structures of the brain.

In case that life events are encoded in the brain's resting state and its information is stored in its neural activity as "autobiographical self" (Damasio 1999) which can be seen as synonymous with the ego, as Solms and Turnbull (2002) state, therapeutic methods such as psychoanalysis or psychodynamics would be able to change positively the condition of the stored information and therefore the condition of the resting state or the connectivity of the networks in the brain.

When we ask how and what psychodynamic psychotherapy changes in the brain, Solms and Turnbull (2002, p. 287) already formulated a summary of the relevant scientific investigations:

"In the first place, they show that the functional activity of the brain is indeed *altered by* psychotherapy. Second, they show that *specific* changes are correlated with the therapeutic outcome. Third, and most pertinently, they show that *these outcome-specific changes are essentially localized to the prefrontal lobes*." The authors emphasize that internalization takes place and the mutative power of which "is probably largely confined to certain critical periods of frontal-lobe development (in the first few years of life), but which may perhaps be artificially rekindled by the regressive nature of the transference relationship" (p. 290). But this is not more than a hypothesis which should be investigated in the future carefully and in detail.

Kandel's (2001) experiments have demonstrated that brain structures and synaptic connections are dynamic. The crucial point is that synapses can be modified by the environment, including learning and memory processes and even psychotherapy. Mundo (2006) points out that those recent results from neuroimaging studies, using fMRI and PET, have found that dynamic psychotherapy has measurable effects on the brain and may modify brain function and metabolism in specific brain areas. The neurobiological correlates to elements of psychotherapy such as attachment, empathy, memory, learning, emotional regulation, and fear for extinction seem to be emerging, but a precise description of the changes and influences that occur during psychotherapy is currently not possible in the way we wish it. Therefore we have to be careful and have to wait until future imaging and neurobiological investigation will elucidate these processes.

## 21.2 Essentials of Neuropsychodynamic Treatment

An essential element common to all psychotherapies studies is the *relationship* between patient and therapist. Thus, when neurobiological changes take place in psychodynamic psychotherapy, they are also caused by the therapeutic relationship, because this relationship is the basis for an effective psychotherapy.

The relationship is characterized by:

- Lack of fitting "Passung" patient-analyst match
- Transference
- Coutertransference
- Resistance

### 21.2.1 Lack of "Fitting" (Patient-Analyst Match)

Sometimes the psychotherapist's and patient's communication does not fit together. The constellation can be unfavorable from the outset. Refusing feelings can be from one side or both sides. Occasionally, the intensity is so violent and also complicated with a particularly unfavorable real relationship constellation that a fertile therapeutic process does not take place. In such a case, one also speaks of the lack of fitting (patient-analyst match) of the two persons involved in psychodynamic therapy. Boeker (2011) points out that in the phenomenon of fitting of the therapist and the patient, this occurs beyond the verbalization and the specific techniques of psychotherapeutic procedures. It is important not to confuse the "lack of fitting" with unconscious resistance. The resistance can be worked through in the psychoanalytic and psychodynamic psychotherapy. It can then help the patient to progress better in the therapeutic process. Thus, it is important to note that not every problematic relationship between the patient and the therapist is based on resistance but can also be a lack of the fitting.

### 21.2.2 Transference

In psychodynamic and psychoanalytic therapy, one of the most important elements is the significance of transference. Some childhood patterns of the psychic organization can continue in the adult age: "the past is repeating itself in the present" (Gabbard 2014, p. 18). Characteristics of the person of the past are attributed to the treating person and also to the current partner; the feelings associated with the historical person are also felt in the current encounter. On the basis of the transference, patients experience relations from the past unconsciously, instead of remembering them.

The knowledge about the environment that the individual gains is manifested in the relationship between a recognizing subject and a recognized object (Latour 2014, p. 150). As a rule, the investment energy (cathexis), according to the concept of Freud (1926), is strong in this relation when the object recognized is a person important to the individual. As a result, the patient's relationship with the therapist is usually associated with quantitatively high investment energy (cathexis). But a special quality is also added; although this is also the case in many other constellations of relationships, it is of particular importance in psychodynamic therapy. The particular quality is, in fact, composed of the elements of the present real relationship as regards hopes, expectations, sympathy, etc. and at the same time of elements derived from earlier, often early childish desires and experiences with primary reference persons.

Both elements intermingle in the experience of the patient, whereby early relationship experiences are established in the memory of emotional experience in a way that they can modify all perceptions on a present relationship. Crucial is that this influence, which takes place through the early experiences, is unconscious. This is called transference (Freud 1912) (see Chap. 11).

## 21.3 Neuropsychodynamic Perspective of Transference

Considering that the internal representations of objects are the core elements which activate the transference, we follow the description of Gabbard (2014, p. 19): "Representations exist as a network of neurons that can be activated in tandem. Thus representations are like potentials waiting to be activated when aspects of a therapist remind the patient of qualities similar to the figure represented in the patient's neural network."

In psychodynamic psychotherapy, much of the joint work of therapist and patient is about the awareness of the elements of early relationship experiences. These are usually unconscious and often embodied in the neural network from a preverbal development stage. The advantage is that they can be reactivated by the transference to the therapist. This delivers the chance for the patient to become aware and to re-experience the feelings associated with these experiences. But this is only the first step in a psychotherapeutic process. Then, in a lengthy therapeutic working through, the therapist helps the patient to recognize the influences of the old relationship constellations in the transference to him again and again. These should be assigned to the actual reference person of the past. It is crucial and particularly important to gradually reduce the associated emotional intensities with their special energies of investment. In this way, the patient learns to let go the unreal modifications for the present and to experience the respective relationship constellations to the partner and other people, to whom the patient has special investment energy, real and appropriate. This psychotherapeutic process alleviates illness-causing conflicts more and more. In a neuropsychodynamic understanding, we assume that the potentials of the network of neurons, which represent the embodied memory of early relationships, can be reduced; and additionally the neural connection between old representations of objects will be less activated in the encounter with other people in the presence.

### 21.3.1 Case Report

Because of a so-called burnout, the 48-year-old Mrs. S. goes to the psychotherapist; however, he notes that she suffers from symptoms of chronic depression. Her mood is depressed; she cannot rejoice, feels without drive, and suffers from sleeping problems. She does not feel understood by her family and laments various changing physical impairments, which has been years, but now getting particularly bad. In her report it is noticeable that she speaks quietly and monotonously, weak and helpless, and as someone who is seeking maternal protection.

She reveals that she expects the psychotherapist to take the burden off her shoulders and give her life joyfulness. From the biographical history, it can be learned that her mother was suffering from cancer when the patient was 1 or 2 years old. That was why the mother was often for weeks in therapy in hospitals and health resorts far from home. She finally died when the patient was 7 years old. The girl then grew up with the grandparents and partly with the new wife of her father.

#### 21.3.1.1 Comment

As far as the transference is concerned, it is already clear in the first encounter which becomes evident more and more in the psychotherapy. Her loss experience is so lasting that she wishes and expects unconsciously from the therapist to understand her pain, to comfort her, and to restore the constellation as it was before the loss experience in her childhood. This goes hand in hand with idealizing unconscious hopes about the therapist's abilities to undo the former loss in her childhood. As is often the case, there are similar unconscious desires to her husband, about whom she complains that he is less and less concerned about her lately.

The partner had unconsciously taken over the protective and often dominant role in the relationship, whereby the already asymmetrical relationship was further strengthened. If the therapist is able to work effectively with the transference to him, the established relationship constellation with the husband also has a chance to change, namely, the special quality of the unconscious elements of the early development period would no longer influence the present relationship. Both partners would get the opportunity to shape their relationship symmetrically according to reality. This should also usually be accompanied by a development process of the depressed patient's husband. The therapeutic work on the transference thus comes to mean that the patient can actually remove her unconscious, unreal wishes and hopes, i.e., the husband replacing the lost mother, undoing the once-lost life events, and continually comforting her. The constant disappointment of the overpowered husband is no longer necessary, and the source of the constant conflicts that have contributed to the depression can disappear.

## 21.4 Countertransference

On the other hand in the interactional psychodynamic process between patients and psychotherapists, there are also the attitudes and emotional reactions which represent the counterpart to transference, namely, the therapist's countertransference (Freud 1912). Initially, one understood the complementary unconscious imprints from the psychotherapist's own childhood. However, we speak today of a general countertransference (Laplanche and Pontalis 1972; Kernberg 1985) in a broader sense, since these are often mixed with present attitudes, such as sympathy, aversion, and other emotional reactions. This extended understanding is meant the totality of unconscious responses that can be manifested in attitudes, actions, and feelings of a psychotherapist and also of an entire therapeutic team. This applies to the treatment of all mental illnesses, including psychodynamic therapy in psychoses.

Kernberg (1985, p. 183/184) points out the variability of the weighting of both components in the different psychiatric disorders: one could describe a whole spectrum of countertransference reactions, the countertransference forms in neuroses at one end up to the countertransference in psychotics as another extreme. In the course of this series, the respective ratio of the reality and transference proportions of the patient and the therapist is changed. On the one hand, the psychotherapist is conscious of his countertransference, which happens occasionally with the help of the supervision. The patient should be accompanied by a therapist who is as neutral as possible and unaffected by personal feelings. This is often not realistic. Therefore, on the other hand, the conscious countertransference is still very important. It is a subtle and reliable tool for the detection of unconscious psychological attitudes in the patient. In addition, the countertransference can also be used as a *therapeutic tool.*

For the neuropsychodynamic psychotherapist who works with psychotic patients, e.g., with schizophrenics, the common countertransference is impatience, fear, anger, and feeling of distance. Often there is a countertransference obstacle, because the therapist has to bear the danger of psychotic disintegration and fragmentation of the self of his patient. So he should learn what we call "chaos ability" (Hartwich 2007). It is normal that the therapist wants to protect himself from own self-fragmentation; therefore he usually builds up a countertransference resistance. When he is aware of that in the psychodynamic setting with the patient, then he can better and earlier realize the patient's danger of self-fragmentation.

### 21.4.1 Example

In the depressive patient described above, the therapist will feel compassion at the first encounter and lean toward her in order to understand her better as she speaks so quietly. The help-seeking facial expression will appeal to his maternal protective feelings. Their unconscious desire to find a healer of their problems with the corresponding idealization will make the therapist feel that he is irreplaceable and that he will be pleased with the assigned role of the powerful. In the course of time, when his "healing power" is still not enough to alleviate the symptoms, and the constant offerings are "beaten out," with complaints, e.g., it has never been as bad as today, the therapist's patience is put to a severe test. Gradually, his countertransference feelings change, and he becomes annoyed more and more. The aggressive fundamental tone of his feelings becomes louder in him, but his anger is tied by a sense of guilt caused by the patient's vulnerability. It is now the time to reflect on the countertransference feelings. It is a question of making aware the source of the aggression of the therapist in order to become neutral and "therapeutic." On the other hand, the experience of his countertransference feelings helps him to feel the unconscious aggressiveness of the patient as an essential component of her illness and to work through this part of her unconscious feelings. This is really a great chance for the patient to change her behavior. If the therapist does not reflect his countertransference, a similar pattern of interaction between the patient and her husband, as observed in the patient's marriage, would occur.

### 21.4.2 Neuropsychodynamic Perspective of Countertransference

Neuropsychodynamic therapists should have *in mind the imagination* of abnormal neuronal events of their patient's brain:

- For example, hyperconnectivity in the cortical midline structures of the brain and DMN-CEN imbalance in schizophrenics (see Chap. 10).
- They do have in mind numerous neurophysiological impairments during manic-depressive episodes and also after remission. These dysfunctions are associated with structural and functional changes in cortical and limbic brain regions (see Chap. 12).
- They have in mind the neural aberrations which are found in depressive patients (see Chap. 11).
- Abnormal neural events in anxiety disorders (Chap. 13), obsessive-compulsive disorders (Chap. 14), anorexia nervosa (Chap. 16), and personality disorders (Chap. 18).

When psychotherapists do have the abovementioned neural conditions in mind, they can get a benefit from *imagining simultaneously* their patient's neural aberrations referring to empathy and countertransference. The chance grows to understand and to consider the problems of the patients additionally from a neural point of view, when they are suffering from symptoms. This kind of considering both, not only the phenomenal but also the neuronal perspective at the same time, will cause a *change in the psychotherapist's paradigm* with the result that, e.g., his countertransference resistance reduces. Especially while treating psychotic patients, the fear of the therapist's own self-fragmentation will decrease. If this is the case, more psychotherapists could attempt to treat severe psychiatric patients neuropsychodynamically.

### 21.4.3 Case Report (with the Permission of the Psychoanalyst Wolfgang Hering)

A female schizophrenic patient in advanced psychoanalytic treatment became very annoyed in the session. The therapist could not calm her. She stood up between the therapist and her chair, stared out of the window, and repeated for a long time stereotypically: "I'm sorry, I'm sorry, I'm sorry…." There was no chance to clarify and to understand her demand. In short, the therapist could not at all come into contact with her. The emotional excitement was growing and became dangerous. Then the psychoanalyst came to the idea of humming a well-known melody (Händel Rinaldo, aria of the Almira: Lascia ch'io pianga). While he was humming, she stopped her stereotypy, turned the view from the window to therapist and said: "You sing." Suddenly, there was again a feeling of relationship. Therapist: "Yes, it's the song of a prisoner"—pause—"caught as you now." The further conversation was more relaxed. Later, she explained, her behavior was about her helplessness not to be able to protect the therapist from "the aliens" who were going to influence him.

#### 21.4.3.1 Comment

We suppose that the patient's overly increased hyperconnectivity in her cortical midline structures and the abnormally changed balance of the connectivity between DMN and CEN were modulated positively by the therapist's humming of the well-known melody. With the melody, a new spatial-temporal pattern was, as stimuli, delivered which may have had a modifying influence of the spatial-temporal pattern of the patient's resting state activity and the consolidation of the spatial-temporal structure of the self.

## 21.5 Compensation as a Strategy to Combine Neuronal Aspects with Psychodynamic Therapy

Compensation would mean enhancing and increasing neuronal mechanisms which are able to compensate aberrant connectivity, e.g., with creative therapeutic methods (Arieti 1976; Hartwich and Fryrear 2002). When a person is in a state of creativity, then one can find an enhanced neuronal activity in the temporal and parietal regions, sometimes with a sudden increase of high frequency of activity (summarized by Kandel 2012). Therapists may be able to strengthen these neuronal activities with therapeutic methods, which stimulate and unfold creativity with the help of expressive media such as painting, sculpture, music, movement, and poetry. We postulate that such creative therapies could modify aberrant changes in connectivity in a positive way. In the future one could work out more specific nuances in creative therapies on the basis of neuronal compensation, which refers to the interaction and relation between the phenomenal and neuronal level.

Another compensatory mechanism could use the self-related processing (Northoff 2011, 2014; Northoff et al. 2009). Neuroimaging studies have demonstrated that those words and pictures that are highly related to the individual's self are considered to be more emotional than those that show rather low degree of self-relatedness. On the neuronal level, a meta-analysis (Northoff et al. 2006) underlines that there is a concentration of neuronal activation in the cortical midline structures, the premotoric and bilateral parietal cortex. Huang et al. (2016) showed that the MPFC region (medial prefrontal cortex) is closely linked to a self-related processing.

In psychopathology the mirror phenomenon is well known as "signe du miroir." The first description goes back to the French psychiatrists Delmas (1929) and Abély (1930). Some schizophrenic patients look into the mirror for hours to "find themselves," and this can be interpreted as an autotherapeutic (Scharfetter 1986) habit to stabilize their self, when it is in danger to deteriorate. On the phenomenal level and even on the neuronal level, they may unconsciously use the activation which is given by the self-related processing. This corresponds with empirical studies about video mirroring on schizophrenics, which experimentally show an increase of ego strength (Hartwich and Lehmkuhl 1979). One could enhance the neuronal activation which is given by the self-related processing using audiovisual mirroring systematically. There may be a positive influence to the abnormal resting state activity and their imbalance in interaction with improving the spatial-temporal structure of the self also on the phenomenal level.

## References


Abély P. Le signe du miroir dans les psychoses et plus spécialement dans la démence précoce. Ann Med Psychol. 1930;88:28–36.

Arieti S. Creativity. The magic synthesis. New York: Basic Books; 1976.

Boeker H. Psychotherapie der depression. Bern: Huber/Hogrefe AG; 2011.

Damasio A. The feeling of what happens. London: Heinemann; 1999.

Delmas FA. Le signe du miroir dans la démence précoce. Ann Med Psychol. 1929;87:227–33.

Freud S. The dynamics of transference, Vol. 12. Standard Edition. New York: Norton; 1912. P. 97–108.

Freud S. Neurosis and psychosis, Vol. 19. Standard Edition. London: Hogarth; 1926. P. 385–91.

Gabbard GO. Psychodynamic psychiatry in clinical practice. 5th ed. Washington DC: American Psychiatric Publishing; 2014.

Hartwich P. Psychodynamisch orientierte Therapieverfahren bei Schizophrenien. In: Hartwich P, Barocka A, editors. Schizophrene Erkrankungen. Sternenfels: Wissenschaft & Praxis; 2007. p. 33–98.

Hartwich P, Fryrear JL. Creativity, the third therapeutic principle in psychiatry. Sternenfels: Wissenschaft & Praxis; 2002.

Hartwich P, Lehmkuhl G. Audiovisual self confrontation in schizophrenia. Arch Psychiatr Nervenkr. 1979;227:341–51.

Huang Z, Obara N, Davis H, Pokorny J, Northoff G. The temporal structure of resting-state brain activity in the medial prefrontal cortex predicts self-consciousness. Neuropsychologia. 2016;82(2016):161–70.

Kandel ER. Psychotherapy and the single synapse: the impact of psychiatric thought on neurobiological research. J Neuropsychiatry Clin Neurosci. 2001;13:290–300.
Kandel ER. Das Zeitalter der Erkenntnis. Die Erforschung des Unbewussten in Kunst, Geist und Gehirn von der Wiener Moderne bis heute. München: Siedler/Random House GmbH; 2012.
Kernberg OF. Objektbeziehungen und Praxis der Psychoanalyse. 2nd ed. Stuttgart: Klett-Cotta; 1985.
Laplanche J, Pontalis JB. Das Vokabular der Psychoanalyse. Frankfurt: Suhrkamp; 1972.
Latour B. Existenzweisen. Berlin: Suhrkamp; 2014.
Mundo E. Neurobiology of dynamic psychotherapy: an integration possible? J Am Acad Psychoanal Dyn Psychiatry. 2006;34(4):679–91. https://doi.org/10.1521/jaap.2006.34.4.679.
Northoff G, Heinzel A, de Greck M, Bermpol F, Dobrowolny H, Panksepp J. Self-referential processing in our brain–a metanalysis of imaging studies on the self. Neuroimage. 2006;31:440–57.
Northoff G, et al. Differential parametric modulation of the self-relatedness and emotions in different brain regions. Hum Brain Mapp. 2009;30(2):369–82.
Northoff G. Neuropsychoanalysis in practice. New York: Oxford University Press; 2011.
Northoff G. Unlocking the brain. Consciousness, vol. 2. Oxford: Oxford University Press; 2014.
Northoff G. Neuro-philosophy and the healthy mind: learning from the unwell brain. New York: Norton; 2016.
Sartre JP. Being and nothingness. An essay on phenomenological ontology. Translated and with an introduction by Barnes HE. New York: Philosophical Library Inc.; 1956.
Scharfetter C. Schizophrene Menschen. 2nd ed. München: Urban & Schwarzenberg; 1986.
Solms M, Turnbull O. The brain and the inner world. An introduction to the neuroscience of subjective experience. London: Karnac; 2002.

# Acute Psychiatric Wards

# 22

Peter Hartwich and Heinz Boeker

**Abstract**

Acute psychiatric wards are protected emergency areas in a psychiatric hospital. Treated are all patients who are in danger to harm themselves or others because of their psychiatric illness or who want to commit suicide.

In some *schizophrenic patients*, whose attention is greatly reduced to themselves, one can assume that they focus strongly on their spatiotemporal Self, so the hyper-connectivity of their CMS is overly emphasized. At the same time, because of a shift in the CMS-CEN balance with an overload of the CEN, they are hardly aware of the environment.

Other *schizophrenic and schizoaffective patients* are particularly sensitive to external stimuli because they cannot filter and mitigate them. Consequently, the neuronal aberrations that amplified the symptoms of the disease become worse.

We see the same with severe *depressive patients* with suicidal ideas that they can hardly be influenced by the environment. But other depressives are overly irritated by the extreme external conditions. When *manic* patients are admitted in a locked ward against their will, they create a high level of tension. Patients with *personality disorders* and drug addicts usually are very disturbed by the overly arousal.

We describe how to deal with aggressiveness and hyperarousal from the side of the patient and the therapeutic team and especially with suicidal patients. The treatment of chronic suicidal tendencies creates a problem what Gabbard described as "countertransference hate." We describe examples of group psychotherapy for those patients and supervision sessions for the therapeutic team.

P. Hartwich (✉)
Department of Psychiatry, Psychotherapy, Psychosomatics, General Hospital of Frankfurt am Main, Teaching Hospital University Frankfurt, Frankfurt, Germany
e-mail: Prof.Hartwich@casytec.de

H. Boeker
Department of Psychiatry, Psychotherapy, and Psychosomatics, Psychiatric University Hospital Zurich, University of Zurich, Zurich, Switzerland
e-mail: heinz.boeker@bli.uzh.ch



H. Boeker et al. (eds.), *Neuropsychodynamic Psychiatry*,
https://doi.org/10.1007/978-3-319-75112-2_22

Acute psychiatric wards are usually protected wards and emergency areas in a psychiatric hospital. These are locked wards where patients are admitted, who have to be protected from injuring themselves or from harming other people. After the acute state, those patients usually are transferred to on open ward.

The diagnoses of the patients, who are treated in an acute psychiatric ward, are not limited to special groups. These locked wards are for all patients who are in danger to harm themselves or others because of their psychiatric illness or who want to commit suicide because of their psychiatric illness.

So one finds nearly all psychiatric diagnoses: organic mental disorders, substance-related disorders (drug dependence, opiate, cocaine, crack, psychedelics, alcohol, etc.), schizophrenia, schizoaffective disorders, mood disorders (depression, mania, and bipolar disorders), borderline and other personality disorders, dementia, etc. In short, all psychiatric diagnoses and those who are not yet diagnosed may suffer from organic diseases or even those who are malingerers.

## 22.1 Self-Brain-Environment and Acute Psychiatric Wards

We assume that the disturbances of the *self-brain-environment relationship* are the basis for psychiatric disorders. Consequently, when conditions of the environment drastically change, they influence individuals very intensely. This happens when the habitual room is changed to an unknown place. In addition, there are extreme situations, such as overly loud music in a rock concert, overly loud children's crying in a schoolyard, excessive traffic noise from cars and motorcycles, and other noise sources made by humans. This also affects many healthy people because they are irritated by the overload of stimuli when it goes on for a longer time. It comes to a hyperarousal in which the inhibition functions of the brain become overloaded. The brain can then no longer shield and filter out the external stimuli. Such overloads from the environment affect the brain and Self in such a way that the neural systems and neural connectivity can be altered pathologically.

On the other hand, there are situations where the environmental stimuli are reduced to a minimum. They are extreme situations in which people are in total darkness and without any acoustic external stimuli. After a few hours or days, when they cannot orient themselves to space and time, it comes to a state that may be similar to a psychosis. Here, too, the probability is that neuronal aberrations occur in healthy people, which are, for example, similar to hyper-connectivity in the CMS and the disbalance between DMS and CEN (see Chap. 9).

For all psychiatric patients, not only those suffering from psychoses and psychosis-related conditions, *admission to an acute care unit in a psychiatric hospital* means an unusual environmental condition. These are characterized by a place that is alien, filled with people who may be dangerous and who are sometimes very noisy. The foreign environment, into which a psychiatric emergency patient is brought, is often a closed ward in which some patients are held against their own will.

The burden and strain of the relationship between the environment and the Self can become extreme, so that the continuity of the spatiotemporal structure of the Self is in danger to interrupt. It is obvious that psychiatric patients, in whom the spatio-temporal continuity of the Self is disturbed, experience additional irritation in such an environment. This can lead to an increase in anxiety and subsequent amplification of symptoms. The situation of the *hyperarousal* can cause the disease to get worse.

How do patients react to this extreme environment, influencing the Self and the brain not only during the admission but also after some days of staying?

### 22.1.1 Schizophrenia and Schizoaffective Psychoses

There are patients who, after a few hours or a few days, are able to adapt to the new environment and to experience the protection of the ward positively. Although the relatives are often frightened and think that the patient will get worse, the patients themselves feel safe. These include, for example, a *schizophrenic*, whose attention is greatly reduced to himself. On the neural level, one can assume that they focus strongly on their spatiotemporal Self, so the hyper-connectivity of their CMS (Northoff 2016; Whitfield-Gabrieli et al. 2009) is overly emphasized. At the same time, because of a shift in the CMS-CEN balance (Robinson et al. 2015) with an overload of the CEN, they are hardly aware of the environment.

Other patients from the *schizophrenic and schizoaffective group* are particularly sensitive to external stimuli because they cannot filter and mitigate them. The filter mechanism against external stimuli is not sufficiently effective. It is also assumed that compensation and counter-regulation of the brain are not sufficient to protect them. Consequently, the neuronal aberrations are amplified on the neuronal level of the brain, and on the phenomenal level, the symptoms of the disease become worse.

### 22.1.2 Depressive Patients with Suicidal Ideas

There are severe *depressive patients* with suicidal ideas, in which the emphasis on experiencing their own self (CMS, see Chap. 11 in this book) is so strong that they can hardly be influenced by the environment. But other patients with depression are overly irritated by the extreme environment. On the neural level of the brain, a number of neuronal aberrations are to be found, for example, a resting-state hyperactivity in the anterior midline regions and hypoactivity in the posterior midline regions. These may result from a disturbance in the DMN (default mode network) (see Chap. 11 in this book for detailed information). We assume that the disturbance may increase in the stressful environment of a psychiatric hospital.

### 22.1.3 Bipolar Patients

When *manic* patients are admitted in a locked ward against their will, they usually behave inconsiderately, without any distance to other people, and they create a high level of tension in the ward. The manic symptoms are usually strengthened. Therefore they disturb other patients, which again increases the tension and irritation in the ward's rooms. On the neuronal level, investigations of bipolar patients show that the amygdala is a key region within a frontal limbic network of emotion regulation, and it displays increasing resting-state activity as well as hyper-reactivity in response to emotional stimuli (see Chap. 12 in this book for more detailed information.)

Patients with *personality disorders* usually are very disturbed by the overly tension and arousal on emergency wards. They react very differently, some with an increase of the psychopathological symptoms and some others with withdrawal.

Cocaine *addicts* and other addicts of stimulant drugs are so irritated that they want to open the door to flee.

One of the *treatment* methods in acute psychiatric wards is high doses of psychopharmaceuticals which are usually given to almost all patients. This makes it possible to reduce the tension caused by the environment in the psychiatric ward with acute psychiatric patients and to make individuals more tolerant toward the pathologizing environment. In individual cases, there are protected resting rooms, in which very restless patients can be isolated in order to calm them down. More importantly, there should be professionals, nurses, doctors, and psychologists, who have a lot of empathy to take care of the patients.

In this chapter we will describe psychiatric patients who are:

1. Aggressive against other people
2. In danger to commit suicide or harm themselves
3. Severe ill psychotic patients, who are treated in a group psychotherapy in a locked ward of a psychiatric hospital

## 22.2 On Dealing with Aggressiveness

In acute psychoses, aggressive discharges can occur. Delusional experiences, misunderstanding, and misjudging of persons can trigger aggressive behavior, especially in schizophrenic and manic patients.

**Case Report**

Madness, which leads to aggression, is reported by the patient Mrs. G. in the postacute course:

> When I was taken to the hospital, I felt that the intelligence service took me to the torture chamber. At the ward the nurses looked like slaughterers. At that time I was convinced that corpses were stored in the basement under the ward. When the nurse gave me medication, I thought I was being anesthetized and then killed. I took the medication from her hand, hit her in her face, pushed her aside, and tried to flee.

In catatonic stupor too, there may be unintended aggressive breakthroughs, which manifest themselves unexpectedly.

**Case Report**

Sudden aggressiveness in catatonic schizophrenia: Mr. K. sits in the room immersed in himself without any visible contact with other patients. Suddenly he utters a cry. The previously rigid nearly paralyzed patient takes the chair on which he has sat on and thus hits the table and a shelf. As the nursing staff and the physician try to calm him down, he takes a shelf loosened from the wall and lashes out with it. A few weeks later he reported embarrassedly: "*I felt as if my energy was being sucked. I felt my body dissolved and I could not move properly. It was until I slammed with the objects, than I had again the strength and the power of my movements.*"

Even in patients suffering from affective psychoses, aggressive actions can occur in the manic episode. Frequently, the patients feel superior to other people and are impatient. Necessary limitations are not always tolerated by those patients.

**Case Report**

Aggressiveness in the manic phase of an affective psychosis: Mr. Q. is used to take extended baths in manic phases when he is irritated. He then hears loud music or sings operatic aria to hoarseness. Since the patient is of a particularly strong stature, the nurse team tolerates this behavior, also for fear that he might otherwise react aggressively, as is known from previous inpatient treatments. Only when the bathroom and floor of the ward are under water, the nurse team decides to intervene: while talking to Mr. Q., he snatches the scrubber from the cleaning lady, who is also present, provocatively leans on it, and cries "*Get me out of here, if you can.*" Only the doctor who has been called can persuade the patient to leave the bathroom.

### 22.2.1 The Patient's Side

The examples of aggressive action show that aggressiveness can often be understood: it may be an attempt to escape a threat that is subjectively realistically experienced in delusion. Aggressive behavior can have a structure-cohesive effect with given self-fragmentation. Thus, aggressiveness can counteract the threatening disintegration by ego vitalization and promote the experience of self-integrity. Aggressive utterances and actions can serve to improve the ego demarcation and restore a sustainable distance.

## 22.2.2 The Side of the Therapeutic Team

Dealing with aggressive patients is a particular challenge for the members of the therapeutic team: if a patient is acting aggressively due to illness, he delegates the control of the impulses to the outside. Despite the possibility of understanding aggressive behavior within the context of the psychodynamic connections, a limiting intervention is usually required. If de-escalating verbal strategies fail, action must also be taken. This results in different consequences:

- The inevitable confrontation with the need to limit aggressive behavior provokes a consequent counterreaction. This can be experienced as a kind of one's own aggressiveness, contrary to the image of the professional helper. This puts the professional ideal in question.
- An aggressive behavior of patients can cause a strong anxiety effect and lead to uncertainty in the treatment of the patient (Grube 2003).
- The patient strikes with his aggressiveness the psychological vulnerability of individual team members. This vulnerability may result from their own biographical development. This can lead to considerable insults. If this happening remains unconscious, the response to the aggression can be excessive and inadequate.

The primary objective of the management should be to control the aggressiveness with appropriate means. If strong emotions are aroused by the team members and if they remain unconscious, there is a risk of an excessive reaction. Then it is no longer about the necessary control of aggressiveness but unconsciously about the establishment of a "dominance versus subordination" hierarchical structure. Frequently, the result is a new aggressive act of the patient, which in turn reacts to the exaggerated reaction of the therapeutic team. There is a risk that this new aggressive action will lead to further escalations. The original linear sequence - after aggression follows control - has now become an aggressive circle, with a considerable risk of escalation and uncontrolled aggression on both sides (Maier 1999).

## 22.2.3 What Can the Therapeutic Team Do?

The experienced therapist is often, but not always, able to anticipate the risk of aggressive behavior of patients with psychosis. This should be discussed explicitly in case of discussions in order to sensitize the members of the therapeutic team. If this sensitization is successful, many situations can be defused before the manifestation of concrete aggressive action. After each aggressive action, it is also necessary for team discussions and supervision to work out characteristics of the pre-aggressive situation, which contain signals of an aggressive action. There is then the chance to use de-escalation strategies in the future before aggressive actions can manifest themselves. Also of great importance is to become aware, to express, and to name one's own feelings after an aggressive action. In particular, if a team member or a patient was a victim of a physical attack, a detailed working through the traumatizing event is also required. When fear or counter-aggression becomes conscious, there is an opportunity to get control over it and reduce the risk of entry into the spiral of "dominance versus subordination" hierarchical structure. The management to ward off dangers should be practiced again and again (Grube 2001). The consequent safety results in greater serenity. A conversation with the patient about his aggressive actions should take place as soon as possible in order to better understand the psychodynamic basis of the actions and also to strengthen the intrapsychic control mechanisms. In some stages it is not avoidable that body-oriented approaches may be helpful in directing the aggressive potential into constructive, persistent, and self-limiting behaviors.

For body-oriented protection measures, a special training for the staff is necessary. The main point is that the patients and the staff are not injured. It is about the protection of all, especially the patients. For this purpose, a particular concept has to be developed for each psychiatric institution, and it is significantly important to find a good trainer who is physically, sportily, and psychologically experienced. It is not that the staff is trained once or twice, but it has to be done continuously; also the team supervision should be held continuously.

## 22.3 The Danger to Commit Suicide

"Suicidal persons are probably the most frequent causes of psychiatric emergencies. The suicidal person is difficult to identify with certainty; he is often unmanageable in an outpatient setting and resistant to hospitalization, and he is subject to recurrent crises in management" (Kaplan and Sadock 1981, p. 704). Often it is necessary to get a suicidal patient into safe custody in an acute psychiatric locked ward, even if it is sometimes against the patient's wishes.

### 22.3.1 The Pre-suicidal Syndrome

Almost all patients with severe depressive episodes can become suicidal in the acute disease stage. To find out in what extent the suicidal risk is, the pre suicidal syndrome (Ringel 1953, 1984) should be considered here. It usually consists of the sequence of three characteristic developmental stages:

1. Restriction of thinking and feeling: situational, dynamic, interpersonal
2. Turning back of aggression against oneself: fainting rage, aggression against one's own person
3. Death fantasies: increasing fantasy activity around the process of self-killing

Additionally, a fourth subject should be discussed with a possible suicidal patient: it is especially important to find out if the patient already has *a plan* to carry out a suicide.

The lack of emotional investment energy of self and objects restricts the sustainability of relationships. Thus the promise not to commit suicide is not always reliable under these circumstances, because a trustful relationship with the therapist is not firmly built in those cases and the emotional anchorage of the promise cannot be achieved in the depth of the depression.

In some cases, a severe nihilistic delusion, which Henseler (1974) describes as a negatively distorted representation of omnipotence, can contribute to the motive of an extended suicide. Thus, in rare instances, the depressive does not only want to kill himself but also those close to him, for example, his children, because he experiences them as an associated part of his own self.

The suicidal aggressiveness, which is in many cases psychodynamically seen as aggression which is turning back against oneself, should be perceived sensitively both in the outpatient and in the inpatient setting. When the therapist communicates with the patient, this crucial point should always be questioned actively. If a high degree of auto-aggressivity is confirmed, the therapist should place emphasis on treatment in a protected ward or in a protected environment with professionals. The decision is not always easy because the doctor sometimes decides not only against the patient's wishes but also against the wishes of the family and friends. Despite all precautions, suicide cannot be prevented in all cases.

#### 22.3.1.1 Case Study: Suicidal Action

Suicide of a bipolar patient with comorbidity of a neurological disease (multiple sclerosis):

The 54-year-old woman suffers from a bipolar affective disorder that has existed for 25 years. Despite phase prophylactics, episodes with psychotic symptoms occur nearly every 2 years, and she has to be hospitalized. In addition to psychosis, she suffers from multiple sclerosis with increasing cerebral impairment, which manifests itself in the form of creeping progression of cognitive deficits.

In previous years the patient was a passionate member of a shooting club. A few years ago, the marriage was divorced, and the now grown-up son was about to leave the parental home. In the current inpatient stay, she was dominated by a depressive-dysphoric mood, which seemed to be well compensated in the course of therapy. Suddenly she committed suicide. For this she went home without the therapist's permission and shot herself there with one of her revolvers, which she had secretly kept. In the months before, she had expressed several times life-threatening thoughts and alarmed her relatives and her therapists. The relatives then contacted the clinic. After the event, we learned that she had just said goodbye to her son and some close friends. These persons unfortunately did not inform us this time but obviously tolerated her decision to commit suicide.

This is an example of the higher rate of suicide in the comorbidity of psychosis and organic diseases, such as multiple sclerosis, Parkinson's disease, Huntington's disease, and AIDS. These are organic states of illness, in which there is subjectively and in many cases also objectively no improvement, but rather a progression of the patient's pathology is expected in the further course. If such an organic development is mixed with an affective disorder, it is less likely that a patient commits suicide in the depth of the depression but rather in the interval period.

### 22.3.2 The Subjective Side of the Patient

A long decision-making process had taken place for weeks and months. On the one hand, the patient suffered from the high discrepancy between her once good relationship with the husband, her challenging job, and her creative power and on the other hand her present loss of these abilities, including the loss of personally valuable relations. In addition she suffered from the fear of further cerebral decay. This condition was accentuated by the psychosis, especially by the negative view during depression. There had long been a distressing despair. From our point of view, the patient ended her life because she was not anymore able to accept the increasing decay, loss, and dependence on other people. So this decision was her last active personal action.

### 22.3.3 How About the Therapeutic Team?

Among those who were involved in the therapy both during the inpatient period and also in interval outpatient treatment and have been willing to help her, a fainting impotence was spreading. In the discussion, many thoughts, which had previously only thought of, were now spoken aloud: whether one had done enough and whether one had failed. The question also came up, as to whether it really was her real intention to commit the suicide. How often has the team experienced that patients do not irrevocably make such a decision but that they go through ambivalent conditions? Also the question of guilt raised in the team supervision.

Gabbard (2000, p. 222) points out "*that a completed suicide is a reminder of the limitations inherent in our craft. The natural tendency, whether in hospital practice or in psychotherapy, is for clinicians to go to great length to prevent suicide. To implement reasonable measures to prevent patients from taking their life is certainly good judgment from a clinical standpoint, responsible behavior from an ethical standpoint and sound defensive medicine from a medicolegal standpoint. However, when the role of savior becomes all-consuming, the results may be countertherapeutic.*"

Hoff and Venzlaff (2009) explicitly emphasize that the experience of modern psychiatry shows that too much control and the attitude that one can avoid all suicide if one only has enough professionalism have just opposite challenges. Hoff and Venzlaff (2009, p. 857) comment on the difficult decision of how many or how few preventive measures should be carried out for the affected person in a closed psychiatric department. The decisive factor is always the appropriateness of a suicide preventive measure. The suicide prevention measures, such as isolation, fixation, taking all personal belongings, or highly dosed sedation, which are usually practiced in a strongly custodial-oriented psychiatry, are not only ethically questionable but can even be regarded as a medical malpractice in individual cases, especially when other accompanying interventions are lacking, as they are supposed to increase the already existing self-devaluation tendencies and anxieties of the patients and are thus suitable for triggering or strengthening a vicious circle.

For those who are responsible, the dignity and intimacy of suicidal patients in need of protection must be taken into account both in the structural and in the personal treatment. There is thus a scope, a balance of considerations, and often a conflict between the two aspects, respect for dignity and intimacy on the one hand and extreme control on the other. Clothing, for example, is more of an intimate sphere. Belts are a frequent subject, since they can be used for strangulation. On the other hand, a patient should be allowed to have a belt construction on, in order not to injure his shame and dignity. Ultimately, many measures are a compromise in balancing, how high the threat is, how far the dignity of the concerned person is to be respected, and to what extent the nursing staff should take protection. It also plays an important role in the extent to which a good therapeutic relationship can be established and permanently strengthened. It is important to bear in mind that, in the long run, too much supervision causes the patient to feel completely defenseless and to no longer possess self-determination, because that can enhance suicidal tendencies. In monitoring rooms with appropriate inspection windows, visual inspection is always in the ideal case possible; in reality, however, there are always distractions. Here, too, the individual case with the current extent of the suicidal threat is to be concerned. The balancing between "total control," which can increase suicidal behavior, and low control in which suicide can be possible must be taken into consideration. It is also difficult to evaluate because suicidal behavior has many facets and many different interindividual and intraindividual degrees of intensity, which are sometimes difficult to detect even for an experienced psychiatrist.

In case a psychiatrist is accused and has to appear in court, there are some rules which go back to Ankermann (1984, p. 272). He emphasizes, especially on the basis of the personal assessment, the therapist should make his decisions, which measures are to be taken, and the extent to which the patient can be granted freedom. In this respect, as for the therapist, who has created a relationship of trust with the patient, and who is therefore more likely to judge him by his experience than an outside person, there is a margin of discretion which should not be too narrow. But if the physician fails to take a concrete evidence of a patient's possible self-harm, then he has exceeded these limits of patient's granted freedom.

### 22.3.4 The Problem of Uncertain Prediction

There will always be patients who want to take their own lives and ultimately do so with all their conviction. Although there are many studies on pre-suicidal risk factors with the statement of statistical probabilities, these include affective disorders, other psychoses, additional organic diseases, loss of social bonds, difficult life events, suicide attempts in case history, etc. (Schneider 2003; Harris and Barraclough 1994, 1997; Isometsä et al. 1995; Wolfersdorf et al. 2016). However, the real prediction of a suicide risk for the individual patient is very difficult if the patient does not express himself accordingly. Gabbard (2000, p. 223) writes: "Clinicians cannot read minds and must not berate themselves for their perceived failures when not clear verbal or nonverbal indications of suicidality were present."

### 22.3.5 Countertransference and Suicidality

From the psychodynamic point of view, the countertransference in the suicidal patient should be considered with special care and should be worked through in the supervision sessions of the team. Occasionally, countertransference feelings such as anger, impotence, and injury to the therapist's pride may become a serious obstacle to the patient's therapy. The impending suicide or the real suicide of a patient is a nuisance to the therapist, which can lead to expectations and accusations of the environment. In the treatment of prolonged and chronic suicidal tendencies, what Gabbard (2000, p. 225) described as "countertransference hate" occasionally comes to mind: "... treaters will often harbor an unconscious wish for the patient to die so that the torment will end." As a result, the inability to recognize and respond to their own unconscious aggressive wishes occasionally results in an expression of the opposing emotions; it can happen that the therapist does not respect or "forget" the patient. Gabbard (2000, p. 226) continues, "Countertransference hatred must be accepted as part of the experience of treating suicidal patients."

When the therapist's countertransference anger or hatred is split off, he denies and projects into the patient. The therapist's unconscious destructive wishes are also added to the patient's own suicidality. In addition, Milch (1994) points to the danger of the own suicidal impulses in the treater, which are reversed unconsciously to the contrary and lead him to actions which can have deleterious results because of the jointly shared death wishes. These statements, derived from the psychodynamic view, seem to contradict the psychiatrist's and psychotherapist's understanding in his professional role. In the dilemma of the therapist between professional roles and the abovementioned countertransference, a personal or team supervision can be helpful.

### 22.3.6 Workshops and Psychodynamic Team Groups

Hartwich and Langegger (2003) report the experience that *workshops*, where case studies are worked through, can help employees of all professional groups dealing with suicidal patients. For example, a colleague who works in her own practice reported about a young man who had suffered from a chronic psychosis: when his mother had gone for a holiday, he committed suicide. She, the therapist, was supposed to "take care" of the patient during this time. By working through in the group, the therapist's individual feelings of guilt were transferred into a collective experience. For the group leaders, it was important to create space and an understanding atmosphere for the evolving emotional experience. The group has discussed the problematic feelings of the colleague and transferred them from the individual experience to a collective experience of those who care for psychotic patients. It was important to make it clear that they themselves had very similar experiences.

There was given enough time for listening to the subject and for reflection, without interruption of conversations, and the participants came to a series of very empathic contributions. Own experiences of similar events were communicated. The emotional inner stirrings emanating from the reporting colleague were recorded, shared, and enriched with their own experiences. The psychopathological situation of the patient, the family environment, and the more exact circumstances of the suicide were reconstructed in detail in an unfolding emotionally severe group process. The following aspects, which were discussed in group discussion, are briefly outlined: in somatic medicine it is accepted that there exist diseases leading to death. In the case of mentally ill patients, on the other hand, it is difficult to accept this analogously.

## 22.4 Seriously Sick Psychotic Patients, Who Are Treated in a Group Psychotherapy on a Locked Ward of a Psychiatric Hospital

We used to modify our group therapeutic experiences for treatment of patients with severe illnesses in the closed ward. All patients got pharmacological antipsychotic treatment with an additional group therapy offer. As a rule, there are patients with schizophrenia and schizoaffective psychoses, often with additional problems regarding alcohol and other drugs as well as organic comorbidity. The leaders of the group are a psychiatrist and a psychologist, who are in psychoanalytical training. The group takes place twice a week for about 30 minutes.

- *Example of a Group Session*
- Ms. B., 45 years old, suffers from a catatonic schizophrenia; she sits stiffly on her chair and is today able to speak, and she holds a monolog to herself about death and from the leaning tower of Pisa. She looks at the therapist and says, "I am the fish on the hook." She allows no questions and no contact, and the others do not exist for her at all.
- Mr. F., 65 years old, suffers from a schizoaffective psychosis and regressive changes of the brain volume. In the state of maniform irritation, he stands up and lays himself on the couch, which stands in the room, and then he curses: "I am fine, I can no longer bear the whole lamentation." He gets up again and verbally attacks another patient, who then leaves the room.
- Mrs. M's thoughts and speech sped up; her thoughts are constantly diverted by the intervention of strange ideas. Her logorrhea is partly whispering and partly overly loud in the direction to the therapists.
- Mr. H. roars into the group: "It is too restless here," and he changes the seat; after that he goes out and comes back into the room again.

- *In the Supervision Session*
- The two therapists reported after the group session the feeling of being somewhat confused: nothing seemed possible, helplessness spread out, and they felt the threatening fear and the danger of losing themselves. They said that they have become very uneasy and dissatisfied and have had the impression that nothing really "reasonable" has been achieved. According to Bion (1961), such patients are not capable of grouping; they seem not to be able to build a group cohesion (Yalom 1974). But as the philosopher Ortega y Gasset (1964) realized, whether we want or not, we are always in group situations, and Slater (1970) described the psychotherapeutic group as a microcosmos.
- In this context with those severe psychotic patients, we speak of a different quality of cohesion, which we call *para-cohesion* (Hartwich). From the psychoanalytic point of view, it is an encounter with the "primary process" (Freud 1915), with fragments without recognizable cohesion, embodied by the individual patients in the session. *The fragments in the group are mirroring the fragmentation in the individual patient's self.*
- The group therapist has the chance to experience the fragmentation, the chaos, and the primary process on closer examination, which is only possible with some experience, and to have a closer look at what the patients with schizophrenia have experienced in their illness. Therapists are increasingly learning to have more "chaos-bearing ability," which is a good training for psychodynamic therapy of psychotics (Hartwich 2007). As Battegay (1976) emphasizes, severe psychotic patients in the group session press for an exclusive contact with the therapist and do not talk to each other.
- The patients were caught up in their private and isolating conviction of ideas (the leaning tower of Pisa, the fish on the hook, etc.), and they tried to capture their own fragments and build a border with them very rigidly and autistically. This has a presently important function, since otherwise they would merge with the "outer objects." The lack of communication, which is not a noncommunication, thus also serves to protect against flooding by fusion. If the threat from outside had been intensified by perceived proximity and resonance to what others have said, for example, the death thoughts, then the patient had to leave the group room temporarily. The group meeting became a place of fragments, a collective mirror of what is lived in fragmentation in the individual self. The therapist's countertransference could be reflected. They could come nearer to the patient's psychotic experience, and if they succeed in carrying self-fragmentation, they have a cohesion-forming function as group leaders. The therapists represent a crystallization nucleus for the further group process, in the vicinity of which the self-fragments can constellate themselves by embodying safety. Thus in later sessions, it was also possible to feel more calmness as a basic phenomenon, which in such groups is an indication that psychotic fantasies can be tolerated.
- In this way, we create the prerequisite for group psychotherapy on open wards and later in outpatient settings. Our experience is about "slow open groups" (Foulkes 1964). As regards the course of the disease and its psychopathological condition, the view has been established that the time after reduction of the acute

symptoms as well as a stable psychopharmacological treatment is among the most favorable prerequisites (e.g., Kanas et al. 1980; Hartwich 1982; Gabbard 2014) for successful group psychotherapy in schizophrenic and schizoaffective psychoses (Schwarz 1986, 2001).

*Some important experiences of the patients are outlined here briefly*:

- The individual patient meets compassionates and connoisseurs.
- A group cohesion can strengthen its own cohesion efforts.
- A positive anticipatory movement when a patient experiences a progress and improvement in other psychotics.
- Omega projections (Schindler 1968, 1969) are less endangered in psychosis groups because of the member's group dynamic position, which changes faster than in groups with other diagnoses.
- Greater range of tolerance toward psychotic experience.
- Chance to learn and to observe the early warning signs.
- Chance to work through the recalling of early traumatizations.
- Help with the linguistic formulation of otherwise unspeakable (transitional subject, Benedetti 1992).
- Handling of psychopharmaceuticals is governed by "experts among themselves" and has greater conviction than in the dual relationship with the doctor.

Hartwich and Schumacher (1985) have presented an empirical study showing that the relapse rate after the 4th and 5th year compared to the comparable control patients was significantly lower in the observation period of 5 years. Gabbard (2014) points out that group psychotherapy is often seen as a "second-class" treatment modality versus individual therapy. In fact, many review articles in which treatment results are compared to group therapy with those of individual psychotherapy do not support this prejudice. Group therapies are performed for patients with almost all psychiatric diagnoses: affective disorders, alcohol dependence, other seeker disease, post-traumatic stress disorders, borderline disorder, psychiatric age disorders, bipolar disorders, personality disorders, anxiety disorders, etc. In psychodynamic group psychotherapy, we prefer homogeneity referring to the diagnoses inconsideration of the strength of the individual structural level. Therefore, e.g., we avoid mixing psychotic patients with borderline personality disorders or drug-dependent patients with other diagnoses.

## 22.5 Neurobiological Findings

Neurobiological investigations about aggressiveness to other people or to the patient himself may depend on the diagnoses. So in Chap. 10 one will find the neurobiological findings about schizophrenia and other psychoses in detail. This is also true for group psychotherapy with schizophrenic and schizoaffective patients who are in an acute emergency condition and who have to be treated on a protected ward of a psychiatric hospital. Also in Chap. 11 about depressive syndromes, all the modern investigations about this topic are mentioned in detail. Therefore it should not be repeated here.

Wolfersdorf et al. (2016) summarize neurobiological investigations referring to the problem of suicide: one of the hypotheses is that a reduced serotonergic innervation of the ventral prefrontal cortex may promote impulsiveness and autoaggression. Neurobiochemically a presynaptic serotonergic deficit with compensatory increase of the 5-HT2 receptors in the prefrontal cortex and a reduced CSF-5-HIAA is assumed. Also, a decrease in serotonergic activity may be associated with a decrease in cholesterol in the synaptic membranes in the central nervous system. But Wolfersdorf et al. (2016) also emphasize that the described psychological and neurobiological manifestations do not explain the question of the cause of suicidal behavior.

## References

Ankermann E. Urteil des Oberlandesgerichts Düsseldorf. Zur BGB §§ 276, 823 Abs. 1: Sorgfaltspflicht der behandelnden Ärzte in einer Nervenheilanstalt gegenüber infolge zeitweiser Wahnideen suizidgefährdeten Patienten. Spektrum 13/6; 1984. pp. 266–75.

Battegay R. Der Mensch in der Gruppe, Bd 1. 5th ed. Bern: Huber; 1976.

Benedetti G. Psychotherapie als existenzielle Herausforderung. Göttingen: Vandenhoeck & Ruprecht; 1992.

Bion WR. Experiences in groups. London: Tavistock Publications; 1961.

Foulkes SH. Therapeutic group analysis. London: Allan & Unwin; 1964.

Freud S. The unconscious S.E. 14:161 Das Unbewusste.G.W.Bd. 10; 1915. pp. 264–303.

Gabbard GO. Psychodynamic psychiatry in clinical practice. 3rd ed. Washington: American Psychiatric Press; 2000.

Gabbard GO. Psychodynamic psychiatry in clinical practice. 5th ed. Washington DC: American Psychiatric Press; 2014.

Grube M. Aggressivität bei psychiatrischen Patienten – Einflussmöglichkeiten durch ein Selbstschutztraining. Nervenarzt. 2001;72:867–71.

Grube M. Emotionale Reaktionen von Mitarbeitern im Umgang mit aggressiven psychiatrisch Erkrankten. Psychiatr Prax. 2003;30:187–91.

Harris EC, Barraclough BM. Suicide as an outcome for medical disorders. Medicine. 1994;73:281–96.

Harris EC, Barraclough BM. Suicide as an outcome for mental disorders: a metaanalysis. Br J Psychiatry. 1997;170:205–28.

Hartwich P. Gruppentherapie bei Schizophrenen in der Nachsorgeambulanz. In: Helmchen H, Linden M, Rüger U, editors. Psychotherapie in der Psychiatrie. Berlin: Springer; 1982. p. 110–5.

Hartwich P. Psychodynamisch orientierte Therapieverfahren bei Schizophrenien. In: Hartwich P, Barocka A, editors. Schizophrene Erkrankungen. Sternenfels: Wissenschaft & Praxis; 2007. p. 33–98.

Hartwich P, Langegger F. Supervision bei Suizidalität und erfolgtem Suizid. In: Hartwich P, Haas S, editors. Suizidalität. Diagnostik und Therapie. Sternenfels: Wissenschaft & Praxis; 2003. p. 117–20.

Hartwich P, Schumacher E. Zum Stellenwert der Gruppenpsychotherapie in der Nachsorge Schizophrener. Eine 5-Jahres-Verlaufsstudie. Nervenarzt. 1985;56:365–72.

Henseler H. Narzisstische Krisen, zur Psychodynamik des Selbstmords. Hamburg: Reinbeck; 1974.

Hoff P, Venzlaff U. Psychiatrische Begutachtung von Suizidhandlungen. In: Foerster K, Dreßing H, editors. Psychiatrische Begutachtung. 5th ed. München: Urban & Fischer; 2009.
Isometsä ET, Heikkinen ME, Marttunnen MJ. The last appointment before suicide: is suicide intent communicated? Am J Psychiatry. 1995;152:919–92.
Kanas N, Rogers M, Kreth E, et al. The effectiveness of group psychotherapy during the first three weeks of hospitalisation: a controlled study. J Nerv Ment Dis. 1980;168:487–92.
Kaplan HI, Sadock BJ. Modern synopsis of comprehensive textbook of psychiatry/III. 3rd ed. Baltimore: Williams & Wilkins; 1981.
Maier GJ. Psychological issues in treatment: transference and countertransference. In: Tardiff K, editor. Medical management of the violent patient. New York Basel: Marcel Dekker; 1999.
Milch W. Gegenübertragungsprobleme bei suizidalen Patienten unter stationärer psychiatrischer Behandlung. Psychiatr Prax. 1994;21:221–5.
Northoff G. Neuro-philosophy and the healthy mind. Learning from the unwell brain. New York: Norton; 2016.
Ortega y Gasset J. Sozialisierung des Menschen. Vertreibung des Menschen aus der Kunst. München: dtv; 1964. p. 40–4.
Ringel E. Der Selbstmord – Abschluss einer krankhaften psychischen Entwicklung. Wien: Maudrich; 1953.
Ringel E. Suizid. In: Battegay R, et al., editors. Handwörterbuch der Psychiatrie. Stuttgart: Enke; 1984.
Robinson JJD, Nils-Frederic Wagner NF, Northoff G. Is the sense of agency in schizophrenia influenced by resting-state variation in self-referential regions of the brain? Schizophr Bull. 2015;42(2):270–6. https://doi.org/10.1093/schbul/sbv/102.
Schindler R. Dynamische Prozesse in der Gruppenpsychotherapie. Z Gruppenpsychother Gruppendyn. 1968;1:31–7.
Schindler R. Das Verhältnis von Soziometrie und Rangordnungsdynamik. Z Gruppenpsychother Gruppendyn. 1969;2:9–20.
Schneider B. Risikofaktoren für Suizid. In: Hartwich P, Haas S, editors. Suizidalität. Diagnostik und Therapie. Sternenfels: Wissenschaft & Praxis; 2003.
Schwarz F. Übertragung und Gegenübertragung in der analytischen Gruppenpsychotherapie mit psychotischen Patienten. In: Sandner D, editor. Analytische Gruppentherapie mit Schizophrenen. Göttingen: Vandenhoeck & Ruprecht; 1986. p. 117–32.
Schwarz F. Gruppenprozesse und Gruppentherapie. In: Schwarz F, Maier C, editors. Psychotherapie der Psychosen. Stuttgart: Thieme; 2001. p. 102–9.
Slater PA. Mikrokosmos: Eine Studie über Gruppendynamik. Frankfurt: Fischer; 1970.
Whitfield-Gabrieli S, et al. Hyperactivity and hyperconnectivity of the default network in schizophrenia and in first-degree relatives of persons with schizophrenia. Natl Acad Sci. 2009;106(4):1279–84.
Wolfersdorf M, Prurucker M, Schneider B. Suizidales Syndrom. In: Boeker H, et al., editors. Neuropsychodynamischen Psychiatrie. Berlin: Springer; 2016. p. 395–408.
Yalom ID. Gruppenpsychotherapie. München: Kindler; 1974.

# Psychoanalytic Treatment of Borderline Patients in a Day Hospital Setting

23

Heinz Weiss and Margerete Schött

**Abstract**

The authors present the relevance of day hospital settings for the psychoanalytic treatment of borderline patients. Referring to specific areas of conflict which are central to borderline pathology (identity problems, dealing with separation and dependency, claustro-agoraphobic anxieties, taking refuge to 'psychic retreats'), they argue that the specific elements of a day hospital setting may be more favourable in dealing with these problems than a classical psychiatric inpatient treatment. Using clinical vignettes, the significance of focusing on the emotional experiences in the here and now of the interactions, maintaining the borders of the setting and the thorough working through of the transference and counter-transference is emphasised. Possibilities and limits of day hospital care are discussed with reference to the existing literature and future research strategies.

## 23.1 Introduction

Borderline personality disturbance is characterised by a vacillation of feelings of self-worth, instabilities of personal relationships and affects, difficulties in impulse control, self-harm, chronic feelings of futility and emptiness often associated with suicidal ideation and (para)suicidal actions, intolerance of separation as well as occasional dissociative states and brief psychotic episodes (American Psychiatric

H. Weiss (✉)
Robert-Bosch-Krankenhaus, Abt. Psychosomatische Medizin, Stuttgart, Germany

Sigmund-Freud-Institut, Frankfurt, Germany
e-mail: heinz.weiss@rbk.de

M. Schött
Sigmund-Freud-Institut, Frankfurt, Germany
e-mail: schoett@sigmund-freud-institut.de



H. Boeker et al. (eds.), *Neuropsychodynamic Psychiatry*,
https://doi.org/10.1007/978-3-319-75112-2_23

Association (APA) 2013). The prevalence in the population is biased towards young adulthood and female gender by about 2% (in various European and American studies 0.5–5.9%; see Zanarini and Hörz 2011). Borderline patients use up to 20% of psychiatric services available to a population. Approximately 8–10% commit suicide (American Psychiatric Association (APA) 2013).

Borderline patients pose a particular challenge for psychoanalytic therapy because of their instability, their tendency to put others under pressure and to easily break up relationships. While some authors suggested a modification of the classic treatment techniques (Rudolf 2004, 2006; Kernberg 1999; Clarkin et al. 2005, 2006; Dammann et al. 2000), others prefer a more direct working with the complex and at times confusing transference manifestations without major changes to technique (Rosenfeld 1981, 1987; Segal 1991; Steiner 1993; Weiß 2009). There is agreement that understanding the countertransference and working through of the enactments in the here and now of the therapeutic situation, as well as the transition from concrete to symbolic thinking, assume particular importance. By using these techniques, the pressure to act out internal tension decreases, and the patient regains more space to tolerate contradictory feelings and to think about internal states.

The development of a detailed understanding of 'early' defence mechanisms, such as projective identification and pathological splitting, has been central to the understanding of borderline pathology. These mechanisms make it possible to keep 'good' and 'bad' experiences apart and to project unbearable parts of the self into others. This can provide a short-term relief but in the long term leads to feelings of confusion, imprisonment and pronounced agora-claustrophobic anxieties. Some of these patients slip into their external objects as if into a 'shell', because they lack a sense of internal stability, but as a consequence they feel confused and imprisoned. On the other hand, if they distance themselves too far away from the other, they experience panic attacks and states of emptiness and abandonment as if they had lost parts of their self.

Kernberg (1975) has described 'diffusion of identity' as a core symptom of borderline pathology, and Henry Rey (1979, 1994) regards the 'claustro-agoraphobic syndrome' as a basic element of schizoid dysfunction. More recent psychoanalytic approaches have highlighted problems in mentalising (Fonagy et al. 2002) and symbol formation (Green 1990; Segal 1991), as well as the tendency of some of these patients to construct highly complex defensive organisations (Rosenfeld 1971; Meltzer 1968). Such pathological organisations can manifest clinically as states of psychic retreat (Steiner 1993) and often lead to considerable limitations in the capacity to build relationships.

## 23.2 Aetiology and Neurophysiological Findings

Aetiologically genetic and epigenetic factors (see Maier and Hawellek 2011) as well as the role of a neglectful, frequently traumatic early environment seem to be established. Borderline patients report significant loss and traumatic experiences in early childhood more than any other diagnostic group. Frequently the atmosphere in the family of origin was characterised by a fear of violence or sexual transgression by the caregivers who were often themselves psychologically unstable (Osofsky 2011; Dulz and Jensen 2011). Attachment studies in children and adolescent borderline patients have shown the predominance of unstable relationship patterns with 'insecure-enmeshed' or 'disorganised' attachment patterns (see Buchheim 2011) and the absence of a mirroring relationship matrix (Gergely and Watson 1996).

Recent investigations in experimental and magnetic resonance scanning have indicated that there might be neurophysiological correlates of different aspects of borderline pathology. It seems that frontal lobe/limbic system dysfunctions are associated with hypersensitivity for negatively charged emotional stimuli as well as with a restricted capacity for emotion regulation (Ruocco et al. 2013). Altered activity patterns in the area of the medial frontal cortex, the amygdala as well as the basal ganglia were observed (Kamphausen et al. 2013), while regulation disturbances in the frontal cingulate gyrus and other areas of the prefrontal cortex may play a part as well (Mauchnik and Schmahl 2010; Lang et al. 2012). Imbalance in serotonin and dopamine neurotransmissions has been discussed as a possible basis of poor behaviour and impulse control, but the results have not been securely verified. Changes in pain perception have been repeatedly reported, whereby the perception of physical and psychological pain has overlapping brain areas (Schmahl et al. 2006; Ducasse et al. 2014; Simons et al. 2014). There appear to be neuronal correlates of dissociative states (Krause-Utz et al. 2014) as well as for neural processing of rejection the sensitivity of being rejected (Domsalla et al. 2014) in borderline patients.

## 23.3 Therapeutic Approaches

There is little evidence for the long-term therapeutic efficacy of neuroleptic, antidepressant or anxiolytic medication (Stoffers and Lieb 2011; Lieb et al. 2014). On the contrary, there is an increased risk of possible misuse, for the purpose of suicide or attempted suicide or in the development of drug dependency. By contrast, the efficacy of various psychotherapeutic approaches could be shown in randomised controlled trials, for example, for dialectic behavioural therapy (Linehan 1996; Bohus 2011). The efficacy of psychoanalytically based approaches has been shown in treatments using the concept of mentalisation (Bateman and Fonagy 2011) as well as Kernberg's transference focussed therapy (Kernberg 1999; Clarkin et al. 2005, 2006; Yeomans and Diamond 2011) which uses classic techniques of working in the here and now of the therapeutic transference with an early focus on negative transference and adherence to the therapeutic setting, as well as the working through of primitive defence mechanisms.

In this context the question of which *treatment setting* is most suitable has come up. In this paper the possibilities and limitations of a psychoanalytic outpatient day hospital treatment for borderline patients will be discussed. This will be done with particular reference to the concepts of *containment* (Bion 1962), *enactment*, (Joseph 1971, 1985, 1989; Jacobs 1986) and *pathological organisations of the personality* (Steiner 1993, 2006, 2014).

## 23.4 Clinical Experiences in a Day Hospital Setting

Borderline patients can be seen as people who live in a precarious balance between inner and outer reality. Steiner (1993) formulated the concept of the *borderline position* and showed that these patients on the one hand feel threatened by feelings of fragmentation and persecution and on the other by feelings of loss and agonising guilt. In order to protect themselves from both, they seek out borderline states, which he clinically described as 'psychic retreats'.

Such psychic retreats can become idealised as 'a secure haven', a 'lonely island' or an enchanted resort. Often though, the depressive quality of the retreat becomes more obvious as it turns into a bleak desert, a dungeon or a prison. They can also be portrayed as *interpersonal groups* like a business organisation, religious sects or a mafia-like gang, be it in dreams, in phantasies or in the patient's actual relationships. Rosenfeld (1971) and Meltzer (1968) have shown how such intrapsychic organisations take over the control of the needy side of the self by promising security and control at the cost of emotional contact and development.

In the *day hospital setting*, these relationships are frequently projected onto fellow patients or the treatment staff, so that the patients enact their tendency to live out their internal relationships in one way or another. At the start of treatment, it is often these *enactments* which provide the sole avenue of communicating inner conflict. It is important for the therapeutic work that these communications are recognised as such at an early stage in order to work on them. The prerequisite for this is that the relevant countertransference feelings are taken up in the individual and group therapies as well as by the whole team so that they can be understood and worked through. In order to achieve this, daily team meetings and intensive supervision are essential to assure the stability of the setting.

The same accounts for the distortion of internal and external reality, which can affect self-perception as well as object relationships. Some patients seem to depend on those *misrepresentations* (Bion 1962; Money-Kyrle 1968) to maintain their psychic equilibrium. The following case vignette is an illustration of this.

### 23.4.1 Case Vignette 1

Mr. A, a 39-year-old married employee, complained that his bosses and his colleagues misjudged him and did not value his performance adequately. For this reason he had had several breakdowns with burst of anger and massive physical symptoms. Eventually he resigned from his work as the financial manager of a development project. In his biography he described himself as the 'emotional head' of weak, needy parents who were longing for harmony. He said that they could not give him much but relied on his tolerance and understanding.

In the day hospital, he attempted to establish a special relationship with the therapists but was exceptionally vulnerable and testy in response to comments by his fellow patients, who questioned his superiority. He tried to justify his behaviour with many rationalisations but had to concede that he got into conflict with anybody who did not agree with his views.

After one of these disputes, he got into a fierce power struggle with a driver who hassled him. He attempted to thwart the other driver without realising that this endangered him and his family. When this situation nearly escalated into a duel between him and the other driver in the next car park, he registered that his young son had cried and screamed with anxiety all along in the back of the car.

This experience mirrored the transference situation. A weak, needy part was at the mercy of an omnipotent part of his self, which initially did not allow any support or help. In the same way, the therapists initially had the choice to admire his rationalisations or take on the role of a helpless companion. Only when this narcissistic misrepresentation of various 'overtaking manoeuvres' and the experience of being 'thwarted' became clear within the therapeutic setting, could he relinquish his pseudo co-operation. He got into a temporary state of helplessness and despair, akin to his son on the back seat. Only now could he give up the control over the steering wheel and the accelerator and look for a way together with the therapists and the fellow patients.

These kinds of contortions of psychic reality can evoke helplessness, anger and confusion in the treatment team. They reflect the difficulties of the borderline patient in dealing with the experience of dependency without feeling at someone's mercy, humiliated or persecuted. Not infrequently, these patients linger around the perimeter of the clinic, arrive too early or too late and threaten to end treatment, just to return moments later to beg for help and support.

### 23.4.2 Case Vignette 2

Ms. B, a young art student, could only linger in the glassed entrance hall of the clinic when she was not in a therapy session. She idealised the aesthetic ambiance of this room, where she could see trees, water and sky as if she were 'in nature'. In individual therapy sessions, she spoke in a melodic singsong while experiencing group therapy as a threat. She was afraid that her fellow patients would drive her 'mad' and behaved like a shy doe, which retreated to the unattainability of 'nature' when the contact with humans became too close.

In the second week of treatment, she already threatened to break off therapy. She voiced this intention with such forcefulness that the nurse asked the individual therapist to have a closing session with her. In the subsequent session, the therapist took up her paranoid feeling that therapy might make her mad. This reduced her paranoia enough to make her decide to stay in treatment. During the following session, she began to talk about the rather recent suicide of her father. Midway she suddenly began to cry, lost control and asked for a paper handkerchief. She was still disturbed when she left the clinic in the afternoon. However, when she arrived at home, she started an argument with her boyfriend and intimated suicidal threats, as he had not instantly ended a telephone conversation on her arrival to concentrate on her.

Through her crying in the session, she had in a way 'arrived' in the day hospital no longer lingering in the borderline area of the clinic, but the loss of control had been so shameful and threatening that she had to pick a quarrel at home to forestall a breakdown. In this way she had unconsciously identified with her father who had committed suicide in similar circumstances.

The distress of this patient was that she could not bear to be either inside or outside the day hospital. This is a problem which is frequently encountered with borderline patients. For this reason, the periphery and the boundary, i.e. the borderline area, of the day hospital setting assumes a particular significance.

Therefore, the *holding function of the caregiving staff* with the usual morning and evening ward rounds as well as communal group activities assumes an important function in protecting the space and time limits of the setting (see Küchenhoff 1998). When these boundaries are not adhered to, the internal space, which allows for the therapeutic process of the individual patient to develop, may become fragile and break down.

Narcissistic and borderline patients find dependency on another person difficult to bear without feeling humiliated or imprisoned. At the same time, they find it difficult to be alone without feeling confused or empty. Getting too close makes them feel confined, and they desperately try to free themselves from this enclosed state. But by distancing themselves, they develop panic anxiety of abandonment.

This *agoraphobic-claustrophobic situation* was described by Rey (1979, 1994) as characteristic of the desperate attempt of the borderline patient to find a secure place (see Weiß 2015). In their early childhood development, they lacked a transitional space, referred to as a 'marsupial space' by Rey, in which internal and external reality can gradually be differentiated. Instead they feel locked into the internal space of their early caregiver or traumatically expelled thereof. It is for this reason that borderline patients find it sometimes difficult to tolerate inpatient treatment.

Because of its permeable structure, the day hospital can provide an acceptable solution for this dilemma. It does not lock patients into too narrow a space but at the same time offers structure and boundaries, i.e. they are not left on their own.

Clinical experience and research findings on early child development (Mahler et al. 1975; Fonagy et al. 2002) have shown that *rapprochement and separation* play a significant role in early personality development. By exposing the patients daily to these experiences, the day hospital setting provides a chance for development and change. Empirical findings do indeed indicate that a day hospital setting is in the long-term more conducive for this particular group of patients than, for instance, inpatient psychiatric treatment (Bateman and Fonagy 1999, 2001; Chiesa et al. 2004).

With the concepts of 'containment' (Bion 1962), the 'marsupial' (Rey 1979) and 'transitional space' (Winnicott 1953), psychoanalytic authors have given an idea of the complex processes playing a part in these changes and developments. In the therapeutic work, it is particularly important to create a space where 'learning through experience' (Bion 1962) is facilitated. This 'learning', however, is an *internal learning*, and the experiences are *emotional experiences* for which symbols have to be found in order to be able to think them. To do this, a good enough, reliable external setting has to be provided.

Similarly this applies to borderline patients with pronounced hypochondriac anxieties, somatic and dissociative symptoms. For those patients the body is often the only 'container' in which they can localise their emotional experiences. In the internal spaces of the body, these experiences are condensed and trapped, sometimes in the hope that they can be treated with medication or be surgically removed in the same magical way as they had got there.

Somatic manipulation, somatising and dissociative symptoms as well as hypochondriac anxieties can be understood as desperate attempts to maintain a fragile psychic balance. The day hospital offers an interesting concept to transform such states and to give them a different space. Like the body, the hospital has entrances and exits, internal spaces and transitional spaces. And unlike in the enclosed space of a ward, the patient has to leave the 'body' of the day hospital each day and return to it the next morning. In this way the setting is less of a retreat to protect the individual from the 'evil' external world and provide a permanent refuge.

Rosenfeld (1997) came up with the idea that in some patients physical symptoms become a kind of refuge, which binds disturbing phantasies and feelings. He spoke of 'psychotic islands', which can disturb the psychic equilibrium if taken out of their encapsulation. In order to take such a step, the patient needs a space in which to accommodate such experiences.

To date little is known about the efficacy of a day care setting on patients with complex dissociative and somatic symptomatologies. What is known, however, is that such patients evoke disturbing countertransference reactions in the treatment staff team. In contrast to a locked ward or an inpatient setting, the day clinic offers an *open space* with more freedom and less control. Correspondingly, the patients have more responsibility about the use of this space, their contribution to the treatment, how they deal with somatic symptoms, self-harm tendencies, etc.

Therapeutically it seems important to adopt an open, exploratory attitude, directly taking up the therapeutic relationship and the current emotional experiences without restricting the therapeutic space by premature explanations or a controlling, symptom-centred approach. In our own practice, we have therefore largely dispensed with 'contracts', 'controls', 'therapeutic aims', etc. to facilitate the exploration of the transference situation and not define the relationship with the patient in the different therapies (individual as well as group therapy, creative and body work-oriented therapies within the psychoanalytic setting) from the beginning. This is relevant not just for the general setting of the day hospital but also for important topics such as eating behaviour, self-harm or suicidal ideation. This mode of working is encouraging the patient to contribute these topics openly in therapy in order to address them and explore them without the fear of immediate consequences (see Weiß et al. 2008).

This approach is based on the conviction that contracts and controls are often not so much about patient security than about dealing with disturbing countertransference feelings in the therapist. Not uncommonly this can lead to mutual acting out as will be illustrated in the next case vignette, because the therapist's decisions and sanctions got out of hand and could hardly be contained.

### 23.4.3 Case Vignette 3

Mrs. C, a very intelligent 35-year-old patient, had lived on the edge of society for many years because of her emotional difficulties. After the collapse of a low-frequency, supportive therapy, she had retreated and had not dared to seek further therapeutic help.

This first therapy had failed because the therapist reacted increasingly helplessly to her vehement, pressurised transference by making contracts and sanctions, reducing the length of the sessions as well as the frequency. When the patient subsequently phoned more and more frequently outside her regular treatment session, he ended her treatment, as he did not know how to deal with this.

Consequently the patient got into a desperate state and climbed over the balcony into the consulting room to get her case notes. The therapist felt threatened and reported her to the police, which led to increasing self-incrimination and an acting out of her guilt feelings, ending in a suspended sentence of half a year.

In the first few days of the day hospital treatment, the patient presented with intrusive and provocative behaviour. Her unconscious aim seemed to be that she wanted to be 'kicked out'. For the fourth group therapy session, she turned up on her own as her fellow patients refused to come to the group if she was present. In this way she had created a situation in which one had to decide in her favour or in favour of the others. She was close to breaking off therapy and was experienced as 'unbearable' by some of the therapy team.

The attempt to explore this 'unbearable' feeling eventually led to a reconstruction of a phantasy, whereby the patient felt violently excluded from her family and more or less convinced that she could only gain access by violence to what had been taken from her in childhood.

It was only when this phantasy, which had been so concretely acted out in the group, was worked through and interpreted that the desperate anxiety and neediness underlying her behaviour could be felt. She now explained in retrospect that she had had to get her case file because 'you always leave part of yourself with the therapist' and she would not have been able to separate from him otherwise.

This example not only illustrates *the risk of an escalating acting out* but also demonstrates the *concreteness* of experience of many borderline patients. On the basis of her conviction that the end of treatment meant the loss of parts of herself, Ms. C had to actually get into the consulting room of her therapist to get back these 'parts' in the shape of her case notes. But in the day clinic, her intrusive behaviour was not condoned in such a way, and she managed to redefine her interaction with the group. This enabled her to continue treatment, and subsequently she went into an outpatient psychoanalytic treatment which she concluded successfully after a couple of years.

This kind of approach, which does not introduce premature rules but aims at understanding the patient's communications by *working through the countertransference*, requires a certain tolerance of experiences of uncertainty, anxiety and guilt in the staff team. This does not necessarily involve higher risk but can beneficially contribute in handing back the responsibility for the treatment to the patient. The relatively low dropout rate (in our clinic between 6 and 16% over 10 years) seems to confirm the efficacy of this approach and is possibly an indication that the day hospital setting is particularly effective in enhancing these specific therapeutic principles, which foster psychic change.

The possibility of bringing the stresses of daily life into the therapy setting and conversely checking out what was worked on in therapy in everyday life forms part of the efficacy of this approach. This mutual exchange between the therapy setting and the setting of social relationships provides the opportunity to validate the experiences of therapy and allow new possibilities to emerge in dealing with conflict. In this way, the day hospital setting is more like a high-frequency outpatient treatment than an inpatient treatment.

This structural similarity of daily coming together and leaving, and the high responsibility the patient has for the treatment, usually has the effect of easing *the transition into a subsequent outpatient therapy*. But this transition is nevertheless a critical area in the day hospital treatment. If it is not possible to transfer the patient into an appropriate outpatient treatment in good time following the intensive day patient care, patients frequently deteriorate with tendencies to retreat, breaking up relationships and a breakdown of professional and social integration.

The findings of Chiesa and Fonagy (2000, 2003) confirm this. Our experience with borderline patients who have gone through the day hospital treatment shows that they only achieve significant lasting reduction of psychic stress if they were able to take up a long-term, high-frequency outpatient treatment closely following on from the day hospital treatment.

The improvement and the working through of this transition is therefore a paramount aim in the day hospital treatment. This can, for instance, be done by the therapeutic work around the daily separation. Early referral of the patients to available therapists is another way of securing continuity. Conversely, arrangements can be made with the therapists who take on the patient that the patient can return short term to the day hospital at times of crises or when there is a break in treatment. This will be illustrated by the final case vignette.

### 23.4.4 Case Vignette 4

Mr. D, a 42-year-old patient, was admitted to the day hospital because of severe dissociative episodes and loss of impulse control, which endangered the relationship with his wife and children. He had been pensioned off early because he suffered various physical illnesses, including chronic inflammable bowel disease, a heart infarct and hydrocephalus which had been relieved by a shunt. Again and again the physical symptoms and the labile psychic states, self-harm and hypochondriac states were mixed up in an obscure way.

Although the initial prognosis was unpromising, the patient made good progress over several years. A prerequisite for this was taking up a long-term outpatient treatment following his second day hospital stay. There was close co-operation between the colleague and the day hospital regarding referral back to the day hospital, which allowed the patient to see outpatient and day patient treatment as unified at times of crisis. After more than 5 years and a longer period of separation, the patient was again with his family and could maintain personal and social relationships and gradually reduce the many psychoactive medications he had been prescribed. There had not been any problems with impulse control and dissociative episodes for several years, and the extent of physical complaints had significantly reduced so that he was rarely admitted to specialist services.

In his childhood and adolescence, this patient had experienced massive traumas including the suicide of a violent and alcohol-dependent father. He repeatedly ran away from home and at the age of 13–15 years together with a friend hid away in a closed down part of the sewage system of his home town. In this hideaway he felt relatively safe and, fearing the outbreaks of violence from his father, only secretly returned home at night to get food and clothing. During adolescence he led a peripatetic borderline life with many broken relationships until he met his wife, who in his view was 'the only good experience' in his life. Tragically, it was exactly this good experience which he felt compelled to repeatedly attack, which led to feelings of agonising guilt and desperate anxiety of loss.

The day hospital represented a 'refuge' in which he was relatively protected from impulsive outbursts and catastrophic anxieties, akin to the closed down part of the sewage system in which he had spent part of his youth. In this setting he could gently approach the many problems of his life. Gradually, these 'excursions' expanded, and he could better face his internal and external conflicts as he experienced his mourning and guilt feelings as less devastating and persecutory. In this way a process of reparation began which allowed him to recover and to reconcile his damaged internal objects.

Defective *processes of symbolisation and reparation* are of central importance to borderline pathology. If reparation fails, this may lead to repetitive cycles which perpetuate the damage to the internal objects. Sometimes self-harm can be understood as a desperate attempt at reparation, which is why it should be interpreted rather than sanctioned (see Weiß 2012).

## 23.5 Summary, Outlook and Perspectives

Up to now there are only few empirical studies on the effectiveness of hospital psychoanalytic settings for the treatment of patients with borderline pathology. However, there are indications that at least for a subgroup of borderline patients this approach is more beneficial than a psychiatric inpatient treatment. It combines psychotherapy in individual as well as group settings in a containing framework, which provides security without confining the patient too much. Even problems of chronic suicidal ideation as well as self-harm can be treated in the day care setting.

Psychoanalytic day care differs from psychiatric outpatient care in providing high-frequency individual psychotherapy. It focuses on the actual emotional experience and the related phantasies, the systematic working through of primitive defence mechanisms and coping strategies within the therapeutic relationship without directly guiding the patient's behaviour. Another part is the systematic reflection of the setting by continuing external and internal supervision (of the individual therapist as well as of the whole treatment team).

It is because of its open structure that the day care fosters the patient's personal responsibility and the interaction with the social environment. Thereby a mutual exchange between everyday life and the therapeutic space is encouraged which can become a focus of therapeutic work (see Küchenhoff 1998). In this way a pathological splitting between the 'bad' outside world, often experienced as cold and rejecting, and the idealised refuge of the day centre as a 'psychic retreat' (Steiner 1993) is less likely to occur.

Instead the examination of the daily arrival and departure promotes the working through of areas of conflict (identity and dependency), which are central to many borderline patients and closely related to their agoraphobic-claustrophobic anxieties.

The close co-operation of the care staff in maintaining the therapeutic setting is of supreme importance, particularly at times of cancelled and missed sessions, appointments outside the day hospital as well as in managing the acting out, self-harm and suicidal threats which can put the team under a lot of pressure. In this context the co-operation with inpatient services can be helpful.

Similarly, the co-operation with other medical specialities may be needed, for instance, gynaecology and obstetrics in the support of patients during pregnancy, and mother and baby units to support patients during pregnancy and after giving birth.

The development of day care units within different medical institutions (including intensive care) and the close network of consultancy and liaison services as well as the integration of specialist services in psychosomatic-psychotherapy day centres often have beneficial effects. For borderline patients these services can provide longer-term support while waiting for a vacancy with a therapist.

The prompt transfer from day hospital care to long-term therapy is a critical issue, as it crucially affects the sustainability of the treatment effects and the avoidance of rehospitalisation. It is of utmost importance that long-waiting periods are avoided and a 'smooth' transition from the day hospital to further psychotherapy is ensured. Therefore close links with therapists and training institutions are maintained, and there are service provisions by the day hospital to bridge this gap and ensure continuity of treatment (including intermittent treatment). For, as experience shows, the waiting period for outpatient psychotherapy treatment is longest for those patients who are psychologically most burdened and therefore most in need.

Prospective randomised studies by Bateman and Fonagy (1999, 2001) have demonstrated the relative advantage of day hospital care compared to conventional psychiatric inpatient care. Chiesa and Fonagy (2000, 2003; see Chiesa et al. 2004) could show that a subsequent outpatient group therapy had more beneficial effects than individual inpatient treatment.

The naturalistic study by Zeeck et al. (2009) indicates that the severity or diagnosis is not relevant as to the decision whether to refer to day patient or inpatient psychotherapy. The short-term outcomes were similar in both settings. Agarwalla and Küchenhoff (2004) have published encouraging results regarding long-term outcomes.

In summary, psychoanalytic day hospital therapy, in conjunction with inpatient and outpatient treatments, seems to be a promising treatment approach for borderline patients. Working out specific indication criteria, clarification of setting-specific effectiveness as well as the evaluation of the different elements of the treatment and the analysis of treatment drop-outs are left to further quantitative and qualitative studies.

## References


Agarwalla P, Küchenhoff J. Teilstationäre Psychotherapie. Ergebnisse, Katamnese, Einflussfaktoren. Psychotherapeut. 2004;49:261–71.

American Psychiatric Association (APA). Diagnostic and statistical manual of mental disorders DSM-5. Washington DC: American Psychiatric Association; 2013. p. 2013.

Bateman A, Fonagy P. Effectiveness of partial hospitalization in the treatment of Borderline personality disorder: a randomized controlled trial. Am J Psychiatry. 1999;156:1563–9.

Bateman A, Fonagy P. Treatment of borderline personlity disorder with psychoanalytically oriented partial hospitalization: an 18th month follow-up. Am J Psychiatry. 2001;158:36–42.

Bateman A, Fonagy P. Borderline-Persönlichkeitsstörung und Mentalisierungsbasierte Therapie (MBT). In: Kernberg OF, Sachsse U, editors. Handbuch der Borderline-Störungen. 2nd ed. Stuttgart: Schattauer; 2011. p. 566–75.

Bion WR. Learning from experience. London: Heinemann; 1962.

Bohus M. Dialektisch-Behaviorale Therapie für Borderline-Störungen. In: Kernberg OF, Sachsse U, editors. Handbuch der Borderline-Störungen. 2nd ed. Stuttgart: Schattauer; 2011. p. 619–39.

Buchheim A. Borderline-Persönlichkeitsstörungen und Bindungserfahrungen. In: Dulz B, Herpertz SC, Kenberg OF, Sachsse U, editors. Handbuch der Borderline-Störungen. 2nd ed. Stuttgart: Schattauer; 2011. p. 158–67.

Chiesa M, Fonagy P. Cassel personality disorder study. Br J Psychiatry. 2000;176:485–91.

Chiesa M, Fonagy P. Psychosocial treatment for severe personality disorder. 36-month follow-up. Br J Psychiatry. 2003;183:356–62.

Chiesa M, Fonagy P, Holmes J, Drahorad C. Residential versus community treatment for severe personality disorders: a comparative study of three treatment programs. Am J Psychiatry. 2004;161:1463–70.

Clarkin JF, Levy KN, Schiavi JM. Transference focused psychotherapy: development of a psychodynamic treatment for severe personality disorders. Clin Neurosci Res. 2005;4:379–85.

Clarkin JF, Yeomans FE, Kernberg OF. Psychotherapy for borderline personality: focusing on object relations. Washington DC: American Psychiatry Publishing; 2006.

Dammann G, Buchheim P, Clarkin JF, Kernberg OF. Einführung in eine übertragungsfokussierte, manualisierte psychodynamische Therapie der Borderline-Störung. In: Kerberg OF, Dulz B, Sachsse U, editors. Handbuch der Borderline-Störungen. Stuttgart: Schattauer; 2000. p. 461–81.

Domsalla M, Koppe G, Niedtfeld I, Vollstädt-Klein S, Schmahl C, Bohus M, Lis S. Cerebral processing of social rejection in patients with borderline personality disorder. Soc Cogn Affect Neurosci. 2014;9(11):1789–97.

Ducasse D, Courtet P, Olié E. Physical and social pains in borderline disorder and neuroanatomical correlates: a systematic review. Curr Psychiatry Rep. 2014;16(5):1–12.

Dulz B, Jensen M. Aspekte einer Traumaätiologie der Borderline-Persönlichkeitsstörung – psychoanalytisch-psychodynamische Überlegungen und empirische Daten. In: Kernberg OF, Sachsse U, editors. Handbuch der Borderline-Störungen. 2nd ed. Stuttgart: Schattauer; 2011. p. 203–24.

Fonagy P, Gergely G, Jurist EL, Target M. Affektregulierung, Mentalisierung und die Entwicklung des Selbst, vol. 2004. Stuttgart: Klett-Cotta; 2002.

Gergely G, Watson J. The social biofeedback model of parental mirroring. Int J Psychoanal. 1996;77:1181–212.
Green A. Geheime Verrücktheit. Grenzfälle der psychoanalytischen Praxis, vol. 2000. Gießen: Psychosozial-Verlag; 1990.
Jacobs TJ. On countertransference enactments. J Am Psychoanal Assoc. 1986;34:289–307.
Joseph B. A clinical contribution to the analysis of a perversion. Int J Psychoanal. 1971;52:441–9.
Joseph B. Transference: the total situation. Int J Psychoanal. 1985;66:447–54.
Joseph B. In: Feldman M, Bott Spillius E, editors. Psychic equilibrium and psychic change: selected papers of Betty Joseph. London: Routledge; 1989.
Kamphausen S, Schröder P, Maier S, Bader K, Feige B, Kaller CP, Tüscher O. Medial prefrontal dysfunction and prolonged amygdala response during instructed fear processing in borderline personality disorder. Curr Psychiatry Rep. 2013;14(4):307–18.
Kernberg OF. Borderline conditions and pathological narcissism. New York: Jason Aronson; 1975.
Kernberg OF. Psychoanalysis, psychoanalytic psychotherapy and supportive psychotherapy: contemporary controversies. Int J Psychoanal. 1999;80:1075–91.
Krause-Utz A, Winter D, Niedtfeld I, Schmahl C. The latest neuroimaging findings in borderline personality disorder. Curr Psychiatry Rep. 2014;16(3):1–13.
Küchenhoff J. Teilstationäre Psychotherapie. Theorie und Praxis. Stuttgart: Schattauer; 1998.
Lang S, Kotchoubey B, Frick C, Spitzer C, Grabe HJ, Barnow S. Cognitive reappraisal in trauma-exposed women with borderline disorder. NeuroImage. 2012;59(2):1727–34.
Lieb K, Stoffers J, Dulz B. Pharmakologische Behandlung von Borderline-Persönlichkeitsstörungen. Nervenheilkunde. 2014;10:720–2.
Linehan MM. Dialektisch-Behaviorale Therapie der Borderline-Persönlichkeitsstörung. München: CIP Medien; 1996.
Mahler MS, Pine F, Bergman A. The Psychological birth of the human infant. New York: Hutchinson; 1975.
Maier W, Hawellek B. Genetik. In: Dulz B, Herpertz SC, Kernberg OF, Sachsse U, editors. Handbuch der Borderline-Störungen. 2nd ed. Stuttgart: Schattauer; 2011. p. 69–74.
Mauchnik J, Schmahl C. The latest neuroimaging findings in borderline personality disorder. Curr Psychiatry Rep. 2010;12(1):46–55.
Meltzer D. Terror, persecation and dread. Int J Psychoanal. 1968;49:396–401.
Money-Kyrle R. Cognitive development. In: Meltzer D, O'Shaughnessy E, editors. The collected papers of Roger Money-Kyrle. Perthshire: Clunie Press; 1968. p. 416–33.
Osofsky JD. Aspekte der frühen Entwicklung als Verständnisgrundlage der Borderline-Persönlchkeitsorganisation. In: Dulz B, Herpertz SC, Kernberg OF, Sachsse U, editors. Handbuch der Borderline-Störungen. 2nd ed. Stuttgart: Schattauer; 2011. p. 148–57.
Rey H. Schizoid phenomena in the borderline. In: Le Boit C, editor. Advances in the psychotherapy of the borderline patient. New York: Jason Aronson; 1979.
Rey H. Universals of psychoanalysis in the treatment of psychotic and borderline states. London: Free Association Books; 1994.
Rosenfeld HA. A clinical approach to the psychoanalytic theory of the life and death instincts: an investigation into the aggressive aspects of narcissism. Int J Psychoanal. 1971;52:169–78.
Rosenfeld HA. Zur Psychopathologie und psychoanalytischen Behandlung einiger Borderline-Patienten. Psyche – Z Psychoanal. 1981;35:338–52.
Rosenfeld HA. Impasse and interpretation. London: Tavistock; 1987.
Rosenfeld HA. The relationship between psychosomatic symptoms and latent psychotic states. In: De Masi F, editor. Herbert Rosenfeld at work. The Italian seminars. London: Karnac; 1997. p. 24–44.
Rudolf G. Strukturbezogene Psychotherapie. Stuttgart: Schattauer; 2004.
Rudolf G. Psychoanalytische Therapie struktureller Störungen. Behandlung 'as usual' oder strukturbezogene Modifikation. In: Springer A, Gerlach A, Schlösser A-M, editors. Störungen der Persönlichkeit. Gießen: Psychosozial-Verlag; 2006. p. 93–112.

Ruocco AC, Amirthavasagam S, Choi-kain LW, Mcmain SF. Neural correlates of negative emotionality in borderline personality disorder: an actionlikeklihood-estimation meta-analysis. Biol Psychiatry. 2013;73:153–60.

Schmahl C, Bohus M, Esposito F, Treede RD, Di Salle F, Greffrath W, Seifritz E. Neural correlates of antinoception in borderline personality disorder. Arch Gen Psychiatry. 2006;63(6):659–66.

Segal H. Dream, phantasy and art. London: Routledge; 1991.

Simons LE, Moulton EA, Linnman C, Carpino E, Becerra L, Borsook D. The human amygdala and pain: evidence from neuroimaging. Hum Brain Mapp. 2014;35(2):527–38.

Steiner J. Psychic retreats. Pathological organizations in psychotic, neurotic and borderline patients. London: Routledge; 1993.

Steiner J. Narzißtische Einbrüche: Sehen und Gesehenwerden. In: Weiß H, Frank C, editors. Scham und Verlegenheit bei pathologischen Persönlichkeitsorganisationen. Stuttgart: Klett-Cotta; 2006.

Steiner J. Seelische Rückzugsorte verlassen. In: Weiß H, Frank C, editors. Therapeutische Schritte zur Aufgabe der borderline-position. Stuttgart: Klett-Cotta; 2014.

Stoffers J, Lieb K. Pharmakotherapie der Borderline-Persönlichkeitsstörung. In: Dulz B, Herpertz SC, Kernberg OF, Sachsse U, editors. Handbuch der Borderline-Störungen. 2nd ed. Stuttgart: Schattauer; 2011. p. 854–64.

Weiß H. Das Labyrinth der Borderline-Kommunikation. Klinische Zugänge zum Erleben von Raum und Zeit. Stuttgart: Klett-Cotta; 2009.

Weiß H. Wiedergutmachung beim Borderline-Patienten. Jahrb Psychoanal. 2012;65:59–80.

Weiß H. Überlegungen zum agora-klaustrophoben Dilemma des Borderline-Patienten. Psyche – Z Psychoanal. 2015;69:916–35.

Weiß H, Horn E, Kidess A, Roman A, Winkler R. Das mobbende innere Objekt – der kleinianische Ansatz in einem teilstationären psychotherapeutischen Setting. In: Dreyer A, Schmidt MG, editors. Niederfrequente psychoanalytische Psychotherapie. Stuttgart: Klett-Cotta; 2008. p. 246–65.

Winnicott DW. Transitional objects and transitional phenomena: a study of the first not-me possession. Int J Psychoanal. 1953;34:89–97.

Yeomans FE, Diamond D. Übertragungsfokussierte Psychotherapie (Transference-focused Psychotherapy, TFP) und Borderline-Persönlichkeitsstörung. In: Kernberg OF, Sachsse U, editors. Handbuch der Borderline-Störungen. 2nd ed. Stuttgart: Schattauer; 2011. p. 543–58.

Zanarini MC, Hörz S. Epidemiologie und Langzeitverlauf der Borderline-Persönlichkeitsstörung. In: Dulz B, Herpertz SC, Kernberg OF, Sachsse U, editors. Handbuch der Borderline-Störungen. 2nd ed. Stuttgart: Schattauer; 2011. p. 44–56.

Zeeck A, Hartmann A, Küchenhoff J, Weiß H, Sammet I, Gaus E, Semm E, Harms D, Eisenberg A, Rahm R, Wietersheim JV. Differentielle Indikationsstellung stationärer und tagesklinischer psychotherapie: die DINSTAP-Studie. Psychother Psych Med. 2009;59:354–63.

# Creative Therapies and Neuropsychodynamics

# 24

Peter Hartwich

**Abstract**

Creative therapy methods in psychiatry are procedures in which psychiatric patients work with artistic equipment. In doing so, a special quality of drive is used for creative activities. This can be unfolded in activities, such as creating poetry, painting, music, sculpturing, etc. Neurobiological research has shown that the amplification of certain neuronal brain activities as well as a reduction of cerebral inhibitory functions is effective by the moment of creative actions.

We show how some creative therapy methods are effective in psychiatric disorders. Especially the principle of the binding force in self-fragmentation in psychoses has a therapeutic power. It can be used neuropsychotherapeutically when the risk of self-fragmentation occurs. The resulting improvement of the coherence of the spatiotemporal self is exemplified by different methods:

1. Painting with the help of a computer painting program: The advantage for the patients is that the tools with which a picture is painted can be offered with clear rules and structure, especially for patients who have lost this kind of ability. The electronic distance and the possibility to "rub out" parts of the painting are close to the *defense mechanism of undoing*; storing the picture is like *repression*. Thus, psychotic patients can learn again defense mechanisms in a subtly adapted way.
2. Sculpture with stone: The creativity of the sculptor manifests in the slowness of the design process. The stone material offers a particularly *high degree of structure*. The meditative rhythmic beating with the hammer and the working of the

P. Hartwich
Department of Psychiatry, Psychotherapy, Psychosomatics, General Hospital of Frankfurt am Main, Teaching Hospital University Frankfurt, Frankfurt, Germany
e-mail: Prof.Hartwich@casytec.de



H. Boeker et al. (eds.), *Neuropsychodynamic Psychiatry*,
https://doi.org/10.1007/978-3-319-75112-2_24

whole body enfold a highly creative binding force which reduces the fragmentation of the self.

3. Video mirroring the image of the self: It is impressive how strongly patients respond to their own audiovisual mirror image. They experience a high investment energy in the self-reference which improves self-coherence and the strengthening of self-demarcation in psychotic patients.

The creative art therapist is the one who awakens and catalyzes the creativity in the mentally ill persons in order to use it for the healing process. This poses the questions: What is creativity? And what are the neuronal correlations? Searching for an answer, one encounter concepts associated with creativity: originality, ingenuity, flexibility, discovery, extraordinary, and playfulness.

## 24.1 What Is Creativity, and Who Is a Creative Person?

Guilford (1950) combines creativity with its manifestations: the creative power unfolds itself in activities such as discovery, invention, and divergent thinking; he points out that the most important features of divergent thinking are flexibility, originality, and fluency. Decisive is the connection with constructively ordering features; creativity is therefore composed of both, the unstructured creativity and the constructively ordered. When we ask what characterizes a creative person, it is someone who is able to combine both abilities the freedom of originality and something that gives order, in a balanced relationship. These two features, the unfolding and the structure, should come together in a personality, who is able to truly create a creative product. Hicklin (1979) sees in the creative product that something is made visible, what has hitherto remained hidden.

Jung (1922) describes the creativeness as an autonomous complex that means that the creative complex in man has autonomy on its own. He sees a quality in the creative energy and dynamic that can go so far as to dominate the actions of a person. Matussek (1976, 1979) assumes that creative expressions are original, inventive, flexibly discovered, and uniquely new. He points out that creative potential may exist in every person and can be activated in every life situation. Creativity is not just a feature of especially talented people but exists in nearly everyone. Navratil (1965), referring to the mentally ill and their artistic work, says that every human being is creative. Arieti (1976) distinguishes two different levels: "ordinary" creativity that almost everyone has and "great" creativity, reserved for special people, as, e.g., Dante, Galileo, Shakespeare, and Newton, and which is responsible for human's great achievements (Fig. 24.1).

All in all, creativity is characterized by an experience that many people know as an energy-driven movement when they pursue an intensive creative activity in writing, making music, poetry, painting, or sculpturing. By doing so people seem to "forget" everything else in their environment in a state inflamed with passion.

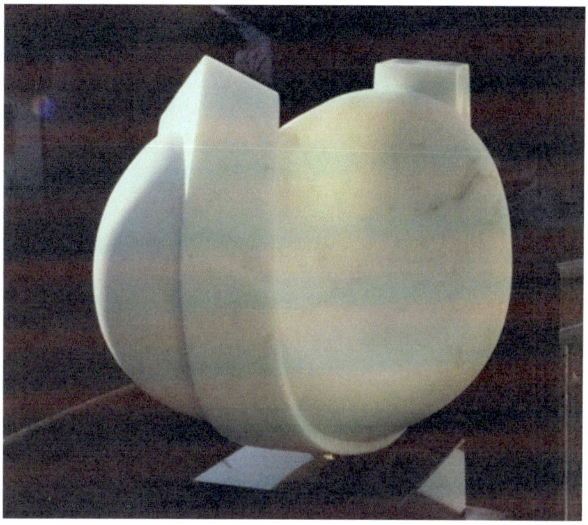

**Fig. 24.1** Balance of unstructured dynamics and structure (Carrara-marble, sculptor Peter Hartwich 2001)

## 24.2 Creativity and Neuronal Findings

First of all, the question should be whether this particular kind of the dynamic power of the unfolding creativity in healthy and mentally ill people is correlated to neuronal findings. This is not meant to decipher the phenomenon of the creativity as such but rather the question as to which neural brain activities are to be found in the state of creative acts.

Kandel (2012) summarizes that brain research studies, which are looking for biological explanations, are still rare. At first, Kandel points out that creative people have a controllable connection between the unconscious and the conscious, which is mediated by brain top-down processes, in the sense of a regression at the service of the ego. For the hitherto explored brain biological components of some aspects of creativity, reference is made to association fields of the cerebral cortex. There could be indications that creativity occupies the right hemisphere of the cerebral cortex and, in particular, the right temporal gyrus and the right parietal cortex. If test persons had to solve linguistic problems that required creative insights, there would be an increase in activity in this region of the right temporal lobe. Shortly before, a sudden increase in high-frequency brain activity occurs in the same region. The conclusion seems to be that interaction between the left and right prefrontal cortices can promote or inhibit originality and creativity. This suggests the hypothesis that creativity involves a reduction of inhibition.

If one assumes that the inhibition functions are diminished in some psychotic disorders, the neurobiological findings provide a parallel to what Benedetti (1979, p. 1052) emphasizes: in the state of a psychotic illness, a person can experience "an eruption of creativity."

But we should be careful. There are now a large number of experimental studies, in which some different aspects of subdivided types of creativity were investigated neurobiologically and some of the results seem to be controversial.

Dietrich and Kanso (2010) report that the interest into the neuronal underpinnings of creative behavior has produced many self-contradictory data. The review of the literature with a total of 72 experiments reported in 63 articles. Electroencephalographic studies of divergent thinking yield highly variegated results. Neuroimaging studies of this paradigm also indicate no reliable changes above and beyond diffuse prefrontal activation. They emphasize: "*Neuroelectric and imaging studies of insight are more consistent, reflecting changes in anterior cingulate cortex and prefrontal areas. Taken together, creative thinking does not appear to critically depend on any single mental process or brain region, and it is not especially associated with right brains, defocused attention, low arousal, or alpha synchronization, as sometimes hypothesized. To make creativity tractable in the brain, it must be further subdivided into different types that can be meaningfully associated with specific neurocognitive processes.*"

Abraham (2013) points out also that the field of creative neurocognition is a rapidly growing area of research. The problem is the heterogeneity, which associates with the creativity construct. The methodological and conceptual problems pose considerable limitations. Therefore they suggest three issues in order to avoid the problems. One should be aware whether the creativity neuroimaging paradigms are clear enough to find out what makes creative cognition different from normative cognition and how types of creativity can be distinguished.

Wenfu Li et al. (2015) used in their study voxel-based morphometry to identify the brain regions underlying individual differences in trait creativity, as measured by the Williams creativity aptitude test, in a large sample ($n = 246$). They found that creative individuals had higher gray matter volume in the right posterior middle temporal gyrus (pMTG), which might be related to semantic processing during novelty seeking (e.g., novel association, conceptual integration, and metaphor understanding).

Beaty et al. 2014 using functional magnetic resonance imaging (fMRI) explored dynamic interactions between brain regions during a divergent thinking task. Multivariate pattern analysis revealed a distributed network associated with divergent thinking, including several core hubs of the default (posterior cingulate) and executive (dorsolateral prefrontal cortex) networks. "*Moreover, temporal connectivity analysis revealed increased coupling between default and salience network regions (bilateral insula) at the beginning of the task, followed by increased coupling between default and executive network regions at later stages. Such dynamic coupling suggests that divergent thinking involves cooperation between brain networks linked to cognitive control and spontaneous thought, which may reflect focused internal attention and the top-down control of spontaneous cognition during creative idea production.*" And "*divergent thinking ability was associated with increased resting-state functional connectivity between the inferior prefrontal cortex and the DMN.*"

When we look at the studies on creativity and its neuroimaging studies today, it remains a problem to carefully examine the many aspects of the creativity of the people defined as creative. For therapeutic purposes, the question would arise which neural changes could be provoked by creative therapy methods. It then depends on whether the therapeutically provoking changes have a positive influence in psychotic patients. One might consider that the aberrant neuronal findings in schizophrenia and depressive disorders can be modulated positively by creative therapy methods. This has been plausible in clinical observation but has not yet been verified at the neuronal level. Here, we must remain modest today and wait to see if such an expectation can be met in the future.

## 24.3 Critical to the Question of the Connection Between Creativity and Psychiatric Disorders in Artists

In the context of creativity, bipolar disorders are often cited by artists or their families, as is the case with Kandel (2012). It is often emanated from observation people in maniform states. At the beginning of manically elevated moods, combined with the high feeling of an unlimited range of cognitive options, artistic creativity is strongly inspired. As a result, being bipolar and being creative are seen in a close context by some people, even by psychiatrists. Critically it is objectionable; it seems almost fashionable and chic to be a little "bipolar." The extent to which the given diagnoses of the artists (e.g., van Gogh) and writers in retrospect, especially those who have lived long before our current ICD or DSM diagnostics, are correct is questionable. There are many psychical constellations of a creative nature, based on particular gifts, as well as many abnormal conditions, sometimes due to the supply of substances suitable for this purpose, so that the creativity of man is not at all to be reduced to variants with mental disorders.

Accordingly, only a partial aspect is to be seen here. And it makes sense to compare the neurobiological findings of brain activity of manic state in bipolar patients with creative people in the creative process. It is to examine the questions of similarity or dissimilarity and to determine to what extent the reduction of inhibitions results in an aspect of creativity. This can presumably only be true if sufficient control (ability to inhibit) has also been maintained. This is however not the case with many manic states, if the lack of self-criticism and the feeling of irresistibility with increasing drive upturn take over.

## 24.4 Creative Therapy Methods and Possible Neurobiological References

Benedetti (1999) has summarized the psychodynamics of the psychoses: "*creativity is not to be found where the psychic decay manifests itself, but where the patient takes a step forward in the overcoming of his suffering by a symptom of his suffering. ... From the contributions of both the patient and the therapist, the power of gestalt change arises. The transformative power of the symptom, which creates a plus situation instead of a minus situation, makes possible an energy flow, instead of a deadening of the communication a new development emerges*" (Benedetti 1999, p. 50).

When we use creative methods in the treatment of mentally ill patients, we use a principle which stimulates and unfolds creativity as special force that is associated with a bonding strength of the ego. Once, creativity can open up new areas by loosening neurotic thought structures and revitalizing emotions. In the case of psychotic patients, creativity can also be used as a binding force. I am talking about the "bonding-power of creativity" (Hartwich 2010, 2012), which can again bring into line fragments of the fragmented self or stop further fragmentation in psychoses. In the positive sense, this helps the patient to free himself from his symptoms, especially paraconstructions (see Chap. 10).

One of the most important questions will be whether the described psychodynamic changes in the sense of a reduction of the symptoms caused by creativity also correspond to the accompanying neurobiological events. This means that, for example, an existing hyperconnectivity in the midline areas of a schizophrenic brain is modified by intensifying the activity in the temporal lobe or in other areas of the brain and possibly reducing inhibition functions through the creative act. If this can be confirmed experimentally in the future, we would come to a genuine neuropsychodynamically underpinned treatment basis. However, today we must assume that we are still far from being able to establish such direct links. The neuroscientific findings on psychoses are presented in detail in Chap. 9 and are not to be repeated here (see Chaps. 10 and 12).

## 24.5 Art Therapy and Empirical Research

Today's art therapy for psychiatric patients is on the shoulders of our psychiatric ancestors, who have made a breakthrough in this field. These include Lombroso (1890), Morgenthaler (1921), Prinzhorn (1922), Jung (1922), Volmat (1956), Jakab (1956), Rennert (1962), Navratil (1965), and others. If neurobiological aspects are involved in the discussion, the question must also be examined as to whether the effectiveness of art therapy can be tested empirically or whether creative procedures can ultimately not be made operational and measurable without an intolerable reductionism.Interestingly, this problem has already been formulated very early, namely, by Prinzhorn, when he described the resonance of the professional world to his book *Bildnerei der Geisteskranken*: "*… such researches are no longer highly regarded because they do not go back to the exactly measurable*" (Prinzhorn 1922).

So it remains a challenge, with our current technical means, to push the limits of the measurable in this area further forward. However, the area of creativity, which cannot be operationalized, should remain respectfully unaffected. Art therapeutic approaches, especially for psychiatric patients, which can fulfill the stated claim, will be here illustrated by three examples. It is also about the special modifications to psychoreactive and psychosomatic illnesses to work out:

- Painting by means of a computer program
- Sculpture with stone
- Video mirroring (the image of the self)

### 24.5.1 Painting with the Help of a Computer Painting Program

The psychiatric tradition of art therapeutic methods with psychiatric patients, which has already been systematically described by Morgenthaler (1921) and Prinzhorn (1922), consists of painting with pens, brushes, and colors on paper, cardboard, and wood. In some diseases, the development of the painted image can be associated with such a high intensity of the reactivation of feelings, for example, in the case of imaging a sexual abuse, the limit of what is bearable in the patients is exceeded. In the case of psychotic patients, the missing structure on an empty sheet of paper can also cause anxiety that has an unfavorable effect on a healing process.

We (Hartwich and Brandecker 1997) had also noticed that some borderline cases in psychotherapeutic treatment switched into psychotic states, while we were working in our usual painting therapy with conventional art materials. Such a dangerous breakdown of ego boundaries led us to look for new treatment methods. So we did look for a creative art therapy method for patients with a weak personality structure.

Computer painting on the screen—with its strict rules, its distance between patient and picture, and the possibility of keeping and protecting defense mechanisms—seemed to be a useful art therapy development and worthwhile to explore further. But there is a good deal of prejudice about computer techniques being used in psychotherapy on mentally ill patients. Some people even talk fearfully about an artificial brain or an artificial relationship. But today young people grow up with computers, and also handicapped people are trained with the help of computer programs. Why shouldn't computers also be useful in therapy with psychiatric patients? Hartwich and Brandecker (1997) decided to try out a computer painting program within the creative art therapy methods with inpatients. The advantage for the patients is that the painting tools with which an image is made on the screen of the monitor have high-demand ability, can be offered with clear rules, and can thus specify a structure. As far as the overly intensity of the unfolding emotions in the creative process is concerned, the electronic distance behind the screen, which is not painted on a piece of paper, helps. It is easy to "rub out" the painted. This is close to the *defense mechanism of the undoing*. If the design exceeds the bearable and, for example, leads to psychotic state, the painting process can be stored at any point so that it can be reloaded and resumed fractionally at a later session. This therapeutic approach is comparable to *the defense mechanism of repression*. Thus, patients can learn again defense mechanisms in a subtly adapted alternation between detection and protection. This is also helpful for many psychotic illnesses. Federn (1956) already emphasized, to stabilize psychotherapeutically the ego strength in psychotic patients. For documentation, the entire process of painting can be recorded, stored at any point, and repeated for the purpose of mirroring. The indication for conventional or computer-assisted painting therapy is not competing with usual painting therapy. But the respective use depends on the structural level of the personality structure of the psychotic or personality disorder. The weaker the psychic structure is, the more rules giving and structuring therapy procedures are required (Figs. 24.2 and 24.3).

- Painting tools with high-demand ability
- Clear rules and structure specification
- Electronic distance
- "Rub out" as a defense mechanism of undoing
- Save as defense mechanism of repression
- Balance between revealing and protection

**Fig. 24.2** Painting with the help of a computer painting program (schizophrenia) (Hartwich 2007, by courtesy of the publisher Wissenschaft & Praxis)

**Fig. 24.3** Computer painting using letters and colors (Hartwich 2007, by courtesy of the publisher Wissenschaft & Praxis)

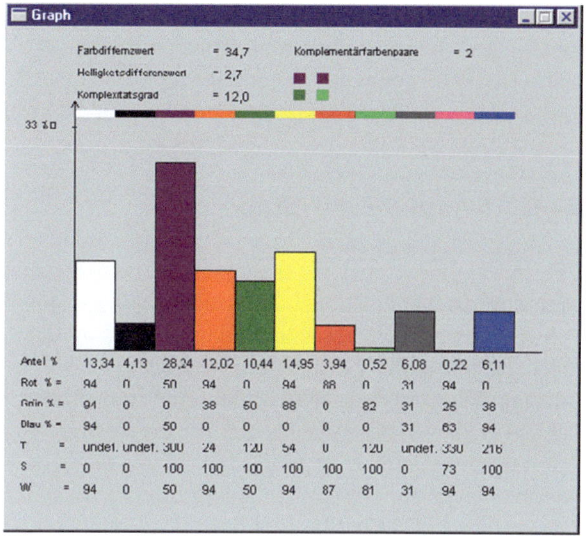


**Fig. 24.4** Example of the image analysis program (Hartwich 2002, by courtesy of the publisher Wissenschaft & Praxiss)

Since the computer images are numerically digitized in the form of pixels and a ratio scale level is present, they can be processed mathematically well. As a result, we have developed an image analysis program in which color criteria and formal criteria can be recorded (Hartwich and Brandecker 1999). In the case of the complexity regarding an image, there are many or a few elements that make up the image. In the spectrum analysis (Fourier transform), each pixel is represented in three color portions corresponding to the RBG values (red-green-blue) and converted into a frequency amplitude diagram. Both intraindividual changes could be parallelized with psychopathology and psychodynamics; also interindividual differences among the disease groups were presented (Hartwich 2002, 2010) (Fig. 24.4).

## 24.5.2 Neural Dimension of Computer Painting

Compared to conventional painting therapy, one of the decisive differences is that clear-cut rules and a firm structure specification are offered when using a computer painting program. For most patients with neurotic and psychoreactive disorders, this would mean a narrowing of their unfolding creative possibilities. In the case of psychotic illnesses with ego disintegration and self-fragmentation, the structure giving is a help in the sense of a prerequisite that creativity can be developed and unfolded at all. This may allow the abovementioned interaction between left and right anterior temporal gyri and the right parietal and prefrontal cortices, where originality and creativity may be promoted or inhibited and may be raised in terms of the activity level. This would imply that the decrease of inhibition, which may pose a risk to the psychotic structure, is compensated by the therapeutic structuring. The gained activity could modulate positively the hyperconnectivity in the midline structures of the psychotic's brain. One day we may be able to compare the neural reaction of colors of the paintings and may find the relation to measurements of our pixel diagram of computer pictures. Especially complementary contrasts of colors, which are painted next to each other, e.g., purple and yellow or red and green, are interesting to be investigated, because they were found to be favored by schizophrenics (5% niv. Hartwich 2002) compared to normal people and other diagnoses. The neural reaction to those strong complementary contrasts should be investigated in more detail.

### 24.5.3 Sculpture with Stone

The three-dimensional art therapy work will be described by means of marble sculpture. In our history of mankind, shaping a stone has an ancient tradition, in which a sculptor creates a three-dimensional product with the use of his focused muscular power that he and art viewers can also feel by touching the stone to comprehend. Thus, the creativity of the sculptor manifests itself differently from painting and music but has a special quality: the slowness of the design process; the stone can only be worked over long time, weeks sometimes months, depending on the hardness and size. The stone material offers a particularly *high degree of structure*. The sculpting of the hardness of the object is only possible gradually, layer by layer; there are no quick solutions, such as painting and drawing. A creative product of a sculpture can only arise in a respectful treatment of the stone. Even severely disturbed psychiatric patients benefit from the described structuring. On the other hand, if they were given a structureless clay or a blank sheet of paper, they would be overwhelmed, since their structural weakness in the acute stage cannot yet afford the necessary order and structure (Fig. 24.5).

- Slowness of the creative process.
- The stone has a particularly high degree of structure.
- Meditative rhythmic beating with the hammer.
- Physical working of the whole body.
- Creative binding force reduces fragmentation of the self.

**Fig. 24.5** Creative sculpturing in marble

Duration and intensity of work with the stone, the use of the whole body, and the meditative rhythmic process of beating with the hammer on the chisel lead to a high libidinal energetic investment. The object becomes a self-object (Winnicott 1953). For psychodynamically working therapists, it is fascinating to witness that this self-object occurs in the patients' dreams. One should not interpret this at first but waits as the positive force of the self-object unfolds. The creative process, which extends over several weeks, unfolds a considerable creative binding force, which helps to mitigate the fragmentation of the self (Kohut 1973; Kohut and Wolf 1980) in a psychotic patient. Strengthening the self-cohesion is the prerequisite for letting go of the symptoms.

### 24.5.4 The Neural Dimension of Sculpturing with Stone Material

From a neuronal perspective, the long-lasting therapeutic creativity of the patients, which lasts several weeks, could increase activity in the temporal lobe and high-frequency brain activity, which may be associated with high-frequency gamma activity in the EEG. These functions could modulate the pathological abnormalities, such as hyperconnectivity in schizophrenia, as well as change brain activity in the medial and right frontal brains in depressants. The described chance of decreasing inhibitory functions in the case of creativity would be controlled in this type of therapy by the firm structure of the stone material with the inherent slowness of the processing in such a way that the balance between the decrease of the inhibition and control is maintained, which leads to a higher level of activity than previously possible in the course of the disease.

### 24.5.5 Video Mirroring: The Image of the Self

When looking at family photos or holiday films, it is noticeable that each person first looks at the image of himself and is more affected by it than by the image of other persons. The decisive factor here is that the extent of the investment energy of the observer depends on the intensity of the self-reference. This observation has been used therapeutically in the use of the method of video mirroring on psychiatric patients (e.g., Hartwich and Lehmkuhl 1979, 1981). Video mirroring is used in neurotic, anorexia nervosa, schizophrenic, schizoaffective, and affective psychoses patients (Hartwich 1986, 1993; Deister 1984). First, an interview with a patient is recorded audiovisually; after this suitable short (¼ to 1 min) sequences are cut out. With these records, patients are systematically mirrored. Their reactions are again audiovisually documented with a second video camera system. It is impressive to observe how strongly patients respond to their own mirror image, which appears audiovisually, that is, optically and acoustically in its own language. The crucial experience is that attention, interest, and emotional movements are much stronger on one's own reflection than when other people are to be seen. Psychodynamically it is a particularly pronounced form and

**Fig. 24.6** One of the first examples of mirroring

quality of investment energy (cathexis), which is connected with the consciousness of the own self. This is used for therapeutic purposes (Fig. 24.6).

- High investment energy in the self-reference
- Audiovisual mirroring of one's own facial expressions, gestures, and speech
- Awareness of the self and investment energy (cathexis)

## 24.5.6 Mirror Phenomenon and Neurobiology

H. W. Diamond (see Burrows and Schumacher, 1979) in England was the first psychiatrist who took photographs of mentally ill patients with the just-invented technique of photography. He confronted his patients with their own portraits. In his lecture in front of the Royal Society in London on May 22, 1856, he described the effect of mirroring with the self-portrait as a treatment method of psychiatric patients: "*There are still points of view, under which the value of the portrayal of mental illnesses are particularly prominent, namely, the effect which they exert on the patients themselves. ... Such photographs are often viewed with interest and pleasure, and the best effect is obtained by those images which show the progress and the recovery of a severe period of their madness*" (Diamond 1856, cited by Burrows and Schumacher 1979, p. 156).

In psychoanalysis, Kohut (1973) has dealt with the mirror phenomenon. In the development of childhood, the transition takes place from autoerotism to narcissism which is supported by mirroring by reference persons. This is a development step from the level of the fragmented self (self-cores) to the level of the coherent self that leads to the development of the perception of the self as a physical and mental unity, which is spatially connected and also temporally continuous (Kohut 1973, p. 143).

With regard to neurobiological findings, we report those of schizophrenic diseases (see more detailed in Chap. 9). Northoff (2012) assumes, among other things, a disruption of the connection between the anterior and posterior median regions (hyperconnectivity) is correlated to the disintegration or fragmentation of the self in schizophrenia.

Seeing one's own image in the mirror and experiencing the audiovisual recording of oneself (optically in the movement of the face and acoustically in the hearing of one's own speech) are accompanied by intense emotional reactions, which differ from the perception of neutral objects. Similar phenomena have already been investigated by means of neurobiological experiments. For example, Northoff et al. (2009) reports from a Chinese research group who have examined patients with a disorder of consciousness, namely, with fMRI and EEG. The patients were presented with acoustic stimuli in the form of their own names or by names of unknown persons. Even comatose patients responded with a strong activity in the region of the speech area and the amygdala (responsible for the emotional processing of stimuli), when listening to their own name, but not when hearing foreign names. Northoff et al. (2006) compared the previous neuroimaging studies, which relate to their own self, with several other different areas: verbal, emotional, social, facial expression, etc. In the different studies, the different modalities and areas were found to be common that the relation to the own self did enhance the neural activity of the subcortical-cortical midline system. Neuroimaging studies have shown that words and images that are related to the individual self are more emotionally valued than those who have less self-reference (Northoff and Bermpohl 2004; Northoff et al. 2006) In the cortical midline structures, the neuronal activity is enhanced when the stimuli are experienced as self-referenced.

Northoff explains how he understands self-reference:

> Another example is the way we perceive pictures of ourselves or close friends versus pictures of completely unknown people or pictures of our childhood houses versus pictures of unknown houses. Such comparisons are possible in different sensory modalities. Self-relatedness is here understood and presupposed in a rather cognitive sense. This implies self-awareness, meaning that one becomes aware of one's self, one sees the stimulus that is related to one's own self as distinguished from the stimuli that are not related to the self. (Northoff 2014, p. 581)

### 24.5.7 Improving Self-Coherence Through Mirroring

When the video mirroring with well-selected sequence is used therapeutically, the person on the monitor sees himself as a unit in mimic and gesture, as well as a speaking person, whose thoughts are formulated in an appropriate way. It is impressive to see how psychotic patients feel strongly positively stimulated and how they experience the strengthening of their ego coherence. There are spontaneous remarks like "*I seem quite normal, much more secure than I thought.*"

The method in more detail: Selected sequences from a previously recorded interview are shown; then, in the presence of the therapist, the self-picture remains still on the monitor and serves as an anchor for the psychotherapeutic conversation. The patient experiences himself as a physical mind entity, whose psychomotor expression usually acts on him more favorably than corresponds to his current self-assessment. In this approach it is important to emphasize that the psychotic patient carries inside the memory of the self-fragmentation experience and the uncertainty of the synchronization of thoughts, words, and feelings. On the monitor, however, he encounters the connection of his own cognitive and affective possibilities and their psychomotor expression. Here we see one of the main mechanisms of video mirroring in schizophrenics.

In the experimental study of this phenomenon (video mirroring) by Hartwich and Lehmkuhl (1979), the influence of schizophrenic ego disturbances was systematically investigated. The result was that after a short-term worsening at the beginning, which was accompanied by an intensification of anxiety, the ego demarcation (Scharfetter 1986) was measured as significantly enhanced over the whole time of the video-mirroring application (several weeks). Affects and thoughts were more synchronously experienced, and language and mimetic expression behavior became livelier. In another study (Hartwich 1982) of investigated schizophrenics, a subgroup was extracted by means of type analysis, in which the effects were not measurable, while another subgroup noticed a more lively and intense affectivity as well as a decrease in the ego disturbances.

- Self-fragmentation and improvement of coherence
- Strengthening of self-demarcation
- Synchronization of affects and thoughts

### 24.5.8 What May Happen Neurobiologically While Using Video Mirroring on Schizophrenics?

We start out from the idea of Northoff (2011) from an abnormally strong communication between anterior and posterior midline regions (hyperconnectivity) in schizophrenics as prerequisite of disintegration or fragmentation of the consciousness of the self, the "broken self in schizophrenia" as he says. Then we should ask how the neurobiological findings of the "*self-related processing*" should be related.

The method of systematic video mirroring appears to be particularly suitable for intensifying the self-related processing. It is therefore appropriate to take into account the amplifications of the neuronal activity in the subcortical-cortical central nervous system. It could be probable that the systematic video mirroring of the individual in schizophrenics could modify positively their hyperconnectivity in the central region by the neuronal activations provoked in the region of the subcortical and cortical central nervous system. This would mean that an influence would be possible in the sense of a reduction in the hyperconnectivity in the midline structures. In the sense of the neuropsychodynamic approach, this would correspond to the improved self-coherence experience of the schizophrenics.

## References


Abraham A. The promises and perils of the neuroscience of creativity. Front Hum Neurosci. 2013;7:246.

Arieti S. Creativity. The magic synthesis. New York: Basic Books; 1976.

Beaty RE, Benedek M, Wilkins RW, et al. Creativity and the default network: a functional connectivity analysis of the creative brain at rest. Neuropsychologia. 2014;64:92–8. https://doi.org/10.1016/j.neuropsychologia.2014.09.019.

Benedetti G. Psychopathologie und Kunst. In: Condrau G (Hrsg) Die Psychologie des 20. Jahrhunderts, Bd XV: Transzendenz, Imagination und Kreativität. Zürich: Kindler; 1979. pp. 1045–54.

Benedetti G. Das Symptom als kreative Leistung. In: Thomashoff H-O, Naber D, editors. Psyche und Kunst. Stuttgart: Schattauer; 1999. p. 49–56.

Burrows A, Schumacher J. Über die Anwendung der Photographie auf die physiognomonischen und seelischen Erscheinungen der Geisteskrankheit. Vortrag vor der Royal Society am 22 Mai 1856. In: Doktor Diamonts Bildnisse von Geisteskranken. Frankfurt: Syndikat Autoren- und Verlagsgesellschaft; 1979. p. 155–8.

Deister A. Audiovisuelle Selbstkonfrontation bei Schizophrenie. Med. Diss RWTH Aachen. 1984.

Dietrich A, Kanso R. A review of EEG, ERP, and neuroimaging studies of creativity and insight. Psychol Bull. 2010;136(5):822–48.

Federn P. Ich-Psychologie und die Psychosen. Bern: Huber; 1956.

Guilford JP. Kreativität. In: Ulmann G, editor. Kreativitätsforschung. Köln: Kiepenheuer & Wisch; 1950. p. 25–43.

Hartwich P. Experimentelle Untersuchungen zur audiovisuellen Selbstkonfrontation bei Schizophrenen. In: Kügelgen B, editor. Video und Medizin. Erlangen: Perimed; 1982.

Hartwich P. Audiovisuelle Verfahren. In: Müller C, editor. Lexikon der Psychiatrie. Berlin: Springer; 1986. p. 74–6.

Hartwich P. Videospiegelung in der Behandlung schizophrener Psychosen. In: Ronge J, Kügelgen B, editors. Perspektiven des Videos in der klinischen Psychiatrie und Psychotherapie. Berlin: Springer; 1993.

Hartwich P. Creative therapeutic methods and quantification. In: Hartwich P, Fryrear JL, editors. Creativity. . The third therapeutic principle in psychiatry. Sternenfels: Wissenschaft & Praxis; 2002.

Hartwich P. Psychodynamisch/somatopsychodynamisch orientierte Therapieverfahren bei Schizophrenen. In: Hartwich P, Barocka A, editors. Schizophrene Erkrankungen. Prophylaxe, Diagnostik und Therapie. Wissenschaft & Praxis: Sternenfels; 2007. p. 33–98.

Hartwich P. Bildnerisches Gestalten in der Kunsttherapie mit Psychosekranken. In: Sinapius P, Wendland Baumeister M, Niemann A, Bolle R, editors. Wissenschaftliche Grundlagen der

Kunsttherapie, Bd III: Bildtheorie und Bildpraxis in der Kunsttherapie. Frankfurt: Peter Lang; 2010. p. 195–210.
Hartwich P. Bildhauerei mit psychotisch Kranken. Die Bedeutung von Kreativität und Parakonstruktion. Forum der Psychoanalytischen Psychosentherapie, Bd 28. Göttingen: Vandenhoeck & Ruprecht; 2012. pp. 56–70.
Hartwich P, Brandecker R. Computer-based art therapy with inpatients: acute and chronic schizophrenics and borderline cases. Arts Psychother. 1997;24:367–73.
Hartwich P, Brandecker R. Quantifizierung bildnerischer Gestaltungselemente in der Computermaltherapie bei Schizophrenen. In: Hartwich P, editor. Videotechnik in Psychiatrie und Psychotherapie. Sternenfels: Wissenschaft & Praxis; 1999.
Hartwich P, Lehmkuhl G. Audiovisual self-confrontation in schizophrenia. Arch Psychiatr Nervenkr. 1979;227:341–51.
Hartwich P, Lehmkuhl G. Experimentelle Einzelfalluntersuchung zur schizophrenen Affektivität. Z Psychother Psychosom Med Psychol. 1981;31:83–6.
Hicklin A. Das Schöpferische als Zentralproblem der Psychotherapie. In: Condrau G (Hrsg) Psychologie des 20. Jahrhunderts, Bd XV: Transzendenz, Imagination und Kreativität. Zürich: Kindler; 1979. pp. 1063–68
Jakab I. Zeichnungen und Gemälde der Geisteskranken, ihre psychiatrische und künstlerische Analyse. Berlin: Henschel; 1956.
Jung CG Über die Beziehung der analytischen Psychologie zum dichterischen Kunstwerk. In: GW Bd XV. Über das Phänomen des Geistes in Kunst und Wissenschaft. Olten: Walter; 1922. pp. 75–96.
Kandel E. Das Zeitalter der Erkenntnis. Die Erforschung des Unbewussten in Kunst, Geist und Gehirn von der Wiener Moderne bis heute. München: Siedler; 2012.
Kohut H. Narzißmus. Frankfurt: Suhrkamp; 1973.
Kohut H, Wolf ES. Die Störungen des Selbst und ihre Behandlung. In: Peters UH (Hrsg) Die Psychologie des 20. Jahrhunderts, Bd X: Ergebnisse für die Medizin (2). Zürich: Kindler; 1980. pp. 667–82.
Li W, Li X, Huang L, et al. Brain structure links trait creativity to openness to experience. Soc Cogn Affect Neurosci. 2015;10(2):191–8. https://doi.org/10.1093/scan/nsu041.
Lombroso C. Der geniale Mensch. Übers. v. Fraenkl MO. Verlagsanstalt und Druckerei Actien Gesellschaft, Hamburg (italien. 1872). 1890.
Matussek P. Kreativität als Chance. München: Piper; 1976.
Matussek P. Kreativität. In: Condrau G (Hrsg) Psychologie des 20. Jahrhunderts. Bd:XV: Transzendenz, Imagination und Kreativität. Zürich: Kindler; 1979. pp. 44–66
Morgenthaler W. Ein Geisteskranker als Künstler. Bern: Bircher; 1921.
Navratil L. Schizophrenie und Kunst, Bd 28. dtv Gesamtausgabe, München. 1965.
Northoff G. Neuropsychoanalysis in practice. New York: Oxford University Press; 2011.
Northoff G. Das disziplinlose Gehirn – Was nun Herr Kant? Auf den Spuren unseres Bewusstseins mit der Neurophilosophie. München: Irisiana; 2012.
Northoff G. Unlocking the brain. Consciousness, vol. 2. New York: Oxford University Press; 2014.
Northoff G, Bermpohl F. Cortical midline structures and the self. Trends Cogn Sci. 2004;8(3):102–7.
Northoff G, Heinzel A, de Greck M, et al. Self-referential processing in our brain – a meta-analysis of imaging studies on the self. NeuroImage. 2006;31(1):440–57.
Northoff G, et al. Differential parametric modulation of the self-relatedness and emotions in different brain regions. Hum Brain Mapp. 2009;30(2):369–28.
Prinzhorn H. Bildnerei der Geisteskranken. Ein Beitrag zur Psychologie und Psychopathologie der Gestaltung. 2nd ed. Berlin: Springer; 1922.
Rennert H. Die Merkmale schizophrener Bildnerei. Jena: VEB G Fischer; 1962.
Scharfetter C. Schizophrene Menschen. 2nd ed. München: Urban & Schwarzenberg; 1986.
Volmat R. L'art psychopathologique. Paris: Presses Universitaires de France; 1956.
Winnicott DW. Transitional objects and transitional phenomena. Int J Psychoanal. 1953;34:89–97.

# Four Psychotherapeutic Settings 25

Heinz Boeker

**Abstract**

Four different psychotherapeutic settings for the outpatient psychotherapy and the respective indication criteria are presented in this chapter.

The neuropsychodynamic approach supports the application of effective principles of psychotherapy and pharmacotherapy in the treatment of psychiatric disorders on the basis of a conceptualization which focuses on the circular relationship between different biological, neurobiological, psychological, and social dimensions of the disorders.

Therapeutic changes may be induced "top-down" as well as "bottom-up". Consequently therapeutic strategies should be developed and modified which aim at reactivating the brain's inherent plastic potency and inducing a reorganization of the neural networks which had developed on the basis of genetic dispositions or by means of epigenetic influences and which had been stabilized by means of their previous use.

Indication criteria, aspects of therapeutical techniques, and adequate therapeutical settings are described. Four settings with different frequencies of the sessions (low, medium, higher frequency, outpatient group psychotherapy), duration, and therapeutical focus may serve as an orientation for the outpatient psychotherapy. Their selection has always to be related to each single patient with regard to the symptomatology, the history of the disorder, the personality, the degree of psychological strain, and the motivation of the patient.

H. Boeker
Department of Psychiatry, Psychotherapy, and Psychosomatics, Psychiatric University Hospital Zurich, University of Zurich, Zurich, Switzerland
e-mail: heinz.boeker@bli.uzh.ch



H. Boeker et al. (eds.), *Neuropsychodynamic Psychiatry*,
https://doi.org/10.1007/978-3-319-75112-2_25

Four psychotherapeutic settings for the outpatient psychotherapy and the respective indication criteria are presented in the following chapter.

The neuropsychodynamic approach supports the application of effective principles of psychotherapy and pharmacotherapy in the treatment of psychiatric disorders on the basis of a conceptualization which focuses on the circular relationship between different biological, neurobiological, psychological, and social dimensions of the disorders. Neuropsychodynamically orientated interventions, resting upon a psychotherapeutic attitude, are in compliance with the knowledge of neuroinformatics which also underline the preconditions of psychotherapeutic interventions and learning in psychotherapy. The basic requirement is a fear-free atmosphere, because it has to be assumed that anxiety adjusts the neuromodulators in an adverse manner. Furthermore, the studies on the neuroplasticity of the brain underline that growing insight as well as repetitive learning may be therapeutically successful. Therapeutic changes may be induced "top-down" as well as "bottom-up" (cf. Boeker 2005; Spitzer 2000). In cognitive behavioural therapy, new associations are generated by training, and the old, unfavourable ones are extinguished. In insight-orientated psychotherapies, also far off associations are made available. Implicit meanings are recognized and modulated, and new associations are created.

Consequently therapeutic strategies should be developed and modified which aim at reactivating the brain's inherent plastic potency and inducing a reorganization of the neural networks which had developed on the basis of genetic dispositions or by means of epigenetic influences and which had been stabilized by means of their previous use. A precondition therefore is the analysis of those factors which had led in the course of the previous development of a patient to the priming and stabilization of the associations and networks in the brain and to the activation of neuroendocrinological processes.

The effects of the somato-psychic-psychosomatic vicious circles of psychiatric disorders withdraw the directly interpreting access (e.g. in the classical psychoanalytical setting). New experiences are enabled in the dialogue between patient and therapist which lead to a reduction and modulation of previous dysfunctional and painful coping strategies (Boeker 2005; Mentzos 1995). Indication criteria, aspects of therapeutical techniques, and adequate therapeutical settings are described in the following.

The psychoanalytical psychotherapy represents a psychotherapeutical method which is closely orientated to psychoanalytical theory and technique but in fact differs from classical psychoanalysis due to modifications of the setting and their impact on the therapeutical process (Boeker 2005). Where and how precise the borderline should be set is in the focus of critical discussions occasionally. In order to reveal the unconscious aspects of intrapsychical and interpersonal dynamics, it is necessary that the psychoanalytical psychotherapist adopts a "reflective function" (Fonagy and Target 1997) which involves empathy, verbalization of intrapsychical processes, and mirroring. This provides a basis for the so-called moments of meeting (Stern 1998), during which something happens or is emerging what cannot be verbalized any longer, but represents an expression of implicit, unconscious, and relational knowledge ("implicit relational knowledge"), thus enabling new experiences (Hoffmann and Schauenburg 2000).

It is often the case that only after remission of the somatic blockade (e.g. in depression) the premorbid structures and basic conflicts may be addressed (Boeker 2005; Rudolf 1996). In the further course of the treatment, psychoanalytical psychotherapy focuses on the high degree of self-expectations (high ego-ideal), self-doubts (turning aggression towards one's own self), feelings of guilt (rigid superego), and tendencies to be dependent on others (idealization, separation anxieties, shame, separation guilt) or to develop an accelerated autonomy (regressive actualization of the grandiose self in mania) taking into account the biographical background, the internalized relationships to significant persons, and the unconscious conflicts. Besides the interpretations of the predominant unconscious dynamics and transference, especially the new experiences in the "here-and-now" of the therapeutical relationship contribute to a reduction and modification of defensive strategies of defence and modification and habitual attachment patterns.

Four settings with different frequencies of the sessions, duration, and therapeutical focus may serve as an orientation for the outpatient psychotherapy (Mentzos 1995; Boeker 2000). Their selection has always to be related to each single patient with regard to the symptomatology, the history of the disorder, the personality, the degree of psychological strain, and the motivation of the patient:

*Setting A*: Low *frequency of sessions* (e.g. one or two times per month over the course of several years)

- Short-session duration (e.g. 20–30 min)
- Constantly accepting relatively distanced therapeutical attitude (enables the new experience of autonomy without guilt and contributes to the development of an emotional attachment without shame)

*Setting B*: Medium frequency of sessions (one session per week mostly)

- Therapeutical focus on actual conflicts (e.g. relationship, profession).
- No interpretation of the transference at the beginning and occasional interpretation of the transference and reconstruction later on.
- Therapeutical attitude: the ambivalence in the relational experience of the patient remains of great importance in the reflexion of the therapist.

*Setting C*: Higher frequency of sessions (2–3 sessions per week)

- In case of depression after reduction or remission of the severe depressive symptoms.
- Working through of the personality structural components of the disorder.
- Interpretation of the transference.
- Evidential experience of the patient: the appreciation of the other maintains in spite of the noticeable aggression and distance in the relationship.

*Setting D*: Outpatient group psychotherapy (1.5 hours per week)

- Coping with and prevention of relapses in the long-term course of the disorder (psychoeducation)
- Increase of social competence in the interactional exchange of experiences between the members of the group therapy
- Multilateral transferences: experience and overcoming social vicious circles in the actual group situation
- Stepwise working through of the ambivalent wish for an "ideal object" (over a longer term of the therapeutical process)

The usually used indication criteria for psychotherapy (psychological strain, motivation, capability to introspection, severity and duration of the illness, personality) may also be considered in the context of neuropsychodynamic psychiatry (see Table 25.1).

**Table 25.1** Indication criteria for a neuropsychodynamic-orientated psychotherapy and selection of the therapeutical setting

| | | Therapeutical setting | | | |
|---|---|---|---|---|---|
| Indication criteria | | A | B | C | D |
| Psychological strain | High | | X | (X) | (X) |
| | Low | X | | | |
| Motivation | Basic precondition, has to be clarified and supported under certain circumstances | X | X | X | X |
| Capability to introspection | High | | X Focus: actual conflicts | X Working through of dysfunctional psychic mechanisms | |
| | Low | X | | | X |
| Severity and duration of the disorder | Long-lasting course of the disorder, severe psychosocial deficits | X | (X) | | X |
| Personality structure | Severe personality disorder with low structural level according to operationalized psychodynamic diagnostics (OPD) | X | X | | (X) |
| | Narcissistic structure (medium structural level according OPD) | | X | (X) | (X) |
| | Higher structural level according OPD with histrionic or anxious-insecure traits | | X | X | |

*OPD* operationalized psychodynamic diagnostics

## References

Boeker H. Depression, manie und schizoaffektive psychosen: psychodynamische theorien, einzelfallorientierte forschung und psychotherapie. Gießen: Psychosozial-Verlag; 2000.

Boeker H. Melancholie, depression und affektive störungen: zur entwicklung der psychoanalytischen depressionsmodelle und deren rezeption in der klinischen psychiatrie. In: Boeker H, editor. Psychoanalyse und psychiatrie: geschichte, krankheitsmodelle und therapiepraxis. Berlin: Springer; 2005. p. 115–57.

Fonagy P, Target M. Attachment and reflective function. Their role in self-organization. Dev Psychopathol. 1997;9:679–700.

Hoffmann N, Schauenburg H, editors. Psychotherapie der depression. Stuttgart: Thieme; 2000.

Mentzos S. Depression und manie: psychodynamik und psychotherapie affektiver störungen. Göttingen: Vandenhoeck & Ruprecht; 1995.

Rudolf G. Psychotherapieforschung bezogen auf die psychotherapeutische Praxis. Psychother Forum. 1996;4:124–34.

Spitzer N. Geist im netz: modelle für lernen, denken und handeln. Berlin: Spektrum, Akademischer Verlag; 2000.

Stern D. The process of therapeutic change involving implicit knowledge: some implications of developmental observation for adult psychotherapy. Infant Ment Health J. 1998;19:300–8.

# Psychotherapy, Psychopharmacotherapy, and Neuromodulation

# 26

Peter Hartwich and Heinz Boeker

**Abstract**

Whether you want it or not, when psychopharmaceuticals are given, there is always involved a psychodynamic component. Thus, the prescription of the drug is embedded in the relationship of the two dissimilar partners: patient and doctor. In the case of many psychoses, the use of psychopharmaceuticals can help to build up a viable therapeutic relationship and can be the prerequisite for psychotherapy. The bottom-up effects by means of psychopharmacotherapy can be completed with the top-down effects which are mediated by means of psychotherapy. This underlines that psychotherapy modulates biological processes indirectly (by activating psychogenetic material). On the other hand, psychopharmaceuticals can change biological reaction patterns directly. Thus psychopharmacotherapy and psychotherapy are to be seen as different—but additional—components of a multimodal treatment concept.

Furthermore, different modalities of targeted neuromodulation (e.g., vagus nerve stimulation (VNS), deep brain stimulation (DBS)) are being actively researched as putative treatment options for treatment-resistant forms of depression.

A special challenge concerns the adequate weighting of the single component in different disorders regarding the individual situation of each single patient.

We show how this therapeutic approach is experienced in depressive patients, schizophrenics and other psychiatric disorders. Important is that the therapist

P. Hartwich (✉)
Department of Psychiatry, Psychotherapy, Psychosomatics, General Hospital of Frankfurt am Main, Teaching Hospital University Frankfurt, Frankfurt, Germany
e-mail: Prof.Hartwich@casytec.de

H. Boeker
Department of Psychiatry, Psychotherapy, and Psychosomatics, Psychiatric University Hospital Zurich, University of Zurich, Zurich, Switzerland
e-mail: heinz.boeker@bli.uzh.ch



H. Boeker et al. (eds.), *Neuropsychodynamic Psychiatry*,
https://doi.org/10.1007/978-3-319-75112-2_26

reflects his inner position and represents a clear attitude in the interaction of psychodynamics and psychopharmaceuticals; we try to explain which factors are effective and how the role of relatives is in relation to the reliable intake of medication. In difficult patients, who omit their medication in a critical situation, it is important that a treatment partnership is established with a family member. The example of group psychotherapy serves to show that "experts among themselves" have a great influence on the willingness to take medication in a neuropsychodynamic context.

## 26.1 Introduction: How Do Drugs Work?

It is assumed that medications, especially psychopharmaceuticals, do not affect illnesses but symptoms. Küchenhoff (2016) emphasizes however that this view is too simplistic. Pharmacological interventions do not affect symptoms but certain neurophysiological systems, which are anchored in brain regions. They alter the qualities of the limbic system, they increase the filter function of the thalamus, etc. The neurobiological knowledge becomes more and more precise with regard to the effects on synapses and brain function. Küchenhoff also points out that psychopharmaceuticals do not cure depression but that the basis of the experience of the depressive patient is changed.

Whether you want it or not, when psychopharmaceuticals are given, there is always involved a psychodynamic component. Why? In the prescription of a medication are always two people involved, doctor and patient. Thus, the prescription of the drug is embedded in the relationship of the two dissimilar partners. In this respect, different facets can unfold. For example, in the case of psychoses, the use of psychopharmaceuticals can help to build up a viable therapeutic relationship at all, as mentioned, e.g., by Küchenhoff (2016), and can be the prerequisite for psychodynamic psychotherapy.

## 26.2 Dream and Interaction with the Therapist

### 26.2.1 Case Study

A 43-year-old female patient has now been treated for the third time in a psychiatric hospital because of a paranoid-hallucinatory schizophrenic psychosis. In the meantime, she was treated as an outpatient by a psychiatrist. As maintenance therapy she took a depot neuroleptic as well as a daily neuroleptic medication, the latter being taken as needed. With pharmacotherapy and psychotherapy, she was so well balanced that she was able to continue her job as a saleswoman in the fashion industry.

Now she was treated in a psychiatric hospital, had violent psychotic symptoms, and was given a combination of a potent, a low-potent antipsychotic medication and a tranquilizer. The psychotic symptoms improved very quickly, almost within 14 days.

*She tells a dream: She is in a room with pale light; the walls are made of bricks. She is afraid that the walls could not hold. Some pressure from the outside causes the wall to wobble. The walls get cracks. Her fear grows that the walls do not hold. The stones threaten to break in. She also hears a menacing noise. With great fear she wakes up and the dream is finished.*

She feels confused and expresses this clearly in the description of the dream in terms of facial expression and gesture. When the therapist had seen her insecurity and heard the dream, his first countertransference was "I should give her help and protection in her fear. She shows her fragility." The therapist takes the dream as an illustration of the risk of self-fragmentation or re-fragmentation, since her last severe relapse was only a short time ago.

Countertransference: the therapist feels her fear of self-fragmenting. However, in the stage after the just experienced florid psychotic symptoms, it was positive to see that she was able to distinguish between dream and psychosis. The therapist decided to follow his need for protection first in enhancing the positive side, according to Benedetti (1987).

Therapist: "I think it's good that you can already distinguish between what you have experienced in the psychotic state and what you dream of. There you are with yourself again."

Patient: "Yes, but I am very unsure."

How should the therapist handle the dream now? It was a message. There might be an object-level interpretation offer, such as the question "Is there anyone who threatens you?" or "What do you think you are not able cope with?" This form of working with the dream appeared to the therapist to be inappropriate. It seemed rather appropriate to conceive what she said as a message, as if she wanted to communicate something to the therapist.

Therapist: "Are you afraid that something psychotic will break in and you cannot endure it?" (He wanted to remain in her own dream picture on a subject-level).

Patient: "Yes, with all that I have already experienced." (So she went into the concrete understanding of the dream message.)

Therapist: "Do you think the drugs are enough?"

Patient: "I think I should have a little more."The interview ended with the therapist and the patient agreeing on a certain dose increase for a limited time.

## 26.3 Discussion of the Case

She is one of the patients who have had several years of experience with her schizophrenic psychosis and also with antipsychotics. Consequently, a "third element of the therapeutic encounter" is added here, as Boeker (2016) emphasizes. Thus, the drug becomes self- and/or object fantasies. Both patient and therapist transfer fantasies into the drug. This is about the consolidation of the spatiotemporal structure of the self. The patient has learned to strengthen the principle of reality, which is emphasized by the prescription of the drug, as Jessner (2016) and Meißel (2001) also describe. Decisive is the respect for the patient's autonomy. The therapist offers an increase in the dose, and the patient has the freedom to decide to agree or not. Frequently, psychopharmaceuticals can contribute to building a viable therapeutic relationship at all; this strategy is also emphasized by Küchenhoff (2016).

There are patients who suffer from, e.g., anorexia nervosa or mania and who feel injured in their autonomy when the doctor prescribes them medication. In addition, they fear that the drugs could change their personality. This point of view is emphasized by Muench (2016). From an interesting psychoanalytical point of view, he states that for the manic patient, the drug could become an object with an influence on the regulation of the self-esteem and thus its loss: "With the drug one is nothing and without everything! In this case, the medication is denied because one does not want to be nothing. It is different in the depression, with the drug one is something and without it one is nothing, it is taken, because then one is at least something" (Muench 2016, p. 105).

## 26.4 Therapist and Psychopharmaceuticals

Medication is in principle prescribed by the physician. In consideration of the genetic disposition and other somatic facts, as well as the treatment success with psychopharmaceuticals documented today in practice and in the literature, the psychiatrically trained physician represents the importance of the chemical substances. It is also his duty to educate patients and family members about a number of important studies about side effects (such as Aderhold et al. 2014), e.g. dealing with structural abnormalities in the brain in people with the diagnosis of schizophrenia. By taking antipsychotics, depending on the dosage level, the structural changes in schizophrenia can be further enhanced. The consequence is that cognitive impairments can occur more intensely. In addition to the important positive aspects, however, attention must also be paid to other side effects. The doctor has to clarify as early as possible about side effects in antipsychotics, e.g., late dyskinesia. If he fails to do the clarification, it can not only have legal consequences but also can disrupt the doctor-patient relationship.

Often, patients would like to hear that their illness is a purely psychogenic one, and that this can be healed quickly with psychotherapeutic methods alone. If, however, a patient is prescribed psychopharmaceuticals, a somatic accent of the understanding of his illness is documented. Taking medication such as antidepressants and antipsychotics over a long period of time can be an admission for the patient to bear his destiny with consequence.

The same is often the case for relatives. In many other cases, it is hoped that the antipsychotics will be given and taken to overcome the psychiatric disease only by chemical means. In this attitude, the belief that there may be a "spice" or "plant" which can cure the disease can sometimes be included. Since chlorpromazine was introduced as the first neuroleptic in 1952, the attitude toward the therapy of psychoses has changed considerably for all participants, including for the physician. In this way, he is able to influence the psychopathological behavior and experience with chemical substances, which is often possible in the positive sense. The distance regulation in schizophrenic patients could also be stabilized, as Meißel (2001) emphasizes. This can give strength, hope, and confidence to the physician, since he becomes a medical practitioner with a medicine prescription which does not differ from the other disciplines of medicine.

As a further point, it should be considered that some patients experience the medical treatment of psychopharmaceuticals in a certain way as a "charismatic use." The doctor should then become aware that the grandiose projections do not mean his person but a superior healer principle, which may be "behind him." Hippocrates, too, is said to have understood himself as the servant of the god Asclepius. If the doctor is aware of the omnipotence projection on the part of the patient in the action dialogue of the drug prescription, he will be able to use this inherent force therapeutically after a corresponding reflection of his grandiose self (Kohut 1973). In the case of longer outpatient psychotherapy of psychiatric patients, the treatment is occasionally divided. This means that a psychotherapist only performs the psychotherapeutic sessions with the patient and another doctor prescribes the psychopharmaceuticals. From our point of view, it is important to pay attention to the danger of splitting, which can reinforce the internal tendencies of splitting in many mental illnesses, e.g., borderline patients, other personality disorders with hysterical features, and schizophrenia.

The splitting seduces the psychotherapist and patient into an illusion, namely, one could get along without medication, and the taking of the psychopharmaceuticals is associated with a negative connotation. These psychologists' and doctors' attitudes, which are exclusively psychotherapeutic, come then when one's own grandiosity (Kohut 1973) leads to an unrealistic overestimation of the psychotherapeutic possibilities. Frequently, the patient is also in harmony with this unconscious overestimation since he can suppress the somatic side of his illness and thus the inherent severity. It is not to be ignored that the common illusion with the common fascinating power can temporarily have a positive therapeutic effect. It then reaches its limits when severe disease stages become manifest.

As regards transference and countertransference, Gabbard (2014) points out that some patients reject drugs: "By rejecting help offered to them, these patients may be unconsciously seeking revenge against their parents. When such patients sense that they are making their doctor miserable, they often feel a secret triumph" (p. 152). If the patient is noncompliant, countertransference anger is often provoked. Gabbard (2014) points out that "an ego-syntonic grandiose psychosis was the most powerful discriminating factor that distinguished schizophrenic patients who did not comply from those who did. Clearly, the noncompliant patients preferred their experience of psychotic grandiosity" (p. 154). Another reason to reject medication is the denial of the illness.

## 26.5 Patients and Psychopharmaceuticals

In the first few weeks or months, many patients with a new disease are ambivalent of the intake of medication. Initially, antipsychotics are often experienced as an unpleasant operation. Many patients fear to lose their ego-demarcation and the control of themselves by the influence of psychopharmaceuticals. Zeek (2016) stresses that in the case of anorexia nervosa and the bulimia, the prescription of medication is experienced as a loss of autonomy, since the patients are afraid that they can no longer control themselves. Since self-control plays a major role in their psychodynamics. The result is the medication is secretly omitted. This factor is too little considered in scientific studies on the efficacy of psychopharmaceuticals in the anorexia nervosa and other psychiatric illnesses. In depressive patients, thymoleptics can have an additional negative effect on their diminished self-esteem. Vollmoeller (2003) points out that in some depressives, the prescription of psychopharmaceuticals is misunderstood as a signal that they have given up. However, as soon as the thymoleptics have an antidepressant effect, self-esteem is also improved. Then the positive significance of psychopharmaceuticals is also perceived subjectively.

In some patients, the use of antipsychotics relieves their feelings of guilt. They no longer feel responsible for their destiny as a psychiatric ill person. Some patients experience the psychopharmaceuticals as evidence of the biological basis of their disease and can even counteract this with their relatives.

The feature of constancy is an important factor in the long-term treatment of many psychiatric disorders. The doctor tells the patient that this is to avoid relapses. Many patients are internally ambivalent. On the one hand, they want to protect themselves from relapses, but on the other hand they do not want to be reminded that they have a mental illness that may reappear. From this, many patients derive several reasons to omit psychopharmaceuticals:

- In the case of relative recovery, the medication can be experienced in the same way as an opponent in which the disease is symbolized, which one no longer wants to see.
- Libido and potency are reduced by many antipsychotics and some thymoleptics, which can lead to impairment of the sexual relationship and reduction of one's own quality of life. Then the patient often leaves out the medication without consultation with the doctor. First, he experiences an improvement in sexuality. He registers this positively and feels no reason to take the drugs again soon. Unfortunately, after a few weeks of beginnings of psychopathological symptoms, the relapse occurs.
- In many patients, especially women, weight gain plays an important role. The attractiveness is often diminished by taking psychopharmaceuticals. This leads to the desire to omit the drugs. Here, it is important that the doctor respects the patient's wishes and, together with her, negotiates the dosage or alteration of the drug. This increases the autonomy and self-responsibility of the patient.

Frequently, the psychiatrist's everyday life comes to the following situation, which is typical not only for schizophrenic but also for schizoaffective psychoses. A psychotic patient is obviously quite well put on psychopharmaceuticals, and he is better. Suddenly he reports more of psychotic symptoms.

### 26.5.1 What Does the Psychiatrist Usually do?

The routine reflex of us psychiatrists is to increase the medication. But we should look more closely at this. It can have two faces (Hartwich and Grube 2014):

1. The psychosis has been re-intensified; then the routine reflex of increasing the medication is correct.
2. The patient who is stabilized can better talk about his psychotic experiences. He is now able to communicate his bizarre experiences, which are not new. It is therefore a stage in which patients are able to report more about their prior symptoms. Here the increase in medication is not indicated. On the contrary, if the dose is increased, the patient will not tell you anything more. His need to express himself, to communicate his experience, and to find an anchor in the therapist, who holds him psychodynamically, would be terminated abruptly.

## 26.6 Relatives and Psychopharmaceuticals

If a family member experiences a severe psychiatric illness, most relatives experience the use of psychopharmaceuticals as a relief. There are, however, also relatives who regard the taking of psychopharmaceuticals as an admission that it is something morbid in the behavior of the family. This can go so far that the mother or the spouse turns against the taking of medication and instead demands only psychological treatment. In rare cases even the partner is involved in the psychotic experience and reacts similarly to a folie à deux. In such cases, a common defense strategy of the patient and the relative, where the drugs are regarded as a common opponent, occurs. This opponent emphasizes the fact of the illness, which one gladly displaces from the consciousness. If such a chemical treatment is rejected, then one also denies the fact of a serious illness and the risk of a relapse.

The helpless family member wants to do something for the sick partner. He can now become active and rebel against chemistry, against an institution, or against medical treatment. He then feels particularly close and helpful to his ill partner. In the interactional relationship between the sick and the healthy family member, the feeling of connectedness is enhanced. In some cases, the sick partner will, consciously or unconsciously, provoke such behavior in the healthy person, in order to make him more helpful to him.

There are cases where relatives at first project too much expectation into the effects of medication. Later on it comes to the contrary, namely, an enthusiastic devaluation, for example, they say: "It is not enough, nothing is done right." With relative groups and family counseling, the relatives manage to take care of the medication and become a "treatment partner."

### 26.6.1 Treatment Partnership

The need for medication is accepted. Together in the family, the concern for the drug intake is borne. In the case where in the outpatient treatment the sick partner shows new symptoms at home, which require a variation of the dosage, the knowledgeable relatives will ensure that the appropriate measures take place as early as possible. This will be a preventive support for the sick partner. In many cases, such a favorable constellation can greatly promote the willingness for psychodynamic treatment.

In the case of chronic depression and longer-lasting partnership, taking care of medication may also have a substitute function. If interaction with the partner is blocked because of lack of drive and withdrawal, then monitoring and allocation become the vehicles of communication.

## 26.7 Group Therapy and Psychopharmaceuticals

In regular outpatient and inpatient group psychotherapy with psychiatric patients, a remarkable phenomenon occurs with regard to the attitude toward medication: an intensifying effect occurs. In group psychotherapy, it is always impressive with how well the patients become experts in dealing with their disease. Psychiatric patients, who have developed positive experiences with psychopharmaceuticals over the course of time following initial skepticism, pass on this knowledge to the other group members, especially new members. In the group psychotherapy sessions, it is then shown that adherence to a well-balanced medication of older and longer illness-experienced members is strongly advocated. Consequently, those who are indecisive or skeptical are gradually taking this position. Within the group, we can also learn what we called the "early warning principle" (Hartwich and Schumacher 1985). The extent to which the flare of a psychotic relapse can be detected at an early stage is often discussed in detail in group sessions. In particular, the following symptoms usually are carefully discussed:

- Sleep disorders
- Restlessness
- Delusional ideas
- Withdrawal
- Individual symptom accentuation

Experienced psychiatric patients in the group report that they are able to adapt the dose of their maintenance drug or their supplementary medication somewhat more prophylactically in such early warning symptoms. This is not a general self-medication, but it is about the temporal bridging until the visit to the treating psychiatrist. Experienced patients carry on-demand medicines, for example, on vacation. They are then able to act on the first signs themselves. For example, in sleep disorders, one of the most common early warning symptoms, low-dose antipsychotics and, in some cases, sleeping pills are taken.

## 26.8 Top-Down and Bottom-Up Effects of Psychotherapy and Pharmacotherapy in the Treatment of Depression

The former dichotomy in the assessment of the effects of psychotherapy and pharmacotherapy (according the motto "psychotherapy has influences on the psyche; pharmacotherapy has influences on the body") could be overcome in the meanwhile, not at last because of the results of multidimensional studies in depression. Due to neuronal plasticity, it can be presupposed that the neuronal network structure may be modified during the whole life. Because of the circular relationships between neurobiological, psychological, and social factors, it is possible to treat mental disorders by means of therapeutical interventions which either have an influence on subjective interpretations and validation (in order to modify negative self-concepts or dysfunctional relationship patterns) or which have a direct influence on the functionality of those neuronal networks, leading to the continuation of conflictuous processing and validation of one's own actions and the resulting disturbed social relationships. Thus psychopharmacotherapy and psychotherapy form different brick stones of a multimodal treatment concept. A special challenge consists in the adequate weighting of the single components in different disorders regarding the individual problematics of each single patient.

Psychopharmacological drugs change the functionality of the brain by intervening in those normally occurring partial processes of the intercellular communication in a more or less specific way and with different sustainability. The high density of neuronal networks and their interdependency cause each intervention in a specific partial process to contribute to a complex chain of secondary and further reactions and counterreactions. Individual differences in single partial systems of the central nervous system may result in different effects and side effects of the same substance from person to person. Last but not least, the subjective validation (i.e., positive or negative expectation) has a strong impact whether a drug is effective and which of the drug-induced neuronal processes are supported or suppressed. Especially in case of treatment with antidepressants, the importance of the so-called placebo effect may not be underestimated: In comparison studies with active substances, the placebo effect explains up to 80 percent of the efficacy (cf. Stassen et al. 2007; Szegedi et al. 2009).

It may be assumed that the serotonergic system functions as a globalizing transmitter system due to its widespread and multiple connections in the whole brain and its tonic activity. Thus, it contributes to harmonizing of the activities which are generated in different regional networks of the brain (cf. Boeker 2011a, b, 2017).

Also the axons of the noradrenergic system are widespread and have multiple connections. The original neurons of the noradrenergic system are located in the core nuclei (nucleus coeruleus) and in noradrenergic core regions of the brain stem. Contrary to the serotonergic system, its fire rate is not tonic but phasic; accordingly activations are induced when something new is percepted. By means of this, the production and depletion of astrocytic growth factors increase, which themselves serve as triggers for neuroplastic reactions. Accordingly noradrenalin has an indirect effect on the secondary cascade of neurons, and there it supports the reorganization and channeling processes.

It may be hypothesized that the effect of antidepressants is in all probability no specific causal therapy of the assumed biological factor. Rather one may start from the assumption that their effect is not based directly on the increase of the synaptic transmitter concentration but on secondary adaptation reactions of the receptors on the increased transmitter supply (so-called downregulation). By means of antidepressants, adaptive processes of neuromodulation are induced (e.g., induction of the cortisol level, increase of tCREB (cf. Aldenhoff 2000)). Obviously antidepressants are able to modulate parts of the disastrous, negative vicious circles in the somatic dead end of the feedback loops of the regulation of the mood drive systems, emotion regulation, or the emotional-cognitive interaction (cf. Boeker 2011a, b).

Among others results of psychopharmacotherapy studies point to this system-theoretical or systembiological view of the unspecific effect of antidepressants (Stassen et al. 2007; Szegedi et al. 2009). It was shown that the beginning of the effect of antidepressants is independent from the substances, each of which was the applied. All of the antidepressants, which were investigated until now, showed a clinical effect on day 14, which differed significantly from the placebo effect. Because of the similar time course of recovery of all investigated antidepressants, it may be concluded that they do not have a specific antidepressant effect but that they may induce a trigger effect, which finally contributes to the adaptation of the neurobiological system.

In view of the effectiveness of psychotherapy in depressed patients, it also can be hypothesized that new patterns of using the neuronal structure and networks are established and that the existing neuroplasticity of the brain can be activated by means of psychotherapy. In this way priming and reorganization processes are induced and stabilized on the level of those neuronal networks, which were involved in the thinking, feeling, and acting of the respective patient till then (cf. Boeker 2011a, b, 2017). By means of a complementary drug treatment, the neuroplasticity of the brain may be supported and reactivated additionally. Through early and adequate psychotherapeutical interventions, the chance increases in all likelihood that the specific priming processes in the course of a mental disorder may be reduced and that the reduction of neuroplasticity and reorganization capability of neuronal networks of the respective patient can be avoided.

In the following the question is discussed, how psychotherapy influences the neuronal functionality in depressed patients. To answer this question, studies are consulted, which compared the neuronal activity and the neuronal metabolism under pharmacotherapy and psychotherapy. Goldapple et al. (2004) found changes of the regional glucose metabolism (by means of PET) during successful treatment with cognitive-behavioral therapy and with the serotonin reuptake inhibitor paroxetine. Psychotherapy (CBT) led to an increased metabolism in the hippocampus and dorsal cingulum (BA 24) and to a decreased metabolisms in the dorsal (BA 9/46), ventral (BA 47/11), and medial (BA 9/10/11) frontal cortex.

In contrast a frontal increase and a decrease of the metabolism in the hippocampus subgenual cingulum were found during the treatment with the antidepressant (paroxetine). Accordingly therapy-specific, inverse effects can be assumed in the treatment of depressed patients by means of psychotherapy and pharmacotherapy. Cognitive-behavioral therapy led to an increase of metabolism in the hippocampus and to a decrease of metabolism in the frontal cortex, whereas the antidepressant led to a decrease of metabolism in the hippocampus and an increase of metabolism in the frontal cortex. "Top-down effects" are mediated by means of psychotherapy and "bottom-up effects" by means of psychopharmacotherapy.

Seminowicz et al. (2004) studied the effects of different therapeutical interventions in patients with major depression on the level of the limbic-cortical networks. Depressive phenotypes were characterized on the neuronal system level: antidepressant responders were characterized by the activation of limbic-cortical networks (LPFC-subgenual cingulum-OMPFC-hippocampus), on the contrary, antidepressant nonresponders showed additional disturbances of limbic-cortical networks (anterior thalamus-AC-subgenual cingulum-OMPFC-hippocampus).

CBT responders were characterized by an activation of limited limbic-cortical networks (hippocampus-LPFC) and cortical-cortical networks (OMPFC, OF11—MF10). On the basis of these depressive phenotypes, the authors suggested neuronal-based algorithms for the development of individual treatment of depressed patients.

Functional neuronal correlates of the response and the delayed response using antidepressants had been already described before by Mayberg et al. (1999, 2000). Those patients, who had responded to the antidepressive treatment, did not only show a mere "correcteur" of the diagnosed functional disturbances before beginning of the treatment but were characterized by a complex adaptational process, which was accompanied by a normalization of the cortical hypometabolism and changes of specific subcortical and subcortical regions of the brain, in which no metabolic disturbances had existed before (limbic-striatal decrease of glucose metabolism in the subgenual cingulum, hippocampus, insula, and pallidum; dorsal cortical increase in PFC and parietal, anterior, and posterior cingulum; increase of glucose metabolism in the brain stem).

A delayed response was characterized by a lack of changes of the cerebral glucose metabolism in the subgenual cingulum and the PFC. The adaptations of the early metabolic changes were lacking in the further course (e.g., in week 6).

As a summary top-down effects of effective psychotherapeutical treatment (CBT) and bottom-up-effects of psychopharmacological antidepressant treatment (e.g., with paroxetine) could be differentiated by means of changes in the regional glucose metabolism (cf. Goldapple et al. 2004). In the comparison of pharmacological and psychotherapeutical treatment of depression, therapy-specific, inverse effects were found: an increase of the glucose-metabolism in the hippocampus under psychotherapy (CBT) and a decrease of glucose metabolism in the frontal cortex. On the contrary, glucose metabolism in the hippocampus is decreased during an antidepressant treatment (e.g., with paroxetine) and is increased in the frontal cortex.

Meanwhile, the anterior cingulate (Cg 25) is acknowledged as a key region for therapy response and prediction in depression treatment; neurophysiological adaptations in the region of the ACC were found during antidepressant treatment and cognitive-behavioral treatment and in view of the placebo effect as well (Ressler and Mayberg 2007).

Neuronal adaptation processes were also found under psychodynamic and psychoanalytical psychotherapy. Viinamäki et al. (1998) found changes in the serotonin receptor density, which were similar to those under psychopharmacological treatment, after a 1-year psychodynamic psychotherapy.

In a further study of Lehto et al. (2008) in depressed patients (8 patients with atypical depression and 11 patients with "non-atypical" depression), a significant symptom reduction was found in all patients. What was especially remarkable, changes in the serotonin receptor density in the midbrain were found in patients with atypical depression but not in those with "non-atypical" depression. Last but not least, this result is interesting because atypical depression represents the only psychiatric disorder, for which operationalized definition of the bonding dimension, especially dependency on others, is considered. The result underlines the neuronal effects of psychodynamic psychotherapy, especially in the context of working through of conflictuous relationship patterns (dependency) in patients with atypical depression.

Furthermore, a modulation of neuronal activation patterns in limbic-frontal networks was found in depressed patients who were treated with psychoanalytical psychotherapy (so-called HANSE study, Buchheim et al. 2012).

An integrative, neuropsychodynamic approach considers that the respective therapeutical interventions represent different referential systems and epistemological dimensions. The reflection of the possibilities and limitations of the respective method is connected with therapeutical consequences (e.g., possible change or combination of interventions). Nevertheless, in case of a combination therapy, the cooperation of the therapists is of great importance.

This kind of an integrative perspective in the treatment of depression considers among others the personality and comorbid personality disorders. The psychopharmacological treatment is involved in action dialogues in the framework of the therapeutical relationship (e.g., using a drug underlines the asymmetry of the therapeutical relationship, may activate conflicts of dependency and autonomy, or may be experienced as a "transitional object").

The results of psychotherapy research stress the complementarity of psychotherapy and psychopharmacotherapy and, furthermore, the possibility of an alternate application in different subgroups of depressed patients. The differential indication has always been related to the single, individual patient.

The comparison studies (pharmacotherapy versus psychotherapy) indicate that psychotherapy has equal effects in mild and moderate depression and, last but not least, has also a lower rate of therapy breaks. Psychopharmacotherapy is more effective in severe depression (e.g., in psychotic depression) in the short-time course. Combination therapy is more effective in severe depression in the long-time course.

The planning of the treatment presupposes a necessary overcoming of traditional one-sided etiological concepts, e.g., the implicit assumptions of the former triadic model: the indication for pharmacotherapy is given in "endogenous" depression, and the indication for psychotherapy exists in "psycho-reactive" depression. Beyond these former dichotomies rather it can be started from the knowledge that psychotherapy also causes neurobiological changes, whereas psychopharmacotherapy of depression mostly includes further therapeutical dimensions (supportive and psychoeducative elements). A neuropsychodynamic approach to depression here (Boeker et al. 2015) focuses on the understanding of the circular effects of biological-neurobiological, psychodynamic, and social factors of depression ("depression as a somato-psychic-psychosomatic disorder of emotion regulation"; cf. Boeker 1999, 2011a, b). This contributes essentially to a more holistic concept which takes the neurobiological effects of pharmacotherapy and psychotherapy as well into account. Different steps for the development of such an integrative concept were taken during the last decades:

1. Genetics/Epigenetics: interaction of indirect and direct biological changes (Gabbard 2000, 2014). This approach underlines that psychotherapy modulates biological processes indirectly (by activating genetic material). On the other hand, antidepressants change biological reaction patterns directly.
2. Correlation between traumatic experiences in childhood and youth and changes of neuronal functions: different etiologically relevant factors which occur during different times of the pathogenesis (Mentzos 1991). Psychotherapy focuses on the experience-based vulnerability, while antidepressants focus on biological disturbances.
3. Common final pathway of the effects of psychotherapy and psychopharmacotherapy: overlap of neurobiological mechanisms of psychotherapy and antidepressants (Mayberg et al. 2000; Goldapple et al. 2004).
4. Psychopathological model of the structural dynamics of the effects of psychopharmacotherapy (Janzarik 1988): Janzarik described in his psychopathological model the effects of drugs on the dynamics of the personality, which enables a new consolidation of the personality structure. With this he underlines that drugs do not intrude upon the personality structure, but that psychopharmacotherapy may principally only have an impact on the "dynamic derailment" (Janzarik 1988, p. 26).
5. Psychodynamic model (Mentzos 1991, 1995): Mentzos described in his psychodynamic concept of psychotherapy of psychoses the effects of pharmacotherapy especially in the context of a necessary reduction of the emotional excitation. The consequence of this reduction is that psychotic mechanisms of defense are no longer necessary and moreover that a chemically induced ego restriction limits the perceptive faculty, i.e., psychopharmacotherapy contributes to a chemical substitute of defense mechanisms (cf. Küchenhoff 2010). The ego gets less permeable resulting in a recompensation, which then enables psychotherapeutical interventions again.
6. Neuropsychodynamic model (Boeker et al. 2015): Considering multiple psychodynamic and conlictuous elements on the one hand side and the phenomena of depressive inhibition on the other hand side, this neuropsychodynamic model of depression especially underlines the importance of the increased resting state activity on a neurobiological process level. The reduction of the hyperarousal (by means of psychotherapy and by means of psychopharmacotherapy as well) finally enables the development of functional mechanisms of compensation.

## 26.9 Neuromodulation

Neuropsychodynamic psychiatry and neurostimulation: how does that fit together?

Neuropsychodynamic psychiatrist should keep in mind the severity and chronic course of many psychiatric disorders and the chances and limitations of the existing therapies as well. That is, many patients with MDD profit little if anything even after many years of treatment. Between 10 and 30% of depressed patients taking antidepressants are partially or totally resistant to treatments (Rush et al. 2006).

Even electroconvulsive therapy (ECT), which is very effective in severe, especially psychotic depression and in catatonia and also may be very helpful as maintenance therapy, has a limited effect in these depressed patients who are resistant to treatments.

Four about two decades, different modalities of targeted neuromodulation (e.g., vagus nerve stimulation (VNS), deep brain stimulation (DBS)) are being actively researched as putative treatment options for treatment-resistant forms of depression. Recently, promising results on the use of DBS in small series of patients with treatment-resistant depression have been reported for four different stimulation targets: the subgenual cingulate gyrus (Lozano et al. 2008), the interior limb of the capsula interna (Mallone Jr et al. 2009), the nucleus accumbens (NAcc, Bewernick et al. 2010), and the medial forebrain bundle (Schlaepfer et al. 2013).

Focusing on the role of the brain's reward regions in motivated behavior, depression, and antidepressant treatment, converging evidence has identified the NAcc and its dopaminergic inputs from the ventral tegmental area (VTA) of the midbrain as one of the important anatomical substrates for drug reward as well as for natural rewards (Nestler and Carleson Jr 2006). This reward pathway evolved to promote activities that are essential to the survival of the species and consists of core structures: the NAcc, the VTA, the ventromedial and lateral nuclei of the hypothalamus, and the amygdala (cf. Schlaepfer et al. 2013). These core structures are interconnected through the medial forebrain bundle (MFB, cf. Zellner et al. 2011).

It had been speculated that the reward system might be partly dysfunctional in major depression (Blood et al. 2010; Martin-Soelch 2009).

Schlaepfer et al. (2013) showed that DBS to the human reward system in closer proximity to the VTA is efficacious in decreasing ratings of depression by bilaterally stimulating the superolateral branch of the medial forebrain bundle (slMFB), a structure with proven convergence onto the prefrontal cortex. During intraoperative test stimulation, all patients showed strikingly similar signs of appetitive motivation (e.g., orientation, reaction, initialization of eye contact, engaging in conversation) and mood improvement. No indication of hypomania or anxiety was observed. A rapid reduction of depression ratings was found in six of seven patients after 2 days of stimulation, and after 1week, four of seven patients had reached the response criterion (50% reduction in MADRAS). At last observation (12–33 weeks), six of seven patients were responders, among them, four were classified as remitters.

The most prominent adverse events were blurred vision and strabismus, which occurred in all patients at higher amplitudes when specific electrode contacts were activated. Oculomotor problems were a limiting factor for parameter changes.

The study of Schlaepfer et al. (2013) demonstrated that DBS to the NAcc has acute anti-anhedonic and longer-term antidepressant effects.

Because the slMFB is a structure that cannot be identified with conventional MRI, it has to be individually mapped by means of deterministic diffusion tensor imaging (with the use of an area lateral of the VTA as seed region, which allows displaying the projections of the slMFB through the NAcc to the prefrontal cortex).

Unexpectedly, a rapid onset of antidepressant response was found, as described above. Comparably instant effects in major depression had only been demonstrated in studies with the use of ketamine (Berman et al. 2000; Zarate et al. 2006) before. In contrast to the time-limited effect of ketamine (and sleep deprivation), the effects of slMFB DBS seem to be enduring. It remains open so far whether slMFB DBS exerts its antidepressant effect in combination with other treatments or if it would be effective alone.

**Conclusion**

The question "What helps whom?" cannot be answered sufficiently until now, because of the lack of valid diagnostic, therapeutic, and prognostic parameters. Nevertheless, top-down and bottom-up mechanisms represent essential fundaments for tailoring of individualized treatments (e.g., in depression). The reduction of the increased resting state activity in depression, through pharmacotherapy or psychotherapy, for instance, enables the development of more functional coping mechanisms. Aiming at developing an integrative perspective, the keen elements of neuropsychiatric psychiatry and its circular concept of mental disorders may enable to "think together" what often is separated from one another: psychopharmacotherapy, psychotherapy, and neuromodulation. Further therapy research in all these fields, including individualized research paradigms, may open the door for the development of new illness models and more efficacious therapeutic strategies. We are convinced that these future developments will underline the significance of the therapeutic relationship.

## References


Aderhold V, Weinmann S, Hägele C, Heinz A. Frontale Hirnvolumenminderung durch Antipsychotika? Nervenarzt. 2014;86:302–23.

Aldenhoff J. Biologische Veränderungen bei der Psychotherapie der depression. Nervenarzt. 2000;50:415–9.

Benedetti G. Psychotherapeutische Behandlungsmethoden. In: Kisker KP, et al., editors. Schizophrenien. Psychiatrie der Gegenwart, vol. 4. Berlin: Springer; 1987. p. 285–323.

Berman RM, Capiello A, Anand A, Oren DA, Heninger GR, Charney DS, et al. Antidepressant effects of ketamine in depressed patients. Biol Psychiatry. 2000;47:351–4.

Bewernick BH, Hurlemann R, Matusch A, Kayser S, Gubert C, Hadrysiewilz B, et al. Nucleus accumbens deep brain stimulation decreases ratings of depression and anxiety in treatment-resistant depression. Biol Psychiatry. 2010;67:110–6.

Blood AJ, Iosefescu DV, Makris N, Perlis RH, Kennedy EN, Daugherty DD, et al. Microstructural abnormalities in subcortical reward circuitry of subjects of subjects with major depressive disorder. PLoS One. 2010;5:e13945.

Boeker H. Selbstbild und Objektbeziehungen bei Depressionen: Untersuchungen mit der Repertory Grid-Technik und dem Gießen-Test an 139 PatientInnen mit depressiven Erkrankungen. In: Monographien aus dem Gesamtgebiete der Psychiatrie. Darmstadt: Steinkopff-Springer; 1999.

Boeker H. Psychotherapie der Depression. Bern: Huber; 2011a.

Boeker H. Ergebnisse der Therapie- und Psychotherapieforschung. In: Psychotherapie der Depression. Bern: Huber; 2011b. p. 195–212.

Boeker H. Psychotherapie und Pharmakotherapie bei depressiv Erkrankten. In: Küchenhoff J, editor. Psychoanalyse und Psychopharmakologie. Stuttgart: Kohlhammer; 2016. p. 131–48.

Boeker H. Psychodynamische Psychotherapie der Depression. Theorie und Praxis. Giessen: Psychosozial-Verlag; 2017.

Boeker H, Hartwich P, Northoff G, editors. Neuropsychodynamische Psychiatrie. Heidelberg: Springer; 2015.

Buchheim A, Viviani R, Kessler H, Kächele H, Cierpka M, Roth G, George C, Kernberg OF, Bruns G, Taubner S. Changes in prefrontal-limbic function in major depression after 15 months of long-term psychotherapy. PLoS One. 2012;7(3):e33745.

Gabbard GO. A neurobiologically informed perspective on psychotherapy. Br J Psychiatry. 2000;177:117–22.

Gabbard GO. Psychodynamic psychiatry in clinical practice. 5th ed. Washington DC: American Psychiatric Publishing; 2014.

Goldapple K, Segal Z, Garson C, Lau M, Bieling P, Kennedy S, Mayberg H. Modulation of cortical-limbic pathways in major depression. Arch Gen Psychiatry. 2004;61:34–41.

Hartwich P, Grube M. Psychotherapie bei Psychosen. Neuropsychodynamisches Handeln in Klinik und Praxis. 3rd ed. Berlin: Springer; 2014.

Hartwich P, Schumacher E. Zum Stellenwert der Gruppenpsychotherapie in der Nachsorge Schizophrener. Eine 5-Jahres-Verlaufsstudie. Nervenarzt. 1985;56:365–72.

Janzarik W. Strukturdynamische Grundlagen der Psychiatrie. Stuttgart: Enke; 1988.

Jessner M. Antipsychotika und Unbewusstes. In: Küchenhoff J, editor. Psychoanalyse und Psychopharmakologie. Stuttgart: Kohlhammer; 2016. p. 149–66.

Kohut H. Narzißmus. Frankfurt: Suhrkamp; 1973.

Küchenhoff J. Zum Verhältnis von Psychopharmakologie und Psychoanalyse – am Beispiel der Depressionsbehandlung. Psyche – Z Psychoanal. 2010;64:890–916.

Küchenhoff J. Die analyse der therapeutischen Beziehung und die Psychopharmakotherapie. In: Küchenhoff J, editor. Psychoanalyse und Psychopharmakologie. Stuttgart: Kohlhammer; 2016. p. 13–33.

Lehto SM, Tolmunen T, Joensuu M, Saarinen PI, Valkonen-Korhonen M, Vnninen R, Ahola P, Tihonen J, Kuikka J, Lehtonen J. Changes in midbrain serotonin transporter availability in atypically depressed subjects after one year of psychotherapy. Prog Neuro-Psychopharmacol Biol Psychiatry. 2008;32(1):229–37.

Lozano AM, Mayberg HS, Giacobbe P, Hamami C, Craddock RC, Kennedy SH, et al. Subcallosal cingulate gyrus deep brain stimulation of treatment-resistant depression. Biol Psychiatry. 2008;64:46–467.

Mallone DA Jr, Dougherty DD, Rezai AR, Carpenter LC, Friehs GM, Escander EN, et al. Deep brain stimulation of the ventral capsule/ventral striatum for treatment-resistant depression. Biol Psychiatry. 2009;65:267–75.

Martin-Soelch C. Is depression associated with dysfunction of the central reward system? Biochem Soc Trans. 2009;37:313–7.

Mayberg HS, Liotti M, Brannan SK, McGinnis S, Mahurin RK, Jerabek PA, Silva JA, Tekell JL, Martin CC, Lancaster JL, Fox PT. Reciprocal limbic-cortical function and negative mood: converging PET findings in depression and normal sadness. Am J Psychiatry. 1999;156:675–82.

Mayberg HS, Brannan SK, Tekell JL, Silva JA, Mahurin RK, McGinnis S, Jerabek PA. Regional metabolic effects of fluoxetine in major depression: serial changes and relationship to clinical response. Biol Psychiatry. 2000;48:830–43.

Meißel T. Psychodynamik der Medikation bei schizophrenen Psychosen. In: Schwarz F, Maier C, editors. Psychotherapie der Psychosen. Stuttgart: Thieme; 2001. p. 48–52.

Mentzos S. Psychodynamische Modelle in der Psychiatrie. Göttingen: Vandenhoeck and Ruprecht; 1991.

Mentzos S. Depression und Manie; Psychodynamik und Therapie affektiver Störungen. Göttingen: Vandenhoeck & Ruprecht; 1995.

Mentzos S. Psychoanalyse der Psychosen. Psychother Dialog. 2002;3:223–34.
Muench A. Psychodynamische Modelle der Medikamentenapplikation und ihre Synergien mit der Psychotherapie. In: Küchenhoff J, editor. Psychoanalyse und Psychopharmakologie. Stuttgart: Kohlhammer; 2016. p. 91–108.
Nestler EJ, Carleson WA Jr. The mesolimbic dopamine reward circuit in depression. Biol Psychiatry. 2006;59:1151–9.
Ressler KJ, Mayberg HS. Targeting abnormal neural circuits in mood and anxiety disorders: from the laboratory to the clinic. Nat Neurosci. 2007;10(9):1116–24.
Rush AJ, Trivedi MH, Wisniewski SR, Nierenberg AA, Stewart JW, Warden D, et al. Acute and longer-term outcomes in depressed outpatients requiring one or several treatment steps: a STAR*D report. Am J Psychiatry. 2006;163:1905–17.
Schlaepfer TE, Bewernick BH, Sayser S, Mädler B, Koenen V. Rapid effects of deep brain stimulation for treatment-resistant Major Depression. Biol Psychiatry. 2013;73:1204–12.
Seminowicz DA, Mayberg HS, McIntos AR, Goldapple K, Kennedy S, Segal Z, Rafi-Tari S. Limbic-frontal circuitry in major depression: a path modeling metanalysis. NeuroImage. 2004;22:409–18.
Stassen HH, Angst J, Hell D, Scharfetter C, Szegedi A. Is there a common resilience mechanism underlying antidepressant drug response? Evidence from 2848 patients. J Clin Psychiatry. 2007;68:1195–205.
Szegedi A, Jansen WT, van Willigenburg AP, van der Meulen E, Stassen HH, Thase ME. Early improvement in the first 2 weeks as a predictor of treatment outcome in patients with major depressive disorder: a meta-analysis including 6562 patients. J Clin Psychiatry. 2009;70:344–53.
Viinamäki H, Kuikka J, Tiihonen J, Lehtonen J. Change in monoamine transporter density related to clinical recovery: a case-control study. Nord J Psychiatry. 1998;52:39–44.
Vollmoeller W. Psychodynamik in der medikamentösen Therapie. In: Vollmoeller W, editor. Integrative Behandlung in Psychiatrie und Psychotherapie. Stuttgart: Schattauer; 2003.
Zarate CA Jr, Singh JB, Carlson PJ, Brutsche NE, Ameli R, Luckenbaugh DA, et al. A randomized trial of an N-methyl-D-aspartate antagonist in treatment-resistant major depression. Arch Gen Psychiatry. 2006;63:856–64.
Zeek A. SSRI in der psychodynamischen Psychotherapie der Bulimia nervosa. In: Küchenhoff J, editor. Psychoanalyse und Psychopharmakologie. Stuttgart: Kohlhammer; 2016. p. 187–98.
Zellner MR, Watt DF, Solms M, Panksepp J. Affective neuroscientific and neuropsychoanalytic approaches to two intractable psychiatric problems: why depression feels so bad and what addicts really want. Neurosci Rev. 2011;35:2000–8.

# Psychotherapy Research in the Context of Neuroscience

# 27

Heinz Boeker and Georg Northoff

**Abstract**

The chapter focuses on the essentials of psychotherapy research in the context of neuroscience and the question of how the mental correlates of psychotherapy, especially of psychodynamic psychotherapy, can be investigated.

The psychoanalytic approach aims at enabling the patient to "remember, repeat and work through" concerning the explicit memory. Moreover, the relation between analyst and patient establishes a new affective configuration which enables a reconstruction of the implicit memory. The resulting psychic changes correspond to neuronal changes.

The variables that need to be controlled and measured in individualized neuroimaging will be discussed.

Two main methodological problems can be distinguished: The design problem addresses the issue of how to account for functionally related variables in an experimentally independent way. The translation problem raises the question of how to bridge the gaps between different levels of the concepts presupposed in individualized neuroimaging (e.g. the personal level of the therapist and the client, the neuronal level of the brain).

This chapter is based on an article which was published in *Frontiers in Human Neuroscience*: Boeker et al. (2013).

H. Boeker (✉)
Department of Psychiatry, Psychotherapy, and Psychosomatics, Psychiatric University Hospital Zurich, University of Zurich, Zurich, Switzerland
e-mail: heinz.boeker@bli.uzh.ch

G. Northoff
Mind, Brain Imaging, and Neuroethics,
Institute of Mental Health Research, University of Ottawa, Ottawa, ON, Canada
e-mail: georg.northoff@theroyal.ca; www.georgnorthoff.com



H. Boeker et al. (eds.), *Neuropsychodynamic Psychiatry*,
https://doi.org/10.1007/978-3-319-75112-2_27

An overview of the individualized paradigms which have been used until now will be given, including the Operationalized Psychodynamic Diagnostics (OPD-2) and the Maladaptive Interpersonal Patterns Q-Sort (MIPQS). The development of a new paradigm which will be used in fMRS experiments, the "Interpersonal Relations Picture Set" (IRPS), will be described. Further perspectives and limitations of this new approach concerning the design and the translation problem will be discussed.

## 27.1 Introduction

The recently emerged dialogue between psychoanalysis and neuroscience (Beutel et al. 2003; Kandel 1999; Northoff 2007; Northoff et al. 2007; Solms et al. 1998) led to several empirical hypotheses and investigations of psychodynamic concepts like defence mechanisms (Northoff 2007; Boeker et al. 2006, 2013), self (Milrod 2002), memories (Gabbard 2000; Mancia 2006; Peres et al. 2008), dreams (Andrade 2007; Solms 1995/2000) and empathy (Gallese et al. 2007). While these originally psychodynamic concepts are currently investigated in the neuroscientific context, the neuronal basis of one core element of psychoanalysis and psychodynamic psychotherapy has not been elucidated yet. Though neurobiological changes in some single cases undergoing psychodynamic psychotherapy have been reported (Lai et al. 2007; Lehto et al. 2008; Overbeck et al. 2004; Saarinen et al. 2005; Viinamäki et al. 1998), systematic and well-controlled brain imaging studies of the neuronal effects of psychodynamic psychotherapy are still lacking.

In contrast to psychodynamic psychotherapy, the neuronal effects of other forms of psychotherapy like cognitive behavioural therapy (CBT) and interpersonal therapy (IPT) have been studied in brain imaging more often (see Beauregard 2007a, b; Frewen et al. 2008; Linden 2006; Roffman et al. 2005 for reviews). These studies demonstrated neuronal modulation in various brain regions encompassing subcortical and medial and lateral cortical regions during predominantly cognitive-emotion regulation tasks before and after CBT or IPT. Interpretation of these findings is however constrained by various methodological problems; these include issues like objectification and quantification of the effects of psychotherapy in behavioural and subjective parameters; selection of the activation task in functional imaging; appropriate control groups; physiological, behavioural and psychological variables indicating task-specific effects of neuronal stimulation; distinction between the target symptom and its possible underlying psychodynamic processes; etc. (see Frewen et al. 2008 for a detailed discussion).

While brain imaging studies of both CBT and IPT are already confronted with numerous methodological problems, the situation is even more difficult in the case of psychodynamic psychotherapy. For instance, the therapeutic relationship, including transference and countertransference, plays a much more essential role in psychodynamic psychotherapy than in CBT and IPT; this makes it necessary to include the client-therapist relationship as an intervening variable in neuronal analysis.

Another problem is the conceptualization of the psychodynamic phenomena like ego, defence mechanisms, etc. and their translation into psychological variables for subsequent experimental testing in functional brain imaging. The neuropsychoanalyst who wants to study the neuronal effects of psychodynamic psychotherapy is thus confronted with numerous and highly complex input variables that he needs to account and control in order to make reliable and valid investigation of the output, the neuronal effects, possible.

## 27.2 Essentials of Psychoanalytic Process and Change

To this end it is necessary to describe and characterize the essential and specific aspects that account for the process and change of a patient during a psychoanalysis or psychoanalytic psychotherapy. This could be the basis for the development of meaningful research designs and paradigms. The main questions in this respect are:

- What is the process and what is changed within a psychoanalysis or psychoanalytic psychotherapy?
- Which are the mechanisms, techniques and actions that enable psychoanalytic process and change?

Patients mostly seek psychotherapy because of distress, i.e. they suffer from psychic symptoms, from dysfunctional behaviours and/or from disturbances in their psychosocial environment (interpersonal problems, in relationships, at work, etc.) with the intention to reduce and resolve the distress. Often patients also aim to achieve a greater self-understanding. Others wish to be supported emotionally and personally or receive guidance and instructions from the therapist for resolving their problems.

The specific aspects concerning the therapeutic process and change in psychoanalysis and psychoanalytic psychotherapy address not only symptoms and dysfunctional behaviours. Another objective is to find out what may lie behind the symptoms and dysfunctional behaviours. This is connected to the fundamental psychoanalytic concept that conscious, so-called "manifest" symptoms, thoughts and actions of the patient imply an unconscious "latent" meaning and motivation. Within a psychoanalytic perspective, conscious symptoms and disturbances are assumed to be the result of mechanisms of defence and formations of compromise, which deal with multiple preconscious or unconscious factors. Such dynamics have a strong impact on how one thinks, feels and behaves. Preconscious or unconscious factors may constitute intrapsychic conflicts or dilemmas, wishes, expectations, fantasies or structural psychic functions (super-ego, ego and id, self- and object representations, capacities to regulate affects, impulses, self-esteem, relationships with others, etc.).

Consequently, a basic psychoanalytic approach to enable a therapeutic process is to generate and foster a patient's insight into and understanding of these preconscious or unconscious aspects and parts of him- or herself. This is to make conscious what had been unconscious before, which is part of what Freud (1933) wrote:

"Where id was, there shall ego be". A fundamental psychoanalytic technique during sessions is to ask patients to report about "what comes into their heads, even if they think it is unimportant, irrelevant or nonsensical", which was called by Freud "free association". Another approach is the patient's report of dreams and the associations to them.

On the psychoanalysts' part, the correspondent technical approach is a special form of listening ("evenly suspended attention"), the use of clarification and interpretation and the formulation of hypotheses on how the patient functions mentally to establish links to unconscious conflicts and aspects, something the patient cannot perceive on his/her own and/or accept as being connected with his/her conscious thinking and current-day functioning. Reclining on the couch, clients might remember better and feel more freely to associate. However, against the patient's free association, the building of links with unconscious aspects and gaining insight in oneself, resistances and transference come into play—which both can build the grounds for interpretations of the analyst. The psychoanalytic approach aims at enabling the patient to "remember, repeat and work through" (Freud 1914) what has been experienced in the past, repressed or internalized. Psychic difficulties are not yet resolved to understand the underlying mechanisms of connection. Classical interpretation and insight may be the start of a reorganization of thoughts—the former preconscious may become conscious.

Another focus of psychoanalysis is on the patient's childhood experiences and relations with significant others (mother, father, siblings, etc.) and the impact these important childhood relations had and still have on a person's life. This can be seen in current relationships of the patient in the here-and-now with important persons or the analyst. Beutel (2009) gives a summary of effects that early childhood interpersonal experiences have on cerebral development through genetic expression and the development of neuronal connections.

It was Freud's conception of the transference of the patient and later on by his followers that of countertransference of the analyst—first seen as obstacles to the therapy process—that were further developed to the second basic approach in psychoanalysis and psychoanalytic psychotherapy: the focus is here on what is happening in the therapeutic relationship on the basis of transference and countertransference. Both are established in the therapeutic relationship and have to be analysed and understood. Dysfunctional, maladaptive relationship patterns, fears and wishes in relationships of the patient tend to be repeated in the relation to the analyst. The relation towards the analyst (and the analyst itself) constitutes the groundwork for the patient's internal structure of expectations in relationships. The analytic setting fosters the evolution of these inner conceptions. Within the transference situation, unconscious processes can be actualized. Experiencing a secure attachment with the analyst, the patient may be enabled to become aware and reconstruct his/her memories and relationships (that may have structured him/her) and eventually work them through. Andrade (2005) stresses the effect of positive transference as the basis of therapeutic action. The relationship between patient and analyst promotes an identification that is based upon introjection (of a good object) and empathy and can construct a new affective organization. According to Andrade (2005), the affective nucleus fosters cognitive development. Interpretation—as the classic method of psychoanalysis—is related to explicit memory (as part of the cortex) only and does not effectively deal with implicit memory (as part of subcortical areas), which can only be seen through repetitive transference (cp. Beutel 2009). These implicit memories are unconscious affective structures that can be emptied of their quotas of affect (Andrade 2005). The relation between analyst and patient establishes a new affective configuration.

The Boston Change Process Study Group (2007) depicts early childhood memories (e.g. attachment patterns within the second year of life) as implicit relation knowledge. This internal configuration constitutes the intercourse with others, which becomes evident in subsequent object relations. The Boston Change Process Study Group (2007) defines the intrapsychic as interpersonal experience that is implicitly incorporated. To link therapeutic change with neuroscience, Andrade (2005) deduces that "inadequate object relations can lead to neurophysiological changes and that adequate analytic relations lead to psychic changes that correspond to neuronal changes" (p. 684). As described before, introjection may be the neurochemical basis of psychic change, since new neuronal circuits—as a result of the secretion of neurotransmitters—develop. Also, Beutel (2009) describes the neuronal plasticity that evolves after mechanisms of learning and their repetition. In psychotherapy, these progressions take time and need affective involvement (Beutel 2009).

Within a psychotherapeutic environment, Sterba (1934) described the "therapeutic division of the ego" into an experiencing and an observing ego. During psychoanalysis the patient's ego is at the same time remembering or working through and also analysing this process. The conscious, reasonable, non-neurotic parts of the patient's ego can be distinguished from the unconscious, conflict-motivated, irrational portions of the ego (Sterba 1934). This potential of the neocortex (analysing subcortical activities) may be linked to neurosciences in the way Andrade (2005) explains the difference between unconscious implicit and conscious explicit memories (cp. Beauregard 2007a, b).

Beutel and Huber (2008) state in their review that psychoanalysis has an effect on the brain and argue that the division of psychological psychotherapy and biological psychiatry regarding a patient's treatment has become outdated. The authors stress that today the main theorem of psychoanalysis (a major part of psychic activity remains unconscious) is a convention in neuroscience. Within psychoanalysis, patients learn to deal with their reactivation of patterns in their transference. In this sense, the reconstruction of object-related, psychic configurations resembles the therapeutic effect Andrade (2005) explained.

From a neuroscientific perspective, the problem of showing these effects within methodical borders remains a challenge for the future. For instance, Zwiebel (2007) depicts the difficultness to fully understand the functioning of the analyst. Analysts oscillate between so-called personal and technical poles when treating their patients (Zwiebel 2007). Thus, the therapeutic process can be understood to be on a micro-psychological level that can hardly be quantified (cp. Beutel and Huber 2008).

How can the complexity of input variables be dealt with in order to enable future brain imaging studies in psychodynamic psychotherapy? The general aim of this paper is to discuss the variables that need to be controlled and measured in studying the neuronal effects of psychodynamic psychotherapy. We will discuss the various variables and the methodological problems which can be subsumed under two main headings, the design problem and the translation problem. The design problem addresses the issue of how to account for functionally related variables in an experimentally independent way. For instance, the activation tasks employed in brain imaging should somehow mirror and simulate those functional processes that are assumed to mediate the therapeutic effects of psychodynamic psychotherapy. Experimentally, however, we need to measure and account for both variables in an independent way without any confusion between them. The translation problem raises the question of how to bridge the gaps between the different levels of the concepts presupposed in such investigation. The gap between the personal level of the therapist and the client, on the one hand, and the neuronal level of the brain, on the other, needs to be bridged. There is also a gap between the behavioural effects of psychodynamic psychotherapy the therapist can observe and the subjective effects the client himself experiences. Finally, the gap between the psychodynamic/neuropsychodynamic level of the psychodynamic psychotherapy, the psychological level of the activation task in brain imaging and the neuronal level of the parameters to be measured needs to be bridged. The development of bridges for the various gaps is crucial in developing an experimentally sound design that allows for valid and reliable measurement and interpretation of the data. We will discuss both problems here, the design problem and the translation problem in their various facets which will be illustrated by a specific example, the example of introjection (see below for exact definition).

## 27.3 Design Problem

The design problem deals with the issue how to account for functionally related variables in an experimentally independent way. Relevant inputs that enter such study designs include the psychotherapist, the client, the therapeutic relationship and the investigator, i.e. the experimentator (see Fig. 27.1 and Table 27.1). This discussion of the relevant inputs will shed some light on which and how their variables can be controlled and accounted for in experimental design.

### 27.3.1 The Psychotherapist as "Input"

What does the psychotherapist put into psychodynamic psychotherapy? First and foremost, he puts in his own personality, his cognitions, his affects and ultimately his own life history. In the further course of the interaction between the patient and the therapist, it is the psychotherapist's perspective on the patient's thoughts, feelings and behaviour which essentially contributes to the development of the therapeutic relationship. Recent research demonstrated that the psychotherapist himself,

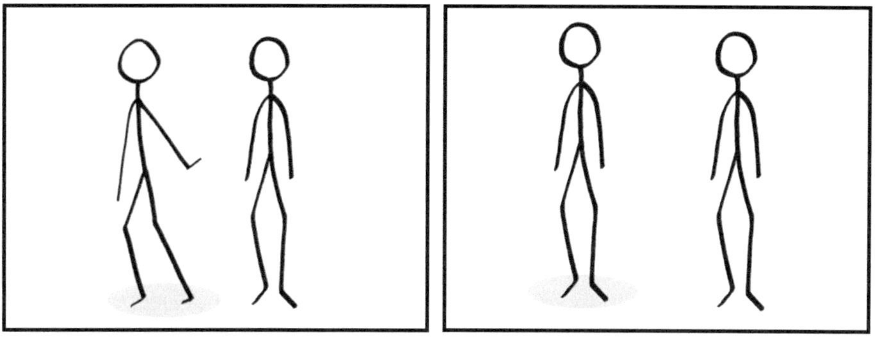

**Fig. 27.1** The design problem

**Table 27.1** Input, empirical variables and experimental measures

| Input | Empirical variables | Experimental measures |
|---|---|---|
| *Psychotherapist* | Personality, empathy | Scales for personality and empathy |
| | Psychotherapeutic intervention as input | Psychotherapeutic identity |
| | Psychotherapeutic output | Psychodynamic, subjective and behavioural measures |
| | Psychodynamic process mediating between psychotherapeutic input and output | Measurement of psychodynamic process with STIPO, OPD, etc. |
| *Client/patient* | Personality and psychodynamic structure as input | Measurement of psychodynamic process with STIPO, OPD, etc. |
| | Behavioural and subjective input in the gestalt of symptoms | Likert scales, reaction times and other behavioural parameters |
| | Therapeutically induced changes in subjective and behavioural output | Psychophysiological measures like skin conductance, etc. |
| *Patient-psychotherapist match* | Quality of therapeutic relationship | Scales for measurement of fit of match between client and therapist and thus of therapeutic relationship with Helping Alliance Questionnaire, Vanderbilt Psychotherapy Process Scale or Working Alliance Inventory |
| *Investigator* | Concept and hypothesis of brain function | Localization versus integration |
| | Behavioural task as activation paradigm and input | Neurophysiological, methodological, psychodynamic, symptomatic and experiential demands |
| | Changes in neuronal activity as output | Method of measurement (fMRI, PET, etc.) |

as a personality with all his/her affects, cannot remain abstinent in psychotherapy as originally envisioned by Freud. If the patient experiences the analyst as an "impenetrable object", it can lead to serious difficulties in the analytic process, e.g. the patient will transfer his/her projections onto the therapist, which in turn can trigger unconscious "hardening" by the therapist (Skogstad 2013). Kohut pointed out that the capacity to show empathy is a major factor in how the relationship between therapist and client can develop which in turn has a strong impact on potential therapeutic effects (see Kohut 1959). A recent study investigated cognitive and emotional aspects of empathy in psychotherapists (Hassenstab et al. 2007). When compared to control subjects, psychotherapists showed higher empathy scores when making inferences based on language mirroring cognitive aspects. Affective aspects of empathy did not differ between both groups though psychotherapists showed better emotion regulation with less personal distress in response to the distress of others. Though preliminary because of the low number of cases ($n = 19$), this psychological study supports the crucial importance of empathy in psychotherapists. Certainly though further studies are necessary to reveal the exact role of empathy and its distinct aspects (sensory, cognitive, affective; see also Zanocco et al. 2006) in psychotherapeutic interaction. Furthermore, one may investigate whether the neuronal network implicated in empathy (insula, anterior cingulate, thalamus, temporoparietal junction, amygdala; see Frewen et al. 2008) may show a higher neuronal reactivity in psychotherapists when compared to non-psychotherapists. Ideally, one would include neuronal and psychophysiological (skin conductance, heart rate, etc.; see Marci and Riess 2005) measures of the psychotherapist's emphatic abilities as confounding variables, i.e. as regressor or covariate, in the measurement and analyses of the client's neuronal changes during psychodynamic psychotherapy.

Why consider the personality and empathic abilities of the psychotherapist as a confounding variable? Imagine, for instance, a psychotherapist with strong tendencies to identify with the patient. This, of course, enables the therapist to understand the patient and also, a client who has great difficulty internalizing significant others might well benefit from such an empathic psychotherapist and a supportive approach. On the other hand, however, it might hinder his empathic abilities and may also be problematic when he encounters, for instance, a depressed client who has internalized highly ambivalent object relationships. This case might be problematic for therapeutic interaction (e.g. when the transference is directed by these ambivalent aspects of the patient's internalized relationships). Sandler et al. (2011) also point out that for a successful psychotherapy beyond the actual transference relationship, which enables the transference neurosis, a different form of relationship—a "working alliance"—is required. This should also enable the patient to maintain an analytic attitude even if the transference conflicts are intense. Psychotherapeutic effects might thus not only depend on the personality and psychic structures of the psychotherapist himself but also on the specific constellation between therapist and client including their respective attachment styles (see, for instance, Schauenburg et al. 2006). This makes it clear that experimentally, we do not only need to include personality scales for both the client and the therapist but measures for attachment styles on both sides.

Another variable the psychotherapist himself puts in are of course the psychotherapeutic interventions he uses to induce psychotherapeutic change; the factor accounting for the psychotherapeutic intervention may be conceptualized as "psychotherapeutic input" which describes the intervention the psychotherapist uses to induce therapeutic change in the client. Freud (1937) tackles the desired changes in psychotherapy and appropriate therapeutic interventions, when he raises the question of the "natural end of the analysis". He emphasizes that therapeutic interventions should be aimed at overcoming the patient's inner resistances, and thus the symptoms he is suffering from will disappear. It is a question of undoing "ego changes", which are created by mobilizing ego defences against dangerously experienced drive derivative in the course of development. This makes up the analytical process. The therapist may, for instance, maintain a state of analytic abstinence together with an evenly suspended attention as a basis for interpreting unconscious conflicts, the transference or dreams. Or he may choose to focus on working with imagination letting the client imagine various kinds of scenarios to put traumatic events into a broader context. Contrary to long-term psychoanalytical psychotherapy, the therapist may focus—within the framework of short-term psychodynamic psychotherapy—on so-called core conflictual relationship themes (CCRT) or interpersonal conflicts in the actual relationship of the patient (Luborsky and Crits-Christoph 1989; Luborsky et al. 1985; Roth and Fonagy 1996).This must be accounted for in a quantified and objective way, for instance, by the recently developed questionnaire of psychotherapeutic identity that asks for various issues of the psychotherapists' education, experience, style and values (see Sandell et al. 2002).

In addition to psychotherapeutic input and psychodynamic process, we also need to account for the psychotherapeutic output, the effects. There have been various studies showing the therapeutic efficacy of psychodynamic psychotherapy (see, for instance, Haase et al. 2008; Leichsenring and Leibing 2007; Leichsenring and Rabung 2008; Taylor 2008). Recently developed instruments like the Operationalized Psychodynamic Diagnostics System (Boeker and Richter 2008; Boeker et al. 2008; Cierpka et al. 2007; OPD-Taskforce 2008) enable an operationalized psychodynamic diagnostic approach based on a multiaxial system (consisting of four psychodynamic axes and one descriptive axis). Furthermore, OPD enables the definition of relevant therapeutic foci and the measurement of therapeutic changes (Rudolf et al. 2004). The Structured Interview for Personality Organization (Clarkin et al. 2004) was developed according to the psychodynamic concept of Kernberg (1996). The STIPO allows the evaluation of an individual's personality organization with respect to the following dimensions: identity consolidation, quality of object relations, use of primitive defences, quality of aggression, adaptive coping versus character rigidity and moral values. The psychotherapeutic output is accounted here only on a psychodynamic level; this is problematic because the measure that measures something, the psychotherapeutic output, should be different from what it shall measure, the independent variable in the experimental design (which though remains constitutively dependent on it). Therefore, what is needed additionally are some dependent variables of psychotherapeutic change and their underlying psychodynamic processes on a different level, the subjective and behavioural level.

One might argue that the neuronal effects themselves may well serve as dependent though different measure of psychotherapeutic outcome. This however is to confuse different evidences. The neuronal effects are supposed to evidence the effects of psychotherapy on the neuronal level, while they are not supposed to reflect evidence of the psychotherapeutic effects by themselves. We cannot measure and evidence psychotherapeutic effects by neuronal measures that are supposed to mediate them if we want to avoid circularity. Hence, to reliably link neuronal effects to psychotherapeutic effects, we need a measure of psychotherapeutic effects that is neither psychodynamic, thereby avoiding identity with the output, nor neuronal in order to avoid identity with the process that is supposed to mediate its effects. As such a measure, Beutel (2009) suggests changes in the known memory systems—declarative (explicit) and procedural (implicit)—that (memory) in turn can be localized in specific brain structures. He discussed that the repression, which has been overcome by analytic interventions, can lead to the repressed being recalled and then being reproduced and detected by memory tests. The findings of Nader et al. confirm the well-known fact in memory research that memory performance is affected by the constellations of encoding and retrieval situations and may distort the memory of content ("false memory"). In contrast to these findings, we assume that the influence of the memory is insignificant in the constellation of the encoding and the retrieval situation, because it retrieves meaningful biographical information. Thus memory systems can on the one hand reflect the effects of psychotherapy; on the other hand, they can be localized in specific areas of the brain itself. However, this requires a careful conceptualization of such experiments: first, the confounding variables should be detected (e.g. influencing memory performance by the current emotional state of the patient/subject) and controlled and second, a careful selection of test instruments should be made. Only then can the memory performance be a measure that maps evidence of the effects of psychotherapy, on the one hand, and locates and maps the neuronal level on the other hand.

### 27.3.2 The Patient as "Input"

First and foremost, the client comes with a specific psychodynamic constellation and his particular personality, his psychodynamic and personality input. For instance, a certain mechanism may predominate to such a degree that it becomes pathological (e.g. introjection in the "introjective type" of depressed patients (see Boeker et al. 2000; Blatt 1974; Taylor and Richardson 2005). Consequently, more mature mechanisms cannot be used. The psychodynamic constellation of the client needs to be objectified and verified, and several instruments like the OPD, the STIPO and the KAPP (Weinryb et al. 1991a, b) have been developed for this purpose. For instance, KAPP is a rating instrument based on psychoanalytical theory to assess relatively stable modes of mental functioning as they appear in self-perception of the own personality and interpersonal relations. KAPP has 18 subscales covering interpersonal relations, personality functioning like frustration tolerance, impulse control, personality organization and others. In addition, one should also include measures of the personality like the Temperament and Character Inventory (TCI) that measures various dimensions of reward (reward dependence, novelty seeking, etc.) and self (self-directedness, self-transcendence, etc.).

However, the client does not come to the psychotherapist because of his specific psychodynamic constellation. He comes because he encounters some behavioural and subjective problems which outside observers may call symptoms. These symptoms are the aim and targets of the psychotherapeutic intervention. For instance, a client with high degree of introjection does not come because of his abnormally high introjection but because he may be severely depressed, and it is his depressive symptoms that are the target of psychotherapeutic intervention. What we need to account for experimentally is thus the behavioural and subjective problems encountered by the client, i.e. his symptoms. They may, for instance, be measured subjectively with scales like the Beck Depression Inventory (BDI) or the Beck Hopelessness Scale (BHS) where the client himself rates and evaluates his subjective and behavioural problems. Or the client's problems may be rated objectively by somebody else using, for instance, the Hamilton Depression Rating Scale (HDRS). In order to avoid confusion between psychotherapeutic intervention and symptom measurement, objective scales shall be accounted by a person that is different and independent of the psychotherapist himself since otherwise some bias and contamination by the latter cannot be excluded. Most importantly, what is needed here in the future is a clear empirical linkage between specific psychodynamic processes and particular symptoms, i.e. behavioural and subjective abnormalities. For instance, introjection or anaclitic needs have often been associated with depression (see Blatt 1974). Referring to the psychotherapeutical context of introjection, Blatt's distinction between introjective and anaclitic depression is of special importance. Patients suffering from anaclitic depression are primarily preoccupied with issues of interpersonal relatedness (e.g. trust, caring, intimacy and sexuality) and use primarily avoidant defences (e.g. denial and repression) to cope with psychological conflict and stress. In contrast, patients suffering from introjective depression are primarily concerned with establishing and maintaining a viable sense of self, ranging from a basic sense of despair to concerns about autonomy and control and to issues of self-worth, and use primarily counteractive defences (e.g. projection, doing and undoing, intellectualization, reaction formation and overcompensation). Interestingly, this differentiation is significantly related to different kinds of outcome in long-term intensive treatment of seriously disturbed young adults and different responses to two forms of therapy—psychoanalysis and psychotherapy (cf. Blatt 1993). What is needed are studies to show the correlation between both psychodynamic and symptomatic measures entailing what we call psychodynamic-symptomatic specificity.

Finally, we need to account for the change in the client as induced by the psychotherapy. These changes may be measured in behavioural and psychodynamic terms as discussed above and should also be accounted for in subjective terms. For instance, one hypothesis is that introjection may be accounted for by what we call self-related processing (Boeker and Richter 2008; Northoff 2008). If so one would expect increased self-relatedness in depressed patients when compared to healthy subjects which is indeed the case as demonstrated recently (Northoff 2007; Grimm et al. 2009). Psychodynamic psychotherapy should lead to a decrease in the self-focus in depressed patients which ideally should be accompanied by decreased introjection. If so, the subjective experience of self-relatedness may be taken as marker of subjective change induced by psychodynamic psychotherapy. This may be accompanied ideally by behavioural markers like reaction time measures during tasks implicating self-relatedness. Most importantly, the subjective and behavioural measure of self-relatedness should be sensitive to both, the psychodynamic processes, as induced by psychotherapeutic intervention, and the symptoms, i.e. the clients' behavioural and subjective input. This means that self-relatedness should serve as a dependent variable of both introjection and depressed symptoms and that the latter two should also be linked in functional regard. All three, self-relatedness, introjection and depressed symptoms, are thus closely linked to each other in functional and hence constitutional regard, while experimentally they should be kept distinct and separate. We are thus again confronted with the discrepancy between clinical and experimental levels encountering the constitution of clinical symptoms by various interdependent functions which though experimentally need to be kept apart and thus independent of each other.

### 27.3.3 The Therapeutic Relationship as "Input"

Over the past decades, the psychoanalytical situation was reconceptualized as a dyadic system in which the psychoanalytic psychotherapist is both participant and observer. The broadened definition of countertransference and the influence of object relations theory and various intersubjective perspectives have led to increased emphasis on the relationship between psychotherapist and patient. Many new terms have been coined to emphasize various facets of the "two-personness" of analysis including the therapeutic alliance and the "real" relationship (cf. Vaughan and Roose 2000). The most far-reaching attempt to distinguish transference-countertransference from "reality" aspects of the dyad has occurred in the context of the growing emphasis on patient-therapist match.

Kantrowitz et al. (1989) defined match in the following way: Match is "a broader field of phenomena in which counter-transference is included as one of many types of match. The individual history, characteristics, attitudes and values of each analyst and patient predispose them respectively to certain counter-transference and transference reactions. Match, however, can also refer to observable styles, attitudes and personal characteristics which are rooted in residual and unanalyzed conflicts, shared or triggered in any patient-analyst pair" (Kantrowitz et al. 1989, p. 895). Different types of facilitating and impeding matches are distinguished from one another which based on similarity and complementarity very much resemble the concordant and complementary transference-countertransference paradigms delineated by Racker (1968) within an object relations model. The importance of interactive, non-verbal affective communication that shapes the behaviour and response of the patient and the therapist also needs to be pointed out as one central factor constituting the match (cf. Kantrowitz 1995).

Some psychotherapy studies have focused on the question of what constitutes a good match. Luborsky et al. (1988) observed that from ten pretreatment demographic variables (age, marital status, having children, religion and level of religious activity, education, cognitive style, etc.), only match in marital status was found to be significantly predictive of positive outcome (see Gruenbaum 1983; Hollander-Goldfein et al. 1989; Garfield and Bergin 1978 for other studies in this direction).

Recently, instruments to measure the fit or match between therapist and client have been developed. The Helping Alliance Questionnaire (Luborsky 1984) investigates the subjective evaluation of the therapeutic relationship from the perspective of both the client and the therapist so that the correlation between both may reflect the fit or "match". Another instrument is the Vanderbilt Psychotherapy Process Scale (O'Malley et al. 1983) that allows an evaluation of the client-therapist relationship by means of an external observer, for instance, a video recording. It includes dimensions like patient involvement, therapist-offered relationship and exploratory process. Finally, the Working Alliance Inventory (Horvath and Greenberg 1989) (see also Bordin 1975/1976) includes 36 items to the dimensions' goal, task and bond that can be evaluated by the client, the therapist and an external observer.

Taken together, there is still a need for psychotherapeutic research that collects data from both participants in dyadic situations. To date there are only very few studies attempting to operationalize different factors of the therapeutic relationship and developing adequate paradigms using neuroimaging approaches (see Kaechele and Buchheim 2008).

### 27.3.4 The Investigator as "Input"

The investigator targets the brain; more specifically he aims to reveal the neuronal effects of psychodynamic psychotherapy. By developing his hypothesis about possible neuronal effects, he must presuppose (either implicitly or explicitly) a specific concept and theory of brain function. For instance, presupposing strict localizationism and modularity, he may hypothesize that neuronal activity in a specific region like the often observed abnormality in the subgenual anterior cingulate cortex (Mayberg 2003) may be changed and normalized by psychodynamic psychotherapeutic intervention in depression. This hypothesis is based upon similar observations in CBT and pharmacotherapy (see Goldapple et al. 2004; Kennedy et al. 2007). However, these and almost any other brain imaging study on the neuronal effects of psychotherapy do show a wide variety of different regions showing neuronal changes. This puts the presupposition of strict localization into doubt and may make a different concept and theory of brain function.

Alternatively to localizationism, one may assume neuronal integration. Neuronal integration describes the coordination and adjustment of neuronal activity across multiple brain regions. The interaction between distant and remote brain areas is considered necessary for a complex function to occur, such as emotion or cognition (Friston 2003; Price and Friston 2002). Neuronal integration focusing on the interaction between two or more brain regions must be distinguished from neuronal segregation (Friston 2003; Price and Friston 2002). Here a particular cognitive or emotional function or processing capacity is ascribed to neuronal activity in a single area that is both necessary and sufficient; one can subsequently speak of neuronal specialization and localization. We assume that, for instance, mechanisms as complex emotional-cognitive interactions cannot be localized in specialized or segregated brain regions. Instead, we consider specific psychodynamic mechanisms to require interaction between different brain regions and thus neuronal integration.

For neuronal integration to be possible, distant and remote brain regions have to be linked together which is provided by connectivity. Connectivity describes the relation between neuronal activities in different brain areas. There is anatomical connectivity for which we will use the term connections in order to clearly distinguish it from functional connectivity. In addition, Friston and Price (2001) distinguish between functional and effective connectivity: functional connectivity describes the "correlation between remote neurophysiological events" which might be due to either direct interaction between the events or other factors mediating both events. A correlation can either indicate a direct influence of one brain area on another or their indirect linkage via other factors. In the first case, the correlation is due to the interaction itself, whereas in the second the correlation might be due to other rather indirect factors like stimuli based on common inputs. In contrast, effective connectivity describes the direct interaction between brain areas; it "refers explicitly to the (direct) influence that one neuronal system exerts over another, either at a synaptic or population level" (Friston and Price 2001). Here, effective connectivity is considered on the population level because this corresponds best to the level of different brain regions investigated here. For example, the prefrontal cortex might modulate its effective connectivity with subcortical regions, thereby influencing specific functions like interoceptive processing.

Based upon connectivity, neuronal activity between distant and remote brain regions has to be adjusted, coordinated and harmonized. Coordination and adjustment of neuronal activity might not be arbitrarily but guided by certain principles of neuronal integration (Northoff et al. 2004). These principles describe functional mechanisms according to which the neuronal activity between remote and distant brain regions is organized and coordinated. Such principles of neuronal integration might, for instance, include reciprocal modulation, modulation by functional unity, top-down modulation and modulation by reversal (see Boeker et al. 2006; Northoff 2008 for details). As hypothesized by us, each of these principles may be associated with a specific psychodynamic mechanism.

One may want to argue that the debate about the presupposed theory and concept of brain function is of mere theoretical interest while remaining empirically irrelevant. This however is to neglect that the experimental measure of neuronal change strongly depends upon the concepts we as investigators put into the investigation itself. If we, for instance, hypothesize a single or specific regions to be effected by psychodynamic psychotherapy, we only measure and analyse our data with regard to such localizationism. Thus, we neither measure nor analyse brain function in orientation on, for instance, the above-mentioned principles of neuronal integration that require different methods of analysis. This may be necessary in depression where the specific abnormality may not consist in one particular region but an abnormal reciprocal modulation between medial and lateral prefrontal cortex with both regions no longer activating in a converse, i.e. opposite and reciprocal, way (see Boeker et al. 2006; Northoff et al. 2004; Grimm et al. 2006). Hence, by concentrating on changes in single regions, we may miss neuronal changes that are induced by psychotherapeutic effects like the normalization of, for instance, reciprocal modulation between the medial and lateral prefrontal cortex. This demonstrates that the very concept of the brain the investigator himself most often implicitly presupposes may strongly impact what and how he measures brain function and which neuronal variables can and will be linked to psychotherapeutic change.

Another crucial input by the investigator is the behavioural task he employs in brain imaging to induce changes in neuronal activity. Brain imaging may be performed in resting state and/or during an activation state with the latter requiring a specific behavioural task. The choice and selection of this behavioural task is of vital importance. Functionally, the behavioural task should be linked to the psychodynamic processes targeted by the psychotherapist in psychodynamic psychotherapy as well as to the client's symptoms, his subjective and behavioural complaints. Psychotherapeutic intervention may then assumed to contribute to "normalize" abnormal reciprocal modulation in depression. This implies for the experimental designs that the behavioural task used in scanning should recruit those neuronal processes and mechanisms that supposedly mediate both the psychodynamic interventions and the client's symptoms. In addition to such psychodynamic and symptomatic requirements, the behavioural task needs to meet experimental demands. Such experimental demands include careful control conditions, behavioural and subjective measurement of the effects of the task itself, empirical linkage to the targeted neuronal processes and mechanisms, etc. The main problem here is to reconcile psychodynamic and symptomatic requirements with experimental demands. The unit of interest on both the psychodynamic and symptomatic level includes usually a mixture of several psychological, subjective and behavioural variables which though on the experimental level need to be carefully controlled and spaced apart. Since the development of the behavioural task, the activation paradigm, is vital, we will discuss this issue in more detail in the next section.

## 27.4 Translation Problem

The above description of the design problem in its various facets reveals that different levels of investigation are involved. The translation problem deals with the methods and strategy how we can bridge the gap between the different levels. To simplify things, we want to discuss in the following four examples with each showing one pair of different levels. We will contrast personal and neuronal levels, psychodynamic and processual levels and first- versus third-person levels (see Fig. 27.2).

### 27.4.1 Personal and Neuronal Levels: Persons vs. Brains

The psychotherapist and the client are individual subjects and must therefore be characterized as persons. The brain, in contrast, is not a person but rather an object.

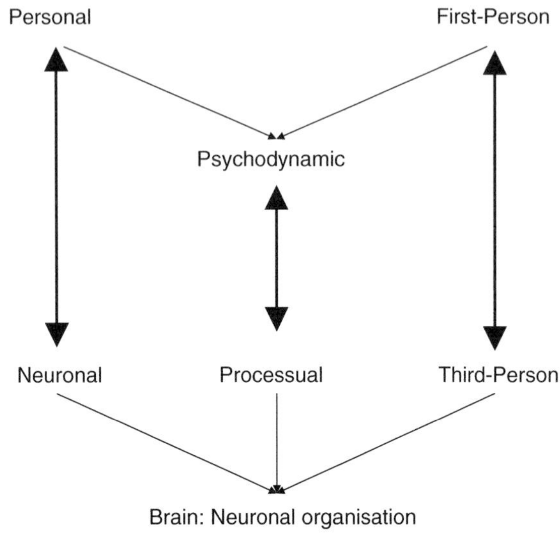


**Fig. 27.2** Different levels and the translation problem

Though this seems obvious, it has major implications in both conceptual and empirical regard. Let us consider first the conceptual implications. Bennett and Hacker (2003) warn not to confuse individual subjects with their brains because that means to neglect the basic difference between persons and objects, they speak of what they call mereological fallacy where the whole, i.e. the person, is confused with one of its parts, the brain. This means, for instance, that one cannot say that the brain thinks, feels or acts since these attributes belong only to persons.

What is treated in psychodynamic psychotherapy is not the brain but the person. We may treat the person in a neurophysiologically constrained way by considering neuronal processes and mechanisms in our psychotherapeutic interventions, but this concerns only the neuronal processes that supposedly mediate the therapeutic outcome. Thus to argue that we treat the client's brain rather than himself as a person is not only to confuse the person and brain but also to neglect the difference between neuronal processes/mechanisms and psychotherapeutic output. Neuronal processes and mechanisms concern the brain and may be regarded a necessary though not sufficient condition of psychotherapeutic output since other factors like interpersonal constellations, the cultural environment, etc. have to be considered too.

The psychotherapeutic output, in contrast, concerns the level of the person that of course is somehow related to the brain but should at least conceptually not be identified with it. Hence, psychodynamic psychotherapy targets the person rather than the brain though its effects may, at least in part, be mediated neurally and thus by the brain. This implies that we should not aim to map the psychodynamic concepts in a one-to-one way with neuronal activity in particular brain regions or networks and thus to strive for what is described by the concept of "neuronal correlates". This is so because that would mean to neglect the various other factors or variables that are implied and included in psychodynamic concepts as we saw in the specific case of psychotherapeutic intervention. Instead of the concept of neuronal correlates, one may therefore want to preferably use the concept of "neuronal mechanisms" that, unlike the concept of neuronal correlates, does not presuppose one-to-one mapping between psychodynamic concepts and neuronal activity. As such, neuronal mechanisms are supposed to underlie (rather than correlate with) psychodynamic concepts and thus psychotherapeutic interventions which leaves open conceptual and empirical space for including variables other than purely neuronal ones.

## 27.4.2 Personal and Neuronal Levels: Generality vs. Individuality

Another important empirical implication of the conceptual difference between persons and objects is the difference between individual and general levels. Persons concern individual subjects each with major idiosyncrasies both psychodynamically and neutrally (as it is focused on in the neuropsychodynamic approach). The focus of psychodynamic psychotherapy is always on the individual, its specific subjective and personal contents as derived from its life history. Psychoanalysis gives us a conceptual framework to link these individual contents, as they are experienced from the inside of the experiencing person itself, to general structures of the psyche of persons, as they are observed from the outside by the observer. Neuroscience, in contrast, concerns the brain as we can observe it from the outside; thereby however the individual person's specifics get lost because the experimental approach averages across different individual subjects. The difference between individuality and generality marks a principal difference between psychoanalysis and neuroscience which is nicely expressed by David Milrod in the following quote:

> Neuroscientists strive to explain fundamental phenomena such as perception, consciousness, emotion, memory, etc. including the subtleties of their integration, and in this way build up an understanding of the basic functioning of the organism. In recent years they have included a study of the self as it integrates with consciousness, emotion, and awareness of the object. They try to discover the general truth, and often use phylogenetic concepts in arriving at an understanding of what a particular concept (e.g., the self) is, what it interrelates with, and what it influences, and is influenced by. In short, they concern themselves with the universal and objective. Psychoanalysis, which has historically focused on the individual and has been more interested in ontology, has as its goal the understanding of protracted intrapsychic, interpersonal, and subjective functioning of the individual. It was in order to better understand that functioning that psychoanalysts had to deal with the self and its representation. In dealing with the self, the psychoanalyst is more likely to focus on the contents of the self and its representation, the state of stability or fragility it may possess, and under what circumstances these characteristics shift back and forth in the course of treatment. In other words, they focus on those elements that make each individual different from one another. (Milrod 2002, p. 22–23)

How can we bridge the principal difference between the individual level of persons and the general level of brains? One way is to investigate only single cases and to focus on case studies (see, for instance, Solms and Lechevalier 2002 with regard to lesions patients as well as Lai et al. 2007; Lehto et al. 2008; Overbeck et al. 2004;

Rudolf et al. 2004 for single case studies of psychodynamic psychotherapy and brain imaging). This however precludes a deeper insight into the neuronal processes and mechanisms that may eventually mediate psychotherapeutic output. What we need to develop are experimental designs and analyses that allow to take the individualized data as starting point and then to take and preserve these individual features as starting point for group analyses without averaging and generalizing them out into a group mean.

One may, for instance, imagine that the regions of interest in the individual subjects are taken as starting point for averaging and group analysis. The individual regions of interest may not only be determined in oriented on anatomical constraints but also psychodynamic constraints like the predominance of a certain psychodynamic mechanism. Another possibility is to group the individual subjects according to their subjective or psychodynamic profiles as revealed in empirical investigation of subjective experience. For instance, subjects with "high scores of introjection" may then be grouped together and compared with those showing "low scores of introjection". One of the major methodological challenges in the future is thus to develop experimental designs and ways of analyses that allow to link individual and general features on the neuronal level in the same way Freud achieved it on the psychological level in such an ingenious way.

### 27.4.3 Personal and Neuronal Levels: Content vs. Organization

Another issue in this regard is the difference between neuronal contents and neuronal organization. Psychodynamic concepts may mirror the general organization of psychological activity which then may be manifested and realized in specific psychological contents of that individual person. This parallels to the neuronal level. We mentioned above that one may search for principles of neuronal integration rather than specific regions and networks. Specific regions and networks mirror what may be called neuronal contents, and these are the targets in, for instance, the search for the neuronal correlates of consciousness (NCC) presupposing mere correlation and one-to-one mapping strategies. The principles of neuronal integration refer rather to the organization of neuronal activity and hence to what we call neuronal organization.

If one now searches for psychodynamic concepts in specific neuronal regions and networks, one may attempt to link structures of psychological organization with neuronal contents. This however may be doomed to fail because one then confuses the level of organization, as presupposed on the psychological level, with the level of contents, as implied by the neuronal level, with both remaining unable to match or correspond on a one-to-one basis. Instead, one may rather link psychodynamic concepts to the neuronal organization with both presupposing and implying analogous structures. This however remains rather speculative at this point (see Northoff 2011, for a first attempt in this sense with regard to the self) since especially the principles and structures of neuronal organization, as distinguished from neuronal contents, remain to be explored.

### 27.4.4 Psychodynamic Level Versus Process Level

One of the main issues is the translation of psychodynamic concepts into processes that then can be psychologically and neurally investigated. Consider again the example of introjection as psychodynamic mechanism.

Introjection is considered a psychodynamic mechanism that, based on Mentzos (1995), can generally be determined as a form of appropriating and relating objects to the subjects in a personal way that is called internalization. Internalization includes three different mechanisms, identification, introjection and incorporation. These mechanisms of identification depend on different structural levels of the ego functions and of the personality. Incorporation describes that the subject incorporates and integrates objects into itself so that the object becomes part of the subject itself with the former being indistinguishable from the latter. The subject may also introject the object.

What distinguishes introjection from identification and incorporation (cf. Meissner 1978)? The separate reality of the object is acknowledged by the subject in introjection, but the object relations are highly ambivalent including aggressive and narcissistic conflicts and feelings of anxiety, which are defended by projective mechanisms. In contrast, identifications depend on differentiated, continuous object relations and enable a selective internalization of partial aspects of the object. Ambivalent emotions may be tolerated and expressed. Incorporations, introjections and identifications are important steps and components of the maturation process. Disturbances of the maturation process may lead—in a psychological developmental and psychoanalytical perspective—to the development of pathological defence mechanisms and the reactivation of early modes of internalization and object relationships (e.g. introjection in depression and borderline; see Boeker et al. 2006).

Introjection allows the distinction between subject and object by the subject; however, the price for acknowledging their difference consists in ambivalence with subsequent affects and anxieties. Metaphorically speaking, the object becomes strongly affectively coloured by the subject while at the same time retaining its separate reality and reality for the subject. By means of affective involvement, the object is thus subjectivized and related to the subject, or, as one could say with Mentzos (1995), something objective (object) is transformed into something subjective (object): the parentheses are included because the object becomes only coloured by subjectivity while retaining its status as object, whereas in projection, as the opposite of introjection, one would probably speak of objective subject. The result of this process of introjecting may be what is called an introject, the internal representation of an object. An object can be internalized and introjected and thus become an introject only if it has a special meaning and personal significance to the subject which usually is reflected in strong emotional involvement with the respective emotional feelings. If, for instance, somebody has a rather close but ambivalent and therefore a strongly emotionally loaded relationship to her/his mother, the mother as object may become internalized and introjected to resolve the ambiguity in the relationship resulting in the mother being an introject for the subject. If, in contrast, the relationship to the mother is positive and free of ambiguity, it is possible to identify with the mother selectively as well as being separated from her.

This short description of introjection points out some cardinal psychological processes like relating objects to the person's self which has recently been described as self-related processing (Northoff et al. 2006a, b). Moreover, it is clear that emotion processing is involved and closely linked to self-relatedness. Furthermore, the ability to relate to other people that involves empathy is crucial in introjecting. At the same time, however, self-awareness is also involved since otherwise the introjecting person remains unable to distinguish itself as subject from the object, i.e. from other persons. These psychological processes may then be regarded as starting point to develop a neuropsychodynamic hypothesis of introjection (cf. Chap. 10). Accordingly, translation from the level of the psychodynamic concept, i.e. introjection, to the process level, i.e. self-related and emotion processing, etc., is needed to develop neuropsychodynamic hypothesis let alone appropriate experimental designs. The psychological processes that may eventually be involved in psychodynamic concepts may then be used as guiding thread for where to look in the brain and what kind of principles of neuronal organization may be involved.

### 27.4.5 First-Person Level Versus Third-Person Level

Systematic examination and evaluation of subjective experience must preserve its richness and complexity, on the one hand, and objectively quantify its main characteristics on the other. Objectification and quantification of subjective first-person data allow for scientific investigation and consequently for establishing what can be called a "science of experience" (Gabbard 2000). Based on a "science of experience", a "science of psychodynamic processes" needs to be developed. The "science of psychodynamic processes" should place great emphasis on patients' mental life or inner experience in order to preserve the richness and complexity of subjective experience and clinical description. At the same time, these subjective features must be objectified to provide reliable and quantifiable data. This can be achieved by asking the subjects to complete rating scales. For example, visual analogue scales (Weinryb et al. 1991a, b), with regard to personal identity or idiographic instruments like the Repertory Grid Test (Boeker et al. 2000) which enables the evaluation of idiosyncratic experiences and views by means of a semi-quantitative measurement, might be applied to let the subjects themselves evaluate their experiences. One might also apply structured interviews with valid and reliable instruments for evaluation of the subjects' relevant psychodynamic features by an experienced investigator. General instruments include the Karolinska scale that assesses different psychodynamically relevant dimensions of a person's structure (Weinryb et al. 1991a, b). Another instrument is the Operationalized Psychodynamic Diagnostics (OPD) (OPD-Taskforce 2008) which develops four psychodynamically relevant axes (illness experience and expectations, relationship, conflict and structure) and one descriptive axis (psychic and psychosomatic disturbances according to ICD and DSM, cf. Chap. 4).

One of the main methodological challenges in investigating the neuronal processes underlying mechanisms is to link these first-person data about psychodynamic processes to third-person observation of neuronal states (cf. Chap. 3). Being based upon subjective experience, psychoanalysis relies on first-person data or more precisely on data obtained by introspection that presupposed what may be called second-person perspective (which in the following we will subsume under the concept of first-person perspective). This contrasts with neuroscience which requires third-person observation of neuronal states. Due to the neglect of first-person subjective experience, neuronal states as third-person data can be quantified and objectified. This, in contrast, remains impossible in the case of first-person data which are rather qualitative and subjective. If, however, the neuronal processes of mechanisms are to be investigated, subjective experience and neuronal states (i.e. first- and third-person data) have to be linked to each other in a systematic way. For this purpose, we have created an appropriate methodological strategy, first-person neuroscience, which aims at systematically linking first- and third-person data (see Northoff 2007) that also conceptualizes many investigations in current brain imaging that correlate subjective experiential variables (as, for instance, in visual analogues scales) with neuronal measures of brain function (see, for instance, Grimm et al. 2009).

We define "first-person neuroscience" as a methodological strategy to systematically link first-person subjective experience to third-person observation of neuronal states. The development of such methods distinguishes first-person neuroscience from neuroscience as it is commonly practised which most often relies on third-person observation of neuronal states more or less independently of subjective experience. The main challenge in establishing first-person neuroscience consists in linking the individual contents of subjective experience to neuronal states. How can we link subjective experience to neuronal states?

Linkage between subjective experience and neuronal states requires two steps: First, subjective experience needs to be evaluated systematically including objectification and quantification of subjective data. Such "science of experience" is a necessary precondition for any linkage between subjective experience and neuronal states. Second, the systematically objectified and quantified subjective data then enable the linkage to analogous data about neuronal states. For this, special methodological strategies need to be developed—this is the core of what we call "first-person neuroscience" (cf. Chap. 1). The above-described discussion of how to translate the psychodynamic concept of introjection, which experimentally is accounted for by first-person data, into a behavioural task as activation paradigm that yields third-person data about the brain, can be regarded as example of how to link first- and third-person perspectives and may therefore be regarded an instance of first-person neuroscience.

## 27.5 Individualized Paradigms in Neuroimaging: State of the Art

Over the past years, a number of papers have attempted to link neuroscience and psychoanalysis. And neurobiological changes in some (single) cases undergoing psychodynamic psychotherapy have been reported, mostly using SPECT (Lai et al. 2007;

Lehto et al. 2008; Saarinen et al. 2005; Viinamäki et al. 1998) with few using fMRI in studies of obsessive-compulsive disorder, panic disorder and somatoform disorder (Overbeck et al. 2004; Beutel et al. 2010; De Greck et al. 2011). To date, studies examining the functional neuroanatomy of psychotherapy in depressed patients have applied interpersonal therapy or cognitive behavioural therapy (Linden 2006; Roffman et al. 2005). Buchheim et al. were the first to conduct an fMRI study with depressed patients treated with psychodynamic psychotherapy (Buchheim et al. 2006), using two fMRI paradigms (Kessler et al. 2011; Taubner et al. 2012). The first paradigm (OPD) confronted patients with their maladaptive interpersonal relationship themes, presenting them sentences in the scanner derived from an OPD interview. These four sentences were individual for each person and represented the core dysfunctional relationship theme of each one. During the control condition (traffic), patients recalled a stressful traffic situation they had experienced inducing negative emotions and recalling autobiographical memories. Conditions were separated by a "relaxation" condition. The second paradigm (AAP) used attachment-related pictures from the Adult Attachment Projective Picture System (AAP) eliciting mental engagement with attachment-related experiences such as loss, illness, danger and separation. During an AAP interview, patients described the scene in the picture including what characters were thinking and what could happen next. From this interview, three sentences representing the attachment pattern of each patient were extracted. The same pictures were used for the neutral condition, describing only the environment.

One might argue that the paradigms described above have an enhanced cognitive load so one should also take the effect of cognition on emotion processing into account. Furthermore patients didn't immediately rate the pictures/sentences to evaluate the subjective experience.

Methodological shortcomings in the above-mentioned studies are expression of the complex endeavour that the investigation of neuronal mechanisms of subjective experience implies. Developing valid experimental designs taking into account the very individual dimension of experience is an arduous methodological challenge, as it has been illustrated here before. In the following section, our attempt to create a new experimental design for the investigation of neuronal mechanisms in depressed patients during psychodynamic psychotherapy will be depicted.

Coherent with the above-described methodological strategy named "first-person neuroscience", the experimental design aims at systematically linking subjective experience to the observation of neuronal states. This necessitates the objective evaluation of subjective experience in a first step. In a second step, the so-created data has to be linked to observations on the neuronal level.

To begin with, we will outline our choices of how to evaluate subjective experience, giving a brief description of the OPD-2 axis "interpersonal relations" as well as the Maladaptive Interpersonal Patterns Q-Sort (MIPQS). We will then describe the development of the "Interpersonal Relations Picture Set" (IRPS) and a new neuroimaging experiment based on the ISPS.

The OPD-2 axis "interpersonal relations" and the Maladaptive Interpersonal Patterns Q-Sort (MIPQS)

An adapted instrument to attempt an objective evaluation of subjective experience is the Operationalized Psychodynamic Diagnostics (OPD-2) (OPD-Taskforce 2008). It operationalizes psychodynamic dimensions in different diagnostic axes. Regarding depressive disorders, the interpersonal relations axis is particularly relevant, as depressive disorders go along with various impairments in interpersonal and social functioning. For example, depressive patients tend to have deficits in emotional expression and emotion recognition in others. They also tend to have difficulties with affective modulation in the basic interpersonal communication of emotions and feelings. Broadly speaking, psychodynamic psychotherapy sets a specific focus on these affective processes using phenomena of transference and countertransference. The interpersonal relation axis of the OPD offers a classificatory system describing different patterns in interpersonal behaviour. This diagnostic axis enables clinicians to assess and precisely describe specific maladaptive interpersonal patterns in patients with depression.

Recently, the Maladaptive Interpersonal Patterns Q-Sort (MIPQS), a self-report version of the OPD interpersonal relation axis, has been developed. Using a card sorting procedure, the MIPQS allows the establishment of a subjective and hierarchized profile of typical interpersonal behaviour. One of its substantial advantages compared to the OPD interpersonal relation axis is its facile use. The MIPQS exists in two versions. The MIPQS-A concerns the self-evaluation of the participant concerning typical interpersonal behaviour. In contrast, the MIPQS-B enables the clinician to evaluate the participant's interpersonal behavioural patterns from his point of view and based on the interview situation. We will concentrate here on the MIPQS-A. The card sorting procedure of the MIPQS-A (named MIPQS in the following) comprises two steps. In a first step, the participant rates the relevance of behavioural patterns (described by 32 items) in his own behaviour towards significant others. In a second step, the participant rates the relevance of described behavioural patterns (equally 32 items) in the behaviour of others towards himself. The description of patterns in interpersonal behaviour was adopted from rating items of the OPD interpersonal relation axis. These items are theoretically close to interpersonal circumflex models such as used in interpersonal psychology (e.g. the SASB/Structural Analysis of Social Behaviour model, Benjamin et al. 2006). The SASB model describes different qualities of interpersonal behaviour by means of two orthogonal and bipolar dimensions: control (dominant versus submissive) and affiliation (friendly versus hostile). The MIPQS items have been empirically tested and can be located in the circumflex model comprising these two dimensions. Every one of the 32 MIPQS items consists of two easy-to-understand descriptions of a specific interpersonal behaviour pattern like "I tend to ignore others or give them the cold shoulder" and "Others tend to ignore me or give me a cold shoulder," respectively. They are printed on separate cards. The sorting procedure includes the depositing of the 32 cards on finally 9 columns ranging from "most typical" to "most untypical" with "unimportant" as the fifth column. Furthermore, the instrument offers a sequential ranking of all items from 1 (most untypical) to 32 (most typical).

Based on the items of the MIPQS, we have developed graphic illustrations forming the so-called Interpersonal Relations Picture Set (IRPS), which we use in fMRI experiments. In the following, the different steps of its development will be described.

### 27.5.1 The Interpersonal Relations Picture Set (IRPS)

The MIPQS provides 32 items describing patterns of interpersonal behaviour (Fuchs et al. 2017). The IRPS comprises pictures illustrating interpersonal situations such as described by the 32 MIPQS items. These pictures were developed step by step. At first, there had been an attempt to illustrate the interpersonal situations by means of multicoloured symbols. However, this approach was abandoned because of the high level of abstraction of these symbols. Consecutively, a collection of pictures illustrating the different situations by means of stick figures was composed. The pen drawings were scanned for further processing in a widely used image editor (Seashore©). Each picture shows two or more black stick figures on white background. The figures vary in size but do not show any gender specification, facial expression, clothes or other specific characteristics. Some figures, e.g. the figure taking a neutral body position, occur repeatedly in different pictures. Different interpersonal situations are expressed only in the specific posture of figures as well as their positions towards each other.

### 27.5.2 Validation of the Pictures

In order to confirm the relation of every picture to the assigned item of the MIPQS, an online survey was performed. It was sent to a mailing list of the university. The survey included an introduction, an instruction concerning the questionnaire and a request of age, gender, educational level and history of interpersonal relationships. Then the ISPS pictures were presented in randomized order. Participants were asked to judge the pictures in two conditions. Firstly, in the "discrimination task", pictures were presented combined with the matching description as well as four randomly selected descriptions out of the remaining 31 MIPQS items. Furthermore, a response that none of the offered descriptions fitted the picture was added. This form of questioning was designed as multiple-choice task and only one answer was allowed. Participants were asked to distinguish the matching description from the offered choice. Secondly, in the "relevance task", participants were asked to rate the level of relevance of each picture. A nine-point Likert scale was combined with the picture and its matching description for this task. The sample of the online survey was randomly split into two groups to avoid a repetition effect. Hence, no item was presented twice in the survey. Statistical analysis was conducted using the SPSS® software package (IBM). The data for all variables collected were subjected to descriptive statistics according to the respective scale level. Due to the purpose of direct interpretability of the result values of both tasks, the evaluation of results was limited on descriptive statistics indicating the proportion of correct responses for multiple-choice tasks in percentage and the median for nine-point Likert scales.

In a further step, the correspondence between the picture set and the MIPQS was tested when participants performed them separately. A number of patients suffering from major depressive disorder (MDD) were recruited from the Department of Psychiatry, Psychotherapy and Psychosomatics of the University Hospital of Psychiatry Zurich. Patients with neurological or other physical illnesses, disorders of personality or alcohol or substance abuse were excluded. Diagnosis was made according to DSM-IV (1994). Clinical symptoms were assessed with the Beck Depression Inventory II and the Hamilton Depression Rating Scale (Hamilton 1960). After execution of the MIPQS, patients were asked to perform the relevance task (see above) using the experimental control software Presentation® (Neurobehavioral Systems). Ratings for each item were related to the number of the column chosen by the patient in the MIPQS using Spearman rank correlations.

### 27.5.3 Course of the Experiment

We investigate changes in depressed patients during 1 year of psychodynamic psychotherapy concerning psychodynamic, behavioural and neuronal parameters. Different instruments are employed for the evaluation of these changes, such as the OPD interview, an fMRI examination and a series of questionnaires (MIPQS (Maladaptive Interpersonal Patterns Q-Sort), OPD-SF ("OPD Strukturfragebogen", OPD questionnaire concerning psychic structure), BDI (Beck Depression Inventory), BAI (Beck Anxiety Inventory), BHI (Beck Hopelessness Inventory), FKBS ("Fragebogen zu Konfliktbewältigungsstrategien", coping with conflicts questionnaire), IIP-D (Inventory of Interpersonal Problems), HCSC ("Heidelberger Umstrukturierungsskala", Heidelberg's structural change scale)). A specific focus is set on changes in interpersonal behavioural patterns such as evaluated by the MIPQS. The IRPS will be employed during the fMRI examinations. The procedure for the fMRI experiment will be roughly described hereafter.

In a first step, participants rate the MIPQS and the IRPS. Six pictures representing the most typical interpersonal situations for each participant are selected. Participants are then invited to develop a personal narrative for every one of the six target pictures. Ideally, this determines the individual meaning of each of these stimuli. After, participants proceed to fMRI examinations. Instructions include a structured description of the task and a trial run. Stimuli are presented in a block design in randomized order. Four experimental conditions are used in the scanning procedure. In a first condition ("typical"), the six target pictures are presented. A second condition ("untypical") consists of pictures rated beforehand by the participants as being not typically representative for their interpersonal behaviour. The control condition ("neutral") includes pictures showing a number from two to four stick figures in a frontal, neutral position. Finally, a resting condition ("rest") showing a black fix-cross on white background is included. In order to evaluate subjective experience during the "typical" and "untypical" condition, a nine-point Likert scale is presented after every picture during the scanning procedure. Participants are asked to rate the level of personal involvement experienced while watching the pictures. In the "neutral" condition, subjects are asked to report the number of stick figures presented in every picture. Stimuli are presented using Presentation®, and all feedbacks are given using a trackball response pad (Current Designs®). The experimental design was optimized for further analysis of effective connectivity.

## 27.6 Discussion

We discussed the methodological problems in designing a brain imaging study to measure neuronal effects of psychodynamic psychotherapy. Two main problems, the design problem and the translation problem, were encountered. The design problem points to the many inputs including the psychotherapist himself, the client and the investigator, which each by itself may need to be included as distinct experimental variables in the study design. The translation problem refers to the different levels involved in such project such as the personal versus the neuronal level, the psychodynamic versus the process level and the first-person level versus the third-person level. Thereby the personal versus the neuronal level is of particular interest in that it includes conceptually and empirically relevant distinctions like persons vs. brains, generality vs .individuality and organization vs. content.

In conclusion, we have illustrated our attempt to develop an individualized neuroimaging paradigm. As mentioned before, the choice and selection of the behavioural task employed during fMRI examinations is of vital importance. We have chosen to base the behavioural task on a validated instrument (MIPQS) describing a central dimension in psychodynamic psychotherapy: changes in interpersonal behaviour and associated feelings. We hence try to isolate a specific mechanism relevant in the psychodynamic treatment of depression and operationalize it in an fMRI experiment. The experiment incorporates a number of specific principles that are, in our view, of great importance for this type of experimental design.

First, focus is set on the very individual dimension of experience as well as their emotional implications. This is reflected in the individual choice and subjective determination of meaning of stimuli (IRPS pictures) used during neuroimaging.

Second, the association of the picture to autobiographical experience should strengthen the affective reaction of the participant when the IRPS pictures are presented in the scanner. During the scanner procedure, participants rate their subjective emotional arousal induced by the IRSP pictures. Having a subjective rating of this kind enhances the validity of the experiment.

Third, the use of visual stimuli in form of pictures may reduce the cognitive demand on participants during fMRI examinations compared to tasks involving stimuli using words or sentences.

Fourth, the fMRI experiment comprises a valid control condition. The control condition consists in presentation of (a) IRPS pictures that the patient rated as non-relevant for himself and (b) pictures showing stick figures in a neutral position.

Results of fMRI exams will be linked to results from other diagnostic instruments such as the OPD and a series of other questionnaires. By the choice of these instruments, we tempt to take into account the complexity of subjective experience.

There are several factors that have been pointed out earlier to be relevant for the design of neuroimaging paradigms in psychodynamic/neuropsychodynamic research that are not taken into account in our paradigm. This includes factors resumed under the "design problem" as well as those evoked concerning the "translational problem". For example, we have not considered the psychotherapist or the therapeutic relationship as "input" in depth. Our design includes one questionnaire possibly giving a hint on the matching of therapist and patient (IIP-D), but it does not include personality or attachment style ratings for the therapist. Ideally, this should also be taken into account. To give another example, we also need to carefully consider our hypothesis concerning neuronal activation during fMRI exams and take into account considerations illustrated earlier with reference to "the investigator as input".

Our experimental paradigm does not aspire to satisfy all of the requirements that experimental designs in psychodynamic psychotherapy research using neuroimaging should ideally fulfil and which were described earlier in this article. In this vast and complex research domain, the development of adequate and valid experimental designs stays a defiant methodological challenge. Our experimental paradigm represents a further step into this direction. It aspires to create an experimental design that does reflect—even though in a limited way—the complexity of subjective experience.

**Conclusion**

Taken together, this demonstrates that brain imaging studies of the neuronal effects of psychodynamic psychotherapy are confronted with a rather high degree of complexity raising various conceptual, empirical and experimental problems. The discussion of these problems shall not discourage future investigators; instead it shall provide them with some suggestions for guidance through the jungle of complexity. Though any such investigation requires multi-professional efforts and emphatic collaboration, we are sure the merits are highly rewarding. The complexity of investigating the neuronal effects of psychodynamic psychotherapy mirrors in an almost paradigmatic way the complexity of our brain so that neuropsychodynamic findings entail insight and a better understanding of the general principles of neuronal organization and our brain's very human nature. The answer to this question is twofold. On the one hand, we do think that by revealing the neuronal mechanisms underlying psychotherapeutic processes, we may be able to develop more specific protocols of psychotherapy in orientation on the respective neuronal functions associated with the respective region. For instance, taking a rather simplistic example, if the level of neuronal activity in the amygdala may be involved in psychotherapeutic processes and even be predictive of psychotherapeutic outcome, it may be an indicator that the involvement of emotions may have been crucial in psychotherapeutic success. This may be the case even if the therapy was not emotion-focused but rather oriented in psychodynamic mechanisms. Such results may then be considered as evidence for the central involvement of emotional processes in the psychodynamic processes which may then lead to further refinement and specification of psychotherapeutic protocols. This raises not only the question for the linkage between emotions and psychodynamic processes but also how we can more explicitly emotions in relation to, for instance, defence mechanisms in psychotherapy. Such orientation on neuronal functions may then lead to the development of neurally based psychodynamic psychotherapy in the future and may therefore be empirically, i.e. neurally, more plausible and compatible with respect to the brain and its mode of function than the current purely clinically and observationally based approaches.

On the other hand, revealing the neuronal mechanisms underlying psychotherapeutic processes may also contribute in the reverse direction, by giving us a better understanding of the psychological, i.e. psychodynamic, mechanisms associated with certain patterns of neuronal activity across different regions. Hence, it is not only that psychodynamic psychotherapy may benefit from brain imaging but also the other way in that the latter may also be complemented by the former.

Taken together, we thus assume bilateral exchange and contribution between psychodynamic psychotherapy, neuropsychodynamic psychiatry and brain imaging. This may ultimately, as we hope, lead to the development of diagnostic and therapeutic predictive markers with especially the latter predicting what subjects may benefit from what kind of psychotherapy in general and the kind of focus in psychodynamic/neuropsychodynamic psychotherapy in particular.

## References

American Psychiatric Association. Diagnostic and statistical manual of mental disorders (DSM IV). 4th ed. Washington, D.C.: American Psychiatric Association; 1994.

Andrade VM. Affect and therapeutic action of psychoanalysis. Int J Psychoanal. 2005;86:677–97.

Andrade VM. Dreaming as a primordial state of the mind: the clinical relevance of structural faults in the body ego as revealed in dreaming. Int J Psychoanal. 2007;88:55–74.

Beauregard M. Mind does really matter: evidence from neuro-imaging studies of emotional self-regulation, psychotherapy and placebo effect. Prog Neurobiol. 2007a;81:218–36.

Beauregard M. Mind does really matter: evidence from neuro-imaging studies of emotional self-regulation, psychotherapy, and placebo effect. Prog Neurobiol. 2007b;81(4):218–36.

Bennett MR, Hacker PMS. Philosophical foundations of neuroscience. Oxford: Blackwell; 2003.

Beutel ME, Huber M. Functional neuro-imaging – can it contribute to our understanding of processes of change. Neuropsychoanalysis. 2008;10:5–16.

Beutel ME, Stern E, Silbersweig DA. The emerging dialogue between psychoanalysis and neuroscience: neuro-imaging perspectives. J Am Psychoanal Assoc. 2003;51(3):773–801.

Beutel M. Neurowissenschaften und psychodynamische Psychotherapie. Z Psychiatr Psychol Psychother. 2009;57(2):87–96.

Beutel ME, Stark R, Pan H, Silbersweig D, Dietrich S. Changes of brain activation pre- post short-term psychodynamic inpatient psychotherapy: an fMRI study of panic disorder patients. Psychiatry Res Neuroimaging. 2010;184:96–104.

Benjamin LS, Rothweiler JC, Critchfield KL. The use of structural analysis of social behavior (SASB) as an assessment tool. Annu Rev Clin Psychol. 2006;2:83–109.
Blatt SJ. Levels of object representation in anaclitic and introjective depression. Psychoanal Study Child. 1974;29:107–57.
Blatt SJ. Different kinds of folks may need different kinds of strokes: the effect of patients' characteristics on therapeutic process and outcome. Psychother Res. 1993;3(4):245–59.
Boeker H, Richter A. Commentary On: "Functional neuro-imaging – can it contribute to our understanding of processes of change?" Neuropsychoanalysis and the process of change: questions still to be answered. Neuropsychoanalysis. 2008;10(1):23–5.
Boeker H, Hell D, Budischewski K, Eppel A, Härtling F, Rinnert H, von Schmeling C, Will H, Schoeneich F, Northoff G. Personality and object relations in patients with affective disorders: Idiographic research by means of the repertory grid-technique. J Affect Disord. 2000;60:53–60.
Boeker H, Kleiser M, Lehman D, Jaenke L, Bogerts B, Northoff G. Executive dysfunction, self, and ego pathology in schizophrenia: an exploratory study of neuropsychology and personality. Compr Psychiatry. 2006;47(1):7–19.
Boeker H, Himmighoffen H, Straub M, Schopper C, Endrass J, Kuechenhoff B, Weber S, Hell D. Deliberate self-harm in female patients with affective disorders: investigation of personality structure and affect regulation by means of operationalized psychodynamic diagnostics. J Nerv Ment Dis. 2008;196(10):743–51.
Boeker H, Richter A, Himmighoffen H, Ernst J, Bohleber L, Hofmann E, Vetter J, Northoff G. Essentials of psychoanalytic process and change: how can we investigate the neural effects of psychodynamic psychotherapy in individualised neuro-imaging? Front Hum Neurosci. 2013;7:355.
Bordin, E. The working alliance: basis for a general theory of psychotherapy. Symposium of the American Psychological Association, Washington, DC; 1975.
Bordin E. The generalizability of the psychoanalytic concept of the working alliance. Psychother Theory Res Pract. 1976;16:252–60.
Boston Change Process Study Group. The foundational level of psychodynamic meaning: Implicit processes in rela- tion to conflict, defense and the dynamic unconscious. Int J Psychoanal. 2007;88:843–60.
Buchheim A, Erk S, George C, Kachele H, Ruchsow M, Spitzer M, Kircher T, Walter H. Measuring attachment representation in an fMRI environment: a pilot study. Psychopathology. 2006;39(3):144–52.
Cierpka M, Grande T, Rudolf G, von der Tann M, Staschf M. The operationalized psychodynamic diagnostics system: clinical relevance, reliability and validity. Psychopathology. 2007;40(4):209–20.
Clarkin J, Caligor E, Stern B, Kernberg OF. Structured interview for personality organisation (Stipo). New York: Personality Disorders Institute, Weill Medical College of Cornell University; 2004.
de Greck M, Rotte M, Paus R, Moritz D, Thiemann R, Proesch U, Bruer U, Moerth S, Tempelmann C, Bogerts B, Northoff G. Is our self based on reward? Self-Relatedness recruits neural activity in the reward system. NeuroImage. 2008;39(4):2066–75.
De Greck M, Scheidt L, Bölter AF, Frommer J, Ulrich C, Stockum E, Enzi B, Tempelmann C, Hoffmann T, Northoff G. Multimodal psychodynamic psychotherapy induces normalization of reward activity in somatoform disorder. World J Biol Psychiatry. 2011;12(4):296–308.
Freud S. Die endliche und die unendliche Analyse. Studienausgabe, Ergänzungsband. Frankfurt: Fischer; 1937.
Freud S. Erinnern, Wiederholen und Durcharbeiten (Weitere Ratschläge zur Technik der Psychoanalyse, II). Internationale Zeitschrift für ärztliche Psychoanalyse. 1914;2:485–91. Remembering, repeating and working-through. SE, 12: 147–156
Freud S. New introductory lectures on psycho-analysis. The standard edition of the complete psychological works of Sigmund Freud, Volume XXII (1932-1936): new introductory lectures on psycho-analysis and other works, 1–182; 1933. p. 79.

Frewen PA, Dozois DJ, Lanius RA. Neuro-imaging studies of psychological interventions for mood and anxiety disorders: empirical and methodological review. Clin Psychol Rev. 2008;28(2): 228–46.
Friston K. Learning and inference in the brain. Neural Netw. 2003;16(9):1325–52.
Friston KJ, Price CJ. Dynamic representations and generative models of brain function. Brain Res Bull. 2001;54(3):275–85.
Fuchs S, Bohleber LM, Ernst J, Soguel-dit-Piquard J, Boeker H, Richter A. One look is worth a thousand words: new picture stimuli of interpersonal situations. J Soc Neurosc. 2017;18:1–9. https://doi.org/10.1080/17470919.2017.1327457.
Gabbard GO. A neurobiologically informed perspective on psychotherapy. Br J Psychiatry. 2000;177:117–22.
Gallese V, Eagle MN, Migone P. Intentional attunement: mirror neurons and the neural underpinnings of interpersonal relations. J Am Psychoanal Assoc. 2007;55(1):131–76.
Garfield S, Bergin AE. Handbook of psychotherapy and behavior change: an empirical analysis. New York: Wiley; 1978.
Goldapple K, Segal Z, Garson C, Lau M, Bieling P, Kennedy S, Mayberg H. Modulation of cortical-limbic pathways in major depression: treatment-specific effects of cognitive behavior therapy. Arch Gen Psychiatry. 2004;61(1):34–41.
Grimm S, Schmidt CF, Bermpohl F, Heinzel A, Dahlem Y, Wyss M, Hell D, Boesiger P, Boeker H, Northoff G. Segregated neural representation of distinct emotion dimensions in the prefrontal cortex-an fMRI study. NeuroImage. 2006;30(1):325–40.
Grimm S, Ernst J, Boesiger P, Schuepbach D, Hell D, Boeker H, Northoff G. Increased self-focus in major depressive disorder is related to neural abnormalities in subcortical-cortical midline structures. Hum Brain Mapp. 2009;30(8):2617–1627.
Gruenbaum H. A study of therapists' choice of a therapist. Am J Psychiatr. 1983;140:1336–9.
Haase M, Frommer J, Franke GH, Hoffmann T, Schulze-Muetzel J, Jager S, Grabe HJ, Spitzer C, Schmitz N. From symptom relief to interpersonal change: treatment outcome and effectiveness in inpatient psychotherapy. Psychother Res. 2008;18(5):615–24.
Hamilton M. A rating scale for depression. J Neurol Neurosurg Psychiatry. 1960;23:56–62.
Hassenstab J, Dziobek I, Rogers K, Wolf OT, Convit A. Knowing what others know, feeling what others feel: a controlled study of empathy in psychotherapists. J Nerv Ment Dis. 2007;195(4):277–81.
Hollander-Goldfein B, Fosshage JL, Bahr JM. Determinants of patients' choice of therapist. Psychotherapy. 1989;26:448–61.
Horvath A, Greenberg LS. Development and validation of the working alliance inventory. J Consell Psychol. 1989;36:225–33.
Kaechele H, Buchheim A. Neuro-Psychoanalyse-Studie und einige Widerspiegelungen im Erleben der Beteiligten Patienten und Psychoanalytiker. Bad Homburg: DPV Herbsttagung; 2008.
Kandel ER. Biology and the future of psychoanalysis: a new intellectual framework for psychiatry revisited. Am J Psychiatr. 1999;156(4):505–24.
Kantrowitz JL. The beneficial aspects of the patient-analyst match. Int J Psychoanal. 1995;76:299–313.
Kantrowitz JL, Katz AL, Greenman DA, Morris H, Paolitto F, Sashin J, Solomon L. The patient-analyst match and the outcome of psychoanalysis: a pilot study. J Am Psychoanal Assoc. 1989;37(4):893–919.
Kennedy SH, Konarski JZ, Segal ZV, Lau MA, Bieling PJ, McIntyre RS, Mayberg HS. Differences in brain glucose metabolism between responders to Cbt and venlafaxine in a 16-week randomized controlled trial. Am J Psychiatr. 2007;164(5):778–88.
Kernberg O. A psychoanalytic theory of personality disorders. In: Clarkin J, Lenzenweger MF, editors. Major theories of personality disorders. New York: Guilford Press; 1996.
Kessler H, Taubner S, Buchheim A, Münte TF, Stasch M, Kächele H, Roth G, Heinecke A, Erhard P, Cierpka M, Wiswede D. Individualized and clinically derived stimuli activate limbic structures in depression: an fMRI study. PLoS ONE. 2011;6(1):e15712.

Kohut H. Introspection, empathy, and psychoanalysis; an examination of the relationship between mode of observation and theory. J Am Psychoanal Assoc. 1959;7(3):459–83.
Lai C, Daini S, Calcagni ML, Bruno I, De Risio S. Neural correlates of psychodynamic psychotherapy in borderline disorders - a pilot investigation. Psychother Psychosom. 2007;76(6):403–5.
Lehto SM, Tolmunen T, Kuikka J, Valkonen-Korhonen M, Joensuu M, Saarinen PI, Vanninen R, Ahola P, Tiihonen J, Lehtonen J. Midbrain serotonin and striatum dopamine transporter binding in double depression: a one-year follow-up study. Neurosci Lett. 2008;441(3):291–5.
Leichsenring F, Leibing E. Psychodynamic psychotherapy: a systematic review of techniques, indications and empirical evidence. Psychol Psychother. 2007;80:217–28.
Leichsenring F, Rabung S. Effectiveness of long-term psychodynamic psychotherapy: a meta-analysis. J Am Med Assoc. 2008;300(13):1551–65.
Linden DE. How psychotherapy changes the brain--the contribution of functional neuro-imaging. Mol Psychiatry. 2006;11(6):528–38.
Luborsky L. Principles of psychoanalytic psychotherapy. A manual for supportive expressive psychotherapy. New York: Basic Books; 1984.
Luborsky L, Crits-Christoph P. A relationship pattern measure: the core conflictual relationship theme. Psychiatry. 1989;52(3):250–9.
Luborsky L, McLellan AT, Woody GE, O'Brien CP, Auerbach A. Therapist success and its determinants. Arch Gen Psychiatry. 1985;42:602–11.
Luborsky L, Crits-Christoph P, Mintz J, Auerbach A. Who will benefit from psychotherapy: predicting therapeutic outcomes. New York: Basic Books; 1988.
Mancia M. Implicit memory and early unrepressed unconscious: their role in the therapeutic process (how the neurosciences can contribute to psychoanalysis). Int J Psychoanal. 2006;87:83–103.
Marci C, Riess H. The clinical relevance of psychophysiology: support for the psychobiology of empathy and psychodynamic process. Am J Psychother. 2005;59(3):213–26.
Mayberg HS. Modulating dysfunctional limbic-cortical circuits in depression: towards development of brain-based algorithms for diagnosis and optimised treatment. Br Med Bull. 2003;65:193–207.
Meissner M. Internalisation and object relations. J Am Psychoanal Assoc. 1978;27:345–60.
Mentzos S. Dream sequences. On the psychodynamic aspects of the dramaturgy of dreams. Psyche. 1995;49(7):653–71.
Milrod D. The concept of the self and the self representation. Neuropsychoanalysis. 2002;4(1):7–23.
Northoff G. Psychopathology and pathophysiology of the self in depression - neuropsychiatric hypothesis. J Affect Disord. 2007;104(1-3):1–14.
Northoff G. Neuropsychiatry. An old discipline in a new gestalt bridging biological psychiatry, neuropsychology, and cognitive neurology. Eur Arch Psychiatry Clin Neurosci. 2008;258(4):226–38.
Northoff G. The self and its brain. In: Neuropsychoanalysis in practice. Oxford: Oxford University Press; 2011. p. 212–35.
Northoff G, Heinzel A, Bermpohl F, Niese R, Pfennig A, Pascual-Leone A, Schlaug G. Reciprocal modulation and attenuation in the prefrontal cortex: an fMRI study on emotional-cognitive interaction. Hum Brain Mapp. 2004;21(3):202–12.
Northoff G, Heinzel A, de Greck M, Bermpohl F, Dobrowolny H, Panksepp J. Self-referential processing in our brain - a meta-analysis of imaging studies on the self. NeuroImage. 2006a;31(1):440–57.
Northoff G, Böker H, Bogerts B. Subjektives Erleben und neuronale Integration im Gehirn: Benötigen wir eine Erste-Person Neurowissenschaft? Fortschr Neurol Psychiat. 2006b;74:627–33.
Northoff G, Bermpohl F, Schoeneich F, Boeker H. How does our brain constitute defense mechanisms? first-person neuroscience and psychoanalysis. Psychother Psychosom. 2007;76:141–53.
O'Malley S, Suh CS, Strupp HH. The Vanderbilt psychotherapy process scale: a report on the scale development and a process-outcome study. J Consult Clin Psychol. 1983;51:581–6.

OPD-Taskforce. Operationalized psychodynamic diagnosis Opd-2. In: Manual of diagnosis and treatment planning. Seattle: Hogrefe & Huber; 2008.
Overbeck G, Michal M, Russ MO, Lanfermann H, Roder CH. Convergence of psychotherapeutic and neurobiological outcome measure in a patient with Ocd. Psychother Psychosom Med Psychol. 2004;54(2):73–81.
Peres JF, McFarlane A, Nasello AG, Moores KA. Traumatic memories: bridging the gap between functional neuro-imaging and psychotherapy. Aust N Z J Psychiatry. 2008;42(6):478–88.
Price CJ, Friston KJ. Degeneracy and cognitive anatomy. Trends Cogn Sci. 2002;6(10):416–21.
Racker H. Transference and counter-transference. New York: International Universities Press; 1968.
Roffman JL, Marci CD, Glick DM, Dougherty DD, Rauch SL. Neuro-imaging and the functional neuroanatomy of psychotherapy. Psychol Med. 2005;35(10):1385–98.
Roth A, Fonagy P. What works for whom? New York: Guilford; 1996.
Rudolf G, Grande T, Jakobson T. Struktur Und Konflikt. Gibt Es Strukturspezifische Konflikte? In: Dahlbender R, Buchheim P, Schüssler G, editors. OPD – Lernen an der Praxis. Bern: Huber; 2004. p. 195–205.
Saarinen PI, Lehtonen J, Joensuu M, Tolmunen T, Ahola P, Vanninen R, Kuikka J, Tiihonen J. An outcome of psychodynamic psychotherapy: a case study of the change in serotonin transporter binding and the activation of the dream screen. Am J Psychother. 2005;59(1):61–73.
Sandell R, Broberg J, Schubert J, Blomberg J, Lazar A. Psychotherapeutische Identität (Thid). Ein Fragebogen zu Ausbildung, Erfahrung, Stil und Werten. Deutsche Fassung Von Günther Klug, Dorothea Huber & Horst Kächele. Linköping, Stockholm; 2002.
Sandler J, Dare C, Holder A. Die Grundbegriffe der psychoanalytischen Therapie. Stuttgart: Klett-Cotta; 2011.
Schauenburg H, Dinger U, Buchheim A. Attachment patterns in psychotherapists. Z Psychosomatische Med Psychother. 2006;52(4):358–72.
Skogstad W. Impervious and intrusive: the impenetrable object in transference and countertransference. Int J Psychoanal. 2013;94:221–38.
Solms M. New findings on the neurological organization of dreaming: implications for psychoanalysis. Psychoanal Q. 1995;64(1):43–67.
Solms M, Lechevalier B. Neurosciences and psychoanalysis. Int J Psychoanal. 2002;83:233–7.
Solms M, Turnbull OH, Kaplan-Solms K, Miller P. Rotated drawing: the range of performance and anatomical correlates in a series of 16 patients. Brain Cogn. 1998;38(3):358–68.
Sterba RF. The fate of the ego in analytic therapy. Int J Psychoanal. 1934;15:117–26.
Taubner S, Buchheim A, Rudyk R, Kächele H, Bruns G. How does neurobiological research influence psychoanalytic treatments? Clinical observations and reflections from a study on the interface of clinical psychoanalysis and neuroscience. Am J Psychoanal. 2012;72(3):269–86.
Taylor D. Psychoanalytic and psychodynamic therapies for depression: the evidence base. Adv Psychiatr Treat. 2008;14:401–13.
Taylor D, Richardson P. The psychoanalytic/psychodynamic approach to depressive disorders. In: Gabbard G, Beck JS, Holmes J, editors. Oxford textbook of psychotherapy. Oxford: Oxford University Press; 2005.
Vaughan SC, Roose SP. Patient-therapist match: revelation or resistance? J Am Psychoanal Assoc. 2000;48(3):885–900.
Viinamäki H, Kuikka J, Tiihonen J, Lehtonen J. Change in monoamine transporter density related to clinical recovery: a case-control study. Nord J Psychiatry. 1998;52:39–44.
Weinryb RM, Rossel RJ, Asberg M. The Karolinska psychodynamic profile. I. Validity and dimensionality. Acta Psychiatr Scand. 1991a;83(1):64–72.
Weinryb RM, Rossel RJ, Asberg M. The Karolinska psychodynamic profile. II. Interdisciplinary and cross-cultural reliability. Acta Psychiatr Scand. 1991b;83(1):73–6.
Zanocco G, De Marchi A, Pozzi F. Sensory empathy and enactment. Int J Psychoanal. 2006;87:146–58.
Zwiebel R. Von der Angst, Psychoanalytiker zu sein: Das Durcharbeiten der phobischen position. Stuttgart: Klett-Cotta; 2007.

# Concepts of Empirical and Clinical Research in Psychoanalysis and Neuropsychoanalysis

28

Marianne Leuzinger-Bohleber

**Abstract**

Psychoanalysis has a long tradition where different concepts have been developed. The beginnings of psychoanalysis focused on psychic mechanisms of various disorders ranging from melancholia to the various forms of neuroses.

Importantly, Freud himself aimed to develop a psychology on a mechanistic basis. Initially, being a neuroanatomist, he considered the brain and, in his 1895 writing of a scientific psychology, wanted to link it to the psychic level. That, following Freud, failed, for which reason he gave up such neuropsychodynamic approach.

The present chapter sketches Freud's and subsequent developments in psychoanalysis. It finishes with the most recent development of neuropsychoanalysis where it highlights the different approaches including the cognitive-affective approach (Solms, Panksepp) and relational-spatiotemporal as highlighted in this book.

## 28.1 Introduction

What kind of a science is psychoanalysis really? What did Freud mean when he defined psychoanalysis as a special "science of the unconscious"? As a young man, Freud was very interested, as is known, in philosophy and in the humanities before he turned with a remarkably strong emotional reaction to the natural sciences.

* The first part of this chapter is based on the research lecture at the celebration of the centenary of the International Psychoanalytical Association in Berlin, 2010.

** This chapter is based on former publications (Leuzinger-Bohleber 2009, 2015).

M. Leuzinger-Bohleber
Sigmund-Freud-Institut, Frankfurt, Germany
e-mail: m.leuzinger-bohleber@gmx.de



H. Boeker et al. (eds.), *Neuropsychodynamic Psychiatry*,
https://doi.org/10.1007/978-3-319-75112-2_28

He worked at that time on research in medicine and neurology in the laboratory of Ernst Brücke's Institute of Physiology, where he became acquainted with a strict positivistic understanding of science, that attracted him throughout his whole life. As we know, Freud later, however, turned away from the neurology of his time since he recognized the boundaries of the methodological possibilities concerning research of the psyche in this discipline. With *The Interpretation of Dreams*, the founding work of psychoanalysis, he defined this as "pure psychology." He further understood himself, however, to be a physician who observed very exactly as a natural scientist. His wish of a precise, "empirical" examination of hypothesis and theories protected Freud, as Whitebook (2010) notes, from his own predilection to wild speculation. Thus, Freud as a "philosophical physician" could establish a new "science of the unconscious."

Concerning the history of the institution, this understanding of psychoanalysis has been a key to its success. It is well known that Freud even in 1909 considered integrating psychoanalysis into the medical organization "medical psychology and psychotherapy" of August Forel or even into the Orden for ethics and culture. Fortunately, he decided during the Sylvester night, 1910, to found his own independent organization, the IPA (see Falzeder 2010). By this decision the independence of psychoanalysis as a scientific discipline with its own research methodology and institution was protected. Afterward, Freud (1926) always emphasized that psychoanalysis did not deserve to be "swallowed by the medical faculty" "but could instead as 'the psychology of the unconscious' (Tiefenpsychologie), the discipline of the unconscious, become indispensible to all sciences that have to do with the emergence of human culture and its great institutions as art, religion and social systems" (p. 283).

In the century of its history, the specificity of psychoanalytic science became more and more precise. Psychoanalysis developed a differentiated, independent method of research for the examination of its specific object of research, of unconscious conflicts and fantasies.

It has additionally, as all other current disciplines, its own criteria of quality and truth which it has to represent with transparency and self-confidence in scientific dialogue, in order, as any science, to be criticized from outside.

In Sect. 28.2 of this chapter, I would like to present our view for discussion, that it is important for psychoanalysis in our current media-influenced, "knowledge society", to authentically present in the public in new forms that it has its own elaborated, empirical-clinical research and treatment methods that connects it in countless studies with various forms of extra-clinical, e.g., empirical-quantitative, experimental (e.g., neuroscientific), but also interdisciplinary, socially critical research. I am starting with some remarks on psychoanalysis in the contemporary "knowledge society" (1) compared with some episodes during its 100 years of history (2) and then illustrate the specificity as well as the richness of contemporary psychoanalytical research (focusing on the situation in Germany and without being able to give a complete overview) (3).

## 28.2 Psychoanalysis: A Special Scientific Discipline in the Politicized, Commercialized, and Media-Influenced World of Science, Part of the "Knowledge Society"

Western societies have used a great part of their resources in the last 300 years for the acquisition, expansion, and examination of their knowledge. The "industrial society" has changed to a "knowledge society" in the last century. If psychoanalysis wants to remain in this world of science, then it must realize the extreme changes in this field and to attempt to understand its influence on the reality of psychoanalytic research.

(a) The first component of the change in science has to do with *differentiation*. As Hermann von Helmholtz ascertained 100 years ago, each single researcher is increasingly forced to dedicate himself to more and more *specific* methods with more and more *narrow* questions. For this reason the age of the universal geniuses belongs to the past: modern scientists are, for the most part, highly specialized experts with a limited knowledge about adjacent disciplines (Helmholtz 1867 quoted by Weingart 2002, p. 703). They are dependent upon networking on an international, intergenerational, and interdisciplinary level. In connection with this process of differentiation, also the criteria of "science" and "scientific truth" in the respective disciplines have changed, and this is becoming also more specific: not only in the natural sciences but also in the humanities. The concept of a unified science, of "science," relying on the experimental design, on the double-blind experiment in classical physics has proven to be a myth: we live in the times of the "plurality of science" (see also Hampe 2003; Guggenheim et al. 2017; Leuzinger-Bohleber 2015).

(b) A second characteristic of these changes has to do with the *relationship of science and society:* modern scientific disciplines—and thus also psychoanalysis—are in permanent, accelerated, and globalized competition at different levels with one another. Thus, for example, the practical relevance of its research results is permanently evaluated by society's foundations and political interest groups that, for example, increasingly gain influence over the financing of research projects. In this sense, science loses more and more its self-determination. Science becomes politicized—politics more scientific.

(c) A *third characteristic* is connected with this: because politics and society expect more quick results from science concerning recommendations for the solution of societal problems, less and less peace and quiet is left for basic research, from which relatively certain knowledge for practical application was derived. This leads to a paradox situation: on the one hand, ever fewer "normal citizens" and politicians have confidence in their own judgment on complex issues without consulting scientists, but on the other hand, it has become common knowledge that also scientific experts do not have "objective" truths, that so-called scientific knowledge is to be regarded critically. Moreover, it also carries new risks, as the catastrophes of Tschernobyl, the financial crisis or now the so-called refugee crisis, have suddenly shown. This leads to a new source of insecurity and diffuse fears. Which scientific expert is given the most confidence is dependent on his media-transmitted credibility, which now become a relevant factor in society that is competed for in politics and in the public.

(d) A *fourth factor is the roll of the media.* Scientific knowledge is usually taken note of, when it—correspondingly simple and dramatic but credible—finds its way into the media. "It is paradox—the more independent science and the media are, the tighter their coupling. And as the media gain importance, science is losing the monopoly of judging scientific knowledge. The abstract criterion of truth is no longer sufficient in the public debate because the media add the criterion of public acceptance. This does not mean that scientific verification is being replaced, but it is being supplemented by other measures… The loss of distance (between science and the media, LB) will not lead to the end of communication of truths. Trust and confidence remain both constitutive and rare values in communication, and the more society depends on reliable knowledge, the more these are required. The main characterization of today's society is the competition for trust. Once achieved, this is invaluable and science should be keen to preserve it. Therefore, it is only the efforts needed to produce trust and confidence that have become greater" (Weingart, loc cit. p. 706) (emphasis LB).

## 28.3 Remarks to the 100-Year-Old History of Research of Psychoanalysis

What influence did and do the just mentioned changes have on psychoanalysis specifically? It is my opinion that psychoanalysis as a science that relies on the intimacy of the psychoanalytic situation is quite severely hit by the mentioned paradox and dilemma of these changes. As a science of the unconscious, it seems to me to be especially dependent upon if and how it is successful in gaining and keeping the confidence of the world of science, of the public, of politicians and funders, but also of potential patients, candidates in training and the health system. In the last century, it has experienced that the wind of the Zeitgeist has blown from very many directions, as Bohleber (2010) has discussed in respect to German psychoanalysis. This has, although seldom reflected, had its effect on the understanding of research of psychoanalysis and on its concrete research projects, its questions, designs, and goals. In this framework just a few remarks may be allowed.

Freud's lifelong hope, that is, due to the development of the modern natural sciences, the time would come in which the insights of psychoanalysis that have been won with pure psychological, clinical-empirical methods of observation could be also "objectively" examined with the "hard" methods of natural science, seems to often become reality today through the dialogue with the modern neurosciences. Forty years ago, however, Jürgen Habermas (1968) called, as is known, this Freudian longing the "scientistic misunderstanding" (Szientistisches Selbstmissverständnis) of psychoanalysis. He characterized psychoanalysis as following an *emancipatory interest in insight,* in contrast to behavior therapy, that has a *technical interest.* This distinction met with a positive response from a whole generation and psychoanalysis, of course, due to other factors, was at its zenith as it has never been before or after. Psychoanalysis experienced, on the whole, as a critical hermeneutic method of individual and social contradictions, of unconscious sources of psychic and psychosomatic suffering, an exclusive social acceptance in these years that at times verged on idealization. Although there were always attacks and controversies, psychoanalysis as a method of treatment and as a critical theory of culture did not have to worry about its existence during this period.

The social acceptance of that time formed also the understanding of science and research of psychoanalysis in those decades. Shortly summarized, in the 1970s and 1980s besides the genuine clinical psychoanalytic research, this concerned above all hermeneutic-oriented and social psychological approaches, analysis of culture and an interdisciplinary exchange with philosophy and sociology and the sciences of literature, humanities and pedagogy, as well as film and art. Empirical and especially quantitative research in psychoanalysis and the dialogue with the natural sciences were considered by many to be naïve and not fitting for psychoanalysis, even to the point of being harmful. This problematic way of communication had long-lasting consequences: To mention just one example, Siri Hustveth (2010) writes in her bestseller *The Shaking Woman or a History of My Neves* "laconically": "Although American psychiatry was once heavily influenced by psychoanalysis, the two disciplines have grown further and further apart, especially since the 1970s. Many psychiatrists have little or no knowledge of psychoanalysis, which has become increasingly marginalized in the culture. Large numbers of American psychiatrists now leave most of the talk to social workers and stick to writing prescriptions. Pharmacology dominates. Nevertheless, there are still many psychoanalysts practicing around the world, and it's a discipline I've been fascinated by since I was sixteen and first read Freud" (Hustveth 2010, p. 17).

As Thomas Kuhn describes in his analysis of the history of science, different paradigms often exist side by side within a scientific discipline. However, one of them usually dominates—the one that fits best to the Zeitgeist. It seems to me that the just mentioned understanding of psychoanalysis as a critical hermeneutics of the 1970s and 1980s is still currently represented in French psychoanalysis and partly in the Latin-American IPA societies (see, e.g., Green 2003; Widlöcher 2003; Ahumada and Doria-Medina 2009; Bernardi 2003; Vinocur de Fischbein 2009; Duarte Guimaraes Filho 2009), while in the Anglo-Sachsen and German-speaking psychoanalysis, the discussion or perhaps even the adjustment to an empirical-quantitative research paradigm has been pushed to the fore (see among others Fonagy 2015). In these countries the Zeitgeist has changed: in times of "evidence-based medicine" and of medical guidelines, the impression can at times arise, that also for psychoanalysis there exists only one form of research, namely, empirical-quantitative psychoanalytic research, in the sense of the classical natural sciences, of "science." This is—by closer inspection—a strange reoccurrence of an outdated and problematical idea of a "unified science" (Einheitswissenschaft) (see, e.g., Hampe 2003; Guggenheim et al. 2017), an unconscious simplification of the complexities of research in the before-mentioned knowledge—society, which, as is my impression, also involves certain dangers for psychoanalysis.

I (LB) would like to shortly illustrate this point by means of a diagram of clinical and extra-clinical research in psychoanalysis, which I have developed in other papers (see Leuzinger-Bohleber et al 2003a, b; Leuzinger-Bohleber 2009, 2015; Leuzinger-Bohleber & Kaechele 2015). In order not to flounder in abstraction, I will refer in my plea for the creative use of a broad spectrum of current psychoanalytic research strategies, to some of my ongoing research projects. In all these projects, I attempt to encounter the actual Zeitgeist without uncritically submitting ourselves to it and without renouncing the autonomy and specificity of psychoanalysis as a scientific discipline.

## 28.4 Clinical and Extra-clinical Research in Psychoanalysis

Today we can differentiate between two different groups of psychoanalytic research, the *clinical* and *extra-clinical*. By *clinical research* we mean the genuine psychoanalytic research in the psychoanalytic situation itself. Ulrich Moser describes it as on-line research, while the *extra-clinical research* (the off-line research) takes place after the psychoanalytic sessions and embraces a variety of different research strategies as will be shortly sketched

But first to *clinical research*, it takes place in the intimacy of the psychoanalytic situation and can be described as a circular process of discovery in which—together with the patient—idiosyncratic observations of unconscious fantasies and conflicts are successively visualized, symbolized, and finally put into words at different levels of abstraction, an understanding that molds our processes of perception in subsequent clinical situations, even though we enter into each new session with the basic, genuine psychoanalytic attitude that has been described as "not knowing." The circular processes of discovery take place first above all unconsciously and in the realm of implicit private theories. Only a small part hereof is accessible to conscious reflection by the psychoanalyst.

The insights that are won in this clinical research are presented in and outside the psychoanalytic community for critical discussion. In agreement with many current psychoanalysts, clinical research is for me the *central core of psychoanalytic research in general.* It is connected with a characteristic psychoanalytic idea of experience and linked to epistemic values (Erkenntniswerte) (compare Toulmin 1977; Hampe 2004; Guggenheim et al. 2017). Clinical, psychoanalytic research deals with the understanding of unconscious construction of meaning, of personal and biographical uniqueness, as in the exact analysis of the complex weavings of various determinants in the microworld of the patient Moser (2013) and for that reason can be characterized, as mentioned, as critical hermeneutics.

The professionalism of the psychoanalyst makes a stance of free-floating attention (gleichschwebenden Aufmerksamkeit) of his own countertransference, the scenic observation of "embodied enactments" of the patient (see also Argelander

1967; Leuzinger-Bohleber 2015; Leuzinger-Bohleber et al. 2008), Freudian slips, dreams, etc., for the successive understanding of the actual unconscious psychodynamic of the analysand. The typical groping, psychoanalytic process of search for "unconscious truths" can only be carried out with the analysand and is regarded as one of the marked characteristics of psychoanalysis—for example, in opposition to the top-down procedure of behavior therapy. As Jonathan Lear (1995) so impressively described it, psychoanalysis is distinguished as the most democratic of current therapeutic procedures. Combined with this is the characteristic "criterion of truth" of psychoanalytic interpretation: if a certain interpretation of unconscious fantasies or conflicts is "true" can only be decided *together* with the patient, i.e., by the common observation of his (unconscious and conscious) reactions to an interpretation.

As is known, we owe our specific psychoanalytic, clinical-empirical method of research the intensive and detailed "field observations" with single patients in the analytic situation, the most part of all insights that we have won in the last 100 years of our scientific history—for example, the genesis and treatment of chronically depressed patients. Christina von Braun (2010) also sees in clinical research of psychoanalysis the unique chance to recognize and critically reflect the deeper cultural changes by the ubiquitous exploitation mentality of global and "emotional capitalism" (Illouz 2006) on the unconscious of modern man in the analytic relationship that is not only highly relevant for the affected individual but also for an analysis of culture.

But still, let there be no misunderstanding: Peter Fonagy is right when he points out that not every clinician is automatically a researcher. A methodologically systematic procedure, that—through exact description and lucid considerations—makes clinical observations accessible to the understanding and the critique of a third party, is a precondition, that *a gain in knowledge in this form is not only a professional skill but also a clinical science.* Psychoanalysis has at its disposal, as does hardly any other clinical discipline, a differentiated culture of intervision and supervision—closely modeled on psychoanalytic practice—in which the clinical processes of research and gains in insight can be critically discussed. However, there is much room for improvement. Many problems are well known, for example, the chance selection of clinical case reports that only *illustrate* theoretical concepts instead of *verifying* them and critically developing them. Moreover, psychoanalytic concepts are too seldom compared with the results of extra-clinical research, something I would like to deal with later.

*We urgently need good clinical research in order not only to hold our standing in the world of psychotherapy but also to continually develop our professional treatment skills* (compare Boesky 2005; Chiesa 2005; Colombo and Michels 2007; Knoblauch 2005; see also Altmann de Litvan 2014).

Thus we have developed, for example, in the LAC depression study, *our own form of clinical research:* in weekly "clinical conferences," we discuss the treatment sessions that have been partially taped and systematically document our discussion. Based on this joint clinical research, narrative case reports that have been "expert-validated" are developed that belong to the most important results of this study. These case studies convey psychoanalytic insights about the specific psychodynamics of chronic depression, its complex individual and cultural determinants, as well as the details of treatment to the psychoanalytic and non-psychoanalytic community (see, e.g., Leuzinger-Bohleber 2015; Leuzinger-Bohleber et al. submitted).

## 28.5 Psychoanalytic Conceptual Research

This above sketch of new forms of clinical research that are always part of a creative and original research on concepts, a field of research that likewise is as old as psychoanalysis itself. The creative development and enhancement of concepts always distinguished the innovative minds of psychoanalysis and lends our discipline a great attraction for intellectuals, writers, artists, and researchers of other disciplines.

A new characterization of psychoanalytic conceptual research was finally laid out by Joseph Sandler and Anna Ursula Dreher in 1990s, setting themselves apart from other forms of psychoanalytic research. In the Research Subcommittee for Conceptual Research that was initiated by the then IPA President Daniel Widlöchen 2002 with the wish of building more bridges between the conceptual traditions in the different IPA regions, we attempted to further delineate and differentiate the research on concepts during 8 years, as well as to clarify criteria of quality for this specific psychoanalytic research and other involved epistemological questions (compare illustration 1).

In the following administrations of the IPA, this theme has been renewed, and with great effort, the existing psychoanalytic concepts have been integrated in new ways in order to counteract the risk of theoretic fragmentation. The Project Committee for Conceptual Integration (Chair Werner Bohleber) dedicated itself to this work (see, e.g., Bohleber et al. 2017)

### 28.5.1 Extra-clinical Research

The results of not only the clinical psychoanalytic but also of the conceptual research can then in the next step become the subject of other extra-clinical studies (see illustration 1). We distinguish between empirical, experimental, and interdisciplinary studies.

#### 28.5.1.1 Extra-clinical Empirical Studies: An Example of Psychoanalytic Psychotherapy Research[1]

As an example of extra-clinical empirical studies, I would like to shortly discuss psychoanalytic psychotherapy research because it is indispensable in the "knowledge

[1]The literature in the following section is selected illustrating some important historical and contemporary studies. For a more complete overview see Leuzinger-Bohleber and Kaechele (2015).

society" for political and public reasons, in order to prove the effectiveness of psychoanalytic treatment also by the criteria of evidence-based medicine.

Wallerstein (2001) traces these attempts back to their beginnings in 1917 and defines different generations of psychotherapy researchers. He mentions above all a number of American studies, that I—without making a claim to be all-exclusive—will supplement with some European studies (for a more complete overview, see Leuzinger-Bohleber and Kaechele 2015).

*First generation (1968–1971)*, for the most part, retrospective studies that verified with unspecific criteria of success that most psychoanalytic treatment was successful. (Coriat 1917; Fenichel 1930; Jones 1936; Alexander 1937; Knight 1941; Hamburg et al. 1967; Feldman 1968).

*Second generation (1959–1985)*, in which two different groups of studies were carried out:

(a) Prospective, aggregated comparisons of different, exactly defined groups of psychoanalytic treatment. These studies relied on more sophisticated research methods and operationalized, for example, the criteria of success for the expected success of the therapy. Also they could verify that approximately 80% of all psychoanalytic treatment was successful (Bachrach et al. 1985; Erle 1979).
(b) Individual studies that resulted from a methodological uneasiness that individual differences between the patients should not be mixed with group examinations, but to place the main focus on the individual consideration of the single treatment of different patients, as is fitting in psychoanalytic procedure, in which it always has to do with the understanding of unconscious structures of meaning. For this reason they used, for example, in their interviews, also careful psychoanalytic methods, such as psychoanalytic follow-up interviews (Pfeffer 1961; Norman et al. 1976; Schlessinger and Robbins 1974; later follow-up studies at the Anna Freud Center by Target and Fonagy (see Fonagy 2015), DPV Follow-Up- Study by Leuzinger-Bohleber et al. 2003a, b). These studies verified not only the effectiveness of psychoanalytic therapy but also developed a number of unexpected, clinically interesting questions, for example, that with reference to the reduction of symptoms and to other therapy goals, some treatments proved to be effective but that these patients had not gone through a psychoanalytic process in a narrower sense.

*Third generation (1945–1986)*: In these systematic and formal psychoanalytic studies of psychotherapy, examinations of results and of the process were combined, i.e., statistical comparisons were made between the groups but in combination with systematic single-case studies, that, for example, followed the fates of single patients over a longer period of time (Bachrach et al. 1991; Kantrowitz 1986). An example of this third generation of psychoanalytic psychotherapy research is exemplified by the Psychotherapy Research Project of the Menninger Foundation that led to a wealth of insights on the results of psychoanalytic and supportive psychoanalytic therapies and on details concerning treatment techniques. Impressive is, for example, the careful longitudinal study of 42 patients over the course of several decades that Wallerstein published with the moving title "Forty-two Lives in Treatment" (Wallerstein 1986; Wallerstein et al. 1956).

*Fourth generation (the current) (1970–present)* combines not only research of results and therapeutic processes but, thanks to new techniques (video/audio recordings), links microanalysis of therapeutic processes with research on results (beginning with early analysis of tape recordings by Dahl et al. 1988; Strupp et al. 1988; Beenen 1997, Leuzinger-Bohleber 1987, 1989; Varvin 1997; Grande et al. 1997; Huber et al. 1997; Sandell 1997; Leuzinger-Bohleber et al. 2003a, b; compare Fonagy's (2015) excellent overview, as well as the third edition of the Open Door Review (see Leuzinger-Bohleber and Kaechele 2015); as well as new studies of long-term therapies complied by Leichsenring and Rabung (2008), see also Leuzinger-Bohleber et al. (submitted). Important has been the comparison with other therapeutic approaches, e.g., cognitive behavioral therapies (CBT) in the last decades, for CBT researchers have published the outcomes of numerous studies on short-term therapies. Therefore it was important that, e.g., Driessen et al. (2010) showed that psychodynamic short-term treatments are equally effective as cognitive behavioral ones. Positive long-term effects of psychoanalysis have also been shown in several studies, but due to the enormous costs and methodological challenges of such studies, there are still relatively few research groups engaged in the investigation of psychoanalyses and psychoanalytic long-term therapies under naturalistic conditions (see Leuzinger-Bohleber et al. submitted; see also, e.g., Shedler 2010; Leichsenring and Rabung 2011). In all these contemporary studies, it is a great challenge to combine clinical and extra-clinical, quantitative and qualitative, as well as process and outcome research (see, e.g., Leuzinger-Bohleber 2015).

However it is too little known therefore, above all, by clinicians how many psychoanalytic research groups are currently involved in extra-clinical studies. Fonagy (2015) gives a comprehensive survey of the worldwide studies on the effectiveness of psychoanalytic short-term therapies. In the third edition of the Open Door Review (available on the IPA website), the numerous studies in contemporary psychoanalysis have been summarized (see Leuzinger-Bohleber and Kaechele 2015)

#### 28.5.1.2 Experimental Psychoanalytic Studies

It is self-evident that it is impossible to test psychoanalytic processes directly in an experimental design. However, over the last decades, different research groups are successfully working on an examination also experimentally of single psychoanalytic concepts, for example, on the preconscious and the unconscious processing of information in memory and in dreams (to mention just a few of them, the workgroup of Howard Shevrin and his group (see, e.g., Shevrin 2000); Steven Ellman and his group in NY (see, e.g., Ellman 2010); by Wolfgang Leuschner, Stephan, Hau, Tamara Fischmann at the SFI (Hau 2008) to the concept of embodied memory from Pfeifer and his research group in Zürich (Leuzinger-Bohleber and Pfeifer 2002); as well as other studies of facial interaction with the help of the FACs from Rainer Krause in Saarbrücken (e.g., Krause 1997)).

In the last years, as is well known, the dialogue with the neuroscientists has opened new doors for psychoanalysis, probably a reason, why, for example, in the

Society for Neuropsychoanalysis newly founded by Mark Solms and in other institutions, currently a wealth of experimental FmRI und EEG studies involving psychoanalytic questions have been carried out: to only mention a few, studies at the Anna Freud Center (Peter Fonagy), at Yale University (Linda Mayer among others), at Columbia University (Brad Peterson, Andrew Gerber, Steven Roose), or in Germany at the University of Mainz (Manfred Beutel), at dem Wissenschafts Hanse Kollege (Horst Kächele, Anna Buchheim, Manfred Cierpka, Gerhard Roth, Jürgen Bruns among others), at the Psychiatric University Clinic in Zürich (Heinz Böker and Georg Northof), and also from us at the SFI, Lehtonen in Kuopio, and many other groups (compare publications in Neuro-Psychoanalysis, Pincus 2000, or, e.g., Mancia 2006).

#### 28.5.1.3 Interdisciplinary Research (Compare Fig. 28.1)

We would like to finally at least mention that the interdisciplinary dialogue with the neurosciences in these experimental studies is not only decisive for the acceptance of psychoanalysis in the modern world of science but also the creative exchange, for example, with attachment research, empirical developmental research, and the embodied cognitive science. Just as important is the interdisciplinary research in cooperation with literature and cultural studies, social psychology, philosophy, the media and communication sciences, as well as ethno-psychoanalysis.

At the same time, the political and public awareness of science demands from such specialized research projects, as from the abovementioned example of the study on the effectiveness of psychotherapy for chronic depression, that the new found insights, for example, of the lasting therapeutic change, be carried out in an interdisciplinary dialogue involving culture critique of the societal roots of the illness. We are convinced that psychoanalysis as a specific treatment and research

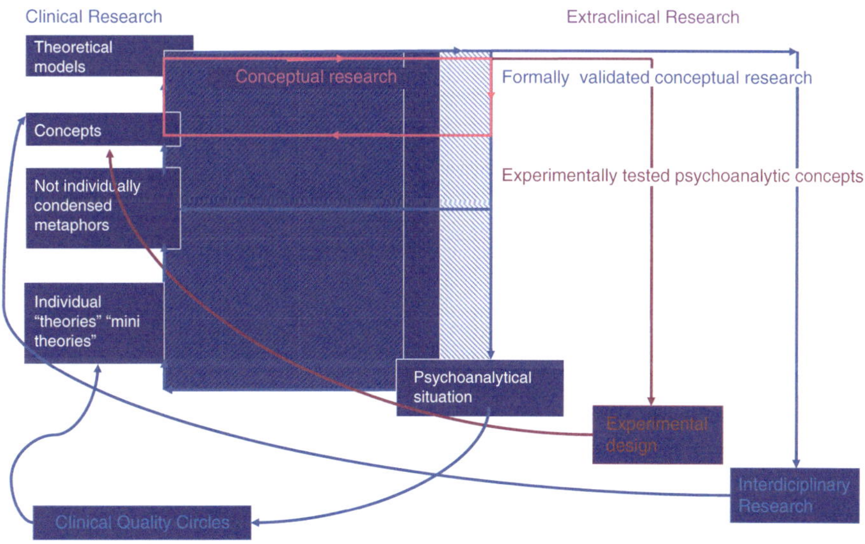


**Fig. 28.1** Psychoanalytical Research

method must take on topics again and again that are of societal relevance in order to communicate the indispensable nature of its research results to the world of the media. We think, for example, of the field of early prevention, ADHS, migration, youth violence, right-wing radicalism, nationalism, and anti-Semitism; the return of fundamentalism, religion and violence; the trauma research (e.g., as one focus working with refugees, see, e.g., Leuzinger-Bohleber et al. 2017); as well as the short- and long-term influence of new media and technologies on processes of psychic development and of modern conflicts in the realms of sexuality and object relations.

## 28.6 Summary

To summarize:

(a) Already Freud hoped that psychoanalysis by means of "objective research results" could win the acceptance in the scientific community of medicine and natural sciences. On the other hand, it was only through the insistence on its own autonomy and specifity—as a method and institution—that psychoanalysis as a scientific discipline could secure its survival and its productive unfolding in the last 100 years.
(b) In the first century of its history, psychoanalysis developed a differentiated, specific method of research for the examination of its own specific research object, of unconscious fantasies and conflicts, that it connected in diverse studies with a variety of forms of extra-clinical research.
(c) Contemporary psychoanalytic research takes place in an extreme field of tension. On the one pole exists the danger of retreating to the psychoanalytic ivory tower and refuting the dialogue with the nonpsychoanalytic community—on the other pole, the over-adaptation to, for psychoanalysis, inadequate understanding of science and therefore a loss of identity and independence. This field of tension cannot be resolved but can only be critically reflected upon and productively shaped again and again in an interdisciplinary and intergenerational dialogue. This critical reflection may also be seen as a safeguard against submission to the dominating "Zeitgeist." As it is well known, the gold of contemporary science may well be the iron of the future.
(d) The future of psychoanalysis will be dependent upon which innovative and creative insights can be found in its rich spectrum of different fields of research in the clinical, conceptual, empirical, experimental, and interdisciplinary research and be transferred into the scientific and nonscientific community. As the Nobel Prize laureate, Eric Kandel has pointed out in different of his papers: the contemporary neurosciences have opened a door for psychoanalysis which, as is demonstrated in this volume, could be creatively used for psychoanalytic research.
(e) In today's political, economical, and media-influenced "knowledge society" in which scientific experts compete at all levels for authenticity and credibility, it

has in a new way become a question of survival for psychoanalysis—if it can assert itself as a specific, irreplaceable, effective, and productive clinical method of treatment and as a theory of culture. Only when it becomes publicly visible that psychoanalysis still, through its specific research method, has developed unique and effective forms of short-term and long-term treatments and has interesting and innovative explanations to offer for the complex phenomenon of society will it then time and again exert its attractiveness as a "specific science of the unconscious."

## References


Ahumada JL, Doria-Medina R. Über Forschung. Ein kontrapunktischer Dialog. In: Leuzinger-Bohleber M, Canestri J, Target M, editors. Früh e Entwicklungen und ihre Störungen. Frankfurt: Brandes & Apsel; 2009. p. 204–18.

Alexander F. The neurotic personality of our time, by Karen Horney. New York, 1937. Psychoanal Q. 1937;6:536–40.

Altmann de Litvan M, editor. Time for change. London: Karnac; 2014.

Argelander H. Das Erstinterview in der Psychotherapie. 3rd ed. Darmstadt: Wissenschaftliche Buchgesellschaft; 1967. p. 1987.

Bachrach HM, Weber JJ, Solomon M. Factors associated with the outcome of psychoanalysis (clinical and methodological considerations): report of the Columbia Psychoanalytic Center Research Project. Int Rev Psychoanal. 1985;12:379–89.

Bachrach HM, Galatzer-Levy R, Skolnikoff A. On the efficacy of psychoanalysis. J Am Psychoanal Assoc. 1991;39:871–916.

Beenen F. Die Amsterdamer PEP-Studie. In: Leuzinger-Bohleber M, Stuhr U, editors. Psychoanalysen im Rückblick. Gießen: Psychosozial-Verlag; 1997. p. 336–47.

Bernardi R. What kind of evidence makes the analyst change his or her theoretical and technical ideas? In: Leuzinger-Bohleber M, Dreher AU, Canestri J, editors. Pluralism and unity? Methods ot research in psychoanalysis. London: International Psychoanalytical Association; 2003. p. 125–37.

Boesky D. Psychoanalytic controversies contextualized. J Am Psychoanal Assoc. 2005;53:835–63.

Bohleber W. Die Entwicklung der Psychoanalyse in Deutschland nach 1945. Vortrag auf der Tagung der DPG und DPV: 100 Jahre Internationale Psychoanalytische Vereinigung (IPV) – 100 Jahre institutionalisierte Psychoanalyse in Deutschland, Berlin, 7 Mar 2010; 2010.

Bohleber W, Jimenez JP, Scarfone D, Varvin S, Zysman S. Unconscious fantasy: an attempt at conceptual integration. In: Leuzinger-Bohleber M, Arnold S, Solms M, editors. The unconscious. A bridge between psychoanalysis and cognitive neuroscience. New York: Routledge; 2017. p. 39 65.

von Braun C. Haben Institutionen ein Unbewusstes? Vortrag auf der Tagung der DPG und DPV: 100 Jahre Internationale Psychoanalytische Vereinigung (IPV) – 100 Jahre institutionalisierte Psychoanalyse in Deutschland, Berlin, 5 Mar 2010; 2010.

Chiesa M. Can psychoanalytic research integrate and improve knowledge for clinical practice? Some reflections and an example. Scand Psychoanal Rev. 2005;28:31–9.

Colombo D, Michels R. Can (should) case reports be written for research use? Psychoanal Inq. 2007;27:640–9.

Coriat I. Some statistical results of the psychoanalytic treatment of the psychoneuroses. Psychoanal Rev. 1917;4:209–16.

Dahl H, Kächele H, Thomä H, editors. Psychoanalytic process research strategies. Berlin: Springer; 1988.

Driessen E, Cuijpers P, de Maat SC, Abbass AA, de Jonghe F, Dekker JJ. The efficacy of short-term psychodynamic psychotherapy for depression: a meta-analysis. Clin Psychol Rev. 2010;30:25–36.

Duarte Guimaraes Filho P. Klinisch-konzeptuelle Forschung im Bereich des Aufbaus heutigen psychoanalytischen Wissens. In: Leuzinger-Bohleber M, Canestri J, Target M, editors. Frühe Entwicklungen und ihre Störungen. Frankfurt: Brandes & Apsel; 2009. p. 237–52.

Ellman S. When theories touch. A historical and theoretical integration of psychoanalytic thought. London: Karnac; 2010.

Erle JB. An approach to the study of analyzability and analyses: the course of forty consecutive cases selected for supervised analysis. Psychoanal Q. 1979;48:198–228.

Falzeder E. Die Gründungsgeschichte der IPV und der Berliner Ortsgruppe. Vortrag auf der Tagung der DPG und DPV: 100 Jahre Internationale Psychoanalytische Vereinigung (IPV) – 100 Jahre institutionalisierte Psychoanalyse in Deutschland, Berlin, 6 Mar 2010; 2010.

Feldman F. Results of psychoanalysis in clinic case assignments. J Am Psychoanal Assoc. 1968;16:274–300.

Fenichel O. Statistischer Bericht über die therapeutische Tätigkeit 1920-1930. In: Zehn Jahre Berliner Psychoanalytisches Institut. Wien: Verlag Internationale Psychoanalyse; 1930. p. 13–9.

Fonagy P. Epistemological and methodological issues on process and outcome research. In: Leuzinger-Bohleber M, Kaechele H, editors. Open door review. 3rd ed. London: International Psychoanalytical Association; 2015.

Freud S. Die Frage der Laienanalyse, vol. 14. Leipzig: Internationaler Psychoanalytischer Verlag; 1926. p. 207–86.

Grande T, Rudolf G, Oberbracht C. Die Praxisstudie Analytische Langzeittherapie. In: Leuzinger-Bohleber M, Stuhr U, editors. Psychoanalysen im Rückblick. Gießen: Psychosozial-Verlag; 1997. p. 415–31.

Green A. The pluralism of sciences and psychoanalytic thinking. In: Leuzinger-Bohleber M, Dreher AU, Canestri J, editors. Pluralism and unity? Methods of research in psychoanalysis. London: International Psychoanalytical Association; 2003. p. 26–45.

Guggenheim Z, Hampe M, Schneider P, Strassberg D. Im Medium des Unbewußten: Zur Theorie der Psychoanalyse. Stuttgart: Kohlhammer; 2017.

Habermas J. Erkenntnis und Interesse. Frankfurt: Suhrkamp; 1968.

Hamburg DA, Bibring GL, Fisher C, Stanton AH, Wallerstein RS, Weinstock HI, Haggard E. Report of ad hoc committee on central fact-gathering data of the American Psychoanalytic Association. J Am Psychoanal Assoc. 1967;15:841–61.

Hampe M. Pluralism of sciences and the unity of reason. In: Leuzinger-Bohleber M, Dreher AU, Canestri J, editors. Pluralism or unity? Methods of research in psychoanalysis. London: International Psychoanalytical Association; 2003. p. 45–63.

Hampe M. Pluralität der Wissenschaften und Einheit der Vernunft – Einige philosophische Anmerkungen zur Psychoanalyse. In: Leuzinger-Bohleber M, Deserno H, Hau S, editors. Psychoanalyse als Profession und Wissenschaft. Stuttgart: Kohlhammer; 2004. p. 17–32.

Hau S. Unsichtbares sichtbar machen. Forschungsprobleme in der Psychoanalyse. Göttingen: Vandenhoeck u. Ruprecht; 2008.

Huber D, Klug G, von Rad M. Münchener Psychotherapie-Studie (MPS). In: Leuzinger-Bohleber M, Stuhr U, editors. Psychoanalysen im Rückblick. Methoden, Ergebnisse und Perspektiven der neueren Katamneseforschung. Gießen: Psychosozial-Verlag; 1997.

Hustveth S. Die zitternde Frau. Eine Geschichte meiner Nerven. Berlin: Rowohlt; 2010.

Illouz E. Gefühle in Zeiten des Kapitalismus. Adorno-Vorlesungen 2004. Frankfurt: Suhrkamp; 2006.

Jones E. The future of psycho-analysis. Int J Psychoanal. 1936;17:269–77.

Kantrowitz JL. The role of the patient-analyst "match" in the outcome of psychoanalysis. Annu Psychoanal. 1986;14:273–97.

Knight RP. Evaluation of the results of psychoanalytic psychotherapy. Am J Psychiatr. 1941;98:434–46.

Knoblauch SH. What are we trying to do when we write about the psychoanalytic interaction? The relevance of theory and research to clinical responsiveness: reply to commentaries. Psychoanal Dialogues. 2005;15:883–96.

Krause R. Allgemeine Psychoanalytische Krankheitslehre, vol. 1. Stuttgart: Kohlhammer; 1997.

Lear J. The shrink is in. In: Leuzinger-Bohleber M, Stuhr U, editors. Psychoanalysen im Rückblick. Gießen: Psychosozial-Verlag; 1995. p. 92–106.

Leichsenring F, Rabung S. Effectiveness of long-term psychodynamic psychotherapy: a meta-analysis. J Am Med Assoc. 2008;300:1551–65.

Leichsenring F, Rabung S. Long-term psychodynamic psychotherapy in complex mental disorders: update of a meta-analysis. Br J Psychiatry. 2011;199:15–22.

Leuzinger-Bohleber M. Veränderung kognitiver Prozesse in Psychoanalysen. Bd. 1: Eine hypothesengenerierende Einzelfallstudie. Berlin: Springer; 1987.

Leuzinger-Bohleber M. Veränderung kognitiver Prozesse in Psychoanalysen. Bd. 2: Fünf aggregierte Einzelfallstudien. Berlin: Springer; 1989.

Leuzinger-Bohleber M. Finding the body in the mind – embodied memories, trauma, and depression. London: International Psychoanalytical Association/Karnac; 2015.

Leuzinger-Bohleber M, Kaechele H. Open door review. 3rd ed. London: International Psychoanalytical Association; 2015.

Leuzinger-Bohleber M, Pfeifer R. Remembering a depressive primary object? Memory in dialogue between psychoanalysis an cognitive science. Int J Psychoanal. 2002;83:3–33.

Leuzinger-Bohleber M, Dreher AU, Canestri J, editors. Pluralism and unity? Methods of research in psychoanalysis. London: International Psychoanalytical Association; 2003a.

Leuzinger-Bohleber M, Hautzinger M, Fiedler G, Keller W, Bahrke U, Kallenbach L, Kaufhold J, Ernst M, Negele A, Schoett M, Küchenhoff H, Günther F, Rüger B, Beutel M. Outcome of psychoanalytic and cognitive-behavioral long-term-therapy with chronically depressed patients. A controlled trial with preferential and randomized allocation. Can J Psychiatry. 2018a; in press.

Leuzinger-Bohleber M, Kaufhold J, Kallenbach L, Negele A, Ernst M, Keller W, Fiedler G, Hautzinger M, Beutel M. Does sustained symptomatic improvement of chronically depressed patients need structural change in long-term psychotherapies? Findings from the LAC depression study comparing the outcomes of cognitive-behavioral and psychoanalytic long-term treatments. Int J Psychoanal. 2018b; in press.

Leuzinger-Bohleber M, Rüger B, Stuhr U, Beutel M. How to study the 'quality of psychoanalytic treatments' and their long-term effects on patients' well-being: a representative, multi-perspective follow-up study. Int J Psychoanal. 2003b;84:263–90.

Leuzinger-Bohleber M, Tahiri M, Hettich N. STEP-BY-STEP: Ein Pilotprojekt zur Unterstützung von Geflüchteten in der Hessischen Erstaufnahmeeinrichtung Michaelisdorf in Darmstadt. Psychotherapeut. 2017;62(1):341–7. https://doi.org/10.1007/s00278-017-0208-6.

Leuzinger-Bohleber M, Henningsen P, Pfeifer R. Die psychoanalytische Konzeptforschung zum Trauma und die Gedächtnisforschung der embodied cognitive science. In: Leuzinger-Bohleber M, Roth G, Buchheim A, editors. Psychoanalyse, Neurobiologie, Trauma. Stuttgart: Schattauer; 2008. p. 157 71.

Leuzinger-Bohleber M, Hautzinger M, Fiedler G, Keller W, Bahrke U, Kallenbach L, Kaufhold J, Ernst M, Negele A, Schoett M, Kuechenhoff H, Günther F, Rüger B, Beutel M. Outcome of psychoanalytic and cognitive-behavioral longterm-therapy with chronic depressed patients. A controlled trial with preferential and randomized allocation. Can J Psychiatry, in print.

Mancia M, editor. Psychoanalysis and neuroscience. New York: Springer; 2006.

Moser U. Was ist eine Mikrowelt? Psyche. 2013;67(5):401–31.

Norman HF, Blacker KH, Oremland JD, Barrett WG. The fate of the transference neurosis after termination of a satisfactory analysis. J Am Psychoanal Assoc. 1976;24:471–98.

Pfeffer AZ. Research in psychoanalysis. J Am Psychoanal Assoc. 1961;9:562–70.

Pincus D. Mind and brain sciences in the 21st century, by Robert L. Solso. Cambridge, 1997. Psychoanal Psychol. 2000;17:600–7.

Sandell R. Langzeitwirkung von Psychotherapie und Psychoanalyse. In: Leuzinger-Bohleber M, Stuhr U, editors. Psychoanalysen im Rückblick. Gießen: Psychosozial-Verlag; 1997. p. 348–65.

Schlessinger N, Robbins F. Assessment and follow-up in psychoanalysis. J Am Psychoanal Assoc. 1974;22:542–67.

Shedler J. The efficacy of psychodynamic psychotherapy. Am Psychol. 2010;65(2):98–109.

Shevrin H. The investigation of unconscious conflict, unconscious affect, and signal anxiety. In: Velmans M, editor. Investigating phenomenal consciousness: new methodologies and maps. New York: John Benjamins; 2000.

Strupp HH, Schacht TE, Henry WP. Problem – treatment – outcome congruence: a principle whose time has come. In: Dahl H, Kächele H, Thomä H, editors. Psychoanalytic process research strategies. Berlin: Springer; 1988. p. 1–14.

Toulmin S. Kritik der kollektiven Vernunft. Frankfurt: Suhrkamp; 1977. p. 1983.

Varvin S. Die Oslo-Studie. Eine Prozeß-Ergebnis-Studie der Psychoanalyse – Werkstattbericht. In: Leuzinger-Bohleber M, Stuhr U, editors. Psychoanalysen im Rückblick. Gießen: Psychosozial-Verlag; 1997. p. 407–14.

Vinocur de Fischbein S. Plädoyer für die interdisziplinäre konzeptuelle und klinische Erforschung von Traumnarrativen. In: Leuzinger-Bohleber M, Canestri J, Target M, editors. Frühe Entwicklungen und ihre Störungen. Frankfurt: Brandes & Apsel; 2009. p. 252–85.

Von Helmholtz H. Handbuch der physiologischen Optik, vol. 9. Leipzig: Voss; 1867.

Wallerstein RS. Forty-two lives in treatment: a study of psychoanalysis and psychotherapy. New York: Guilford Press; 1986.

Wallerstein RS. The generations of psychotherapy research: an overview. Psychoanal Psychol. 2001;18:243–67.

Wallerstein RS, Robbins L, Sargent H, Luborsky L. The psychotherapy research project of the Menninger Foundation. Bull Menn Clin. 1956;20:221–80.

Weingart P. The moment of truth for science. The consequences of the 'knowledge society' for society and science. EMBO Rep. 2002;3:703–6.

Whitebook J. Sigmund Freud – a philosophical physician. Lecture at the 11th Joseph Sandler Research Conference: persisting shadows of early and later trauma, Frankfurt; 2010.

Widlöcher D. Foreword. In: Leuzinger-Bohleber M, Dreher AU, Canestri J, editors. Pluralism and unity? Methods of research in psychoanalysis. London: International Psychoanalytical Association; 2003. p. xix–xxiv.

# Neuroscientifically Inspired Psychoanalysis: Chronic Depression as a Paradigmatical Example

29

Marianne Leuzinger-Bohleber and Tamara Fischmann

**Abstract**

What can neuropsychoanalysis contribute to psychodynamic therapy? This is the central question in the current chapter.

We therefore introduce our own brain imaging study on chronic depression and psychodynamic therapy. A central aspect of psychodynamic therapy is the access to the own memories, specifically, autobiographical memories as distinguished from semantic memories as they are targeted in cognitive-behavioural therapy.

After reviewing some recent findings, we report a single case on chronic depression and how that is related to psychodynamic therapy and brain imaging. The focus is especially put on dreams and memories with both serving and indexing inner transformations in the structure of the ego.

Hence, we consider the treatment of autobiographical memories that allows to integrate trauma into the own self as well as the underlying neurobiological pathways in the brain as crucial to future neuropsychodynamic therapy.

## 29.1 Personal Introductory Remarks

The editors of this book asked us to talk about "neuropsychodynamic psychotherapy". But as I (MLB) have discussed in my epistemological chapter (see Leuzinger-Bohleber, Chap. 28 in this volume), I am sceptical about an integration of neuroscientific and psychoanalytical concepts in clinical practice already at this stage of the interdisciplinary dialogue. To mention just one aspect: The intention to directly apply neuroscientific knowledge in a specific psychotherapeutic situation

M. Leuzinger-Bohleber (✉) · T. Fischmann
Sigmund-Freud-Institut, Frankfurt, Germany
e-mail: Leuzinger-Bohleber@sigmund-freud-institut.de; M.Leuzinger-Bohleber@gmx.de
Dr.Fischmann@sigmund-freud-institut.de



H. Boeker et al. (eds.), *Neuropsychodynamic Psychiatry*,
https://doi.org/10.1007/978-3-319-75112-2_29

is, in my perspective, naive and even harmful because it simplifies and neglects the rich clinical and technical knowledge which has been collected during the 100 years of psychoanalytical (clinical) research. How to treat a specific (depressed) patient in a specific psychoanalytical situation can best be reflected in the dialogue with other experienced psychoanalysts or psychotherapists: one reason why psychoanalysis has developed such an intensive culture of supervision and intervision (see graph in Chap. 28). Psychoanalysis is not such an exception in this respect: how to operate a tumour in the heart is best learnt and critically discussed with other experienced heart surgeons—not with a neurologist.

At the same time: new interdisciplinary knowledge is, of course, very important for the development of psychoanalysis as a scientific discipline and its concepts and understanding of the human mind. As we have discussed in other papers: psychoanalytic models have to prove to be "externally coherent" (C. Strenger) with contemporary scientific knowledge in order to remain a "science" (a "Wissenschaft") and not a closed up religion or "only" a professional trade or art. In other words, the interdisciplinary dialogue (e.g. with the neurosciences) may influence a basic attitude of the analyst as a "clinical researcher" in contrast to a "believer" which only wants to replicate "old findings" and theories again and again. A "clinical researcher" is driven by curiosity and self-critical thinking, hoping to discover new and innovative aspects of unconscious fantasies and conflicts of his patients in the psychoanalytical situation.

Such basic attitudes have, of course, not only a theoretical relevance but an (indirect) influence on the quality of the psychoanalytical practice as well. An intensive interdisciplinary dialogue is always challenging and strengthens—as a confrontation with the foreign (alien) scientific discipline—a professional self-critical attitude of a researcher in the clinical psychoanalytical situation. It also inspires the models in the mind of the analyst, his implicit (and explicit) theories, as has been discussed in psychoanalysis during the last years (see, e.g. Bohleber et al. 2007). The constant modification and widening of (implicit) theories and models of the analyst, of course—indirectly—influence his attitudes, understandings and emotions in a certain clinical situation (see graph in Chap. 28). To put it in simplified terms, although these interdisciplinary inspired models don't directly "tell the analyst what to do in a certain clinical situation with a specific patient", but they influence (and broaden)—in an indirect way—the mind of the analyst, which means his basic attitudes, his perceptions, fantasies, thoughts and even emotions. These (unconscious) cognitive-affective processes enable him to see structures in the clinical material which he/she has not been able to see before. The interdisciplinary dialogue—in this sense— widens his horizon, his professional perspectives and even his clinical intuition (see, e.g. Hampe 2016; Leuzinger-Bohleber 2015). Therefore we prefer to talk about a "neuroscientifically inspired psychoanalysis".

In the following chapter, we would like to illustrate these epistemological considerations:

(a) We refer to a neurobiologically based integrated memory model by Lane et al. (2015) in order to understand the clinical transformation processes of a severely traumatized, chronic depressed patient of the LAC depression study. This model is on a relatively high level of abstraction but enables to capture specificities in the psychoanalytical approach to chronic depression in contrast, e.g. to CBT.

(b) We also refer to the psychoanalytical model of depression developed by Bleichmar (2010) which is formulated on medium level of abstraction integrating knowledge from many different psychoanalytical schools. Some of them have integrated neuroscientific knowledge into their psychoanalytical models (e.g. on trauma and depression). By doing this we are trying to make use of the richness of psychoanalytical theories and clinical experiences in the times of "theoretical and clinical pluralism" (see Leuzinger-Bohleber 2015a, b).

(c) Finally we focus on the dream-generating model by Moser and von Zeppelin which was discussed in Chap. 8. This model on the generation of dreams integrates systematically psychoanalytical and neuroscientific knowledge even developing own terms for capturing these new integrations. As we have discussed in Chap. 8, we derived categories from this model in order to systematically and empirically investigate changes of manifest dreams of severely traumatized, depressed patient (see page 146 ff).

## 29.2 An Integrative Memory Model for Comparing Different Psychotherapeutic Approaches: Memory Reconsolidation, Emotional Arousal and the Process of Change in Psychotherapy—New Insights from Brain Science

Lane et al. (2015) proposed an integrative, neurobiologically based model of change in psychotherapy which allows to compare diverging, different pathways of psychodynamic/psychoanalytical therapy (PAT) to behavioural, cognitive-behavioural therapy (CBT) or emotion-focused psychotherapy. The model has three interactive components—autobiographical (event) memories, semantic structures and emotional responses—supported by emerging evidence for cognitive neurosciences on implicit and explicit emotions, implicit and explicit memory, emotion-memory consolidations and the relationship between autobiographical and semantic memory. According to the model, essential ingredients of therapeutic change include (1) reactivating old memories, (2) engaging in new emotional experiences that are incorporated into these reactivated memories via the process of reconsolidation and (3) reinforcing the integrated memory structures by practising a new way of behaving and experiencing the world in a variety of contexts.

The model allows to differentiate between mechanisms of change, e.g. in CBT and PAT. CBT focuses on identifying irrational thoughts that induce distressing emotions and changing the thoughts to bring about a different emotional experience (Butler et al. 2006). As described in Leuzinger-Bohleber et al (in press), CBT in the LAC study (see below) is based in Beck's cognitive therapy for depression (Beck et al. 1979) and Hautzinger's (2013) approach, which aims to reduce depressive feelings by having clients identify and reevaluate their integrative thoughts, assuming that the depressed feeling results from maladaptive thinking. According to Lane's model, these interpretations are driven by the semantic structures that derive from prior experience. New evaluative structures, once in place, enable one to experience the original eliciting circumstances or stimulus in the context of an altered emotional state that then permits updating through reconsolidation. This is one reason why "homework" seems so important to effect change in CBT (Lane et al. p. 15).

Although CBT traditionally focuses on emotion such as depression as an outcome, the model by Lane et al. highlights emotional arousal as a mediator of therapeutic success.

> Although CBT does not empathize the exploration of past memories that originally lead to development of the maladaptive response, it clearly uses exploration of similar, albeit more recent, experiences that have elicited distressing reactions. To the extent that these experiences share common characteristics with the original memories, they will also be subject to reconsolidation through the corrective experiences (p. 15)

To summarize: CBT focuses mainly on semantic memories.

In contrast all psychoanalytic schools are sharing the common fundamental ingredient of transference and a developmental perspective.

> Although there are many different schools of thought within psychoanalysis, the common fundamental ingredients of transference and a developmental perspective are important to consider in light of the integrated memory model. Time and cost considerations aside, the technique of meeting three, four or five times per week for several years creates a special opportunity to activate old memories and observe their influence on present-day construals and emotional experiences with an emotional intensity and vividness that is difficult or impossible with other methods (Freud 1914/1958). As such, this approach has the potential to offer something not available with other modalities that can have pervasive effects on a person's functioning in a wide variety of social, occupational, and avocational settings. New learning can involve improvement in function above and beyond symptom reduction, such as better self-esteem, greater ability to tolerate and manage stress, improved flexibility in social relations, a greater capacity for intimacy and the construction of a coherent life narrative that exceed what would be expected based on symptomatic improvement alone. (Shedler 2010; Lane et al. 2015, p. 16/17)

Therefore for PAT "access to memory structure is most often via old episodic memories", mainly autobiographical memories. These are also connected with (old) rules and expectations that derive from them which lead to a change of emotional responsing. "Reconsolidation revises the original memory by incorporating aspects of the new event, as well as the expectations and rules that will be applied to new situations" (p. 17).

In our view the integrated model by Lane et al. describes neurobiological mechanisms in memory functioning of so-called structural change which has been discussed extensively in literature as a central aim in psychoanalysis in contrast to CBT, which mainly focuses on symptomatic change (see, e.g. Leuzinger-Bohleber et al. 2003) (see Sect. 29.3). Structural change, as defined by increasing consciousness of focal themes of intrapsychic conflict and of deficiencies of mental structure, appears to be relevant for psychotherapeutic change in general and particularly relevant for change in long-term psychoanalytical psychotherapies.

As specified in the neurobiologically grounded model of psychotherapeutic change by Lane et al. (2015), therapeutic change is achieved in different forms of psychotherapies by different points of entry into the integrated memory structure (p. 14). The characteristic point of entry of psychodynamic therapies and psychoanalysis aiming at structural changes is episodic memories. The technique of meeting several times a week for several years creates a specific opportunity to activate old memories particularly on traumatic, autobiographical experiences of the patients. These activations make it possible to observe in detail the specific influence on present-day construals and emotional experiences in the transference with an emotional intensity and vividness which, as mentioned above, is difficult or impossible with other psychotherapeutic methods. It enables to change problematic implicit emotional procedures through insight based on observations in the transference and thus interrupts the automatic behavioural enactment. The associated "underlying" emotions become conscious, a presupposition for changing behaviour. In psychoanalytic terms, the emotionally intensive working through of the unbearable traumatic emotions and episodic memories in a holding and containing relationship to the therapists allowed the chronic depressed patients (as Mr. X, see Sect. 29.4. and Chaps. 8, 28) in PAT a psychic integration of the traumatic experiences in their (conscious) idiosyncratic autobiographic memory. Thus, the traumatizations could be integrated into a more mature self and identity which no longer were unconsciously determined by past traumatizations (see, e.g. Bohleber and Leuzinger-Bohleber 2016; Negele et al. 2016). Such processes need a high emotional intensity in the therapeutic relationship and take time (see, e.g. Monsen et al. 1995; Leuzinger-Bohleber et al. 2003; Sandell and Shedler 2010; Rudolf 2012).

## 29.3 Trauma and Depression[1]: Some Psychoanalytical Conceptual Remarks

Current research in depression postulates a multifactorial model of explanation: genetic vulnerability, experiences of early relationships, environmental conditions, traumatization and acute individual, institutional and social stress situations combine to produce severe depressive conditions (cf., among others, Holmes 2013; Schulte-Körner and Allgaier 2008, see also Chaps. 6, 28 in this volume). To this multifactorial explanatory attempt, psychoanalysis contributes specific knowledge on the unconscious, individual roots of depressive conditions: in every analysand we find complex, idiosyncratic, unconscious determinants which, in a specific form of depression, flow into a specific life situation—every depression shows its, distinctive, individual face. "There is no unitary concept of depression…" (McQueen 2009, p. 225). Depression is no closed, clearly circumscribed and static category, but an entirely unique, individual process. Thus, in psychoanalytical literature, we discover a wealth of conceptual works on the emergence and treatment of depression.

[1] The following section is based on a previous publication by Leuzinger-Bohleber (2015).

For the sake of brevity, we refer here to Hugo Bleichmar's (2013) tables for a short overview of contemporary psychoanalytic explanatory models of depression in various psychoanalytic schools and thus taking into account the current pluralism in theories in contemporary psychoanalysis (see Leuzinger-Bohleber 2015a). In these, Bleichmar summarizes a model which sketches the multiple, though not exclusive paths of genesis which can lead to a chronic depression (see Fig. 29.1).

For Bleichmar (1996, p. 941), Freud's (1916–1917) work *Mourning and Melancholia* still represents one of the fundamental texts for a psychoanalytic understanding of depression. Freud characterized depression as a reaction to a real or imagined loss of an object. Hence, he defines depression as a reaction which is not only connected to a "real" loss of an object, to an idea, a self-image, etc., but one which also depends on the way in which the loss is linked to unconscious fantasies and thoughts. In *Inhibitions, Symptoms and Anxiety*, Freud (1926) emphasizes the insatiable yearning of depressive persons following the loss of an object: the satisfaction of drives, the desire for commitment, narcissistic needs as well as the yearning for well-being can no longer be fulfilled by way of a real or fantasized object. Thus, a central feeling of helplessness and hopelessness dominates in depression. The self experiences itself as powerless and impotent. Those emotions directed at the desired object are deactivated: apathy, inhibition and passivity belong to the results (cf., e.g. Bibring 1953; Bohleber 2005, 2010; Haynal 1977, 1993; Jacobson 1971; Joffe and Sandler 1965; Klein 1935, 1940; Kohut 1971; Leuzinger-Bohleber et al. 2010; in press; Mentzos 1995; Steiner 2005; Stone 1986; Taylor 2010). Rado (1928, 1951) noted severe rage and aggression as an attempt to retrieve the lost object. He described, furthermore, that severe self-accusation has the function of attenuating feelings of guilt and of regaining the love of the superego through self-punishment (cf. also Abraham 1911, 1924; Blatt 2004; Bohleber 2012; Campos et al. 2011; Freud 1916–17; Jacobson 1971; Kernberg 2006; Klein 1935, 1940;

| Scales | Clinically significant traumatization LAC, N = 367 |
|---|---|
| Emotional Abuse | 222 (60,5%) |
| Physical Abuse | 97 (26,4%) |
| Sexual Abuse | 91 (24,8%) |
| Emotional Neglect | 189 (51,5%) |
| Physical Neglect | 117 (31,9%) |
| **Trauma Overall** | **278 (75,7%)** |

**Fig. 29.1** Child Trauma Questionnaire (CTQ)

Kohut 1971, 1977; Rado 1928, 1951; Steiner 2005; Steiner and Schafer 2011; Taylor 2010). Similarly, Bleichmar (1996, p. 942) emphasizes the connection between aggression, guilt feelings and depression and, in so doing, distinguishes between different forms of psychodynamic processes which, at high intensity of pain and when lasting for extended periods of time, can lead to an extreme defence of psychic life in general, to an inner state of "non-existence" (cf. Bowlby 1980; Kennel 2013; Ogden 1982; Spitz 1946; Steiner and Schafer 2011; Weiss 2012) (cf. pathways in the table to the above right).

By contrast, Kohut (1971) and others explain that, frequently, it is not guilt feelings that comprise the central motifs underlying a depression, but shame and narcissistic suffering. He refers to tragic disturbed self-regulation as determined by an immature ego-self, ideal-object and super-ego, as the decisive components involved in the emergence of depression. Various pathological developments lead to different forms of depression (e.g. to mania, to anaclitic depression or to guilt depression) (cf. dynamic above right in Fig. 29.1).

A further consequence of extreme feelings of helplessness and powerlessness are phobias and anxieties: the representations of the self as weak and impotent lead to a psychic state in which everything is perceived as dangerous and can overcome the weak ego. In this connection Melanie Klein (1935, 1940) underlined that the fear of persecution vitiates psychic functioning, the development of the ego, the object relations and, finally, reality testing, which leads to depression (dynamic above right in Fig. 29.1) (see also Mentzos 1995). John Steiner (2005) describes the "psychic retreat" as a chronified psychic state of the depressed with a pathological organization of the inner objects and the self. The projection of the needy and aggressive parts of the self leads to its impoverishment and to a secondary dependence on the real objects.

The identification with a depressive parent may also underlie a depression (cf. also Freud 1965; Hellman 1978; Leuzinger-Bohleber 2001, 2013b; Markson 1993; Morrison 1983). In addition, all conditions that lead to deficits in the ego (inner conflicts, traumatic realities, deficits in the parents' egos, deficient ability to mentalize, etc.) complicate the possibilities of entering satisfying object relations, etc. and thus increase the probability of suffering from a depression (cf., e.g. Fonagy 2010; McGinley and Varchevker 2010) (path to the bottom right of Fig. 29.1).

Hence, there are several ways that could lead to the central depressive sense of hopelessness and powerlessness: neither of the paths has shown itself as an obligatory condition. Each is determined by various factors or areas of the respective pathology.

Bleichmar (1996) also mentions the influence of traumatic, external realities in the genesis of depression (cf. also Balint 1968; Baranger et al. 1988; Brown and Harris 1978; Winnicott 1965). However, as discussed in the following, the link between trauma and depression appears far more dramatic and causal than has been represented, to date, in most of psychoanalytic literature on depression: although single authors have made mention of this in recent literature, the central role which traumatization plays in the genesis of severe depression, in my view, continues to be underestimated (Blum 2007; Bohleber 2005, 2012; Bokanowski 2005; Bose 1995; Bremner 2002; Denis 1992; Hovens et al. 2010; Leuzinger-Bohleber 2010, 2013a; Leuzinger-Bohleber et al. 2013; in press; Lubbe 2011; Skalew 2006; Taylor 2010; Varvin 2003).

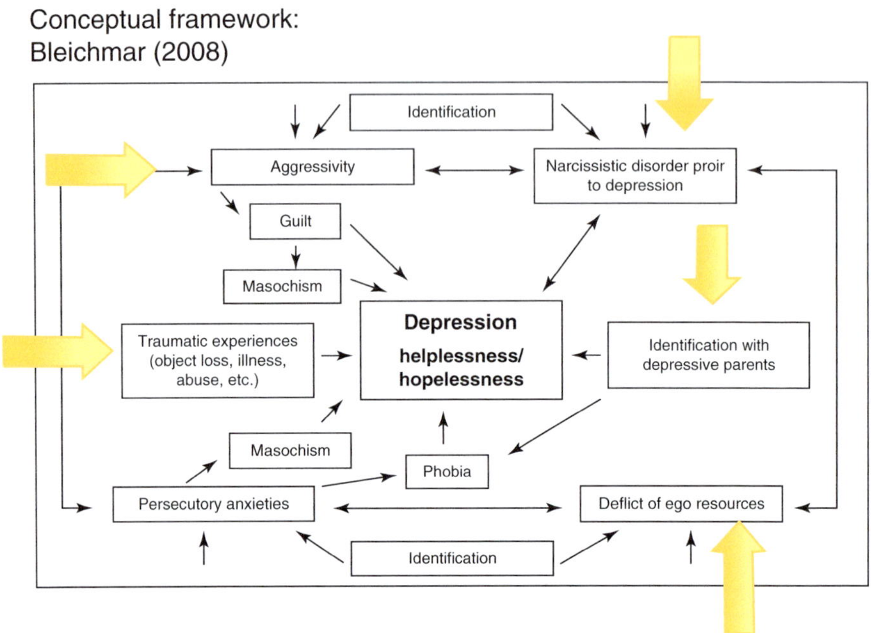


## 29.4 Selected Findings from Neurobiological, Epigenetic Studies and from Basic Research

And yet not only psychotherapy studies but also studies from various neighbouring disciplines discuss the connection between trauma and depression (overview, among others, in Boeker and Seifritz 2012; Schore 2012; Holmes 2013). Here are just a few, selected examples on the matter:

Hill (2009) emphasizes, for example, in his general article, that numerous studies have meanwhile proven that the probability of suffering from a depression as an adult is increased by the early loss of the parents or an experience of early emotional neglect (Hill 2009, p. 200; Bifulco et al. 1987; Hill et al. 2001). Fergusson and Mullen (1999) have similarly shown in a substantial survey of literature on the subject that in the case of childhood victims of sexual abuse the risk of suffering from depression in adulthood increases considerably. According to Lynne Murray (2009), this is the most important result of the major Cambridge longitudinal study, which shows that children of depressive mothers are at greater risk, beginning with problematic neurobiological reaction patterns on the HPA axis, through to depressively coloured cognitions about the self and other social problems and psychosomatic sicknesses (Cf. also Ammaniti et al. (in press); Feldman 2012; Kernberg 2012; Mayes 2012, Schechter 2012).

Finally, also worthy of mention are studies of twins showing that unipolar depression reveals itself as "moderately hereditary" (Hill 2009, p. 202; Kendler et al. 2006). However, especially relevant for psychoanalysts is the fact that first epigenetic studies tend to show that genetic vulnerability only leads to a depressive sickness in cases in which the individual experiences an early traumatization at the same time. Thus, in 2003, Caspi and his research team published a fascinating paper in science, "Influence of life stress on depression: Moderation by a polymorphism in the 5-HTT gene" (Caspi et al. 2003). In a prospective longitudinal study, the researchers tested in a representative birth cohort why stressful experiences lead to depression in some people but not in others. A functional polymorphism in the promoter region of the serotonin transporter (5-HTT) gene was found to moderate the influence of stressful life events on depression. Individuals with one or two copies of the short allele of the 5-HTT promoter polymorphism exhibited more depressive symptoms, diagnosable depression and suicidality in relation to stressful life events than individuals homozygous for the long allele. Their epidemiological study thus provided evidence of a gene-by-environment interaction, in which an individual's response to environmental insults is moderated by his or her genetic make-up. However, this finding could not be replicated in a recent meta-analysis; it thus remains questionable as to whether the serotonin transporter genotype and negative (traumatic) life events really predict depressive illnesses (Risch et al. 2009).

Steven Suomi (2010), a student of Harlow, was also able to prove the influence of early separation trauma in rhesus monkeys at a neuromolecular level. An earlier separation from the mother animal releases depression, aggression and anxiety as well as massively disturbed social behaviour. However, without an early separation, the monkeys developed normally in spite of the proven genetic (neuromolecular) vulnerability (cf. also Medina 2010). What remains relevant for psychoanalysts is the proof that the triggering of the 5-HTT allele could be stopped in cases in which the monkeys were returned again to a caring mother animal after a lapse of a few days: analogous to the classical study of hospitalization studies by René Spitz, the psychotoxic effects of traumatization were attenuated in cases in which separation did not last too long, and an empathetic replacement object existed.

Goldberg (2009) concludes his overview of recent studies in this field:

> These interactions between gene and environment, between behaviour and genotype are important in the way they provide explanations of how the many different features that make-up the 'depressive diathesis' arise. However, they have a much wider significance. They provide a possible pathway by which changing interpersonal and cultural factors across the generations can be cause as well as effect of genotype, and through which changes in human culture might possibly operate as an accelerator of evolutionary processes. In summary, we see that adverse environmental conditions are especially harmful to some particular genotypes, leaving the remainder of the population relatively resilient; research in this area is expanding very fast—and we may expect many more advances in the years to come… (p. 244–5)

Kernberg (2000) draws similar preliminary consequences from these studies:

> Early separation trigger depressive affects, a chain reaction of rage, dispair, and despondency ad thus corresponding neuro-hormonal correlates, as we now know, in humans as well as other primates (Suomi 2010). The link between felt emotion and neurochemical response begins to connect the psychoanalytic theory of internalized object relations with biological research into the genetic and neurobiological determinants of aggressive and depressive affect (p. 2).

And Bosch and Wetter (2012) add to this perspective:

> What was especially interesting was the result that patients with childhood traumas (early loss of parents, experiences of violence, sexual abuse, neglect) clearly profit from psychotherapy considerably more than patients who have not suffered from trauma. With these patients, only psychotherapy was not only more effective than medicinal mono-therapy, but also the combination of both processes (psychotherapy/medicinal treatment [ML-B]) only led to minimally improved results … The increase in knowledge over the last two decades has revealed complex connections between hormones, genes and environmental influences on the human psyche, while at the same time opens the foundation for individualized, therapeutic interventions (p. 376)

Finally, the epigenetic findings concerning the transgenerational transmission of depression and trauma are most relevant for psychoanalysts. Holmes (2013), after discussing the major findings of epigenetic research concerning mood disorders, summarizes: "… (the developmental environment), through epigenetic mechanisms, inscribes itself in an individual's biology, which is then transmitted to future generations" (p. 76) (see also Boeker 2013; Peterson 2013; Reinhold and Markowitsch 2010, p. 22).

Thus the new epigenetic research adds a new dimension to the psychoanalytical knowledge, although the results of the epigenetic studies are still the subject of controversial discussion (see, e.g. Rutter 2009, p. 1288).

Ultimately, interesting insights may also be gleaned from basic research on the unconscious continued effect of traumatization through so-called embodied memories, such as in the field of the biologically oriented "embodied cognitive science". "Embodied memories" are evoked by structural analogies of current conflict situations (e.g. following separation from a love object), through the sensomotoric coordination in the here and now, a current interactive situation of analogous emotional, cognitive and psychophysiological reactions that are akin to the original traumatic experience. As discussed in various papers, the concept of "embodiment" proves itself as exceptionally helpful in recognizing traces of traumatization suffered by the patient's body in the transference situation and to use it as key for her psychoanalytical understanding (cf. Leuzinger-Bohleber and Pfeifer 2002, 2011; Leuzinger-Bohleber 2015, see also Chap. 6 in this volume). Thus, it was decisive for a chronically depressive analysand in her third psychoanalysis that the extremely burdensome symptoms that she had hitherto understood as "psychotic" or "bipolar" could be induced by way of analogous sensomotoric coordination, such as at the beginning of her traumatic polio disease. The mutual understanding of these "embodied memories" finally enabled her to attenuate her "psychotic" breakdowns and dissociative states and even to partially overcome them—an enormous relief for herself but also for her marriage and the relationship to her children (cf. Leuzinger-Bohleber 2008; Leuzinger-Bohleber and Pfeifer 2011).

The few examples taken from studies of affiliated disciplines must suffice here for highlighting the influence of earlier traumas in the genesis of severely depressive sickness and to also take account of their transgenerational transmission, as well as "embodied memories" to the unprocessed extreme experience of childhood in new conceptualizations.

## 29.5 The LAC Study Comparing Outcomes of Psychoanalytic and Cognitive-Behavioural Long-Term Therapies with Chronic Depressed: An Research Example

One of the unexpected findings of the LAC study is the frequency of severely traumatized, chronic depressive patients in currently ongoing comparative therapy studies on the short-term and long-term outcomes of psychoanalytic, as compared to cognitive-behavioural, long-term therapies. This is a large-scale multicentric study carried out in Frankfurt am Main, Berlin, Hamburg and Mainz by a large group of researchers (Chairs: M. Leuzinger-Bohleber, M. Hautzinger, M. Beutel, W. Keller, G. Fiedler, B. Rüger; see http://www.sigmund-freud-institut.de). The design has been published in TRIALS (Beutel et al. 2012). In this study we have interviewed more than 500 chronically depressive patients. Two hundred and fifty two were included in the study following strictly defined inclusion criteria. The patients had been diagnosed according to DSM-IV by independent evaluators in (psychiatric) SKID interviews. In the first year of treatment, the patients (and the analysts) are investigated by different instruments (questionnaires, interviews, etc.) every 3 months and, in the following 3 years, every 6 months. Follow-ups after 3 and 5 years were planned (for more details, see http://www.sigmund-freud-institut.de).

We will publish the first results of the study in English in 2018 (see Leuzinger-Bohleber et al. in press, a and b). In this paper we would like to focus on one finding from the extra-clinical as well as from the clinical parts of the study concerning the topic: chronic depression and trauma (see Chaps. 6, 8, 28 in this volume). As already mentioned, we stumbled across the unexpected result of how many of the chronic depressed patients in our sample are severely traumatized.

The patients filled out the Child Trauma Questionnaire which contains questions concerning the following five subscales (see http://www.sigmund-freud-institut.de) (see Fig. 29.1).

Seventy-six percent of severely traumatized patients in our sample had, according to the Child Trauma Questionnaire, experienced severe traumatization (self-rating of the patients). These findings were confirmed by the clinical psychoanalytical research, by the ratings of the treating psychoanalysts from over 100 ongoing psychoanalytical long term therapies forming part of the LAC studies in Frankfurt. The analysts filled out a questionnaire with similar categories as those of the Child Trauma Questionnaire. In the evaluations of this subsample of analysts, even 84% of patients had been severely traumatized in their life histories. Negele (2016) using another sample of the LAC study found that even more chronic depressed patients were diagnosed as severely traumatized by their treating psychoanalysts (see Negele et al. 2016).

In order to illustrate the relationship between chronic depression and trauma, as well as psychoanalytical technique which is inspired by neuropsychoanalytical knowledge, we would like to shortly summarize one of the psychoanalyses of the LAC study.

## 29.6 "As far as I can Remember: I Always Have Been Depressed..." (Mr X.) Assessment Interviews and Short Biography[2] (Case Study: M. Leuzinger-Bohleber)

The patient has come to the outpatient department of the Sigmund Freud Institute. He reports that he has in fact always been depressive but that since two years this state becomes steadily worse. He mentions that he can hardly concentrate at work and suffers from severe sleep and eating disturbances—"I feel empty inside, like a dark hole—I can't imagine to be important for anybody, not even for my wife…". Before the interview the patient has not left his dark room for several weeks.

On enquiry he reports that his wife has had three miscarriages, the last one 2 years ago. "This time she had been 3 months pregnant—therefore we had already imagined a life with the baby. I was very sad…." He reports that this event strengthened his depressive mood but that it had occurred before, actually had always existed.

The patient is the younger one of two sons. His mother is a housewife and his father is a natural scientist. During the assessment he does not have a single memory of the first years of his life, only of the time at primary school when his mother increasingly slipped into alcoholism. "It was horrible. We had to hide alcoholic drinks from her or even had to take them away from her. She was angry, cross, uncontrolled, yelled at us or was just drunken. My father was very helpless and took refuge in work. The whole issue was up to me and my brother—it overshadowed my whole childhood and adolescence. It wasn't possible to bring friends home. If this however once happened, I was horribly ashamed, when they found out what was going on at home…." Fortunately the patient had a close and good relationship to his older brother (+2 years). "The whole situation might have been even worse for him. He has even bigger problems than me. He isn't married and still lives with my mother and he suffers from a severe chronic illness. He is also a natural scientist. We are very similar. He is also very secluded, has hardly any friends and can hardly express himself…."

The patient describes extensively his social isolation and severe feelings of inferiority since his first years in school and particularly during adolescence. "I could not imagine ever managing to get a woman—I always found myself unattractive, too inhibited and withdrew into my own world. I play the guitar since I was 14, mostly just for me alone, in the past sometimes in a group." After his A levels, he completed the basic military service. Also during this time he suffered because he hardly built any friendly contacts.

Afterwards he studied a field in natural sciences (like his brother by the way) and successfully completed the university time. As a student he found a job as a research assistant in a project and became a development engineer in a big industrial company. When he was 30 years old, he met his future wife who is a natural scientist

[2] Mr. X. agreed to publish material from his psychoanalysis. In order to protect confidentiality some biographical and socioeconomical data are actively changed although without destroying the "narrative truth".

too. It was her who gave him the advice to start a psychotherapy when she noticed that his state got worse and worse.

The patient was highly motivated for the treatment. Already during the assessment interviews, he showed a surprisingly differentiated capacity of self-reflection and was able to report dreams. *He again and again dreams of standing on a high object, e.g. a bookshelf, which suddenly starts to quiver under him. It increasingly quivers until he eventually falls down and wakes up in a panic*. These associations lead to his mother's constantly increasing addiction, which literally made the "ground under him quiver" when he was a child and pushed the whole family into the abyss. The patient responds with another dream to the attempt to understand this dream in the first assessment interviews and his hopes connected with a therapeutic treatment: "I'm getting around by bicycle, wanting to reach any destination...then I suddenly stand in front of an abyss. I see my brother and a companion down in the valley. I take a big ladder and want to climb down to them into the abyss. But when I'm on the ladder, it starts to shake– it shakes and shakes so that I get scared to fall down. My brother's companion shouts to me that this would not work out and that I should ascend the ladder again and take the path down further right... I follow his advice and really find my way down..." The patient tells me that this is the first dream in which he is not alone and from which he does not wake up in panic and instead finds a solution based on the advice of someone else. Therein we see his hope to find a way out of his loneliness and to leave behind the "shaking ladder" with the help of the psychoanalytical relationship with his analyst, prognostically an indicator for a beginning of a positive transference.

*Short Summary of the Psychoanalytic Process with the Focus on the Changes in the Manifest Dream Contents*

Of course (MLB) I can only summarize fragments of the psychoanalytical process (4 sessions a week, 5 years). In order to make this summary a bit less arbitrary, we discussed the material in the so-called weekly "clinical conferences" of the LAC study, following the expert validation in the "Three level of observation model", developed by the clinical observation group in the IPA (see, e.g. Altmann 2015). My chosen focus for systematically studying transformation processes in this psychoanalysis had been the changes in the manifest and latent dreams.

*Nightmares and Early Enactments of the Trauma in the Transference*

During the initial phase of psychoanalysis, the patient seems highly motivated for the treatment, has hardly cancelled any session and tried to get an alternative date when I had to cancel a session. The conflicts at work escalate during the first months of the treatment. He often feels unable to concentrate or work creatively. Additionally due to reorganizations at his company he has to switch the project, leave the previous team and fit in a new work group. Besides the conflicts with a married colleague he has fallen in love with escalate. One day the colleague's husband comes to see him and blames him that his wife developed a severe neurodermatitis due to her relationship to him. For this reason she had to be hospitalized. The patient reacts with feelings of guilt and an aggravation of his depressive states: Once again—due to his unconscious omnipotent conviction to be able to destroy his love object—he feels guilty about a wife's mental illness (his mother, his wife after the miscarriage, now his colleague). It seems to be mentally easier for the patient to blame himself and see himself as the source of "the women's" misfortune instead of bearing the massive feelings of helplessness and admitting that he has fallen in love with his colleague because he has turned away from his wife after the traumatic loss of their unborn child and unconsciously blames her for the miscarriage and childlessness. The patient is only capable to experience his depression; he has not access to his own aggressive impulses.

For a long time, it is difficult for the patient to freely associate. It is often difficult to reach him emotionally—many analytic sessions deal with depressive self-accusations, massive inferiority feelings and the unconscious belief "to be a strange loner, who no one likes, much less isn't loved…." As one strategy to cope with his extreme anxieties to give up his control, he is working with dreams: The patient often remembers dreams, often nightmares, and fills whole sessions with them. The dream subject in these narrations is often flooded by waves, stuck in narrow channels, tortured or persecuted. After a conflict with the married woman at work, mentioned above, he dreamt:

> I was condemned to death and wanted to hang myself in the death cell…. I woke up in panic…

Dreams like this one (and many others during this time of psychoanalysis) may illustrate the enormous feelings of helplessness and archaic forms of guilt feelings. These feelings are in the centre of Mr. X's depression: a basic feeling of being the passive victim of strokes of fate which he can do little to counter, as he was confronted during the 12 years of growing up with a severely ill mother (see graphic of Bleichmar mentioned above). Besides, due to his pathological early object relationship in the first years of life (severely depressed mother, missing father), he probably hardly had been able to integrate his early sadistic and destructive impulses and phantasies and to regulate his unbearable affects.

*Approaching Aggression and Containing Objects*

After 6 months of psychoanalysis, the manifest dream contents start to change. As in the dream, mentioned above, which was told during the assessment interviews (with his brother), now the dreamer is not alone in the dream anymore. Other persons are present. The dreamer also seems to have at least some chances to decide "autonomously", as in the following example:

> It has been a very short dream. I dreamt that I am in a room together with a friend. He explains to me that I will be executed. I can choose between two possibilities: either to have my head being cut off or to be killed by a cord hanging around my neck in a fixed position which would be then strangulated by a running horse… I awoke in panic….

The process of regaining some control and activity in the dreams continues. In the 8th month of psychoanalysis, after Mr. X. had spent one night in the sleeping laboratory, he reports the following dream:

> I was in a party with my wife. I had to go to the toilette and found one in rooms in another floor. I could not close the door and had the feeling that someone is following me. There had been two men—an older and a younger one. The first one had a strange outlook. He had thick glasses on his nose and stared at me continuously. This was threatening. I tried to chase him away with the toilet brush.... And afterwards I woke up because I really was kicking with my foot.

The associations lead to experiences of helplessness—in the sleeping laboratory where he felt unprotected but also in the psychoanalytical sessions. Memories arise of scenes of complete loneliness during adolescence. We talk about his wish "to kick someone" instead of being "a good guy"—"Well—in the dream you are defending yourself with a toilet brush in order to protect yourself... this seems to be an important, active step in contrast to the situation in the laboratory last night or here on the couch—and probably also to many situations as an adolescent where you felt like a passive victim..." (comment of the analyst).

*Differentiation Between Self- and Object Representations and Positive Transference*

Then a series of dreams occur in which his ill brother appears. *Once he discovers that his brother is tortured, he flees despite his feelings of guilt for not helping his brother.*

Among other things it becomes possible to recognize massive feelings of guilt thinking of improving his job situation which for him would feel as if he also overtakes his brother in his job. This is the reason why he often despises colleagues and also his wife (and his analyst) who (in his eyes) concentrate on their career. "If you give up making a career, you leave this domain to your brother. According to your fantasies, you should not also be successful in your career: eventually your brother doesn't have a wife, is ill and still lives at home. You are feeling responsible for his life and don´t realize what a burden it might be for you to have such a handicapped brother..." He reports the following dream in the next psychoanalytic session:

> In the dream I was with my brother. We both were sitting in the car. He was driving but drove faster and faster and did not control the car. I said: I am now counting 1, 2, 3 and then I am taking over the wheel. And I really was able to prevent a catastrophy...

This is an impressive example for the successive differentiation between the self-representation and object representation (the brother) as well as the struggling of the dream subject to gain activity and control in order "to prevent catastrophes..."

*Mourning, Revenge, Self-Destructiveness and Creativity*

In the third year of psychoanalysis, the patient feels able to work most of the time and is partially able to solve conflicts with colleagues at work in an open and productive manner. He is able to reflect his tendency towards social withdrawal in order to work against it. He has partly regained his creativity and joy at work although he still suffers from depressive breakdowns. Therefore he expects that a continuation of the treatment will among other things help him to have the courage to change his job in which he often feels intellectually under challenged or to work his way up in his company. He feels that for this he needs more analytic work, especially in order to no longer be determined by his fear of rivalries and competitive situations and his massive inhibition of aggression.

Further processing that his marriage will remain childless is still painful. The female colleague mentioned above has just born a child. This causes intensive grief and a new crisis for the patient connected with intensive envy also towards the analyst who has children.

The massive fears to destroy his objects through his archaic revenge impulses and aggressive fantasies become now accessible in the transference and open a window for the psychoanalytical understanding. For a long time, they still provoke massive fears and defences in the patient.

This is also shown by the dreams of this time. *In a dream people are destroyed by hurricanes. He is in a panic and does not dare to watch how the people are lifted up in the air by the hurricane and are smashed afterwards.* This dream shows us his unconscious belief that his aggressive impulses could destroy his love objects and lead to an irrepressible catastrophe (like his mother's alcoholic disease). The psychoanalytic understanding of these archaic fears and omnipotent fantasies and the increased psychic integration of his aggressive and libidinal impulses are often in the centre of the psychoanalytical work in the transference during these months. Closely connected to this is the understanding of the psychic function of his rigid ego ideal and superego structures. It seems favourable for the prognosis that the patient increasingly shows some signs of humour and partially revised his self-image as "low-maintenance analysand". He partially endures intense conflicts in the analytic sessions although they still threatened him. Thus there seems to be some indicators that the psychoanalytic treatment of this severely traumatized patient leads to a regaining of trust in a "helping internal and external object", connected with a felt reduction of his loneliness as well as an improvement of his narcissistic self-regulation.

After a session, in which we are talking about his terrible loneliness as an adolescent again and that he constantly felt like an "odd bird" (ein schräger Vogel), he starts the next session:

Mrs. Bohleber—you will not believe it. I really have dreamt of a "schräger Vogel" (an odd (oblique) bird):

"There was a large, black "odd bird" which was hopping on the street: It looked really interesting, original und loveable. Then suddenly a sport car stopped. Very fancy young people got out of the car and looked at the bird. They wanted to take the bird with them but I protested and did not allow them to take it away: "The odd bird does belong to me…", I said.

The associations lead to the just mentioned self-perception as an "odd bird" in his adolescence. In his adolescence he "always felt different than the other, strange, not loved. I felt I didn't belong…". At that time he tried to make a virtue out of necessity: "I am just different from them, so what…." Thus during these months in psychoanalysis, the patient deals with his own part in his social isolation, e.g. his arrogant scorn for the fun-loving colleagues during military service or his angry withdrawal from his colleagues at work who forgot to invite him to have lunch with them. Working through this complex of problems leads to a change of his position in the outside reality as well as in the transference.

Let me illustrate the summarized changes in the manifest dreams as well as in the psychoanalytical work with dreams by one session in the third year of psychoanalysis. According to the clinical evaluation of the already mentioned group of psychoanalytical experts,[3] it illustrates some of the transformations which the analysand already has gone through but at the same time that the working through of his severe traumatizations was still in progress, which meant that psychoanalysis had to continue.

*Section of an Analytic Dream Session at the End of the Third Year of Analysis Indicating Psychoanalytical Transformations*

P. started the session right away. "I had two dreams last night. One was quite funny: I played with the famous jazz guitarist Ralf Towner. I went quite well and made fun. I didn't fail and the neck of the guitar was also not soft[4] (laughs). The guitarist played along my improvisations and held back. Of course I knew that he is better than me, but this did not play a role—it was just good fun…"

The associations imply that Ralf Towner gave him much room. He reports that indeed at the beginning he was nervous, but later it was just good fun. He played a relatively easy melody and did not overextend himself on it as usual. Usually his perfectionism often plays tricks on him and blocks him, e.g. when playing a trio with his neighbour they would have chosen a too difficult piece of music.

A following long passage deals with "my perfectionisms" which especially during the depressive periods removed joy from him. He often thought in advance: "It isn't worth it anyway…".

A.: "Probably this was the voice of "the black dog."[5]… but in the dream the famous guitarist was interested in you…"

P.: "Yes, he gave me much room and did not show that he can make everything better…. Thus I wasn't under such a pressure to perform…"

A.: "And you are allowed to present yourself and to develop your ambition…"

P.: "That's true—I was probably ambitious … in the dream…"

Thereupon he is concerned with the issue of ambition for a relatively long time—an issue we talked a lot about during the last sessions because he adjudged his wife's and colleagues' ambition. The guitarist who is about 70 years older reminds him of his grandfather who was interested in him in a comprehensible manner. He remembers that when he visited his mother and brother at the weekend they intensively talked about the grandfather. His brother used a meaningful phrase: When we visited one of the grandparents, we just have been there—when we visited the other ones we just were visitors…

---

[3] See summary below.

[4] The patient refers to another "funny" dream. Before the dream he had a conflict with his wife which wasn't treated openly. Instead the conflict led to an erectile dysfunction. Then he dreamed that he played on a guitar which had a very soft neck…

[5] Mathew Johnson: "......." has created the metaphor of the "Black Dog" for his depression. Mr. X. has read the book and sometimes refers to it during the psychoanalytic sessions.

A.: “And the guitarist is also simply there, as I during our psychoanalytic sessions.”

P.: “He backs off and just let me do…and this is just good the way it is…”

We again talk about how much energy it cost him to survive with his alcoholic mother. Therefore he did not experience enough that his parents were interested in him and that his performance was also well without perfectionism. Once again we talk about his job situation and that he thinks about maybe becoming an audio engineer. This could be a possible profession that allows him to combine his creativity and perfectionism (In an earlier session he talked about the musical artist Ray Kurzweil). He reports that he wants to create a studio at home; however “this might perhaps not bring that many benefits…it is just fun…I have wanted this for a long time…”

P.: “Somehow the dream was calming… In normal life I often block myself by a high performance…”

A.: “And in your dream you simply play with the world class—thereby ambition does play a role, but this does not exclude fun and pleasure…”

As the manifest dream content as well as the associations may illustrate, Mr. X. has refound his sense of humour. In the dream this is an indicator for a growing capability of the dreaming self to create a certain distance to feelings and thoughts and to create a sense of control and self-agency. Looking back to the transformation processes during psychoanalysis, the humorous creation of the “odd bird” in the dream mentioned above was a “turning point” in the treatment: The analysand had refound his sense of humour an indicator for a line of identifications with a “good inner object”. In the terminology of Moser, the model of a depressed self and object no longer dominated completely the psychic reality of the analysand (the dream scene) and the generation of dreams but was replaced by “embodied memories” of more adequate, less traumatic early object relationship experiences (probably with his analyst building up on early embodied experiences with his mother before she became addicted as well as with his father and grandfather) in which positive affects as well as a “mobility of the self” (self-agency) have been present.

The same characteristics can be found in the dream just mentioned: The dream self can liberate itself from the traumatic paralysis: supported by a “good object” (the famous guitarist, the listening analyst), he can show his capabilities and does not punish himself in a depressive modality anymore (as becoming impotent in the conflict with his wife, mentioned above). He no longer seems to be inhibited completely in the fulfilment of his exhibitionistic wishes.

## 29.7 Summary

In this chapter we tried to illustrate what we understand as a *neuroscientifically inspired psychoanalysis*:

(a) We referred to an integrated memory model which integrates the state of art of neurobiologically based memory research (by Lane et al. 2015). We referred to this model in order to demonstrate the *specific entry of psychoanalysis* into the integrated memory structure. The chronic depressed patient, e.g. Mr. X., discovered unconscious (embodied) memories of his idiosyncratic early traumatizations which mainly determined the chronic depression in the transference. The high emotional intensity was—in accordance with Lane et al., a presupposition for understanding the unbearable quality of the trauma in the presence of a new human relationship: the holding and containing relationship to the analyst. These processes were a presupposition to understand the specific embodied memories of the traumatizations of Mr. X. and—finally—a psychic integration of the traumatic experiences in his (conscious) idiosyncratic autobiographic memory. Thus, the traumatizations could be integrated into a more mature self and identity which no longer were unconsciously determined by past traumatizations.

(b) The (systematic but narrative) summary of the clinical material aimed to illustrate the different pathways which may lead into a chronic depression following the integrative, psychoanalytical model by Bleichmar (2010). The richness of such a conceptual approach might also be a specificity for contemporary (neuroscientifically inspired) psychoanalysis.

(c) The focus of the summary of the clinical transformations was changes of the manifest dream contents which could be investigated by a systematic empirical method which was illustrated in Chap. 8. This content analysis is based on a model by Moser and von Zeppelin, integrating psychoanalytical and interdisciplinary (e.g. neuroscientific) knowledge on dreams.

Within the limited space of this chapter, we tried to illustrate that a neuroscientifically inspired psychoanalysis indeed may open new doors to the understanding of unconscious fantasies and conflicts as they are, e.g. represented in dreams. This knowledge may also open new research perspectives for investigating structural changes in psychoanalyses—an important topic in contemporary comparative psychotherapy research.

## References

Bohleber W. Psychoanalytiker bei der Arbeit-ihre Praxis, ihre Theorien. Psyche. 2007;61:831–6.

Hautzinger M. Cognitive Behavior Therapy of Depression (7th edition). Stuttgart: Beltz Edition. 2013.

Lane R, Ryan L, Nadel L, Greenberg L. Memory consolidation, emotional arousal, and the process of change in psychotherapy. New insights from brain science. Behav Brain Sci. 2015;38:1–64.

Leuzinger-Bohleber M, Hautzinger M, Fiedler G, Keller W, Bahrke U, Kallenbach L, Kaufhold J, Ernst M, Negele A, Schoett M, Küchenhoff H, Günther F, Rüger B, Beutel M. Outcome of psychoanalytic and cognitive-behavioral long-term-therapy with chronically depressed patients. A controlled trial with preferential and randomized allocation. Can J Psychiatr. 2018a; in press.

Leuzinger-Bohleber M, Kaufhold J, Kallenbach L, Negele A, Ernst M, Keller W, Fiedler G, Hautzinger M, Beutel M. Does sustained symptomatic improvement of chronically depressed patients need structural change in long-term psychotherapies? Findings from the LAC depression study comparing the outcomes of cognitive-behavioral and psychoanalytic long-term treatments. Int J Psychoanal. 2018b; in press.

# Neurophilosophy and Neuroethics: Template for Neuropsychoanalysis?

# 30

Georg Northoff

**Abstract**

The encounter of neuroscience has led to intense debate in philosophy on how its originally mind-based concepts are related to the brain and its neural function. While some proponents in especially the Anglo-American world suggest reductive replacement of philosophy by neuroscience as manifest in what they call neurophilosophy, the opponents claim for a more non-reductive form of neurophilosophy where both philosophy and neuroscience are closely intertwined but distinct. I here sketch the field of such non-reductive neurophilosophy by distinguishing different domains, empirical (neuroscientific investigation of originally philosophical concepts) theoretical (methodological and conceptual issues), and practical (neuroethical questions) neurophilosophy. In conclusion, a non-reductive neurophilosophy opens the door for a truly transdisciplinary exchange between philosophy and neuroscience which will lead to novel questions and approaches in both disciplines. This, in turn, will also make possible intensive dialogue between psychoanalysis and neuroscience, i.e. neuropsychoanalysis.

## 30.1 Introduction: Concept of Neurophilosophy

Recent neuroscientific progress has led to the extension of neuroscience to apply and include also concepts like consciousness, free will, self, etc. that were originally discussed in philosophy. This has led to the recent emergence of a new field, neurophilosophy. The term "neurophilosophy" is often used either implicitly or explicitly for the characterization of an investigation of philosophical theories in relation to neuroscientific hypothesis. According to Breidbach, "neurophilosophy" has already

G. Northoff
Mind, Brain Imaging, and Neuroethics,
Institute of Mental Health Research, University of Ottawa, Ottawa, ON, Canada
e-mail: georg.northoff@theroyal.ca; www.georgnorthoff.com

 
H. Boeker et al. (eds.), *Neuropsychodynamic Psychiatry*,
https://doi.org/10.1007/978-3-319-75112-2_30

been implicitly practiced at the turn of the last century by, for example, W. Wundt (Breidbach 1997, p. 393–4). Another neurophilosopher though not named as such was Schopenhauer who was probably the first philosopher to introduce the concept of the brain in the philosophical context. The French philosopher M. Merleau-Ponty may also be considered a neurophilosopher since in his *Phenomenology of perception* he explicitly introduces the brain and its neural organization and links it to perception and other originally philosophical concepts. Other important developments in this regard were the paper about naturalized epistemology by Quine (1969) and the book about the self and its brain by Popper and Eccles (1989). Though these approaches differ widely, they at least share the presupposition that the brain may be important to consider in explaining our possible knowledge and the concept of mind.

Yet, it was P. Churchland who explicitly introduced the term "neurophilosophy" (Churchland 1986). Her concept of neurophilosophy set a certain standard in defining neurophilosophy by possible reduction and elimination of originally philosophical concepts by neuroscientific concepts and facts. She thus did not only consider the brain to be relevant for knowledge and the concept of mind but claimed much stronger that the latter can be reduced to the former. This had important implications since then the term neurophilosophy is almost exclusively reserved for reductive-eliminative approaches. Neurophilosophy in this sense is considered to be the "application of neuroscientific concepts to traditional philosophical questions" (Bickle et al. 2006, p. 1). Since "neurophilosophy" in this sense aims at revealing the neural correlates of originally philosophical terms (like, e.g. free will, personal identity, consciousness, etc.), one may also speak of a "neuroscience of philosophy" or "empirical neurophilosophy". "Empirical neurophilosophy" focuses on the investigation of the neural (and psychological) conditions of originally philosophical concepts like free will, self, action, consciousness, etc.

While neurophilosophy often completely identified or equated with empirical neurophilosophy these days, especially in the Anglo-American world, European continental authors (Walter 1998; Northoff 2001, 2004) point out a wider notion and concept of neurophilosophy. Such wider concept of neurophilosophy can then also include European authors like M. Merleau-Ponty and A. Schopenhauer who argue against a reductive-eliminative approach. Such different concepts of neurophilosophy being either narrow, as dominating in the Anglo-American world, or wide as in the European continental tradition raise the question for methodological and conceptual issues in neurophilosophy which may be subsumed under the concept of "theoretical neurophilosophy" (Northoff 2001, 2004).

Theoretical neurophilosophy focuses predominantly on the development of a definition and methodological principles and strategies for the linkage between philosophical theory and neuroscientific hypothesis. These methodological principles may differ from the ones that are presupposed in philosophy and neuroscience, respectively, as well as from the ones that are applied in the linkage of philosophical concepts with concepts from other sciences (like physics or chemistry). The core feature of theoretical neurophilosophy is the investigation and definition of the specific neurophilosophical methodology as distinguished from neuroscientific and philosophical methodology.

The specific methodological feature of neurophilosophy consists in the hybrid nature of neurophilosophical concepts (see also Bennett and Hacker 2003). Neurophilosophical concepts like consciousness, free will, etc. are on the one hand neuroscientific concepts that are measured in orientation on empirical-experimental standards thus presupposing facts, while on the other they are philosophical concepts that are measured in orientation on logical-conceptual standards. Since both empirical-experimental and logical-conceptual measures are integral components of neurophilosophical concepts, they must be linked to each other in neurophilosophical investigation. This requires special methodological strategies that are different from both neuroscience and philosophy that both investigate only one component, i.e. either facts or concepts. Hence, neurophilosophical methodology may be characterized by what may be called "concept-fact linkage" that must be considered truly transdisciplinary rather than intradisciplinary (Fig. 30.1).

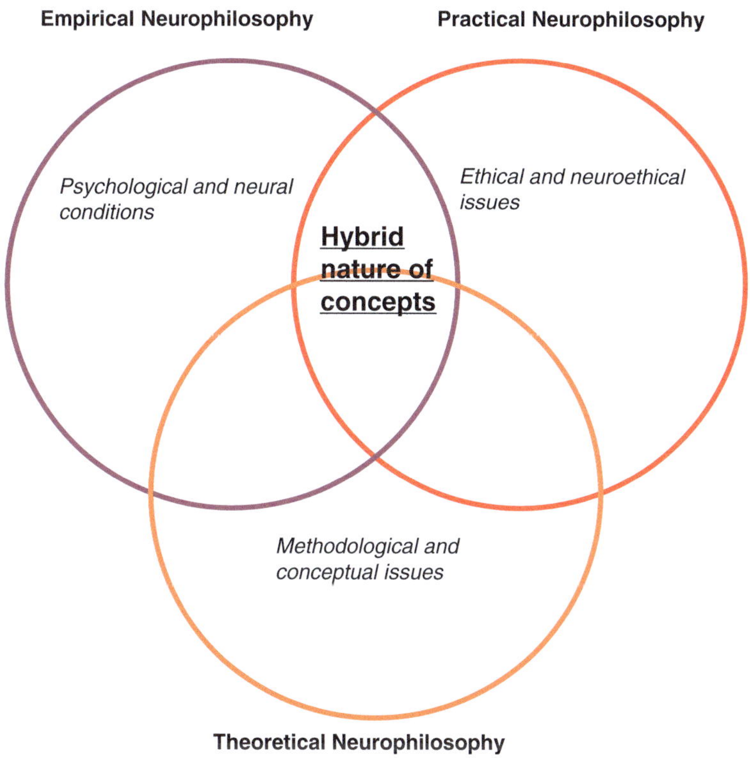


**Fig. 30.1** Distinction between empirical, practical, and theoretical neurophilosophy. The figure shows the three main different domains of neurophilosophy, empirical, theoretical, and practical. Despite their differences in focus and content, they all share the hybrid nature of concepts as including both theoretical, i.e. conceptual-logical, and empirical, i.e. observational, features. That makes linkage between concepts and facts necessary for which we need to develop a specific methodological strategy, concept-fact linkage

The concept of theoretical neurophilosophy is closely related to the one of "philosophy of neuroscience" as it is reflected in the recent literature (see Bechtel et al. 2001; Bickle et al. 2006): Like philosophy of psychology and philosophy of physics, the "philosophy of neuroscience" represents an "attempt to address foundational issues in neuroscience" (see Bechtel et al. 2001, p. 7). For example, the question of the sort of explanation in neuroscience is raised, like whether neuroscientific explanation is in accordance with the deductive-nomological model as suggested by Hempel. Another central question concerns the problem of "naturalization". Can neuroscience apply the same strategies for "naturalization" of philosophical terms as other disciplines (e.g. physics and chemistry)? Are the general methodological principles for "naturalization" valid in neuroscience too or is there a need to develop special strategies for neuroscience in particular?

The question for "concept-fact linkage" does though not concern a specific philosophical problem in neuroscience but rather a specific philosophical problem in neurophilosophy. One may consequently speak of a "philosophy of neurophilosophy" rather than a "philosophy of neuroscience". Theoretical neurophilosophy as conceptualized here is understood in a rather broad sense and is supposed to include both "philosophy of neurophilosophy" rather than a "philosophy of neuroscience".

In addition to empirical and theoretical neurophilosophy, one may also distinguish "practical neurophilosophy" which these days is coined "neuroethics" (Roskies 2002; Moreno 2003; Northoff et al. 2006). Neuroethics can broadly and preliminarily be defined by the drawing of relationships between neuroscientific observations and ethical concepts. Neuroethics is concerned with ethical issues in clinical and scientific neuroscience like informed consent and enhancement that arise from neuroscientific progress. Most importantly, neuroethics also investigates the neural mechanisms and conditions underlying ethical concepts like moral judgement, free will, etc.

## 30.2 Empirical Neurophilosophy: Application of Neuroscience to Philosophy

Empirical neurophilosophy is concerned with the investigation of the neural conditions and mechanisms underlying originally philosophical concepts like consciousness and self. Empirical neurophilosophy may best be illustrated by briefly discussing two core examples, consciousness and self. I here want to briefly discuss two very influential theories of consciousness, the global workspace theory and the concept of re-entrant circuits, which will be followed by a brief discussion of the concept of the neural self.

### 30.2.1 Neural Conditions of Consciousness

The global workspace theory by Baars (2005) is a theory about the cognitive architecture that is necessary to constitute consciousness (see also Baars and Franklin 2007). Conscious states evoke widespread activity in and synchronization (as, e.g. gamma oscillations) across various regions including many cortical and subcortical regions, while unconscious states can be characterized by a more restricted and limited network. This leads to the assumption that consciousness may require access to and integration between various functions and hence what is called a global workspace. The function of consciousness is then to provide global access with the consecutive integration and coordination between different functions that are associated with the different regions (see Baars 2005, p. 51–2). Within this global workspace, Baars (2005, p. 49–51) distinguishes between content systems and context systems. Content systems are those that mediate specific contents like the visual ventral stream, while context systems provide the context and must be associated with frontoparietal regions. The context system is supposed to "observe" the content system, and for such observation to be possible, one must assume some kind of observer, a self, more specifically what Baars et al. (2003) calls the "observing self" which he consecutively associates with the brain's resting-state activity in frontoparietal networks.

Baar's global workspace theory is considered a theory of consciousness, i.e. how consciousness is constituted and what its function is for the brain. Most importantly, Baars' concept of global workspace is supposed to account for how phenomenal states like conscious states can possibly be constituted and thus why we are able at all to have conscious subjective experience in first-person perspective (FPP) rather than mere unconscious neural processing. By coordinating and integrating different kinds of information, the global workspace is supposed to make the constitution of conscious contents as distinguished from unconscious ones possible—such coordination and integration allows for the transformation from unconscious into conscious contents and thus for what I call neuronal-phenomenal transformation (Northoff 2001, 2004).

Another influential theory of consciousness is the concept of re-entrant circuits. Edelman (2003) introduced the concept of re-entrant circuits that feedback back anatomically and functionally from one region to the very same region via some other regions. The best instance of such re-entrant circuits is the connection between the prefrontal cortex, the basal ganglia, and the thalamus. There are efferent connections from the prefrontal cortex to the striatum, from there to the subthalamic nucleus and the external globus pallidus which both project then to the internal globus pallidus, and from there one can observe efferent connections to the various nuclei of the thalamus which in turn project back to the prefrontal cortex. Hence the thalamus is the critical relay station in completing re-entrant circuitry.

Edelman (2003) argues that such re-entrant circuits allow to integrate information from different brain regions by connecting them to each other and that it is by such integration and coordination that consciousness is constituted. More specifically, these re-entrant circuits allow the circulation of information in metastable loops within a temporal window of about 500 ms. This constitutes what Edelman (2003, p. 5521–5523) calls the "re-entrant dynamic core" that consists in functional clusters of neural activity. Each functional cluster or dynamic core is supposed to constitute a unitary scene which corresponds to what phenomenologically is described as qualia, i.e. a specific phenomenal state as discriminated from others.

### 30.2.2 Concept of the Neural Self

The question of the self has been one of the most salient problems throughout the history of philosophy and more recently also in psychology and neuroscience (Gallagher 2000; Gallagher and Frith 2003; Metzinger and Gallese 2003; Northoff 2004; Northoff et al. 2006; Legrand and Ruby 2009). For example, William James distinguished between a physical self, a mental self, and a spiritual self. These distinctions seem to reappear in recent concepts of self as discussed in neuroscience (Panksepp 1998a, b; Damasio 1999; Gallagher 2000; Stuss et al. 2001; Churchland 2002; Kelley et al. 2002; Lambie and Marcel 2002; LeDoux 2002; Turk et al. 2002; Damasio 2003a, b; Gallagher and Frith 2003; Keenan et al. 2003; Kircher and David 2003; Panksepp 2003; Turk et al. 2002; Vogeley and Fink 2003; Dalgleish 2004; Marcel and Lambie 2004; Northoff and Bermpohl 2004). Damasio (1999) and Panksepp (1998a, b, 2003) suggest a "proto-self" in the sensory and motor domains, respectively, which resembles William James' description of the physical self. Similarly, what has been described as "minimal self" (Gallagher 2000; Gallagher and Frith 2003) or "core or mental self" (Damasio 1999) might correspond more or less to James' concept of mental self. Finally, Damasio's (1999) "autobiographical self" and Gallagher's (2000) and Gallagher and Frith (2003) "narrative self" strongly rely on linking past, present, and future events with some resemblances to James' spiritual self.

These distinct selves are now related to distinct brain regions. For instance, the "proto-self" outlining one's body in strongly affective and sensory-motor terms is associated with subcortical regions like the PAG, the colliculi, and the tectum (Panksepp and Northoff 2008). The "core or mental self" building upon the "proto-self" in mental terms is associated more with the thalamus and cortical regions like the ventromedial prefrontal cortex (see, for instance, Damasio 1999, 2003a, b). Finally, the "autobiographical or extended self" that allows one to reflect upon one's "proto-self" and "core or mental self" is associated with cortical regions like the hippocampus and the cingulate cortex.

Recent imaging studies in humans show various cortical regions, predominantly the so-called cortical midline structures (CMS), to be involved in what is called self-related processing (SRP) (see Northoff and Bermpohl 2004; Northoff et al. 2006; Uddin et al. 2007) that are integrated with subcortical processes to yield an integrated subcortical-cortical midline system (SCMS). The lowest regions of this distributed SCMS network include the periaqueductal grey (PAG), the superior colliculi (SC), and the adjacent mesencephalic locomotor region (MLR) as well as preoptic areas, the hypothalamus, and dorsomedial thalamus (DMT) (Holstege et al. 1996; Panksepp 1998a, b). While cortical regions include the ventro- and dorsomedial prefrontal cortex (VMPFC, DMPFC), the pre- and supragenual anterior cingulate cortex (PACC, SACC) and the posterior cingulate cortex (PCC), and the medial parietal cortex (MPC) (Northoff et al. 2006).

The assumption of an integrated subcortical-cortical midline system (SCMS) is consistent with various lines of research showing that core self-related functioning involves both cortical and subcortical regions (Northoff and Panksepp 2008; Panksepp 1998a; Phan et al. 2004). Many imaging studies in humans have focused on cortical regions, in part because of the involvement of strong cognitive components such as evaluative judgements in the respective tasks (see Northoff et al. 2006). This is well reflected in the experimental paradigms of most imaging studies of the self. They compare the evaluation of self-related stimuli with the evaluation of non-self-related stimuli. This however raises the question whether the neural activity in the SCMS is associated with the cognitive functions implicated in the evaluation of stimuli as self-related or in the self-relatedness of the stimuli themselves.

Many regions of the SCMS show high neural activity in the resting-state activity independent of stimulus processing. High resting-state activity has been associated with the so-called default-mode network (DMN) (Raichle et al. 2001; Buckner et al. 2008) which has been observed in both humans (Damoiseaux et al. 2006; Fox and Raichle 2007; Fransson 2005; Raichle et al. 2001) and chimpanzee/macaques (Rilling et al. 2007; Vincent et al. 2007). Since the initial discovery of the DMN (Raichle et al. 2001), also regions outside the DMN have been reported to show high resting-state activity including various subcortical and cortical medial regions (Buckner et al. 2008; Northoff and Panksepp 2008). Hence, there seems to be a strong neuroanatomical overlap between the DMN and the SCMS, the latter too showing high resting-state activity.

Is such high resting-state activity in the SCMS related to the self? High resting-state activity is most likely related to the observation of predominant deactivation, so-called negative BOLD responses, in these regions in functional magnetic resonance imaging (fMRI) (see Fox and Raichle 2007; Raichle et al. 2001; Northoff et al. 2007). Hence high resting-state activity in this so-called default-mode network is supposed to be associated with introspection of one's self-specific contents as attention towards its internal contents and their consecutive representation as physical or mental. The self is consecutively characterized by higher-order cognitive function, e.g. by attention, and meta-representation of specific contents as self-related in awareness (Wicker et al. 2003). While the self itself is considered innate and related to intrinsic resting-state activity in the DMN/SCMS, representation of the self is associated with different degrees of stimulus-induced deactivation in the very same regions.

## 30.3 Theoretical Neurophilosophy: Relevance of Neurophilosophy to Neuroscience

Theoretical neurophilosophy is concerned with the discussion of concepts and methodological strategies and how they relate to the empirical observations made in neuroscience. I here want to briefly discuss the methodological strategies how to neuroscientifically investigate consciousness in dependence on presupposing a certain concept of consciousness. This will be followed by the discussion of the concept of self.

### 30.3.1 Research of Consciousness: Building-Bloc Approach Versus Unified Field Model

The building-bloc approach pursues the strategy to investigate consciousness in one particular sensory modality (or one domain) like visual consciousness. The specific functional mechanisms revealed in one particular case like the conscious experience of the colour red might then be applied to the consciousness of other colours and other sensory modalities and domains like taste, smells, sounds, and consciousness generally. The prime example of such building-bloc in current science is visual consciousness which is studied in patients with blind sight, binocular switching, and analysis of the temporal course of the visual stimulus in the brains' pathways. Strategically, the building-bloc approach starts with one particular example or building-bloc of consciousness like visual consciousness, aims to apply those neuronal mechanisms to other forms of consciousness, e.g. other building-blocs, and then ultimately hopes to explain consciousness in general by putting all building-blocs together.

Conceptually, the building-bloc approach presupposes that consciousness is, as Searle says (2004, p. 106), atomistic in that it consists in "more-or-less independent conscious units" which might correspond to different experiences like "the experience of red, the taste of beer, the sound of middle C". According to Searle, this is however not in accordance with phenomenology. It is not that we have an isolated consciousness of red independent of all other colours or an isolated taste of beer independent of its colour, its smell, etc.: "the unconscious subject would suddenly have a conscious experience of red and nothing else" (Searle 2004, p. 108). Instead, we experience the colour red in dependence on the surrounding colours, and we experience the taste of the beer in relation to its smell, its colour, etc.—the experience of red and the taste of the beer surface as aspects or parts of an always already unified field of consciousness. The object's red colour thus does not induce consciousness per se but rather modifies an already existing "qualitative subjective unity", e.g. the conscious field: "To put the point very crudely, a conscious experience of red can only occur in a brain that is already conscious. We should think of perception not as creating consciousness but as modifying a preexisting conscious field. Again, consider dreams. Like many people, I dream in colour. When I see the colour red in a dream, I do not have a perceptual input that creates a building-bloc of red. Rather the mechanisms in the brain that create the whole unified field of dream consciousness create my experience of red as part of the field" (Searle 2004, p. 108).

Consciousness may then no longer be conceptualized as the mere collection of possible building-blocs reflecting consciousness in the different sensory modalities but rather as unified field with "qualitative subjective unity" that may be manifest and expressed in different surface features including different sensory modalities. This implies a different research strategy. Instead of the building-bloc approach, Searle (2004, p. 108; see also Searle 1999, p. 2074–5) suggests what he calls a "unified field model". The "unified field model" does not consider consciousness to be created and caused anew each time it occurs; instead, our single conscious experiences are considered rather as modifications or surface features of an already preexisting consciousness: "But I like to think it this way: you are not creating a new consciousness; you are modifying the preexisting conscious field. On the unified-field model we should think of perceptual inputs not as creating building-blocs of consciousness but as producing bumps and valleys in the conscious field that has to exist a priori to our having the perceptions" (Searle 2004, p. 108; see also Noe and Thompson 2004).

### 30.3.2 Concept of Self: Self as "Special Addition" or "Formal Notion"

The characterization of the self in the above-mentioned sense presupposes that the self is considered something specific and distinct from everything else, i.e. from all other functions. The self may then be regarded what I call a "special addition" that may be necessary to integrate and coordinate neural activity across the brain. Churchland (2002), for instance, considers the self is as some "special addition" that is needed to organize and coordinate the various functions: "Rather, the self is something like a squadron of capacities flying in loose formation. Depending on context, it is one or another of these capacities, or their exercise, to which we refer when we speak of the self. Some of these capacities involve explicit memory, some involve detection of changes in glucose or CO2 levels, others involve imagery in diverse modalities or emotions of diverse valence. The fundamental capacity, however, probably consists in coordinating needs, goals, perception, and memory with motor control" (Churchland 2002, p. 63).

Churchland assumes that for this to be possible, there must be what she calls "self-representational capacities" that operate on all levels including "representing the internal milieu and viscera via chemical and neural pathways aimed largely at the brainstem and hypothalamus, representing musculoskeletal structures via the somatic sensory system, representing autobiographical events via medial temporal lobe structures, deferring gratification and controlling impulses via prefrontal lobe and limbic structures, and representing the sequence of actions to take next, as well as representing where one is in space-time and the social order" (Churchland 2002, p. 309). The most characteristic feature of these self-representational capacities is that they represent the brain's activity during these various processes: "A brain can also have models of its own processes. If some neuronal activity represents a motor command to reach for an apple, other neuronal activity represents the fact that a specific command has been issued. If some neuronal activity represents a light touch on the left ear, higher-order neuronal activity may represent the integration of many lower-order representations (light touch on the left ear and buzzing sound to left, which means that there is a mosquito, etc.). The brain not only represents the sensations of one's limbs: it specifically represents the sight and feel of that limb as belonging to oneself (there is a mosquito on my left ear). Yet further neuronal activity may represent that representation as a mental states (I know I feel a mosquito on my left ear)" (Churchland 2002, p. 64). According to Churchland (2002, p. 77), these inner models of one's body and experience may be mediated by what she, in orientation on Grush, calls emulators.

How does such neurophilosophical concept of self differ from the philosophical concept of the self by one of the most important philosophers, Descartes? Descartes assumed a special substance which he supposed to underlie the self which, due to his famous insight "I think there I am", he characterized as non-physical and hence as mental. Current neuroscience and neurophilosophy diverge from this in two aspects. First, the concept of self is multiplied with, for instance, "proto-self", "core self", and "autobiographical or extended self", describing distinct aspects. Hence the concept of self is no longer unified and homogeneous as in Descartes but diverge and heterogenous. Second, the assumption of a mind as underlying substance is replaced by the brain making any neuroscientific approach to the self impossible. This however points out a crucial underlying similarity. Current neuroscientific approaches to the self are still very much in the tradition of Descartes in assuming a neural mechanisms (or substance) as "special addition" to underlie the self. In both approaches, the self is considered a "special addition" where it is characterized by contents be they mental (Descartes, Philosophy) or neuronal (Churchland, Neurophilosophy).

There are though alternative concepts of self that focus more on the self as formal notion rather than as content. For instance, the philosopher Searle (2004, p. 200–206) characterizes the self as form or mode when he describes as "purely formal notion": "The self as I am describing it is a purely formal notion; it does not involve having a particular type of reason or a particular type of perception. Rather, it is a formal notion involving the capacity to organize its intentionality under constraints of rationality in such a way as to undertake voluntary actions, intentional actions, where the reasons are not causally sufficient to fix the action" (Searle 2004, p. 204). The self is devoid of any content and is rather some organizational principle that allows to constitute the relation to the environment (see also Northoff et al. 2006 for a somewhat similar approach from the neuroscientific perspective). As such the self must be considered a "special addition" to all other functions: "nonetheless there is a formal or logical requirement that we postulate a self as something in addition to the experiences in order that we can make sense of the character of our experiences" (Searle 2004, p. 205).

## 30.4 Practical Neurophilosophy: Neuroethics and the Relevance of Ethical Concerns in Neuroscience

Practical neurophilosophy or neuroethics is concerned with the relationship between neuroscientific findings and ethical concepts like free will, moral judgement, informed consent, etc. Neuroethics focuses on the one hand on the investigation of the psychological and neural conditions of these ethical concepts and on the other on the ethical problems arising from neuroscientific progress. Roskies (2002) distinguishes correspondingly between ethics of neuroscience and neuroscience of ethics. The ethics of neuroscience deals with ethical problems in neuroscience and thus

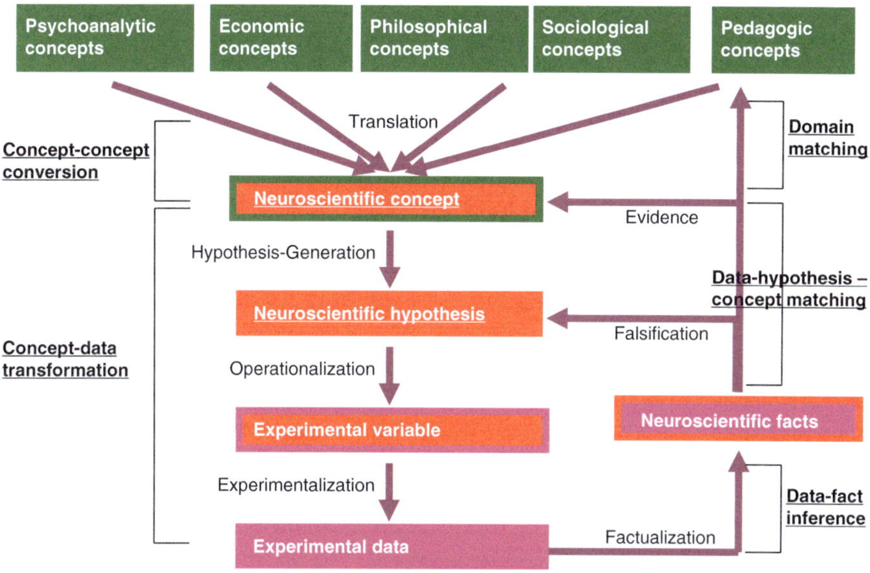


**Fig. 30.2** Methodological iterativity between concepts and facts. The figure shows the linkage between concepts from different disciplines including philosophy and psychoanalysis (upper part with green boxes). These concepts are then linked to neuroscientific concepts and hypotheses and subsequently to experimental testing. This results in empirical or neuroscientific data (lower right) which, through steps, can be compared with the original theoretical concept. If both empirical data and original concept match and fit with each other, one can say that the original philosophical or psychoanalytic concept is empirically plausible; if there is not matching and fit with each other, the concept remains empirically implausible. Hence, empirical plausibility between concepts and facts replaces reduction of concepts to facts

with issues of validity of informed consent in psychiatric patients, enhancement of cognitive functions by neuroscientific interventions, coincidental findings in neuroimaging, etc., while the neuroscience of ethics investigates the neural mechanisms that may possibly underlie ethical concepts like informed consent, moral judgement, free will, etc. One central problem is the question for informed consent, the ability to approve, refuse, or remain ambivalent about the participation in clinical or scientific investigation (Fig. 30.2).

### 30.4.1 Psychological (and Neural) Conditions of Informed Consent

Informed consent is crucial in clinical scientific studies of healthy subjects and among other psychiatric patients. This has led to extensive research of the factors and functions determining the kind of consent psychiatric patients give, i.e. whether they accept, refuse, or remain ambivalent about study participation (see Northoff 2006; Candilis et al. 2006; Carpenter Jr et al. 2000; Koren et al. 2005; Charland 1998; Breden and Vollmann 2004; Appelbaum et al. 1999; Grisso et al. 1997; Jeste et al. 2006). Giving informed consent requires the capacity to make a decision, i.e. decision-making (Candilis et al. 2006, 2008; Carpenter Jr et al. 2000; Koren et al. 2005; Jeste et al. 2006; Palmer and Jeste 2006; Stroup et al. 2005; Palmer et al. 2004; Howe et al. 2005; Roberts et al. 2002; Moser et al. 2002). Decision-making is a complex process that involves both cognitive and affective functions as pointed out by recent neuroscientific research (Damasio 1999; Northoff et al. 2006; Northoff 2006).

Cognitive functions in decision-making concern attention, working memory, executive functions, and others (see Candilis et al. 2006, 2008; Carpenter Jr et al. 2000; Koren et al. 2005; Jeste et al. 2006; Palmer and Jeste 2006; Stroup et al. 2005; Palmer et al. 2004; Howe et al. 2005; Roberts et al. 2002; Moser et al. 2002). More specifically, the capacity to make a decision in informed consent, the so-called decisional capacity, involves cognitive functions like understanding, appreciation, and reasoning. These cognitive functions and thus the decisional capacity have recently been systematically investigated with the MacArthur Competence Assessment Tool for Clinical Research (MacCAT-CR) (Appelbaum et al. 1999; Grisso et al. 1997). The MacCAT-CR focuses predominantly on cognitive functions like understanding, appreciation, reasoning, and expressing a choice that are supposedly implicated in the decisional capacity for informed consent.

In addition to cognitive functions, decision-making in general and decisional capacity in informed consent in particular involve empathy and emotions (see Damasio 1999; Northoff et al. 2006; Northoff 2006; Breden and Vollmann 2004). Empathy describes the ability to share another person's cognitive and emotional inner life (see de Vignemont and Singer 2006; Hein and Singer 2008; Singer et al. 2006; Fan et al. 2011; Dziobek et al. 2008; Decety and Lamm 2006). Empathic sharing between the consenting subject and the investigator may be crucial in determining decisional capacity and ultimately the consent itself. This is possible only when the consenting subject is able to recognize the investigator's emotions thus requiring emotion recognition (Bowers et al. 1992).

### 30.4.2 Ethical Problems in Informed Consent

While these issues concern the psychological (and possibly also neural) conditions of the ethical concept of informed consent, they also raise ethical issues. Consider particularly the possible involvement of empathy and emotion recognition in informed consent.

This may touch upon the selection of subjects to be examined in clinical trials and for the validity of informed consent in general. One may only select those subjects with low cognitive empathic abilities and emotion recognition for study participation in order to gain higher approval rate in informed consent. This bias in selecting subjects for study participation may ethically be problematic though because it artificially minimizes the risk of refusal (or ambivalence).

Another ethical problem concerns the impact of the investigator. The outcome of informed consent, approval, refusal, or ambivalence, is usually considered to be solely dependent upon the subject itself while the investigator, his emotions and empathic abilities, remains absent. If however empathy and emotion play a major role in determining the outcome of the informed consent, the emotional and empathic abstinence of the investigator can no longer be maintained. Analogous to the subjects, the empathic and emotional abilities of the investigator itself may determine the outcome of the informed consent which may be probed in future studies. If the investigator has indeed an impact, we may need to develop what may be called an investigator-dependent concept of informed consent as distinguished from current concepts that remain rather investigator-independent.

The inclusion of the investigator may also raise another problem. A high empathic investigator may have a positive impact on subjects' informed consent. If so, one may only select high empathic investigators to obtain the consent from the subjects in order to reduce approval rates in scientific investigation. Such bias on the side of the investigator, i.e. selection of only high empathic investigators, may though be ethically as problematic as the above-mentioned bias in the selection of only low empathic subjects.

These results are also of particular relevance for the validity of informed consent in psychiatric patients. Psychiatric disorders like schizophrenia, depression, and personality disorders are well known to show dysfunction in both cognitive and affective components of empathy. This in turn may strongly impact their ability to give valid informed consent. From an ethical point of view, our results therefore question the hitherto predominance of cognitive functions in judging the validity of subjects' informed consent. Severe empathic and emotional deficits may confound the subjects' informed consent to a great extent even if their cognitive skills are preserved as it is, for instance, the case in personality disorders and neurosis.

### 30.4.3 Neurophilosophy: Template for Neuropsychoanalysis?

Why is all that relevant for neuropsychoanalysis? Similar to neurophilosophy, neuropsychoanalysis is "situated" right at the interface between the conceptual and empirical realms. Similar to neurophilosophy, we cannot reduce psychodynamic concepts to neuronal findings as the former cannot be found in the latter as in specific brain images. Reductive neuropsychoanalysis thus remains impossible in the same way reductive neurophilosophy suffers from a birth defect.

However, as in the case of neurophilosophy, there is nevertheless some close nonarbitrary or non-contingent relation between psychodynamic concepts and neuronal mechanisms. We need to investigate how neuronal activity transforms into psychodynamic features. Though investigation of such neuronal-psychodynamic transformation is still at the beginning, it will be, as I am sure, the centrepiece of a future neuropsychoanalysis in very much the same as the quest for neuromental transformation is the centrepiece of neurophilosophy.

Finally, one may distinguish between different branches of neuropsychoanalysis. Empirical neuropsychoanalysis (as analogous to empirical neurophilosophy) may focus on the empirical mechanisms of neuro-psychodynamic transformation—the focus here is on empirical studies of their relationship, while theoretical neuropsychoanalysis may focus more on conceptual issues, the definition of its concepts and terms as they are to be distinguished from both psychodynamic and neuroscientific concepts. Finally, there is also a practical and ethical aspect to neuropsychoanalysis as it may, as we hope, enter therapeutic practice.

## References

Appelbaum PS, Grisso T, Frank E, O'Donnell S, Kupfer DJ. Competence of depressed patients for consent to research. Am J Psychiatry. 1999;156(9):1380–4.

Baars BJ. Global workspace theory of consciousness: toward a cognitive neuroscience of human experience. Prog Brain Res. 2005;150:45–53.

Baars BJ, Franklin S. An architectural model of conscious and unconscious brain functions: Global Workspace Theory and IDA. Neural Netw. 2007;20(9):955–61.

Baars BJ, Ramsoy TZ, Laureys S. Brain, conscious experience and the observing self. Trends Neurosci. 2003;26(12):671–5.

Bechtel W, Mandik P, Mundale J. Philosophy meets the neurosciences. In: Bechtel W, Mandik P, Mundale J, Stufflebeam RS, editors. Philosophy and the neurosciences: a reader. Oxford: Blackwell; 2001.

Bennett MR, Hacker PMS. Philosophical foundations of neuroscience. Oxford: Blackwell Publishing; 2003.

Bickle J, Mandik P, Landreth A. The philosophy of neuroscience. Journal [serial on the Internet]. 2006. Available from: http://plato.stanford.edu/entries/neuroscience/.

Bowers D, Blonder L, Heilman K. The Florida affect battery. Center of Neuropsychological Studies; 1992.

Breden TM, Vollmann J. The cognitive based approach of capacity assessment in psychiatry: a philosophical critique of the MacCAT-T. Health Care Anal. 2004;12(4):273–83. discussion 65-72

Breidbach O. Die Materialisierung des Ichs - Eine Geschichte der Hirnforschung im 19. und 20. Jahrhundert. Frankfurt: Suhrkamp; 1997.

Buckner RL, Andrews-Hanna JR, Schacter DL. The brain's default network: anatomy, function, and relevance to disease. Ann N Y Acad Sci. 2008;1124:1–38.

Candilis PJ, Geppert CM, Fletcher KE, Lidz CW, Appelbaum PS. Willingness of subjects with thought disorder to participate in research. Schizophr Bull. 2006;32(1):159–65.

Candilis PJ, Fletcher KE, Geppert CM, Lidz CW, Appelbaum PS. A direct comparison of research decision-making capacity: schizophrenia/schizoaffective, medically ill, and non-ill subjects. Schizophr Res. 2008;99(1-3):350–8.

Carpenter Jr WT, Gold JM, Lahti AC, Queern CA, Conley RR, Bartko JJ, Kovnick J, Appelbaum PS. Decisional capacity for informed consent in schizophrenia research. Arch Gen Psychiatry. 2000;57(6):533–8.

Charland LC. Appreciation and emotion: theoretical reflections on the MacArthur Treatment Competence Study. Kennedy Inst Ethics J. 1998;8(4):359–76.

Churchland P. Neurophilosophy: toward a unified science of the mind-brain. Cambridge: MIT Press; 1986.

Churchland PS. Self-representation in nervous systems. Science. 2002;296(5566):308–10.

Dalgleish T. The emotional brain. Nat Rev. 2004;5(7):583–9.

Damasio AR. How the brain creates the mind. Sci Am. 1999;281(6):112–7.

Damasio A. Feelings of emotion and the self. Ann N Y Acad Sci. 2003a;1001:253–61.

Damasio A. Mental self: the person within. Nature. 2003b;423(6937):227.

Damoiseaux JS, Rombouts SA, Barkhof F, Scheltens P, Stam CJ, Smith SM, Beckmann CF. Consistent resting-state networks across healthy subjects. Proc Natl Acad Sci U S A. 2006;103(37):13848–53.

Decety J, Lamm C. Human empathy through the lens of social neuroscience. Sci World J. 2006;6:1146–63.
Dziobek I, Rogers K, Fleck S, Bahnemann M, Heekeren HR, Wolf OT, Convit A. Dissociation of cognitive and emotional empathy in adults with Asperger syndrome using the Multifaceted Empathy Test (MET). J Autism Dev Disord. 2008;38(3):464–73.
Edelman GM. Naturalizing consciousness: a theoretical framework. Proc Natl Acad Sci U S A. 2003;100(9):5520–4.
Fan Y, Duncan NW, de Greck M, Northoff G. Is there a core neural network in empathy? An fMRI based quantitative meta-analysis. Neurosci Biobehav Rev. 2011;35(3):903–11. https://doi.org/10.1016/j.neubiorev.2010.10.009.
Fox MD, Raichle ME. Spontaneous fluctuations in brain activity observed with functional magnetic resonance imaging. Nat Rev. 2007;8(9):700–11.
Fransson P. Spontaneous low-frequency BOLD signal fluctuations: an fMRI investigation of the resting-state default mode of brain function hypothesis. Hum Brain Mapp. 2005;26(1):15–29.
Gallagher S. Philosophical conception of the self: implications for cognitive science. Trends Cogn Sci. 2000;4:14–21.
Gallagher HL, Frith CD. Functional imaging of theory of mind. Trends Cogn Sci. 2003;7(2):77–83.
Grisso T, Appelbaum PS, Hill-Fotouhi C. The MacCAT-T: a clinical tool to assess patients' capacities to make treatment decisions. Psychiatr Serv. 1997;48(11):1415–9.
Hein G, Singer T. I feel how you feel but not always: the empathic brain and its modulation. Curr Opin Neurobiol. 2008;18(2):153–8.
Holstege G, Bandler R, Saper CB. The emotional motor system. Prog Brain Res. 1996;107:3–6.
Howe V, Foister K, Jenkins K, Skene L, Copolov D, Keks N. Competence to give informed consent in acute psychosis is associated with symptoms rather than diagnosis. Schizophr Res. 2005;77(2-3):211–4.
Jeste DV, Depp CA, Palmer BW. Magnitude of impairment in decisional capacity in people with schizophrenia compared to normal subjects: an overview. Schizophr Bull. 2006;32(1):121–8.
Keenan JP, Wheeler M, Platek SM, Lardi G, Lassonde M. Self-face processing in a callosotomy patient. Eur J Neurosci. 2003;18(8):2391–5.
Kelley WM, Macrae CN, Wyland CL, Caglar S, Inati S, Heatherton TF. Finding the self? An event-related fMRI study. J Cogn Neurosci. 2002;14(5):785–94.
Kircher T, David A. The self in neuroscience and psychiatry. Cambridge: Cambridge University Press; 2003.
Koren D, Poyurovsky M, Seidman LJ, Goldsmith M, Wenger S, Klein EM. The neuropsychological basis of competence to consent in first-episode schizophrenia: a pilot metacognitive study. Biol Psychiatry. 2005;57(6):609–16.
Lambie JA, Marcel AJ. Consciousness and the varieties of emotion experience: a theoretical framework. Psychol Rev. 2002;109(2):219–59.
LeDoux J. Synaptic self: how our brains become who we are. New York: Viking; 2002.
Legrand D, Ruby P. What is self-specific? Theoretical investigation and critical review of neuroimaging results. Psychol Rev. 2009;116(1):252–82.
Marcel AJ, Lambie JA. How many selves in emotion experience? Reply to Dalgleish and power. Psychol Rev. 2004;111(3):820–6.
Metzinger T, Gallese V. The emergence of a shared action ontology: building blocks for a theory. Conscious Cogn. 2003;12(4):549–71.
Moreno JD. Neuroethics: an agenda for neuroscience and society. Nat Rev. 2003;4(2):149–53.
Moser DJ, Schultz SK, Arndt S, Benjamin ML, Fleming FW, Brems CS, Paulsen JS, Appelbaum PS, Andreasen NC. Capacity to provide informed consent for participation in schizophrenia and HIV research. Am J Psychiatry. 2002;159(7):1201–7.
Noe A, Thompson E. Are there neural correlates of consciousness? J Conscious Stud. 2004;11:3–28.
Northoff G. Personale Identität und operative Eingriffe in das Gehirn. Paderborn: Mentis; 2001.
Northoff G. Philosophy of the brain. Amsterdam: John Benjamins Publishing; 2004.

Northoff G. Neuroscience of decision making and informed consent: an investigation in neuroethics. J Med Ethics. 2006;32(2):70–3.
Northoff G, Bermpohl F. Cortical midline structures and the self. Trends Cogn Sci. 2004; 8(3):102–7.
Northoff G, Panksepp J. The trans-species concept of self and the subcortical-cortical midline system. Trends Cogn Sci. 2008;12(7):259–64.
Northoff G, Heinzel A, de Greck M, Bermpohl F, Dobrowolny H, Panksepp J. Self-referential processing in our brain–a meta-analysis of imaging studies on the self. NeuroImage. 2006;31(1):440–57.
Northoff G, Walter M, Schulte RF, Beck J, Dydak U, Henning A, Boeker H, Grimm S, Boesiger P. GABA concentrations in the human anterior cingulate cortex predict negative BOLD responses in fMRI. Nat Neurosci. 2007;10(12):1515–7.
Palmer BW, Jeste DV. Relationship of individual cognitive abilities to specific components of decisional capacity among middle-aged and older patients with schizophrenia. Schizophr Bull. 2006;32(1):98–106.
Palmer BW, Dunn LB, Appelbaum PS, Jeste DV. Correlates of treatment-related decision-making capacity among middle-aged and older patients with schizophrenia. Arch Gen Psychiatry. 2004;61(3):230–6.
Panksepp J. Affective neuroscience: the foundations of human and animal emotions. New York: Oxford University Press; 1998a.
Panksepp J. The preconscious substrates of consciousness: affective states and the evolutionary origins of the self. J Conscious Stud. 1998b;5:566–82.
Panksepp J. At the interface of the affective, behavioral, and cognitive neurosciences: decoding the emotional feelings of the brain. Brain Cogn. 2003;52(1):4–14.
Panksepp J, Northoff G. The trans-species core SELF: the emergence of active cultural and neuro-ecological agents through self-related processing within subcortical-cortical midline networks. Conscious Cogn. 2008;18(1):193–215.
Phan KL, Taylor SF, Welsh RC, Ho SH, Britton JC, Liberzon I. Neural correlates of individual ratings of emotional salience: a trial-related fMRI study. NeuroImage. 2004;21(2):768–80.
Popper K, Eccles J. Das Ich und sein Gehirn. München: Piper; 1989.
Quine WVO. Epistemology naturalized. In: Quine WVO, editor. Ontological relativity and other essays. New York: Columbia University Press; 1969.
Raichle ME, MacLeod AM, Snyder AZ, Powers WJ, Gusnard DA, Shulman GL. A default mode of brain function. Proc Natl Acad Sci U S A. 2001;98(2):676–82.
Rilling JK, Barks SK, Parr LA, Preuss TM, Faber TL, Pagnoni G, Bremner JD, Votaw JR. A comparison of resting-state brain activity in humans and chimpanzees. Proc Natl Acad Sci U S A. 2007;104(43):17146–51.
Roberts LW, Warner TD, Brody JL, Roberts B, Lauriello J, Lyketsos C. Patient and psychiatrist ratings of hypothetical schizophrenia research protocols: assessment of harm potential and factors influencing participation decisions. Am J Psychiatry. 2002;159(4):573–84.
Roskies A. Neuroethics for the new millenium. Neuron. 2002;35(1):21–3.
Searle JR. The future of philosophy. Philos Trans R Soc Lond. 1999;354(1392):2069–80.
Searle J. Mind: a brief introduction. Oxford: Oxford University Press; 2004.
Singer T, Seymour B, O'Doherty JP, Stephan KE, Dolan RJ, Frith CD. Empathic neural responses are modulated by the perceived fairness of others. Nature. 2006;439(7075):466–9.
Stroup S, Appelbaum P, Swartz M, Patel M, Davis S, Jeste D, Kim S, Keefe R, Manschreck T, McEvoy J, Lieberman J. Decision-making capacity for research participation among individuals in the CATIE schizophrenia trial. Schizophr Res. 2005;80(1):1–8.
Stuss DT, Gallup GG Jr, Alexander MP. The frontal lobes are necessary for theory of mind. Brain. 2001;124(Pt 2):279–86.
Turk DJ, Heatherton TF, Kelley WM, Funnell MG, Gazzaniga MS, Macrae CN. Mike or me? Self-recognition in a split-brain patient. Nat Neurosci. 2002;5(9):841–2.
Uddin LQ, Iacoboni M, Lange C, Keenan JP. The self and social cognition: the role of cortical midline structures and mirror neurons. Trends Cogn Sci. 2007;11(4):153–7.

de Vignemont F, Singer T. The empathic brain: how, when and why? Trends Cogn Sci. 2006;10(10):435–41.
Vincent JL, Patel GH, Fox MD, Snyder AZ, Baker JT, Van Essen DC, Zempel JM, Snyder LH, Corbetta M, Raichle ME. Intrinsic functional architecture in the anaesthetized monkey brain. Nature. 2007;447(7140):83–6.
Vogeley K, Fink GR. Neural correlates of the first-person-perspective. Trends Cogn Sci. 2003;7(1):38–42.
Walter H. Neurophilosophie der Willensfreiheit: von libertarischen Illusionen zum Konzept natürlicher Autonomie. Paderborn: Mentis; 1998.
Wicker B, Ruby P, Royet JP, Fonlupt P. A relation between rest and the self in the brain? Brain Res Brain Res Rev. 2003;43(2):224–30.

# Outlook: Neuropsychodynamic Psychiatry and the Impact of Therapeutic Relationships

31

Heinz Boeker, Peter Hartwich, and Georg Northoff

**Abstract**

The neuropsychodynamic approach has far-reaching implications for how mental phenomena, like the self, the conscious, the unconscious and their underlying psychological functions, are viewed. Mental phenomena are considered in a relational perspective, taking the relationships between the self, brain and environment as a starting point. What has historically been attributed to the mind, and in more modern times frequently to the brain as basic content, can, we believe, only be understood as part of this self-brain-environment relationship. This relationship is of a primary nature, and what is individually described and perceived as the self, the brain or the environment is precisely a result of this basic, fundamental relationship.

Psychiatric disorders may be characterised by shifts or an imbalance in the self-brain-environment relationship. The chapters in this book have taken on the challenge of endeavouring to bridge the principal differences between a person's individual level and the brain's general level. Neuroscientific results are examined in light of the essentials of psychodynamic psychiatry, and their epistemological and clinical significance is highlighted in the relevant chapters.

H. Boeker (✉)
Department of Psychiatry, Psychotherapy, and Psychosomatics, Psychiatric University Hospital Zurich, University of Zurich, Zurich, Switzerland
e-mail: heinz.boeker@bli.uzh.ch

P. Hartwich
Department of Psychiatry, Psychotherapy, Psychosomatics, General Hospital of Frankfurt am Main, Teaching Hospital University Frankfurt, Frankfurt, Germany
e-mail: Prof.Hartwich@casytec.de

G. Northoff
Mind, Brain Imaging, and Neuroethics,
Institute of Mental Health Research, University of Ottawa, Ottawa, ON, Canada
e-mail: georg.northoff@theroyal.ca; www.georgnorthoff.com



H. Boeker et al. (eds.), *Neuropsychodynamic Psychiatry*,
https://doi.org/10.1007/978-3-319-75112-2_31

Neuropsychodynamic psychiatry is presented in this book as a diagnostic and therapeutic approach for our present day. At the same time, neurodynamic psychiatry is also a scientific model for the future: it focuses on explaining, understanding, researching, diagnosing and treating psychopathological phenomena and includes dilemmata and distortions in intrapsychic structures and internalised object relationships. This means that the functionality and dysfunctionality of psychic and neuronal mechanisms can be taken into account and put into the context of neuronal correlations.

From a methodological perspective, we are confronted with the claim that neuropsychodynamic psychiatry should be bilingual, understanding and speaking the language of the brain and that of the mind: we feel that neuropsychodynamic psychiatry can contribute to overcoming deep-rooted dichotomies surrounding somatopsychic-psychosomatic phenomena.

The authors hope that neuropsychodynamic psychiatry, as outlined in this book, will help scientific and clinical "relationship work" to come together.

## 31.1 How Can We Understand Psychiatric Disorders and Treat the Individual Patient in a Better Way?

Throughout the long history of psychiatry, many different answers to this question have been proposed: social factors in the environment, neuronal activation patterns in the brain, psychodynamics and the phenomenological contents of experience are often mentioned when trying to explain these disorders and their often initially baffling symptoms. The authors of this book have deviated from the traditional focus on certain contents, an approach currently favoured particularly in Anglo-American countries, and have instead turned their attention to relationships. The most fundamental relationship is that between the self, the brain and the environment: psychiatric disorders are relationship disorders or alterations in the self-brain-environment relationship.

This approach has far-reaching implications for how mental phenomena, like the self, the conscious, the unconscious and their underlying psychological functions, are viewed. Mental phenomena are considered in a relational perspective, taking the relationships between the self, brain and environment as a starting point. What has historically been attributed to the mind, and in more modern times frequently to the brain as basic content, can, we believe, only be understood as part of this self-brain-environment relationship. This relationship is of a primary nature, and what is individually described and perceived as the self, the brain or the environment is precisely a result of this basic, fundamental relationship.

Psychiatric disorders may be characterised by shifts or an imbalance in the self-brain-environment relationship. For instance, in depression, there is an abnormal shift towards the self at the expense of the environment (see Chap. 10), whereas in acute psychosis, a reversed dysbalance towards the environment can be found (see Chap. 9). In a nutshell, we consider psychiatric disorders to be disorders of the self-brain-environment relationship, that is, the quintessence neuropsychodynamic psychiatry as outlined in our book.

Neuroscientific results are examined in light of the essentials of psychodynamic psychiatry, and their epistemological and clinical significance is highlighted in the relevant chapters. In the course of the some 100 years of psychodynamic psychiatric history, a great amount of knowledge has been gathered, based on extensive clinical experience and often validated in empirical studies. Therefore, the "wheel" of psychodynamic psychiatry does not need to be invented again, but rather what is known about psychodynamic psychiatry should be connected with both subjective modes of experience and neuroscientific approaches and findings and then applied to basic questions about neuroscience. Linking the brain with subjective experience and psychodynamics can give us a new view of both the brain and psychodynamic concepts. We can see neuronal processes, such as the intrinsic activity of the brain, in a new light, and this can be validated empirically using sometimes new experimental approaches in the future. However, some of the traditional definitions of psychodynamic concepts, such as cathexis, should also be looked at in a different context, for instance, in the neuronal context of the brain.

This approach in neuropsychodynamic psychiatry is multidimensional and constantly requires translation between the mind's language and the brain's language. Neuropsychodynamic psychiatry is presented in this book as a diagnostic and therapeutic approach for our present day. At the same time, neurodynamic psychiatry is also a scientific model for the future: it focuses on explaining, understanding, researching, diagnosing and treating psychopathological phenomena and includes dilemmata and distortions in intrapsychic structures and internalised object relationships. This means that the functionality and dysfunctionality of psychic and neuronal mechanisms can be taken into account and put into the context of neuronal correlations.

What is of great importance here is the concept of the context, which is used both methodologically and empirically and ontologically. For instance, neuronal processes have been put into the context of psychodynamic mechanisms; for example, a direct connection has been made between the neuronal dysbalance between direct and indirect cortico-subcortical loops in obsessive-compulsive disorder and maintaining and strengthening autonomy (see Chap. 13). At the same time, we consider the contextualisation of the brain through the environment and the self (and the contextualisation of these two through the brain) to be an empirical and ontological fundamental feature of the brain—the brain is a relationship organ, and its neuronal states are relationship states. Thus, new findings should always be put into the respective context, the context of neuronal processes and mechanisms or the context of interpersonal constellations and the cultural and social environment.

The chapters in this book have taken on the challenge of endeavouring to bridge the principal differences between a person's individual level and the brain's general level. Therefore, neuropsychodynamic psychiatry should not be equated with the biopsychosocial model of psychiatric disorders. Even though this model admittedly may contribute considerably to overcoming previous dualistic perspectives, a tendency towards biologically linear thinking can still be observed. If biological, mental and social contents are simply juxtaposed, the central element is lost, that is, the relationship between the different contents or dimensions. Currently, these relationships, or rather interactions between the different dimensions, are not being given sufficient consideration either in daily clinical routine or in scientific studies. This failure to fully understand and put the biopsychosocial model into practice may well be seen then in the caricature of a clinician looking at a patient's computer tomogram in a psychiatric clinic ("narrow ventricle, here brain loss, so it looks like a psychosis"), instead of talking to the schizophrenic patient and trying to build a relationship with him/her. From a scientific point of view, failing to distinguish in the brain the purely objective from the subjective self shows that the relationship between doctor and patient has been neglected.

Neuropsychodynamic psychiatry is a mechanism-based approach focussing particularly on the *dimension of the self*, which is of great significance for psychotherapeutic processes. The relationship between the self, the brain and the environment is constantly changing; it is variable and, in a healthy person, adapts to the environment, which is generated by neuropsychodynamic mechanisms. Exactly the same neuropsychodynamic mechanisms can act in the opposite way and become maladaptive in the case of psychiatric illness.

Early on, Freud spoke about the importance of mental (and ultimately also of neuronal) mechanisms and processes: "In mourning, it is the world which has become poor and empty; in melancholia it is the ego itself" (Freud 1914, p. 246). Therefore, it is important to understand the functionality and dysfunctionality of psychic phenomena (in the sense of Mentzos 2011) on the basis of neuropsychodynamic mechanisms and apply this understanding to psychotherapeutic interventions. In this context, our book describes different energy processes as expressions of neuropsychodynamic mechanisms: cathexis/decathexis in schizophrenia (Hartwich and Northoff), paraconstructions and expansion of the concept of cathexis to include clinical subforms of schizophrenia (Hartwich) and high-energy standstill in obsessive-compulsive disorders (Hartwich, Boeker and Northoff).

## 31.2 What Challenges Will We Face in the Future?

There will undoubtedly be clinical, methodological and scientific *challenges*. From a clinical viewpoint, one of the future challenges will be to convey and implement scientific know-how as an element of the psychotherapeutic attitude in the therapeutic relationship. Specific questions remain unanswered, such as the therapeutic consequences of dysfunction in cortical midline structures, of heightened arousal (hyperactivity in the central nervous system) in depressed patients, of hypoactivity in the dorsal lateral prefrontal cortex (DLPFC) and of disturbed self-reference and intensified body experience in depressed patients. These kinds of questions may also be considered, for instance, in the treatment of schizophrenia in view of the functionality of psychotic symptoms, and the discerning encounter with existential compensation mechanisms and neuropsychodynamically oriented psychotherapy as a preventive method and aid in strengthening the personality.

From a methodological perspective, we are confronted with the claim that neuropsychodynamic psychiatry should be bilingual, understanding and speaking the language of the brain and that of the mind: we feel that neuropsychodynamic psychiatry can contribute to overcoming deep-rooted dichotomies surrounding somatopsychic-psychosomatic phenomena. It aims at an *integrative understanding*, by relating different concepts with their respective contexts. This is often confusing at first glance, but it makes sense, particularly in view of the *relationship changes* observed in psychiatric disorders. Can biography and brain fit together? We think so, because it is the basis of the self-brain-environment relationship.

The studies carried out so far and those planned in the near future involve individual or group studies and help us draw conclusions for the individual patients we are currently treating. However, there will always be imponderabilities in these conclusions if we are not able to methodologically understand the different weightings of the individual components in their interrelationships for the individual in a current situation and over a longer period of time. Because of this, we must assume that there are ultimately numerous and probably even an infinite number of variations, which we can only approach by determining the respective weighting in the relationship interaction network. The neuropsychodynamically oriented therapist will take the psychiatric patient's disturbed self-brain-environment relationship as a starting point, consider the scientific essentials and experimental results and then, using his experience and, to some extent, his intuition, will apply them to his patient's individual fate and use them in the therapy.

From a scientific perspective, future methodological challenges will be to develop experimental designs and analytical methods enabling individual and general features to be correlated on a neuronal level. This can lead to novel experimental approaches in the field of the brain, which we, Northoff and Boeker, have investigated in depression using neuroimaging (2010). Important work for the future will be understanding the brain's intrinsic activity, which has always been linked to the self and the environment. Thus, it is our task to investigate the relationship of the brain and its intrinsic activity to the self and the environment.

## 31.3 How Is This Relationship Encoded in the Brain's Intrinsic Activity? Why Can the Brain Only Develop Its Intrinsic Activity in Relation to Itself and the Environment, Unlike a Computer, Which Does Not Have a Relationship Either with Itself or with the Environment?

In short, brain science will have to do some "relationship work" in the future and will be joined by your future clinicians and therapists who will also have to do the same. Ultimately, they will have to individually calibrate and rebalance the dysfunctional imbalance between the self, the brain and the environment in their patients. The better the neuroscientist does his relationship work, and the better he/she understands relationship processes and mechanisms of the brain concerning the self and the outside world, the better and more specifically the clinician and therapist can do their "relationship work". Scientific and clinical relationship work come together, then, and in our opinion cannot be separated from one another. We hope that neuropsychodynamic psychiatry, as outlined in this book, will help scientific and clinical "relationship work" to come together, now in the present and particularly later; however the future may be.

## References

Freud S. Mourning and melancholia. Standard Edition, Vol. XIV. London: Hogarth; 1914–1916. p. 237–260.

Mentzos S. Lehrbuch der Psychodynamik. Die Funktion der Dysfunktionalitat psychischer Störungen. 5th ed. Göttingen: Vandenhoeck & Ruprecht; 2011.

Boeker H, Northoff G. Die Entkopplung des Selbst in der Depression: Empirische Befunde und neuropsychodynamische Hypothesen. Psyche Z Psychoanal. 2010;64:934–76.

# Index




H. Boeker et al. (eds.), *Neuropsychodynamic Psychiatry*,
https://doi.org/10.1007/978-3-319-75112-2

MIX
Papier aus verantwortungsvollen Quellen
Paper from responsible sources
FSC® C105338

If you have any concerns about our products,
you can contact us on
**ProductSafety@springernature.com**

In case Publisher is established outside the EU,
the EU authorized representative is:
**Springer Nature Customer Service Center GmbH**
**Europaplatz 3, 69115 Heidelberg, Germany**

Printed by Libri Plureos GmbH
in Hamburg, Germany